1991

Practical Handbook of
Biochemistry and Molecular Biology

Edited by

Gerald D. Fasman, Ph.D.

Graduate Department of Biochemistry
Brandeis University
Waltham, Massachusetts

CRC Press
Boca Raton   Ann Arbor   Boston

Library of Congress Cataloging-in-Publication Data

Practical handbook of biochemistry and molecular biology/editor,
  Gerald D. Fasman.
      p.   cm.
    Includes selections from Handbook of biochemistry and molecular
biology. Cleveland : CRC Press, c1975-1977.
    Bibliography: p.
    Includes index.
    ISBN 0-8493-3705-4
    1. Biochemistry--Handbooks, manuals, etc.   2. Molecular biology-
-Handbooks, manuals, etc.   I. Fasman, Gerald D.
QP514.2.P73   1989
574.1'92--dc19

88-38230
CIP

This book represents information obtained from authentic and highly regarded sources. Reprinted material is quoted with permission, and sources are indicated. A wide variety of references are listed. Every reasonable effort has been made to give reliable data and information, but the author and the publisher cannot assume responsibility for the validity of all materials or for the consequences of their use.

Direct all inquiries to CRC Press, Inc., 2000 Corporate Blvd., N.W., Boca Raton, Florida, 33431.

© 1989 by CRC Press, Inc.
Second Printing, 1990
Third Printing, 1990

International Standard Book Number 0-8493-3705-4

Library of Congress Card Number
Printed in the United States

PREFACE

New methodologies and databases for biochemistry and molecular biology are constantly being added to current sources. This *Handbook* updates previous handbooks in a format which is easy to use in the laboratory.

New information, in areas such as restriction enzymes, is always being published, so it is impossible to be completely up-to-date. However, the tables in this *Handbook* contain the most relevant data available at the time of publication.

Gerald D. Fasman
Editor

THE EDITOR

Gerald D. Fasman, Ph.D., is the Rosenfield Professor of Biochemistry, Graduate Department of Chemistry, Brandeis University, Waltham, Massachusetts.

Dr. Fasman graduated from the University of Alberta in 1948 with a B.S. Honors Degree in Chemistry, and he received his Ph.D. in Organic Chemistry in 1952 from the California Institute of Techology, Pasadena, California. Dr. Fasman did postdoctoral studies at Cambridge University, England, Eidg. Technische Hochschule, Zurich, Switzerland, and the Weizmann Institute of Science, Rehovoth, Israel. Prior to moving to Brandeis University, he spent several years at the Children's Cancer Research Foundation at the Harvard Medical School. He has been an Established Investigator of the American Heart Association, a National Science Foundation Senior Postdoctoral Fellow in Japan, and has been awarded two John Simon Guggenheim Fellowships.

Dr. Fasman is a member of the American Chemical Society, a Fellow of the American Association for the Advancement of Science, member of Sigma Xi, the Biophysical Society, the American Society of Biological Chemists, the Chemical Society (London), the New York Academy of Science, the Protein Society, and a Fellow of the American Institute of Chemists. He has published 250 research papers.

ACKNOWLEDGMENT

With exceptions as noted, materials in this volume were derived from the multi-volume *CRC Handbook of Biochemistry and Molecular Biology*, edited by Gerald D. Fasman:

Section 1. Proteins: derived from *Proteins,* Volumes I and II (pages 3 to 102 from Volume I and pages 103 to 366 from Volume II) as well as new material, beginning on page 367.

Section 2. Nucleic Acids: derived from *Nucleic Acids,* Volumes I and II (pages 385 to 402 and 403 to 448, respectively), with added new material on pages 449 to 511.

Section 3. Lipids: contains selected tables from *Lipids, Carbohydrates and Steroids.*

Section 4. Physical-Chemical Data: drawn from Volumes I and II (pages 531 to 565 and 566 to 576, respectively) of *Physical and Chemical Data.*

TABLE OF CONTENTS

SECTION 2. NUCLEOSIDES, NUCLEOTIDES, AND NUCLEIC ACIDS

SECTION 3. LIPIDS

SECTION 4. PHYSICAL-CHEMICAL DATA

Section 1
Amino Acids and Proteins

AMINO ACIDS

DATA ON THE NATURALLY OCCURRING AMINO ACIDS

Elizabeth Dodd Mooz

The amino acids included in these tables are those for which reliable evidence exists for their occurrence in nature. These tables are intended as a guide to the primary literature in which the isolation and characterization of the amino acids are reported. Originally, it was planned to include more factual data on the chemical and physical properties of these compounds; however, the many different conditions employed by various authors in measuring these properties (i.e., chromatography and spectral data) made them difficult to arrange into useful tables. The rotation values are as given in the references cited; unfortunately, in some cases there is no information given on temperature, solvent, or concentration.

The investigator employing the data in these tables is urged to refer to the original articles in order to evaluate for himself the reliability of the information reported. These references are intended to be informative to the reader rather than to give credit to individual scientists who published the original reports. Thus not all published material is cited.

The compounds listed in Sections A to N are known to be of the L configuration. Section O contains some of the D amino acids which occur naturally. This last section is not intended to be complete since most properties of the D amino acids correspond to those of their L enantiomorphs. Therefore, emphasis was placed on including those D amino acids whose L isomers have not been found in nature. The reader will find additional information on the D amino acids in the review by Corrigan[263] and in the book by Meister.[1]

Compilation of data for these tables was completed in December 1974. Appreciation is expressed to Doctors L. Fowden, John F. Thompson, Peter Müller, and M. Bodanszky who were helpful in supplying recent references and to Dr. David Pruess who made review material available to me prior to its publication. A special word of thanks to Dr. Alton Meister who made available reprints of journal articles which I was not able to obtain.

DATA ON THE NATURALLY OCCURRING AMINO ACIDS (continued)

A. L-Monoamino, Monocarboxylic Acids

No.	Amino acid (synonym)	Source	Structure	Formula (mol wt)	Melting point °C[a]	$[\alpha]_D^{\ b}$	pK$_a$	References Isolation and purification	References Chromatography	References Chemistry	References Spectral data
1	Alanine (α-aminopropionic acid)	Silk fibroin	$CH_3CH(NH_2)COOH$	$C_3H_7NO_2$ (89.09)	297°	+1.8^{25} (c 2, H$_2$O) (1) ≠14.6^{25} (c 2, 5 N HCl) (1)	2.34 9.69	2	3	4	4
2	β-Alanine (β-aminopropionic acid)	*Iris tingitana*	$NH_2CH_2CH_2COOH$	$C_3H_7NO_2$ (89.09)	196° (dec)	—	3.55 10.24	5	5	5	—
3	α-Aminobutyric acid	Yeast protein	$CH_3CH_2CH(NH_2)COOH$	$C_4H_9NO_2$ (103.12)	292° (dec)	+20.5^{25} (c 1–2, 5 N HCl) (290) +9.3^{25} (c 1–2, H$_2$O) (290) +42^{25} (c 1–2, gl acetic) (290)	2.29 9.83	6	7	6	—
4	γ-Aminobutyric acid (piperidinic acid)	Bacteria	$NH_2CH_2CH_2CH_2COOH$	$C_4H_9NO_2$ (103.12)	203° (dec)	—	4.03 10.56 (290)	8–10	9, 10	11	—
5	1-Aminocyclopropane-1-carboxylic acid	Pears and apples	(structure)	$C_4H_7NO_2$ (101.11)	—	—	—	11	11	12	12
6	2-Amino-3-formyl-3-pentenoic acid	*Bankera fulgineoalba* (a mushroom)	$CH_3CH=CCH(NH_2)COOH$ with CHO	$C_6H_9NO_3$ (143.15)	—	—	—	13	13	13	13
7	α-Aminoheptanoic acid	*Claviceps purpurea*	$CH_3(CH_2)_4CH(NH_2)COOH$	$C_7H_{15}NO_2$ (145.21)	—	—	—	14	14	—	—

DATA ON THE NATURALLY OCCURRING AMINO ACIDS (continued)

No.	Amino acid (synonym)	Source	Structure	Formula (mol wt)	Melting point °C[a]	$[\alpha]_D$[b]	pK_a	References Isolation and purification	References Chromatography	References Chemistry	References Spectral data
7a	2-Amino-4,5-hexadienoic acid	*Amanita solitaria*	$H_2C=C=CHCH_2-CH(NH_2)COOH$	$C_6H_9NO_2$ (127.16)	200° (dec) (14a)	—	—	14a	—	—	14a
8	2-Amino-4-hexenoic acid	Ilamycin	$CH_3CH=CHCH_2CH(NH_2)COOH$	$C_6H_{11}NO_2$ (129.17)	—	—	—	15	15	—	—
8a	2-Amino-4-hydroxy-hept-6-ynoic acid	*Euphoria longan*	$HC\equiv C-CH_2CH(OH)CH_2CH(NH_2)COOH$	$C_7H_{11}NO_3$ (157.19)	—	$-27°$ (c 2, H_2O) $-8°$ (c 1, 5 N HCl) (15a)	—	15a	15a	—	15a
8b	2-Amino-6-hydroxy-4-methyl-4-hexenoic acid	*Aesculus Californica* seeds	HOH_2C / H_3C C=C $CH_2CH(NH_2)COOH$ (with H)	$C_7H_{13}NO_3$ (159.21)	—	$-31°$ (c 2.2, H_2O) $+2°$ (c 1.1, 5 N HCl) (23b)	—	23b	23b	—	15b
8c	2-Amino-4-hydroxy-5-methyl hexenoic acid	*Euphoria longan*	$HC\equiv C$ CHCH$_2$CH(NH$_2$)COOH (HOH_2C)	$C_7H_{11}NO_3$ (157.19)	—	$-27°$ (c 2, H_2O) $-13°$ (c 1, 5 N HCl) (15a)	—	15a	—	15a	15a
8d	2-Amino-3-hydroxy-methyl-3-pentenoic acid	*Bankera fuligineoalba*	CH_2OH / $H_3CCH=CCH(NH_2)COOH$	$C_6H_{11}NO_3$ (145.18)	160–161° (dec) (13)	$+182^{25}°$ (c 0.8, H_2O) $+201^{25}$ (c 0.8, 0.3 N HCl) (13)	—	13	13	13	13
9	α-Aminoisobutyric acid	*Iris tingitana*, muscle protein	$(CH_3)_2C(NH_2)COOH$	$C_4H_9NO_2$ (103.12)	200° (dec)	—	2.36 10.21 (290)	16	16	—	—
10	β-Aminoisobutyric acid	*Iris tingitana*	$NH_2CH_2CHCOOH$ (CH_3)	$C_4H_9NO_2$ (103.12)	179° (17)	-21^{26} (c 0.43, H_2O) (17)	—	17	17	17	17

DATA ON THE NATURALLY OCCURRING AMINO ACIDS (continued)

No.	Amino acid (synonym)	Source	Structure	Formula (mol wt)	Melting point °C[a]	$[\alpha]_D$[b]	pK_a	References Isolation and purification	Chromatography	Chemistry	Spectral data
10a	2-Amino-4-methoxy-*trans*-3-butenoic acid	*Pseudomonas aeruginosa*	H$_3$CO–CH=CH–CH(NH$_2$)COOH	C$_5$H$_9$NO$_3$ (131.15)	—	—	—	17a	17b	—	17a, 17b
11	γ-Amino-α-methylene butyric acid	*Arachis hypogaea* (groundnut plants)	NH$_2$CH$_2$CH$_2$CCOOH, =CH$_2$	C$_5$H$_9$NO$_2$ (115.13)	152° (18)	—	—	18	18	18	—
12	2-Amino-4-methyl-hexanoic acid (homoisoleucine)	*Aesculus California* seeds	CH$_3$CH$_2$CHCH$_2$CH(NH$_2$)COOH, CH$_3$	C$_7$H$_{15}$NO$_2$ (145.21)	—	$-2.2°$ (c 1, H$_2$O) (19) $+24.2°$ (c 0.87, 5 N HCl) (19)	—	19	19	19	19
13	2-Amino-4-methyl-4-hexenoic acid	*Aesculus California* seeds	H$_3$C–CH=C(CH$_3$)–CH$_2$CH(NH$_2$)COOH	C$_7$H$_{13}$NO$_2$ (143.19)	—	$-61.2°$ (c 2.4, H$_2$O) (19) -36 (c 1.2, 6 N HCl) (19)	—	19	19	19	19
13a	2-Amino-4-methyl-5-hexenoic acid	*Streptomyces* species	CH$_2$=CHCH(CH$_3$)CH$_2$CH(NH$_2$)COOH	C$_7$H$_{13}$NO$_2$ (143.21)	260° (dec) (19a)	$-9.6°$ (c 1.78, H$_2$O) $+5.7°$ (c 0.7, 1 N HCl) (99a)	—	19a	19a	—	19a
14	2-Amino-5-methyl-4-hexenoic acid	*Leucocortinarius bulbiger*	CH$_3$C=CHCH$_2$CH(NH$_2$)COOH, CH$_3$	C$_7$H$_{13}$NO$_2$ (143.19)	260—270° (dec) (22a)	$-45.9°^3$ (c 0.47, H$_2$O) $-7°^3$ (c 0.4, 1 N HCl) (22a)	—	22, 22a	22a	—	22a

DATA ON THE NATURALLY OCCURRING AMINO ACIDS (continued)

No.	Amino acid (synonym)	Source	Structure	Formula (mol wt)	Melting point °C[a]	$[\alpha]_D$[b]	pK_a	References Isolation and purification	Chromatography	Chemistry	Spectral data
14a	2-Amino-4-methyl-5-hexenoic acid	*Euphoria longan*	$HC{\equiv}C{\diagdown}CHCH_2CH(NH_2)COOH$, $H_3C{\diagup}$	$C_7H_{11}NO_2$ (141.19)	—	$-33^{2\circ}$ (c 2, H_2O) $-27^{2\circ}$ (c 1, 5 N HCl) (15a)	—	15a	—	15a	15a
15	α-Amino-octanoic acid	*Aspergillus atypique*	$CH_3(CH_2)_5CH(NH_2)COOH$	$C_8H_{17}NO_2$ (159.23)	—	—	—	23	23	23	23
15a	2-Amino-4-pentynoic acid	*Streptomyces* sp. #8–4	$HC{\equiv}CCH_2CH(NH_2)COOH$	$C_5H_7NO_2$ (113.13)	241–242° (dec) (23a)	-31.1^{25} (c 1, H_2O) -5.5^{25} (c 1, 5 N HCl) (23a)	—	23a	23a	23a	23a
15a'	*cis*-α-(Carboxycyclopropyl)glycine	*Aesculus parviflora*	$HOOCHC{\diagdown}{\diagup}CHCH(NH_2)COOH$, CH_2	$C_6H_9NO_4$ (159.16)	—	$+25^{2\circ}$ (c 1, H_2O) $+58$ (c 0.5, 5 N HCl) (23a')	—	23a'	23a'	—	23a'
15b	*trans*-α-(Carboxycyclopropyl)glycine	*Blighia sapida*	$HOOCHC{\diagdown}{\diagup}CHCH(NH_2)COOH$, CH_2	$C_6H_9NO_4$ (159.16)	—	$+107^{2\circ}$ (c 2, H_2O) $+146^{2\circ}$ (c 1, 5 NHCl) (23a')	—	23a'	23a'	—	23a'
15b'	*trans*-α-(2-Carboxymethylcyclopropyl)glycine	*Blighia unijugata*	$HC{\diagdown}{\,CH_2\,}{\diagup}CHCH(NH_2)COOH$, CH_2COOH	$C_7H_{11}NO_4$ (173.19)	—	$+12^{2\circ}$ (c 1, H_2O) $+45^{2\circ}$ (c 0.5, 5 N HCl) (99a)	—	99a	99a	—	99a

DATA ON THE NATURALLY OCCURRING AMINO ACIDS (continued)

No.	Amino acid (synonym)	Structure	Source	Formula (mol wt)	Melting point °C[a]	$[\alpha]_D$[b]	pK_a	Isolation and purification	Chromatography	Chemistry	Spectral data
										References	
15c	γ-Glutamyl-2-amino-4-methyl-hex-4-enoic acid	H_3C, H, C=C, H_3C, CH_2—CH—COOH, NH, O=C—CH_2—CH_2—$CH(NH_2)$COOH	*Aesculus California* seeds	$C_{12}H_{20}N_2O_5$ (272.34)	—	+17° (c 3, H_2O) (23b)	—	23b	23b	—	—
		B. L-Monoamino, Dicarboxylic Acids									
16	Glycine (α-aminoacetic acid)	$CH_2(NH_2)COOH$	Gelatin hydrolyzate	$C_2H_6NO_2$ (75.07)	290° (dec) (1)	—	2.35 9.78 (290)	24	3	25	25
17	Hypoglycin A [α-amino-β-(2-methylene cyclopropyl)-propionic acid]	$H_2C=C$, CH_2, CH, CH_2, NH_2CHCOOH	*Blighia sapida*	$C_7H_{11}NO_2$ (141.18)	280—284° (26)	+9.2 (c 1, H_2O) (26)	—	26	26	27	27
18	Isoleucine (α-amino-β-methyl-valeric acid)	CH_3CH_2, CHCH(NH_2)COOH, CH_3	Sugar beet molasses	$C_6H_{13}NO_2$ (131.17)	284° (1)	$+39.5^{25}$ (c 1, 5 N HCl) (290) $+12.4^{25}$ (c 1, H_2O) (290)	2.36 9.68	3	9	29	1
19	Leucine (α-aminoisocaproic acid)	$(CH_3)_2CHCH_2CH(NH_2)COOH$	Muscle fiber, wool	$C_6H_{13}NO_2$ (131.17)	337° (1)	-11^{25} (c 2, H_2O) $+16^{25}$ c 2,5 N HCl) (1)	2.36 960 (1)	30	3	31	31
19a	N-Methyl-γ-methyl-*alloisoleucine*	$CH_3CH(CH_3)CH(CH_3)CH(NHCH_3)COOH$	Etamycin	$C_8H_{17}NO_2$ (159.26)	—	—	—	31a	31a	—	—

DATA ON THE NATURALLY OCCURRING AMINO ACIDS (continued)

No.	Amino acid (synonym)	Source	Structure	Formula (mol wt)	Melting point °C[a]	$[\alpha]_D$[b]	pK$_a$	Isolation and purification	Chromatography	Chemistry	Spectral data
19b	β-Methyl-β-(methylene-cyclopropyl) alanine	*Aesculus Californica* seeds	$H_2C=C\underset{CH_2}{\diagup}CH(CH_3)CH(NH_2)COOH$	$C_8H_{13}NO_2$ (155.22)	—	1.5^{20} (c 2, H$_2$O) $+45^{20}$ (c 1, 5 N HCl) (23b)	—	23b	23b	—	15b
20	α-(Methylene-cyclopropyl) glycine	*Litchi chinensis*	$H_2C=C\underset{CH_2}{\diagup}CHCH(NH_2)COOH$	$C_6H_9NO_2$ (127.15)	—	$+43^{22.5}$ (c 0.5, 5 N HCl) (32)	—	32	32	32	32
21	β-(Methylene-cyclopropyl)-β-methylalanine	*Aesculus California*	$H_2C=C\underset{CH_2}{\diagup}CHCHCH(NH_2)COOH$ CH_3	$C_8H_{13}NO_2$ (155.19)	—	$+1.5^{20}$ (c 2, H$_2$O) $+45^{20}$ (c 1, 5 N HCl)	—	19	19	19	19, 21
21a	β-Methylenenor-leucine	*Amanita vaginata*	$H_3C-CH_2-CH_2$ \ $C-CH(NH_2)COOH$ / H_2C	$C_7H_{13}NO_2$ (143.21)	171° (35a)	$+158^{20}$ (c 0.51, 1 N HCl) $+149^{20}$ (c 0.56, H$_2$O) (35a)	—	—	35a	35a	—
22	Valine (α-aminoisovaleric acid)	Casein	$(CH_3)_2CHCH(NH_2)COOH$	$C_5H_{11}NO_2$ (117.15)	292–295° (1)	$+28.3^{25}$ (c 1.2, 5 N HCl) $+5.63^{25}$ (c 1–2, H$_2$O) (290)	2.32 9.62(1)	35	3	35, 36	36
23	α-Aminoadipic acid	*Pisum sativum*	$HOOC(CH_2)_3(CH(NH_2)COOH$	$C_6H_{11}NO_4$ (161.18)	195° (37)	$+3.2^{25}$ (c 2, H$_2$O) (290) $+23^{22}$ (c 2, 6 N HCl) (37)	2.14 4.21 9.77 (290)	37	37	37	—
24	3-Aminoglutaric acid	*Chondria armata*	$HOOCCH_2CH(NH_2)CH_2COOH$	$C_5H_9NO_4$ (162.13)	280–282° (38)	$\pm0^c$ (c 2, 5 N HCl) (38)	—	38	38	38	38

DATA ON THE NATURALLY OCCURRING AMINO ACIDS (continued)

No.	Amino acid (synonym)	Source	Structure	Formula (mol wt)	Melting point °C[a]	$[\alpha]_D$[b]	pK_a	References Isolation and purification	References Chromatography	References Chemistry	References Spectral data
25	α-Aminopimelic acid	*Asplenium septentrionale*	$HOOC(CH_2)_4CH(NH_2)COOH$	$C_7H_{13}NO_4$ (175.19)	204° (39)	—	—	39	39	39	—
26	Asparagine (α-aminosuccinamic acid)	Asparagus	$H_2NCOCH_2CH(NH_2)COOH$	$C_4H_8N_2O_3$ (132.12)	236° (1)	-5.6^{25} (c 2, H_2O) (290) +33.2 (3 N HCl) (1)	2.02 8.80 (1)	40	3	41	42
27	Aspartic acid (α-aminosuccinic acid)	Conglutin, legumin	$HOOCCH_2CH(NH_2)COOH$	$C_4H_7NO_4$ (133.10)	270° (1)	$+5.05^{25}$ (c 2, H_2O) $+25.4^{25}$ (c 2, 5N HCl) (1)	1.88 3.65 9.60 (1)	43	3	41	41
28	Ethylasparagine	*Ecballium elaterium, Bryonia dioica*	$CH_3CH_2NHCOCH_2CH(NH_2)COOH$	$C_6H_{12}N_2O_3$ (160.19)	—	—	—	45	45	45	45
29	γ-Ethylidene glutamic acid	Mimosa	$CH_3CH{=}HOOCCCH_2CH(NH_2)COOH$	$C_7H_{11}NO_4$ (173.18)	—	$+21^{20}$ (c 2.8, H_2O) (47) $+38.3^{20}$ (c 1.4, 6 N HCl) (47)	—	46	46	—	46
30	N-Fumarylalanine	*Penicillium recticulosum*	$CH_3CHCOOH$ NHCOCH$=$CHCOOH	$C_7H_9NO_5$ (187.16)	229° (48)	—	—	48	—	48	—
31	Glutamic acid (α-aminoglutaric acid)	Gluten-fibrin hydrolyzates	$HOOC(CH_2)_2CH(NH_2)COOH$	$C_5H_9NO_4$ (147.13)	249° (1)	$+12^{25}$ (c 2, H_2O) $+31.8^{28}$ (c 2, 5 N HCl) (1)	2.19 4.25 9.67 (1)	49	3	50	50
32	Glutamine (α-aminoglutaramic acid)	Beet juice	$H_2NCO(CH_2)_2CH(NH_2)COOH$	$C_5H_{10}N_2O_3$ (146.15)	185° (1)	$+6.3^{25}$ (c 2, H_2O) $+31.8^{25}$ (c 2, 1 N HCl) (1)	2.17 9.13 (1)	51	3	50	50

DATA ON THE NATURALLY OCCURRING AMINO ACIDS (continued)

No.	Amino acid (synonym)	Source	Structure	Formula (mol wt)	Melting point °C[a]	$[\alpha]_D$[b]	pKa	Isolation and purification	Chromatography	Chemistry	Spectral data
									References		
33	N^5-Isopropylglutamine	*Lunaria annua*	$HNCO(CH_2)_2CH(NH_2)COOH$ H_3CCHCH_3	$C_8H_{16}N_2O_3$ (188.23)	—	$+7.1^{22}$ (c 1.7, H_2O) (53)	—	53	53	53	—
34	N^4-Methylasparagine	—	$CH_3NHCOCH_2CH(NH_2)COOH$	$C_5H_{10}N_2O_3$ (146.15)	241–244° (54)	-4.2 (c 5.5, H_2O) (54)	—	—	54	—	—
35	β-Methylaspartic acid	*Clostridium tetanomorphum*	$HOOCCH(CH_3)CH(NH_2)COOH$	$C_5H_9NO_4$ (147.13)	—	-10 (c 0.4, H_2O) $+12.4$ (c 3, 1 N HCl) $+13.3$ (c 3, 5 N HCl) (55)	3.5 9.9 (290)	55	55	55	—
36	γ-Methylglutamic acid	*Phyllitis scolopendrium*	$HOOCCH(CH_3)CH_2\ CH(NH_2)COOH$	$C_6H_{11}NO_4$ (141.17)	—	—	—	56	56	—	—
37	γ-Methyleneglutamic acid	*Archis hypogaea*	$HOOCCCH_2CH(NH_2)COOH$ $=CH_2$	$C_6H_9NO_4$ (159.15)	196° (57)	—	—	57	57	57	—
38	γ-Methyleneglutamine	*Archis hypogaea*	$H_2NCOCCH_2CH(NH_2)COOH$ $=CH_2$	$C_6H_{10}N_2O_3$ (158.17)	173–182° (57)	—	—	57	57	57	—
39	Theanine (α-amino-γ-N-ethylglutaramic acid)	*Xerocomus badius*	$HOOCCH(NH_2)(CH_2)_2CONHCH_2CH_3$	$C_7H_{14}N_2O_3$ (174)	—	—	—	59	59	59	59
39a	β-N-Acetyl-α-,β-diaminopropionic acid (β-acetamido-L-alanine)	*Acacia armata* seeds	$CH_3CONHCH_2CH(NH_2)COOH$	$C_5H_{10}N_2O_3$ (146.17)	—	-87^{20} (c 8, H_2O) -35^{20} (c 4, 6 N HCl) (59a)	—	59a	59a	59a	59a

DATA ON THE NATURALLY OCCURRING AMINO ACIDS (continued)

C. L-Diamino, Monocarboxylic Acids

No.	Amino acid (synonym)	Source	Structure	Formula (mol wt)	Melting point °C[a]	$[\alpha]_D$[b]	pK_a	References Isolation and purification	Chromatography	Chemistry	Spectral data
40	N-Acetylornithine	*Asplenium* species	$CH_3CONH(CH_2)_3CH(NH_2)COOH$	$C_7H_{14}N_2O_3$ (174.11)	200° (dec) (60)	—	—	60	60	60	—
41	α-Amino-γ-N-acetylamino-butyric acid	Latex of *Euphorbia pulcherrima*	$CH_3CONH(CH_2)_2CH(NH_2)COOH$	$C_6H_{12}N_2O_3$ (160.18)	220–222° (dec) (61)	—	4.45 (33)	61	61	—	—
42	N-ε-(2-Amino-2-carboxyethyl)-lysine	Alkali-treated protein	$HOOCCH(NH_2)CH_2NH(CH_2)_4CH(NH_2)COOH$	$C_9H_{19}N_3O_4$ (233.28)	—	—	2.2 6.5 8.8 9.9 (62)	62	62	62	—
43	N-δ-(2-Amino-2-carboxyethyl)-ornithine	Alkali-treated wool	$HOOCCH(NH_2)CH_2NH(CH_2)_3CH(NH_2)COOH$	$C_8H_{17}N_3O_4$ (273.72)	—	—	—	63	63	63	—
44	2-Amino-3-dimethylamino-propionic acid	*Streptomyces neocaliberis*	$(CH_3)_2NCH_2CH(NH_2)COOH$	$C_5H_{12}N_2O_2$ (117.15)	—	-17.8^{25} (c 1, H₂O) +18.1 (c 1, HCl pH 3) (64)	—	64	64	64	64
45	α-Amino-β-methyl-aminopropionic acid	*Cycas circinalis*	$CH_3NHCH_2CH(NH_2)COOH$	$C_4H_{11}N_2O_2$ (105.15)	165–167° (65)	—	—	65	65	65	65
45a	α-Amino-β-oxalyl-aminopropionic acid	*Crotalaria*	$HOOC–CH(NH_2)CH_2NHCOCOOH$	$C_5H_8N_2O_5$ (176.15)	—	—	—	—	65a	—	—
46	Canaline	*Canavalia ensiformis*	$H_2NO(CH_2)_2CH(NH_2)COOH$	$C_4H_{10}N_2O_3$ (134.14)	—	—	2.40 3.70 9.20 (20)	66	—	67	—
46a	Threo-α,β-diamino-butyric acid	Amphomycin hydrolyzate	$CH_3CH(NH_2)CH(NH_2)COOH$	$C_4H_{10}N_2O_2$ (118.16)	213–214° (dec) (67a)	$+27.1^{25}$ (c 2, 5 N HCl) (67a)	—	67b	67b	67a	67b

DATA ON THE NATURALLY OCCURRING AMINO ACIDS (continued)

No.	Amino acid (synonym)	Source	Structure	Formula (mol wt)	Melting point °C[a]	$[\alpha]_D$[b]	pKa	Isolation and purification	Chromatography	Chemistry	Spectral data
									References		
47	α,γ-Diaminobutyric acid (γ-aminobutyrine)	Glumamycin	$H_2NCH_2CH_2CH(NH_2)COOH$	$C_4H_{10}N_2O_2$ (134.14)	—	$+7.2^{25}$ (c 2, H_2O) $+14.6^{18}$ (c 3.67, H_2O)[g] (290)	1.85 8.28 10.50 (20)	68	68	68	—
48	3,5-Diaminohexanoic acid	*Clostridium stricklandii*	$CH_3CH(NH_2)CH_2CH(NH_2)CH_2COOH$	$C_6H_{14}N_2O_2$ (146.19)	204—208° (69)	—	—	69	69	69	69
48a	2,6-Diamino-7-hydroxyazelaic acid	*Bacillus brevis* (edeine A and B)	$HOOCCH_2CH(OH)CH(NH_2)(CH_2)_3CH(NH_2)COOH$	$C_9H_{18}N_2O_{15}$ (234.29)	—	—	—	69a	69a	69a	—
49	α,β-Diaminopropionic acid (β-aminoalanine)	Mimosa	$H_2NCH_2CH(NH_2)COOH$	$C_3H_8N_2O_2$ (104.11)	—	—	1.23 6.73 9.56 (20)	70	70	70	—
49a	$N\varepsilon,N\varepsilon$-Dimethyllysine	Human urine	$(CH_3)_2N(CH_2)_4CH(NH_2)COOH$	$C_8H_{18}N_2O_2$ (174.28)	214—216° (dec) (70a)	—	—	70a	70a	—	70a
49b	N^5-1-Iminoethylornithine	*Streptomyces* broth	$HN{=}C(CH_3)NH(CH_2)_3CH(NH_2)COOH$	$C_7H_{15}N_3O_2$ (173.25)	226—229° (70b)	$+20.6^{25}$ (c 1, 5 N HCl) (70b)	1.97 8.86 11.83 (70b)	70b	70b	—	70b
50	Lathyrus factor (β-N-(γ-glutamyl)-aminopropionitrile)	*Lathyrus pusillus*	$CH(CH_2)_2NHCO(CH_2)_2CH(NH_2)COOH$	$C_8H_{13}N_3O_3$ (199.22)	193—194° (72)	$+28^{18}$ (c 1, 6 N HCl) (72)	2.2 9.14	71	72	72	72
51	Lysine (α,ε-diaminocaproic acid)	Casein	$H_2NCH_2(CH_2)_3CH(NH_2)COOH$	$C_6H_{14}N_2O_2$ (146.19)	224—225° (dec) (73)	$+14.6^{20}$ (H_2O) (73)	2.16 9.18 10.79 (290)	74	3	75	—
52	β-Lysine (isolysine; β,ε-diaminocaproic acid)	Viomycin	$HOOCCH_2CH(NH_2)CH_2CH_2CH_2(NH_2)$	$C_6H_{14}N_2O_2$ (146.19)	240—241° (76)	—	—	76	76	76	76

DATA ON THE NATURALLY OCCURRING AMINO ACIDS (continued)

No.	Amino acid (synonym)	Source	Structure	Formula (mol wt)	Melting point °C[a]	$[\alpha]_D$[b]	pK_a	References Isolation and purification	Chromatography	Chemistry	Spectral data
53	Lysopine N^2-(D-1-carboxyethyl)-lysine	Calf thymus histone	$H_2N(CH_2)_4CHCOOH$ / NH / $H_3CCHCOOH$	$C_9H_{18}N_2O_4$ (204.25)	157–160° (77)	+18 (c 1.4, H_2O) (77)	—	78	78	77	77
54	ε-N-Methyllysine	Calf thymus histone	$CH_3NH(CH_2)_4CH(NH_2)COOH$	$C_7H_{16}N_2O_2$ (160.23)	—	—	—	79	79	—	79
55	Ornithine (α,δ-diaminovaleric acid)	Asplenium nidus	$H_2N(CH_2)_3CH(NH_2)COOH$	$C_5H_{12}N_2O_2$ (132.16)	—	+12.1 (c 2, H_2O) +28.4 (c 2, 5 N HCl) (1)	1.71 8.69 10.76 (290)	60	60	81	81
55a	4-Oxalysine	Streptomyces	$H_2N-(CH_2)_2O(CH_2)_2CH(NH_2)COOH$	$C_5N_1N_2O_3$ (148.19)	—	—	—	81a	81a	—	81a
56	β-N-Oxalyl-α,β-diaminopropionic acid	Lathyrus sativus	$HOOCCONHCH_2CH(NH_2)COOH$	$C_5H_6N_2O_5$ (176.13)	206° (dec) (82)	-36.9^{27} (c 0.66, 4 N HCl) (82)	1.95 2.95 9.25 (82)	82	82	82	82
56a	β-Putreanine [N-(4-aminobutyl)-3-aminopropionic acid]	Bovine brain	$NH_2(CH_2)_4NH(CH_2)_2COOH$	$C_7H_{16}N_2O_2$ (160.25)	250–251° (dec) (82a)	—	3.2 9.4 11.2 (82a)	82a	82a	—	82a

D. L-Diamino, Dicarboxylic Acids

No.	Amino acid (synonym)	Source	Structure	Formula (mol wt)	Melting point °C[a]	$[\alpha]_D$[b]	pK_a	References Isolation and purification	Chromatography	Chemistry	Spectral data
57	Acetylenic dicarboxylic acid diamide	Streptomyces chibaensis	$CCONH_2$ ‖ $CCONH_2$	$C_4H_4N_2O_2$ (112.09)	216–218° (dec)	—	—	83	83	83	83
58	α,ε-Diaminopimelic acid	Pine pollen	$HOOCCH(NH_2)(CH_2)_3CH(NH_2)COOH$	$C_7H_{14}N_2O_4$ (190.20)	—	$+8.1^{25}$ (c 5, H_2O) $+45^{26}$ (c 1, 1 N HCl) $+45.1^{24}$ (c 2.6, 5 N HCl) (290)	1.8 2.2 9.9 8.8 (290)	84	84	85	—

DATA ON THE NATURALLY OCCURRING AMINO ACIDS (continued)

No.	Amino acid (synonym)	Source	Structure	Formula (mol wt)	Melting point °C[a]	$[\alpha]_D$[b]	pK_a	Isolation and purification	Chromatography	Chemistry	Spectral data
									References		
59	2,3-Diaminosuccinic acid	*Streptomyces rimosus*	$HOOCCH(NH_2)CH(NH_2)COOH$	$C_4H_8N_2O_4$ (148.10)	240–290° (dec)	–	–	86	86	86	86
			E. L-Keto, Hydroxy, and Hydroxy Substituted Amino Acids								
60	O-Acetylhomoserine	*Pisum*	$H_3CCO(CH_2)_3CH(NH_2)COOH$ (with $\overset{O}{\overset{\|}{C}}$)	$C_6H_{11}NO_4$ (161.17)	–	–	–	87	87	87	–
60a	Threo-α-amino-β,γ-dihydroxybutyric acid	*Streptomyces*	$HOCH_2CH(OH)CH(NH_2)COOH$	$C_4H_9NO_4$ (135.14)	210° (dec)	-13.3^{25} (c 1, H_2O) -1.1^{25} (c 1, 2.2 N HCl)(87a)	–	87a	1	.	–
61	2-Amino-4,5-dihydroxy pentanoic acid	*Lunaria annua*	$HOCH_2CH(OH)CH_2CH(NH_2)COOH$	$C_5H_{11}NO_4$ (149.15)	–	–	–	88	88	88	–
61a	2-Amino-3-formyl-3-pentenoic acid	*Bankera fuligineoalba*	$H_3CCH{=}CCH(NH_2)COOH$ (with CHO)	$C_6H_9NO_3$ (143.16)	–	–	–	88a	88a	88a	88a
62	α-Amino-γ-hydroxy-adipic acid	*Vibrio comma*	$HOOCCH_2CH(OH)CH_2CH(NH_2)COOH$	$C_6H_{11}NO_5$ (177.17)	–		–	89	–	–	–
63	2-Amino-6-hydroxy-aminohexanoic acid	*Mycobacterium phlei*	$HOCH(NH_2)(CH_2)_3CH(NH_2)COOH$	$C_6H_{14}N_2O_3$ (162.19)	–	$+6.3°$ (c 5, H_2O) $+23.9°$ (c 5.1, 1 N HCl)(90)	–	90	90	90	–
64	α-Amino-γ-hydroxy-butyric acid	*Escherichia coli* mutants	$HOCH_2CH_2CH(NH_2)COOH$	$C_4H_9NO_3$ (119.12)	199° (91)	–	–	91	91	91	–
65	γ-Amino-β-hydroxy-butyric acid	*Escherichia coli* mutants	$CH_3CH(NH_2)CH(OH)COOH$	$C_4H_9NO_3$ (119.12)	–	–	–	–	92	–	–

DATA ON THE NATURALLY OCCURRING AMINO ACIDS (continued)

No.	Amino acid (synonym)	Source	Structure	Formula (mol wt)	Melting point °C[a]	$[\alpha]_D$[b]	pK_a	References Isolation and purification	Chromatography	Chemistry	Spectral data
66	2-Amino-6-hydroxy-4-methyl-4-hexenoic acid	*Aesculus Californica*	HOCH$_2$ H / C=C / H$_3$C, CH$_2$CH(NH$_2$)COOH	C$_7$H$_{13}$NO$_3$ (159.19)	—	$-30^{2.0}$ (c 2.2, H$_2$O) $+2^{2.0}$ (c 1.1, 5 N HCl)(19)	—	19	19	19	19, 21
67	2-Amino-3-hydroxy-methyl-3-pentenoic acid	*Bankera fuligineoalba*	CH$_2$OH / CH$_3$CH=CCH(NH$_2$)COOH	C$_6$H$_{11}$NO$_3$ (145.17)	160–161° (13)	$+182^{2.5}$ (c 0.8, H$_2$O) $+201^{2.5}$ (c 0.8, 0.3 N HCl (13)	—	13	13	13	13
68	α-Amino-γ-hydroxy-pimelic acid	*Asplenium septentrionale*	HOOC(CH$_2$)$_2$CH(OH)CH$_2$CH(NH$_2$)COOH	C$_7$H$_{13}$NO$_5$ (191.19)	—	—	—	96	96	—	—
69	α-Amino-δ-hydroxy-valeric acid	*Canavalia ensiformis*	HOCH$_2$(CH$_2$)$_2$CH(NH$_2$)COOH	C$_5$H$_{11}$NO$_3$ (133.15)	216° (dec) (97)	$+6^{2.5}$ (c 2.65, H$_2$O) $+2.4^{2.5}_{365}$ (c 2.65, H$_2$O) (97)	—	97	97	97	97
70	α-Amino-β-ketobutyric acid	Mikramycin A	O‖ CH$_3$CCH(NH$_2$)COOH	C$_4$H$_7$NO$_3$ (117.11)	—	—	—	98	—	—	—
71	α-Amino-β-methyl-γ,δ-dihydroxyisocaproic acid	Phalloidin	COOH HCNH$_2$ HCCH$_3$ HOCCH$_3$ CH$_2$OH	C$_7$H$_{15}$NO$_4$ (177.21)	208–210° (99)	—	—	99	99	99	99
71a	2-Amino-5-methyl-6-hydroxyhex-4-enoic acid	*Blighia unijugata*	HOH$_2$C, H$_3$C / C=CHCH$_2$CH(NH$_2$)COOH	C$_7$H$_{13}$NO$_3$ (159.21)	—	—	—	99a	99a	—	99a

DATA ON THE NATURALLY OCCURRING AMINO ACIDS (continued)

No.	Amino acid (synonym)	Source	Structure	Formula (mol wt)	Melting point °C[a]	$[\alpha]_D$[b]	pK_a	References			
								Isolation and purification	Chromatography	Chemistry	Spectral data
72	O-Butylhomoserine	Soil bacterium	$CH_3(CH_2)_3OCH_2CH_2CH(NH_2)COOH$	$C_8H_{17}NO_3$ (175.23)	267° (100)	—	—	100	100	100	100
72a	Dihydrorhizobitoxine [O-(2-amino-3-hydroxypropyl)-homoserine]	*Rhizobium japonicum*	$HOCH_2CH(NH_2)CH_2-O-(CH_2)_2CH(NH_2)COOH$	$C_7H_{16}N_2O_4$ (192.25)	—	—	7.2 8.6 (100b)	100a	—	—	100
73	β,γ-Dihydroxyglutamic acid	*Rheum rhaponticum*	$HOOCCH(OH)CH(OH)CH(NH_2)COOH$	$C_5H_9NO_6$ (179.13)	—	—	—	101	101	101	—
74	β,γ-Dihydroxyisoleucine	Thiostrepton	CH_3 (branch) $CH_3CH_2C(OH)CH(OH)CH(NH_2)COOH$	$C_6H_{15}NO_4$ (165.20)	—	—	—	102	—	—	—
75	γ,δ-Dihydroxyleucine	Phalloin	CH_3 (branch) $HOCH_2C(OH)CH_2CH(NH_2)COOH$	$C_6H_{13}NO_4$ (163.18)	—	—	—	103	—	103	103
75a	δ,ε-Dihydroxynorleucine	Bovine tendon	$HOOC-CH(NH_2)(CH_2)_2CH(OH)CH_2OH$	$C_6H_{13}NO_4$ (163.20)	—	—	—	103a	—	103a	—
76	O-Ethylhomoserine	Soil bacterium	$CH_3CH_2OCH_2CH_2CH(NH_2)COOH$	$C_6H_{13}NO_3$ (147.18)	262° (100)	$-14.3°$ (c 2.5, H₂O) (100)	—	100	100	100	100
76a	β-Guanido-γ-hydroxyvaline	Viomycin	$\begin{array}{l} NH \\ \| \\ H_2NCNHCH_2 \\ \quad CHCH(NH_2)COOH \\ HOH_2C \end{array}$	$C_6H_{14}N_4O_3$ (190.24)	182°(dec) (104a)	—	—	—	—	—	104a
77	Homoserine (α-amino-γ-hydroxybutyric acid)	*Pisum sativum*	$HOCH_2CH_2CH(NH_2)COOH$	$C_4H_9NO_3$ (119.12)	—	-8.8^{25} (c 1–2, H₂O) $+18.3^{26}$ (c 2, 2 N HCl)(290)	2.71 9.62 (290)	105	105	—	—

DATA ON THE NATURALLY OCCURRING AMINO ACIDS (continued)

No.	Amino acid (synonym)	Source	Structure	Formula (mol wt)	Melting point °C[a]	$[\alpha]_D$[b]	pK_a	References			
								Isolation and purification	Chromatography	Chemistry	Spectral data
78	α-Hydroxyalanine	Peptides of ergot	OH | $CH_3C(NH_2)COOH$	$C_3H_7NO_3$ (105.10)	—	—	—	106	—	—	—
79	α-Hydroxy-γ-amino-butyric acid	*E. coli* mutants	$H_2NCH_2CH_2CH(OH)COOH$	$C_4H_9NO_3$ (119.12)	—	—	—	91	91	—	—
80	β-Hydroxy-γ-amino-butyric acid	Mammalian brain	$H_2NCH_2CH(OH)CH_2COOH$	$C_4H_9NO_3$ (119.12)	—	—	—	—	108	—	—
81	α-Hydroxy-ε-amino-caproic acid	*Neurospora crassa*	$N_2NCH_2(CH_2)_3CH(OH)COOH$	$C_6H_{13}NO_3$ (147.18)	—	—	—	109	109	—	—
82	β-Hydroxyasparagine	Human urine	$H_2NCOCH(OH)CH(NH_2)COOH$	$C_4H_8O_4N_2$ (148.12)	238–240° (dec) (110)	—	2.09 8.29 (20)	110	110	110	110
83	β-Hydroxyaspartic acid	*Azotobacter*	$HOOCCH(OH)CH(NH_2)COOH$	$C_4H_7NO_5$ (149.10)	—	+41.4 (c 2.42, H_2O) +53.0 (c 2.46, 1 N HCl) (290)[d]	1.91 3.51 9.11 (20)	111	111	—	—
83a	N-(2-Hydroxyethyl)-alanine	Rumen protozoa	$HO(CH_2)_2NHCH(CH_3)COOH$	$C_5H_{11}NO_3$ (133.17)	—	—	—	111a	—	—	111a
84	N⁴-(2-Hydroxyethyl)-asparagine	*Bryonia dioica*	$HO(CH_2)_2NHCOCH_2CH(NH_2)COOH$	$C_6H_{12}N_2O_4$ (176.20)	199–200° (112)	−2.9° (c 5, H_2O) (112)	—	112	112	112	112
85	N⁵-(2-Hydroxyethyl)-glutamine	*Lunaria annua*	$HO(CH_2)_2NHCO(CH_2)_2CH(NH_2)COOH$	$C_7H_{14}N_2O_4$ (190.23)	—	+5.8° (c 1.8, H_2O) (88)	—	88	88	88	—
86	β-Hydroxyglutamic acid	*Mycobacterium tuberculosis*	$HOOCCH_2CH(OH)CH(NH_2)COOH$	$C_5H_9NO_5$ (163.13)	187° (dec) (290)	+8.6° (H_2O) +30.8²° (c 2, 20% HCl) (290)[d]	—	114	114	—	—

DATA ON THE NATURALLY OCCURRING AMINO ACIDS (continued)

No.	Amino acid (synonym)	Source	Structure	Formula (mol wt)	Melting point °C[a]	$[\alpha]_D$[b]	pK$_a$	Isolation and purification	Chromatography	Chemistry	Spectral data
									References		
87	γ-Hydroxyglutamic acid	*Linaria vulgaris*	HOOCCH(OH)CH$_2$CH(NH$_2$)COOH	C$_5$H$_9$NO$_5$ (163.13)	—	—	—	115	115	115	—
88	γ-Hydroxyglutamine	*Phlox decusata*	H$_2$NCOCH(OH)CH$_2$CH(NH$_2$)COOH	C$_5$H$_{10}$N$_2$O$_4$ (162.15)	163—164° (dec) (116)	—	—	116	116	116	—
89	ε-Hydroxylamino-norleucine (α-amino-ε-hydroxy-aminohexanoic acid)	*Mycobacterium phlei*	HONHCH$_2$(CH$_2$)$_3$CH(NH$_2$)COOH	C$_5$H$_{12}$N$_2$O$_3$ (148.17)	223—225° (dec) (117)	+6.3²° (c 5, H$_2$O) +23.9¹⁸ (c 5.1, 1 *N* HCl) (117)	—	117	117	117	—
89a	4-Hydroxyiso-leucine	*Trigonella foenumgraecum*	CH$_3$CHOH 　　＼ 　　CHCH(NH$_2$)COOH 　　／ 　CH$_3$	C$_6$H$_{13}$NO$_3$ (147.20)	—	+31²° (c 1, H$_2$O) (117a)	—	117a	117a	—	117a
90	δ-Hydroxy-γ-keto-norvaline	*Streptomyces akiyoshiensis* novo	HOCH$_2$COCH$_2$CH(NH$_2$)COOH	C$_5$H$_9$NO$_4$ (147.13)	—	−8.2¹⁷ (c 3.4, H$_2$O) (118)	2.0 9.1	118	118	118	118
91	δ-Hydroxyleucenine (δ-ketoleucine)	Phalloidin	CH$_3$ 　｜ HOCH$_2$C=CHCH(NH$_2$)COOH	C$_6$H$_{11}$NO$_3$ (145.17)	—	—	—	119	119	119	119
92	β-Hydroxyleucine	Antibiotic from *Paecilomyces* strain	(CH$_3$)$_2$CHCH(OH)CH(NH$_2$)COOH	C$_6$H$_{13}$NO$_3$ (147.17)	—	—	—	16	16	16	—
93	δ-Hydroxyleucine	*Paecilomyces*	HOH$_2$C 　　＼ 　　CHCH$_2$CH(NH$_2$)COOH 　　／ 　H$_3$C	C$_6$H$_{13}$NO$_3$ (147.17)	—	—	—	121	121	121	121
94	Threo-β-hydroxy-leucine	*Deutzia gracilis*	HOOCCH(NH$_2$)CH(OH)CH(CH$_3$)CH$_3$	C$_6$H$_{13}$NO$_3$ (147.17)	—	—	—	122	122	122	122

DATA ON THE NATURALLY OCCURRING AMINO ACIDS (continued)

No.	Amino acid (synonym)	Source	Structure	Formula (mol wt)	Melting point °C[a]	$[\alpha]_D$[b]	pKa	Isolation and purification	Chromatography	Chemistry	Spectral data
95	α-Hydroxylysine (α,ε-diamino-α-hydroxycaproic acid)	*Salvia officinalis*	OH $H_2N(CH_2)_4C(NH_2)COOH$	$C_6H_{14}N_2O_3$ (162.19)	—	—	2.13 8.62 9.67 (20)	123	123	—	—
96	δ-Hydroxylysine (α,ε-diamino-δ-hydroxycaproic acid)	Fish gelatin	$H_2NCH_2CH(OH)(CH_2)_2CH(NH_2)COOH$	$C_6H_{14}N_2O_3$ (162.19)	—	—	—	124	111	—	—
96a	β-Hydroxynorvaline	*Streptomyces* species	$CH_3CH_2CH(OH)CH(NH_2)COOH$	$C_5H_{11}NO_3$ 133.17	244° (dec) (124a)	—	—	124a	124a	—	124a
97	γ-Hydroxynorvaline	*Lathyrus odoratus*	$CH_3CH(OH)CH_2CH(NH_2)COOH$	$C_5H_{11}NO_3$ (133.15)	—	+22° (c 5, H₂O) +32 (c 2.5, gl acetic) (126)	—	126	126	126	—
98	γ-Hydroxyornithine	*Vicia sativa*	$H_2NCH_2CH(OH)CH_2CH(NH_2)COOH$	$C_5H_{12}N_2O_3$ (148.16)	—	—	—	127	127	—	—
99	α-Hydroxyvaline	Ergot	OH $(CH_3)_2CHC(NH_2)COOH$	$C_5H_{11}NO_3$ (133.15)	—	—	2.55 9.77 (20)	106	—	—	—
100	γ-Hydroxyvaline	*Kalanchoe daigremontiana*	H_3C HOH_2C >CHCH(NH₂)COOH	$C_5H_{11}NO_3$ (133.15)	228° (dec) (129)	+10° (H₂O) (129)	—	129	129	—	—
100a	Hypusine	Bovine brain	$H_2N(CH_2)_2CH(OH)CH_2NH(CH_2)_4CH(NH_2)COOH$	$C_{10}H_{23}N_3O_3$ (233.27)	234—238° (dec) (129a)	—	—	129a	—	129a	129a
100b	Isoserine	*Bacillus brevis* (edeine A and B)	$H_2NCH_2CH(OH)COOH$	$C_3H_7NO_3$ (105.11)	—	—	—	129b	129b	129b	—
101	4-Ketonorleucine (2-amino-4-ketohexanoic acid)	*Citrobacter freundii*	$HOOCCH(NH_2)CH_2COCH_2CH_3$	$C_6H_{11}NO_3$ (145.17)	142—143° (130)	—	—	131	131	130	130

DATA ON THE NATURALLY OCCURRING AMINO ACIDS (continued)

No.	Amino acid (synonym)	Source	Structure	Formula (mol wt)	Melting point °Ca	$[\alpha]_D$ b	pK_2	References Isolation and purification	References Chromatography	References Chemistry	References Spectral data
102	γ-Methyl-γ-hydroxy glutamic acid	*Phyllitis scolopendrium*	CH_3 $HOOCC(OH)CH_2CH(NH_2)COOH$	$C_6H_{11}NO_5$ (157.17)	—	—	—	56	56	—	—
103	Pantonine (α-amino-β,β-dimethyl-γ-hydroxybutyric acid)	*Escherichia coli*	$HOCH_2C(CH_3)_2CH(NH_2)COOH$	$C_6H_{13}NO_3$ (147.18)	—	—	—	132	132	132	—
103a	*Threo*-β-phenylserine	*Canthium eurysides*	$HO-CH-CH(NH_2)COOH$ (phenyl)	$C_9H_{11}NO_3$ (181.21)	—	—	—	—	—	—	252a
103b	Pinnatanine [N^5-(2-hydroxymethylbutadienyl)-*allo*-γ-hydroxyglutamine]	*Staphylea pinnata*	HOH_2C $H_2C=HC$ $C=CHNH-COCH(OH)CH_2CH(NH_2)COOH$	$C_{10}H_{16}N_2O_5$ (244.28)	165° (dec) (132a)	$+3.2^{27}$ (c 0.5, H_2O) (132a)	9.1 (132a)	132a	132a	—	132a
104	O-Propylhomoserine	Soil bacterium	$CH_3(CH_2)_2O(CH_2)_2CH(NH_2)COOH$	$C_7H_{15}NO_3$ (161.21)	265° (100)	-11^{30} (c2, H_2O) (100)	—	100	100	100	100
104a	Rhizobitoxine [2-amino-4-(2-amino-3-hydroxypropoxy)-but-3-enoic acid]	*Rhizobium japonicum*	$HOH_2CCH(NH_2)CH_2OCH=CHCH_2(NH_2)COOH$	$C_7H_{14}N_2O_4$ (191.24)	—	—	—	132b	—	—	132c
105	Serine (α-amino-β-hydroxypropionic acid)	Silk fibroin	$HOCH_2CH(NH_2)COOH$	$C_3H_7NO_3$ (105.09)	228° (dec) (1)	-7.5^{25} (c2, H_2O) $+15.1^{25}$ (c 2, 5N HCl) (290)	2.19 9.21 (290)	134	3	134	134
106	O-Succinylhomoserine	*Escherichia coli*	$HOOC(CH_2)_2CO(CH_2)_2CH(NH_2)COOH$	$C_9H_{15}NO_6$ (219.20)	180—181° (135)	—	4.4 9.5 (135)	135	135	135	—

DATA ON THE NATURALLY OCCURRING AMINO ACIDS (continued)

No.	Amino acid (synonym)	Source	Structure	Formula (mol wt)	Melting point °C[a]	$[\alpha]_D$[b]	pK_a	Isolation and purification	Chromatography	Chemistry	Spectral data
107	Tabtoxinine (α,ε-diamino-β-hydroxypimelic acid)	*Pseudomonas tabaci*	$HOOCCH(NH_2)(CH_2)_2CH(OH)CH(NH_2)COOH$	$C_7H_{14}N_2O_5$ (186.20)	—	—	—	136	137	137	—
108	Threonine (α-amino-β-hydroxybutyric acid)	Fibrin hydrolyzate	$CH_3CH(OH)CH(NH_2)COOH$	$C_4H_9NO_3$ (119.12)	253° (1)	-28^{25} (c 1–2, H_2O) -15^{25} (c 1–2, 5 N HCl) (290)	2.09 9.10 (290)	138	3	138, 139	139

F. L-Aromatic Amino Acids

No.	Amino acid (synonym)	Source	Structure	Formula (mol wt)	Melting point °C[a]	$[\alpha]_D$[b]	pK_a	Isolation and purification	Chromatography	Chemistry	Spectral data
109	α-Amino-β-phenylbutyric acid	*Streptomyces bottropensis*	$-CH(CH_3)CH(NH_2)COOH$	$C_{10}H_{13}NO_2$ (187)	176–177° (140)	—	—	140	140	140	140
109a	β-Amino-β-phenylpropionic acid	*Roccella canariensis* hydrolyzate	$-CH(NH_2)CH_2COOH$	$C_9H_{11}NO_2$ (165.21)	—	—	—	—	—	—	140a
110	3-Carboxy-4-hydroxyphenylalanine (m-carboxytyrosine)	*Reseda odorata*	$-CH_2CH(NH_2)COOH$	$C_{10}H_{11}NO_5$ (165.15)	—	-7.7^{25} (c 0.9, 1 N NaOH) -29.9^{24} (c 0.6, 0.2 M PO_4, pH 7) (297)	2 3.4 9.3 12–13 (297)	141	141	141	141
111	m-Carboxyphenylalanine	Iris bulbs	$-CH_2CH(NH_2)COOH$	$C_{10}H_{11}NO_4$ (191.26)	—	—	1.5 3.9 (142)	142	142	142	142
111a	2,3-Dihydroxy-N-benzoylserine	*Escherichia coli*	$-CONHCH(CH_2OH)COOH$	$C_{10}H_{11}NO_6$ (241.22)	193–194° (142a)	—	—	142a	142a	—	142a

DATA ON THE NATURALLY OCCURRING AMINO ACIDS (continued)

No.	Amino acid (synonym)	Source	Structure	Formula (mol wt)	Melting point °C[a]	$[\alpha]_D$[b]	pK_a	References			
								Isolation and purification	Chromatography	Chemistry	Spectral data
112	2,4-Dihydroxy-6-methylphenylalanine	*Agrostemma githago*	OH, CH₂CH(NH₂)COOH, CH₃, HO (ring)	$C_{10}H_{13}NO_4$ (211.23)	252° (144)	+19.7[a] (1 N HCl) (144)	–	144	144	144	144
113	3,4-Dihydroxy-phenylalanine (DOPA)	*Vicia faba*	OH, CH₂CH(NH₂)COOH, HO (ring)	$C_9H_{11}NO_4$ (209.21)	–	–14.3 (1 N HCl) (146)	2.32[e] 8.68[e] 9.88[e]	145	1	146	–
114	3,5-Dihydroxy-phenylglycine	*Euphorbia helioscopia*	OH, CH(NH₂)COOH, HO, OH (ring)	$C_8H_9NO_4$ (183.16)	230–232° (147)	–	–	147	147	147	147
114a	γ-Glutaminyl-4-hydroxybenzene	*Agaricus bisporus*	NHCO(CH₂)₂CH(NH₂)COOH, HO (ring)	$C_{11}H_{14}N_2O_4$ (238.27)	225–226° (147a)	+42.5[55] (c 0.67, 0.1 N NaOH) (147a)	–	147a	–	–	147a
114b	3-Hydroxymethyl-phenylalanine	*Caesalpinice tinctoria*	HOH₂C, CH₂CH(NH₂)COOH (ring)	$C_{10}H_{13}NO_3$ (195.24)	–	–	–	147b	147b	–	147b
114c	4-Hydroxy-3-hydroxymethyl-phenylalanine	*Caesalpinia tinctoria*	HOH₂C, CH₂ — CH(NH₂)COOH, HO (ring)	$C_{10}H_{13}NO_4$ (211.24)	–	–36° (c 1, H₂O) –4° (c 0.67, 1 N NaOH) (147b)	–	147b	147b	–	147b
115	3-Hydroxy-kynurenine	Human urine	COCH₂CH(NH₂)COOH, NH₂, OH (ring)	$C_{10}H_{12}N_2O_4$ (224.21)	–	–	–	148	148	–	–

DATA ON THE NATURALLY OCCURRING AMINO ACIDS (continued)

No.	Amino acid (synonym)	Source	Structure	Formula (mol wt)	Melting point °C[a]	$[\alpha]_D$[b]	pK_a	Isolation and purification	Chromatography	Chemistry	Spectral data
115a	p-Hydroxymethyl-phenylalanine	*Escherichia coli*	HOH$_2$C–⟨C$_6$H$_4$⟩–CH$_2$CH(NH$_2$)COOH	C$_{10}$H$_{13}$NO$_3$ (195.24)	231—233° (dec) (148a)	-32.5^{20} (148a)	—	—	148b	148b	148a
116	m-Hydroxyphenyl-glycine	*Euphorbia helioscopia*	⟨C$_6$H$_4$(OH)⟩–CH(NH$_2$)COOH	C$_8$H$_9$NO$_3$ (177.16)	212—214° (147)	—	—	147	147	147	147
117	Kynurenine (β-arthraniloyl-α-aminopropionic acid)	Rabbit urine	⟨C$_6$H$_4$(NH$_2$)⟩–COCH$_2$CH(NH$_2$)COOH	C$_{10}$H$_{12}$N$_2$O$_3$ (208.21)	191° (dec) (290)	-30.5^{25} (c 1, H$_2$O) (290)	—	148	—	148	148
118	O-Methyltyrosine (β-(p-methoxy-phenyl)-alanine)	Puromycin	H$_3$CO–⟨C$_6$H$_4$⟩–CH$_2$CH(NH$_2$)COOH	C$_{10}$H$_{13}$NO$_3$ (195.22)	191° (150)	$-5.9,_{46}$ (HCl) (150) $-3.2,_{46}$ (1 N NaOH) (150)	—	149	—	150	—
119	Phenylalanine	*Lupinus luteus*	⟨C$_6$H$_5$⟩–CH$_2$CH(NH$_2$)COOH	C$_9$H$_{11}$NO$_2$ (165.19)	284° (1)	-34.5^{25} (c 1–2, H$_2$O) (290) -4.5^{25} (c 1–2, 5 N HCl) (290)	2.16 9.18 (290)	151	3	152	152
120	Tyrosine (α-amino-β-hydroxyphenyl propionic acid)	Casein, alkaline hydrolyzate	HO–⟨C$_6$H$_4$⟩–CH$_2$CH(NH$_2$)COOH	C$_9$H$_{11}$NO$_3$ (181.19)	344° (1)	-10^{25} (c 2, 5 N HCl)(1)	2.20 9.11 (1) 10.13 (290)	153	3	154	154
120a	β-Tyrosine	*Bacillus brevis* (edeine A and B)	HO–⟨C$_6$H$_4$⟩–CH(NH$_2$)CH$_2$COOH	C$_9$H$_{11}$NO$_3$ (181.21)	—	—	—	129b	129b	129b	—

DATA ON THE NATURALLY OCCURRING AMINO ACIDS (continued)

No.	Amino acid (synonym)	Source	Structure	Formula (mol wt)	Melting point °C[a]	$[\alpha]_D$[b]	pK$_a$	Isolation and purification	Chromatography	Chemistry	Spectral data
									References		
121	m-Tyrosine	*Euphorbia myrsinites* L	(OH-phenyl)CH$_2$CH(NH$_2$)COOH	C$_9$H$_{11}$NO$_3$ (181.19)	272–274° (155)	-14.5^{22} (70% EtOH) $+8.9$(70% EtOH, 2 N HCl) (155)	—	155	155	155	155
121a	N-Acetylarginine	Cattle brain	H$_2$N—C—NH(CH$_2$)$_3$CH(NHCOCH$_3$)COOH ‖ NH	C$_8$H$_{16}$N$_4$O$_3$ (204.27)	270° (155a)	—	—	155a	155a	155a	155a

G. L-Ureido and Guanido Amino Acids

No.	Amino acid (synonym)	Source	Structure	Formula (mol wt)	Melting point °C[a]	$[\alpha]_D$[b]	pK$_a$	Isolation and purification	Chromatography	Chemistry	Spectral data
122	Albizziine (2-amino-3-ureido-propionic acid)	*Mimosaceae*	HOOCCH(NH$_2$)CH$_2$NHCONH$_2$	C$_4$H$_9$NO$_3$ (119.12)	—	-66^{24} (c 2, H$_2$O) (157)	—	156	157	157	157
123	Arginine (amino-δ-guanidino-valeric acid)	*Lupinus luteus*	H$_2$NCNH(CH$_2$)$_3$CH(NH$_2$)COOH ‖ NH	C$_6$H$_{14}$N$_4$O$_2$	238° (1)	$+12.5^{25}$ (c 2, H$_2$O) $+27.6^{25}$ (c 2, 5 N HCl)(1)	1.82 8.99 12.48 (290)	158	3	159	159
124	Canavanine (α-amino-(O-guanidyl)-γ-hydroxybutyric acid)	*Canavalia ensiformis*	H$_2$NCNHOCH$_2$CH$_2$CH(NH$_2$)COOH ‖ NH	C$_5$H$_{12}$N$_4$O$_3$ (176.19)	172° (160)	$+18.6^{18.5}$ (c 7.8, H$_2$O) (160)	2.50 6.60 9.25 (33)	161	160	160	162
125	Canavanosuccinic acid	*Canavalia ensiformis*	COOH | HOOCCH$_2$CHNHCNHOCH$_2$CH$_2$CH(NH$_2$)COOH ‖ NH	C$_9$H$_{16}$N$_4$O$_7$ (292.27)	—	—	—	163	163	—	—

DATA ON THE NATURALLY OCCURRING AMINO ACIDS (continued)

No.	Amino acid (synonym)	Source	Structure	Formula (mol wt)	Melting point °C[a]	$[\alpha]_D$[b]	pK_a	References Isolation and purification	Chromatography	Chemistry	Spectral data
126	Citrulline	Watermelon	$H_2NCONH(CH_2)_3CH(NH_2)COOH$	$C_6H_{13}N_3O_3$	202° (164)	$+4^{35}$ (c 2, H₂O) $+24.2^{25}$ (c 2, N HCl) $+10.8$ (c 1.1, 0.1 N NaOH) (290)	2.43 9.41 (290)	164	165	164	—
127	Desaminocanavanine	Canavanine	$HN{=}CNHO(CH_2)_2CHCOOH$ NH	$C_5H_9N_3O_3$	256–257° (166)	$+26.61^{21}$ (H₂O) (166)	—	166	—	166	—
127a	N^G,N^G-Dimethyl-arginine	Bovine brain	H_3C NH ‖ $NCNH(CH_2)_3CH(NH_2)COOH$ H_3C	$C_8H_{18}N_4O_2$ (202.30)	198–201° (70a)	—	—	166a	166a	—	166a
127b	N^G,N'^G-Dimethyl-arginine	Bovine brain	NCH₃ ‖ $H_3C{-}NH{-}CNH(CH_2)_3CH(NH_2)COOH$	$C_8H_{18}N_4O_2$ (202.30)	237–239° (dec) (70a)	—	—	166a	166a	70a	166a
128	Gigartinine [α-amino-δ-(guanylureido)-valeric acid]	*Gymnogongrus flabelliformis*	NH ‖ $H_2NCNHCONH(CH_2)_3CH(NH_2)COOH$	$C_7H_{15}N_5O_3$ (217.25)	—	—	—	167	167	167	167
129	Homoarginine	*Lathyrus* species	NH ‖ $H_2NCNH(CH_2)_4CH(NH_2)COOH$	$C_7H_{16}N_4O_2$ (188.25)	—	$+42.4$ (c 0.452, 1.02 N HCl)[f] (168)	—	168	168	168	168
130	Homocitrulline	Human urine	$H_2NCONH(CH_2)_4CH(NH_2)COOH$	$C_7H_{14}N_3O_3$ (189.22)	—	—	—	169	169	—	—
131	γ-Hydroxyarginine	*Vicia sativa*	NH ‖ $H_2NCNHCH_2CH(OH)CH_2CH(NH_2)COOH$	$C_6H_{14}N_4O_3$ (190.20)	—	—	—	170	170	170	170

DATA ON THE NATURALLY OCCURRING AMINO ACIDS (continued)

No.	Amino acid (synonym)	Source	Structure	Formula (mol wt)	Melting point °C[a]	$[\alpha]_D{}^b$	pK$_a$	Isolation and purification	Chromatography	Chemistry	Spectral data
									References		
131a	N^5-Hydroxyarginine	*Bacillus species*	$\overset{\overset{\text{NH}}{\parallel}}{}$ H$_2$NCNOH(CH$_2$)$_3$CH(NH$_2$)COOH	C$_6$H$_{14}$N$_4$O$_3$ (190.24)	206–212° (dec) (170a)	+21^{25} (c 1, 5 N HCl) (170a)	–	170a	170a	170a	170a
132	γ-Hydroxyhomoarginine (α-amino-ε-guanidino-γ-hydroxy-hexanoic acid)	*Lathyrus tingitanus*	$\overset{\overset{\text{NH}}{\parallel}}{}$ H$_2$NCNH(CH$_2$)$_2$CHOHCH$_2$CH(NH$_2$)COOH	C$_7$H$_{16}$N$_4$O$_3$ (204.23)	–	–	–	171	171	171	171
133	Indospicine (α-amino-ε-amidino caproic acid)	*Endigofera spicata*	$\overset{\overset{\text{NH}}{\parallel}}{}$ H$_2$NC(CH$_2$)$_4$CH(NH$_2$)COOH	C$_7$H$_{15}$N$_3$O$_2$ (173.23)	131–134° (173)	+18^{22} (c 1.1, 5 N HCl)(173)	–	173	173	173	–
133a	ω-N-Methylarginine (guanidinomethyl-arginine)	Bovine brain	$\overset{\overset{\text{NH}}{\parallel}}{}$ H$_3$CNHCNH(CH$_2$)$_3$CH(NH$_2$)COOH	C$_7$H$_{16}$N$_4$O$_2$ (188.27)	–	–	–	166a	166a	–	166a

H. L-Amino Acids Containing Other Nitrogenous Groups

No.	Amino acid (synonym)	Source	Structure	Formula (mol wt)	Melting point °C[a]	$[\alpha]_D{}^b$	pK$_a$	Isolation and purification	Chromatography	Chemistry	Spectral data
134	Alanosine [2-amino-3-(N-nitrosohydroxy-amino) propionic acid]	*Streptomyces alanosinicus*	$\overset{\overset{\text{NO}}{\parallel}}{}$ HONCH$_2$CH(NH$_2$)COOH	C$_3$H$_7$N$_3$O$_4$ (149.11)	–	+8 (1 N HCl) –46 (0.1 N NaOH) (174)	(174)	174	–	174	174
135	Azaserine (O-diazoacetylserine)	*Streptomyces*	$\overset{\overset{\text{O}}{\parallel}}{}$ N$_2$CHCOCH$_2$CH(NH$_2$)COOH	C$_5$H$_7$N$_3$O$_4$ (173.14)	146–162° (175)	–0.5$^{27.5}$ (c 8.46, H$_2$O) (175)	8.55 (34)	175	–	175	175
136	β-Cyanoalanine	*Vicia sativa*	N≡CCH$_2$CH(NH$_2$)COOH	C$_4$H$_6$N$_2$O$_2$ (114.11)	214.5° (176)	–2.9^{26} (c 1.4, 1 N acetic acid) (177)	1.7 7.4 (177)	176	176	177	177

DATA ON THE NATURALLY OCCURRING AMINO ACIDS (continued)

No.	Amino acid (synonym)	Source	Structure	Formula (mol wt)	Melting point °C[a]	$[\alpha]_D$[b]	pK_a	References Isolation and purification	References Chromatography	References Chemistry	References Spectral data
136a	γ-Cyano-α-amino-butyric acid	*Chromobacterium violaceum*	$N{\equiv}(CH_2)_2CH(NH_2)COOH$	$C_5H_8N_2O_2$ (128.15)	221–223° (177a)	+32.1[21] (c 0.38, 1 N HOAc) (177a)	—	177a	177a	177a	177a
137	ε-Diazo-δ-ketonorleucine	*Streptomyces*	$N_2CHCO(CH_2)_2\ CH(NH_2)COOH$	$C_6H_9N_3O_3$ (171.17)	145–155° (178)	21[26] (c 5.4, H₂O) (178)	—	178	178	178	178
138	Hadacidin (*N*-formyl-*N*-hydroxyaminoacetic acid)	*Penicillium frequentans*	CH₂COOH | NOH | CHO	$C_3H_5NO_4$ (119.08)	205–210° (179)	—	—	179	—	179	179

I. L-Heterocyclic Amino Acids

No.	Amino acid (synonym)	Source	Structure	Formula (mol wt)	Melting point °C[a]	$[\alpha]_D$[b]	pK_a	References Isolation and purification	References Chromatography	References Chemistry	References Spectral data
138a	2-Alanyl-3-isoxazolin-5-one	Pea seedlings	(structure)	$C_6H_8N_2O_4$ (172.16)	203–205° (dec) (180b)	—	—	180a	180b	180b	180b
139	Allohydroxyproline	*Santalum album*	(structure)	$C_5H_9NO_3$ (131.13)	248° (180)	−57[25] (c 0.65, H₂O) (180)	—	180	180	180	—
140	Allokainic acid (3-carboxymethyl-4-isopropenyl proline)	*Digenea simplex*	(structure)	$C_{10}H_{15}NO_4$ (213.24)	—	+8[26] (H₂O) (184, 185)	—	184, 185	—	184, 185	184, 185

DATA ON THE NATURALLY OCCURRING AMINO ACIDS (continued)

No.	Amino acid (synonym)	Source	Structure	Formula (mol wt)	Melting point °C[a]	$[\alpha]_D$[b]	pK$_a$	Isolation and purification	Chromatography	Chemistry	Spectral data
									References		
140a	1-Amino-2-nitro-cyclopentane carboxylic acid	*Aspergillus wentii*		$C_6H_{10}N_2O_4$ (174.18)	150° (dec) (185a)	–	–	185a	185a	185a	185a
141	4-Aminopipecolic acid	*Strophanthus scandens*		$C_6H_{11}N_2O_2$ (143.18)	–	–	–	186	186	–	–
141a	*cis*-3-Aminoproline	*Morchella esculenta*		$C_5H_{10}N_2O_2$ (130.17)	215° (dec) (186a)	+5.8²⁰ (c 2, H₂O) +23.0²⁰ (c 2, 5 N HCl) (186a)	–	186a	186a	–	186a
141b	Anticapsin	*Streptomyces griseoplanus*		$C_9H_{13}NO_4$ (199.23)	240° (dec) (186b)	+125²⁵ (c 1, H₂O) (186b)	4.3 10.1 (186b)	186b	186b	–	186b
142	Ascorbigen	Cabbage		$C_{15}H_{15}NO_6$ (305.30)	–	–	–	187	188	189	–

DATA ON THE NATURALLY OCCURRING AMINO ACIDS (continued)

No.	Amino acid (synonym)	Source	Structure	Formula (mol wt)	Melting point °C[a]	$[\alpha]_D^b$	pK$_a$	Isolation and purification	Chromatography	Chemistry	Spectral data
143	Azetidine-2-carboxylic acid	*Convallaria majalis*		$C_4H_7NO_2$ (101.11)	—	—	—	190	190	190	190
143a	Aziridinomycin (3-methyl-2-hydro-azirine carboxylic acid)	*Streptomyces aureus*		$C_4H_5NO_2$ (99.10)	—	—	—	189a	189a	—	189a
144	Baikiain 1,2,3,6-tetrahydropyridine-α-carboxylic acid	*Baikiaea plurijuga*		$C_6H_9NO_2$ (127.15)	273–274° (183)	—	—	181	182	183	183
144a	N-Carbamoyl-2-(p-hydroxyphenyl)glycine	*Vicia faba*		$C_9H_{10}N_2O_4$ (210.21)	194–195° (dec) (190a)	—	—	190a	—	—	190a
144b	3-Carboxy-6,7-dihydroxy-1,2,3,4-tetrahydroisoquinoline	*Mucuna Mutisiana*		$C_{10}H_{11}NO_4$ (209.22)	286–288° (190b)	-114.9^{25} (c 1.65, 20% HCl) (190b)	—	190b	—	190b	190b
144c	Clavicipitic acid	*Claviceps* (ergot fungus)		$C_{16}H_{18}N_2O_2$ (270.36)	262° (dec) (190c)	—	—	190c	—	—	190c

DATA ON THE NATURALLY OCCURRING AMINO ACIDS (continued)

No.	Amino acid (synonym)	Source	Structure	Formula (mol wt)	Melting point °C[a]	$[\alpha]_D$[b]	pK_a	Isolation and purification	Chromatography	Chemistry	Spectral data
									References		
145	Cucurbitine (3-amino-3-carboxy-pyrrolidine)	*Cucurbita moschata*		$C_5H_{10}NO_2$ (116.14)	—	-19.76^{27} (c 9.3, H_2O) (191)	—	191	191	191	191
145a	N-Dihydrojas-monoylisoleucine	*Gibberella fujikuroi*		$C_{18}H_{31}NO_4$ (325.50)	140—141° (191a)	—	—	—	191a	191a	191a
145b	2,5-Dihydro-phenylalanine (1,4-cyclohexa-diene-1-alanine)	*Streptomycete* X-13, 185		$C_9H_{13}NO_2$ (167.23)	206—208° (191b)	-33.7^{25} (c 1, 5 N HCl) (191b)	—	191b	—	—	191b
145c	2-N,6-N-Di-(2,3-dihydroxy-benzoyl)lysine	*Azobacter vinelandii*		$C_{20}H_{22}N_2O_8$ (418.44)	—	—	4.8 9 (191c)	191c	191c	—	191c

DATA ON THE NATURALLY OCCURRING AMINO ACIDS (continued)

No.	Amino acid (synonym)	Source	Structure	Formula (mol wt)	Melting point °C[a]	$[\alpha]_D^b$	pK_a	References — Isolation and purification	Chromatography	Chemistry	Spectral data
145d	cis-3,4-trans-3,4-Dihydroxyproline	Diatom cell walls	(structure)	$C_5H_9NO_4$ (147.15)	262° (dec) (191d)	-61.2^{20} (c 0.5, H_2O) (191d)	—	191d	191d	—	191e
145e	β-(2,6-Dihydroxy-pyrimidin-1-yl)alanine	Pea seedlings	(structure)	$C_7H_9N_3O_4$ (199.19)	230° (dec) (191f)	—	—	191f	—	191f	191f
145f	4,6-Dihydroxy-quinoline-2-carboxylic acid	Tobacco leaves	(structure)	$C_{10}H_7NO_4$ (205.18)	287° (dec) (191g)	—	—	191g	191g	191g	191g
146	Dihydrozanthurenic acid (8-hydroxy-1,2,3,4-tetrahydro-4-ketoquinaldic acid)	*Lepidoptera*	(structure)	$C_{10}H_9NO_4$ (207.19)	185—190° (192)	$-45°$ (c 0.9, MeOH) $+18^{20}$ (c 0.9, MeOH-HCl) (192)	—	192	192	—	192
147	Domoic acid [2-carboxy-3-carboxymethyl-4-1-methyl-2-carboxy-1,3-hexadienyl)-pyrrolidine]	*Chondria armata*	(structure)	$C_{15}H_{21}NO_6$ (311.35)	217° (193)	-109.6^{12} (c 1.314, H_2O) (193)	2.20 3.72 4.93 9.82 (193)	193	193	193	193

DATA ON THE NATURALLY OCCURRING AMINO ACIDS (continued)

No.	Amino acid (synonym)	Source	Structure	Formula (mol wt)	Melting point °C[a]	$[\alpha]_D$ [b]	pK$_a$	Isolation and purification	Chromatography	Chemistry	Spectral data
									References		
148	Echinine (2-*tert*-pentenyl-5,7-diisopentenyl-tryptophan)	*Aspergillus glaucus*		$C_{26}H_{36}N_2O_2$ (408.50)	169–172° (194)	—	—	194	—	194	194
148a	Enduracididine [α-amino-β-(2-iminoimidazolidinyl)-propionic acid]	Enduracidin hydrolyzate		$C_6H_{12}N_4O_2$ (172.22)	—	+63.3^{32} (1 *M* HCl) +57.6^{32} (1 *M* NaOH) (194a)	2.5 8.3 12 (94a)	—	194a	194	194a
148a'	Furanomycin [α-amino-(2,5-dihydro-5-methyl)furan-2-acetic acid]	*Streptomyces* L-803		$C_7H_{11}NO_3$ (157.19)	220–223° (dec) (194a')	+136.1^{27} (c 1, H$_2$O) (194a')	2.4 9.1 (194a')	194a'	194a'	194a'	194a'
148a''	Furosine [ε-*N*-(2-furoyl-methyl)-lypine]	Heated milk	$COCH_2\ NH(CH_2)_4\ CH(NH_2)\ COOH$	$C_{12}H_{18}N_2O_4$ (254.32)	—	—	—	194a''	194b	—	194a''
148b	γ-Glutaminyl-3,4-benzoquinone	*Agaricus bisporus*	$NHCO(CH_2)_2\ CH(NH_2)COOH$	$C_{11}H_{12}N_2O_5$ (252.25)	—	—	—	194b	194b	—	194b
149	Guvacine	*Areca cathecu*		$C_6H_9NO_2$ (127.15)	—	—	—	195, 196	—	—	—

DATA ON THE NATURALLY OCCURRING AMINO ACIDS (continued)

No.	Amino acid (synonym)	Source	Structure	Formula (mol wt)	Melting point °C[a]	$[\alpha]_D$[b]	pK$_a$	Isolation and purification	Chromatography	Chemistry	Spectral data
150	Histidine	Protamine from sturgeon sperm		C$_6$H$_9$N$_3$O$_2$ (155.16)	277° (1)	−38.5^{35} (H$_2$O) +11.8 (5 N HCl)(1)	1.82 6.00 9.17 (1)	197	3	198	198
150a	β-Hydroxyhistidine	Bleomycin A$_2$ (antibiotic)		C$_6$H$_9$N$_3$O$_3$ (171.18)	205° (dec) (198a)	+40^8 (c 1, H$_2$O) (198a)	<2.0 198a 5.5 8.8 (198a)	198a	—	198a	—
151	4-Hydroxy-4 methyl-proline	Apples		C$_6$H$_{11}$NO$_2$ (145.16)	—	—	—	199	199	—	200
152	Hydroxyminaline	*Penicillium aspergillus*		C$_5$H$_9$NO$_3$ (127.10)	—	—	—	201	—	201	201
153	4-Hydroxypipecolic acid	*Acacia pentadenia*		C$_6$H$_{11}$NO$_3$ (145.16)	250—270° (202)	−12.5^{21} (H$_2$O) +0.34^{21} (1 N HCl) −18.5^{21} (1 N NaOH) (202)	—	202	202	202	203
154	5-Hydroxypipecolic acid	*Rhapis excelsa*		C$_6$H$_{11}$NO$_3$ (145.16)	—	—	—	204	204	—	203
154a	5-Hydroxy-piperidazine-3-carboxylic acid	Monamycin hydrolyzate		C$_5$H$_{11}$N$_2$O$_3$ (146.17)	201—202° (DNP derv.) (204a)	+21.4^{24} (c 0.39, H$_2$O) (204b)	—	204b	204b	204b	204a, 204b

DATA ON THE NATURALLY OCCURRING AMINO ACIDS (continued)

No.	Amino acid (synonym)	Source	Structure	Formula (mol wt)	Melting point °C[a]	$[\alpha]_D$[b]	pK_a	References — Isolation and purification	Chromatography	Chemistry	Spectral data
155	3-Hydroxyproline (3-hydroxypyrrolidine-2-carboxylic acid)	Telomycin	(structure)	$C_5H_9NO_3$ (131.13)	225–235° (206)	-17.4^{20} (c 1, H_2O) $+13.3^{20}$ (c 0.5, 1 N HCl) (206a)	–	206	206	206	206a
156	4-Hydroxyproline (4-hydroxypyrrolidine-2-carboxylic acid)	Gelatin hydrolyzate	(structure)	$C_5H_9NO_3$ (131.13)	273–4° (290)	-76.0^{25} (c 2, H_2O) -50.5^{15} (c 2, 5 N HCl) (290)	1.82 9.66 (290)	207	3	208	208
157	2-Hydroxytryptophan	Phalloidin	$CH_2CH(NH_2)COOH$ (structure)	$C_{11}H_{12}N_2O_3$ (220.22)	257° (210)	$+40.8$ (c 4.2, 1 N NaOH) (210)	–	209	–	210	–
158	5-Hydroxytryptophan	Chromobacterium violaceum	$CH_2CH(NH_2)COOH$ (structure)	$C_{11}H_{12}N_2O_3$ (220.22)	273° (dec) (290)	-32.5^{22} (c 1, H_2O) $+16^{22}$ (c 1, 4 N HCl) (290)	–	211	211	–	–
159	Ibotenic acid (α-amino-3-keto-4-isoxazoline-5-acetic acid)	Amanita strobiliformis	$HC=CCH(NH_2)COOH$ (structure)	$C_5H_6N_2O_4$ (114.10)	177–178° (212)	$\pm0^c$ (212)	5.1 8.2 (212)	212	–	212	212
160	4-Imidazoleacetic acid	Polyporus sulfureus	$CHCH_2COOH$ (structure)	$C_5H_7N_2O_2$ (125.11)	–	–	2.96 7.35 (20)	213	–	–	–
161	N-(Indole-3-acetyl)-aspartic acid	Magnolia	$CH_2CONHCH$ (structure)	$C_{14}H_{14}N_2O_5$ (290.29)	–	–	–	214	214	–	–

DATA ON THE NATURALLY OCCURRING AMINO ACIDS (continued)

No.	Amino acid (synonym)	Source	Structure	Formula (mol wt)	Melting point °C[a]	$[\alpha]_D$[b]	pK_a	Isolation and purification	Chromatography	Chemistry	Spectral data
162	Indole-3-acetyl-ε-lysine	*Pseudomonas savastanoi*	$CH_2CONH(CH_2)_4CH(NH_2)COOH$ (indole)	$C_{16}H_{21}N_3O_3$ (303.36)	259–261° (215)	$+22^3$ (2 N HCl) (215)	—	215	215	215	215
162a	N-Jasmonoyliso-leucine	*Gibberella fujikuroi*	$CONHCH(COOH)CH(CH_3)(C_2H_5)$	$C_{18}H_{29}NO_4$ (323.48)	147–149° (191a)	—	—	—	191a	191a	191a
163	Kainic Acid (3-carboxymethyl-4-isopropenyl proline)	*Digenea simplex*	$H_2C{=}C{-}CH{-}CHCH_2COOH$ / H_3C H_2C $CHCOOH$ N H	$C_{10}H_{15}NO_4$ (213.25)	251° (216)	-15^{17} (H₂O) (185, 216)	—	216	216	—	—
164	4-Keto-5-methyl-proline	Actinomycin	$O{=}C{-}CH_2$ / H_3CHC $CHCOOH$ N H	$C_6H_9NO_3$ (143.18)	215° (217)	—	—	217	217	217	217
165	4-Ketopipecolic acid	Staphylomycin	H_2C CH_2 $CHCOOH$ N H (with C=O)	$C_6H_9NO_3$ (143.15)	—	—	—	218	—	—	—
166	4-Ketoproline	Actinomycin	$O{=}C{-}CH_2$ / H_2C $CHCOOH$ N H	$C_5H_5N_5O_2$ (181.00)	—	—	—	219	—	219	—
167	Lathytine (tingitanine)	*Lathyrus tingitanus*	H_2NC $CHCH_2CH(NH_2)COOH$ (ring)	$C_7H_{10}N_4O_2$ (182.00)	215° (220)	-55.9^{21} (H₂O) (220)	2.4 4.1 9.0 (220)	220	—	—	220
167a	exo-3,4-Methano-proline	*Aesculus parviflora*	CH_2 / HC—CH / H_2C—N—CHCOOH H	$C_6H_9NO_2$ (127.16)	—	-132^{20} (c 2, H₂O) -104^{20} (c 1, 5 N HCl) (23a)	—	23a	23a	—	23a

DATA ON THE NATURALLY OCCURRING AMINO ACIDS (continued)

No.	Amino acid (synonym)	Source	Structure	Formula (mol wt)	Melting point °C[a]	$[\alpha]_D$[b]	pK$_a$	Isolation and purification	Chromatography	Chemistry	Spectral data
168	4-Methyleneproline	*Eriobotrya japonica*		$C_6H_9NO_2$ (127.15)	225° (221)	±0[c]	—	221	221	222	—
168a	1-Methyl-6-hydroxy-1,2,3,4-tetrahydroisoquinoline-3-carboxylic acid	*Euphorbia myrisinites*		$C_{11}H_{13}NO_3$ (207.25)	—	—	—	222a	222a	—	222a
169	4-Methylproline	Apples		$C_6H_{11}NO_2$ (129.16)	232–234° (224)	−52 (c 0.3, H$_2$O) (224)	—	223	223	224	224
170	N-Methylproline (hygric acid)	Apples		$C_6H_{11}NO_2$	—	—	—	223	223	223	—
170a	N-Methylstreptolidine	Streptothricin		$C_7H_{14}N_4O_3$ (202.25)	—	—	—	224a	224a	—	—
171	β-Methyltryptophan	Telomycin		$C_{12}H_{14}N_2O_2$ (218.25)	—	—	—	226	—	226	226
172	Mimosine	*Mimosa pudica*		$C_8H_{10}N_2O_4$ (198.18)	228–229° 227	-21^{22} (H$_2$O) $+10^{22}$ (1% HCl)(227)	—	227a	—	227	227

DATA ON THE NATURALLY OCCURRING AMINO ACIDS (continued)

No.	Amino acid (synonym)	Source	Structure	Formula (mol wt)	Melting point °C[a]	$[\alpha]_D$[b]	pK$_a$	Isolation and purification	Chromatography	Chemistry	Spectral data
									References		
173	Minaline (pyrrole-2-carboxylic acid)	Diastase		$C_5H_5NO_2$ (111.10)	180° (228)	—	—	228	—	228	228
174	Muscazone	*Amanita muscaria*		$C_5H_6N_2O_5$ (174.12)	190° (229)	—	—	229	—	230	231
174a	Nicotianamine [1-(3'-(γ-amino-α-γ-dicarboxypropyl-amino)-propyl-azetidine-2-carboxylic acid]	Tobacco leaves		$C_{12}H_{21}N_3O_6$ (303.36)	240° (dec) (231a)	-60.5^{33} (c 2.7, H$_2$O) (231a)	—	231a	—	231a	231a
175	β-3-Oxindolyl-alanine	Phalloidin		$C_{11}H_{12}N_2O_3$ (220.24)	249–253° (209)	$+39.2^{0}$ (1 N NaOH) (209)	—	209	—	209	—
176	Pipecolic acid	Apples		$C_6H_{11}NO_2$ (129.17)	260° (dec) (290)	-25.4^{18} (c 5, H$_2$O) -13.3^{25} (c 2.5 N HCl)(290)	—	233	233	233	—
176a	Piperidazine-3-carboxylic acid	Monamycin hydrolyzate		$C_5H_{10}N_2O_2$ (130.17)	153–155° (DNP derv.) (204b)	$+307^{25}$ (c 0.18, CH$_3$OH) (204b)	—	—	—	—	204a

DATA ON THE NATURALLY OCCURRING AMINO ACIDS (continued)

No.	Amino acid (synonym)	Source	Structure	Formula (mol wt)	Melting point °C[a]	$[\alpha]_D$[b]	pK_a	References Isolation and purification	References Chromatography	References Chemistry	References Spectral data
177	Proline (pyrolidine-2-carboxylic acid)	Casein hydrolyzate	H_2C—CH_2 / H_2C $CHCOOH$ / N–H	$C_5H_9NO_2$ (115.13)	222° (1)	-86.2^{25} (c 1–2, H_2O) -60.4 (c 1–2, 5 N HCl) (1)	1.95 10.64 (290)	234	3	235	235
178	β-Pyrazol-1-ylalanine	*Citrullus vulgaris*	$HC{=}CH$ / $HC{=}N$—$NCH_2CH(NH_2)COOH$	$C_6H_9N_3O_2$ (155.17)	236–238° (236)	-7.3^{0} (c 3.4, H_2O) (236)	2.2 (236)	236	236	236	236
178a	Pyridosine [ε-(1,4-dihydro-6-methyl-3-hydroxy-4-oxo-1-pyridyl)-lysine]	Heated milk	OH—N$[-NH(CH_2)_4CH(NH_2)COOH]$ ring, CH_3, O	$C_{12}H_{19}N_3O_4$ (269.34)	—	—	—	236a	—	—	236a
178b	Pyrrolidone carboxylic acid (5-oxoproline)	Human plasma	H_2C—CH_2 / C $CHCOOH$ / O N–H	$C_5H_7NO_3$ (129.13)	—	—	—	236b	236b	—	—
179	Roseanine	Roseothricin	CH_2NH_2 / $—CHC(OH)COOH$ / $N{=}CH_2$ / $H_2N{\cdot}N$	$C_6H_{12}N_4O_3$ (188.21)	—	$+56.8^{22}$ (c 2.35, H_2O)[c] (238)	—	237	238	238	—
179a	Stendomycidine	Stendomycin (from *Streptomyces*)	CH_2 / HN $CHCH(NH_2)COOH$ / $CH_3N{=}C$—N—CH_3	$C_8H_{16}N_4O_2$ (200.28)	—	—	—	238a	238a	238a	238a, 238b

DATA ON THE NATURALLY OCCURRING AMINO ACIDS (continued)

No.	Amino acid (synonym)	Source	Structure	Formula (mol wt)	Melting point °C[a]	$[\alpha]_D^b$	pK_a	Isolation and purification	Chromatography	Chemistry	Spectral data
180	Stizolobic acid [β-(3-carboxypyran-5-yl)alanine]	*Stizolobium hassjoo*	HOOCCH(NH₂)CH₂C	$C_9H_9NO_6$ (227.18)	231—233° (239)	—	—	239	239	239	239
181	Stizolobinic acid [β-(6-carboxy-α-pyran-3-yl)alanine]	*Stizolobium hassjoo*	HOOCCH(NH₂)CH₂	$C_9H_9NO_6$ (207.18)	—	—	—	239	239	240	240
181a	Streptolidine	Streptothricin	HOOCHC—CHCH(OH)CH₂NH₂	$C_6H_{12}N_4O_3$ (188.22)	215° (dec) (224a)	+55.3²⁵ (c 1.01, H₂O) (224a)	—	224a	224a	224a	224a
182	Tricholomic acid (α-amino-3-keto-5-isoxazolidine acetic acid)	*Tricholoma muscarium*	H₂C—CHCH(NH₂)COOH	$C_5H_8N_2O_4$ (160.13)	207° (242)	—	6.0 8.6 (242)	242	—	242	242
183	Tryptophan (α-amino-β-3-indolepropionic acid)	Casein	CH₂CH(NH₂)COOH	$C_{11}H_{12}N_2O_2$ (204.24)	282° (1)	−33.7²⁵ (c 1—2, H₂O) +2.8²⁵ (c 1—2, 1 N HCl) (1)	2.43 9.44 (290)	243	3	244	244
183a	Tuberactidine (2-imino-4-hydroxyhexahydro-6-pyrimidinyl glycine)	*Streptomyces griseoverticillatus*	CH(NH₂)COOH	$C_6H_{12}N_4O_3$ (188.22)	182° (dec) (244a)	−25.8¹⁵ (c 0.5, H₂O) (244a)	—	244a	—	—	244a

DATA ON THE NATURALLY OCCURRING AMINO ACIDS (continued)

No.	Amino acid (synonym)	Source	Structure	Formula (mol wt)	Melting point °C[a]	$[\alpha]_D$[b]	pK$_a$	References Isolation and purification	Chromatography	Chemistry	Spectral data
184	Viomycidine (guanidine-1-pyrroline-5-carboxylic acid)	Viomycin	H_2C–C–NHC(NH_2)=NH / H_2C–C / N / CH–COOH	$C_6H_{14}N_4O_2$ (170.19)	181-182° (dec) (246a)	$-151^{3\cdot2}$ (c 1.25, H_2O) $-38^{2\cdot2}$ (c 0.8, HCl) (246a)	1.3 5.5 12.6 (246)	76, 246a	246, 246a	246, 246a	246a
185	Willardine [3-(1-uracyl)-alanine]	*Mimosa*	HOC=N–C=O / HC=CH–N–$CH_2CH(NH_2)COOH$	$C_7H_9N_3O_4$ (199.18)	—	-12.1^{22} (c 1.2, 1 N HCl) (248)	—	247	247	247, 248	248

J. L-N-Substituted Amino Acids

No.	Amino acid (synonym)	Source	Structure	Formula (mol wt)	Melting point °C[a]	$[\alpha]_D$[b]	pK$_a$	References Isolation and purification	Chromatography	Chemistry	Spectral data
186	Abrine (N-methyl-tryptophan)	*Abrus precatorius*	indole–$CH_2CHCOOH$ / $NHCH_3$	$C_{12}H_{14}N_2O_2$ (218.41)	297° (dec) (249)	—	—	250	—	249	—
186a	N-Acetylalanine	Human brain	CH_3–$CH(NHCOCH_3)COOH$	$C_5H_9NO_3$ (131.15)	—	—	—	250a	250a	250a	—
186a'	N-Acetylglutamic acid	Mammalian liver	$HOOC$–$(CH_2)_2$–$CH(NHCOCH_3)COOH$	$C_7H_{11}NO_5$ (189.19)	—	—	—	250a'	250a'	—	—
186b	N-Acetylaspartic acid	Extract of cat brain	$COOH$–CH_2–$CH(NHCOCH_3)COOH$	$C_6H_9NO_5$ (175.16)	123–125° (250b)	-11.8^{27} (2% methyl cellosolve) (250b)	—	250b	250b	—	250b
187	2,3-Dihydro-3,3-dimethyl-1 H-pyrrolo[1,2-α] indole-9-alanine		indole–$CH_2CH(NH_2)COOH$ / $(CH_3)_2CCH$=CH_2	$C_{16}H_{20}N_2O_2$ (272.34)	170–175° (dec) (251)	—	—	251	251	251	251
188	β,N-Dimethylleucine	Etamycin	$(CH_3)_2CHCH(CH_3)CHCOOH$ / $NHCH_3$	$C_8H_{17}NO_2$ (159.23)	315–316° (dec) (252)	+33.15 (c 2, H_2O) $+39.2^{2\cdot9}$ (c 2.2, 5 N HCl)(252)	—	252	252	252	—

DATA ON THE NATURALLY OCCURRING AMINO ACIDS (continued)

No.	Amino acid (synonym)	Source	Structure	Melting point °C[a]	$[\alpha]_D$[b]	pKa	Isolation and purification	Chromatography	Chemistry	Spectral data
								References		
188a	N,N-Dimethyl-phenylalanine	*Canthium eurysides*	$C_{11}H_{15}NO_2$ (193.27)	—	—	—	—	—	—	252a
189	γ-Formyl-N-methyl-norvaline	Ilamycin	$C_7H_{13}NO_3$ (159.19)	—	—	1.8 10.2 (253)	253	—	253	253
190	Fusarinine [δ-N-(cis-5-hydroxy-3-methylpent-2-enoyl)-δ-N-hydroxy-ornithine]	*Fusarium*	$C_{11}H_{20}N_2O_4$ (244.30)	—	—	—	254	254	254	254
191	Homarine (N-methylpicolinic acid)	*Arenicola marina*		—	—	—	255	255	—	—
192	4-Hydroxy-N-methylproline	*Afrormosia elata* heartwood	$C_6H_{11}NO_3$ (145.16)	238–240° (dec) (256)	−86.6 (c 1.5, H₂O) (256)	—	256	—	256	256
193	Merodesmosine	Elastin	$C_{18}H_{34}N_4O_6$ (420)	—	—	—	257	257	257	257
194	N-Methylalanine	Dichapetalum cymosum (Gifblaar)	$C_4H_9NO_2$ (103.12)	—	—	—	258	258	258	258

DATA ON THE NATURALLY OCCURRING AMINO ACIDS (continued)

No.	Amino acid (synonym)	Source	Structure	Formula (mol wt)	Melting point °C[a]	$[\alpha]_D$[b]	pK_a	References — Isolation and purification	References — Chromatography	References — Chemistry	References — Spectral data
195	1-Methylhistidine	Cat urine	HC=C—CH$_2$CH(NH$_2$)COOH / N—C—N—CH$_3$, H	C$_7$H$_{11}$N$_3$O$_2$ (169.18)	245–247° (259)	−25.81[8] (c 3.9, H$_2$O) (290)	1.69 6.48 (imidazole) 8.85 (290)	259	259	—	—
196	3-Methylhistidine	Human urine	C=C—CH$_2$CH(NH$_2$)COOH / H$_3$C—N—N, C—N, H	C$_7$H$_{11}$N$_3$O$_2$ (169.18)	—	−26.5[26] (c 2.1, H$_2$O) (290) +13.5[27] (c 1.9, 1 N HCl)(260)	—	260	260	260	260
197	N-Methylisoleucine	Enniatin A	H$_3$CCH$_2$ CHCHCOOH / H$_3$C NHCH$_3$	C$_7$H$_{15}$NO$_2$ (145.21)	—	+28.6[22] (c 1.034, H$_2$O) +44.8[22] (c 1.162, 5 N HCl)(261)	—	261	261	261	—
198	N-Methylleucine	Enniatin A	CH$_3$CH(CH$_3$)CH$_2$CHCOOH NHCH$_3$	C$_7$H$_{15}$NO$_2$ (145.21)	—	+21.4[15] (c 0.77, H$_2$O) +31.3[15] (c 0.86, 5 N HCl)(262)	—	262	262	262	—
199	N-Methyl-β-methyl-leucine	Etamycin	(CH$_3$)$_2$CHCH(CH$_3$)CHCOOH NHCH$_3$	C$_8$H$_{17}$NO$_2$ (159.23)	—	+26[30] (c 1.8, H$_2$O) +33.2[30] (c 1.9, 5 N HCl)(252)	—	252	252	252	—
200	N-Methyl-O-methyl-serine	*Myobacterium butyricum*	CH$_3$OCH$_2$CHCOOH NHCH$_3$	C$_5$H$_{11}$NO$_3$ (133.14)	203–205° (dec) (264)	—	—	264	264	264	264
201	N-Methylphenyl-glycine (α-phenylsarcosine)	Etamycin	C$_6$H$_5$—CHCOOH NHCH$_3$	C$_9$H$_{11}$NO$_2$ (166.21)	—	+118[31] (c 4.8, 1 N, HCl)(252)	—	252	252	252	—

DATA ON THE NATURALLY OCCURRING AMINO ACIDS (continued)

No.	Amino acid (synonym)	Source	Structure	Formula (mol wt)	Melting point °C[a]	$[\alpha]_D$[b]	pK_a	Isolation and purification	Chromatography	Chemistry	Spectral data
									References		
201a	N-Methylthreonine	Stendomycin hydrolyzate	$CH_3CH(OH)CH(NHCH_3)COOH$	$C_5H_{11}NO_3$ (133.17)	—	-17^{25} (c 2, 5 N HCl) (264a)	—	—	—	—	264a
202	N-Methyltyrosine (surinamine)	*Andira*	HO—⟨C₆H₄⟩—CHCOOH / NHCH₃	$C_{10}H_{13}NO_3$ (195.22)	280–300° (267)	-18.6 (267)	—	266	—	267	—
203	N-Methylvaline	Actinomycin	$(CH_3)_2CHCHCOOH$ / NHCH₃	$C_6H_{13}NO_2$ (131.18)	—	$+17.5$ (H₂O) $+30.9$ (5 N HCl) (269)	—	268	268	269	—
204	Saccharopine (N^6-(2-glutaryl)-lysine)	*Saccharomyces*	$HOOCCH(NH_2)(CH_2)_4NHCHCOOH$ / CH_2CH_2COOH	$C_{11}H_{20}N_2O_6$ (276.30)	—	$+33.6^{23}$ (c 1, 0.5 N HCl) (271)	2.6 4.1 9.2 10.3 (270)	270	270	271	271
205	Sarcosine (N-methylglycine)	*Cladonia silvatica*	CH_3NHCH_2COOH	$C_3H_7NO_2$ (89.10)	210 (dec) (290)	—	2.21 10.20 (290)	272	272	—	—

K. L-Sulfur and Selenium Containing-Amino Acids

No.	Amino acid (synonym)	Source	Structure	Formula (mol wt)	Melting point °C[a]	$[\alpha]_D$[b]	pK_a	Isolation and purification	Chromatography	Chemistry	Spectral data
206	N-Acetyldjenkolic acid	*Acacia farnesiana*	CH₃ / C=O / NH — $HOOCCHCH_2SCH_2SCH_2CH(NH_2)COOH$	$C_9H_{16}N_2O_5S$ (264.32)	170° (273)	-49.0^{25} (c 2, 1% HCl) -60.2^{25} (c i, 1 N HCl)(273)	—	273	273	273	273
207	Alliin	*Allium sativum*	O↑ / $CH_2=CHCH_2SCH_2CH(NH_2)COOH$	$C_6H_{11}NO_3S$ (177.24)	163–165° (274)	$+62.8^{21}$ (H₂O) (274)	—	274	—	356	—
208	S-Allylcysteine	*Allium sativum*	$CH_2=CHCH_2SCH_2CH(NH_2)COOH$	$C_6H_{11}NO_2S$ (161.24)	218° (275)	-8.7^{20} (275)	—	275	275	275	—

DATA ON THE NATURALLY OCCURRING AMINO ACIDS (continued)

No.	Amino acid (synonym)	Source	Structure	Formula (mol wt)	Melting point °C[a]	$[\alpha]_D$[b]	pK_a	Isolation and purification	Chromatography	Chemistry	Spectral data
									References		
209	S-Allylmercapto-cysteine	*Allium sativum*	$CH_2=CHCH_2SSCH_2CH(NH_2)COOH$	$C_6H_{11}NO_2S_2$ (193.31)	188° (276)	-95.3 ±10^{3.5} (c 0.19, 6 N HCl)(276)	—	276	276	276	276
210	α-Amino-β-(2-amino-2-carboxyethyl-mercaptobutyric acid)	Subtilin	$HOOCCH(NH_2)CHCH_3$ — S — $CH_2CH(NH_2)COOH$	$C_7H_{14}N_2O_4S$ (222.28)	—	-34.7^{24} (c 5.4, 1 N HCl)(277)	—	277	277	277	277
211	Cystathionine S-(2-amino-2-carboxyethyl-homocysteine)	Human brain	$HOOCCH(NH_2)CH_2CH_2SCH_2CH_2(NH_2)COOH$	$C_7H_{14}N_2O_4S$ (222.28)	301° (279) 312° (290)	+26.4^{25} (c 0.8, 1 N HCl)(279)	—	279	279	279	279
212	3-Amino-(3-carboxy-propyldimethyl-sulfonium)	Cabbage	$(CH_3)_2S^+CH_2CH_2CH(NH_2)COOH$	$C_6H_{14}NO_2S$ (164.26)	—	—	—	280	280	280	—
212a	α-Amino-γ-[2-(4-carboxy)-thiazolyl butyric acid	*Xeromus subtomentosus* (mushroom)	[structure] $HOOC-\!\!\underset{N}{\overset{S}{\diagup}}\!\!-CH_2-CH_2-CH(NH_2)COOH$	$C_8H_{10}N_2O_4S$ (230.26)	237—238° (280a)	—	3.7 (280a)	280a	280a	280a	280a
213	Carbamyltaurine	Cat urine	$H_2NCONHCH_2CH_2SO_3H$	$C_3H_8N_2O_3S$ (152.17)	—	—	—	281	—	—	—
214	2-[S-(β-Carboxy-β-aminoethyl-tryptophan)]	*Amanita phalloides*	[structure] indole $SCH_2CH(NH_2)COOH$, $CH_2CH(NH_2)COOH$	$C_{14}H_{17}N_3O_4S$ (323.40)	—	—	—	—	119	—	—
215	S-(β-Carboxyethyl)-cysteine	*Albizia julibrissin*	$HOOC(CH_2)_2SCH_2CH(NH_2)COOH$	$C_6H_{11}NO_4S$ (193.17)	218° (282)	-9.33^{20} (c 3, 1 N HCl)(283)	—	283	283	282	283
216	N-(1-Carboxyethyl)-taurine	Red algae	$HO_3S(CH_2)_2NHCH(CH_3)COOH$	$C_5H_{11}NO_5S$ (197.22)	258° (283)	-1.15^{13} (c 5, 1 N NaOH)(284)	—	284	284	284	—

DATA ON THE NATURALLY OCCURRING AMINO ACIDS (continued)

No.	Amino acid (synonym)	Source	Structure	Formula (mol wt)	Melting point °C[a]	$[\alpha]_D$[b]	pKa	Isolation and purification	Chromatography	Chemistry	Spectral data
216a	S-(Carboxymethyl)-homocysteine	Human urine	$HOOCCH_2S(CH_2)_2CH(NH_2)COOH$	$C_6H_{11}NO_4S$ (193.24)	223–225° (dec) (284a)	—	—	284a	—	—	284a
217	S-(2-Carboxy-isopropyl)-cysteine	*Acacia*	CH_3CHCH_2COOH ‑ S ‑ $CH_2CH(NH_2)COOH$	$C_7H_{13}NO_4S$ (207.26)	202° (284)	$+6.6^{22}$ (c 2, H₂O) $+31^{25}$ (c 1.94, 1 N NaOH) (285)	—	285	285	285	285
218	S-(2-Carboxypropyl)-cysteine	Onions	CH_3 ‑ CHCHCOOH ‑ S ‑ $CH_2CH(NH_2)COOH$	$C_7H_{13}NO_4S$ (207.26)	191–194° (285) (286)	-50.1^{21} (H₂O) (286)	—	286, 287	286, 287	286, 287	286, 287
219	Chondrine (1,4-thiazane-5-carboxylic acid 1-oxide)	*Chondria crassicaulis*	(structure: thiazane ring with S→O, H_2C, CH_2, $HOOCHC$, N–H)	$C_5H_9NO_3S$ (163.20)	255–257° (287)	$+20.9^{16}$ (c 2, H₂O) (288) $+30.2^{16}$ (c 2, 6 N HCl)(288)	—	288	288	288	—
220	Cycloallin (3-methyl-1,4-thiazane-5-carboxylic acid oxide)	Onions	(structure: thiazane ring with S→O, H_2C, CH_2, H_3CHC, CHCOOH, N–H)	$C_6H_{11}NO_3S$ (177.24)	—	-17.4^{20} (H₂O) (289)	—	289	289	289	—
221	Cysteic acid (β-sulfo-α-aminopropionic acid)	Sheep's fleece	$HO_3SCH_2CH(NH_2)COOH$	$C_3H_7NO_5S$ (169.17)	289° (dec) (290)	$+8.66$ (c 7.4, H₂O) (290)	1.3 (SO₃H) 1.9 8.7 (290)	291	292	—	—
222	Cysteine (α-amino-β-mercaptopropionic acid)	Cystine	$HSCH_2CH(NH_2)COOH$	$C_3H_7NO_2S$ (121.15)	178° (1)	-16.5^{25} (c 2, H₂O) $+6.5^{25}$ (c 2, 5 N HCl) (290)	1.92 8.35 10.46 (SH) (290)	293	3	294	294

References

DATA ON THE NATURALLY OCCURRING AMINO ACIDS (continued)

No.	Amino acid (synonym)	Source	Structure	Formula (mol wt)	Melting point °C[a]	$[\alpha]_D$[b]	pK_a	Isolation and purification	Chromatography	Chemistry	Spectral data
223	Cysteine sulfinic acid (β-sulfinyl-α-aminopropionic acid)	Rat brain	$HO_2SCH_2CH(NH_2)COOH$	$C_3H_7NO_4S$ (153.17)	—	+11 (H_2O) +24 (c 1, 1 N HCl)	ca 2.1 (290)	295	295	—	—
224	Cystine [β,β'-Dithiodi-(α-aminopropionic acid)]	Urinary calculi	CH_2—S—S—CH_2 $\mid$ $\mid$ CHNH$_2$ CHNH$_2$ $\mid$ $\mid$ COOH COOH	$C_6H_{12}N_4O_4S_2$ (240.29)	261° (1)	-232^{25} (c 1, 5 N HCl)(1)	<1 2.1 8.02 8.71 (290)	296	3	294	294
225	Cystine disulfoxide	Rat tissue	O O $\parallel$ $\parallel$ CH_2—S—S—CH_2 $\mid$ $\mid$ CHNH$_2$ CHNH$_2$ $\mid$ $\mid$ COOH COOH	$C_6H_{12}N_2O_6S_2$ (272.33)	—		—	—	298	298	298
226	S-(1,2-Dicarboxyethyl)-cysteine	Bovine lens	COOH $\mid$ $HOOCCH_2CHSCH_2CH(NH_2)COOH$	$C_7H_{11}NO_6S$ (237.25)	—	—	—	299	299	299	
227	Dichrostachinic acid [S-(β-hydroxy-β-carboxyethane-sulfonyl methyl) cysteine]	Mimosa	$HOOCCH(OH)CH_2SO_2CH_2SCH_2CH(NH_2)COOH$	$C_7H_{13}NO_7S$ (255.27)	201° (300)	$+9.2^{24}$ (c 2.2, 1 N HCl) (300)	—	300	300	300	300
228	Dihydroalliin (S-propylcysteine sulfoxide)	Allium cepa	O $\parallel$ $CH_3CH_2CH_2SCH_2CH(NH_2)COOH$	$C_6H_{13}NO_3S$ (179.26)	—	—	—	301	301	301	
229	Djenkolic acid	Djenkol beans	$SCH_2CH(NH_2)COOH$ CH_2 $SCH_2CH(NH_2)COOH$	$C_7H_{14}N_2O_4S_2$ (254.32)	300–350° (303)	$-65^{0.5}$ (c 1, 1 N HCl)(303)	—	302	—	303	
230	Ethionine	Escherichia coli	$CH_3CH_2SCH_2CH_2CH(NH_2)COOH$	$C_6H_{13}NO_2S$ (163.23)	—	—	—	304	304	—	—

DATA ON THE NATURALLY OCCURRING AMINO ACIDS (continued)

No.	Amino acid (synonym)	Source	Structure	Formula (mol wt)	Melting point °C[a]	$[\alpha]_D$[b]	pK_a	References Isolation and purification	Chromatography	Chemistry	Spectral data
231	Felinine	Cat urine	$HOCH_2CH_2C(CH_3)_2SCH_2CH(NH_2)COOH$	$C_8H_{17}NO_3S$ (207.30)	177° (305)	$+23^{20}$ (c 2.2, H_2O) (305)	—	305	305	305	—
232	N-Formyl methionine	*Escherichia coli*	$CH_3SCH_2CH_2CHCOOH$ $\overset{\displaystyle |}{NH}$ $\overset{\displaystyle |}{CHO}$	$C_6H_{11}NO_3$ (177.22)	—	—	—	306	306	—	—
233	Glucobrassicin [S-β-1-(glucopyranosyl)-3-indolylacetothio-hydroximyl-HO-sulfate]	*Brassica* species		$C_{16}H_{20}N_2O_9S_2$ (447.47)	—	-13.3^{23} (c 3, H_2O) (189)	—	189	189	189	189
234	Guanidotaurine	*Arenicola marina*	$\underset{NH}{\overset{\displaystyle \|}{H_2NCNHCH_2CH_2SO_3H}}$	$C_3H_9N_3O_3S$ (167.20)	228–230° (308)	—	—	308	308	308	—
235	Homocysteine	*Neurospora*	$HSCH_2CH_2CH(NH_2)COOH$	$C_4H_9NO_2S$ (135.18)	—	—	2.22 8.87 10.86 (290)	278	310	—	—
236	Homocystine	Human urine	$CH_2-S-S-CH_2$ $\mid \quad\quad\quad \mid$ $CH_2 \quad\quad CH_2$ $\mid \quad\quad\quad \mid$ $CHNH_2 \quad CHNH_2$ $\mid \quad\quad\quad \mid$ $COOH \quad\; COOH$	$C_8H_{16}N_2O_4S_2$ (268.36)	282–3° (dec) (290)	-16^{21} (c 0.06, H_2O) $+78^{25}$ (c 1–2, 5 N HCl) (290)	1.59 2.54 8.52 9.44 (290)	311	311	—	—
237	Homocysteinecysteine disulfide	Human urine	$CH_2-S-S-CH_2$ $\mid \quad\quad\quad \mid$ $CH_2 \quad\quad CH_2$ $\mid \quad\quad\quad \mid$ $CHNH_2 \quad CHNH_2$ $\mid \quad\quad\quad \mid$ $COOH \quad\; COOH$	$C_7H_{16}N_2O_4S_2$ (254.35)	—	-52.2^{25} (1 N HCl) (312)	—	312	—	312	312
238	Homolanthionine	*Escherichia coli*	$HOOCCH(NH_2)(CH_2)_2S(CH_2)_2CH(NH_2)COOH$	$C_9H_{16}N_2O_4S$ (236.29)	—	$+37.3^{24}$ (c 1, 1 N HCl)(363)	—	313	313	313	—

DATA ON THE NATURALLY OCCURRING AMINO ACIDS (continued)

No.	Amino acid (synonym)	Source	Structure	Formula (mol wt)	Melting point °C[a]	$[\alpha]_D$[b]	pKa	References			
								Isolation and purification	Chromatography	Chemistry	Spectral data
239	Homomethionine (5-methylthionor-valine)	Cabbage	$CH_3S(CH_2)_3CH(NH_2)COOH$	$C_6H_{13}NO_2S$ (163.25)	223–225° (314)	$+21^{25.5}$ (c 0.3, 6 N HCl)(314)	–	314	314	314	314
239a	β-6-(4-Hydroxy-benzothiazolyl)-alanine	Chicken feather pigment	[structure]	$C_{10}H_4N_2O_3S$ (237.27)	–	–	–	314a	314a	–	314a
239′	S-(2-Hydroxy-2-carboxyethyl)-homocysteine	Human urine	$COOHCH(OH)CH_2S(CH_2)_2CH(NH_2)COOH$	$C_7H_{13}NO_5S$ (223.27)	–	–	–	284a	–	–	–
239″	S-(2-Hydroxy-2-carboxyethylthio)-homocysteine	Human urine	$HOOCCHCH_2SS(CH_2)_2CH(NH_2)COOH$ (OH)	$C_7H_{13}NO_5S_2$ (223.27)	–	–	–	377	377	377	–
239b	S-(3-Hydroxy-3-carboxyl-n-propyl thio)-cysteine	Human urine	$HOOC—CH(CH_2)_2—SSCH_2CH(NH_2)COOH$ (OH)	$C_8H_{13}NO_5S_2$ (255.33)	176–177° (dec) (377)	-96.3^{26} (c 2.3, 1 N HCl) (377)	–	377	377	377	377
239c	S-(3-Hydroxy-3-carboxy-n-propyl)cysteine	Human urine	$COOHCH(OH)(CH_2)_2SCH_2CH(NH_2)COOH$	$C_7H_{13}NO_5S$ (223.27)	–	–	–	284a	–	–	–
239c′	S-(3-Hydroxy-3-carboxy-n-propyl-thio)-homocysteine	Human urine	$HOOCCH—(CH_2)_2SS(CH_2)_2CH(NH_2)COOH$ (OH)	$C_9H_{15}NO_5S_2$ (269.36)	187–188° (dec) (377)	$+8.1^{25}$ (c 15.2, 1 N HCl) (377)	–	377	377	377	377
239d	α-Hydroxycysteine-cysteine disulfide	Human urine	$CH_2—S—S—CH_2$ / $CHNH_2$ $COHNH_2$ / $COOH$ $COOH$	$C_6H_{11}N_2O_5S_2$ (256.32)	181° (dec) (378)	-263^{23} (c 2.2, H₂O) (378)	–		378	378	–
240	Hypotaurine (2-aminoethane-sulfinic acid)	Rat brain	$HO_2SCH_2CH_2NH_2$	$C_2H_7NO_2S$ (109.14)	–	–	–	295	295	–	–

DATA ON THE NATURALLY OCCURRING AMINO ACIDS (continued)

No.	Amino acid (synonym)	Source	Structure	Formula (mol wt)	Melting point °C[a]	$[\alpha]_D$[b]	pK_a	References			
								Isolation and purification	Chromatography	Chemistry	Spectral data
241	Isovalthine (isopropylcarboxy-methylcysteine)	Human urine	COOH ‖ $(CH_3)_2CHCHSCH_2CH(NH_2)COOH$	$C_8H_{15}NO_4S$ (221.28)	—	—	—	316	317	317	317
242	Lanthionine [β,β'-thiodi-(α-aminopropionic acid)]	Wool	$HOOCCH(NH_2)CH_2SCH_2CH(NH_2)COOH$	$C_6H_{12}N_2O_4S$ (208.23)	270—304° (318)	—	—	318	—	319	—
242a	β-Mercaptolactate-cysteine disulfide	Human urine	CH_2—S—S—CH_2 CHOH CHNH$_2$ COOH COOH	$C_6H_{11}NO_5S_2$ (241.30)	—	—	—	—	319a	319a	—
243	S-Methylcysteine	*Phaseolus vulgaris*	$CH_3SCH_2CH(NH_2)COOH$	$C_4H_9NO_2S$ (135.19)	220° (320)	-26^{25} (c 2.5) (320)	8.75 (20)	320	320	320	320
243a	S-Methylcysteine sulfoxide	Cabbage	H_3C—SCH_2CH(NH$_2$)COOH (with O)	$C_4H_9NO_3S$ (151.20)	173° (320a)	—	—	—	—	—	—
244	Methionine (α-amino-γ-methylthiobutyric acid)	Casein hydrolyzate	$CH_3SCH_2CH_2CH(NH_2)COOH$	$C_5H_{11}NO_2S$ (149.21)	283° (1)	-10^{25} (H$_2$O) $+23.2^{25}$ (5 N HCl) (1)	2.28 9.21 (1)	321	3	322	322
245	3,3'-(2-Methylethylene-1,2-dithio)-dialanine	*Allium shoenoprasum*	S—$CH_2CH(CH_3)$—S CH_2 CH_2 CHNH$_2$ CHNH$_2$ COOH COOH	$C_8H_{17}N_2O_4S_2$ (282.40)	—	—	—	323	323	323	323
246	β-Methyllanthionine	Yeast	$HOOCCH(NH_2)CH_2SCH(CH_3)CH(NH_2)COOH$	$C_7H_{14}N_2O_4S$ (222.28)	—	—	—	277	—	277	—
247	S-Methylmethionine (α-aminodimethyl-γ-butyrothetin)	Asparagus	$(CH_3)_2S^+CH_2CH_2CH(NH_2)COO^-$	$C_6H_{13}NO_2S$ (164.24)	—	—	—	325	325	280	325

DATA ON THE NATURALLY OCCURRING AMINO ACIDS (continued)

No.	Amino acid (synonym)	Source	Structure	Formula (mol wt)	Melting point °C[a]	$[\alpha]_D$[b]	pK$_a$	Isolation and purification	Chromatography	Chemistry	Spectral data
248	β-Methylselenoalanine (β-methylseleno-cysteine)	*Stanleya pinnata*	CH₃SeCH₂CH(NH₂)COOH	C₄H₉NO₂Se (182.08)	—	—	—	327	327	—	—
249	Neoglucobrassicin	*Brassica napus*	(structure)	C₁₁H₂₁N₂O₁₀S₂ (476.48)	175° (328)	—	—	328	328	328	328
250	S-(Prop-1-enyl)-cysteine	*Allium sativum*	CH₃CH=CHSCH₂CH(NH₂)COOH	C₆H₁₁NO₂S (161.24)	—	—	—	329	329	—	—
251	S-(Prop-1-enyl)-cysteine sulfoxide	Onions	CH₃CH=CHSCH₂CH(NH₂)COOH (O↑)	C₆H₁₁NO₃S (177.22)	146–148° (330)	—	—	330	330	330	330
252	S-n-Propylcysteine	*Allium sativum*	CH₃(CH₂)₂SCH₂CH(NH₂)COOH	C₆H₁₃NO₂S (163.26)	—	—	—	331	331	—	—
253	S-n-Propylcysteine sulfoxide	Onions	CH₃CH₂CH₂SCH₂CH(NH₂)COOH (O↑)	C₆H₁₃NO₃S (179.26)	—	—	—	301	301	—	—
254	Selenocystathionine	*Stanleya pinnata*	HOOCCH(NH₂)CH₂CH₂SeCH₂CH(NH₂)COOH	C₇H₁₄N₂O₄Se (269.07)	—	—	—	327	327	327	—
255	Selenocystine	*Astragalus pectinatus*	CH₂–Se–Se–CH₂ / CHNH₂ CHNH₂ / COOH COOH	C₆H₁₂N₂O₄Se₂ (334.11)	263–265° (334)	—	—	334		334	—
255a	Selenomethionine	*Escherichia coli* (hydrolyzate)	H₃CSe(CH₂)₂CH(NH₂)COOH	C₅H₁₁NO₂Se (196.13)	—	—	—	334a	334a	—	—
256	Selenomethyl-selenocysteine	*Astragalus bisulcatus*	SeCH₂CH(NH₂)COOH	C₃H₇NO₂Se (167.03)	—	—	—	335	335	335	—
257	Taurine (2-amino-ethane-sulfonic acid)	Plant and animal tissue	H₂NCH₂CH₂SO₃H	C₂H₇NO₃S (125.15)	320° (dec) (290)	—	−0.3 9.06 (290)	—	292	—	—

DATA ON THE NATURALLY OCCURRING AMINO ACIDS (continued)

No.	Amino acid (synonym)	Source	Structure	Formula (mol wt)	Melting point °C[a]	$[\alpha]_D$[b]	pKa	Isolation and purification	Chromatography	Chemistry	Spectral data
258	β-(2-Thiazole)-β-alanine	*Bottromycin*	HC—N, HC—S, C—CH(NH₂)CH₂COOH	$C_6H_8N_2O_2S$ (172.22)	197.5–201.5° (337)	—	—	337	337	337	337
259	Thiolhistidine	Erythrocytes and microorganisms	HC=C—CH₂CH(NH₂)COOH, N—NH, C, SH	$C_6H_9N_3O_2S$ (173.23)	—	$-10^{2.5}$ (c 2, 1 N HCl) (339)	1.84 8.47 11.4 (290)	338	—	339	—
260	Thiostreptine	Thiostrepton	CH₃, S—CH, H₃CCH(OH)C(OH)CH—C—N—C—CCOOH, NH₂	$C_9H_{14}N_2O_4S$ (246.32)	—	$-4^{2.5}$ (c 1, 1 N acetic acid)(102)	—	102	102	102	102
261	Tyrosine-*O*-sulfate	Human urine	HO₃SO—⟨benzene⟩—CH₂CH(NH₂)COOH	$C_9H_{11}NO_6S$ (261.26)	—		—	341	342	—	—

L. L-Halogen-Containing Amino Acids

No.	Amino acid (synonym)	Source	Structure	Formula (mol wt)	Melting point °C[a]	$[\alpha]_D$[b]	pKa	Isolation and purification	Chromatography	Chemistry	Spectral data
262	2-Amino-4,4-dichlorobutyric acid	*Streptomyces armentosus*	Cl₂CHCH₂CH(NH₂)COOH	$C_4H_7Cl_2NO_2$ (172.02)	—	$+6.7^{7.5}$ (c 0.74, H₂O) $+26.2^{2.5}$ (c 0.74, 1 N HCl) (64)	—	64	64	64	64
262a	5-Chloropiperidazine-3-carboxylic acid	Monamycin hydrolyzate	CH₂, CHCl, CHCOOH, NH, H₂C—N, H	$C_5H_9ClN_2O_2$ (164.61)	83–85° (DNP deriv) (204a)	$+157^{3.3}$ (c 0.18, CHCl₃) (204b)	—	—	204b	—	204a, 204b
263	3,5-Dibromotyrosine	*Gorgona species*	Br, HO—⟨benzene⟩—CH₂CH(NH₂)COOH, Br	$C_9H_9NO_3Br_2$ (339.01)	245° (344)	—	2.17 6.45 (OH) 7.60 (20)	344	—	344	—

DATA ON THE NATURALLY OCCURRING AMINO ACIDS (continued)

No.	Amino acid (synonym)	Source	Structure	Formula (mol wt)	Melting point °C[a]	$[\alpha]_D$[b]	pK$_a$	References			
								Isolation and purification	Chromatography	Chemistry	Spectral data
263a	2,4-Diiodohistidine	Human urine	I—C=C—CH$_2$CH(NH$_2$)COOH (imidazole, HN–C–N, I)	C$_6$H$_7$I$_2$N$_3$O$_2$ (406.96)	—	—	—	344a	344a	—	—
264	3,3′-Diiodothyronine	Bovine thyroid gland	(diiodo diphenyl ether)CH$_2$CH(NH$_2$)COOH	C$_{15}$H$_{13}$I$_2$NO$_4$ (525.11)	233–234° (dec) (345)	—	—	345	345	345	—
265	3,5-Diiodotyrosine (iodogorgoic acid)	Coral protein	HO–(diiodophenyl)–CH$_2$CH(NH$_2$)COOH	C$_9$H$_9$I$_2$NO$_3$ (433.01)	194° (dec) (290)	+2.9 (1.1 N HCl)(1)	2.12 6.48 (OH) 7.82 (290)	346	347	348	—
266	3-Monobromotyrosine	Sea fans and sponges	HO–(bromophenyl)–CH$_2$CH(NH$_2$)COOH	C$_9$H$_{10}$NO$_3$Br (260.10)	—	—	—	349	349	—	—
266a	3-Monobromo-5-monochlorotyrosine	*Buccinum undatum*	HO–(bromochlorophenyl)–CH$_2$CH(NH$_2$)COOH	C$_9$H$_9$ClNO$_3$Br (294.55)	—	—	—	349a	349a	—	349a
266b	5-Monochlorotyrosine	*Buccinum undatum*	HO–(chlorophenyl)–CH$_2$—CH(NH$_2$)COOH	C$_9$H$_{10}$ClNO$_3$ (215.65)	—	—	—	349b	349b	—	349b
267	2-Monoiodohistidine	Rat thyroid gland	HC=C—CH$_2$CH(NH$_2$)COOH (imidazole, HN–C–N, I)	C$_6$H$_8$IN$_3$O$_2$ (281.02)	—	—	—	350	350	—	—

DATA ON THE NATURALLY OCCURRING AMINO ACIDS (continued)

No.	Amino acid (synonym)	Source	Structure	Formula (mol wt)	Melting point °C[a]	$[\alpha]_D$[b]	pKa	References Isolation and purification	References Chromatography	References Chemistry	References Spectral data
268	Monoiodotyrosine	*Nereocystis luetkeana* (an alga)	HO–⟨ring, I⟩–$CH_2CH(NH_2)COOH$	$C_9H_{10}NO_3$ (307.11)	—	—	—	351	351	—	—
269	Thyroxine	Thyroid gland	HO–⟨ring, I,I⟩–O–⟨ring, I,I⟩–$CH_2CH(NH_2)COOH$	$C_{15}H_{11}I_4NO_4$ (776.88)	236° (dec) (290) 95% EtOH) (290)	+15 (c 5, 1 N HCl, (OH) 10.1 (290)	2.2 6.45	352	353	348	348
270	3,5,3′-Triiodo-thyronine	*Phaseolus vulgaris*	HO–⟨ring, I⟩–O–⟨ring, I,I⟩–$CH_2CH(NH_2)COOH$	$C_{15}H_{12}I_3NO_4$ (650.98)	233–234° (dec) (290)	+23.6[24] (c 5, 1 N HCl-EtOH) (290)	2.2 8.40 (OH) 10.1 (290)	355	347	—	—
			M. L-Phosphorus-Containing Amino Acids								
271	α-Amino-β-phosphonopropionic acid	*Tetrahymena pyriformis, Zoanthus sociatus*	$H_2O_3PCH_2CH(NH_2)COOH$	$C_3H_8NO_5P$ (157.07)	228° (dec) (357)	—	2.2 4.5 8.8 11.0 (357)	358	358	357	—
272	Ciliatine (2-aminoethyl phosphonic acid)	Sea anemone	$H_2NCH_2CH_2PO_3H_2$	$C_2H_8NO_3P$ (125.07)	280–281° (dec) (359)	—	6.4 (359)	359	359	359	359
273	2-Dimethylamino-ethylphosphonic acid	Sea anemone	$(CH_3)_2NCH_2CH_2PO_3H_2$	$C_4H_{12}NO_3P$ (153.13)	249.5° (360)	—	—	360	360	360	360
273a	1-Hydroxy-2-amino ethylphosphonic acid	*Acanthamoeba castellanii* (plasma membrane)	$H_2N-CH_2-CH(OH)PO_3H_2$	$C_2H_8NO_4P$ (141.08)	—	—	—	360a	360a	—	360a
274	Lombricine (2-amino-2-carboxy-ethyl-2-guanidine-ethyl hydrogen phosphate)	Earthworm	$H_2NCNH(CH_2)_2OPOCH_2CH(NH_2)COOH$ with NH (double bond) and OH	$C_6H_{15}N_4O_6P$ (270.21)	223–224° (361)	+14.5[23.5] (c 0.93, H_2O) (362)	8.9 (20)	361	361	362	362

DATA ON THE NATURALLY OCCURRING AMINO ACIDS (continued)

No.	Amino acid (synonym)	Source	Structure	Formula (mol wt)	Melting point °C[a]	$[\alpha]_D$[b]	pK_a	References Isolation and purification	References Chromatography	References Chemistry	References Spectral data
275	2-Methylamino-ethylphosphonic acid	Sea anemone	HNCH₂CH₂PO₃H₂ | CH₃	$C_3H_{10}NO_3P$ (139.10)	291° (360)	—	—	360	360	360	360
276	O-Phosphohomoserine	*Lactobacillus*	H₂O₃POCH₂CH₂CH(NH₂)COOH	$C_4H_{10}NO_6P$ (199.11)	178° (dec) (364)	$+6.25^{22.5}$ (c 2.4, H₂O) (364)	—	364	364	364	—
277	O-Phosphoserine	Casein	H₂O₃POCH₂CH(NH₂)COOH	$C_3H_8NO_6P$ (185.08)	—	$+7.2$ (H₂O) (366)	—	365	366	366	—
278	2-Trimethyl-aminoethyl-betaine phosphonic acid	Sea anemone	$(CH_3)_3{}^+NCH_2CH_2\overset{O}{\underset{O^-}{P}}OH$	$C_5H_{14}NO_3P$ (167.16)	250—252° (360)	—	—	360	360	360	360

N. L-Betaines

No.	Amino acid (synonym)	Source	Structure	Formula (mol wt)	Melting point °C[a]	$[\alpha]_D$[b]	pK_a	References Isolation and purification	References Chromatography	References Chemistry	References Spectral data
279	N-(3-Amino-3-carboxy-propyl)-β-carboxypyridinium betaine	Tobacco leaves	(pyridinium structure)	$C_{10}H_{14}N_2O_5$ (242.24)	241—243° (dec) (368)	$+24^{2.4}$ (c 2, H₂O) (368)	—	368	368	368	368
280	Betonicine (4-hydroxyproline betaine)	*Stachys (Betonica) officinalis*	HOHC—CH₂ / H₂C—⁺N—CHCOO⁻ (CH₃)₂	$C_7H_{13}NO_3$ (159.19)	243—244° (dec) (369)	$-36.6^{1.5}$ (369)	—	369	—	369	—
281	γ-Butyrobetaine (γ-aminobutyric acid betaine)	Rat brain	(CH₃)₃⁺NCH₂CH₂CH₂COO⁻	$C_7H_{15}NO_2$ (144.20)	180—184° (370)	—	—	370	370	—	—
282	Carnitine (γ-amino-β-hydroxybutyric acid betaine)	Vertebrate muscle	(CH₃)₃N⁺CH₂CH(OH)CH₂COO⁻	$C_7H_{15}NO_3$ (161.21)	195—197° (371)	$+23.5^{3.2}$ (c 5, H₂O) (374)	—	372	373	367, 371	—

DATA ON THE NATURALLY OCCURRING AMINO ACIDS (continued)

No.	Amino acid (synonym)	Source	Structure	Formula (mol wt)	Melting point °C[a]	$[\alpha]_D$[b]	pK_a	Isolation and purification	Chromatography	Chemistry	Spectral data
283	Desmosine	Bovine elastin		$C_{24}H_{39}N_5O_8$ (879.43)	—	—	1.70 2.40 8.80 9.85 (375)	375	—	257, 375	375
284	Ergothionine (betaine of thiol histidine)	Ergot		$C_9H_{15}N_3O_2$ (229.29)	290° (340)	$+115^{27.5}$ (c 1, H$_2$O) (290)	—	336	340	340	340
285	Hercynin (histidine betaine)	Mushrooms		$C_9H_{15}N_3O_2$ (203.22)	—	—	—	332	—	333	—
286	Homobetaine (β-alanine betaine)			$C_6H_{13}NO_2$ (131.18)	—			324	—	—	—
287	Homostachydrine (pipecolic acid betaine)	Alfalfa		$C_8H_{14}NO_2$ (156.21)	—	—	—	315	309	—	—
288	3-Hydroxystachydrine (3-hydroxyproline betaine)	*Courbonia virgata*		$C_7H_{13}NO_3$ (159.20)	210–212° (307)	$+10^{30}$ (c 2.9, H$_2$O) (307)	—	307	—	307	—
289	Hypaphorin (tryptophan betaine)	*Erythrina subumbrans*		$C_{14}H_{18}N_2O_2$ (246.29)	—	—	—	58	—	282	—
290	Isodesmosine	Bovine elastin		$C_{24}H_{39}N_5O_8$ (879.43)	—	—	—	375	—	257	—

DATA ON THE NATURALLY OCCURRING AMINO ACIDS (continued)

No.	Amino acid (synonym)	Source	Structure	Formula (mol wt)	Melting point °C[a]	$[\alpha]_D$[b]	pK$_a$	Isolation and purification	Chromatography	Chemistry	Spectral data
291	Laminine (α-amino-ε-trim-ethylamino adipic acid)	*Laminaria angustata*	$(CH_3)_3N^+(CH_2)_4CH(NH_2)COO^-$	$C_9H_{20}N_2O_4$ (220.29)	–	–	–	245	–	245	245
292	Lycin (glycine betaine)	*Lycium barbarum*	$(CH_3)_3N^+CH_2COO^-$	$C_5H_{11}NO_2$ (118.16)	–	–	–	234	–	–	–
293	Miokinine (ornithine betaine)	Human skeletal muscle	$(CH_3)_3N^+(CH_2)_3CHCOO^-$ $N^+(CH_3)_3$	$C_{11}H_{25}N_2O_2$ (217.33)	–	–	–	232	–	–	–
294	Nicotianine (N-3-amino-3-carboxypropyl-β-carboxypyridinium betaine)	Tobacco leaves	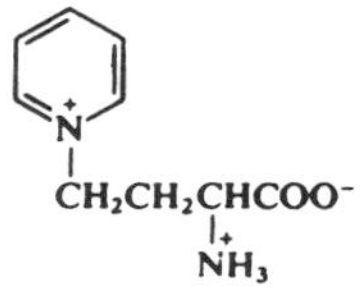 pyridinium ring with $CH_2CH_2CHCOO^-$ / NH_3^+	$C_{10}H_{14}N_2O_3$	241–243° (dec) (156)	+28.4²⁴ (c 5, H_2O) (156)	–	156	156	156	156
295	Stachydrine (proline betaine)	*Stachys tuberifera*	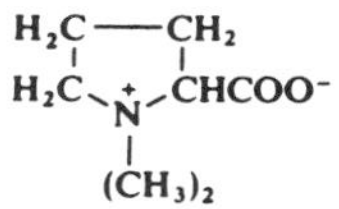 $H_2C{-}CH_2$ / $H_2C{-}N^+{-}CHCOO^-$ / $(CH_3)_2$	$C_7H_{13}NO_2$ (157.22)	–	−20.7²⁵ g (H_2O) (125)	–	128	–	128	–
296	Trigonelline (coffearin)	*Foenum graceum*	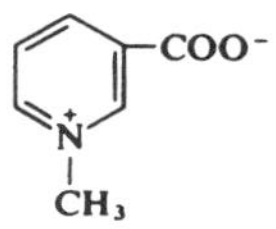 pyridinium ring with COO^- / N^+ / CH_3	$C_7H_7NO_2$ (137.15)	–	–	–	120	107	–	–
296a	ε-N-Trimethyl-δ-hydroxylysine betaine	Diatom cell walls	$(H_3C_3)N^+CH_2CH(OH)(CH_2)_2CH(NH_2)COO^-$	$C_9H_{20}N_2O_3$ (204.31)	243° (dec) (381)	+15.0²⁴ (c 0.84, H_2O) +23.1²⁴ (c 0.84, 2N HCl) (381)	–	381	381	–	–
297	ε-N-Trimethyllysine betaine	Histone of murine ascites cells	$(CH_3)_3N^+(CH_2)_4CH(NH_2)COO^-$	$C_9H_{20}N_2O_2$ (188.28)	225–226° (dec) (70a)	+10.8¹⁸ (c 5, H_2O) (379)	–	44, 379	44, 380	380	379

References (column header span): Chromatography | Chemistry | Spectral data

DATA ON THE NATURALLY OCCURRING AMINO ACIDS (continued)

No.	Amino acid (synonym)	Source	Structure	Formula (mol wt)	Melting point °C[a]	$[\alpha]_D$[b]	pK_a	Isolation and purification	Chromatography	Chemistry	Spectral data
									References		
297a	D-Alloenduracididine [α-amino-β-(2-iminoimidazolidinyl)-propionic acid]	Enduracidin hydrolyzate		$C_6H_{11}N_4O_2$ (172.22)	—	$+8.7^{73}$ (1 M HCl) $+13.3^{23}$ (1 M NaOH) (194)	2.5 8.3 12 (194a)	—	194a	—	194a
297b	D-*Allo*isoleucine	Stendomycin hydrolyzate		$C_6H_{13}NO_2$ (131.20)	—	—	—	—	264a	—	—
				O. D-Amino Acids							
298	D-Allothreonine	*Actinomycetales* species	$CH_3CHOHCH(NH_2)COOH$	$C_4H_9NO_3$ (119.12)	—	—	2.11 9.10 (290)	52	52	—	—
299	1-Amino-D-proline	Flax seed		$C_5H_{10}N_2O_2$ (130.15)	155° (dec) (80)	$+113^{25}$ (c 2, 0.5 N HCl)(80)	—	80	80	80	80
299a	*O*-Carbamyl D-serine	*Streptomyces* strain	$NH_2COOCH_2CH(NH_2)COOH$	$C_4H_8N_2O_4$ (148.14)	238° (dec) (384)	-19.6 (c 2, 1 N HCl) $+2$ (c 2, H$_2$O) (384)	—	384	384	384	—
300	D-(3-Carboxy-4-hydroxyphenyl)glycine	*Reseda luteola*		$C_9H_9NO_5$ (211.18)	<250° (dec)(93)	-121^{22} (c 0.75, 1 N HCl) (93)	—	93	93	93	93
301	D-Cycloserine (oxamycin, D-4-amino-3-isoxazolinone)	*Streptomyces orchidaceus*		$C_3H_6N_2O_2$ (101.09)	156° (dec) (94)	-89^{25} (c 0.6, H$_2$O)(94)	(94)	94	—	94	94

DATA ON THE NATURALLY OCCURRING AMINO ACIDS (continued)

No.	Amino acid (synonym)	Source	Structure	Formula (mol wt)	Melting point °C[a]	$[\alpha]_D$[b]	pK_a	References — Isolation and purification	References — Chromatography	References — Chemistry	References — Spectral data
302	m-Carboxyphenyl-glycine	Iris bulbs	COOH / CH(NH$_2$)COOH (on benzene ring)	$C_9H_9NO_4$ (195.18)	215° (143)	$+112^{25}$ (c 5, 2 N HCl)(326)	4.4 7.3 (326)	143	143	143	143
302a	N-α-Malonyl-D-alanine	Pea seedlings	H$_3$C—CH—COOH / HNCCH$_2$COOH (‖O)	$C_6H_9NO_5$ (175.16)	138–140° (dec)(382)	$+33^{28}$ (c 0.38, H$_2$O) (382)	—	382	382	382	382
303	N-α-Malonyl-D-methionine	Tobacco	CH$_3$SCH$_2$CH$_2$CHCOOH / HNCCH$_2$COOH (‖O)	$C_8H_{13}NO_5S$ (235.30)	—	—	—	95	95	95	—
304	N-α-Malonyl-D-tryptophan	Spinach	CH$_2$CHCOOH / HNCCH$_2$COOH (indole; ‖O)	$C_{14}H_{13}N_2O_6$ (289.27)	—	—	—	104	—	104	—
304a	N-Methyl-D-leucine	Griselimycin hydrolyzate	CH$_3$—CH—CH$_2$—CH—NHCH$_3$ / CH$_3$ COOH	$C_7H_{15}NO_2$ (145.23)	—	-19.6^{20} (c 0.9, H$_2$O) (376)	—	—	—	—	376
305	D-α-Methylserine	Streptomyces	CH$_3$ / HOCH$_2$C(NH$_2$)COOH	$C_4H_9NO_3$ (119.12)	—	—	2.3 9.4 (113)	113	113	113	113
306	D-Octopine [N-α-(1-Carboxy-ethyl-arginine)]	Octopus muscle	H$_2$NCN(CH$_2$)$_3$CHCOOH / ‖NH HNCHCH$_3$ / COOH	$C_9H_{18}N_4O_4$ (246.28)	—	$+20.6^{24}$ (c 1, H$_2$O) 20^{25} (c 2, 5 N HCl) (290)	1.36 2.40 8.76 11.3 (290)	133	—	205	—
307	D-Penicillamine	Penicillin	SH / (CH$_3$)$_2$CCH(NH$_2$)COOH	$C_5H_{11}NO_2S$ (149.22)	—	(225)	1.8 7.9 10.5	225	—	225	—
308	Turicine (D-allohydroxy-proline betaine)	Stachys (Betonica) officinalis	HOHC—CH$_2$ / H$_2$C—N$^+$—CHCOO$^-$ / (CH$_3$)$_2$	$C_7H_{13}NO_3$ (159.19)	249° (241)	$+36.26^{15}$ (241)	—	241	—	241	—

DATA ON THE NATURALLY OCCURRING AMINO ACIDS (continued)

Compiled by Elizabeth Dodd Mooz.

[a]Melting point or decomposition point in degrees, C.
[b]c, grams/100 ml of solution at 20 to 25°C unless specified; wavelength as subscript in millimicrons; temperature as superscript. References are 1, 3, and 290 unless indicated.
[c]The isolated amino acid appears to be a racemic DL mixture.
[d]The naturally occurring isomer is the *erythro*-L-form 290.
[e]Thermodynamic values.
[f]As hydrochloride of amino acid.
[g]For cycloalliin sulfoxide.

REFERENCES

1.   Meister, *Biochemistry of the Amino Acids,* 2nd ed., Academic Press, New York, 1965, 28.
2.   Schutzenberger and Bourgeois, *C. R. Hebd. Seances Acad. Sci.* (Paris), 81, 1191 (1875).
3.   Greenstein and Winitz, *Chemistry of the Amino Acids,* Vol. 2, John Wiley & Sons, New York, 1961, 1382.
4.   Greenstein and Winitz, *Chemistry of the Amino Acids,* Vol. 2, John Wiley & Sons, New York, 1961, 1819.
5.   Morris and Thompson, *Nature,* 190, 718 (1961).
6.   Abderhalden and Bahn, *Hoppe Seyler's Z. Physiol. Chem.,* 245, 246 (1937).
7.   Virtanen and Miettinen, *Biochim. Biophys. Acta,* 12, 181 (1953).
8.   Ackerman, *Hoppe Seyler's Z. Physiol. Chem.,* 69, 273 (1910).
9.   Work, *Bull. Soc. Chim. Biol.,* 31, 138 (1949).
10.   Steward, *Science,* 110, 439 (1949).
11.   Gabriel, *Chem. Ber.,* 23, 1767 (1890).
12.   Vahatalo and Virtanen, *Acta Chem. Scand.,* 11, 741 (1957).
13.   Doyle and Levenberg, *Biochemistry,* 7, 2457 (1968).
14.   Steiner and Hartmann, *Biochem. Z.,* 340, 436 (1964).
14a.   Chilton, Tsou, Kirk, and Benedict, *Tetrahedron Lett.,* p. 6283 (1968).
15.   Takita, *J. Antibiot.* (Tokyo), 17, 264 (1964).
15a.   Sung, Fowden, Millington, and Sheppard, *Phytochemistry,* 8, 1227 (1969).
15b.   Millington and Sheppard, *Phytochemistry,* 7, 1027 (1968).
16.   Kenner and Sheppard, *Nature,* 181, 48 (1958).
17.   Asen, *J. Biol. Chem.,* 234, 343 (1959).
17a.   Scannell, Pruess, Demny, Sello, Williams, and Stempel, *J. Antibiot.* (Tokyo), 25, 122 (1972).
17b.   Sahm, Knobloch, and Wagner, *J. Antibiot.* (Tokyo), 26, 389 (1973).
18.   Fowden and Done, *Biochem. J.,* 55, 548 (1953).
19.   Fowden and Smith, *Phytochemistry,* 7, 809 (1968).
19a.   Kelly, Martin, and Hanka, *Can. J. Chem.,* 47, 2504 (1969).
20.   Perrin, *Dissociation Constants of Organic Bases in Aqueous Solution,* Butterworths, London, 1965.
21.   Millington and Sheppard, *Phytochemistry,* 7, 1027 (1968).
22.   Dardenne, Casimar, and Jadot, *Phytochemistry,* 7, 1401 (1968).
22a.   Dardenne, Casimar, and Jadot, *Phytochemistry,* 7, 1401 (1968).
23.   Staron, Allard, and Xuong, *C. R. Hebd. Seances Acad. Sci.* (Paris), 260, 3502 (1965).
23a.   Scannell, Pruess, Demny, Weiss, Williams, and Stempel, *J. Antibiot.* (Tokyo), 24, 239 (1971).
23a'.   Fowden, Smith, Millington, and Sheppard, *Phytochemistry,* 8, 437 (1969).
23b.   Fowden and Smith, *Phytochemistry,* 7, 809 (1968).
24.   Braconnot, *Ann. Chim. Phys.,* 13, 113 (1820).
25.   Shorey, *J. Am. Chem. Soc.,* 19, 881 (1897).
26.   Hassall and Keyle, *Biochem. J.,* 60, 334 (1955).
27.   Carbon, *J. Am. Chem. Soc.,* 80, 1002 (1958).
28.   Ehrlich, *Chem. Ber.,* 37, 1809 (1904).
29.   Greenstein and Winitz, *Chemistry of the Amino Acids,* Vol. 3, John Wiley & Sons, New York, 1961, 2043.
30.   Proust, *Ann. Chim. Phys.,* 10, 29 (1819).
31.   Greenstein and Winitz, *Chemistry of the Amino Acids,* Vol. 3, John Wiley & Sons, New York, 1961, 2075.
31a.   Walker, Bodanszky, and Perlman, *J. Antibiot.* (Tokyo), 23, 255 (1970).
32.   Gray and Fowden, *Biochem. J.,* 82, 385 (1961).
33.   Kortrum, Vogel, and Andrussow, *Dissociation Constants of Organic Acids in Aqueous Solutions,* Butterworths, London, 1961.
34.   Yukowa, *Handbook of Organic Structural Analysis,* Benjamin, New York, 1965, 584.
35.   Fischer, *Hoppe Seyler's Z. Physiol. Chem.,* 33, 151 (1901).
35a.   Vervier and Casimir, *Phytochemistry,* 9, 2059 (1970).
35b.   Levenberg, *J. Biol. Chem.,* 243, 6009 (1968).
36.   Greenstein and Winitz, *Chemistry of the Amino Acids,* Vol. 3, John Wiley & Sons, New York, 1961, 2368.
37.   Hatanaka and Virtanen, *Acta Chem. Scand.,* 16, 514 (1962).
38.   Takemoto and Sai, *J. Pharm. Soc. Jap.,* 85, 33 (1965).
39.   Virtanen and Berg, *Acta Chem. Scand.,* 8, 1085 (1954).
40.   Vauqelin and Robiquet, *Ann. Chim.,* 57, 88 (1806).
41.   Greenstein and Winitz, *Chemistry of the Amino Acids,* Vol. 3, John Wiley & Sons, New York, 1961, 1856.
42.   Davies and Evans, *J. Chem. Soc.,* p. 480 (1953).
43.   Ritthausen, *J. Prakt. Chem.,* 103, 233 (1868).
44.   Hempel, Lange, and Berkofer, *Naturwissenschaften,* 55, 37 (1968).
45.   Gray and Fowden, *Nature,* 189, 401 (1961).
46.   Gmelin and Larsen, *Biochim. Biophys. Acta,* 136, 572 (1967).
46a.   Nulu and Bell, *Phytochemistry,* 11, 2573 (1972).

47. Fowden, *Biochem. J.*, 98, 57 (1966).
48. Birkinshaw, Raistick, and Smith, *Biochem. J.*, 36, 829 (1942).
49. Ritthausen, *J. Prakt. Chem.*, 99, 454 (1866).
50. Greenstein and Winitz, *Chemistry of the Amino Acids*, Vol. 3, John Wiley & Sons, New York, 1961, 1929.
51. Schulze and Bosshard, *Landwirtsch. Vers. Stn.*, 29, 295 (1883).
52. Ikawa, Snell, and Lederer, *Nature*, 188, 558 (1961).
53. Larsen, *Acta Chem. Scand.*, 19, 1071 (1965).
54. Fowden and Gray, *Amino Acid Pools*, Holden, Ed., Elsevier, Amsterdam, 1962, 46.
55. Barker, Smyth, Wawszkiewicz, Lee, and Wilson, *Arch. Biochem. Biophys.*, 78, 468 (1958).
56. Virtanen and Berg, *Acta Chem. Scand.*, 9, 553 (1955).
56a. Przybylska and Strong, *Phytochemistry*, 7, 471 (1968).
57. Done and Fowden, *Biochem. J.*, 51, 451 (1952).
58. Greshoff, *Chem. Ber.*, 23, 3537 (1890).
59. Casimir, Jadot, and Renard, *Biochim. Biophys. Acta*, 39, 462 (1960).
59a. Seneviratne and Fowden, *Phytochemistry*, 7, 1030 (1968).
60. Virtanena and Linko, *Acta Chem. Scand.*, 9, 531 (1955).
61. Liss, *Phytochemistry*, 1, 87 (1962).
62. Bohak, *J. Biol. Chem.*, 239, 2878 (1964).
63. Ziegler, Melchert, and Lurken, *Nature*, 214, 404 (1967).
64. Argoudelis, Herr, Mason, Pyke, and Zieserl, *Biochemistry*, 6, 165 (1967).
65. Vega and Bell, *Phytochemistry*, 6, 759 (1967).
65a. Bell, *Nature*, 218, 197 (1968).
66. Damodaran and Narayanan, *Biochem. J.*, 34, 1449 (1940).
67. Kitagawa and Monobe, *J. Biochem. (Tokyo)*, 18, 333 (1933).
67a. Bodanszky and Bodanszky, *J. Antibiot. (Tokyo)*, 23, 149 (1970).
67b. Bodanszky, Chaturvedi, Scozzie, Griffith, and Bodanszky, *Antimicrob. Agents Chemother.*, p. 135 (1969).
68. Fujino, Inoue, Ueyanagi, and Miyake, *Bull. Chem. Soc. Jap.*, 34, 740 (1961).
69. Tsai and Stadtman, *Arch. Biochem. Biophys.*, 125, 210 (1968).
69a. Hettinger and Craig, *Biochemistry*, 9, 1224 (1970).
70. Gmelin, Strauss, and Hasenmaier, *Hoppe Seyler's Z. Physiol. Chem.*, 314, 28 (1959).
70a. Kakimoto and Akazawa, *J. Biol. Chem.*, 245, 5751 (1970).
70b. Scannell, Ax, Pruess, Williams, Demm, and Stempel, *J. Antibiot. (Tokyo)*, 25, 179 (1972).
71. Dupuy and Lee, *J. Am. Pharm. Assoc. (Sci. Ed.)*, 43, 61 (1954).
72. Schilling and Strong, *J. Am. Chem. Soc.*, 77, 2843 (1955).
73. Vickery and Leavenworth, *J. Biol. Chem.*, 76, 437 (1928).
74. Drechsel, *Z. Prakt. Chem.*, 39, 425 (1889).
75. Greenstein and Winitz, *Chemistry of the Amino Acids*, Vol. 3, John Wiley & Sons, New York, 1961, 2097.
76. Haskell, Fusari, Frohardt, and Bartz, *J. Am. Chem. Soc.*, 74, 599 (1952).
77. Biemann, Lioret, Asselineau, Lederer, and Polonsky, *Bull. Soc. Chim. Biol.*, 42, 979; *Biochim. Biophys. Acta*, 40, 369 (1960).
78. Lioret, *C. R. Hebd Seances Acad. Sci.* (Paris), 244, 2171 (1957).
79. Murray, *Biochemistry*, 3, 10 (1964).
80. Klosterman, Lamoureux, and Parsons, *Biochemistry*, 6, 170 (1967).
81. Greenstein and Winitz, *Chemistry of the Amino Acids*, Vol. 3, John Wiley & Sons, New York, 1961, 2477.
81a. Stapley, Miller, Mata, and Hendlin, *Antimicrob. Agents Chemother.*, p. 401 (1967).
82. Rao, Adiga, and Sarma, *Biochemistry*, 3, 432 (1964).
82a. Shiba, Kubota, and Kaneto, *Tetrahedron*, 26, 4307 (1970).
83. Suzuki, Nakamura, Okuma, and Tomiyama, *J. Antibiot. (Tokyo)*, 11A(81), 84 (1958).
84. Cummings and Hudgins, *Am. J. Med. Sci.*, 236, 311 (1958).
85. Sorensen and Andersen, *Hoppe Seyler's Z. Physiol. Chem.*, 56, 250 (1908).
86. Hochstein, *J. Org. Chem.*, 24, 679 (1959).
87. Grobbelaar and Steward, *Nature*, 182, 1358 (1958).
87a. Westley, Pruess, Volpe, Demny, and Stempel, *J. Antibiot. (Tokyo)*, 24, 330 (1971).
88. Larsen, *Acta Chem. Scand.*, 21, 1592 (1967).
88a. Doyle and Levenberg, *Biochemistry*, 7, 2457 (1968).
89. Blass and Macheboeuf, *Helv. Chim. Acta*, 29, 1315 (1946).
90. Snow, *J. Chem. Soc.*, 2588, 4080 (1954).
91. Virtanen and Hietala, *Acta Chem. Scand.*, 9, 549 (1955).
92. Umbreit and Heneage, *J. Biol. Chem.*, 201, 15 (1953).
93. Kjaer and Larsen, *Acta Chem. Scand.*, 17, 2397 (1963).
94. Hidy, Hodge, Yound, Harned, Brewer, Phillips, Runge, Staveley, Pohland, Boaz, and Sullivan, *J. Am. Chem. Soc.*, 77, 2345 (1955).
95. Keglevic, Ladesic, and Pokorney, *Arch. Biochem. Biophys.*, 124, 443 (1968).

96. Virtanen, Uksila, and Matikkala, *Acta Chem. Scand.*, 8, 1091 (1954).
97. Thompson, Morris, and Hunt, *J. Biol. Chem.*, 239, 1122 (1964).
98. Okabe, *J. Antibiot. Ser. A*, 13, 412 (1961).
99. Wieland and Haufer, *Justus Liebigs Ann. Chem.*, 619, 35 (1968).
99a. Fowden, MacGibbon, Mellon, and Sheppard, *Phytochemistry*, 11, 1105 (1972).
99a'. Rudzats, Gellert, and Halpern, *Biochem. Biophys. Res. Commun.*, 47, 290 (1972).
100. Murooka and Harada, *Agric. Biol. Chem.*, 31, 1035 (1967).
100a. Giovanelli, Owens, and Mudd, *Biochim. Biophys. Acta*, 227, 671 (1971).
100b. Owens, Thompson, and Fennessey, *Chem. Commun.*, p. 715 (1972).
101. Virtanen and Ettala, *Acta Chem. Scand.*, 11, 182 (1957).
102. Bodanszky, Alicino, Birkhimer, and Williams, *J. Am. Chem. Soc.*, 84, 2004 (1962).
103. Wieland and Schopf, *Justus Liebigs Ann. Chem.*, 626, 174 (1959).
103a. Mechanic and Tanzer, *Biochem. Biophys. Res. Commun.*, 41, 1597 (1970).
104. Good and Andreae, *Plant Physiol.*, 32, 561 (1957).
104a. Takita and Maeda, *J. Antibiot.* (Tokyo), 22, 39 (1969).
105. Saarivirta and Virtanen, *Acta Chem. Scand.*, 19, 1008 (1965).
106. Craig, *J. Biol. Chem.*, 125, 289 (1938).
107. Joshi and Handler, *J. Biol. Chem.*, 235, 2981 (1961).
108. Setsuseo, *J. Osaka Univ.*, 7, 833 (1957).
109. Schweet, Holden, and Lowy, *J. Biol. Chem.*, 211, 517 (1954).
110. Tominaga, Hiwaki, Maekawa, and Yoshida, *J. Biochem.* (Tokyo), 53, 227 (1963).
111. Wilding and Stahlmann, *Phytochemistry*, 1, 241 (1962).
111a. Kemp and Dawson, *Biochim. Biophys. Acta*, 176, 678 (1969).
112. Fowden, *Biochem. J.*, 81, 155 (1961).
113. Flynn, Hinman, Caron, and Woolf, *J. Am. Chem. Soc.*, 75, 5867 (1953).
114. Nagao, *Bull. Fac. Fish Hokkaido Univ.*, 2, 128 (1951).
115. Hatanaka, *Acta Chem. Scand.*, 16, 513 (1962).
116. Brandner and Virtanen, *Acta Chem. Scand.*, 17, 2563 (1963).
117. Snow, *J. Chem. Soc.*, p. 2589 (1954).
117a. Fowden, Pratt, and Smith, *Phytochemistry*, 12, 1707 (1973).
118. Miyake, *Chem. Pharm. Bull.* (Tokyo), 8, 1071 (1960).
119. Wieland and Schön, *Justus Liebigs Ann. Chem.*, 593, 157 (1955).
120. Jahns, *Chem. Ber.*, 18, 2518 (1885); 20, 2840 (1887).
121. Jadot and Casimir, *Biochim. Biophys. Acta*, 48, 400 (1961).
122. Jadot, Casimir, and Alderweireldt, *Biochim. Biophys. Acta*, 78, 500 (1963).
123. Brieskorn and Glasz, *Naturwissenschaften*, 51, 216 (1964).
124. Schryver, *Proc. R. Soc.*, B98, 58 (1925).
124a. Godtfredsen, Vangedal, and Thomas, *Tetrahedron*, 26, 4931 (1970).
125. Steenbock, *J. Biol. Chem.*, 35, 1 (1918).
126. Fowden, *Nature*, 209, 807 (1966).
127. Bell and Tirimanna, *Biochem., J.*, 91, 356 (1964).
128. Planta and Schulze, *Chem. Ber.*, 26, 939 (1893).
129. Pollard, Sondheimer, and Steward, *Nature*, 182, 1356 (1958).
129a. Shiba, Mizote, Kaneko, Nakajima, and Kakimoto, *Biochim. Biophys. Acta*, 244, 523 (1971).
129b. Hettinger and Craig, *Biochemistry*, 9, 1224 (1970).
130. Barry and Roark, *J. Biol. Chem.*, 239, 1541 (1964).
131. Barry, Chen, and Roark, *J. Gen. Microbiol.*, 33, 95 (1963).
132. Ackermann and Kirby, *J. Biol. Chem.*, 175, 483 (1948).
132a. Grove, Daxenbichler, Weisleder, and van Etlen, *Tetrahedron Lett.*, 4477 (1971).
132b. Owens, Guggenheim, and Hilton, *Biochim. Biophys. Acta*, 158, 219 (1968).
132c. Owens, Thompson, Pitcher, and Williams, *Chem. Commun.*, p. 714 (1972).
133. Morizawa, *Acta Sch. Med. Univer. Kioto*, 9, 285 (1927).
134. Cramer, *J. Prakt. Chem.*, 96, 76 (1865).
135. Flavin, Delavier-Klutchko, and Slaughter, *Science*, 143, 50 (1964).
136. Wooley, *J. Biol. Chem.*, 197, 409 (1952).
137. Wooley, *J. Biol. Chem.*, 198, 807 (1953).
138. Rose, McCoy, Meyer, and Carter, *J. Biol. Chem.*, 112, 283 (1935).
139. Greenstein and Winitz, *Chemistry of the Amino Acids*, Vol. 3, John Wiley & Sons, New York, 1961, 2238.
140. Wisvisz, Van der Hoever, and Nijenhuis, *J. Am. Chem. Soc.*, 79, 4522 (1957).
140a. Bohman, *Tetrahedron Lett.*, p. 3065 (1970).
141. Olesen-Larsen, *Biochim. Biophys. Acta*, 93, 200 (1964).
142. Thompson, Morris, Asen, and Irreverre, *J. Biol. Chem.*, 236, 1183 (1961).
142a. O'Brien, Cox, and Gibson, *Biochim. Biophys. Acta*, 177, 321 (1969).

143. Morris, Thompson, Asen, and Irreverre, *J. Am. Chem. Soc.*, 81, 6069 (1959).
144. Schneider, *Biochem. Z.*, 330, 428 (1958).
145. Guggenheim, *Z. Physiol. Chem.*, 88, 276 (1913).
146. Greenstein and Winitz, *Chemistry of the Amino Acids,* Vol. 3, John Wiley & Sons, New York, 1961, 2713.
147. Muller and Schulte, *Z. Naturforsch.*, 23b, 659 (1958).
147a. Weaver, Rajagopalan, Handler, Rosenthal, and Jeffs, *J. Biol. Chem.*, 246, 2010 (1971).
147b. Watson and Fowden, *Phytochemistry*, 12, 617 (1973).
148a. Smith and Sloane, *Biochim. Biophys. Acta*, 148, 414 (1967).
148b. Sloane and Smith, *Biochim. Biophys. Acta*, 158, 394 (1968).
148. Makino, Satoh, Fujik, and Kawaguchi, *Nature*, 170, 977 (1952).
149. Waller, Fryth, Hutchings, and Williams, *J. Am. Chem. Soc.*, 75, 2025 (1953).
150. Behr and Clark, *J. Am. Chem. Soc.*, 54, 1630 (1932).
151. Schultze and Barbieri, *Chem. Ber.*, 14, 1785 (1881).
152. Greenstein and Winitz, *Chemistry of the Amino Acids,* Vol. 3, John Wiley & Sons, New York, 1961, 2156.
153. Liebig, *Justus Liebigs Ann. Chem.*, 57, 127 (1846).
154. Greenstein and Winitz, *Chemistry of the Amino Acids,* Vol. 3, John Wiley & Sons, New York, 1961, 2348.
155. Mothes, Schütte, Müller, Ardenne, and Tümmler, *Z. Naturforsch.*, 196, 1161 (1964).
155a. Ohkusu and Mori, *J. Neurochem.*, 16, 1485 (1969).
156. Noguchi, Sakuma, and Tamaki, *Phytochemistry*, 7, 1861 (1968).
157. Kjaer and Larsen, *Acta Chem. Scand.*, 13, 1565 (1959).
158. Schulze and Steiger, *Chem. Ber.*, 19, 1177 (1886); *Z. Physiol. Chem.*, 11,43 (1886).
159. Greenstein and Winitz, *Chemistry of the Amino Acids,* Vol. 3, John Wiley & Sons, New York, 1961, 1841.
160. Fearon and Bell, *Biochem., J.*, 59, 221 (1955).
161. Kitagawa and Tomiyama, *J. Biochem.* (Tokyo), 11, 265 (1929).
162. Bell, *Biochem. J.*, 75, 618 (1960).
163. Walker, *J. Biol. Chem.*, 204, 139 (1954).
164. Wada, *Biochem. Z.*, 224, 420 (1930).
165. Rogers, *Biochim. Biophys. Acta*, 29, 33 (1958).
166. Kitagawa and Tsukamoto, *J. Biochem.* (Tokyo), 26, 373 (1937).
166a. Nakajima, Matsuoka, and Kakimoto, *Biochim. Biophys. Acta*, 230, 212 (1971).
167. Ito and Hashimoto, *Nature*, 211, 417 (1966).
168. Rao, Ramachandran, and Adiga, *Biochemistry*, 2, 298 (1962).
169. Gerritsen, Vaughn, and Waisman, *Arch. Biochem. Biophys.*, 100, 298 (1963).
170. Bell and Tirimanna, *Nature*, 197, 901 (1963).
170a. Maehr, Blount, Pruess, Yarmchuk, and Kellett, *J. Antibiot.* (Tokyo), 26, 284 (1973).
171. Bell, *Biochem. J.*, 21, 358 (1964).
173. Hegarty and Pound, *Nature*, 217, 354 (1968).
174. Tamoni and Gallo, *Farmaco Ed. Sci.* 21, 269 (1966).
175. Fusari, Bartz, and Elder, *Nature*, 173, 72; *J. Am. Chem. Soc.*, 76, 2878 (1954).
176. Ressler, *J. Biol. Chem.*, 237, 733 (1962).
177. Ressler and Ratzkin, *J. Org. Chem.*, 26, 3356 (1961).
177a. Brysk and Ressler, *J. Biol. Chem.*, 245, 1156 (1970).
178. Dion, Fusari, Jakubowski, Zora, and Bartz, *J. Am. Chem. Soc.*, 78, 3075 (1956).
179. Kaczka, Gitterman, Dulaney, and Folkers, *Biochemistry*, 1, 340 (1962).
180. Radhakrishnan and Giri, *Biochem. J.*, 58, 57 (1954).
180a. Lambein and Van Parijs, *Biochem. Biophys. Res. Commun.*, 32, 474 (1968).
180b. Lambein, Schamp, Vandendriessche, and Van Parijs, *Biochem. Biophys. Res. Commun.*, 37, 375 (1969).
181. King, *J. Chem. Soc.*, p. 3590 (1950).
182. Grobbelaar, *Nature*, 175, 703 (1955).
183. Dobson and Raphael, *J. Chem. Soc.*, p. 3642 (1958).
184. Tanaka, Miyamoto, Honjo, Morimoto, Sugawa, Uchibayshi, Sanno, and Tatsuoka, *Proc. Jap. Acad.*, 33, 47, 53 (1957).
185. Tanaka, Miyamoto, Honjo, Morimoto, Sugawa, Uchibayshi, Sanno, and Tatsuoka, *Chem. Abstr.*, 51, 17881i (1957).
185a. Burrows and Turner, *J. Chem. Soc.*, p. 255 (1966).
186. Schenk and Schütte, *Naturwissenschaften*, 48, 223, (1961).
186a. Hatanaka, *Phytochemistry*, 8, 1305 (1969).
186b. Shah, Neuss, Gorman, and Boeck, *J. Antibiot.* (Tokyo), 23, 613 (1970).
187. Prochazka, *Czech. Chem. Commun.*, 22, 333, 654 (1957).
188. Pironen and Virtanen, *Acta Chem. Scand.*, 16, 1286 (1962).
189. Gmelin and Virtanen, *Ann. Acad. Sci. Fenn. (Med.)*, 107, 3 (1961).
189a. Miller, Tristram, and Wolf, *J. Antibiot.* (Tokyo), 24, 48 (1971).

190. Fowden, *Nature*, 176, 347 (1955).
190a. Eagles, Laird, Matai, Self, and Synge, *Biochem. J.*, 121, 425 (1971).
190b. Bell, Nulu, and Cone, *Phytochemistry*, 10, 2191 (1971).
190c. Robbers and Floss, *Tetrahedron Lett.*, p. 1857 (1969).
191. Fang, Li, Nin, and Tseng, *Sci. Sinica*, 10, 845 (1961).
191a. Cross and Webster, *J. Chem. Soc. (Org.)*, p. 1839 (1970).
191b. Scannell, Pruess, Demny, Williams, and Stempel, *J. Antibiot.* (Tokyo), 23, 618 (1970).
191c. Corbin and Bulen, *Biochemistry*, 8, 757 (1969).
191d. Nakajima and Volcani, *Science*, 164, 1400 (1969).
191e. Karle, Daly, and Witkop, *Science*, 164, 1401 (1969).
191f. Brown and Mangat, *Biochim. Biophys. Acta*, 177, 427 (1969).
191g. Macnicol, *Biochem. J.*, 107, 473 (1968).
192. Brown, *J. Am. Chem. Soc.*, 87, 4202 (1965).
193. Daigo, *J. Pharm. Soc. Jap.*, 79, 353 (1959).
194. Casnati, Quilico, and Ricca, *Gazz. Chim. Ital.*, 93, 349 (1963).
194a. Horii and Kameda, *J. Antibiot.* (Tokyo), 21, 665 (1968).
194a'. Katagiri, Tori, Kimura, Yoshida, Nagasaki, and Minato, *J. Med. Chem.*, 10, 1149 (1967).
194a''. Finot, Bricout, Viani and Mauron, *Experimentia*, 24, 1097 (1968).
194b. Weaver, Rajagopalan, and Handler, *J. Biol. Chem.*, 246, 2015 (1971).
195. Jahns, *Chem. Ber.*, 24, 2615 (1891).
196. Freidenberg, *Chem. Ber.*, 51, 976 (1918).
197. Kossel, *Hoppe Seyler's Z. Physiol. Chem.*, 22, 176 (1896).
198. Greenstein and Winitz, *Chemistry of the Amino Acids*, Vol. 3, John Wiley & Sons, New York, 1961, 1971.
198a. Takita, Yoshioka, Muraoka, Maeda, and Umezawa, *J. Antibiot.* (Tokyo), 24, 795 (1971).
199. Hulme, *Nature*, 174, 1055 (1954).
200. Biemann and Deffner, *Nature*, 191, 380 (1961).
201. Minagawa, *Proc. Imp. Acad.* (Tokyo), 21, 33, 37 (1945).
202. Virtanen and Gmelin, *Acta Chem. Scand.*, 13, 1244 (1959).
203. Schoolery and Virtanen, *Acta Chem. Scand.*, 16, 2457 (1962).
203. Virtanen and Kari, *Acta Chem. Scand.*, 8, 1290 (1954).
204. Virtanen and Kari, *Actu. Chem. Scand.*, 8, 1290 (1954).
204a. Hassall, Morton, Ogihara, and Thomas, *Chem. Commun.*, p. 1079 (1969).
204b. Bevan, Davies, Hassall, Morton, and Phillips, *J. Chem. Soc. (Org.)*, p. 514 (1971).
205. Irvine and Wilson, *J. Biol. Chem.*, 127, 555 (1939).
206. Sheehan and Whitney, *J. Am. Chem. Soc.*, 84, 3980 (1962).
206a. Sung and Fowden, *Phytochemistry*, 7, 2061 (1968).
207. Fischer, *Chem. Ber.*, 35, 2660 (1902).
208. Greenstein and Winitz, *Chemistry of the Amino Acids*, Vol. 3, John Wiley & Sons, New York, 1961, 2018.
209. Wieland and Witkop, *Justus Liebigs Ann. Chem.*, 543, 171 (1940).
210. Kotake, Sakan, and Miwa, *Chem. Ber.*, 85, 690 (1952).
211. Mitoma, Weissbach, and Udenfriend, *Nature*, 175, 994 (1955).
212. Takemoto, Nakajima, and Yokobe, *J. Pharm. Soc. Jap.*, 84, 1232 (1964).
213. List, *Planta Med.*, 6, 424 (1958).
214. von Klämbt, *Naturwissenschaften*, 47, 398 (1960).
215. Hutzinger and Kosuge, *Biochemistry*, 7, 601 (1968).
216. Murakami, Takemoto, and Shimzu, *J. Pharm. Soc. Jap.*, 73, 1026 (1953).
217. Beockman and Staehler, *Naturwissenschaften*, 52, 391 (1965).
218. Vanderhaeghe and Parmetier, *17th Congr. Pure Appl. Chem.*, Butterworths, London, 1959, 56.
219. Brockmann, *Ann. N.Y. Acad. Sci.*, 89, 323 (1960).
220. Bell, *Biochim. Biophys. Acta*, 47, 602 (1961).
221. Gray and Fowden, *Nature*, 193, 1285 (1962).
222. Bethell, Kenner, and Shepperd, *Nature*, 194, 864 (1962).
222a. Müller and Schütte, *Z. Naturforsch.*, 236, 491 (1968).
223. Hulme and Arthington, *Nature*, 173, 588 (1954).
224. Burroughs, Dalby, Kenner, and Sheppard, *Nature*, 189, 394 (1961).
224a. Borders, Sax, Lancaster, Hausmann, Mitscher, Wetzel and Patterson, *Tetrahedron*, 26, 3123 (1970).
225. Chain, *Ann. Rev. Biochem.*, 17, 657 (1948).
226. Sheehan, Drummond, Gardner, Maeda, Mania, Nakamura, Sen, and Stock, *J. Am. Chem. Soc.*, 85, 2867 (1963).
227. Adams and Johnson, *J. Am. Chem. Soc.*, 71, 705 (1949).
227a. Murakoshi, Kuramoto, Ohmiya and Haginiwa, *Chem. Pharm. Bull.* (Japan), 20, 855 (1972).
228. Minagawa, *Proc. Jap. Acad.*, 22, 130 (1946).
229. Eugster, Muller, and Good, *Tetrahedron Lett.*, 23, 1813 (1965).
230. Reiner and Eugster, *Helv. Chim. Acta*, 50, 128 (1967).

231. Fritz, Gagnent, Zbinden, Geigy, and Eugster, *Tetrahedron Lett.*, 25, 2075 (1965).
231a. Noma, Noguchi, and Tamaki, *Tetrahedron Lett.*, p. 2017 (1971).
232. Engeland and Biehler, *Hoppe Seyler's Z. Physiol. Chem.*, 123, 290 (1922).
233. Hulme and Arthington, *Nature*, 170, 659 (1952).
234. Husemann and Marme, *Justus Liebigs Ann. Chem.*, Suppl. 2, 382 (1863).
235. Greenstein and Winitz, *Chemistry of the Amino Acids*, Vol. 3, John Wiley & Sons, New York, 1961, 2178.
236. Noe and Fowden, *Biochem. J.*, 77, 543 (1960).
236a. Finot, Viani, Bricout, and Mauron, *Experimentia*, 25, 134 (1969).
236b. Wolfersberger and Tabachnik, *Experimenta*, 29, 346 (1973).
237. Nakanishi, Ito, and Hirata, *J. Am. Chem. Soc.*, 76, 2845 (1954).
238. Carter, Sweeley, Daniels, McNary, Schaffner, West, Tamelen, Dyer, and Whaley, *J. Am. Chem. Soc.*, 83, 4296 (1961).
238a. Bodanszky, Marconi, and Bodanszky, *J. Antiobiot.*, (Tokyo), 22, 40 (1969).
238b. Marconi and Bodanszky, *J. Antibiot.* (Tokyo), 23, 120 (1970).
239. Hattori and Komamine, *Nature*, 183, 1116 (1959).
240. Senoh, Imamato, Maeno, Tokyama, Sakan, Komamine, and Hattori, *Tetrahedron Lett.*, 46, 3431 (1964).
241. Schulze and Trier, *Hoppe Seyler's Z. Physiol. Chem.*, 76, 258; 79, 235 (1912).
242. Takemoto and Nakajima, *J. Pharm. Soc. Jap.*, 84, 1230 (1964).
243. Hopkins and Cole, *J. Physiol.* (Lond.), 27, 418 (1901).
244. Greenstein and Winitz, *Chemistry of the Amino Acids*, Vol. 3, John Wiley & Sons, New York, 1961, 2316.
224a. Nakamiya, Shiba, Kaneko, Sakakibara, Take, and Abe, *Tetrahedron Lett.*, p. 3497 (1970).
245. Takemoto, Diago, and Takagi, *J. Pharm. Soc. Jap.*, 84, 1176 (1964).
246. Dyer, Hayes, Miller, and Nassar, *J. Am. Chem. Soc.*, 86, 5363 (1964).
246a. Büchi and Raleigh, *J. Org. Chem.*, 36, 873 (1971).
247. Gmelin, *Hoppe Seyler's Z. Physiol. Chem.*, 316, 164 (1959).
248. Kjaer, Knudsen, and Larsen, *Acta Chem. Scand.*, 15, 1193 (1961).
249. Gordon and Jackson, *J. Biol. Chem.*, 110, 151 (1935).
250. Ghatak and Kaul, *Chem. Zentralbl.*, p. 3730 (1932).
250a. Auditore and Wade, *J. Neurochem.*. 18, 2389 (1971).
250a'. Hall, Metzenberg, and Cohen, *J. Biol. Chem.*, 230, 1013 (1958).
250b. Tallan, Moore, and Stein, *J. Biol. Chem.*, 219, 257 (1956).
251. Takita, Naganawa, Maeda, and Umezawa, *J. Antibiot.* (Tokyo), 17, 90 (9164).
252. Sheehan, Zachau, and Lawson, *J. Am. Chem. Soc.*, 80, 3349 (1958).
252a. Bouloin, Ottinger, Pais, and Chiurdoglu, *Bull. Soc. Chim. Belges*, 78, 583 (1969).
253. Takita, *J. Antibiot.* (Tokyo), 16, 175 (1963).
254. Emery, *Biochemistry*, 4, 1410 (1965).
255. Ackerman, *Hoppe Seyler's Z. Physiol. Chem.*, 302, 80 (1955).
256. Morgan, *Chem. Ind.* (Lond.), p.542 (1964).
257. Starcher, Partridge, and Elsden, *Biochemistry*, 6, 2425 (1967).
258. Eloff and Grobbelaar, *J. S. Afr. Chem. Inst.*, 20, 190 1967).
259. Searle and Westall, *Biochem. J.*, 48, 1 (P), (1951).
260. Tallan, Stein, and Moore, *J. Biol. Chem.*, 206, 825 (1954).
261. Plattner and Nager, *Helv. Chim. Acta*, 31, 665 (1948).
262. Plattner and Nager, *Helv. Chim. Acta*, 31, 2192 (1948).
263. Corrigan, *Science*, 164, 142 (1969).
264. Vilkas, Rojas, and Lederer, *C.R. Hebd. Seances Acad. Sci.* (Paris), 261, 4258 (1965).
264a. Bodanszky, Muramatsu, Bodanszky, Lukin, and Doubler, *J. Antibiot.* (Tokyo), 21, 77 (1968).
265. Thompson, Morris, and Smith, *Ann. Rev. Biochem.*, 38, 137 (1969).
266. Hiller-Bombien, *Arch. Pharm.*, 230, 513 (1892).
267. Winterstein, *Hoppe Seyler's Z. Physiol. Chem.*, 105, 20 (1919).
268. Brockmann and Grobhofer, *Naturwissenschaften*, 36, 376 (1949).
269. Plattner and Nager, *Helv. Chim. Acta*, 31, 2203 (1948).
270. Darling and Larsen, *Acta Chem. Scand.*, 15, 743 (1961).
271. Kjaer and Larsen, *Acta Chem. Scand.*, 15, 750 (1961).
272. Linko, Alfthan, Miettinen, and Virtanen, *Acta Chem. Scand.*, 7, 1310 (1953).
273. Gmelin, Kjaer, and Larsen, *Phytochemistry*, 1, 233 (1962).
274. Stoll and Seebeck, *Helv. Chim. Acta*, 31, 189 (1948).
275. Suzuki, *Chem. Pharm. Bull.* (Tokyo), 9, 251 (1961).
276. Sugii, *Chem. Pharm. Bull.* (Tokyo), 12, 1114 (1964).
277. Alderton, *J. Am. Chem. Soc.*, 75, 2391 (1953).
278. Horowitz, *J. Biol. Chem.*, 171, 255 (1947).
279. Tallan, Moore and Stein, *J. Biol. Chem.*, 230, 707 (1958).
280. McRorie, Sutherland, Lewis, Barton, Glazener, and Shive, *J. Am. Chem. Soc.*, 76, 115 (1954).

280a. Jadot, Casimir, and Warin, *Bull. Soc. Chim. Belges*, 78, 299 (1969).
281. Salkowski, *Chem. Ber.*, 6, 744 (1873).
282. Romburgh and Barger, *J. Chem. Soc.* (Lond.), 99, 2068 (1911).
283. Gmelin, Strauss, and Hasenmaler, *Z. Naturforsch.*, 13b, 252 (1958).
284. Kuriyama, *Nature*, 192, 969 (1961).
284a. Kodama, Yao, Kobayashi, Hirayama, Fujii, Mizuhara, Haraguchi, and Hirosawa, *Physiol. Chem. Phys.*, 1, 72 (1969).
285. Gmelin and Hietala, *Hoppe Seyler's Z. Physiol. Chem.*, 322, 278 (1960).
286. Virtanen and Matikkala, *Hoppe Seyler's Z. Physiol. Chem.*, 322, 8 (1960).
287. Mizuhara and Ohmori, *Arch. Biochem. Biophys.*, 92, 53 (1961).
288. Kuriyama, Takagi, and Murata, *Bull. Fac. Fish Hokkaido Univ.*, 11, 58 (1960).
289. Virtanen and Matikkala, *Acta Chem. Scand.*, 13, 623 (1959).
290. Dawson, Elliott, Elliott, and Jones, *Data for Biochemical Research.* Oxford University Press, Oxford, 1969, 2.
291. Martin and Synge, *Adv. Protein Chem.*, 2, 7 (1945).
292. Dent, *Biochem. J.*, 43, 169 (1948).
293. Baumann, *Hoppe Seyler's Z. Physiol. Chem.*, 8, 299 (1884).
294. Greenstein and Winitz, *Chemistry of the Amino Acids*, Vol. 3, John Wiley & Sons, New York, 1961, 1879.
295. Bergeret and Chatagner, *Biochim. Biophys. Acta*, 14, 297 (1954).
296. Wollaston, *Ann. Chim. Phys.*, 76, 21 (1810).
297. Larsen and Kjaer, *Acta Chem. Scand.*, 16, 142 (1962).
298. Sweetman, *Nature*, 183, 744 (1959).
299. Calam and Waley, *Biochem. J.*, 86, 226 (1963).
300. Gmelin, *Hoppe Seyler's Z. Physiol. Chem.*, 327, 186 (1962).
301. Virtanen and Matikkala, *Acta Chem. Scand.*, 13, 1898 (1959).
302. van Veen and Hijman, *Rec. Trav. Chem. Pays-Bas*, 54, 493 (1935).
303. Armstrong and du Vigneaud, *J. Biol. Chem.*, 168, 373 (1947).
304. Fisher and Mallette, *J. Gen. Physiol.*, 45, 1 (1961).
305. Westall, *Biochem. J.*, 55, 244 (1953).
306. Adams and Capecchi, *Proc. Natl. Acad. Sci. USA*, 55, 147 (1966).
307. Cornforth and Henry, *J. Chem. Soc.*, p. 597 (1952).
308. Thoai and Robin, *Biochim. Biophys. Acta*, 13, 533 (1954).
309. Robertson and Marion, *Can. J. Chem.*, 37, 1043 (1959).
310. Strack, Friedel, and Hambsch, *Hoppe Seyler's Z. Physiol. Chem.*, 305, 166 (1956).
311. Gerritson, Vaughn, and Waisman, *Biochem. Biophys. Res. Commun.*, 9, 493 (1962).
312. Frimpter, *J. Biol. Chem.*, 236, PC51 (1961).
313. Huang, *Biochemistry*, 2, 296 (1963).
314. Sugii, Suketa, and Suzuchi, *Chem. Pharm. Bull.* (Tokyo), 12, 1115 (1964).
314a. Minale, Fattorusso, Cimino, DeStefano, and Nicolaus, *Gazzetta*, 97, 1636 (1967).
315. Wiehler and Marion, *Can. J. Chem.*, 36, 339 (1958).
316. Ohmori and Mizuhara, *Arch. Biochem. Biophys.*, 96, 179 (1962).
317. Ohmori, *Arch. Biochem. Biophys.*, 104, 509 (1964).
318. Horn, Jones, and Ringel, *J. Biol. Chem.*, 138, 141 (1941).
319. du Vigneaud and Brown, *J. Biol. Chem.*, 138, 151 (1941).
319a. Ampola, Bixby, Crawhall, Efron, Parker, Sneddon, and Young, *Biochem. J.*, 107, 16P (1968).
320. Thompson, *Nature*, 178, 593 (1956).
320a. Fujiwara, Itokawa, Uchino, and Inoue, *Experimenta*, 28, 254 (1972).
321. Mueller, *Proc. Soc. Exp. Biol. Med.*, 19, 161 (1922).
322. Greenstein and Winitz, *Chemistry of the Amino Acids*, Vol. 3, John Wiley & Sons, New York, 1961, 2125.
323. Matikkala and Virtanen, *Acta Chem. Scand.*, 17, 1799 (1963).
324. Guggenheim, *Die Biogenen. Amine*, S. Karger, Basel, (1951).
325. Challenger and Hayward, *Biochem. J.*, 58, 10 (1954).
326. Friis and Kjaer, *Acta Chem. Scand.*, 17, 2391 (1963).
327. Shrift and Virupaksha, *Biochim. Biophys. Acta*, 100, 65 (1965).
328. Gmelin and Virtanen, *Suomen Kemistilehti*, B35, 34 (1962); *Acta Chem. Scand.*, 16, 1378 (1962).
329. Matikkala and Virtanen, *Acta Chem. Scand.*, 16, 2461 (1962).
330. Spare and Virtanen, *Acta Chem. Scand.*, 17, 641 (1963).
331. Virtanen, Hatanaka, and Berlin, *Suomen Kemistilehti*, B35, 52 (1962).
332. Kutscher, *Zentralbl. Physiol.*, 24, 775 (1910).
333. Barger and Ewins, *Biochem.*, 7, 204, (1913).
334. Horn and Jones, *J. Biol. Chem.*, 139, 649 (1941).
334a. Tuve and Williams, *J. Biol. Chem.*, 236, 597 (1961).
335. Trelease, DiSomma, and Jacobs, *Science*, 132, 618 (1960).
336. Tanret, *C.R. Hebd. Seances Acad. Sci.* (Paris), 149, 222 (1909).
337. Waisvisz, van der Hoeven, and Rijenhuis, *J. Am. Chem. Soc.*, 79, 4524 (1957).

338. Behre and Benedict, *J. Biol. Chem.,* 82, 11 (1929).
339. Greenstein and Winitz, *Chemistry of the Amino Acids,* Vol. 3, John Wiley & Sons, New York, 1961, 2671.
340. Heath, *Nature,* 166, 106 (1950).
341. Tallan, Bella, Stein, and Moore, *J. Biol. Chem.,* 217, 703 (1955).
342. Bettelheim, *J. Am. Chem. Soc.,* 76, 2838 (1954).
343. Partridge, Elsden, and Thomas, *Nature,* 197, 1297 (1963).
344. Morner, *Hoppe Seyler's Z. Physiol. Chem.,* 88, 138 (1913).
344a. Savoie, Massin, and Savoie, *J. Clin. Invest.,* 52, 116 (1973).
345. Gross and Pitt-Rivers, *Biochem. J.,* 53, 645 (1950).
346. Drechsel, *Z. Biol.,* 33, 96 (1896).
347. Greenstein and Winitz, *Chemistry of the Amino Acids,* Vol. 3, John Wiley & Sons, New York, 1961, 1426.
348. Greenstein and Winitz, *Chemistry of the Amino Acids,* Vol. 3, John Wiley & Sons, New York, 1961, 2259.
349. Low, *J. Mar. Res.,* 10, 239 (1951).
349a. Hunt and Breuer, *Biochim. Biophys. Acta,* 252, 401 (1971).
349b. Hunt, *FEBS Lett.,* 24, 109 (1972).
350. Roche, Lissitzky, and Michel, *C.R. Hebd. Seances Acad. Sci.* (Paris), 232, 2047 (1951).
351. Roche and Yagi, *C. R. Soc. Biol.,* 146, 642 (1952).
352. Kendall, *J. Biol. Chem.,* 39, 125 (1919).
353. Coulson, *J. Sci. Food Agric. Abstr.,* 6, 674 (1955).
354. Butenandt, Weidel, Weicher, and Von Derjugen, *Hoppe Seyler's Z. Physiol. Chem.,* 279, 27 (1937).
355. Fowden, *Physiol. Plant,* 12, 657 (1959).
356. Stoll and Seebeck, *Helv. Chim. Acta,* 34, 481 (1951).
357. Chambers and Isbell, *J. Org. Chem.,* 29, 832 (1964).
358. Kittredge and Hughes, *Biochemistry,* 3, 991 (1964).
359. Kittredge, Roberts, and Simonsen, *Biochemistry,* 1, 624 (1962).
360. Kittredge, Isbell, and Hughes, *Biochemistry,* 6, 289 (1967).
360a. Korn, Dearborn, Fales, and Sokoloski, *J. Biol. Chem.,* 248, 2257 (1973).
361. Thoai and Robin, *Biochim. Biophys. Acta,* 14, 76 (1954).
362. Beatty, Magrath, and Ennor, *J. Am. Chem. Soc.,* 82, 4983 (1960); *J. Biol. Chem.,* 236, 1028 (1961).
363. Weiss and Stekol, *J. Am. Chem. Soc.,* 73, 2497 (1951).
364. Agren, *Acta Chem. Scand.,* 16, 1607 (1962).
365. Lipmann, *Biochem. Z.,* 262, 3 (1933).
366. Greenstein and Winitz, *Chemistry of the Amino Acids,* Vol. 3, John Wiley & Sons, New York, 1961, 2208.
367. Tomita and Sendju, *Hoppe Seyler's Z. Physiol. Chem.,* 169, 263 (1927).
368. Noguchi, Sakuma, and Tamaki, *Arch. Biochem. Biophys.,* 125, 1017 (1968).
369. Küng and Trier, *Hoppe Seyler's Z. Physiol. Chem.,* 85, 209 (1913).
370. Hosein and Proulx, *Nature,* 187, 321 (1960).
371. Carter and Bhattacharyya, *J. Am. Chem. Soc.,* 75, 2503 (1953).
372. Gulewitsch and Krimberg, *Hoppe Seyler's Z. Physiol. Chem.,* 45, 326 (1905).
373. Friedman, McFarlane, Bhattacharyya, and Fraenkel, *Arch. Biochem. Biophys.,* 59, 484 (1955).
374. Carter, Bhattacharyya, Weidman, and Fraenkel, *Arch. Biochem. Biophys.,* 38, 405 (1952).
375. Thomas, Elsden, and Partridge, *Nature,* 200, 651 (1963).
376. Terlain and Thomas, *Bull. Soc. Chim. Fr.,* p. 2349 (1971).
377. Kodama, Ohmori, Suzuki, Mizuhara, Oura, Isshiki, and Uemura, *Physiol. Chem. Phys.,* 3, 81 (1971).
378. Wälti and Hope, *J. Chem. Soc. (Org.),* p. 2326 (1971).
379. Larsen, *Acta Chem. Scand.,* 22, 1369 (1968).
380. Delange, Glazer, and Smith, *J. Biol. Chem.* 244, 1385 (1969).
381. Nakajima and Volcani, *Biochem. Biophys. Res. Commun.,* 39, 28 (1970).
382. Ogawa, Fukuda, and Sasaoka, *Biochim. Biophys. Acta,* 297, 60 (1973).
383. Zygmunt and Martin, *J. Med. Chem.,* 12, 953 (1969).
384. Hagemann, Pénasse, and Teillon, *Biochim. Biophys. Acta,* 17, 240 (1955).

α,β-UNSATURATED AMINO ACIDS

α,β-Unsaturated amino acids with free α-amino groups are not stable; *N*-acylated α,β-unsaturated amino acids are stable compounds. The α,β-unsaturated amino acids listed in this table are present in natural products in which they are stabilized by peptide bond formation. The addition of mercaptans to α,β-unsaturated amino acids and the reversible conversion to keto acids and amides are of biological significance.[1,2,11]

α,β-UNSATURATED AMINO ACIDS

No.	Amino acid/ synonym	Source	Structure	Formula (mol wt)	References Chemistry	References Spectral data
1	Dehydroalanine	Nisin Subtilin	$H_2C{=}C{-}COOH$, with NH_2	$C_3H_5NO_2$ 87.08	1, 2 2	1, 2 2
2	β-Methyldehydro-alanine	Nisin, Subtilin Stendomycin	$H_3C{-}H_2C{=}C{-}COOH$, with NH_2	$C_4H_7NO_2$ 101.10	1, 2 3	1, 2 3
3	Dehydroserine	Viomycin Capreomycin	$HOHC{=}C{-}COOH$, with NH_2	$C_3H_5NO_3$ 103.08	4 4	4 4
4	Dehydroleucine	Albonoursin	$(H_3C)(H_3C)CH{-}CH{=}C{-}COOH$, with NH_2	$C_6H_{11}NO_2$ 129.16	5	
5	Dehydrophenyl-alanine	Albonoursin	$C_6H_5{-}CH{=}C{-}COOH$, with NH_2	$C_9H_9NO_2$ 163.17	5	
6	Dehydrotryptophan	Telomycin	indolyl${-}CH{=}C{-}COOH$, with NH_2	$C_{11}H_{10}N_2O_2$ 202.21	6	6
7	Dehydroarginine	Viomycin Capreomycin	$H_2N{-}C({=}NH){-}NH{-}CH_2{-}CH_2{-}CH{=}C{-}COOH$, with NH_2	$C_6H_{12}N_4O_2$ 172.19	4 4	4 4
8	Dehydroproline	Ostreogrycin A	ring: $H_2C{-}CH$, H_2C, N(H), $C({=})COOH$	$C_5H_7NO_2$ 113.11	7	7
9	Dehydrovaline	Penicillin	$(H_3C)(H_3C)C{=}C{-}COOH$, with NH_2	$C_5H_9NO_2$ 115.13	8	
10	Dehydrocysteine	Micrococcin Thiostrepton	$HSHC{=}C{-}COOH$, with NH_2	$C_3H_5NO_2S$	9, 10	10

Compiled by Erhard Gross.

α,β-UNSATURATED AMINO ACIDS (continued)

REFERENCES

1.   Gross and Morell, *J. Am. Chem. Soc.,* 89, 2791 (1967).
2.   Gross and Morell, *Fed. Eur. Biochem. Soc. Lett.,* 2, 61 (1968).
3.   Bodanszky, Izdebski, and Muramatsu, *J. Am. Chem. Soc.,* 91, 2351 (1969).
4.   Bycroft, Cameron, Croft, Hassanali-Walji, Johnson, and Webb, *Tetrahedron Lett.,* p. 5901 (1968).
5.   Khokhlov and Lokshin, *Tetrahedron Lett.,* p. 1881 (1963).
6.   Sheehan, Mani, Nakamura, Stock, and Maeda, *J. Am. Chem. Soc.,* 90, 462 (1968).
7.   Delpierre, Eastwood, Gream, Kingston, Sarin, Todd, and Williams, *J. Chem. Soc.,* p. 1653 (1966).
8.   Abraham and Newton, in *Antibiotics,* Gottlieb and Shaw, Eds., Springer-Verlag, New York, 1967.
9.   Brookes, Clarke, Majhofer, Mijovic, and Walker, *J. Chem. Soc.,* p. 925 (1960).
10.   Bodanszky, Sheehan, Fried, Williams, and Birkhimer, *J. Am. Chem. Soc.,* 82, 4747 (1960).
11.   Gross, Morell, and Craig, *Proc. Natl. Acad. Sci. U.S.A.,* 62, 953 (1969).

This table originally appeared in Sober, Ed., *Handbook of Biochemistry* and Selected Data for Molecular Biology, 2nd ed., Chemical Rubber Co., Cleveland, 1970.

AMINO ACID ANTAGONISTS

Amino acid	Analog	System	Reference
α-Alanine	α-Aminoethanesulfonic acid	Bacteria	1
		Mouse tumor	2
	Glycine	Bacteria	3
	α-Aminoisobutyric acid	Bacteria	4
	Serine	Bacteria	3
D-Alanine	D-Cycloserine	Bacterial cell wall	5-9, 10
	O-Carbamyl-D-serine	Bacterial cell wall	11
	D-α-Aminobutyric acid	Bacterial cell wall	12
β-Alanine	β-Aminobutyric acid	Yeast	13
	Propionic acid	Bacteria	14
	Asparagine	Yeast	15
	D-Serine	Bacteria	16
Arginine	Canavanine	Yeast, *Neurospora*, Bacteria	17, 18-22
		Carcinosarcoma	23
		Animals	24, 25
		Plants	26, 27
		Tissue Culture	28, 29
	Lysine	Arginase	30
	Ornithine	Arginase	31
	Homoarginine	Bacteria	22, 32
Aspartic acid	Cysteic acid	Bacteria	1, 33, 34
		Bacteria	35
	β-Hydroxyaspartic acid	Bacteria	36, 34, 37
	Diaminosuccinic acid	Bacteria	36
	Aspartophenone	Bacteria, yeast	4
	α-Aminolevulinic acid	Bacteria, yeast	4
	α-Methylaspartic acid	Bacteria	38
	β-Aspartic acid hydrazide	Bacteria	38
	S-Methylcysteine sulfoxide	Bacteria	39
	β-Methylaspartic acid	Bacteria	40
	Hadacidin	Purine biosynthesis	41
Asparagine	2-Amino-2-carboxyethane-sulfonamide	*Neurospora*	42
Cysteine	Allylglycine	Bacteria, yeast	43
α, ε-Diaminopimelic acid	α,α-Diaminosuberic acid	Bacteria	44
	α,α-Diaminosebacic acid	Bacteria	44
	β-Hydroxy-α,ε-diaminopimelic acid	*Escherichia coli*	45
	γ-Methyl-α,ε-diaminopimelic acid	*E. coli*	46
	Cystine	*E. coli*	47
Glutamic acid	Methionine sulfoxide	Bacteria	48, 49
		Glutamine synthesis	50
	γ-Glutamylethylamide	Bacteria	51
	β-Hydroxyglutamic acid	Bacteria	52, 53
	Methionine sulfoximine	Bacteria	54, 55
	α-Methylglutamic acid	Enzymes	56, 57
	γ-Phosphonoglutamic acid	Glutamine synthesis	58
	P-Ethyl-γ-phosphonoglutamic acid	Glutamine synthesis	58
	γ-Fluoroglutamic acid[a]	Glutamine synthesis	59
Glutamine	S-Carbamylcysteine	Bacteria	60
		Ascites cells	61
	O-Carbamylserine	Bacteria	62
	O-Carbazylserine	Bacteria	63
	3-Amino-3-carboxypropane-sulfonamide	*E. coli*	64
	N-Benzylglutamine	*Streptococcus lactis*	65
	Azaserine	Enzymes	66
	6-Diazo-5-oxonorleucine	Enzymes	66
	γ-Glutamylhydrazide	*S. faecalis*	67
Glycine	α-Aminomethanesulfonic acid	Bacteriophage	68
		Vaccinia virus	69
		Bacteria	1
		E. coli	4

[a]Some of the observed inhibition may be due to fluoride ion present in the amino acid preparation or formed during incubation.[208]

AMINO ACID ANTAGONISTS (continued)

Amino acid	Analog	System	Reference
Histidine	D-Histidine	Histidase	70
	Imidazole		70
	2-Thiazolealanine	*E. coli*	71
	1,2,4-Triazolealanine	*E. coli*	71, 72
		Salmonella	73
Isoleucine	Leucine	Bacteria	74
		Rats	75
	Methallylglycine	Bacteria, yeast	4, 43
	ω-Dehydroisoleucine	Bacteria	76
	3-Cyclopentene-1-glycine	Bacteria	77
	Cyclopentene glycine	Bacteria	78
	2-Cyclopentene-1-glycine	Bacteria	79
	O-Methylthreonine	Tumor cells	80
	β-Hydroxyleucine	Bacteria	37
Leucine	D-Leucine	Bacteria	81
	α-Aminoisoamylsulfonic acid	Bacteria	1, 33
		Mouse tumor	2
	Norvaline	Bacteria	4
	Norleucine	Bacteria	1, 4, 82
	Methallylglycine	Yeast, bacteria	4
	α-Amino-β-chlorobutyric acid	Yeast, bacteria	4
	Valine	Bacteria	83
	δ-Chloroleucine	*Neurospora*	84
	Isoleucine	Bacteria	85
	β-Hydroxynorleucine	Bacteria	86
	β-Hydroxyleucine	Bacteria	86
	Cyclopentene alanine	Bacteria	87
	3-Cyclopentene-1-alanine	Bacteria	88
	2-Amino-4-methylhexenoic acid	Bacteria	89
	5′,5′,5′-Trifluoroleucine	*E. coli*	90
	4-Azaleucine	*E. coli*	91
Lysine	α-Amino-ε-hydroxycaproic acid	Rat	92
	Arginine	*Neurospora*	93
	2,6-Diaminoheptanoic acid	Bacteria	94
	Oxalysine	Bacteria	95
	3-Aminomethylcyclohexane glycine	Bacteria	96
	3-Aminocyclohexane alanine	Bacteria	97
	trans-4-Dehydrolysine	Bacteria	98
	S-(β-Aminoethyl)-cysteine	Bacteria	99
	4-Azalysine	Bacteria	
Methionine	2-Amino-5-heptenoic acid (crotylalanine)	*E. coli*	100
	2-Amino-4-hexenoic acid (crotylglycine)	*E. coli*	101
	Methoxinine	Bacteria	102
		Vaccinia virus	69
		Rats	103
	Norleucine	Bacteria	82, 104, 105
		Animal tissues	106
		Casein	107
	Ethionine	Bacteria, animals	28, 102, 105, 108-120
		Amylase	121, 122
		Yeast	123
		Tumors	124
		Pancreatic proteins	125
	Methionine sulfoximine	Bacteria	126
	Threonine	*Neurospora*	127
	Selenomethionine	*Chlorella*	128
		E. coli, yeast	129-132
Ornithine	α-Amino-δ-hydroxyvaleric acid	Bacteria	21
	Canaline	Bacteria	21

AMINO ACID ANTAGONISTS (continued)

Amino acid	Analog	System	Reference
Phenylalanine	α-Amino-β-phenylethanesulfonic acid	Mouse tumor	2
	Tyrosine	Bacteria	133
	β-Phenylserine	Bacteria	134, 135, 136
	Cyclohexylalanine	Rats	137
	o-Aminophenylalanine	E. coli	138
	p-Aminophenylalanine	Bacteria	139, 140
	Fluorophenylalanines	Fungi, bacteria	141-146, 140, 147, 148
		Lysozyme, albumin	149
		Muscle enzymes	150
		Amylase	151
		Hemoglobin	152
		Rats	153
Phenylalanine	Chlorophenylalanines	Fungi	147
	Bromophenylalanines	Fungi	147
	β-2-Thienylalanine	Rat, bacteria, yeast	136, 154-164
		β-Galactosidase	165
	β-3-Thienylalanine	Bacteria, yeast	166
	β-2-Furylalanine	Bacteria, yeast	4
	β-3-Furylalanine	Bacteria, yeast	4
	β-2-Pyrrolealanine	Bacteria, yeast	167
	1-Cyclopentene-1-alanine	Bacteria	87
	1-Cyclohexene-1-alanine	Bacteria	87
	2-Amino-4-methyl-4-hexenoic acid	Bacteria	89
	S-(1,2-Dichlorovinyl)-cysteine	E. coli	168
	β-4-Pyridylalanine	Bacteria	169
	Tryptophan	Bacteria	170
	β-2-Pyridylalanine	Bacteria	171
	β-4-Pyrazolealanine	Bacteria	171
	β-4-Thiazolealanine	Bacteria	171
	p-Nitrophenylalanine	Bacteria	140
Proline	Hydroxyproline	Fungi	172
	3,4-Dehydroproline	Bacteria, beans	173, 174
	Azetidine-2-carboxylic acid	Bacteria, beans	175, 176
		Actinomycin	177, 178
Serine	α-Methylserine	Bacteria	4
	Homoserine	Bacteria	4
	Threonine	Bacteria	179, 180
	Isoserine	Enzymes	181
Threonine	Serine	Bacteria	179, 180, 182
	β-Hydroxynorvaline	Bacteria	86, 183
	β-Hydroxynorleucine	Bacteria	86, 183
Thyroxine	Ethers of 3,5-diiodotyrosine	Tadpoles	184
Tryptophan	Methyltryptophans	Bacteria	141, 185, 186-188
		Bacteriophage	189
	Naphthylalanines	Bacteria	190, 191
		Rat	4
	Indoleacrylic acid	Bacteria	192
	Naphthylacrylic acid	Bacteria	193
	β-(2-Benzothienyl)alanine	Bacteria	194
	Styrylacetic acid	Bacteria	193
	Indole	Bacteriophage	195
	α-Amino-β-3(indazole)-propionic acid (Tryptazan)	Yeast	196
		Enzyme	197
		E. coli	198
	5-Fluorotryptophan	Enzyme	199, 200
	6-Fluorotryptophan	Enzyme	201
	7-Azatryptophan	E. coli	202-204, 205

AMINO ACID ANTAGONISTS (continued)

Amino acid	Analog	System	Reference
Tyrosine	Fluorotyrosines	Fungi	147, 206
		Rat	206
	p-Aminophenylalanine	Fungi	139
	m-Nitrotyrosine	Bacteria	140
	β-(5-Hydroxy-2-pyridyl)-alanine	Bacteria	207
Valine	α-Aminoisobutanesulfonic acid	Bacteria	1, 2, 33
		Vaccinia	69
	α-Aminobutyric acid	Bacteria	182, 4
	Norvaline	Bacteria	4
	Leucine, isoleucine	Bacteria	182, 83
	Methallylglycine	Bacteria, yeast	4
	β-Hydroxyvaline	Bacteria	86, 183
	ω-Dehydroalloisoleucine	Bacteria	76

Courtesy of Herbert M. Kagan, Boston University School of Medicine (from Meister, A., Ed., *Biochemistry of the Amino Acids*, 2nd ed., Volume 1, 1965, 233—238).

PROPERTIES OF THE α-KETO ACID ANALOGS OF AMINO ACIDS

α-Keto acid	α-Amino acid analog	2,4-Dinitrophenylhydrazone Crystallization M.p. (°C)	Solvent[a]	Amino acids after hydrogenation (14)	Reduction by lactic dehydrogenase[b,g]	Decarboxylation by yeast decarboxylase[c]
Pyruvic	Alanine	216	h	Alanine	26,800	1.200
α-Ketoadipamic	α-Aminoadipamic acid (homoglutamine)					
α-Ketoadipic	α-Aminoadipic acid	208	h	α-Aminoadipic acid		
α-Ketobutyric	α-Aminobutyric acid	198	h	α-Aminobutyric acid	21,000	
α-Ketoheptylic	α-Aminoheptylic acid	130	e, l	α-Aminoheptylic acid	483	
α-Keto-ε-hydroxycaproic	α-Amino-ε-hydroxy-caproic acid	183	h	α-Amino-ε-hydroxy-caproic acid	181	25
Mesoxalic	α-Aminomalonic acid	205	hc	α-Aminomalonic acid, glycine		
α-Ketophenylacetic	α-Aminophenyl-acetic acid	193	h	α-Aminophenyl-acetic acid, cyclohexyl-glycine	2.6	0
DL-Oxalosuccinic	α-Aminotri-carballylic acid					
α-Keto-δ-guanidinovaleric	Arginine	216, 267 (1)		Arginine	9.0	0
α-Ketosuccinamic	Asparagine	183		Asparagine, aspartic acid	8,930	
Oxalacetic	Aspartic acid	218	h	Aspartic acid, alanine, β-alanine	12.8	
α-Keto-δ-carbamidovaleric	Citrulline	190	h	Citrulline	4.3	0
β-Cyclohexylpyruvic	β-Cyclohexyl-alanine	189	h	β-Cyclohexyl-alanine	14.6	0
β-Sulfopyruvic	Cysteic acid	210	a	Cysteic acid, alanine	89.6	
β-Mercaptopyruvic	Cysteine	195–200 (2) 161–162 (3)		Alanine	27,000	750
α-Keto-γ-ethiolbutyric	Ethionine	131	h	Ethionine	1,650	121
α-Ketoglutaric	Glutamic acid	220	h	Glutamic acid	9.2	0
α-Ketoglutaric γ-ethyl ester	Glutamic acid γ-ethyl ester				49.0	721
α-Ketoglutaramic	Glutamine			Glutamine, glutamic acid		
Glyoxylic	Glycine	203	h	Glycine	21,100	0
β-Imidazolylpyruvic	Histidine	190–192, 240 (1)	hc, e, l	Histidine		
α-Keto-γ-hydroxybutyric	Homoserine					
DL-α-Keto-β-methylvaleric	DL-Isoleucine (or DL-alloisoleucine)	169	h	Isoleucine		
L-α-Keto-β-methylvaleric	L-Isoleucine (or D-alloisoleucine)	176	h	Isoleucine	5.0	1,000
D-α-Keto-β-methylvaleric	D-Isoleucine (or L-alloisoleucine)	176	h	Isoleucine	1.9	280

[a]h = water; e = ethyl acetate; 1 = ligroin; hc = hydrochloric acid; ac = glacial acetic acid; a = ethanol.
[b]Mole $\times 10^{-8}$ of DPNH oxidized per mg of enzyme per minute at 26° (9, 10).
[c]μl. CO_2 per hour (10).
[e]Originally designated *d* (11). Originally designated *l* (11).
[g]Additional data have been published on the reduction of α-keto acids by lactic dehydrogenase (12).
acids by lactic dehydrogenase (12).

PROPERTIES OF THE α-KETO ACID ANALOGS OF AMINO ACIDS (continued)

| α-Keto acid | α-Amino acid analog | 2,4-Dinitrophenylhydrazone | | | Reduction by lactic dehydrogenase[b,g] | Decarboxylation by yeast decarboxylase[c] |
| | | Crystallization | | Amino acids after hydrogenation (14) | | |
		M.p. (°C)	Solvent[a]			
α-Ketoisocaproic	Leucine	162	h	Leucine	3.2	306
Trimethylpyruvic	*tert*-Leucine	180	h	*tert*-Leucine		
α-Keto-ε-aminocaproic	Lysine	212	h	Lysine, pipecolic acid		
α-Keto-γ-methiolbutyric	Methionine	150	h	Methionine	1,550	125
α-Keto-γ-methylsulfonylbutyric	Methionine sulfone	175	h	Methione sulfone		
α-Keto-δ-nitroguanidinovaleric	Nitroarginine	225	ac	Nitroarginine, arginine	42.6	0
α-Ketocaproic	Norleucine	153	h	Norleucine	560	
α-Ketovaleric	Norvaline	167	h	Norvaline	1,470	
α-Keto-δ-aminovaleric	Ornithine, proline	232–242 (4) 219 (5) 211–212 (6)		Ornithine, proline, pentahomoserine[d]		
Phenylpyruvic	Phenylalanine	162–164, 192–194 (7)		Phenylalanine	755	0
S-Benzyl-β-mercaptopyruvic	*S*-Benzylcysteine	150	a			
β-Hydroxypyruvic	Serine	162	e	Serine, alanine	26,000	0[h]
N-Succinyl-α-amino-ε-ketopimelic (15)	*N*-Succinyl-α,ε-diaminopimelic acid	137–143	h	*N*-Succinyl-α,ε-diaminopimelic acid		
DL-α-Keto-β-hydroxybutyric	DL-Threonine (or DL-allothreonine)	157–158 (8)		Threonine, α-aminobutyric acid	20,000	
β-[3,5-Diiodo-4-(3′,5′-diiodo 4′-hydroxyphenoxy)phenyl]	Thyroxine					
β-Indolylpyruvic	Tryptophan	169 (1)		Tryptophan	670	0
p-Hydroxyphenylpyruvic	Tyrosine	178	h	Tyrosine	345	0
α-Ketoisovaleric	Valine	196	h	Valine	103	922

[d]α-Amino-δ-hydroxy-*n*-valeric acid.

[h]This keto acid has been reported to be decarboxylated by yeast preparations; the reaction is much more rapid at pH 6.5 than at 5 (13).

From Meister, *Biochemistry of the Amino Acids*, 2nd ed., Academic Press, New York, 1965, 162—164. With permission.

REFERENCES

1. **Stumpf and Green,** *J. Biol. Chem.,* 153, 387 (1944).
2. **Schneider and Reinefeld,** *Biochem. Z.,* 318, 507 (1948).
3. **Meister, Fraser, and Tice,** *J. Biol. Chem.,* 206, 561 (1954).
4. **Krebs,** *Enzymologia,* 7, 53 (1939).
5. **Blanchard, Green, Nocito, and Ratner,** *J. Biol. Chem.,* 155, 421 (1944).
6. **Meister,** *J. Biol. Chem.,* 206, 579 (1954).
7. **Fones,** *J. Org. Chem.,* 17, 1534 (1952).
8. **Sprinson and Chargaff,** *J. Biol. Chem.,* 164, 417 (1947).
9. **Meister,** *J. Biol. Chem.,* 197, 309 (1953).
10. **Meister,** *J. Biol. Chem.,* 184, 117 (1950).
11. **Meister,** *J. Biol. Chem.,* 190, 269 (1951).
12. **Czok and Büchler,** *Advanc. Protein. Chem.,* 15, 315 (1960).
13. **Dickens and Williamson,** *Nature,* 178, 1349 (1956).
14. **Meister and Abendschein,** *Anal. Chem.,* 28, 171 (1956).
15. **Gilvarg,** *J. Biol. Chem.,* 236, 1429 (1961).

FAR ULTRAVIOLET ABSORPTION SPECTRA OF AMINO ACIDS

The absorption of a chromophoric amino acid in this spectral region is due to the combined absorptions of the side chain chromophore and of the carboxylate group. Because the carboxylate group is consumed in polymerizing amino acids to polypeptides, an amino acid residue absorbs less intensely than a free amino acid. The magnitude of this difference can be estimated from the spectra of the nonchromophoric amino acids — leucine, proline, alanine, serine and threonine — whose total absorption is due only to the carboxylate group. The variations between the absorptions of these amino acids reflect the variability of carboxylate absorption in slightly different environments.

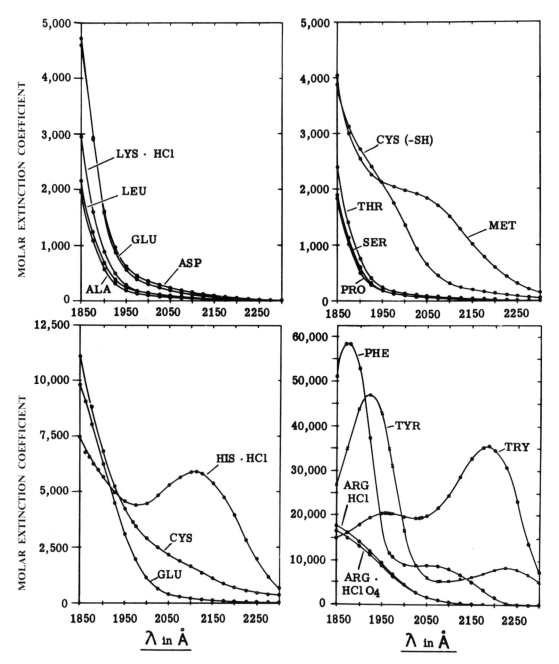

Far ultraviolet spectra of amino acids. All amino acids were in aqueous solution at pH 5, except cystine (Ph 3). The dibasic amino acids were measured as hydrochlorides and the absorbance corrected by subtracting the absorbance contribution of chloride ion. Taken from Wetlaufer, *Adv. Protein Chem.*, 17, 320 (1962). With the permission of Academic Press and R. Sussman-McDiarmid and W. Gratzer.

FAR ULTRAVIOLET ABSORPTION SPECTRA OF AMINO ACIDS

Amino acid	$\lambda_{190.0}$	$\lambda_{197.0}$	$\lambda_{205.0}$	Feature[a] Maxima		Minima		Shoulder	
				λ	ε	λ	ε	λ	ε
In Neutral Water									
Tryptophan	17.60	20.50	19.60	196.7	20.60	203.3	19.40	----	----
				218.6	46.70				
Tyrosine	42.80	35.50	5.60	192.5	47.50	208.0	4.88		
				223.2	8.26				
Phenylalanine	54.50	12.30	9.36	187.7	59.60	202.5	8.96		
				206.0	9.34				
Histidine[b]	5.57	4.35	5.17	211.3	5.86	198.4	4.22		
Cysteine[c]	2.82	1.94	0.730	—	—			195.2	2.18
1/2 Cystine	3.25	1.76	1.05					207.0	0.96
Methionine	2.69	2.11	1.86					204.7	1.89
Arginine[c]	13.1	6.61	1.36						
Acids[d]	1.61	0.460	0.230						
Amides[d]	6.38	2.06	0.400						
Lysine[b]	0.890	0.200	0.110						
Leucine[d]	0.670	0.190	0.100						
Alanine[d]	0.570	0.150	0.070						
Proline[d]	0.540	0.150	0.070						
Serine[d]	0.610	0.160	0.080						
Threonine[d]	0.750	0.180	0.100						
In 0.1 M Sodium Dodecyl Sulfate[d]									
Tryptophan	16.70	19.70	19.00	197.3	19.80	204.1	18.90	----	----
				220.0	46.60				
Tyrosine	39.10	36.60	5.87	193.2	45.10	208.5	4.88		
				223.7	7.86				
Phenylalanine	54.10	11.30	8.45	188.3	57.0	201.5	7.79		
				207.2	8.66				
Histidine[b]	5.88	4.48	5.03	212.0	5.88	199.0	4.25		
Cysteine[c]	2.66	1.79	0.650	—	—			194.5	2.15
1/2 Cystine	2.62	1.61	0.92	—	—				
Methionine	2.67	2.10	1.84	—	—			204.1	1.89
Arginine[c]	12.50	5.70	0.94	—	—				

Compiled by Ruth McDiarmid.

[a]Molar extinctions, $\epsilon \times 10^{-3}$. Wavelength, λ, in millimicrons.

[b]The absorptions of lysine and histidine were determined for the hydrochlorides and corrected for the absorption of the chloride ion. $\epsilon_{Cl^-} = 0.740$ ($\lambda = 190.0$), 0.050 ($\lambda = 197.0$) and 0($\lambda = 205.0$).

[c]The absorptions of cysteine and arginine were determined for the $HClO_4$ salt.

[d]The spectra of the carboxylic acids, the amides, the aliphatic and the hydroxy amino acids and lysine are unchanged from those obtained in neutral water.

This figure and table originally appeared in Sober, Ed., *Handbook of Biochemistry and Selected Data for Molecular Biology*, 2nd ed., Chemical Rubber Co., Cleveland, 1970.

UV ABSORPTION CHARACTERISTICS OF *N*-ACETYL METHYL ESTERS OF THE AROMATIC AMINO ACIDS, CYSTINE AND OF *N*-ACETYLCYSTEINE

	Water[a]		Ethanol[a]			Water[a]		Ethanol[a]	
	λ	ε	λ	ε		λ	ε	λ	ε
Phenylalanine					**Tryptophan**				
Inflection	(208)	10.20	(208)	10.40	Minimum	205	21.40	206	21.30
Inflection	(217)	5.00	(217)	5.30	Maximum	219	35.00	221	37.20
Minimum	240	0.080	244	0.088	Minimum	245	1.900	245	1.560
Maximum	241.2	0.086	242.0	0.093	Maximum	**279.8**	**5.600**	**282.0**	**6.170**
Maximum	246.5	0.115	247.3	0.114	Maximum	288.5	4.750	290.6	5.330
Maximum	251.5	0.157	252.3	0.158					
Maximum	**257.4**	**0.197**	**258.3**	**0.195**	**Cystine**				
Inflection	(260.7)	—	(261.2)	—	Inflection	(250)	0.360	(253)	0.372
Maximum	263.4	0.151	264.2	0.155	Inflection	260	0.280	260	0.320
Maximum	267.1	0.091	267.8	0.096	Inflection	280	0.110	280	0.135
					Inflection	300	0.025	300	0.035
Tyrosine					Inflection	320	0.006	320	0.007
Maximum	193	51.70	—	—					
Minimum	212	7.00	212	6.20	*N*-Acetylcysteine				
Maximum	224	8.80	227	10.20		250	0.015	250	0.020
Minimum	247	0.176	246	0.174		280	0.005	280	0.005
Maximum	**274.6**	**1.420**	**278.4**	**1.790**		320	(nil)	320	(nil)
Inflection	281.9	—	285.7	—					

[a]λ, wavelength in millimicrons; $\varepsilon \times 10^{-3}$, molar extinctions. Inflection denotes unresolved inflection.

Compiled by W. B. Gratzer.

From J. E. Bailey, Ph.D. Thesis, London University (1966).

NUMERICAL VALUES OF THE ABSORBANCES OF THE AROMATIC AMINO ACIDS IN ACID, NEUTRAL, AND ALKALINE SOLUTIONS

Table 1
MOLECULAR ABSORBANCES OF TYROSINE

nm[c]	Neutral[a]	Alkaline[b]	nm[c]	Neutral[a]	Alkaline[b]
230	4980	7752 ± 108	276	1367 ± 0	1206 ± 4
232	3449	8667 ± 38	278	1260 ± 2	1344 ± 5
234	1833 ± 14	9634 ± 19	280	1197 ± 0	1507 ± 5
236	1014 ± 43	10440 ± 20	282	1112 ± 2	1675 ± 5
238	571 ± 36	11000 ± 10	284	845 ± 8	1850 ± 6
240	349 ± 34	11300 ± 20	286	506 ± 7	2024 ± 4
240.5 ↑	—	11340 ± 30	288	248 ± 8	2179 ± 5
242	252 ± 20	11230 ± 40	290	113 ± 0	2300 ± 7
244	209 ± 18	10760 ± 50	292	50 ± 1	2367 ± 5
245.3 ↓	202 ± 20	—	293.2 ↑	—	2381 ± 6
246	205 ± 17	9918 ± 78	294	23 ± 1	2377 ± 8
248	218 ± 15	8734 ± 72	296	13 ± 0	2317 ± 10
250	246 ± 14	7382 ± 56	298	8 ± 1	2195 ± 16
252	287 ± 13	5844 ± 77	300	6 ± 0	2006 ± 23
254	341 ± 14	4471 ± 55	302	5 ± 1	1747 ± 29
256	401 ± 12	3360 ± 46	304	3 ± 0	1445 ± 27
258	485 ± 10	2476 ± 20	306	2 ± 1	1107 ± 35
260	582 ± 9	1883 ± 17	308	1 ± 0	800 ± 27
262	693 ± 13	1467 ± 7	310	1 ± 0	547 ± 21
264	821 ± 13	1204 ± 14	312	—	346 ± 15
266	960 ± 14	1054 ± 16	314	—	206 ± 12
268	1083 ± 13	985 ± 13	316	—	118 ± 9
269.3 ↓	—	974 ± 8	318	—	67 ± 5
270	1197 ± 9	979 ± 9	320	—	32 ± 3
272	1310 ± 9	1019 ± 8	322	—	15 ± 3
274	1394 ± 6	1094 ± 8	324	—	6 ± 2
274.8 ↑	1405 ± 7	—	326	—	1 ± 1

Compiled by Elmer Mihalyi.

[a]0.1 *M* phosphate buffer, pH 7.1.
[b]0.1 *N* KOH.
[c]Maxima, minima, and inflection points are indicated by ·↑, ↓, and ~.

Reprinted with permission from *J. Chem. Eng. Data*, 13, 179 (1969). Copyright 1969 American Chemical Society.

Table 2
MOLECULAR ABSORBANCES OF TRYPTOPHAN

nm[c]	Neutral[a]	Alkaline[b]	nm[c]	Neutral[a]	Alkaline[b]
230	6818	13200	279.0 ↑	5579 ± 14	—
232	4037 ± 60	7470	280	5559 ± 12	5377 ± 43
234	2772 ± 71	4354 ± 81	280.4 ↑	—	5385 ± 34
236	2184 ± 64	2951 ± 50	282	5323 ± 10	5302 ± 34
238	1904 ± 55	2282 ± 29	284	4762 ± 11	4962 ± 42
240	1764 ± 52	1959 ± 30	285.8 ↓	4471 ± 6	—
242.0 ↓	1737 ± 49	1813 ± 25	286	4482 ± 11	4596 ± 22
244	1772 ± 48	1773 ± 29	286.8 ↓	—	4565 ± 27
244.4 ↓	—	1763 ± 29	287.8 ↑	4650 ± 12	—
246	1869 ± 40	1792 ± 27	288	4646 ± 16	4634 ± 19
248	2018 ± 35	1877 ± 23	288.3 ↑	—	4639 ± 28
250	2217 ± 32	2013 ± 25	290	3935 ± 5	4393 ± 32
252	2462 ± 19	2187 ± 37	292	2732 ± 5	3551 ± 46
254	2760 ± 27	2410 ± 38	294	1824 ± 5	2666 ± 27
256	3087 ± 20	2664 ± 25	296	1211 ± 10	1990 ± 24
258	3422 ± 18	2953 ± 39	298	797 ± 4	1472 ± 19
260	3787 ± 17	3261 ± 34	300	510 ± 1	1064 ± 19
262	4142 ± 14	3586 ± 46	302	314 ± 3	755 ± 16
264	4472 ± 10	3895 ± 32	304	184 ± 2	517 ± 10
266	4777 ± 14	4212 ± 48	306	112 ± 4	333 ± 6
268	5020 ± 15	4481 ± 46	308	55 ± 9	217 ± 4
270	5220 ± 8	4742 ± 37	310	27 ± 11	129 ± 5
272	5331 ± 5	4933 ± 45	312	11 ± 8	84 ± 8
272.1 ↑	5344 ± 5	—	314	3 ± 2	53 ± 7
273.6 ↓	5329 ± 10	—	316	—	31 ± 7
274	5341 ± 8	5025 ± 34	318	—	17 ± 4
274.5 ~	—	5062 ± 38	320	—	8 ± 2
276	5431 ± 8	5108 ± 39	322	—	3 ± 4
278	5554 ± 12	5275 ± 46			

Compiled by Elmer Mihalyi.

[a]0.1 *M* phosphate buffer, pH 7.1.

[b]0.1 *N* KOH.

[c]Maxima, minima, and inflection points are indicated by ↑, ↓, and ~.

Reprinted with permission from *J. Chem. Eng. Data*, 13, 179 (1969). Copyright 1969 American Chemical Society.

Table 3
MOLECULAR ABSORBANCES OF PHENYLALANINE

nm[c]	Neutral[a]	nm[c]	Alkaline[b]	nm[c]	Neutral[a]	nm[c]	Alkaline[b]
230	32.8 ± 1.5	230	161.9 ± 1.9	257.6 ↑	195.1 ± 1.5	257	188.4 ± 2.8
232	32.1 ± 1.6	232	99.2 ± 1.9	258	193.4 ± 1.3	258	209.1 ± 0.3
234	35.6 ± 2.1	234	70.7 ± 2.4	259	171.9 ± 1.0	258.2 ↑	209.6 ± 0.2
236	42.8 ± 2.1	236	63.3 ± 2.7	260	147.0 ± 0.6	260	184.2 ± 1.0
238	48.5 ± 2.3	238	62.3 ± 2.6	261.9 ↓	127.7 ± 1.5	260.7 ~	178.6 ± 0.3
240	59.4 ± 2.0	240	68.9 ± 3.2	262	128.1 ± 1.4	262	157.8 ± 0.9
242 ~	72.2 ± 2.3	242	83.0 ± 2.8	263.7 ↑	151.5 ± 0.6	262.7 ↓	105.5 ± 1.3
		243 ~	85.4 ± 2.9	264	148.7 ± 0.4	263.9 ↑	161.2 ± 1.0
244	80.1 ± 2.1	244	89.0 ± 3.0	265	119.8 ± 1.3	264	160.0 ± 2.1
246	102.0 ± 0.6	246	108.9 ± 2.8	266	91.8 ± 1.4	266	114.3 ± 1.6
247.4 ↑	110.7 ± 2.2	247	120.9 ± 1.5	266.8 ~	85.6 ± 1.5	266.5 ↓	109.7 ± 1.8
248	109.8 ± 1.9	248.0 ↑	126.1 ± 1.4			267.7 ↑	117.7 ± 1.8
248.3 ↓	109.5 ± 2.0	248.7 ↓	125.1 ± 1.7	268	74.7 ± 1.0	268	115.0 ± 1.0
250	123.5 ± 2.6	250	132.7 ± 1.8	270	30.0 ± 1.8	270	50.2 ± 2.0
251	143.0 ± 2.8	251	149.3 ± 1.9	272	14.3 ± 1.0	272	18.7 ± 1.1
252	153.9 ± 1.0	252	167.0 ± 1.1	274	5.4 ± 0.3	274	7.4 ± 0.3
252.2 ↑	154.1 ± 1.0	252.9 ↑	171.5 ± 1.3	276	2.2 ± 0.4	276	2.6 ± 0.4
254	139.6 ± 1.0	254	166.3 ± 0.8	278	1.1 ± 0.5	278	0.7 ± 0.3
254.5 ↓	138.5 ± 1.4	254.9 ↓	162.8 ± 1.7	280	0.7 ± 0.3	280	0.4 ± 0.2
256	156.5 ± 2.2	256	168.4 ± 1.9				

Compiled by Elmer Mihalyi.

[a]0.1 M phosphate buffer, pH 7.1.
[b]0.1 N KOH.
[c]Maxima, minima, and inflection points are indicated by ↑, ↓, and ~.

Table 4
ALKALINE[b] VS. NEUTRAL[a] DIFFERENCE SPECTRA
OF TYROSINE, TRYPTOPHAN, AND PHENYLALANINE

nm	Tyrosine	Tryptophan	Phenylalanine	nm	Tyrosine	Tryptophan	Phenylalanine
230	3041	4135	123.9	280	315 ± 1	-191 ± 18	—
232	5440	3213	66.0	282	558 ± 1	-24 ± 4	—
234	7608	1621	35.4	284	994 ± 10	194 ± 3	—
236	9415	732 ± 35	20.7	286	1513 ± 11	110 ± 10	—
238	10490	345 ± 23	13.9	288	1936 ± 1	11 ± 3	—
240	11060	149 ± 21	9.9	290	2196 ± 14	467 ± 8	—
242	11090	45 ± 15	11.1	292	2331 ± 15	802 ± 3	—
244	10660	-40 ± 15	8.8	294	2357 ± 7	830 ± 8	—
246	9844	-104 ± 11	9.0	296	2307 ± 6	755 ± 10	—
248	8567	-172 ± 10	16.4	298	2194 ± 7	652 ± 13	—
250	7205	-233 ± 9	8.3	300	2002 ± 3	527 ± 5	—
252	5671	-298 ± 16	15.3	302	1754 ± 2	413 ± 11	—
254	4344	-371 ± 10	25.8	304	1437 ± 2	300 ± 14	—
256	3127	-435 ± 14	11.3	306	1097 ± 9	205 ± 8	—
258	2142	-490 ± 19	19.1	308	792 ± 14	137 ± 5	—
260	1368 ± 37	-535 ± 13	37.8	310	526 ± 13	88 ± 3	—
262	820 ± 36	-564 ± 8	26.9	312	334 ± 9	55 ± 8	—
264	420 ± 25	-580 ± 10	16.1	314	221 ± 19	22 ± 14	—
266	125 ± 20	-573 ± 12	22.4	316	101 ± 7	16 ± 10	—
268	-78 ± 18	-539 ± 14	42.8	318	62 ± 2	5 ± 2	—
270	-225	-486 ± 12	17.4	320	28 ± 4	0	—
272	-296	-394 ± 12	4.0	322	12 ± 5	—	—
274	-299	-308 ± 7	1.7	324	3 ± 4	—	—
276	-158	-312 ± 16	0.3	326	1 ± 1	—	—
278	89 ± 5	-278 ± 13	0	328	0	—	—

Compiled by Elmer Mihalyi.

[a]0.1 *M* phosphate buffer, pH 7.1.
[b]0.1 *N* KOH.

Reprinted with permission from *J. Chem. Eng. Data,* 13, 179 (1969). Copyright 1969 American Chemical Society.

Table 5
ACID[b] VS. NEUTRAL[a] DIFFERENCE SPECTRA
OF TYROSINE, TRYPTOPHAN, AND PHENYLALANINE

nm	Tyrosine	Tryptophan	Phenylalanine	nm	Tyrosine	Tryptophan	Phenylalanine
230	—	—	46.7	276	−40	128 ± 16	0
232	576	421 ± 49	34.1	278	−45	110 ± 10	—
234	441	610 ± 37	23.3	280	−34	71 ± 11	—
236	346	590 ± 31	16.0	282	−46	− 23 ± 7	—
238	218	512 ± 29	10.8	284	−73	− 92 ± 9	—
240	108	432 ± 23	6.9	286	−71	− 3 ± 9	—
242	40	358 ± 19	3.9	288	−49	− 26 ± 4	—
244	4	305 ± 20	3.3	290	−31	−250 ± 5	—
246	− 13	263 ± 21	1.2	292	−20	−317 ± 9	—
248	− 18	240 ± 16	− 0.3	294	−16	−276 ± 9	—
250	− 20	223 ± 14	2.5	296	−14	−227 ± 5	—
252	− 16	216 ± 17	− 1.8	298	−12	−177 ± 7	—
254	− 12	219 ± 15	− 1.9	300	−10	−131 ± 9	—
256	− 7	222 ± 11	4.0	302	− 9	− 88 ± 6	—
258	− 5	223 ± 14	− 3.3	304	− 7	− 59 ± 8	—
260	− 3	225 ± 14	− 4.3	306	− 6	− 40 ± 10	—
262	0	232 ± 13	2.5	308	− 4	− 24 ± 10	—
264	0	232 ± 18	− 3.1	310	− 3	− 13 ± 10	—
266	− 4	224 ± 15	− 3.4	312	− 2	7 ± 7	—
268	− 8	214 ± 7	− 4.4	314	− 1	− 4 ± 4	—
270	−11	190 ± 9	− 2.0	316	− 1	0	—
272	−13	159 ± 12	− 0.7	318	0	—	—
274	−20	127 ± 15	− 0.3				

Compiled by Elmer Mihalyi.

[a] 0.1 M phosphate buffer, pH 7.1.
[b] 0.1 N HCl.

LUMINESCENCE

Table 1
LUMINESCENCE OF THE AROMATIC AMINO ACIDS

Amino acid	Solvent	pH	Temp.	λ_{ex}[a]	$\lambda_{f,m}$[b]	Q[c]	τ[d] (ns)	Ref.
Phe	H_2O	7	~25	254	282	0.04		1
	H_2O	6	23	260		0.024		2
	H_2O + 0.55% glucose	0	27		285	0.02		3
	H_2O + 0.55% glucose	6	27		285	0.03		3
	H_2O	7	20	248	282	0.025	6.8	4
	H_2O	7	25				6.4	5
Tyr	H_2O	7	~25	254	303	0.21		1
	H_2O	7	~25				3.2	5
	H_2O	6	23	275		0.14		2
	H_2O	7	~25				3.6	6
	H_2O	7	23				2.6	7
	H_2O	7	~25			0.09		8
Trp	H_2O	7	~25	254	348	0.20		1
	H_2O	Alkaline	~25	280		0.51		9
	H_2O	Acid	~25	280		0.085		9
	H_2O	7	~25				3.0	5
	H_2O	7	23				2.6	7
	H_2O	7	~25				2.5	10
	H_2O	1	~25	286	345	0.091		11
	H_2O	7	~25	287	352	0.149		11
	H_2O	11	~25	289	359	0.289		11
	H_2O	8.9	25		342	0.51		12
	H_2O	10.2	25				6.1	10
	H_2O	7	27			0.14		13
	H_2O	7	25			0.12		8
	H_2O	7	25				2.8	13
	H_2O	6	23	280		0.13		2
	H_2O + 0.55% glucose	0	27		350	0.05		3
	H_2O + 0.55% glucose	6	27		350	0.20		3
	H_2O + 0.55% glucose	12	27		355	0.17		3
	H_2O-ethylene glycol (1:1)	4.7	25				2.9	14

Compiled by R. F. Steiner.

[a]The excitation wavelength in nanometers. When not listed it is either not cited in the original reference or else encompasses a range of wavelengths.
[b]The wavelength, in nanometers, of maximum fluorescence intensity.
[c]The absolute quantum yield.
[d]The excited lifetime, in nanoseconds.

REFERENCES

1. Teale and Weber, *Biochem. J.,* 65, 476 (1957).
2. Chen, *Anal. Lett.,* 1, 35 (1967).
3. Bishai, Kuntz, and Augenstein, *Biochim. Biophys. Acta,* 140, 381 (1967).
4. Leroy, Lami, and Laustriat, *Photochem. Photobiol.,* 13, 411 (1971).
5. Gladchenko, Kostko, Pikulik, and Sevchenko, *Dokl. Akad. Nauk Belorrus. SSR,* 9, 647 (1965).

Table 1 (continued)
LUMINESCENCE OF THE AROMATIC AMINO ACIDS

6. Blumberg, Eisinger, and Navon, *Biophys. J.,* 8, A-106 (1968).
7. Chen, Vurek, and Alexander, *Science,* 156, 949 (1967).
8. Borresen, *Acta Chem. Scand.,* 21, 920 (1967).
9. Cowgill, *Arch. Biochem. Biophys.,* 100, 36 (1963).
10. Badley and Teale, *J. Mol. Biol.,* 44, 71 (1969).
11. Bridges and Williams, *Biochem. J.,* 107, 225 (1968).
12. Longworth, *Biopolymers,* 4, 1131 (1966).
13. Eisinger and Navon, *J. Chem. Phys.,* 50, 2069 (1969).
14. Weinryb and Steiner, *Biochemistry,* 7, 2488 (1968).

Table 2
LUMINESCENCE OF DERIVATIVES OF THE AROMATIC AMINO ACIDS

Compound	Solvent	pH	Temp.	λ_{ex}[a]	$\lambda_{f,m}$[b]	Q_{rel}	τ[c] (ns)	Ref.
			Tryptophan[d]					
L-Trp	H_2O	7	23	290		1.00	3.0	10
L-Trp	H_2O-Ethylene glycol (1:1)	4.7	25	290	357	1.00	2.9	1
Acetyl-DL-Trp	H_2O-Ethylene glycol (1:1)	7.5	25	290	361	1.76	5.2	1
Acetyl-Trp	H_2O	5	25	290		1.6	4.8	4
Acetyl-L-Trp	H_2O	7	~25	280	355	1.40		2
Acetyl-L-Trp-amide	H_2O-Ethylene glycol (1:1)	7.5	25	290	356	0.87	3.8	1
Acetyl-Trp-amide	H_2O	7	23	290			3.0	10
Acetyl-Trp-amide	H_2O	5	25	290		1.10	2.6	4
L-Trp-amide	H_2O-Ethylene glycol (1:1)	4.7	25	290	351	0.59	1.4	1
DL-Trp-amide	H_2O	7	25	280	355	1.00		2
Trp-amide	H_2O	7	25	280		0.70		5
Acetyl-Trp-methyl ester	H_2O	7	25	280	350		0.55	5
Trp-methyl ester	H_2O	7	25	280		0.25		5
L-Trp-ethyl ester	H_2O	7	25	280	355	0.16		2
L-Trp-ethyl-ester	H_2O-Ethylene glycol (1:1)	4.7	25	290	349	0.17	0.50	1
L-Trp-Gly	H_2O	7	25	280		0.70		2
L-Trp-Gly	H_2O-Ethylene glycol (1:1)	4.7	25	290	350	0.65	1.60	1
L-Trp-Gly	H_2O	7	23	290			2.2	10
L-Trp-Gly-Gly	H_2O-Ethylene glycol (1:1)	4.7	25	290	348	0.48	1.8	1
Gly-L-Trp	H_2O-Ethylene glycol (1:1)	4.7	25	290	356	0.49	1.4	1
Gly-L-Trp	H_2O	7	25	280	355	0.29		2
Gly-Trp	H_2O	7	25	280		0.25		5

[a] The excitation wavelength in nanometers. When not listed it is either not cited in the original reference, or else encompasses a range of wavelengths.

[b] The wavelength, in nanometers, of maximum fluorescence intensity.

[c] The excited lifetime, in nanoseconds.

[d] Q_{rel} = quantum yield, relative to that of tryptophan.

Table 2 (continued)
LUMINESCENCE OF DERIVATIVES OF THE AROMATIC AMINO ACIDS

Compound	Solvent	pH	Temp.	λ_{ex}[a]	$\lambda_{f,m}$[b]	Q_{rel}[c]	τ[c] (ns)	Ref.
				Tryptophan[d] (continued)				
Gly-L-Trp	H_2O	7	23	290			1.5	10
Gly-Trp	H_2O	9.8	25	280		0.70		5
Gly-Gly-L-Trp	H_2O-Ethylene glycol (1:1)	4.7	25	290	362	0.65	2.0	1
Gly-Gly-Trp	H_2O	7	25	280		0.40		5
Gly-Gly-Trp	H_2O	9.8	25	280		0.50		5
Gly-Gly-Gly-L-Trp	H_2O-Ethylene glycol (1:1)	4.7	25	290	360	0.76	2.4	1
Trp-Gly-Trp-Gly	H_2O-Ethylene glycol (1:1)	4.7	25	290	352	0.69	2.0	1
L-Trp-L-Phe	H_2O-Ethylene glycol (1:1)	4.7	25	290	350	0.63	2.0	1
L-Trp-L-Trp	H_2O	7	~25	280		0.45		2
L-Trp-L-Trp	H_2O	7	23	290			1.6	10
L-Trp-L-Trp	H_2O-Ethylene glycol (1:1)	4.7	25	290	361	0.50	1.5	1
L-Trp-L-Tyr	H_2O-Ethylene glycol (1:1)	4.7	25	290	351	0.75	1.9	1
L-Trp-L-Tyr	H_2O	7	~25	280		0.60		2
Pro-Trp	H_2O	7	25	280		0.25		5
Pro-Trp	H_2O	10.2	25	280		0.95		5
Tryptamine	H_2O	5	25	290		2.10	6.0	4
				Tyrosine[e]				
DL-Tyr	H_2O	7	25	280	320	1.00		2
Tyramine	H_2O	7	25	280	320	0.88		2
L-Tyr-ethyl ester	H_2O	5.5	25	275		0.12		7
Tyr-amide	H_2O	5.5	25	275		0.25		7
N-Methyl-Tyr	H_2O	5.5	25	275		0.38		7
Acetyl-L-Tyr	H_2O	5.5	25	275		0.88		7
Acetyl-Tyr-amide	H_2O	5.5	25	275		0.45		7
L-Tyr-Gly	H_2O	7	25	280	320	0.35		2
L-Tyr-Gly	H_2O	5.5	25	275		0.33		7
L-Tyr-Gly-Gly	H_2O	5.5	25	275		0.22		7
L-Cystinyl-bis-L-Tyr	H_2O	8.5	25	270		0.08		11
Tyr-Cys-S-S-Cys	H_2O	8.5	25	270		0.08		11
Tyr-Ala	H_2O	6	25	270		0.43		3
Tyr-Phe	H_2O	6	25	270		0.38		3
Tyr-Tyr	H_2O	6	25	270		0.38		3
Gly-L-Tyr	H_2O	7	25	280	320	0.33		2
Gly-Tyr	H_2O	5.5	25	280		0.38		8
Gly-Gly-L-Tyr	H_2O	5.5	25	275		0.54		7
Gly-Gly-Gly-L-Tyr	H_2O	5.5	25	275		0.58		7
Gly-L-Tyr-Gly	H_2O	5.5	25	275		0.22		7
Gly-L-Tyr-Gly-amide	H_2O	7	25	280	320	0.17		2

[e] Q_{rel} = quantum yield, relative to that of tyrosine.

Table 2 (continued)
LUMINESCENCE OF DERIVATIVES OF THE AROMATIC AMINO ACIDS

Compound	Solvent	pH	Temp.	λ_{ex}[a]	$\lambda_{f,m}$[b]	Q_{rel}	τ[c] (ns)	Ref.
Tyrosine[e] (continued)								
Gly-Tyr-Gly-amide	H_2O	6	25	270		0.18		3
Leu-L-Tyr	H_2O	7	25	280	320	0.50		2
Leu-Tyr	H_2O	5.5	25	280		0.48		12
Glu-Tyr	H_2O	5.5	25	280		0.45		8
Met-Tyr	H_2O	5.5	25	280		0.45		8
Arg-Tyr	H_2O	5.5	25	280		0.50		8
His-Tyr	H_2O	5.5	25	270		0.50		8
HS-Cys-Tyr	H_2O	8.5	25	270		0.28		11
Phenylalanine[f]								
L-Phe	H_2O-Ethylene glycol (1:1)	4.7	25	250	287	1.00		1
Gly-DL-Phe	H_2O-Ethylene glycol (1:1)	4.7	25	250	287	1.15		1
DL-Ala-DL-Phe	H_2O-Ethylene glycol (1:1)	4.7	25	250	285	1.10		1
L-Lys-L-Phe	H_2O-Ethylene glycol (1:1)	4.7	25	250	287	0.98		1
L-His-L-Phe	H_2O-Ethylene glycol (1:1)	4.7	25	250	288	0.82		1
L-Arg-L-Phe	H_2O-Ethylene glycol (1:1)	4.7	25	250	286	0.77		1
L-Met-L-Phe	H_2O-Ethylene glycol (1:1)	4.7	25	250	287	0.44		1

[f]Q_{rel} = quantum yield, relative to that of phenylalanine.

Compiled by R. F. Steiner.

REFERENCES

1. Weinryb and Steiner, *Biochemistry*, 7, 2488 (1968).
2. Cowgill, *Arch. Biochem. Biophys.*, 100, 36 (1963).
3. Cowgill, *Biochim. Biophys. Acta*, 75, 272 (1963).
4. Kirby and Steiner, *J. Phys. Chem.*, 74, 4480 (1970).
5. Cowgill, *Biochim. Biophys. Acta*, 133, 6 (1967).
6. Edelhoch, Brand, and Wilchek, *Biochemistry*, 6, 547 (1967).
7. Edelhoch, Perlman, and Wilchek, *Biochemistry*, 7, 3893 (1968).
8. Russell and Cowgill, *Biochim. Biophys. Acta*, 154, 231 (1968).
9. Cowgill, *Biochim. Biophys. Acta*, 100, 37 (1967).
10. Chen, *Arch. Biochem. Biophys.*, 158, 605 (1973).
11. Cowgill, *Biochim. Biophys. Acta*, 140, 37 (1967).
12. Cowgill, *Arch. Biochem. Biophys.*, 104, 84 (1964).

Table 3

LUMINESCENCE OF PROTEINS LACKING TRYPTOPHAN

Protein	Solvent	pH	Temp.	λ_{ex}[a]	$\lambda_{f,m}$[b]	Q_{rel}[c]	τ[d] (ns)	Ref.
Angiotensin amide	H_2O	7	25			0.36 (0.075)		2
Insulin (bovine)	H_2O	7	25	277	304	0.19	1.4	1
Insulin B chain	H_2O	7	25			0.26 (0.055)		2
Oxytocin	H_2O	7.8	25			0.26		2, 3
Ribonuclease (bovine)	H_2O	7	25	277	304	0.10 (0.02)	1.9	1

Compiled by R. F. Steiner.

[a]The excitation wavelength in nanometers.
[b]The wavelength of maximum fluorescence.
[c]The quantum yield relative to that of tyrosine (absolute yield in parentheses).
[d]The excited lifetime, in nanoseconds.

REFERENCES

1. **Longworth,** in *Excited States of Proteins and Nucleic Acids,* Steiner and Weinryb, Eds., Plenum, New York, 1971.
2. **Cowgill,** *Biochim. Biophys. Acta,* 133, 6 (1967).
3. **Cowgill,** *Arch Biochem. Biophys.,* 104, 84 (1964).

Table 4
LUMINESCENCE OF PROTEINS CONTAINING TRYPTOPHAN

Protein	Solvent	pH	Temp.	λ_{ex}[a]	$\lambda_{f,m}$[b]	Q[c]	τ[d] (ns)	Ref.
F-Actin	H_2O	7	~25	297	332	0.24		1
Actomyosin	H_2O	7	~25	297	337	0.20		1
Albumin (bovine serum)	H_2O	5.5–7	~25	297	343	0.39		1
	H_2O	7	23		270		4.6	2
	H_2O		~25	280	342	0.15		3
	H_2O	7	25	295	343	0.26		4
	H_2O	7	25		270	0.51		5
	H_2O	7	25	280	342	0.21	4.6	5
Albumin (human serum)	H_2O	5.5–7	~25	297	342	0.31		1
	H_2O		~25	280	339	0.07		3
	H_2O	7	25	295	341	0.155		4
	H_2O					0.24	4.3	5
	H_2O					0.22	4.1	6
	H_2O	5.5	20	295	343	0.21	4.8	7
	H_2O	7	23				4.5	2
	H_2O	7	25	280	339	0.11	3.3	5
Aldolase	H_2O	7	25	280		0.10		4
	H_2O	7	25	280	328	0.10		5
	H_2O	7	25	295	328	0.12		4
β-Amylase	H_2O	7	~25	297	335	0.32		1
Arginase (bovine liver)	H_2O	7	~25	297	337	0.18		1
Avidin	H_2O	7	25	280	338			5
Azurin	H_2O	7	25	280	308	0.10		5
Carbonic anhydrase	H_2O	7	25	280	336	0.17	2.6	5
Carboxypeptidase A	H_2O	7	25	280	327	0.12		4
α-Chymotrypsin	H_2O	7	~25	297	336	0.14		1
	H_2O	7		280	334	0.095		3
	H_2O	7	25	295	332	0.144		4
	H_2O					0.13	3.0	5
	H_2O						3.0	6
	H_2O	7	23	270			3.4	2
	H_2O	7	25	280		0.13		4
Chymotryspinogen A	H_2O	7	~25	297	333	0.127		1
	H_2O	7	~25	280	331	0.07		3
	H_2O	7	25	295	331	0.124		4
	H_2O					0.10	1.6	5
	H_2O						1.6	6
	H_2O						1.9	8
	H_2O	7	23	290			2.9	2
	H_2O	7	25	280		0.11		4
Corticotropin	H_2O	7	25	280	350	0.08		5
Elastase	H_2O	7	25	280	335			5
Endonuclease	H_2O	7	25	280	334			5
γI-Globulin (human)	H_2O	6	25	297	335	0.08		1
	H_2O	7	23				3.2	2
Glutamate dehydrogenase	H_2O	7	25	280	332	0.300	4.6	5
Glucagon	H_2O	7	25	280	345	0.14		5

[a]The excitation wavelength in nanometers.
[b]The wavelength of maximum fluorescence.
[c]The absolute quantum yield.
[d]The excited lifetime, in nanoseconds.

Table 4 (continued)
LUMINESCENCE OF PROTEINS CONTAINING TRYPTOPHAN

Protein	Solvent	pH	Temp.	λ_{ex}[a]	$\lambda_{f,m}$[b]	Q[c]	τ[d] (ns)	Ref.
Glyceraldehyde-3-phosphate dehydrogenase	H_2O	7	25	297	335.5	0.135		1
Growth hormone	H_2O	7	25	280	325	0.15		5
Hemoglobin	H_2O	7	25	280	335	0.001		5
Hyaluronidase (bovine testicle)	H_2O	7	~25	297	336	0.14		1
α-Lactalbumin	H_2O	7	25	295	328	0.06		4
	H_2O	7	25	280		0.05		4
Lactate dehydrogenase	H_2O	7	25	280	345	0.38		5
β-Lactoglobulin A	H_2O	7	25	280	330	0.08		5
β-Lactoglobulin AB (bovine)	H_2O	6–8	~25	297	332	0.12		1
	H_2O					0.15		5
	H_2O	7	25	295	333	0.082		4
	H_2O	7	25	280		0.08		4
Lysozyme (egg white)	H_2O	7.5	25	295	337			9
	H_2O	7	25	295	338	0.079		4
	H_2O		~25	280	341	0.06		3
	H_2O						2.6	5
	H_2O	7	23	270			2.0	2
	H_2O	7	25	280		0.07		4
Apomyoglobin (sperm whale)	H_2O	8.5	~25	297	335.5	0.16		1
	H_2O					0.15	2.9	10
	H_2O	7	15	288	328	0.12	2.8	11
	H_2O					0.16		5
	H_2O	7	23	290			3.0	2
Myosin (rabbit)	H_2O	7	~25	297	338	0.22		1
Ovalbumin	H_2O	7	~25	280	332	0.19	4.5	5
	H_2O	7	25	295	334	0.25		4
Papain	H_2O	9.5–10	~25	297	350	0.16		1
	H_2O	7.5–7.8	~25	297	347	0.16		1
	H_2O	7	25	295	342	0.16		4
	H_2O			280		0.15		12
	H_2O			280		0.13	4.6	13
	H_2O	7	25	280		0.14		4
	H_2O	4.5–5.3	~25	297	340	0.082		1
	H_2O					0.11		5
	H_2O					0.079		4
	H_2O	4.5	23	288		0.08		14
	H_2O			280		0.10	3.0	12
	H_2O			280		0.08	3.4	13
Pepsin	H_2O	5.2–5.5	~25	297	343	0.26		1
	H_2O		~25	280	342	0.13		3
	H_2O					0.22	4.6	15
	H_2O	7	25	295	339	0.185		4
	H_2O					0.31		5
	H_2O	4.5	23	288		0.22		14
	H_2O						4.5	8
Acid phosphatase (wheat bran)	H_2O	7	~25	297	337	0.16		1
Phosphocreatine kinase	H_2O	7	~25	297	333.5	0.11		1

Table 4 (continued)
LUMINESCENCE OF PROTEINS CONTAINING TRYPTOPHAN

Protein	Solvent	pH	Temp.	λ_{ex}[a]	$\lambda_{f,m}$[b]	Q^c	τ^d (ns)	Ref.
Pseudoacetyl cholinesterase (human)	H_2O	7	~25	297	334			1
Pyruvate kinase (rabbit)	H_2O	7	~25	297	339	0.44		1
	H_2O					0.20		5
Staphylococcus aureus nuclease	H_2O	5	~25	270		0.46		18
Subtilisin carlsberg	H_2O	7	~25	280	305			5
Tobacco mosaic virus protein (depolymerized)	H_2O	7.4	~25	297	332	0.40		1
Tobacco mosaic virus protein (polymerized)	H_2O	6.4	~25	297	331.5	0.37		1
Trypsin	H_2O	7	~25	297	334.5	0.11		1
(bovine)	H_2O		~25	280	332	0.08		3
	H_2O	7	25	295	335	0.13		4
	H_2O					0.08		5
	H_2O					0.15		16
	H_2O						2.0	6
	H_2O						2.4	8
	H_2O	7	25	280		0.13		4
Trypsinogen	H_2O	7	25	295	332	0.14		4
	H_2O	7	25	280		0.12		4
Tryptophanyl tRNA synthetase	H_2O	5	~25	270		0.46		18

Compiled by R. F. Steiner.

REFERENCES

1. Burstein, Vedenkina, and Ivkova, *Photochem. Photobiol.*, 18, 263 (1973).
2. Chen, Vurek, and Alexander, *Science*, 156, 949 (1967).
3. Teale, *Biochem. J.*, 76, 381 (1960).
4. Kronman and Holmes, *Photochem. Photobiol.*, 14, 113 (1971).
5. Longworth, in *Excited States of Proteins and Nucleic Acids*, Steiner and Weinryb, Eds., Plenum, New York (1971).
6. Konev, Kostko, Pikulik, and Chernitski, *Biofizika*, 11, 965 (1966).
7. DeLauder and Wahl, *Biochem. Biophys. Res. Commun.*, 42, 398 (1971).
8. Konev, Kostvo, Pikulik, and Volotovski, *Dokl. Akad. Nauk Beloruss. SSR*, 10, 500 (1966).
9. Lehrer, *Biochemistry*, 10, 3254 (1971).
10. Kirby and Steiner, *J. Biol. Chem.*, 245, 6300 (1970).
11. Anderson, Brunori, and Weber, *Biochemistry*, 9, 4723 (1970).
12. Steiner, *Biochemistry*, 10, 771 (1971).
13. Weinryb and Steiner, *Biochemistry*, 9, 135 (1970).
14. Shinitzki and Goldman, *Eur. J. Biochem.*, 3, 139 (1967).
15. Badley and Teale, *J. Mol. Biol.*, 44, 71 (1969).
16. Barenboim, Sokolenko, and Turoverov, *Cytologiya*, 10, 636 (1968).
17. Edelhoch, Condliffe, Lippoldt, and Burger, *J. Biol. Chem.*, 241, 5205 (1966).
18. Edelhoch, Perlman, and Wilchek, *Ann. N. Y. Acad. Sci.*, 158, 391 (1969).

Table 5
COVALENT PROTEIN CONJUGATES

Label	Protein	Groups/molecule	Solvent	pH	Temp.	λ_{ex}[a]	$\lambda_{f,m}$[b]	Q[c]	τ^{d} (ns)	Ref.
Fluorescamine	Bovine plasma albumin	11.2	H_2O	7.4	25	390	490	0.088	10.3	1
		9.3						0.144	10.2	
		3.6						0.103	9.7	
	Ovalbumin	2.2	H_2O	7.4	25	390	490	0.125	9.0	1
		3.9						0.123	9.0	
	Lysozyme	1.3	H_2O	7.4	25	390	490	0.099	8.0	1
		2.2						0.094	7.5	
Fluorescein isothiocyanate	γ-Globulin	0.2	H_2O	9.0	25	490	550	0.5	4.2	2
Anthracene 2-isocyanate	γ-Globulin	1	H_2O	8.0	25	362	460	0.6	29	2
Rhodamine B isothiocyanate	γ-Globulin	1	H_2O	8.0	25	550	585	0.7	3	2
Dansyl	γ-Globulin	2.5	H_2O	8.0	25	345	540–545	0.2	11	2
Dansyl	Bovine plasma albumin	1.2	H_2O	7.4	25	320–360	500	0.6	22	3
Dansyl	Bovine thyroglobulin	2.5	H_2O	7.4	25	320–360	540	0.6	22	3
		16	H_2O	7.4	25	320–360	540	0.6	18	3
		10	H_2O	7.0	25	345			8.1	4
Pyrenebutyrate	Human thyroglobulin	0.84	H_2O	7.0	25	346–347			125	4

Compiled by R. F. Steiner.

[a] The excitation wavelength in nanometers.
[b] The wavelength of maximum fluorescence.
[c] The absolute quantum yield.
[d] The excited lifetime in nanoseconds.

REFERENCES

1. Chen, *Anal. Lett.*, 7, 65 (1974).
2. Chen, *Arch. Biochem. Biophys.*, 133, 263 (1969).
3. Chen, *Arch. Biochem. Biophys.*, 128, 163 (1968).
4. Rawitch, Hudson, and Weber, *J. Biol. Chem.*, 244, 6543 (1969).

SPECIFIC ROTATORY DISPERSION CONSTANTS FOR 0.1 *M* AMINO ACID SOLUTIONS

Amino acid	Solution acidity	Drude parameters[a]				RMS[e] deviation (deg)
		A	λ_a^2	B	λ_b^2	
Alanine	2 *M* HCl	298.5195	0.04220239	294.7037	0.04170058	±0.17
	pH 6.5	360.1463	0.04069301	360.1262	0.04047647	±0.24
	0.25 *M* NaOH	344.4703	0.04989568	343.1793	0.04989547	±0.13
Valine	2 *M* HCl	438.4119	0.03926320	431.6605	0.03866173	±0.46
	pH 6.06	460.4233	0.03465918	459.4160	0.03421852	±0.30
	0.25 *M* NaOH	319.3442	0.03118233	315.3691	0.03050604	±0.39
Leucine	2 *M* HCl	377.4991	0.04044465	374.1618	0.03980577	±0.17
	pH 6.18	274.5246	0.03802343	278.2870	0.03743202	±0.16
	0.25 *M* NaOH	314.6331	0.03777259	312.5013	0.03727019	±0.17
Serine	2 *M* HCl	338.1789	0.04230850	334.9272	0.04186100	±0.24
	pH 5.73	318.8132	0.03560563	321.5245	0.03501944	±0.26
	0.25 *M* NaOH	382.4514	0.03479284	383.6311	0.03466110	±0.14
Aspartic Acid	2 *M* HCl	360.9105	0.04260890	353.2918	0.04253996	±0.22
	pH 2.93[b]	446.1446	0.03954388	444.9421	0.03935718	±0.29
	pH 7.8	374.4645	0.03857365	380.6139	0.03817383	±0.19
	pH 12.1	347.8941	0.03823479	349.8450	0.03755913	±0.12
Glutamic Acid	2 *M* HCl	361.7220	0.03488210	352.7541	0.03430429	±0.57
	pH 3.25[c]	414.0331	0.03695381	410.9175	0.03669701	±0.29
	pH 7.94	393.2113	0.03567558	395.3116	0.03533174	±0.13
	pH 11.6	336.5308	0.03729399	333.6660	0.03702243	±0.24
Asparagine	pH 1.68	398.0964	0.04233022	391.5547	0.04225800	±0.39
	pH 5.2	253.8327	0.03896011	256.2606	0.03865567	±0.12
	pH 10.42	417.5699	0.03106792	421.1466	0.03100037	±0.22
Lysine	2 *M* HCl	1179.298	0.04094295	1172.252	0.04083325	±0.24
	pH 4.83	441.1781	0.03954097	437.5107	0.03937528	±0.16
	pH 9.93	431.4914	0.03917215	427.8984	0.03901931	±0.14
	0.25 *M* NaOH	183.0545	0.03726084	179.6172	0.03679289	±0.14
Ornithine	2 *M* HCl[d]	1350.506	0.04207689	1342.576	0.04200871	±0.21
	pH 5.62	432.2491	0.03778435	428.066	0.03761066	±0.09
	pH 9.80	525.6925	0.04052814	522.2509	0.04041411	±0.24
	0.25 *M* NaOH[d]	486.3102	0.03752583	482.6644	0.03735010	±0.33
Proline	2 *M* HCl	445.3052	0.04125130	461.8143	0.04066942	±0.34
	pH 6.08	403.8729	0.04138662	430.2741	0.04088043	±0.51
	0.25 *M* NaOH	1719.7714	0.01997902	1749.4527	0.02039450	±0.56

Compiled by L. I. Katzin and E. Gulyas from Katzin and Gulyas, *J. Amer. Chem. Soc.*, 86, 1655 (1964).

[a] $[\alpha]_\lambda = \dfrac{A}{\lambda^2 - \lambda_a^2} - \dfrac{B}{\lambda^2 - \lambda_b^2}$ with λ in microns. Parameters valid for 650–270 mμ.

[b] Amino acid 0.0375 *M*.

[c] Amino acid 0.05 *M*.

[d] Amino acid 0.08 *M*.

[e] Root-mean-square deviation between measured and calculated $[\alpha]_\lambda$, for 42 wavelengths between 650 and 270 mμ.

This table originally appeared in Sober, Ed., *Handbook of Biochemistry and Selected Data for Molecular Biology*, 2nd ed., Chemical Rubber Co., Cleveland, 1970, B-78.

AMINO ACID SUFFIXES AND PREFIXES

ABBREVIATIONS

Used	For	Recommended (if different)
ONbzl	*p*-Nitrobenzyl ester	
OIbzl	*p*-Iodobenzyl ester	
Azet	Azetidine-2-COOH	
Pipec	Piperidine-2-COOH	
Thz	Thiazolidine-4-COOH	
GcOH	Glycollic acid	
M.A.	Mixed anhydride	
OEtCl$_2$ SalNH$_2$	*N*-Et-3,5-Cl$_2$-Salicylamide ester	
OPhOH	*o*-Hydrophenyl ester	
Et$_4$P$_2$O$_7$	Tetraethylpyrophosphate	
(PhO$_2$P)$_2$O	Bis-*O*-phenylenepyrophosphite	
OPy	3-OH-Pyridine ester	
O(Bu^{i}NO$_2$MePy)	2-Bu^i-4-NO$_2$-6-Me-pyridine ester	
CMCI	*N*-Cyclohexyl-*N'*-[β-(*N*-methylmorpholinium) ethyl]-carbodiimide *p*-toluenesulfonate	C(NcHx)NMemEt-Tos
HOBztl	l-Hydroxybenzotriazole	
HOBztn	3-OH-4-O-3,4-H$_2$-1,2,3-Benzotriazine	
HCONMe$_2$	Dimethylformamide	
Me$_2$SO	Dimethylsulfoxide	
AcNMe$_2$	Dimethylacetamide	
AcOH	Acetic acid	
H$_4$furan	Tetrahydrofuran	
OP(NMe$_2$)$_3$	Hexamethylphosphatriamide	
F$_6$acetone	Hexafluoroacetone	
Et$_2$PO$_3$H	Diethylphosphite	
DCC	Dicyclohexylcarbodiimide	C(NcHx)$_2$
MePdn	1-Me-Pyrrolidone	
Et$_3$N	Triethylamine	
MeMorph	*N*-Methylmopholine	
OMeSO$_2$Ph	4-(Methylsulfonyl) phenyl ester	
OTcp(2,4,5)	2,4,5-Trichlorophenyl ester	OCl$_3$ph(2,4,5)
OTcp(2,4,6)	2,4,6-Trichlorophenyl ester	OCl$_3$ph(2,4,6)
OPcp	Pentachlorophenyl ester	OCl$_5$ph
ONp	*p*-Nitrophenyl ester	ONph
ODnp(2,4)	2,4-Dinitrophenyl ester	ON$_2$ph(2,4)
ODnp(2,5)	2,5-Dinitrophenyl ester	ON$_2$ph(2,5)
SucNBr	*N*-Bromosuccinimide	
SPh	Thiophenyl ester	
EEDQ	*N*-EtOCO-2-EtO-1,2-N$_2$-quinoline	EE-H$_2$Qnl
Pr^i	Isopropyl	
Dnp	Dinitrophenyl	
F$_3$Ac	Trifluoroacetyl	
OPfp	Pentafluorophenyl ester	OF$_5$ph
NA	Not applicable	

POLY(α-AMINO)ACIDS

POLY(α-AMINO ACIDS), THEIR SOLUBILITY AND SUSCEPTIBILITY TO ENZYMATIC ACTIVITIES[a]

Substrate[b,c]	Soluble in[c]	Enzyme	pH	Reacted (+) or not (−)	Major products[c]	Reference[a]
Ala_n	Cl_2AcOH, F_3AcOH	Aminopeptidase		+		1
(DL-Ala)_n	H_2O, HCO_2H, Cl_2AcOH	Carboxypeptidase		+		1
		Chymotrypsin	7.7, 8.0	slow		2
		Papain	3.7, 7.0	−	Ala, oligomers	2, 3
		Pepsin	2.3, 4.2	+		2, 4
		"Pronase"	7.6, 8.0	+		2
		Trypsin	7.7, 8.0	+		2
(Ala-Pro-Gly)_n[d]		Prolyl hydroxylase	7.8	+		59
[Cys(Aet)]_n	$H_2O < pH\ 5$	Trypsin	7.4	+	[Cys(Aet)]_2, [Cys(Aet)]_3	5
[DL-Phe(NH_2)]_n	H_2O					
Arg_n (7)	H_2O	Histone acetylation system	8.5	+	Acetylated polyarginine	6
		Protein (histone) methylase	7.6	+	Methylated polyarginine	6
		Trypsin	7.8	+	Arg, Arg_2	7
Arg-Gly-(Leu-Pro-Gly)_5	H_2O	Prolyl hydroxylase	7.8	+		53
Arg-Gly-(Pro-Pro-Gly)_5	H_2O	Prolyl hydroxylase	7.8	+		53
Asp_n	$H_2O > pH\ 4$	Acid protease from germinated sorghum	3.6 susp.	+	Asp, oligopeptides	8
		Chymotrypsin	7.7, 8.0	−		2
		Keratinase	9.0	−		9
		Pepsin	2.3, 4.2	+		2, 4, 10
		Penicillium cyaneofulvum protease	4.3	+		11
		"Pronase"	7.6, 8.0	−		2
		Trypsin	7.7, 8.0	−		2
		Yeast protease C	4.2	+	Asp	12
Cys_n	$H_2O > pH\ 9$					
[Tyr(I_2)]_n	$H_2O > pH7$	Carboxypeptidase B	7.5	+	Lys(Me_2)	13
[Lys(Me_2)]_n	$H_2O > pH\ 4$, $HCONMe_2$	Acid protease from germinated sorghum	3.6	+	Glu, oligopeptides	8
Glu_n		Aspergillus fumigatus	5.1	+		14, 15
		Carboxypeptidase-A	5.0, 5.3	+	Glu	16, 17
		Carboxypeptidase-A (DFP-treated)	5.0–7.7	−		18
		Carboxypeptidase B	4.6, 4.9, 5.2	+	Glu	18
		Cathepsin D_2[e]	4.6	+		14
		Chymotrypsin	7.7, 8.0	−		2, 16
		Chymotrypsin	4.5–7.0	+		18
		Chymotrypsin 1	5.3	+		52
		Chymotrypsin 2	5.3	+		52
		Chymotrypsin C	5.3	+		19
		Elastase	5.0–7.0	+	Oligopeptides	18
		Elastase 1	8.8	−		52
		Elastase 2	8.8	−		52
		Esteroproteolytic enzyme from porcine pancrease	4.6–5.6	+	Oligopeptides	20

POLY(α-AMINO ACIDS), THEIR SOLUBILITY AND SUSCEPTIBILITY TO ENZYMATIC ACTIVITIES (continued)

Substrate[b,c]	Soluble in[c]	Enzyme	pH	Reacted (+) or not (−)	Major products[c]	Reference[a]
Glu_n (continued)	H_2O > pH 4, $HCONMe_2$ (continued)	Ficin	4.0–7.5	+	Glu, Glu$_2$	10
		Human thyroid proteinase[e]	4.3–7.7	−		14
		Keratinase	9.0	−		9
		Leucine aminopeptidase	5.0–7.7	+		18
		Leucine aminopeptidase I from *Aspergillus oryzae*	6.5	+	Glu	21
		Leucine aminopeptidase II from *Aspergillus oryzae*	6.5	slow		22
		Leucine aminopeptidase III from *Aspergillus oryzae*	6.0	slow		23
		Pancreatic protease 1	5.3	+		52
		Pancreatic protease 2	5.3	+		52
		Papain	4.0–8.0	+		16, 18, 24
		Penicillium cyaneofulvum protease	4.5	+	Glu$_3$, Glu$_4$	11
		Pepsin	2.0–5.0	+		2, 4, 25
		Pronase	7.6, 8.0	+		2
		Rennin	2.0–5.0	+		10
		Staphylococcus aureus[e]	4.8	+		14
		Subtilisin	3.5–7.5	+		18
		Trypsin	5.0–8.6	−		2, 18
		Yeast proteinase C	4.2	+	Glu	12
(D-Glu)_n	H_2O > pH 4	Chymotrypsin	7.7, 8.0	−		2
		Pepsin	2.3, 4.2	−		2, 4
		"Pronase"	7.6, 8.0	−		2
Glu-Gly(Pro-Pro-Gly)$_5$	H_2O	Prolyl hydroxylase	7.8	+		53
Gly_n	Cl_2 AcOH, F_3 AcOH, Conc. Li$^+$ and NH_4^+ halides	Keratinase	9.0	−		9
(Gly-Ala-Pro)$_4$	H_2O	Prolyl hydroxylase	7.2	+		55
(Gly-Pro-Ala)_n	H_2O	Collagenase	7.0	+	Gly-Pro-Ala	26
		Prolyl hydroxylase	7.8	very slow		59
(Gly-Pro-Gly)$_4$	H_2O if n < 4	Prolyl hydroxylase	7.8	−		59
(Gly-Pro-Pro)$_{\sim16}$	H_2O	Prolyl hydroxylase	8.0	−		59
Har_n		Chymotrypsin	7.7, 8.0	slow	Har$_2$, Har$_3$	61
		Trypsin	8.0	slow		61
His_n	H_2O < pH 6	Chymotrypsin	7.7, 8.0	slow		2
		Lactoperoxidase	6.0	+	Iodinated polyhistidine	27
		Pepsin	2.3, 4.2	−		2
		Takadiastase[e]	5.3	+	His	28
(DL-HSe)_n		Chymotrypsin	7.7, 8.0	slow		2
		Pepsin	2.3, 4.2	slow		2
		"Pronase"	7.6, 8.0	+		2
		Trypsin	7.7, 8.0	−		2
Hyp_n	H_2O	Carboxypeptidase C	5.3	+		29
		Pepsin	2.3	−		4
Lys_n	H_2O	Acid protease from germinated sorghum	3.6	+		8
		Arthrobacter proteinase	8.0	+	Oligopeptides	30
		Carboxypeptidase-A	5.6–9.5	+		31, 32
		Carboxypeptidase-B	7.9–9.3	+	Lys	31, 32
		Cathepsin D₂[e]	4.7	+		14
		Chymotrypsin	7.7, 8.0	+		2, 14, 32, 33
		Elastase,	7.0–10	+		32
		Fibrinolysin (bovine)	7.4	+	Lys$_2$, Lys$_3$, Lys$_4$	28
		Ficin	7.0–12	+	Lys$_3$, Lys$_4$, Lys$_5$	10, 32
		Human thyroid proteinase	4.3–9.7	−		14
		Keratinase	9.0	+	Lys$_2$, Lys$_3$, Lys$_4$	9

POLY(α-AMINO ACIDS), THEIR SOLUBILITY AND SUSCEPTIBILITY TO ENZYMATIC ACTIVITIES (continued)

Substrate[b,c]	Soluble in[c]	Enzyme	pH	Reacted (+) or not (−)	Major products[c]	Reference[a]
Lys$_n$ (continued)	H$_2$O (continued)	Leucine aminopeptidase	4.3–9.7	+	Lys	32
		Leucine aminopeptidase I from *Aspergillus oryzae*	6.5	+		21
		Leucine aminopeptidase II from *Aspergillus oryzae*	6.5	+		22
		Leucine aminopeptidase III from *Aspergillus oryzae*	6.0	slow		23
		Pancreas-protease	7.65	+	Lys, Lys$_2$, Lys$_3$	60
		Papain	7.0–12	+	Oligopeptides	10, 32
		Penicillium cyaneofulvum protease	10.7	+	Oligopeptides	11
		Pepsin	2.3–4.6	−		2, 32, 34
		"Pronase"	7.6, 8.0	+	Lys	2, 14
		Pronase E	7.0, 8.3	+		60
		Staphylococcus aureus[e]	3.2–6.1, 7.1–9.5	−		14
		Subtilisin	6.5–10	+	Lys	32
		Takadiastate	4.85, 5.0	+	Lys	28, 60
		Thrombin	7.8	−		28
(DLys)$_n$	H$_2$O	Trypsin	6.0–10	+	Lys$_2$, Lys$_3$	32, 34, 35
		Carboxypeptidase-B	7.65	−		60
		α-Chymotrypsin	7.65	−		60
		Pancreas powder extract	6.0	+		36
		Pancreas-protease	7.65	−		60
		Pronase E	7.00, 8.3	−		60
		Takadiastase	4.85	−		60
(Lys-Ala-Ala)$_n$	H$_2$O	Trypsin	8.0	+	Lys-Ala-Ala, (Lys-Ala-Ala)$_2$	36
		Elastase	8.6	−		37
		Trypsin	7.5	+	Ala-Ala-Lys	37
Met$_n$ [Lys(Me)]$_n$	CHCl$_3$, Cl$_2$AcOH / H$_2$O	Carboxypeptidase B	7.5	+	Lys(Me)	13
		Trypsin	7.2	+		51
Orn$_n$(38)	H$_2$O	Chymotrypsin	susp.			39
Phe$_n$	33% HBr/AcOH, Hot AcOH	Keratinase	9.0			40
		Acid protease from germinated sorghum	3.6	+		9
Pro$_n$	H$_2$O, AcOH	Aminopeptidase P	8.6	+	Pro	8
		Carboxypeptidase C	5.3	+		41
		Chymotrypsin	7.7, 8.0			42
		Clostridal aminopeptidase	8.6			2
		Dipeptido carboxypeptidase from *E. coli*	8.1			43
		Prolyl hydroxylase	7.5			44
		Leucine aminopeptidase from bovine lens	9.1	very slow		48, 58
		Leucine aminopeptidase I from *Aspergillus oryzae*	6.5	very slow		45
		Leucine aminopeptidase II from *Aspergillus oryzae*	6.5	very slow		21
		Leucine aminopeptidase III from *Aspergillus oryzae*	6.0	very slow		22
		Pepsin	2.3, 4.2			23
		Penicillium cyaneofulvum protease	4.0–10.7			2, 4
		Proline iminopeptidase	7.8–9.5	+	Pro	11
		"Pronase"	7.6, 8.0	+		46, 47
		Prolyl hydroxylase	7.5	−		2
		Trypsin	7.7, 8.0	−		48
		X-Prolyl aminopeptidase	7.7	very slow		2
		Yeast proteinase C	5.0–8.0	−		49
(Pro-Gly-Ala)$_n$	H$_2$O	Collagenase	7.0	−		12
(Pro-Gly-Gly)$_n$	H$_2$O	Collagenase	7.0	+	Pro-Gly, Gly-Pro-Gly, Gly-Pro-Gly-Gly	26

POLY(α-AMINO ACIDS), THEIR SOLUBILITY AND SUSCEPTIBILITY TO ENZYMATIC ACTIVITIES (continued)

Substrate[b,c]	Soluble in[c]	Enzyme	pH	Reacted (+) or not (−)	Major products[c]	Reference[a]
$(Pro-Gly-Pro)_n$	H_2O	Carboxypeptidase C	5.3	−	Pro, Gly-Pro-$(Pro-Gly-Pro)_{n-1}$	29
		Clostridial aminopeptidase	8.6	+	Gly-Pro, Gly-Pro-Pro, Pro-Gly-Pro-Pro	43
		Collagenase	7.0	+	Gly-Pro, $(Pro-Gly-Pro)_{n-1}$-Pro	26
		Dipeptido carboxypeptidase from E. coli	8.1	+		44
		Prolyl hydroxylase	7.2, 7.5, 7.6, 7.8	+	Hydroxylated poly-(Proxylase)	48, 54, 55, 58
$(Pro-Pro-Gly)_n$[f]	H_2O for n < 10, 10% AcOH or 50% EtOH for n ± 20	Prolyl hydroxylase	7.5, 7.8	+	Hydroxylated poly-(Pro-Gly-Pro)	56, 57
Ser_n	Conc. LiBr in H_2O	Acid protease from germinated sorghum	3.6, susp.	−		8
Trp_n	Pyr, $HCONMe_2$, Cl_2 AcOH	Chymotrypsin	susp.	−		40
Tyr_n	H_2O > pH 9, Pyr, $HCONMe_2$, Me_2SO	Chymotrypsin	7.7, 8.0	+		2
		Chymotrypsin	7.3–7.5, 8.3	+		50
		Keratinase	9.0	−		9
		Pepsin	2.3, 4.2	−		2, 4
		Trypsin	7.7, 8.0	−		2
$(Tyr-Ala-Glu)_n$	H_2O > pH 5.5	Acid proteinase from germinated sorghum	3.6	+	Oligopeptides with N-terminal tyrosine	8
Val_n	F_3 AcOH					2

[a] Taken in part from Katchalski et al.[63] The reader is referred to this source if no other reference is given. References 33 and 63 to 82 also deal with properties of polyamino acids.

[b] Only homopolymers and sequence ordered copolymers are listed in this table.

[c] Abbreviations: AcOH, acetic acid; Cys(Aet), S-β-amino-ethyl-cysteine; DL-phe(NH_2), p-amino-DL-phenylalanine; Cl_2AcOH, dichloroacetic acid; Tyr(I_2), 3,5-diiodotyrosine; Lys(Me₂), $N^ε,N^ε$-dimethyllysine; $HCONMe_2$, dimethyl formamide; Me_2SO, dimethyl sulfoxide; HCO_2H, formic acid; Har, homoarginine; Hse, homoserine; Lys(Me), ε,N-methyl lysine; Pyr, pyridine; susp., suspension; F_3 AcOH, trifluoroacetic acid; H_2O, water.

[d] The N-tert-butyloxycarbonyl and methyl ester blocked oligopeptides, n = 2, 3, 4, 5, and 6, were used.

[e] Crude enzymatic preparation has been used.

[f] The nonapeptides, n = 3 with N-tert-pentyloxycarbonyl and benzyl ester blocked N- and C-terminal groups, respectively, were also hydroxylated.

Compiled by Arieh Yaron.

POLY(α-AMINO ACIDS), THEIR SOLUBILITY AND
SUSCEPTIBILITY TO ENZYMATIC ACTIVITIES (continued)

REFERENCES

1. Lindestrom-Lang, *Acta Chem. Scand.,* 12, 851 (1958).
2. Simons and Blout, *Biochim. Biophys. Acta,* 92, 197 (1964).
3. Katchalski, Sela, Silman, and Berger, in *The Proteins,* Vol. II, 2nd ed., Neurath, Ed., Academic Press, New York, 1964, 524.
4. Neumann, Sharon, and Katchalski, *Nature,* 195, 1002 (1962).
5. Lindley, *Nature,* 178, 647 (1956).
6. Kaye and Sheratzky, *Biochim. Biophys. Acta,* 190, 527 (1969).
7. Ariely, Wilchek, and Patchornik, *Biopolymers,* 4, 91 (1966).
8. Garg and Virupaksha, *Eur. J. Biochem.,* 17, 13 (1970).
9. Nickerson and Durand, *Biochim. Biophys. Acta,* 77, 87 (1963).
10. Katchalski, Levin, Neumann, Riesel, and Sharon, *Bull. Res. Counc. Isr. Sect. A,* 10, 159 (1961).
11. Ankel and Martin, *Biochem. J.,* 91, 431 (1964).
12. Hayashi and Hata, *Biochim. Biophys. Acta,* 263, 673 (1972).
13. Seely and Benoiton, *Biochem. Biophys. Res. Commun.,* 37, 771 (1969).
14. Lundblad and Johansson, *Acta Chem. Scand.,* 22, 662 (1968).
15. Martin and Jonsson, *Can. J. Biochem.,* 43, 1745 (1965).
16. Green and Stahmann, *J. Biol. Chem.,* 197, 771 (1952).
17. Avrameas and Uriel, *Biochemistry,* 4, 1750 (1965).
18. Miller, *J. Am. Chem. Soc.,* 86, 3913 (1964).
19. Folk and Schirmer, *J. Biol. Chem.,* 240, 181 (1965).
20. Gjessing and Hartnett, *J. Biol. Chem.,* 237, 2201 (1962).
21. Nakadai, Nasuno, and Iguchi, *Agric. Biol. Chem.,* 37, 757 (1973).
22. Nakadai, Nasuno, and Iguchi, *Agric. Biol. Chem.,* 37, 767 (1973).
23. Nakadai, Nasuno, and Iguchi, *Agric. Biol. Chem.,* 37, 775 (1973).
24. Miller, *J. Am. Chem. Soc.,* 83, 259 (1961).
25. Simons, Fasman, and Blout, *J. Biol. Chem.,* 236, PC64 (1961).
26. Harper, Berger, and Katchalski, *Biopolymers,* 11, 1607 (1972).
27. Holohan, Murphy, Flanagan, Buchanan, and Elmore, *Biochim. Biophys. Acta,* 322, 178 (1973).
28. Rigbi, Ph.D. thesis, Hebrew University, Jerusalem, 1957.
29. Nordwig, *Hoppe-Seyler's Z. Physiol. Chem.,* 349, 1353 (1968).
30. Hofsten and Reinhammar, *Biochim. Biophys. Acta,* 110, 599 (1965).
31. Gladner and Folk, *J. Biol. Chem.,* 231, 393 (1958).
32. Miller, *J. Am. Chem. Soc.,* 86, 3918 (1964).
33. Katchalski, *Adv. Protein Chem.,* 6, 123 (1951).
34. Waley and Watson, *Biochem. J.,* 55, 328 (1953).
35. Katchalski, Grossfeld, and Frankel, *J. Am. Chem. Soc.,* 70, 2094 (1948).
36. Tsuyki, Tsuyuki, and Stahmann, *J. Biol. Chem.,* 222, 479 (1956).
37. Yaron, Tal, and Berger, *Biopolymers,* 11, 2461 (1972).
38. Debabov, Davidov, and Morozkin, *Izvest. Akad. Nauk. SSR SER Khim.,* p. 2153 (1966).
39. Katchalski, Sela, Silman, and Berger, in *The Proteins,* Vol. II, 2nd ed., Neurath, Ed., Academic Press, New York, 1964, 521.
40. Rigbi and Gros, *Bull. Res. Counc. Isr.,* 11A, 44 (1962).
41. Yaron and Mlynar, *Biochem. Biophys. Res. Commun.* 32, 658 (1968); Yaron and Berger, *Methods in Enzymology,* Perlman and Lorand, Eds., Vol. 19, Academic Press, New York, 1970, 521.
42. Nordwig, *Hoppe-Seyler's Z. Physiol. Chem.,* 349, 1353 (1968).
43. Kessler and Yaron, *Biochem. Biophys. Res. Commun.,* 50, 405 (1973); *Methods in Enzymology,* Perlman and Lorand, Eds., Academic Press, New York, in press.
44. Yaron, Mlynar, and Berger, *Biochem. Biophys. Res. Commun.,* 47, 897 (1972).
45. Wiederanders, Lasch, Kirschke, Bohley, Ansorge, and Hanson, *Eur. J. Biochem.,* 36, 504 (1973).
46. Sarid, Berger, and Katchalski, *J. Biol. Chem.,* 234, 1740 (1959).
47. Sarid, Berger, and Katchalski, *J. Biol. Chem.,* 237, 2207 (1962).
48. Kivirikko and Prockop, *J. Biol. Chem.,* 242, 4007 (1967).
49. Dehm and Nordwig, *Eur. J. Biochem.,* 17, 364 (1970).
50. Rigbi, Seliktar, and Katchalski, *Bull. Res. Counc. Isr.,* 6A, 313 (1957).
51. Paik and Kim, *Biochemistry,* 11, 2589 (1972).
52. Uriel and Avrameas, *Biochemistry,* 4, 1740 (1965).

POLY(α-AMINO ACIDS), THEIR SOLUBILITY AND SUSCEPTIBILITY TO ENZYMATIC ACTIVITIES (continued)

53. Kivirikko, Kishida, Sakakibara, and Prockop, *Biochim. Biophys. Acta,* 271, 347 (1972).
54. Prockop, Juva, and Engel, *Hoppe-Seyler's Z. Physiol. Chem.,* 348, 553 (1967).
55. Rhoads and Udenfriend, *Arch. Biochem. Biophys.,* 133, 108 (1969).
56. Kikuchi, Fujimoto, and Tamiya, *Biochem. J.,* 115, 569 (1969).
57. Suzuki and Koyama, *Biochim. Biophys. Acta,* 177, 154 (1969).
58. Hutton, Jr., Marglin, Witkop, Kurtz, Berger, and Udenfriend, *Arch. Biochem. Biophys.,* 125, 779 (1968).
59. Kivirikko and Prockop, *J. Biol. Chem.,* 244, 2755 (1969).
60. Darge, Sass, and Thiemann, *Z. Naturforsch.* 28c, 116 (1973).
61. Rigbi and Elzab, *6th FEBS Meet.,* Madrid, Abstr. No. 730, 1969; Rigbi, Segal, Kliger, and Schwartz, *Bayer Symp. V,* Fritz, Tschesche, Greene, and Truscheit, Eds., Springer-Verlag, Berlin, 1974, 541.
62. Katchalski, Sela, Silman, and Berger, in *The Proteins,* Vol. II, 2nd ed., Neurath, Ed., Academic Press, New York, 1964, 436.
63. Bamford, Elliott, and Hanby, *Synthetic Polypeptides,* Academic Press, New York, 1956.
64. Katchalski and Sela, *Adv. Protein Chem.,* 13, 243 (1958).
65. Szwarc, *Adv. Polymer Sci.,* 4, 1 (1965).
66. Stahmann, *Polyamino Acids, Polypeptides and Proteins,* Wisconsin Press, Madison, 1962.
67. Sela and Katchalski, *Adv. Protein. Chem.,* 14, 391 (1959).
68. Katchalski, *Proc. VI Int. Congr. Biochem.,* p. 80 (1965).
69. Katchalski, *Harvey Lect.,* 59, 243 (1965).
70. Katchalski, in *New Perspectives in Biology,* Sela, Ed., Elsevier, New York, 1964, 67.
71. Kauzmann, *Annu. Rev. Phys. Chem.,* 8, 413 (1957).
72. Scheraga, *Annu. Rev. Phys. Chem.,* 10, 191 (1959).
73. Leach, *Rev. Pure Appl. Chem.,* 9, 1 (1959).
74. Urnes and Doty, *Adv. Protein Chem.,* 16, 401 (1961).
75. Harrap, Gratzer, and Doty, *Annu. Rev. Biochem.,* 30, 269 (1961).
76. Schellman and Schellman, in *The Proteins,* Vol. II, 2nd ed., Neurath, Ed., Academic Press, New York, 1964, 1.
77. Fasman, Tooney, and Shalitin, in *Encyclopedia of Polymer Science and Technology,* Bikales, Ed., John Wiley & Sons, New York, 1965, 837.
78. Fasman, in *Biological Macromolecules, Poly-α-Amino Acids, Protein Models for Conformational Studies,* Timasheff and Fasman, Eds., Marcel Dekker, New York.
79. Goodman, Verdini, Choi, and Masuda, *Top. Stereochem.,* 5, 69 (1970).
80. Scheraga, *Chem. Rev.,* 71, 195 (1971).
81. Johnson, *J. Pharm. Sci.,* 63, 313 (1974).
82. Blaut, Bovey, Goodman, and Lotan, *Peptides, Polypeptides and Proteins,* John Wiley & Sons, New York, 1974.

PROTEINS

SPECIFICITY OF REAGENTS COMMONLY USED TO CHEMICALLY MODIFY PROTEINS

	Amino	Imidazole	Guanidinyl	Indole	Thio Ether	Disulfide	Sulfhydryl	Hydroxyl	Phenol	Carboxyl	Reference
Acetic anhydride	+	±	−	−	−	−	+	±	+	−	1
2-Acetoxy-5-nitrobenzyl chloride	−	−	−	+	−	−	+	−	+	−	2
N-Acetylimidazole	±	−	−	−	−	−	+	−	+	−	3
Acrylonitrile	+	−	−	−	−	−	+	−	−	−	4
Aldehyde/NaBH₄	+	−	−	−	−	−	−	−	−	−	5
Azobenzene-2-sulfenyl bromide	−	−	−	+	−	−	+	−	−	−	6
Bromoacetamido-4-nitrophenol	−	−	−	−	+	−	−	−	−	−	7
D,L-α-Bromo-β-(5-imidazolyl) propionic acid	−	−	−	−	−	−	+	−	−	−	8
N-Bromosuccinimide	−	+	−	+	−	−	+	−	+	−	9
Butanedione	±	−	+	−	−	−	−	−	−	−	10
Carbodiimides, water soluble	−	−	−	−	−	−	−	−	−	+	11
N-Carboxyanhydrides	+	−	−	−	−	−	−	−	−	−	12
NBD chloride	+	−	−	−	−	−	+	−	−	−	13
Citraconic anhydride	+	−	−	−	−	−	+	±	−	−	14
Cyanate	+	−	−	−	−	−	+	+	−	−	15
Cyanogen bromide	−	−	−	−	+	−	+	−	−	−	16
1,2-Cyclohexanedione	+	−	+	−	−	−	−	−	−	−	17
Diacetyl (trimer, dimer)	+	−	+	−	−	−	−	−	−	−	18
Diazoacetates, diazomethane	−	−	−	−	−	−	+	−	−	+	19
Diazonium fluoroborates	+	+	−	−	−	−	−	−	+	−	20
Diazonium salts	+	+	−	±	−	−	+	−	+	−	21
Diazonium 1-H-tetrazole	+	+	−	−	−	−	+	−	+	−	22
1,5-Difluoro-2,4-dinitrobenzene	+	+	−	−	−	−	+	−	+	−	23
4,4'-Difluoro-3,3'-dinitrophenyl sulfone	+	−	−	−	−	−	−	−	+	−	24
Diketene	+	−	−	−	−	−	+	±	+	−	25
Dimethyl adipimidate dimethyl suberimidate	+	−	−	−	−	−	−	−	−	−	26
4,4'-Bis-dimethylaminodiphenyl carbinol	−	−	−	−	−	−	+	−	−	−	27
5-Dimethyaminonapthalenesulfonyl chloride	+	+	−	−	−	−	+	−	+	−	28
Dimethyl(2-methoxy-5-nitrobenzyl) sulfonium bromide	−	−	−	+	−	−	−	−	−	−	29
2,4-Dinitro-5-fluoroaniline	+	+	−	−	−	−	+	−	+	−	30
Dinitrofluorobenzene	+	+	−	−	−	−	+	−	+	−	31
5,5'-Dithiobis(2-nitrobenzoic) acid	−	−	−	−	−	−	+	−	−	−	32
Ethoxyformic anhydride	+	+	−	−	−	−	−	−	+	−	33
Ethyleneimine	−	−	−	−	−	−	+	−	−	−	34
N-Ethylmaleimide	+	−	−	−	−	−	+	−	−	−	35
Ethyl thiotrifluoroacetate	+	−	−	−	−	−	−	−	−	−	36
Fluorescein mercuric acetate	−	−	−	−	−	−	+	−	−	−	37
Formaldehyde	+	+	+	+	−	−	+	−	+	−	19
Glyoxal	+	−	+	−	−	−	−	−	−	−	38
Guanyl-3,5-dimethylpyrazole nitrate	+	−	−	−	−	−	−	−	−	−	39
Haloacetates	+	+	−	−	+	−	+	−	−	+	40
Hydrogen peroxide	−	−	−	+	+	+	+	−	−	−	41

SPECIFICITY OF REAGENTS COMMONLY USED TO CHEMICALLY MODIFY PROTEINS (continued)

	Amino	Imidazole	Guanidinyl	Indole	Thio Ether	Disulfide	Sulfhydryl	Hydroxyl	Phenol	Carboxyl	Reference
2-Hydroxy-5-nitrobenzylbromide	−	−	−	+	−	−	+	−	−	−	42
N-Hydroxysuccinimide esters	+	−	−	−	−	−	−	−	+	−	43
Iodine	−	+	−	+	−	−	+	−	+	−	19
Iodine monochloride	−	+	−	+	−	−	+	−	+	−	44
Iodoacetamide	−	+	−	−	−	−	+	−	−	+	45
O-Iodosobenzoate	−	−	−	−	−	−	+	−	−	−	46
Maleic anhydride	+	+	−	−	−	−	+	−	±	−	47
Mercurials, heavy metals	−	+	−	−	−	−	+	−	−	−	19
P-Mercuribenzoate	−	−	−	−	−	−	+	−	−	−	35
Methanol/HCl	−	−	−	−	−	−	−	−	−	+	48
2-Methoxy-5-nitrotropone	+	−	−	−	−	−	−	−	−	−	49
Methyl acetimidate	+	−	−	−	−	−	−	−	−	−	50
O-Methylisourea	+	−	−	−	−	−	−	−	−	−	51
Methyl 4-mercaptobutyrimidate HCl	−	−	−	−	−	−	+	−	−	−	52
Methyl p-nitrobenzene sulfonate	−	−	−	−	−	−	+	−	−	−	53
O-Nitrophenylsulfenyl chloride	−	−	−	+	−	−	+	−	−	−	54
BNPS-skatole	−	−	−	+	−	−	+	·	−	−	55
Nitrosyldisulfonate, K	−	−	−	−	−	−	−	−	+	−	56
Nitrous acid	+	−	−	−	−	−	+	−	+	−	19
Performic acid	−	−	−	+	+	+	+	−	−	−	57
Phenylglyoxal	+	−	+	−	−	−	−	−	−	−	58
Photooxidation	−	+	−	+	+	−	−	−	+	−	59
Salicylaldehyde	+	−	−	−	−	−	−	−	−	−	60
Sodium borohydride	−	−	−	−	−	+	−	−	−	−	61
Succinic anhydride	+	−	−	−	−	−	+	±	+	−	62
N-Succinimidyl 3-(4-hydroxyphenyl) propionate	+	−	−	−	−	−	−	−	−	−	63
Sulfenyl halides	−	−	−	+	−	−	+	−	−	−	64
Sulfite	−	−	−	−	−	+	+	−	−	−	19
Sulfonyl halides	+	+	−	−	−	−	+	−	+	−	65
Tetranitromethane	−	−	−	+	+	−	+	−	+	−	66
Tetrathionate	−	−	−	−	−	−	+	−	−	−	67
Thiols	−	−	−	−	−	+	−	−	−	−	68
2,4,6-Tribromo-4-methyl cyclohexanedione	−	−	−	+	−	−	−	−	−	−	69
Trinitrobenzenesulfonic acid	+	−	−	−	−	−	+	−	−	−	70

Compiled by James F. Riordan.

REFERENCES

1. Riordan and Vallee, *Methods Enzymol.,* 25, 494 (1972).
2. Horton and Koshland, *Methods Enzymol.,* 25, 468 (1972).
3. Riordan and Vallee, *Methods Enzymol.,* 25, 500 (1972).
4. Riehm and Scheraga, *Biochemistry,* 5, 93 (1966).
5. Means and Feeney, *Biochemistry,* 7, 2192 (1968).
6. Fontana and Scoffone, *Methods Enzymol.,* 25, 482 (1972).
7. Hille and Koshland, *J. Am. Chem. Soc.,* 89, 5945 (1967).
8. Yankeelov and Jolley, *Biochemistry,* 11, 159 (1972).
9. Witkop, *Adv. Protein Chem.,* 16, 261 (1961).
10. Riordan, *Biochemistry,* 12, 3915 (1973).

SPECIFICITY OF REAGENTS COMMONLY USED TO CHEMICALLY MODIFY PROTEINS (continued)

11. Carraway and Koshland, *Methods Enzymol.*, 25, 616 (1972).
12. Sela and Arnon, *Methods Enzymol.*, 25, 553 (1972).
13. Faber et al., *Anal. Biochem.*, 53, 290 (1973).
14. Atassi and Habeeb, *Methods Enzymol.*, 25, 546 (1972).
15. Stark, *Biochemistry*, 4, 588, 1030 (1965).
16. Gross, *Methods Enzymol.*, 11, 238 (1967).
17. Patthy and Smith, *J. Biol. Chem.*, 250, 557 (1975).
18. Yankeelov et al., *J. Am. Chem. Soc.*, 90, 1664 (1968).
19. Herriott, *Adv. Protein Chem.*, 3, 169 (1947).
20. Wofsy et al., *Biochemistry*, 1, 1031 (1962).
21. Riordan and Vallee, *Methods Enzymol.*, 25, 521 (1972).
22. Sokolovsky and Vallee, *Biochemistry*, 5, 3574 (1966).
23. Cuatrecasas et al., *J. Biol. Chem.*, 244, 406 (1968).
24. Zahn and Zuber, *Berufsdermatosen*, 86, 172 (1953).
25. Singhal and Atassi, *Biochemistry*, 10, 1756 (1971).
26. Hunter and Ludwig, *Methods Enzymol.*, 258, 585 (1972).
27. Rohrback, *Anal. Biochem.*, 52, 127 (1973).
28. Gray, *Methods Enzymol.*, 11, 139 (1967).
29. Horton and Tucker, *J. Biol. Chem.*, 245, 3397 (1970).
30. Bergman and Bentov, *J. Org. Chem.*, 26, 1480 (1961).
31. Sanger, *Biochem. J.*, 42, 287 (1948).
32. Ellman, *Arch. Biochem. Biophys.*, 82, 70 (1959).
33. Rosen and Fedoresak, *Biochim. Biophys. Acta*, 130, 401 (1966).
34. Cole, *Methods Enzymol.*, 11, 315 (1967).
35. Riordan and Vallee, *Methods Enzymol.*, 25, 449 (1972).
36. Goldberger, *Methods Enzymol.*, 11, 317 (1967).
37. Karush et al., *Anal. Biochem.*, 9, 100 (1964).
38. Nayaka et al., *Biochim. Biophys. Acta*, 194, 301 (1969).
39. Habeeb, *Biochim. Biophys. Acta*, 34, 294 (1959).
40. Gurd, *Methods Enzymol.*, 11, 532 (1967).
41. Neumann, *Methods Enzymol.*, 25, 393 (1972).
42. Koshland et al., *J. Am. Chem. Soc.*, 86, 1448 (1964).
43. Blumberg et al., *Isr. J. Chem.*, 12, 643 (1974).
44. Koshland et al., *J. Biol. Chem.*, 238, 1343 (1963).
45. Gurd, *Methods Enzymol.*, 25, 424 (1972).
46. Hellerman et al., *J. Am. Chem. Soc.*, 63, 2551 (1941).
47. Butler et al., *Biochem. J.*, 112, 679 (1969).
48. Means and Feeney, *Chemical Modification of Proteins*, Holden-Day, San Francisco, 1971, 139.
49. Tamaoki et al., *J. Biochem.* (Tokyo), 62, 7 (1967).
50. Hunter and Ludwig, *J. Am. Chem. Soc.*, 84, 3491 (1962).
51. Kimmel, *Methods Enzymol.*, 11, 584 (1967).
52. Traut et al., *Biochemistry*, 12, 3266 (1973).
53. Heinrikson, *Biochem. Biophys. Res. Commun.*, 41, 967 (1970).
54. Scoffone et al., *Biochemistry*, 7, 971 (1968).
55. Fontana, *Methods Enzymol.*, 25, 419 (1972).
56. Wiseman and Woodward, *Biochem. Soc. Trans.*, 2, 594 (1974).
57. Moore, *J. Biol. Chem.*, 238, 235 (1963).
58. Takahashi, *J. Biol. Chem.*, 243, 6171 (1968).
59. Westhead, *Methods Enzymol.*, 25, 401 (1972).
60. Williams and Jacobs, *Biochim. Biophys. Acta*, 154, 323 (1968).
61. Light and Sinha, *J. Biol. Chem.*, 242, 1358 (1967).
62. Klapper and Klotz, *Methods Enzymol.*, 25, 531 (1972).
63. Bolton and Hunter, *Biochem. J.*, 133, 529 (1973).
64. Fontana and Scoffone, *Methods Enzymol.*, 25, 482 (1972).
65. Means and Feeney, *Chemical Modification of Proteins*, Holden-Day, San Francisco, 1971, 97.
66. Riordan and Vallee, *Methods Enzymol.*, 25, 515 (1972).
67. Liu, *J. Biol. Chem.*, 242, 4029 (1967).
68. Anfinsen and Haber, *J. Biol. Chem.*, 236, 1361 (1961).
69. Burstein et al., *Isr. J. Chem.*, 5, 65 (1967).
70. Fields, *Methods Enzymol.*, 25, 464 (1972).

ACID HYDROLYSIS OF PROTEINS

The reactions that occur during acid hydrolysis of a protein are complex. They are influenced by the nature and composition of the protein itself, by the presence of impurities, particularly metal ions in the acid used for hydrolysis, and by oxygen. Few systematic studies have been made. Definitive data on the destructions of serine and threonine during acid hydrolysis of proteins are provided by Rees,[1] who studied the stabilities of these amino acids by themselves as well as in mixtures of other amino acids so constituted as to mimic hydrolyates of various proteins. Similar, though less complete data were obtained by Hirs et al.[2] for these amino acids and for cystine and tyrosine. Tryptophan is usually destroyed extensively during acid hydrolysis, particularly in the presence of oxygen. Addition of thioglycollic acid to the medium has been found to provide effective protection[3] − C. H. W. Hirs.

Table 1
DESTRUCTION OF SERINE AND THREONINE

Amino acid mixture	Type of N	Recovery after 24 hours (%)	Amino acid mixture	Type of N	Recovery after 24 hours (%)
Serine alone	Serine-N	89.5	(Insulin)	Serine-N	86.0
Threonine alone	Threonine-N	94.7		Threonine-N	93.8
(Edestin)	Serine-N	89.2	(β-Lactoglobulin)	Serine-N	89.4
	Threonine-N	93.3		Threonine-N	94.0
(Horse Globin)	Serine-N	87.6			
	Threonine-N	95.9			

(From Rees[1])

Table 2
DESTRUCTION OF SERINE, THREONINE, CYSTINE AND TYROSINE

Amino acid in mixture simulating bovine serum albumin hydrolysate	Recovery (%) after various times (hours)			
	Time			
	22	70	23	71
Serine	83.7	61.0	84.9	60.3
Threonine	92.4	81.0	92.7	80.8
Cystine	96.2	83.1	98.0	84.6
Tyrosine	89.7	76.6	94.9	89.2

(From Hirs, et al.[2])

Table 3

**RECOVERY OF TRYPTOPHAN FROM
VARIOUS PROTEINS AFTER ACID
HYDROLYSIS IN THE PRESENCE OF 4%
THIOGLYCOLLIC ACID**

Protein	Recovery (%)
Cytochrome *c*, bovine	96
Ferredoxin, spinach	89
Chymotrypsin, bovine	89
TMV protein	93
Hg-papain	98
Tryptophanase	88

(From Matsubara and Sasaki[3])

REFERENCES

1. **Rees,** *Biochem. J.,* 40, 632 (1946).
2. **Hirs, Stein, and Moore,** *J. Biol. Chem.,* 211, 941 (1954).
3. **Matsubara and Sasaki,** *Biochem. Biophys. Res. Commun.,* 35, 175 (1969).

These tables originally appeared in Sober, Ed., *Handbook of Biochemistry and Selected Data for Molecular Biology,* 2nd. ed., Chemical Rubber Co., Cleveland, 1970.

HYDROLYSIS OF PROTEINS

Table 1
HYDROLYSIS OF PEPTIDES IN ACID SOLUTION[1]

Relative rate of hydrolysis (Gly-Gly = 1)

Peptide[a]	2 N HCl, 99°C (2)	10 N HCl-glacial acetic acid (50 : 50), 37°C (3)	2 N HCl, 104°C (4)	1 N HCl, 104°C (5)	Dowex-50 (5)	0.8 N HCl, 54.5°C (6)
DL-Ala-Gly	0.69	0.62	—	0.62	0.61	0.56
Ala-Leu[b]	0.32	—	—	—	—	—
DL-Ala-DL-Asp	—	—	—	2.15	0.60	—
Ala-Ser	1.1	—	—	—	—	—
Gly-L-Ala	0.37	—	—	—	—	—
Gly-D-Ala	0.40	—	—	—	—	—
Gly-DL-Ala	—	0.62	—	—	—	—
Gly-Asp	1.94	—	—	—	—	—
Gly-Gly	1.0	1.0	1.0	1.0	1.0	1.0
Gly-Leu	0.34	0.40	0.47	0.48	0.49	0.48
Gly-Ser	1.83	—	—	—	—	—
Gly-Tyr	0.43	—	—	—	—	0.52
Gly-Try	—	0.35	—	—	—	0.44
Gly-DL-Val	—	0.31	—	0.34	0.38	—
Gly-Leu-Gly	—	—	0.35 (Gly, Leu)	—	—	—
Gly-Leu-Gly	—	—	0.65 (Leu, Gly)	—	—	—
Leu-Asp	0.86	—	—	—	—	—
Leu-Glu	0.23	—	—	—	—	—
Leu-Gly	0.23	—	0.23	—	—	0.18
DL-Leu-Gly	—	0.23	—	0.22	0.22	—
Leu-Gly-Leu	—	—	0.22 (Leu- Gly)	—	—	—
Leu-Gly-Leu	—	—	1.55 (Gly, Leu)	—	—	—
Leu-D-Leu	0.06	—	—	—	—	—
DL-Leu-DL-Leu	—	0.045	—	—	—	—
Leu-Try	—	0.041	—	—	—	—
Pro-Phe	0.29	—	—	—	—	—
Pro-Tyr	—	—	—	0.116	0.04	—
Ser-Ala	0.74	—	—	—	—	—
Ser-Gly	0.40	—	—	—	—	—
Ser-Ser	0.40	—	—	—	—	—
DL-Val-Gly	—	0.015	—	—	—	—
DL-Val-DL-Ile	—	—	—	0.0086	0.0091	—

Compiled by C. H. W. Hirs.

[a] All optically active amino acids are L isomers unless otherwise stated.
[b] Optical configuration undetermined.

REFERENCES

1. **Hill,** *Advanc. Protein Chem.,* 20, 37 (1965).
2. **Harris, Cole, and Pon,** *Biochem. J.,* 62, 154 (1956).
3. **Synge,** *Biochem. J.,* 39, 351 (1945).
4. **Long and Lillycrop,** *Trans. Faraday Soc.,* 59, 907 (1963).
5. **Whitaker and Deatherage,** *J. Amer. Chem. Soc.,* 77, 3360 (1955).
6. **Lawrence and Moore,** *J. Amer. Chem. Soc.,* 73, 3973 (1951).

Table 2
ENZYMATIC HYDROLYSIS OF PROTEINS AND POLYPEPTIDES

Enzyme	Native protein	Enzyme	Native protein
Trypsin	Human γ-globulin[1]	Pepsin	Horse diphtheria toxin[26]
	Horse diphtheria antitoxin[2]		Pepsin[27]
	Human serum albumin[3]		Botulinum toxin[28]
	Insulin[4]		Catalase[7]
	Trypsin[5]		ACTH[29]
	Enolase[6]		Ribonuclease[30]
	Catalase[7]		Human serum albumin[31]
	Adolase[5]	Chymotrypsin	Human γ-globulin[32]
	Ribonuclease[8]		Growth hormone[33]
	Ribonuclease S[9]		Oxytocin[34]
	Fibrinogen[10]		Insulin[35]
	Myosin[1]		Ribonuclease[36]
	Tropomyosin[12]		Human serum albumin[31]
	Bovine plasma albumin[13]		Chymotrypsin[37]
Papain	Rabbit γ-globulin[14]		Growth hormone[38]
	Human γ-globulin[15]		Ribonuclease[30]
	Human serum globulins[16]		Ribonuclease T$_1$[39]
	Lipovitellin[17]	Leucine amino-peptidase	Insulin[40]
	Thyroglobulin[18]		Oxytocin[34]
Carboxypeptidase A	Insulin[19]		Papain[41]
	Enolase[6]		ACTH[42]
	Hemoglobin[a][20]		Enolase[6]
	ACTH[21]	Subtilisin	Ribonuclease[43]
	Soybean trypsin inhibitor[22]		Ovalbumin[44]
	Tobacco mosaic virus[23]		Human hemoglobin[45]
	Aldolase[24]		Cytochrome c[46]
	Crotoxin[25]	*S. griseus* protease	Taka-amylase[47]
		Collagenase	Collagen[48]

Compiled by C. H. W. Hirs.

[a]Carboxypeptidase A and B.

REFERENCES

1. **Schrohenloher,** *Arch. Biochem. Biophys.,* 101, 456 (1963).
2. **Northrop,** *J. Gen. Physiol.,* 25, 465 (1941–1942); Rothen, *J. Gen. Physiol.,* 25, 487 (1941–1942).
3. **Lapresle, Kaminski, and Tanner,** *J. Immunol.,* 82, 94 (1959); Kaminski and Tanner, *Biochim. Biophys. Acta,* 33, 10 (1959).
4. **Nicol,** *Biochem. J.,* 75, 395 (1960); Carpenter and Baum, *J. Biol. Chem.,* 237, 409 (1962).
5. **Bresler, Glikina, and Fenkel,** *Dokl. Akad. Nauk. SSR,* 96, 565 (1954); Chernikov, *Biokimiya,* 21, 295 (1956); Hess and Wainfon, *J. Amer. Chem. Soc.,* 80, 501 (1958).
6. **Malmstrom,** in *Symposium on Protein Structure,* Neuberger, Ed., John Wiley & Sons, New York, 1958, 338.
7. **Anan,** *J. Biochem. (Tokyo),* 45, 211, 227 (1958).
8. **Ooi, Rupley, and Scheraga,** *Biochemistry,* 1, 432 (1963).
9. **Allende and Richards,** *Biochemistry,* 1, 295 (1962).
10. **Mihalyi and Godfrey,** *Biochim. Biophys. Acta,* 67, 73 (1963).
11. **Mihalyi and Harrington,** *Biochim. Biophys. Acta,* 36, 447 (1959).
12. **de Milstein and Bailey,** *Biochim. Biophys. Acta,* 49, 412 (1961).
13. **Richard, Beck, and Hoch,** *Arch. Biochem. Biophys.,* 90, 309 (1960).
14. **Porter,** *Biochem. J.,* 46, 479 (1950); *Biochem. J.,* 73, 119 (1959); Putnam, Tan, Lynn, Easley, and Shunsuke, *J. Biol. Chem.,* 237, 717 (1962); Fleischman, Porter, and Press, *Biochem. J.,* 88, 220 (1963).
15. **Hsiao and Putnam,** *J. Biol. Chem.,* 236, 122 (1961).
16. **Deutsch, Steihm, and Morton,** *J. Biol. Chem.,* 236, 2216 (1961).
17. **Glick,** *Arch. Biochem. Biophys.,* 100, 192 (1963).
18. **O'Donnell, Baldwin, and Williams,** *Biochim. Biophys. Acta,* 28, 294 (1958).

Table 2 (continued)
ENZYMATIC HYDROLYSIS OF PROTEINS AND POLYPEPTIDES (continued)

19.   Lens, *Biochim. Biophys. Acta*, 3, 367 (1949); Harris, *J. Amer. Chem. Soc.*, 74, 2944 (1952), Harris and Li, *J. Amer. Chem. Soc.*, 74, 2945 (1952); Slobin and Carpenter, *Biochemistry*, 2, 16 (1963).
20.   Antonini, Wyman, Zito, Rossi-Fanelli, and Caputo, *J. Biol. Chem.*, 238, PC60 (1961).
21.   Harris and Li, *J. Biol. Chem.*, 213, 499 (1955).
22.   Davie and Neurath, *J. Biol. Chem.*, 212, 507 (1955).
23.   Harris and Knight, *J. Biol. Chem.*, 214, 215 (1955).
24.   Drechsler, Boyer, and Kowolsky, *J. Biol. Chem.*, 234, 2627 (1959).
25.   Fraenkel-Conrat and Singer, *Arch. Biochem. Biophys.*, 60, 64 (1956).
26.   Pope, *Brit. J. Expt. Pathol.*, 20, 132, 201 (1939); Peterman and Pappenheimer, *J. Phys. Chem.*, 45, 1 (1941).
27.   Perlmann, *Nature*, 173, 406 (1954); Tokuyasu and Funatsu, *J. Biochem. (Tokyo)*, 52, 103 (1962).
28.   Wagman, *Arch. Biochem. Biophys.*, 100, 414 (1963).
29.   Li, Geschwind, Cole, Raacke, Harris, and Dixon, *Nature*, 176, 687 (1955).
30.   Anfinsen, *J. Biol. Chem.*, 221, 405 (1956); Ginsberg and Schachman, *J. Biol. Chem.*, 235, 108 (1960), *J. Biol. Chem.*, 235, 115 (1960).
31.   Kaminski and Tanner, *Biochim. Biophys. Acta*, 33, 10 (1959).
32.   Hanson and Johansson, *Nature*, 187, 600 (1960).
33.   Li, Papkoff, and Hayashida, *Arch. Biochem. Biophys.*, 85, 97 (1959).
34.   Golubow and du Vigneaud, *Proc. Soc. Expt. Biol. Med.*, 112, 218 (1963).
35.   Ginsberg and Schachman, *J. Biol. Chem.*, 235, 108 (1960); *J. Biol. Chem.*, 235, 115 (1960); Butler, Phillips, Stephen, and Creeth, *Biochem. J.*, 46, 44 (1950).
36.   Rupley and Scheraga, *Biochemistry*, 2, 421 (1963).
37.   Gladner and Neurath, *J. Biol. Chem.*, 206, 911 (1954).
38.   Harris, Li, Condliffe, and Pon, *J. Biol. Chem.*, 209, 133 (1954).
39.   Takahashi, *J. Biochem. (Tokyo)*, 52, 72 (1962).
40.   Hill and Smith, *J. Biol. Chem.*, 228, 577 (1957); Smith, Hill, and Borman, *Biochim. Biophys. Acta*, 29, 207 (1958).
41.   Hill and Smith, *J. Biol. Chem.*, 235, 2332 (1960).
42.   White, *J. Amer. Chem. Soc.*, 77, 4691 (1955).
43.   Richards and Vithayathil, *J. Biol. Chem.*, 234, 1459 (1959); Gordillo, Vithayathil, and Richards, *Yale J. Biol. Med.*, 34, 582 (1962).
44.   Ottsen, *Compt. Rend. Trav. Lab. Carlsberg*, 30, 211 (1958).
45.   Ottesen and Schroeder, *Acta Chem. Scand.*, 15, 926 (1961).
46.   Nozaki, Yamanaka, Horro, and Okunuki, *J. Biochem. (Tokyo)*, 44, 453 (1957).
47.   Toda and Akabori, *J. Biochem. (Tokyo)*, 53, 95 (1963).
48.   von Hippel and Harrington, *Biochim. Biophys. Acta*, 36, 427 (1959).

Table 3
ENZYMATIC HYDROLYSIS OF SEVERAL CONJUGATED PROTEINS

Protein	Enzymes	Products isolated
Azaserine labeled enzyme	Papain, pronase, aminopeptidase	C^{14}-labeled azaserinepeptides[1]
Rabbit γ-globulin	Papain	Glycopeptide[2]
Bovine globulin of colostrum	Papain	Glycopeptide[2]
Human γ-globulin	Papain	Glycopeptide[3]
Cytochrome *c*	Pepsin, trypsin	Heme peptides[4]
Ovalbumin	Pepsin, trypsin, chymotrypsin, mold protease	Glycopeptides[5]
Ovalbumin	Pancreatin	Glycopeptide[6]
Ovalbumin	Trypsin, chymotrypsin	Glycopeptides[7]
Chondroitin sulfate complex	Papain	Glycopeptide[8]
Fetuin	Papain, trypsin, chymotrypsin, pepsin, subtilisin	Glycopeptides[9]
Chromatium heme protein	Pepsin	Heme peptide[10]
Phosphorylase	Chymotrypsin	Pyridoxal peptide[11]

Compiled by C. H. W. Hirs.

REFERENCES

1. Dawid, French, and Buchanan, *J. Biol. Chem.*, 238, 2178 (1963).
2. Nolan and Smith, *J. Biol. Chem.*, 237, 453 (1962).
3. Rosevear and Smith, *J. Biol. Chem.*, 236, 425 (1961).
4. Tuppy, in *Symposium on Protein Structure,* Neuberger, Ed., John Wiley, New York, 1958, 66.
5. Johansen, Marshall, and Neuberger, *Biochem. J.,* 78, 518 (1961).
6. Jevons, *Nature,* 181, 1346 (1958).
7. Cunningham, Nuenke, and Nuenke, *Biochim. Biophys. Acta,* 26, 660 (1957).
8. Muir, *Biochem. J.,* 69, 195 (1958); Anderson, Hoffman, and Meyer, *Biochim. Biophys. Acta,* 74, 309 (1963).
9. Spiro, *J. Biol. Chem.,* 237, 382 (1962).
10. Dus, Bartsch, and Kamen, *J. Biol. Chem.,* 237, 3083 (1962).
11. Fischer, Kent, Snyder, and Krebs, *J. Amer. Chem. Soc.,* 80, 2906 (1958).

Table 4
SPECIFICITY OF TRYPSIN TOWARD SYNTHETIC SUBSTRATES

Substrates[a]	Relative rate of hydrolysis	Substrates[a]	Relative rate of hydrolysis
Benzoyl-L-argininamide[1]	100	L-Arginyl-L-leucine[2]	0.036
Glycyl-L-lysinamide[1]	91	L-Arginylglycine[2]	0.029
L-Alanyl-L-lysinamide[1]	200	L-Arginylphenylalanine[2]	0.029
L-α-Aminobutyryl-L-lysinamide[1]	406	L-Arginylglutamic acid	0.017
L-Norleucyl-L-lysinamide[1]	424	Cbz-nitro-L-arginyl-L-leucine[2]	0
β-Alanyl-L-lysinamide[1]	82	Cbz-nitro-L-arginyl-L-phenylalanine	0
L-Valyl-L-lysinamide[1]	324	Cbz-nitro-L-argininamide[2]	0
L-Leucyl-L-lysinamide[1]	155	Benzoyl-L-homoargininamide[3]	0
L-Phenylalanyl-L-lysinamide[1]	124		

Compiled by C. H. W. Hirs.

[a] At 0.05 M substrate, 25°–40°C, pH 7.5–7.8. Cbz, carbobenzoxy.

REFERENCES

1. Izumiya, Yamashita, Uchio, and Kitagawa, *Arch. Biochem. Biophys.,* 90, 170 (1960).
2. Van Orden and Smith, *J. Biol. Chem.,* 208, 751 (1954).
3. Shields, Hill, and Smith, *J. Biol. Chem.,* 234, 1747 (1959).

Table 5
SPECIFICITY OF CHYMOTRYPSIN FOR HYDROLYSIS OF PEPTIDE BONDS IN PROTEINS AND POLYPEPTIDES[a]

Type of bond	Type of bond	Type of bond	Type of bond	Type of bond
-Thr-Asn···Arg-Asn[d]	-Glu-Gln···Ala-Arg[h]	-Lys-His···Ile-Ile[e]	-Leu-Met···Glu-Tyr[c]	-Ala-Thr···Asn-Arg[d]
-Val-Asn···CMC-Ala[d]	-Val-Gln···Ala-Ser[j]	-Lys-His···Lys-Thr[c,f]	-Lys-Met···Ile-Phe[c]	-Gly-Thr···Glu-Gln[h]
-Ala-Asn···Lys-Asn[f]	-Ala-Gln···Lys-His[e]	-Asp-Ile···Asn-Leu[h]	-Ala-Met···Lys-Arg[d]	-Lys-Thr···Gly-Gln[f]
-Lys-Asn···Lys-Gly[c]	-Tyr-Gln···Lys-Met[j]	-Val-Lys···Ala-His[l]		-Ile-Thr···Ser-Leu[j]
	-Ser-Gln···Val-Thr[h]	-Glu-Lys···Gly-Gly[c]	-Thr-Met···Ser[d]	-Arg-Thr···Val-Glu[h]
-Lys-Asn···Val-Ala[e]	-Lys-Gln···[h]		-Lys-Met···Val-Thr[j]	
-Thr-Asn···Val-Lys[b]		-Val-Lys···Gly-His[b]	-Ala-Phe···Pro-Leu[k]	-Val-Thr···Val-Arg[h]
-Lys-Asn···[i]	-Val-Gly···Lys-Lys[k]	-Gly-Lys···Lys-Arg[h]	-Phe-Ser···[i]	-Glu-Val···Arg-Gln[h]
	-Leu-His···Ala-His[b]	-Ile-Lys···Lys-Lys[f]	-Thr-Ser···Ala-Ala[e]	-Pro-Val···Lys-Val[h]
	-Val-His···Ala-Ser[b]			
	-Ala-His···Gly-Lys[j]	-Lys-Lys···Lys[f]	-Tyr-Thr···Ala-Ala[f]	
-Val-CySO₃H···Ser-Leu[g]	-Gly-His···Gly-Lys[b]	-Gly-Met···Asn-Ala[d]	-Val-Thr···Ala-Leu[h]	
-Thr-Glu···Gln-Ala[h]	-Leu-His···Gly-Leu[c,f]	-Ile-Met···Gly-Asn[j]		

Compiled by C. H. W. Hirs.

[a] In the same substrates a total of 32 phenylalanyl bonds, 36 leucyl bonds, 24 tyrosyl bonds, and 6 tryptophanyl bonds were also hydrolyzed.

[b] Human hemoglobin α-chain.[1]
[c] Horse cytochrome c.[2]
[d] Egg-white lysozyme.[3]
[e] Ribonuclease.[4]
[f] Human cytochrome c.[5]
[g] Insulin.[6]
[h] Tobacco mosaic virus.[7]
[i] Papain.[8]
[j] Human hemoglobin, γ-chain.[9]
[k] Ovine corticotropin.[10]

Table 5

SPECIFICITY OF CHYMOTRYPSIN FOR HYDROLYSIS OF PEPTIDE BONDS IN PROTEINS AND POLYPEPTIDES (continued)

REFERENCES

1. **Hill and Konigsberg,** *J. Biol. Chem.,* 237, 3151 (1962).
2. **Margoliash,** *J. Biol. Chem.,* 237, 2161 (1962).
3. **Canfield,** *J. Biol. Chem.,* 238, 2698 (1963).
4. **Hirs, Stein, and Moore,** *J. Biol. Chem.,* 221, 151 (1956); Hirs, Moore, and Stein, *J. Biol. Chem.,* 235, 633 (1960); Smyth, Stein, and Moore, *J. Biol. Chem.,* 237, 1845 (1962).
5. **Matsubara and Smith,** *J. Biol. Chem.,* 238, 2732 (1963).
6. **Sanger and Tuppy,** *Biochem. J.,* 49, 463, 481 (1951); Sanger and Thompson, *Biochem. J.,* 53, 353, 366 (1953).
7. **Anderer, Uhlig, Weber, and Schramm,** *Nature,* 186, 922 (1960).
8. **Light and Smith,** *J. Biol. Chem.,* 237, 2537 (1962).
9. **Schroeder, Shelton, Shelton, Cormick, and Jones,** *Biochemistry,* 2, 992 (1963).
10. **Leonis, Li, and Chung,** *J. Amer. Chem. Soc.,* 81, 419 (1959).

Table 6
PRODUCTS FORMED ON CHYMOTRYPTIC HYDROLYSIS OF α-CHAINS OF HUMAN GLOBIN UNDER TWO DIFFERENT CONDITIONS OF HYDROLYSIS[a]

Peptide substrate, α-chain, human hemoglobin.
Bonds hydrolyzed are indicated by arrows

```
1↓           5        ↓10              ↓15
Val-Leu-Ser-Pro-Ala-Asp-Lys-Thr-Asn-Val-Lys-Ala-Ala-Try-Gly
```

Products on hydrolysis for 6 hr at pH 8, 25°C, with α-chymotrypsin : substrate molar ratio of 1 : 150

Cα14	Val-Leu-Ser-Pro-Ala-Asp-Lys-Thr-Asn-Val-Lys-Ala-Ala-Try	(75% yield)
Cα10	Ser-Pro-Ala-Asp-Lys-Thr-Asn-Val-Lys-Ala-Ala-Try	(20% yield)

Products on hydrolysis for 24 hr at pH 9, 30°C, with α-chymotrypsin : substrate molar ratio of 1 : 150

Cα14	Val-Leu-Ser-Pro-Ala-Asp-Lys-Thr-Asn-Val-Lys-Ala-Ala-Try	(15% yield)
Cα10	Ser-Pro-Ala-Asp-Lys-Thr-Asn-Val-Lys-Ala-Ala-Try	(10% yield)
Cα15	Val-Lys-Ala-Ala-Try	(15% yield)

Peptide, substrate, α-chain, human hemoglobin.
Bonds hydrolyzed are indicated by arrows

```
↓25          ↓30        ↓  35
Tyr-Gly-Ala-Glu-Ala-Leu-Glu-Arg-Met-Phe-Leu-Ser
```

Products on hydrolysis for 6 hr at pH 8, 25°C, with α-chymotrypsin : substrate molar ratio of 1 : 150

Cα9	Gly-Ala-Glu-Ala-Leu-Glu-Arg-Met-Phe	(80% yield)

Products on hydrolysis for 24 hr at pH 9, 30°C, with α-chymotrypsin : substrate molar ratio of 1 : 150

Cα9	Gly-Ala-Glu-Ala-Leu-Glu-Arg-Met-Phe	(32% yield)
Cα16	Glu-Arg-Met-Phe	(50% yield)
Cα29	Gly-Ala-Glu-Ala-Leu	(15% yield)

Compiled by C. H. W. Hirs.

[a]Only portions of the α-chain (substrate) are shown (*1*).

REFERENCES

1. **Hill and Konigsberg,** *J. Biol. Chem.,* 237, 3151 (1962).

Table 7
CONDITIONS FOR CHYMOTRYPTIC HYDROLYSIS OF CERTAIN PROTEINS

Substrate	Substrate (μmoles)	Chymotrypsin[a] (μmoles)	Volume of reaction (ml)	Enzyme substrate molar ratio	Temperature (°C)	pH	Time of hydrolysis (hr)
Performate oxidized ribonuclease[1]	72	0.20	100	1 : 360	25	7.0	24
Oxidized A-chain insulin[2]	22	0.1	5	1 : 220	37	7.5	24
Oxidized B-chain insulin[3]	17	0.1	5	1 : 170	25	7.5–8.0	24
Ovine corticotropin[4]	7.5	0.046	5	1 : 160	40	9.0	24
Human hemoglobin, α-chain[5]	121	0.80	200	1 : 150	30	9.0	24
Human hemoglobin α-chain[5]	60	0.40	90	1 : 150	25	8.0	6
Carboxymethyl lysozyme[6]	70	0.8	100	1 : 87	37	8.0	2
Equine cytochrome c[7]	100	2.4	120	1 : 42	22	7.8	29
Human cytochrome c[8]	70	2.3	90	1 : 33	Room temp.	7.85	26
Oxidized papain[9]	10	0.48	15	1 : 21	39	7.6	6

Compiled by C. H. W. Hirs.

[a]Assumed molecular weight 25,000.

REFERENCES

1. **Hirs, Moore, and Stein,** *J. Biol. Chem.,* 235, 633 (1960).
2. **Sanger and Thompson,** *Biochem. J.,* 53, 353, 366 (1953).
3. **Sanger and Tuppy,** *Biochem. J.,* 49, 463, 481 (1951).
4. **Léonis, Li, and Chung,** *J. Amer. Chem. Soc.,* 81, 419 (1959).
5. **Hill and Konigsberg,** *J. Biol. Chem.,* 237, 3151 (1962).
6. **Canfield,** *J. Biol. Chem.,* 238, 2698 (1963).
7. **Margoliash,** *J. Biol. Chem.,* 237, 2161 (1962).
8. **Matsubara and Smith,** *J. Biol. Chem.,* 238, 2732 (1963).
9. **Light and Smith,** *J. Biol. Chem.,* 237, 2537 (1962).

Table 8
SPECIFICITY OF PEPSIN FOR HYDROLYSIS OF PEPTIDE BONDS IN PROTEINS AND POLYPEPTIDES[a]

Type of bond	Type of bond	Type of bond	Type of bond	Type of bond
-Asn-Arg···Ile-[c]	-Met-Glu···His-[i]	-Leu-Gly···Arg-COO-[f]	-Gly-Ser···Tyr-[c]	-Leu-Tyr···Leu-[g]
		-Leu-Gly···Asp-[h]		
-Val-Asn···Phe-[b]	-Ile-Glu···Leu-[c]	-Gly-Gly···Glu-[f]	-Val-Thr···Ala-[c]	-Ser-Tyr···Ser-[c]
-Phe-Asn···Thr-[d]	-Glu-Glu···Lys-[c]	-Asp-Gly···Leu-[f]	-Val-Thr···Leu-[b]	-Pro-Tyr···Val-[e]
-Gly-Asn···Try-[d]	-Ala-Glu···Phe-[b]	-Arg-Gly···Phe-[g]	-Asn-Thr···Phe-[e]	-Ile-Val···Ala-[e]
-Val-Asp···Glu-[f]	-Lys-Glu···Phe-[f,h]	-Arg-Gly···Tyr-[d]	-Ala-Thr···Val-[c]	-Ala-Val···Asp-[c]
-Ser-Asp···Phe-[c]	-Gly-Glu···Tyr-[b]	-Val-His···Ala-[b]	-Arg-Thr···Val-[c]	-Asn-Val···Asp-[f]
-Ala-Asp···Pro-[c]	-Asp-Glu···Val-[f]	-Glu-His···Phe-[i]	-Val-Thr···Val-[c]	-Ser-Val···CySO$_3$ H-[e,h]
-Val-Asp···Pro-[f,c]	-Val-Gln···Ala-[e,h]	-His-His···Phe-[b]	-Ala-Try···Gly-[b]	-CySH-Val···Leu-[f]
-Leu-CMC···Asn-[d]	-Asn-Gln···His-[g]	-His-Lys···Leu-[b]	-Arg-Try···Try-[d]	-Asn-Val···Lys-[b]
-Try-CMC···Asn-[d]	-Tyr-Gln···Leu-[g]	-Phe-Lys···Leu-[b]	-Ala-Try···Val-[d]	-Ala-Val···Thr-[f]
-Val-Glu···Ala-[g]	-Asn-Gln···Phe-[c]	-Arg-Met···Phe-[b]	-Asn-Try···Val-[d]	-Gln-Val···Try-[c]
-Leu-Glu···Asn-[g]	-Arg-Gln···Phe-[c]	-Ser-Ser···Asp-[d]	-Asn-Tyr···CySO$_3$ H-[g]	-Val-Val···Tyr-[f]
		-Leu-Ser···Ser-[h]		
-Val-Glu···Gln-[g]	-Ser-Gln···Val-[c]	-Val-Ser···Thr-[b]	-Leu-Tyr···Gln-[g]	

Compiled by C. H. W. Hirs.

[a]In the same substrates a total of 22 alanyl bonds, 34 leucyl bonds, and 23 phenylalanyl bonds were also hydrolyzed. Asn, Asparagine; Gln, glutamine; CMC, S-carboxymethyl-cystine.
[b]Human hemoglobin, α-chain.[1]
[c]Tobacco mosaic virus.[2]
[d]Egg-white lysozyme.[3]
[e]Ribonuclease.[4]
[f]Human hemoglobin, β-chain.[1]
[g]Insulin.[5]
[h]Human hemoglobin, γ-chain.[6]
[i]β-MSH.[7]

REFERENCES

1. Hill and Konigsberg, *J. Biol. Chem.*, 237, 351 (1962).
2. Anderer, Uhlig, Weber, and Schramm, *Nature*, 186, 922 (1960).
3. Canfield, *J. Biol. Chem.*, 238, 2698 (1963).
4. Hirs, Moore, and Stein, *J. Biol. Chem.*, 235, 633 (1960).
5. Sanger and Tuppy, *Biochem. J.*, 49, 463, 481 (1951); Sanger and Thompson, *Biochem. J.*, 53, 353, 366 (1953).
6. Schroeder, Shelton, Shelton, Cormick, and Jones, *Biochemistry*, 2, 992 (1963).
7. Harris and Roos, *Biochem. J.*, 71, 434 (1959).

Table 9

SPECIFICITY OF PAPAIN FOR HYDROLYSIS OF PEPTIDE BONDS IN PROTEINS AND POLYPEPTIDES[a]

Type of bond	Type of bond	Type of bond	Type of bond
-Asp-Ala⋯Asn-[d]	-Pro-Asn⋯Ala-[d]	-Glu-Glu⋯Lys-COO-[c]	-Lys-Ile⋯Phe-[d]
-Val-Ala⋯Asn-[b]	-Pro-Asn⋯Leu-[d]	-Val-Gly⋯Ala-[b]	-Ala-Leu⋯Glu-[b]
-Pro-Ala⋯Glu-[b]	-Glu-Asn⋯Phe-[c]	+H₃N-Gly⋯Ala-[b]	+H₃N-Leu⋯Glu-[d]
-Val-Ala⋯Gly-[b]	-Val-Asp⋯Asp-[b]	-Lys-Gly⋯Ile-[d]	-Ala-Leu⋯Ser-[b]
+H₃N-Val-Ala⋯His-[b]	-Asp-Asp⋯Met-[b]	-His-Gly⋯Ser-[b]	-His-Leu⋯Thr-[c]
-Asn-Ala⋯Leu-[b]	-Leu-Glu⋯Asn-[d]	-Ala-Gly⋯Val-[b]	-Asn-Lys⋯Asn-[d]
-Glu-Ala⋯Leu-[b]	-Glu-Glu⋯Asn-[c]	-Leu-His⋯Ala-[b]	-Glu-Lys⋯Gly-[d]
+H₃N-Ser-Ala⋯Leu-[b]	-Pro-Glu⋯Asn-[c]	-Ala-His⋯Leu-[b]	+H₃N-Lys⋯Gly-[d]
-Pro-Ala⋯Val-[b]	-Pro-Glu⋯Glu-[c]	-Val-His⋯Leu-[c]	-Lys-Lys⋯Ile-[d]

Type of bond
-Gly-Lys⋯Lys-[d]
-Pro-Lys⋯Lys-[d]
-Lys-Lys⋯Tyr-[d]
+H₃N-Lys⋯Thr-[d]
-Glu-Phe⋯Thr-[b,c]
+H₃N-Ser⋯Ala-[b]
-Leu-Ser⋯Asp-[b]
-Leu-Ser⋯His-[b]
-Ile-Thr⋯Tyr-[d]
-Thr-Tyr⋯Phe-[b]

Compiled by **C. H. W. Hirs.**

[a]Of the peptides examined, none contained arginine. Bonds formed by this amino acid should be very susceptible to hydrolysis.
[b]Human hemoglobin, α-chain.[1]
[c]Human hemoglobin, β-chain.[2]
[d]Cytochrome c.[3]

REFERENCES

1. **Konigsberg and Hill**, *J. Biol. Chem.*, *237*, 3157 (1962).
2. **Konigsberg, Goldstein, and Hill**, *J. Biol. Chem.*, *238*, 2028 (1963).
3. **Margoliash**, *J. Biol. Chem.*, *237*, 2161 (1962).

Table 10

SPECIFICITY OF SUBTILISIN FOR HYDROLYSIS OF PEPTIDE BONDS IN PROTEINS AND POLYPEPTIDES

Type of bond	Type of bond	Type of bond	Type of bond	Type of bond
-Ala-Ala···Lys-d	-Val-CMC···Ala-d	-Asn-Gln···His-f	-Met-Lys···Arg-d	-Ala-Thr···Asn-d
-CyS-Ala···Ser-f	-Arg-CMC···Asn-d	-Arg-Gln···Phe-b	-Leu-Met···Asp-c	-Asp-Ser···Gly-e
-Val-Ala···Try-d	-Leu-CMC···Asn-d	-Glu-Gly···Gly-e	-Lys-Met···Glu-e	-Arg-Thr···Val-b
-Ser-Arg···Arg-c	-Leu-CySO$_3$ H···Gly-f	-CySO$_3$ H-Gly···Glu-f	-Ala-Met···Lys-d	-Arg-Try···Gly-a
-Glu-Arg···Gly-f	-Val-CySO$_3$ H···Ser-e	-Ser-Gly···Lys-f	-$^+$H$_3$ N-Phe···Val-f	-Ala-Try···Ile-d
-Arg-Arg···Val-b	-Val-CySO$_3$ H···Gly-f	-CySO$_3$ H-Gly···Ser-f	-His-Phe···Arg-a	-Gln-Try···Leu-c
-Met-Asn···Ala-d	-CyS-CyS···Ala-f	-Gln-His···Leu-f	-Gly-Phe···Phe-f	-Asn-Try···Val-d
-CMC-Asn···Asp-d	-Val-CyS···Ser-f	-His-Leu···CySO$_3$ H-f	-Thr-Phe···Thr-e	-Arg-Try···Asn-b
-Val-Asn···Gln-f	-Val-Glu···Ala-f	-Gln-Leu···Glu-f	-Phe-Phe···Tyr-f	-Asn-Tyr···CyS-f
-Ser-Asn···Phe-d	-Leu-Glu···Asn-f	-Ile-Leu···Gln-d	-Asp-Phe···Val-c	-Asp-Tyr···Gly-d
-Ile-Asn···Ser-d	-CySO$_3$ H-Glu···Gly-e	-Ser-Leu···Gly-d	-Gly-Pro···Val-e	-Leu-Tyr···Leu-f
-Arg-Asn···Thr-d	-Thr-Gln···Ala-d	-Ala-Leu···Tyr-f	-CMC-Ser···Ala-d	-Lys-Tyr···Leu-c
-Met-Asn···Thr-COO-c	-Val-Gln···Ala-d	-Ser-Leu···Tyr-f	-Thr-Ser···Asp-c	-Pro-Tyr···Lys-a
-Asn-Asp···Gly-d	-Glu-Gln···Asp-c	-His-Leu···Val-f	-Gly-Ser···His-f	-Gly-Tyr···Ser-d
-Asp-Asp···Ala-b	-Glu-Gln···Cys-S-f	-Tyr-Leu···Val-f	-Tyr-Ser···Lys-c	-Phe-Tyr···Thr-f
-Val-Asp···Asp-b	-Asn-Gln···Glu-f	-Pro-Lys···Ala-COO -f	-Gly-Ser···Thr-d	-$^+$H$_3$ N-Phe-Val···Asn-f
-Gly-Asp···Gly-d	-Ser-Gln···Glu-c		-Ile-Thr···Ala-d	-Leu-Val···CySO$_3$ H-f
				-Pro-Val···CySO$_3$ H-e

Compiled by C. H. W. Hirs.

a β-MSH.[1]
b Tobacco mosaic virus.[2]
c Glucagon.[3]
d Egg-white lysozyme.[4]
e Diisopropyltrypsin peptide.[5]
f Insulin.[6]

REFERENCES

1. **Harris and Roos**, *Biochem. J.*, 71, 434 (1959).
2. **Tsugita, Gish, Young, Fraenkel-Conrat, Knight, and Stanley**, *Proc. Natl. Acad. Sci. (USA)*, 46, 1463 (1960).
3. **Bromer, Staub, Sinn, and Behrens**, *J. Amer. Chem. Soc.*, 79, 2801 (1957); Bromer, Sinn, and Behrens, *J. Amer. Chem. Soc.*, 79, 2807 (1957).
4. **Canfield**, *J. Biol. Chem.*, 238, 2698 (1963).
5. **Dixon, Kauffman, and Neurath**, *J. Amer. Chem. Soc.*, 80, 1260 (1958); Dixon, Kauffman, and Neurath, *J. Biol. Chem.*, 233, 1373 (1958).
6. **Tuppy**, in *Symposium on Protein Structure*, Neuberger, Ed., John Wiley & Sons, New York, 1958, 66; Haugaard and Haugaard, *Compt. Rend. Trav. Lab. Carlsberg*, 29, 350 (1955).

These tables originally appeared in Sober, Ed., *Handbook of Biochemistry and selected data for Molecular Biology*, 2nd ed., Chemical Rubber Co., Cleveland, 1970.

INDEX TO PHYSICAL-CHEMICAL DATA OF PROTEINS

Malcolm H. Smith (deceased)

This index follows earlier compilations, notably by Svedberg[235] and Edsall,[55] and are similarly concerned primarily with information relating to ultracentrifuge studies, particularly sedimentation-velocity studies. The references are taken from the article by Malcolm H. Smith in the 2nd Edition of the Handbook of Biochemistry.

Table 1
GLOBULAR PROTEINS

Protein	References	Protein	References
Acetylcholinesterase, *Electrophorus*	467, 519	D-Amino acid oxidase holo-enzyme, pig kidney	271
Acetyl-CoA synthetase, bovine heart	471	D-Amino acid oxidase, pig kidney	153
Acid proteinase, *Aspergillus*	413	D-Amino acid oxidase, Michaelis complex, pig kidney	270
Acylase I, (Hippuricase) hog kidney	485		
Acyl phosphatase, bovine brain	190	L-Amino acid oxidase, moccasin snake	55, 369
Adenine deaminase, calf intestinal mucosa	25	L-Amino acid oxidase, rat kidney	161, 541
S-Adenine deaminase	133	δ-Aminolevulinate dehydratase, mouse	511
Adrenocorticotropin, pig	55, 223, 320	Amylase, malt	42
Adrenocorticotropin, sheep	55, 319, 323	α-Amylase, *B. subtilis*	67
Adrenocorticotropin, sheep pituitary	137, 231	α-Amylase, pig pancreas	41
Adrenocorticotropin A, pig pituitary	137	α-Amylase, *Pseudomonas*	148
Aequorin, *Aquorea*	452	β-Amylase, sweet potato	55, 375
Alanine dehydrogenase, *B. subtilis*	503	α-Amylase (Zn free), *B. subtilis*	67
Albumin, bovine	185, 188, 354	Anthranilate synthetase, *E. coli*	542
Albumin, bovine serum	55, 188, 408	Anticatalase	72
Albumin, canine serum	253	Apoferritin, Guinea pig	73
Albumin, carp	96	Apoferritin, horse spleen	196
Albumin, cow serum	55, 311, 334, 348, 351, 352, 358, 405	D-Arabinose dehydrogenase, pseudomonad	476
Albumin, dog serum	4	Arachin, *Arachis hypogaea* (ground nut)	108, 111
Albumin esterase, mouse	470	Arginine phosphotransferase, crustacean	61
Albumin, horse serum	55, 235, 297, 336, 356, 402, 403, 404, 411	Ascorbate oxidase apoenzyme, *Cucurbita pepo*	491
Albumin, human serum	55, 177, 300, 351, 410	Ascorbate oxidase, *Cucurbita pepo condensa* (squash)	55, 368
Albumin, rat serum	280, 468	Ascorbate oxidase, *Cucurbita pepo* (inactivated)	491
Alcohol dehydrogenase, horse liver	58, 245		
Alcohol dehydrogenase, yeast	245	Ascorbate oxidase, *Cucurbita pepo* (native)	491
Aldolase	55, 373		
Aldolase, bovine liver, 25°C	184	Ascorbate oxidase, *Cucurbita pepo* (reduced)	491
Aldolase, rabbit liver	500		
Aldolase, rabbit skeletal muscle	243	Aspartate transaminase, ox heart	536
Aldose dehydrogenase, pseudomonad	476	Aspartate transaminase, ox heart (apo-enzyme)	536
Alkaline phosphatase, calf	64		
Alkaline phosphatase, *E. coli*	76	apo-Aspartate transaminase, yeast	490
Amandin	235, 289, 379	Autoprothrombin *c*, bovine	278
Amino acid oxidase, *Crotalus*	261	Autoprothrombin II, bovine	92
D-Amino acid oxidase apo-enzyme, pig kidney	271	Azurin, *Bordetella*	447

Table 1 (continued)
GLOBULAR PROTEINS

Protein	References	Protein	References
Bacillus phlei protein	235, 286	Cocosin	235, 377
Bence-Jones protein	235, 349	Cobalamin protein B, hog gastric	262
Bence-Jones protein, human urine	50	mucosa	
Blue latex protein, *Rhus vernicifera*	175	Colbalamin protein A, hog gastric	262
Blue protein, *Pseudomonas*	234	mucosa	
		Collagenase	212
		Conalbumin, chicken egg white	75
Canavalin, Jack bean	235, 295	Fe-Conalbumin, chicken egg white	75
Carbamate kinase	85	Conarachin	109
Carbamyl phosphate synthetase, frog	149	Concanavalin A, Jack bean	235, 295, 298
liver		Concanavalin B, Jack bean	235, 295
Carbonic anhydrase, human	172	Creatine kinase BB, chicken	535
Carbonic anhydrase, mammalian RBC,	55, 415	Creatine kinase BB, rabbit	535
25°C		Creatine kinase MM, chicken	535
Carbonic anhydrase, ox blood	55, 328	Creatine kinase MM, rabbit	535
Carbonic anhydrase B	140	Crotonase	228
Carbonic anhydrase X_1, human RBC	194	Crotoxin, *Crotalus terrificus*	235, 294
Carbonic anhydrase Y, human RBC	194	(rattlesnake)	
α-Carboxylase, wheat germ	218	Cryptocytochrome *c*, *Pseudomonas*	234
Carboxylesterase, pig kidney	557	Cytochrome, *Chlorobium*	78
Carboxypeptidase	55, 337, 397	Cytochrome, *Rhodopseudomonas*	162
Carboxypeptidase B, pig	69	Cytochrome *a*, mammalian heart	241
Catalase, bovine liver	100, 202, 235,	Cytochrome b_1, *E. coli*	516
	303	Cytochrome b_2, yeast	7, 274
Catalase, horse liver	3	Cytochrome *c*, beef heart	20
Catalase, human blood	34	Cytochrome *c*, bovine heart	55, 56, 321
Catalase, *Micrococcus*	34	Cytochrome *c*, *Chromatium*	9
Catalase, pig blood	165	Cytochrome *c*, horse heart	55, 321, 469
Cathepsin C	88	Cytochrome *c*, human	449
Cathepsin, cod spleen	458	Cytochrome *c*, pig heart	55, 321
Ceruloplasmin, human	185, 188, 350	Cytochrome *c*, rust fungus	167
Ceruloplasmin, human plasma	203	Cytochrome *c*, vertebrate	235
Ceruloplasmin, pig	185	Cytochrome *c*, vertebrate heart	235, 289, 407
Ceruloplasmin, pig blood	179	Cytochrome c_1, bovine heart	283
Chloroperoxidase, *Caldariomyces*	466	Cytochrome c_1, *Rhodopseudomonas*	178
fumago		Cytochrome *f*	44
Chlorocruorin, *Sabella*	6	Cytochrome *h*	121
Choline acetyltransferase, human	473	Cytochrome oxidase, *P. aeruginosa*	99
placenta			
Choline acetyltransferase, rabbit brain	473	DDT dehydrochlorinase	142
Chymopapain	53	Dehydropeptidase II, bovine kidney	214
Chymotrypsin, bovine pancreas	55, 297, 330	Denitrifying enzyme, bacterial	409
	338, 341,	(*Pseudomonas*?)	
	400	Dextran saccharose, *Leuconostoc*	502
Chymotrypsin β, bovine pancreas	55, 327	Diaphorase, pig heart	204
Chymotrypsin inhibitor, potato	461	Dihydrolipoic dehydrogenase, *E. coli*	126
Chymotrypsin α (dimer), bovine	55, 291, 296	Dihydrolipoic dehydrogenase,	493
pancreas		*Spinacea aleracea*	
Chymotrypsin α (monomer), bovine	55, 291	*Diphtheria* antitoxin	55, 362, 412
pancreas		*Diphtheria* toxin	55, 299
Chymotrypsinogen, bovine pancreas	55, 293, 330		
Chymotrypsinogen α, bovine pancreas	55, 263, 327	Edestin	235, 289, 380
Chymotrypsinogen β, bovine pancreas	55, 327	Enolase	55
Citrate cleaving enzyme, rat liver	512	Enolase, rabbit muscle, 1°C	98
Citrate-oxalacetate lyase, *Aerobacter*	23, 514	Erythrocuprin, human	232
Clostridium toxin	55, 419, 420	Erythrocuprein, human RBC	124
Clupein sulphate	285	Excelsin	235, 289, 379

Table 1 (continued)
GLOBULAR PROTEINS

Protein	References	Protein	References
Ferredoxin, *Clostridium*	237	γ-Globulin, bovine	55, 406
Ferri-cytochrome *c*, bovine heart	171	γ-Globulin, cow	55, 358, 365, 370, 408
Ferro-cytochrome *c*, bovine heart	171		
Fibrinogen, cow	55, 358	γ-Globulin, cow serum	55, 372
Fibrinogen, human	55, 418	γ-Globulin, equine serum	188, 366
Ficin	37	γ-Globulin, horse	177, 180, 188
Flavin mononucleotide (FMN), (dimer) 25°C	544	γ-Globulin, horse serum	55, 336, 370
		γ-Globulin, human	185
Fluorokinase	258	γ-Globulin, human	26, 55, 177, 185, 359
Follicle-stimulating hormone, bovine pituitary	531		
		γ-Globulin, human serum	55, 307
Follicle-stimulating hormone, sheep	55, 301	γ-Globulin, pig serum	55, 371
Fucan, *Fucus vesiculosus*	13	γ-Globulin, rabbit	31
Fumarase, hog heart	55, 374	γ-Globulin, rabbit serum	55, 365, 394
		γ_1-Globulin, human plasma	240
Galactosamino-glycan, bovine cornea	555	Glucosamino-glycan, bovine cornea	555
Galactose dehydrogenase, pseudo-monad	476	Glucose oxidase, *P. amagasakiense*	130
β-Galactosidase, *E. coli* ML35	284	Glucose oxidase-microside protein, *Penicillium*	501
β-Galactosidase, *E. coli* ML309	284	Glucose oxidase (Notatin), *Penicillium*	33
Gliadin	70, 235, 322, 398	Glucose-6-phosphate dehydrogenase, human R.B.C.	498
Globulin, *Acacia* seed	43	Glutamate dehydrogenase, bovine liver	55, 305
Globulin, *Arachis* seed	43	Glutamate dehydrogenase, chicken liver	74, 513
Globulin, *Astragalus* seed	43		
Globulin, *Avena* seed	43	Glutamic-aspartic transaminase	107
Globulin, bovine	235, 307, 382	Glyceraldehyde phosphomutase, yeast	243
Globulin, *Cytisus* seed	43	D-Glyceraldehyde-3-phosphate dehydrogenase	60
Globulin, *Dolichos* seed	43		
Globulin, *Ervum* seed	43	D-Glyceraldehyde-3-phosphate dehydrogenase, mammalian	60
Globulin, *Festuca* seed	43		
Globulin, *Genista* seed	43	D-Glyceraldehyde-3-phosphate dehydrogenase, rabbit muscle	60, 243
Globulin, *Glycine* seed	43		
Globulin, *Hordeum vulgare* seed	43	D-Glyceraldehyde-3-phosphate dehydrogenase (+ KCN), rabbit muscle	71
Globulin, horse serum	235, 289, 411		
Globulin, *Lathyrus* seed	43	Glycerokinase, *Candida*	11
Globulin, *Lotus* seed	43	Glycoprotein, bovine plasma	14
Globulin, *Lupinus angustifolius* seed	43, 112	α-Glycoprotein, pleural fluid	22
Globulin, *Medicago* seed	43	Glycoprotein, rat	68
Globulin, *Panicum* seed	43	α-Glycoprotein, fetal calf serum	465
Globulin, *Phaseolus* seed	43	Gonadotrophin, pregnant mare	135
Globulin, *Phleum* seed	43	Growth hormone, pituitary	55, 137, 183, 347, 399
Globulin, pig	235, 307, 382		
Globulin, *Pisum* seed	43		
Globulin, *Secale* seed	43	Haemopexin	209
Globulin, *Trifolium* seed	43	Haptoglobin type *1-1*, human	188, 417
Globulin, *Triticum vulgare* seed	43	Hemagglutinin, castor bean	239
Globulin, *Vicia* seed	43	Hemagglutinin, cold	260
Globulin, *Zea* seed	43	Hemagglutinin, soya bean	181, 256
α-Globulin, barley	43	Hemicellulose A, *Chlorella*	173
α-Globulin, human	188, 343	Hemocyanin, *Achatina*	235, 317
α-Globulin, human plasma	177	Hemocyanin, *Agriolimax*	235, 317
α_2-Globulin	177	Hemocyanin, *Arion*	235, 317
β_1-Globulin	177	Hemocyanin, *Astacus*	235, 364
β-Globulin, human plasma	177	Hemocyanin, *Buccinum*	235, 317, 364
γ-Globulin, barley	43	Hemocyanin, *Busycon*	235, 364

Table 1 (continued)
GLOBULAR PROTEINS

Protein	References	Protein	References
Hemocyanin, *Calocaris*	235, 317	Hemoglobin, *Eumenia*	235, 317
Hemocyanin, *Cancer*	235, 364	Hemoglobin, frog	464
Hemocyanin, *Carcinus*	235, 364	Hemoglobin, *Gasterophilus*	2
Hemocyanin, *Chirodothea*	235, 317	Hemoglobin, *Gasterosteus*	235, 317
Hemocyanin component, *Helix pomatia*	27	Hemoglobin, *Glycera*	235, 317
		Hemoglobin, *Haemopsis*	235, 317
Hemocyanin, *Eledone*	235, 289, 317, 364	Hemoglobin, hedgehog	235, 317
		Hemoglobin, hen	235, 317
Hemocyanin, *Eupagurus*	235, 317	Hemoglobin, *Hirudo*	235, 317
Hemocyanin, *Euscorpius*	235, 317	Hemoglobin, horse	235, 376, 464
Hemocyanin, *Helix pomatia*	27, 235, 289, 364	CO-hemoglobin, horse	55, 287
		Hemoglobin, human	49, 55, 235, 287, 289, 348, 360, 398, 464
Hemocyanin, *Homarus*	235, 289, 364		
Hemocyanin, *Hyas*	235, 317		
Hemocyanin, *Limax*	235, 364	Hemoglobin, *Lampetra*	5, 136
Hemocyanin, *Limnea*	235, 317	Hemoglobin, *Lucioperca*	235, 317
Hemocyanin, *Limulus*	235, 364	Hemoglobin, *Lumbricus*	235, 317
Hemocyanin, *Littorina*	235, 317, 364	Hemoglobin, *Lumbrinereis*	235, 317
Hemocyanin, *Loligo*	235, 364, 508	Hemoglobin, *Marphysa sanguinea*	518
Hemocyanin, *Maia*	234, 317	Hemoglobin, mouse	464
Hemocyanin, *Nephrops*	235, 289, 364	Hemoglobin, *Myxine*	235, 317
Hemocyanin, *Neptunea*	235, 364	Hemoglobin, *Neireis*	235, 317
Hemocyanin, octopus	235, 289, 364	Hemoglobin, *Notomastus*	235, 317
Hemocyanin, *Ommatostrephes sloani pacificus* (Squid)	176	Hemoglobin, *Opsanus*	235, 317
		Hemoglobin, ox	464
Hemocyanin, *Pagurus*	235, 317	Hemoglobin, *Paramecium*	225
Hemocyanin, *Palaemon*	235, 317	Hemoglobin, *Parus*	235, 317
Hemocyanin, *Palinurus*	235, 289, 364	Hemoglobin, *Pectinaria*	235, 317
Hemocyanin, *Paludina vivipara*	27, 235, 317	Hemoglobin, perienteric fluid, *Ascaris*	510
Hemocyanin, *Pandalus*	235, 364		
Hemocyanin, *Rossia*	235, 289, 364	Hemoglobin, *Picus*	235, 317
Hemocyanin, *Sepia*	235, 364	Hemoglobin, pig	464
Hemocyanin, *Sepiola*	235, 317	Hemoglobin, pigeon	235, 317, 464
Hemocyanin, *Squilla*	235, 317	Hemoglobin, *Planorbis*	235, 314, 317
Hemocyanin, *Tonicella*	235, 317	Hemoglobin, *Pleuronectes*	235, 317
Hemoglobin, *Aethelges*	235, 317	Hemoglobin, *Polymnia*	235, 317
Hemoglobin, *Andara (= Arca)*	235, 272, 317	Hemoglobin, *Prionotus*	235, 317
Hemoglobin, *Anguilla*	235, 317	Hemoglobin, *Protopterus*	235, 317
Hemoglobin, *Anguis*	235, 317	Hemoglobin, rabbit	235, 317, 464
Hemoglobin, ape	235, 317	Hemoglobin, *Raja*	235, 317
Hemoglobin, *Arenicola*	235, 381	Hemoglobin, *Rana*	235, 317, 474
Hemoglobin, cat	235, 317	Hemoglobin, *Salmo*	235, 317
Hemoglobin, *Chameleon*	235, 317	Hemoglobin, sheep	464
Hemoglobin, chicken	464	Hemoglobin, *Syrnium*	235, 317
Hemoglobin, *Chironomous*	235, 317	Hemoglobin, *Tautoga*	235, 317
Hemoglobin, *Coluber*	235, 317	Hemoglobin, *Thyone*	235, 317
Hemoglobin, *Corvus*	235, 317	Hemoglobin, *Tubifex*	207
Hemoglobin, cow	235, 317	Hemoglobin, turtle	464
CO-hemoglobin, cow	55, 287, 410	Hemoglobin A, human	472
Hemoglobin, *Cyprinus*	235, 317	Hemoglobin E, human	472
Hemoglobin, *Daphnia*	235, 381	Hemoglobin F, human	472
Hemoglobin, dog	235, 317, 464	Hemoglobin H, human (i)	10
Hemoglobin, dogfish	464	Hemoglobin H, human (ii)	10
Hemoglobin, duck	235, 317, 464	Hemoglobin I, *Bufo*	235, 317
Hemoglobin, *Eisenia*	235, 317	Hemoglobin I, *Chrysemys*	235, 317
Hemoglobin, *Esox*	235, 317	Hemoglobin I, *Lacerta*	235, 317

Table 1 (continued)
GLOBULAR PROTEINS

Protein	References	Protein	References
Hemoglobin I, legume	59	β-Lactoglobulin	55, 334, 336, 340, 355, 430
Hemoglobin I, *Salamandra*	235, 317	β-Lactoglobulin, goat	8
Hemoglobin II, *Bufo*	235, 317	β₂-Lactoglobulin, human	208
Hemoglobin II, *Chrysemys*	235, 317	Lactoperoxidase	247
Hemoglobin II, *Lacerta*	235, 317	Lactotransferrin	16, 160
Hemoglobin II, legume	59	Legumin, *Pisum sativum* (pea)	43
Hemoglobin II, *Salamandra*	235, 317	Lipase, milk	277
Hemoglobin III, *Bufo*	235, 317	Lipoamide dehydrogenase, human liver	539
Hemoglobulin, *Lampetra*	235, 314	α-Lipoprotein, human	188, 205
Hemoglobulin, *Lumbricus*	235, 289	α-Lipoprotein, human serum	205
Hemoglobulin, *Planorbis*	235, 376	Lipoxidase, soya bean	246
δ-Hemolysin, staphylococcal	495	α-Livetin	150
Homogentisate oxygenase, *Pseudomonas fluoresens*	509	β-Livetin	150
		Luciferase, *Cypridina*	66, 250
Hordein	235, 318	Luteinising hormone, pig	55, 361
Hyaluronic acid	255	Luteinising hormone, pig	55, 367
p-Hydroxybenzoate hydroxylase, *Pseudomonas putida*	481	Lysozyme, chicken egg white	55, 227, 312, 315, 316, 324
Hydroxyproline epimerase, *Pseudomonas*	448	Lysozyme, *Papaya*	224
20-β-Hydroxy steroid dehydrogenase *Streptomyces*	497	α-Macroglobulin, rat serum	468
		Malate dehydrogenase, beef heart	63
L-Iditol dehydrogenase	219	Malate dehydrogenase, bovine heart	217
Insulin	39, 55, 287, 308–310, 311, 313, 344, 345, 346, 414, 421, 556	Malate dehydrogenase, *B. subtilis*	496
		Malate dehydrogenase, horse heart	248
		Malate dehydrogenase, ox heart	248
		Malate dehydrogenase, ox heart, 2°C	45
		Malate dehydrogenase, pig heart	248
		Malate-lactate transhydrogenase, *Micrococcus lactilyticus*	534
Interferon	29	Malic dehydrogenase, pig heart	268
Interstitial-cell stimulating hormone (ICSH), human pituitary	455	Megacin	97
Iron-binding protein, cow milk	79	Merino wool protein, SCMKB1	445
Isocitrate lyase, *Pseudomonas*	506	Merino wool protein, SCMKB2	445
Isocitric enzyme	163	Metallothionein, horse kidney	115
Iso-hemagglutinin	188, 300	Metapyrocatechase, *Pseudomonas*	487
		Met-Hemoglobin, horse	55, 357
Kallikrein inactivator, bovine parotid	128	Methylmalonyl racemase, *Propionibacterium*	459
Keratinase, *Streptomyces*	454	Milk protein, mouse	199
		Monoamine oxidase, ox plasma	273
Laccase	166, 174	Monoamine oxidase, plasma	505
Laccase A, *Polyporus*	475	Mucoprotein, urinary	154
Laccase B, *Polyporus*	475	Myeloperoxidase, infected dog uterus	57
Lactalbumin, cow	55, 289	Myogen A, rabbit skeletal muscle	81
Lactalbumin, cow	55	Myoglobin, *Aplysia*	200
Lactalbumin, human	144	Myoglobin, guinea pig	94
Lactate dehydrogenase, ox heart	157, 168	Myoglobin, horse	200
Lactate dehydrogenase, pig heart	105	Myoglobin, horse heart	235, 289, 290
Lactic dehydrogenase H, beef heart	281, 282	Myoglobin, seal	201
Lactic dehydrogenase H, chicken	282	Myoglobin, tortoise	101
Lactic dehydrogenase M, beef heart	282	Myoglobin, Tunny	192
Lactic dehydrogenase M, chicken	282	CO-myoglobin I, mammalian	56
Lactogenic hormone (prolactin), sheep	138	Myokinase	30
Lactoglobulin	235, 289, 339	Myosin	55, 389, 416

Table 1 (continued)
GLOBULAR PROTEINS

Protein	References	Protein	References
NAD glycohydrolase	463	Prolactin, sheep	453
Nerve protein, lobster	155	Protease, *Bacillus thermoproteolyticus*	533
Neuraminase, *Vibrio cholerae*	189	Protease, bacterial	158
Nuclear protein, chicken RBC	440	Protease, *Streptococcus*	170
Nuclear protein, RP2-L, rat carcinoma	443	Protein, *Lupinus* seed	112
		Protein B, lobster nerve	155
Old yellow enzyme	244	Protein X, human blood	49
Old yellow enzyme, brewer's yeast	235, 302	"A" Protein, *E. coli*	95
Ornithine transcarbamylase	197	Proteinase, *B. subtilis*	446
Ovalbumin	55, 235, 348, 398, 408	Proteinase, *Pseudomonas*	102
		Proteinase, *Tetrahymena*	51
Oxytocic hormone	55, 331	Proteolytic inhibitor, blood	82
		Prothrombin, bovine	92
Papain	226	Prothrombin, bovine	92, 188
Paramyosin, *Venus mercenaria*	545	Protochlorophyll-protein complex, bean leaf	18
Penicillase, *Bacillus*	187		
Penicillinase	89	ε-Protoxin, *Clostridium*	456
Pepsin	55, 289, 335, 342, 396, 457, 515	Pseudocholinesterase, horse	106
		Pyridine nucleosidase, bull semen	1
		Pyrophosphatase	55, 206, 401
Pepsinogen	457	Pyrophosphatase, yeast	206
Peptidase A, *Pencillium*	492	Pyruvic acid oxidase	211
Peroxidase, wheat	215		
Peroxidase, wheat germ	215	C-Reactive protein	269
Peroxidase II, horse radish	35, 247	Red protein, bovine milk	47
Phenol oxidase, *Calliphora* larva	116	Red protein, cow milk	86
Phosphoenolpyruvate carboxykinase, pig liver	537	Rhodanese, bovine liver	229
		RHP, *Chromatium*	9
Phosphoglyceric acid mutase, rabbit muscle	54	RHP, *Rhodospirillum rubrum*	9
		Ribonuclease, bovine pancreas	55, 288, 325, 451, 515
Phosphoglyceric acid mutase, yeast	54	Ribonuclease, *B. subtilis*	444
Phosphoglycerate mutase, chicken muscle	462	Ribonuclease, pancreatic	444
		Ribonuclease T₁	252
Phosphoglucomutase	243	Ribonucleoprotein	90
Phospholipase A-I, *Crotalus adamanteus*	488	Ricin	114
		Ricin D	489
Phospholipase A-II, *Crotalus adamanteus*	488	Scarlet fever toxin	55, 333
Phosphorylase a	84	Secretin phosphate, hog intestine	91, 235
Phosphorylase b	83	Somatotrophin, beef	137
Phosphorylase, potato	134	Somatotropin, human	137
Phosphorylase, rabbit heart	276	Somatotrophin, sheep	137
Phosphorylase, rabbit muscle	276	Somatotrophin, whale	137
Phosvitin calcium complex	113	Soya bean protein	267
Phosvitin magnesium complex	113	Succinic dehydrogenase, bacterial	259
Phycocyanin, *Ceramium*	55, 376	Sulfate-binding protein, *Salmonella typhimurium*	532
Phycoerythrin, *Ceramium*	235, 376, 378		
Phytohemagglutinin (lectin), *Phaseolus vulgaris*	538	Taka-amylase A	103, 238
		T.B. bacillus protein	235, 286
Plastocyamin, spinach	118	T Component, equine	188, 366
Polyphenol oxidase, *Camelia sinensis* (tea)	482	Thetin-homocysteinase	52
		Thiogalactoside transacetylase, *E. coli*	479
Pomelin	235, 326	Thrombin, bovine	92, 441
Pre-albumin	188, 350	Thyroglobulin, pig	48, 235, 289, 383
Procarboxypeptidase	123		
Prolactin	55, 332		

Table 1 (continued)
GLOBULAR PROTEINS

Protein	References	Protein	References
Transamidinase, hog kidney	486	Triose-phosphate dehydrogenase, turkey	499
Transferrin, human	139, 188, 350, 477, 480, 483, 484, 494	Triose-phosphate dehydrogenase, yeast	499
		Tropomyosin, rabbit muscle	55, 422
Transferrin, human plasma	484	Trypsin	55, 297, 329
Transferrin, monkey	480	Trypsin inhibitor, bovine pancreas	442
Transferrin, pig	363	Trypsin inhibitor, human	188
Transferrin, porcine	188, 363	α-Trypsin inhibitor, human serum	28
Transferrin, rat	480	Trypsin inhibitor, soya bean	15, 193, 450
Transglutaminase, pig liver, 25°C	543	Trypsinogen, bovine pancreas	55, 292
Triose-phosphate dehydrogenase, bovine	499	Tryptophan synthetase B, *E. coli*	478
		Tryptophanase, *E. coli*	504, 507
Triose-phosphate dehydrogenase, chicken	499	Turnip yellow mosaic virus protein	147
		Tyrosinase, *Psalliota* (mushroom)	145
Triose-phosphate dehydrogenase, *E. coli*	499	Tyrosinase L, *Neurospora crassa*	469
		Tyrosinase S, *Neurospora crassa*	469
Triose-phosphate dehydrogenase, halibut	499	Tryosine transaminase, rat kidney	540
		Tween hydrolyzing enzyme	257
Triose-phosphate dehydrogenase, human	499		
		Urease, Jack bean	235, 304
Triose-phosphate dehydrogenase, lobster	499		
		Venom substrate, bovine plasma	65
Triose-phosphate dehydrogenase, pheasant	499	Vicilin, *Pisum sativum* (pea)	43
		Viper venom coagulant	265
Triose-phosphate dehydrogenase, rabbit	499		
		Wheat germ hemoprotein 550	279
Triose-phosphate dehydrogenase, sturgeon	499		
		Yeast protein	141
		Zein	70

Compiled by Malcolm H. Smith.

Table 2
FIBROUS PROTEINS

Protein	References	Protein	References
F$_1$-Actin, polymerized	529	Histone I, calf thymus	251
F$_2$-Actin, polymerized	529	Histone II, calf thymus	251
G-Actin, cod	38		
G-ADP actin, rabbit skeletal muscle	156	Iridine, *Salmo*	285
α-Amylase, *B. stearothermophilus*	146		
		Lewis blood group substance	122
Casein, mouse milk	198	Lipoeuglobulin III, human	188, 391
α-Casein, cow milk	236	Lipoprotein, human serum (HDL-3)	93
β-Casein, cow milk	236	β-Lipoprotein, human	55, 306
Ceruloplasmin, human	117	α$_1$-Lipoprotein, human	177, 188,
Apo-ceruloplasmin, human	117		338, 388
Chloroplast protein, spinach leaf	36	Lipoprotein, human serum (HDL-2)	93
Clostridium toxin	55, 419	Lipovitellin	254
Collagen, cod skin	275	β-Lipovitellin	129
Collagen, cod swim bladder	275	β-Livetin, hen egg	264
Collagen A, earth worm cuticle	152		
Collagen B, earth worm cuticle	152	α$_2$-Macroglobulin, human	188, 395
α-Crystallin, bovine lens	17, 182	α$_2$-Macroglobulin, human plasma	210
β-Crystallin, bovine lens	17	Mucin, pig submaxillary gland	524
α-Crystallin, bovine lens cortex	182	Mucoid, bovine oestrous cervical	77
Cytochrome *b$_5$*, calf liver	233	Mucoid, bovine pregnancy cervical	77
		Mucoprotein, cartilage	12
Elastin (urea-graded), bovine	24	Mucoprotein, cow milk	104
ligamentum nuchae		Mucoprotein, human	188, 392
Euglobulin, human pathological	185	Mucoprotein, human plasma	222
		Mucoprotein, human urinary	242
Fetuin, bovine	188	Mucoprotein, sheep submaxillary gland	80
Fetuin, calf	186	Mucoprotein, ovine urinary	528
Fetuin, cow fetus	186, 230	Mucoprotein, urinary	154, 213
Fibrinogen, bovine	62	Myosin	55, 110, 389
Fibrinogen, cow	55, 390	Myosin, bovine	525
Fibrinogen, horse	210	Myosin, cod	38, 527
Fibrinogen, human	32, 55, 177	Myosin, dog heart	526
		γ-Myosin	120
Globulin, horse antipneumococcus	235, 307	Myxomyosin	249
Globulin, human "cold-insoluble"	188, 393		
β$_1$-Globulin, human plasma	177	Peptomyosin B	21
Eu-γ-globulin, human plasma	210	Phosvitin	113, 159
Globulin, metal-binding, human plasma	210	Phosvitin, hen serum	159
Gollagen, cod swim bladder	275	Plasminogen, human plasma	46, 216
Gonadotrophin, human urinary	143	α$_3$-Protein, canine	253
Glycoprotein ("ovoglycoprotein"),	521	Pseudo γ-globulin, human plasma	210
egg white			
M$_2$-Glycoprotein, bovine	523	α$_1$-Seromucoid, human plasma	210
α-Glycoprotein, fetal calf serum	465	Sialoprotein, bovine cortical bone	266
α-Glycoprotein, human	188, 385	(EDTA extracted)	
α$_1$-Glycoprotein	188, 386	Sialoprotein (phosphate extracted),	266
		bovine cortical bone	
Haptoglobin, type 1-1, human	188, 387	Silk fibroin	520
Haptoglobin I-(methemoglobin-Hb)^2	87	Spore peptide, *B. megatherium*	191
Haptoglobin II	87		
Haptoglobin II-methemoglobin complex	87	Tropocollagen, calf skin	522
Hemocyanin component, *Paludina*	27	Tropomyosin, adult blowfly	127
vivipara		Tropomyosin, larval blowfly	127
Heparin fraction	132	Tropomyosin, *Pinna nobilis*	119
β-Histone	40	Tropomyosin, rabbit	127
γ-Histone	40		
β-Histone, aggregated	40	Zein	55, 384

Compiled by Malcolm H. Smith.

THE PROTEINS OF BLOOD COAGULATION

Common name	Roman numeral[a]	Associated names	Conversion or activated product	Stage of participation in blood coagulation	Name of disease[b]	References
Fibrinogen	I	—	Fibrin	Late, intrinsic,[e] and extrinsic[f]	Afibrinogenemia or hypofibrinogenemia	1
Prothrombin[g]	II	—	II[a], thrombin, biothrombin	Late, intrinsic, and extrinsic	Prothrombin deficiency	2–4
Tissue thromboplastin	III	Extrinsic thromboplastin	—	Early, extrinsic	—	
Calcium	IV		—	Early, middle, late, intrinsic, and extrinsic	—	
Proaccelerin	V	Labile factor, plasma accelerator globulin, plasma Ac-globulin, thrombogene	Activated proaccelerin,[h] accelerin, Factor VI	Middle, intrinsic, and extrinsic	Parahemophilia	5, 6
Precursor of serum prothrombin conversion accelerator (ProSPCA)[g]	VII	Proconvertin, stable factor, cofactor V, cothromboplastin, kappa factor, serozyme, autoprothrombin I	VII[a], serum prothrombin, conversion accelerator (SPCA), convertin[i]	Middle, intrinsic, and extrinsic	ProSPCA deficiency	7, 8
Antihemophilic factor (AHF)	VIII	Antihemophilic globulin (AHF), platelet cofactor I, antihemophilic globulin A, thromboplastinogen, plasma thromboplastinogen, plasma thromboplastin factor, plasmokinase	Activated AHF[h]	Middle, intrinsic	Classic hemophilia, hemophilia A	9–11
Christmas factor[g]	IX	Plasma thromboplastin component (PTC), autoprothrombin II, auto-prothrombin C, plasma thromboplastin cofactor II, antihemophilic factor B	IX[a], activated Christmas factor, prephase accelerator, PTC[c]	Middle, intrinsic	Christmas disease, hemophilia B	12
Stuart factor[g]	X	Prower factor, venom substrate, autoprothrombin III	X[a], activated Stuart factor, activated venom substrate, autoprothrombin C, thrombokinase	Middle, intrinsic, and extrinsic	Stuart factor deficiency	13, 14
Plasma thromboplastin antecedent (PTA)	XI	Antihemophilic factor C	XI[a], activated PTA, third prothromboplastic factor, activation product	Early, intrinsic	PTA deficiency	15, 16
Hageman factor	XII	Surface factor, clot-promoting factor, contact factor, the fifth plasma thromboplastin precursor	XII[a], activated Hageman factor	Early, intrinsic	Hageman trait	17, 18
Fibrin stabilizing factor (FSF)	XIII	Laki-Lorand factor, fibrinase	XIII[a], activated fibrin stabilizing factor, FSF[c]	Late, intrinsic, and extrinsic	—	19, 20
Prekallikrein	—	Fletcher factor, kallikreinogen	Kallikrein	Early, intrinsic	Fletcher factor disease	21
Plasminogen	—	Profibrinolysin	Plasmin, fibrinolysin	—	—	22

THE PROTEINS OF BLOOD COAGULATION (continued)

Common name	Mode of inheritance	Plasma concentration (μg/ml plasma)	Molecular weight (daltons)	Polypeptide chains	Molecular weight of chains (daltons)	Amino terminus	References
Fibrinogen	Disorder of both sexes, transmitted as autosomal recessive trait	1700–4000	340,000	2 α, 2 β, 2 γ	63,500[d], 56,000[d], 47,000[d]	Ala, Glu[j], Tyr	1
Prothrombin[g]	Disorder of both sexes, transmitted as autosomal recessive trait	70	68,000–72,000	1	—	Ala	2–4
Tissue thromboplastin	—		—	—	—	—	5,6
Calcium	—	90–115					
Proaccelerin	Disorder of both sexes, transmitted as autosomal recessive trait	Trace	310,000–400,000[c]				
Precursor of serum prothrombin conversion accelerator (ProSPCA)[g]	Disorder of both sexes, transmitted as autosomal recessive trait	0.13	45,500[c], 60,000[d]	1	—	Ala	7,8
Antihemophilic factor (AHF)	Disorder primarily of males, transmitted as sex-linked recessive trait	5–10	1.1×10^6 [c,d]	—	200,000[d]	—	9–11
Christmas factor[g]	Disorder primarily of males, transmitted as sex-linked recessive trait	3	54,000[c]	1	—	Tyr	12
Stuart factor[g]	Disorder of both sexes, transmitted as autosomal recessive trait	2–5	55,000[c]	2	39,000, 16,000	Trp, Ala	13,14
Plasma thromboplastin antecedent (PTA)	Disorder of both sexes, transmitted as autosomal recessive trait	Trace	160,000[c,d]	2	80,000	—	15,16
Hageman factor	Disorder of both sexes, transmitted as autosomal recessive or dominant trait	15–47	82,000[c], 80,000[d]	1	—	—	17,18
Fibrin stabilizing factor (FSF)	Disorder of both sexes, transmitted as autosomal recessive trait	Trace	320,000[d]	2 a, 2b	75,000, 88,000	N-AcSer, Glu	19,20
Prekallikrein	—	Trace	90,000[c]	1	—	—	21
Plasminogen	—	Trace	87,000[d]	1	—	Glu	22

Compiled by Walter Kisiel and Earl W. Davie.

THE PROTEINS OF BLOOD COAGULATION (continued)

[a]Roman numerals have been assigned to the well recognized clotting factors by an international nomenclature committee (Wright, *J. Am. Med. Assoc.,* 170, 325, 1959).

[b]Clinical manifestations vary from epitaxes and mild bruising to excessive bleeding following injury and surgery for afibrinogenemia, parahemophilia, ProSPCA deficiency, and Stuart deficiency. Classic hemophilia and Christmas disease are often characterized initially by bleeding in the joints of the knees, ankles, and elbows; spontaneous bleeding is unusual in PTA deficiency, while Hageman trait shows no bleeding tendency.

[c]Protein isolated from bovine plasma.

[d]Protein isolated from human plasma.

[e]Coagulation reactions yielding thrombin due to the interaction of constituents found only in plasma.

[f]Coagulation reactions yielding thrombin due to the interaction of constituents found in plasma and tissue extracts.

[g]Requires vitamin K for biosynthesis.

[h]Activated proaccelerin (accelerin, Factor VI) and activated AHF are thrombin-modified proteins with increased clotting activity. Originally, accelerin was thought to be a new clotting factor and was assigned roman numeral VI.

[i]Proconvertin and convertin have been applied collectively to Factor VII and Stuart factor and their activated forms, respectively.

[j]Pyrrolidone carboxylic acid.

GENERAL REFERENCES

Esnouf and Macfarlane, *Advances in Enzymology,* Nord, Ed., Interscience, N. Y., 1968, 255.

Heimburger and Trobisch, *Angew. Chem.,* 10, 85(1971).

Ratnoff, *Progress in Hemostasis and Thrombosis,* Spaet, Ed., Grune and Stratton, N. Y., 1972, 39.

Davie and Kirby, *Current Topics in Cellular Regulation,* Horecker and Stadtman, Eds., Academic Press, New York, 1972, 51.

Davie and Fujikawa, *Ann. Rev. Biochem.,* 44, 799 (1975).

REFERENCES

1. Doolittle, *Adv. Protein Chem.,* 27, 1 (1973).
2. Cox and Hanahan, *Biochim. Biophys. Acta,* 207, 49 (1970).
3. Pirkle, McIntosh, Theodor, and Vernon, *Thromb. Res.,* 2, 461 (1973).
4. Kisiel and Hanahan, *Biochim. Biophys. Acta,* 304, 103 (1973).
5. Philip, Moran, and Colman, *Biochemistry,* 9, 2212 (1970).
6. Hanahan, Rolfs, and Day, *Biochim. Biophys. Acta,* 286, 205 (1972).
7. Kisiel and Davie, *Biochemistry,* 14, in press (1975).
8. Laake and Ellingsen, *Thromb. Res.,* 5, 539 (1974).
9. Schmer, Kirby, Teller, and Davie, *J. Biol. Chem.,* 247, 2512 (1972).
10. Legaz, Schmer, Counts, and Davie, *J. Biol. Chem.,* 248, 3946 (1972).
11. Shapiro, Anderson, Pizzo, and McKee, *J. Clin. Invest.,* 52, 2198 (1973).
12. Fujikawa, Thompson, Legaz, Meyer, and Davie, *Biochemistry,* 12, 4938 (1973).
13. Jackson and Hanahan, *Biochemistry,* 7, 4506 (1968).
14. Fujikawa, Legaz, and Davie, *Biochemistry,* 11, 4882 (1972).
15. Kato, Legaz, and Davie, unpublished results.
16. Wuepper, in *Inflammation: Mechanism and Control,* Lepow and Ward, Eds., Academic Press, N.Y., 1972, 93.
17. Schoenmakers, Matze, Haanen, and Zilliken, *Biochem. Biophys. Acta,* 101, 166 (1965).
18. Revak, Cochrane, Johnston, and Hugli, *J. Clin. Invest.,* 54, 619 (1974).
19. Schwartz, Pizzo, Hill, and McKee, *J. Biol. Chem.,* 248, 1395 (1973).
20. Takagi and Doolittle, *Biochemistry,* 13, 750 (1974).
21. Takahashi, Nagasawa, and Suzuki, *J. Biochem.,* 71, 471 (1972).
22. Walther, Steinman, Hill, and McKee, *J. Biol. Chem.,* 249, 1173 (1974).

GLYCOPROTEINS

Morris Soodak (deceased)

Glycoproteins may be defined "as conjugated proteins containing as prosthetic group(s) one or more heterosaccharide(s), usually branched, with a relatively low number of sugar residues, lacking a serially repeating unit and bound covalently to the polypeptide chain."

At the present time, 14 types of carbohydrate-protein linkages have been found in living matter:

Amino Acid	Monosaccharide
1. Alanine	Muramic acid
2. Asparagine	N-Acetylglucosamine
3. Cysteine	Glucose
4. Cysteine	Glactose
5. Hydroxylsine	Galactose
6. Hydroxyproline	Arabinose
7. Serine	N-Acetylglucosamine
8. Serine	Galactose
9. Serine	Mannose
10. Serine	Xylose
11. Threonine	N-Acetylgalactosamine
12. Threonine	Fucose
13. Threonine	Glactose
14. Threonine	Mannose

The not uncommon heterogeneity of essentially pure glycoprotein preparations is attributable in part to the variable carbohydrate composition, minor proteolysis in vivo or during resolutions, genetic polymorphism, or all three. This heterogeneity when combined with the difficulties involved in the analysis of the carbohydrate constitutents of the glycoproteins results in a variability of the analytical data found in the literature.

The two-volume work, *Glycoproteins,* edited by A. Gottschalk, 1972, Elsevier Publishing Company, should be consulted for an understanding of these problems.

The tables compiled by Richard J. Winzler for the 2nd Edition of the *Handbook of Biochemistry and Molecular Biology* are reproduced again. Richard Winzler, who was one of the Founding Fathers in this field, died September 28, 1972.

Additions and updating for some proteins are appended to his tables.

Table 1
GLYCOPROTEINS, HUMAN PLASMA*

Protein	Molecular weight	$S_{20,w}$	Carbohydrate content, g/100 g				Reference
			Hexose	Acetyl hexosamine	Sialic acid	Fucose	
1. α_1-Antitrypsin	44,100	3.11	14.7	13.9	12.1	0.7	1
2. Ceruloplasmin	160,000	7.08	3.0	2.4	2.4	0.2	1
3. α_1-Easily precipitable glycoprotein	50,000	3.8	4.8	4.4	3.7	0.4	1
4. Fibrinogen	341,000	7.63	1.0	0.9	0.6	–	1
5. β_{1A}- and β_{1C}-Globulin	–	6.9	1.8	0.6	0.45	0.16	1
6. G_c-Globulin	50,800	3.7	2.0	2.0	0	0.2	1
7. α_{1x}-Glycoprotein	–	3.9	8.0	8.0	6.0	0.7	1
8. α_{2HS}-Glycoprotein [Ba-α_2-glycoprotein]	49,000	3.3	5.2	3.9	4.1	0.2	1
9. β_2-Glycoprotein	–	2.9	6.7	5.8	4.4	0.2	1
10. Haptoglobin	100,000	4.4	7.8	5.3	5.3	0.2	1
11. Hemopexin	80,000	4.8	9.0	7.4	5.8	0.4	1
12. γA-Immunoglobulin	–	7.0	3.2	2.3	1.8	0.22	1
13. γG-Immunoglobulin	156,000	7.0	1.1	1.3	0.3	0.2	1
14. γM-Immunoglobulin	1,000,000	19.0	5.4	4.4	1.3	0.7	1
15. Myeloma globulin	–	–	3.7	3.0	1.0	–	2
16. Inter-α-trypsin inhibitor	–	6.4	3.4	3.5	2.1	0.1	1
17. α_2-Macroglobulin	820,000	19.6	3.6	2.9	1.8	0.1	1, 3
18. α_2-Neuramino glycoprotein	–	3.7	12.0	13.0	17.0	0.6	1
19. Orosomucoid [α_1-acid glycoprotein]	44,100	3.11	14.7	13.9	12.1	0.7	1
20. Plasminogen	143,000	4.28	6.7	5.8	4.4	0.2	1, 4
21. 4.6 S-Postalbumin	–	4.6	4.0	2.8	3.0	0.2	1
22. Prothrombin	–	–	4.6	2.9	4.2	–	5
23. Rheumatoid factor	–	19.0	5.4	3.6	1.8	–	6
24. Transcortin	58,500	3.0–4.1	5.4	4.7	3.2	0.8	7
25. Transferrin	90,000	5.5	2.4	2.0	1.4	0.07	1
26. Tryptophan-poor α_1-glycoprotein	55,000–60,000	3.3	5.5	4.5	3.4	0.3	1
27. Zn-α_2-Glycoprotein	41,000	3.2	7.0	4.0	7.0	0.2	1
28. α_2-β_1-Glycoprotein	60,000		10.7	10.8	7.5	0.5	8
29. Glycine-rich – γ protein		4.3	8.5	–	1.3	–	9
30. Coagulation factor VIII	1,100,000		2.2	3.2	0.9	–	10
31. Coagulation factor XIII	320,000		1.9	1.6	1.2	0.2	11, 12
32. Prothrombin	70,000		2.8	3.8	2.3	–	13, 14
33. Clq of complement	400,000		8.5	1.0	0.5	–	15, 16
34. α-Fetoprotein	70,000		2.2	1.4	1.2	–	17

Compiled by Morris Soodak.

*There are thought to be more than 100 blood plasma proteins and a great many of them are glycoproteins.

Table 1 (continued)
GLYCOPROTEINS, HUMAN PLASMA

REFERENCES

1. Schultze and Heremans, in *Molecular Biology of Human Proteins,* Elsevier, Amşterdam I, 1966, 182.
2. Dawson and Clamp, *Biochem. J.,* 107, 341 (1968).
3. Dunn and Spiro, *J. Biol. Chem.,* 242, 5549 (1967).
4. Slotta and Gonzalez, *Biochemistry,* 3, 285 (1964).
5. Schultze and Schwick, *Clin. Chim. Acta,* 4, 15 (1959).
6. Kunkel, Franklin, and Müller-Eberhard, *J. Clin. Invest.,* 38, 424 (1959).
7. Slaunwhite, Schneider, Wissler, and Sandberg, *Biochemistry,* 5, 3527 (1966).
8. Iwasaki and Schmid, *J. Biol. Chem.,* 245, 1814 (1970).
9. Boenisch and Alper, *Biochim. Biophys. Acta,* 214, 135 (1970).
10. Legaz, Schmer, Counts, and Davie, *J. Biol. Chem.,* 248, 3946 (1973).
11. Bohn, *Ann. N.Y. Acad. Sci.,* 202, 256 (1972).
12. Schwartz, Pizzo, Hill, and McKee, *J. Biol. Chem.,* 246, 5851 (1971).
13. Kisiel and Hanahan, *Biochim. Biophys. Acta,* 304, 103 (1973).
14. Saavedra and Mehl, *J. Biol. Chem.,* 243, 5479 (1968).
15. Calcott and Müller-Eberhard, *Biochemistry,* 11, 3443 (1972).
16. Müller-Eberhard, *Annu. Rev. Biochem.,* 44, 697 (1975).
17. Ruoslahti and Seppälä, *Int. J. Cancer,* 7, 218 (1971).

Table 2
PLASMA GLYCOPROTEINS (OTHER THAN HUMAN)

Protein	Source	Carbohydrate content, g/100 g				Reference
		Neutral sugar	Acetyl hexosamine	Sialic acid	Fucose	
1. Fetuin	Fetal calf	7.6	6.6	8.7	–	1, 2
2. M_2-Glycoprotein	Cow	6.1	7.7	10.3	–	1, 3, 4
3. Gonadotropin-transporting protein	Horse	14.0	13.3	12.0	–	1, 5
4. Corticosteroid-transporting protein	Rabbit	11.4	11.6	8.7	–	6
5. Immunoglobulin IgA	Horse	2.1	1.9	0.9	–	7
6. Immunoglobulin IgA	Cow	0.9	1.8	0.3	–	8
7. Immunoglobulin IgG	Horse	1.1	1.1	0.2	–	7
8. Immunoglobulin IgG	Rabbit	1.2	1.2	0.2	–	9
9. Immunoglobulin light chain	Mouse urine	7.8	7.4	1.8	–	10
10. α_2-Macroglobulin	Fetal calf	2.7	1.6	0.7	–	1, 11
11. α_2-Macroglobulin	Horse serum	4.3	4.3	3.6	–	12
12. α_2-Macroglobulin	Rabbit serum	6.5	4.0	1.6–3.3	–	13
13. α_2-Macroglobulin	Pig serum	8–10	–	2.5–4.0	–	14
14. α_2-Macroglobulin	Rat serum	10.3	–	–	–	15
15. γ-Inhibitor of viral hemagglutination	Horse serum	8.6	3.9	2.5	–	16
16. α_1-Macroglobulin	Rabbit	6.5	4.95	3.3	–	17
17. α-Acid glycoprotein	Cow	12.0	14.5	10–16.2	–	18–20
18. α-Glycoprotein	Dog	11.2	11.1	8.0	–	20, 21
19. α-Glycoprotein	Guinea pig	9.6	8.3	6.7	–	20, 22
20. α-Glycoprotein	Horse	12.3	8.6	6.8	–	20

Table 2 (continued)
PLASMA GLYCOPROTEINS (OTHER THAN HUMAN)

Protein	Source	Carbohydrate content, g/100 g				
		Neutral sugar	Acetyl hexosamine	Sialic acid	Fucose	Reference
21. α-Glycoprotein	Rabbit	12.3	11.2	6.9	–	20
22. α-Glycoprotein	Rat	9.9–15.7	7.9–10.2	5.1–10.0	–	20, 23
23. α_1-Glycoprotein	Sheep	11.1	11.2	13.5	0.5	24
24. Hageman factor	Cow	5.9	5.9	4.4	–	25
25. α_1-Acute phase protein	Rat	4.1	7.3	3.3	–	26
26. Transferrin	Hen	1.2	1.6	0.7	–	27
27. M_1-Glycoprotein	Cow	11.3–18.6	12.6	7.2	0.9	28
28. M_1-Glycoprotein	Pig	13.8–23.1	13.6	7.4	25.3	28
29. M_1-Glycoprotein	Hen	13.8–20.4	15.0	10.1	0.4	28
30. M_2-Glycoprotein	Cow	6.2–10.8	6.5	5.1	0	28
31. M_2-Glycoprotein	Pig	8.3–12.0	6.6	5.0	14.8	28
32. α-Glycoprotein	Sheep	6.2–7.1	6.0–8.1	4.4–4.8	1.1	29, 30
33. Prothrombin	Beef	3.7	2.5	4.2	–	31
34. Freezing-point depression glyco-proteins[3]	Fish	28	29	–	–	32, 33
35. Progesterone binding globulin I	Pig	27.7	31.3	12.7	0.8	34
Progesterone binding globulin II		24.5	25.0	12.2	0.7	
36. Coagulation factor XII	Cow	5.9	6.0	4.4	0.5	35
37. Coagulation factor XI	Cow	4.2	4.9	3.6	–	36
38. Coagulation factor IX	Cow	10.6	6.5	8.7	–	37
39. Coagulation factor VIII	Cow	3.8	5.2	0.6	–	38
40. Coagulation factor X	Cow	2.9	4.4	3.8	–	39, 40
41. Prothrombin	Cow	3.5	2.0	3.7	–	41, 42

Compiled by Morris Soodak.

REFERENCES

1. Eylar, *J. Theor. Biol.,* 10, 89 (1965).
2. Spiro, *J. Biol. Chem.,* 235, 2860 (1960).
3. Bezkorovainy, *Biochemistry,* 2, 10 (1963).
4. Bezkorovainy and Doherty, *Arch. Biochem. Biophys.,* 96, 491 (1962).
5. Bourrillon, Got, and Marcy, *Bull. Soc. Chim. Biol.,* 40, 87 (1958).
6. Chader and Westphal, *Fed. Proc.,* 26, Abstracts-140 (1967).
7. Schultze, *Clin. Chim. Acta,* 4, 610 (1959).
8. Nolan and Smith, *J. Biol. Chem.,* 237, 453 (1962).
9. Fleischman, Porter, and Press, *Biochem. J.,* 88, 220 (1963).
10. Melchers, Lennox, and Facon, *Biochem. Biophys. Res. Commun.,* 24, 244 (1966).
11. Marr, Owen, and Wilson, *Biochim. Biophys. Acta,* 63, 276 (1962).
12. Boretti, di Marco, and Julita, *G. Microbiol.,* 12, 55 (1964).
13. Got, Cheftel, Font, and Moretti, *Biochim. Biophys. Acta,* 136, 320 (1967).
14. Jacquot-Armand and Guinand, *Biochim. Biophys. Acta,* 133, 289 (1967).
15. Fisher and Canning, *Natl. Cancer Inst. Monogr.,* 21, 403 (1966).
16. Pepper, *Biochim. Biophys. Acta,* 156, 327 (1968).
17. Got, Mouray, and Moretti, *Biochim. Biophys. Acta,* 107, 278 (1965).
18. Bezkorovainy, *Biochim. Biophys. Acta,* 101, 336 (1965).

Table 2 (continued)
PLASMA GLYCOPROTEINS (OTHER THAN HUMAN)

19. Bezkorovainy, *Arch. Biochem. Biophys.,* 110, 558 (1965).
20. Weimer and Winzler, *Proc. Soc. Exp. Biol. Med.,* 90, 458 (1955).
21. Athenios, Kukral, and Winzler, *Arch. Biochem. Biophys.,* 106, 338 (1964).
22. Simkin, Skinner, and Seshadri, *Biochem. J.,* 90, 316 (1964).
23. Kawasaki, Koyama, and Yamashina, *J. Biochem.,* 60, 554 (1966).
24. Campbell, Schneider, Howe, and Durand, *Biochim. Biophys. Acta,* 148, 137 (1967).
25. Schoenmakers, Matze, Haanen, and Zilliken, *Biochim. Biophys. Acta,* 93, 433 (1964); 101, 166 (1965).
26. Gordon and Louis, *Biochem. J.,* 113, 481 (1969).
27. Williams, *Biochem. J.,* 108, 57 (1968).
28. Grant, Martin, and Anastassiadis, *J. Biol. Chem.,* 242, 3912 (1967).
29. Samy, *Arch. Biochem. Biophys.,* 121, 703 (1967).
30. Das, *Biochim. Biophys. Acta,* 58, 52 (1962).
31. Magnusson, *Ark. Kemi.,* 23, 285 (1965).
32. DeVries, Vandenheede, and Feeney, *J. Biol. Chem.,* 246, 305 (1971).
33. Glochner, Newman, and Uhlenbruck, *Biochem. Biophys. Res. Commun.,* 66, 701 (1975).
34. Burton, Harding, Aboul-Hosn, MacLaughlin, and Westphal, *Biochemistry,* 13, 3554 (1974).
35. Schoenmakers, Matze, Haanen, and Zilliken, *Biochim. Biophys. Acta,* 101, 166 (1965).
36. Kato, Legaz, and Davie, unpublished results in *Annu. Rev. Biochem.,* 44, 799 (1975), by Davie and Fujikawa.
37. Fujikawa, Thompson, Leger, Meyer, and Davie, *Biochemistry,* 12, 4938 (1973).
38. Legaz, Weinstein, Heldebrandt, and Davie, *Ann. N.Y. Acad. Sci.,* 240, 43 (1975).
39. Fujikawa, Legaz, and Davie, *Biochemistry,* 11, 4882 (1972).
40. Jackson, *Biochemistry,* 11, 4873 (1972).
41. Fujikawa, Coan, Enfield, Titani, Erickson, and Davie, *Proc. Natl. Acad. Sci. U.S.A.,* 71, 427 (1974).
42. Nelsestuen and Suttie, *J. Biol. Chem.,* 247, 6096 (1972).

Table 3

GLYCOPROTEINS, BLOOD GROUP SUBSTANCES, AND MUCINS

Glycoprotein	Source	Carbohydrate content g/100 g				Reference
		Neutral sugar	Acetyl hexosamine	Sialic acid	Fucose	
1. A substance	Ovarian cysts	19.8–26.0	34.4–45.3	1.3–2.9	17.0–19.5	1
2. B substance	Ovarian cysts	33.0–38.1	29.5–34.0	1.9–5.1	16.2–20.8	1
3. AB substance	Ovarian cysts	26.6–31.8	32.0–35.7	1.0–1.7	17.3–17.9	1
4. H substance	Ovarian cysts	22.1–25.1	28.4–32.1	1.5–4.3	16.7–21.6	1
5. Le[a] substance	Ovarian cysts	27.5–31.8	32.6–37.7	3.0–18.0	8.6–13.1	1
6. M and N substances (influenza virus receptor)	Human erythrocytes	13.1–14.8	12.1–18.0	15.7–24.4	0.7–1.1	2–5
7. Infectious mononucleosis antigen	Ovine erythrocytes	–	13.5	7.0	–	6
8. Infectious mononucleosis antigen	Bovine erythrocytes	–	5.5	4.5	–	6
9. Fucose-containing glycoprotein	Human erythrocyte membranes	7.0	4.0	–	3.0	7
10. Cervical mucin	Bovine	27.5	32.9	13.8	5.1	8–10
11. Cervical mucin	Human	46.0	24.9	8.0	11.0	8, 11
12. Submaxillary mucin	Bovine	0.7–3.6	18.3–31.1	22.4–30.8	2.8	12
13. Submaxillary mucin	Ovine	0.45	13.8–15.0	23.8–26.0	0.4	12, 13
14. Submaxillary mucin	Canine	15.9	24.8	9.2	12.1	12
15. Submaxillary mucin	Porcine	11.5–18.4	18.5–25.8	15.3–16.1	6.0–7.6	12, 14, 15
16. Sublingual mucin	Bovine	22.0	27.0	20.9	7.1	12, 16
17. Salivary mucin	Collicallia secretions	18.0	17.3	18.0	–	17
18. Fish skin mucin	Loach	9.8	9.8	22.0	–	18
19. Fish skin mucins	Several species	1.5–10.2	0.7–5.1	0.25–2.0	–	19
20. Bile mucin	Human bile	24.0	17.2	–	–	20
21. Meconium glycoprotein	Human meconium	28.0	29.0	9.5	7.9	4, 5
22. Intestinal mucus	Human – cystic fibrosis patient	9.0	19.7	12.6	6.4	21
23. Mucin	Human colloid breast tumor	20.0	39.6	1.7	–	22

Table 3 (continued)
GLYCOPROTEINS BLOOD GROUP SUBSTANCES, AND MUCINS

Glycoprotein	Source	Carbohydrate content g/100 g				
		Neutral sugar	Acetyl hexosamine	Sialic acid	Fucose	Reference
24. Bronchial sulfated glycoprotein	Human cystic fibrosis patient	29.0	33.5	3.4	16.6	23
25. Jelly coat sulfated glycoprotein	Sea urchin eggs	3.7	3.0	70.0	9.2	24

Compiled by Morris Soodak.

REFERENCES

1. Watkins, in *Glycoproteins*, Gottschalk, Ed., Elsevier, Amsterdam, 1966, 462.
2. Kathan, Johnson, and Winzler, *J. Exp. Med.*, 113, 37 (1961).
3. Kathan and Adamany, *J. Biol. Chem.*, 242, 1716 (1967).
4. Bezkorovainy, Springer, and Hotta, *Biochim. Biophys. Acta*, 115, 501 (1966).
5. Springer, Nagai, and Tegtmeyer, *Biochemistry*, 5, 3254 (1966).
6. Springer and Fletcher, in *Organ Transplantation*, Haymer, Ricken, and Letterer, Eds., Schattauer Verlag, Stuttgart, 1969, 35.
7. Uhlenbruck, Hansen, and Pardoe, *Z. Physiol. Chem.*, 349, 737 (1968).
8. Buddecke, in *Glycoproteins*, Gottschalk, Ed., Elsevier, Amsterdam, 1966, 558.
9. Gibbons, *Biochem. J.*, 73, 209 (1959).
10. Gibbons, *Biochem. J.*, 89, 380 (1963).
11. Gibbons and Roberts, *Ann. N.Y. Acad. Sci.*, 106, 218 (1963).
12. Pigman and Gottschalk, in *Glycoproteins*, Gottschalk, Ed., Elsevier, Amsterdam, 1966, 434.
13. Bhargava and Gottschalk, *Biochim. Biophys. Acta*, 127, 223 (1966).
14. de Salegui and Plonska, *Arch. Biochem. Biophys.*, 129, 49 (1969).
15. Carlson, *J. Biol. Chem.*, 243, 616 (1968).
16. Katzman and Eylar, *Arch. Biochem. Biophys.*, 117, 623 (1966).
17. Howe, Lee, and Rose, *Arch. Biochem. Biophys.*, 95, 512 (1961).
18. Turumi and Saito, *Tohoku J. Exp. Med.*, 58, 247 (1953).
19. Wessler and Werner, *Acta Chem. Scand.*, 11, 1240 (1957).
20. Tiba, *Tohoku J. Exp. Med.*, 52, 103 (1950).
21. Johansen, *Biochem. J.*, 87, 63 (1963).
22. Adams, *Biochem. J.*, 94, 368 (1965).
23. Roussel, Lamblin, Degand, Walker-Nasir, and Jeanloz, *J. Biol. Chem.*, 250, 2114 (1975).
24. Hotta, Hamazaki, Kurokawa, and Isaka, *J. Biol. Chem.*, 245, 5434 (1970).

Table 4
GLYCOPROTEINS, TISSUES

Glycoprotein isolated from	Source	Carbohydrate content, g/100 g				Reference
		Neutral sugar	Acetyl hexosamine	Sialic acid	Fucose	
1. Aorta	Cow	5.5	5.8	7.1	0.8	1, 2, 3
2. Aorta	Human	4.0	4.2	2.2	–	4
3. Bone	Cow	11.3	10.7	20.7	0.7	5–8
4. Bone	Rabbit	7.8	10.5	5.8	1.0	9
5. Cornea	Cow	1.69	1.55	0.22	0–1.2	10
6. Fetal skin	Cow	5.5	5.2	5.9	–	1, 11
7. Vitreous body	Cow	4.3	2.6	0.4	–	1, 12
8. Sarcolemma	Frog	4.1	1.1	1.0	–	13
9. Spleen (apohemosiderin)	Horse	2.4	1.2	0.5	0.3	14
10. Granulation tissue	Rat	12.2	5.8	12.9	0.9	1, 15
11. Albuminoid	Human platelets	5.1	0.9	0.8	–	16
12. Sericin A	Bombyx mori cocoons	0.3	0.8	–	–	17
13. Sericin B	Bombyx mori cocoons	0.8	2.26	–	–	17
14. β-Parotin	Bovine paratid gland	2.5	0.6	–	–	18
15. Intrinsic factor	Hog stomach	18.6	–	–	–	19
16. Collagen	Earthworm	12.0	2.5	0	–	20, 21
17. Collagen	Calf	1.5	0.4	0	–	20, 22
18. Collagen	Dog basement membrane	0.42	0.03	–	–	23
19. Collagen	Bovine cornea	1.9	4.4	0.5	–	24
20. Sperm binding protein	Sea urchin eggs	3.3	3.4	–	–	25
21. Membrane glycoprotein	Pig platelets	17.7	18.4	9.6	–	26
22. Aorta (intima)	Pig	2.5	2.0	1.2	0.6	27
23. Brain G. P. –350 soluble and membrane-bound	Calf	9.4	6.6	2.4	1.0	28, 29
24. Brain	Pigeon					30
G. P. 10–BI		11.1	20.0	0.6	–	
G. P. 10–BII		16.7	22.0	1.0	–	
25. Brain-sulfated glycopeptide	Cow	31.2–37.5	17.5–27.7	13.5–25.0	3.3–5.8	31
26. Brain-tubulin	Pig	0.38	0.57	0.26	0.1	32, 33
27. Rhodopsin	Cow	1.7	2.2	–	–	34
28. Anterior lens capsule	Calf	9.6	2.0	0.3	0.25	35
29. Anterior lens capsule	Cow	11.3	0.9	0.1	0.25	35
30. Anterior lens capsule	Dog	10.4	1.2	0.4	0.6	36
31. Posterior lens capsule	Calf	7.6	2.8	0.6	0.24	35
32. Posterior lens capsule	Cow	10.0	1.2	0.2	0.3	35
33. Glomerular basement membrane	Cow	6.3	2.1	–	0.2	37
34. Glomerular basement membrane	Man	5.5	1.5	0.7	0.2	38
35. Glomerular basement membrane	Dog	6.2	1.8	2.0	0.75	36

Table 4 (continued)
GLYCOPROTEINS, TISSUES

Glycoprotein isolated from	Source	Carbohydrate content, g/100 g				Reference
		Neutral sugar	Acetyl hexosamine	Sialic acid	Fucose	
36. Glycoprotein component VII	Bovine, glomerular membrane	3.7	3.9	1.5	0.2	39
37. Collagen	Carp, swim bladder	0.5	0.1	0.02	0.01	40
38. Collagen	Shark, elastoidin	0.9	0.004	—	—	41
39. Collagen	Bovine Achilles tendon	0.8	0.6	0.06	0.04	40
40. Collagen	Metridium, body wall	7.4	0.5	—	0.3	42
41. Collagen	Lumbricus, whole cuticle	13.1	1.7	—	1.33	43
42. Collagen	Loligo, cephalic cartridge	2.7	0.2	—	0.05	43

Compiled by Morris Soodak.

REFERENCES

1. Buddecke, in *Glycoproteins,* Gottschalk, Ed., Elsevier, Amsterdam, London, New York, 1966, 558.
2. Berenson and Fishkin, *Arch. Biochem. Biophys.,* 97, 18 (1962).
3. Radhakrishnamurthy, Fishkin, Hubbell, and Berenson, *Arch. Biochem. Biophys.,* 104, 19 (1964).
4. Barnes and Partridge, *Biochem. J.,* 109, 883 (1968).
5. Williams and Peacocke, *Biochim. Biophys. Acta,* 101, 327 (1965).
6. Andrews and Herring, *Biochim. Biophys. Acta,* 101, 239 (1965).
7. Herring and Kent, *Biochem. J.,* 89, 405 (1963).
8. Andrews, Herring, and Kent, *Biochem. J.,* 104, 705 (1967).
9. Burckard, Havez, and Dautrevaux, *Bull. Soc. Chim. Biol.,* 48, 851 (1966).
10. Robert and Dische, *Biochem. Biophys. Res. Commun.,* 10, 209 (1963).
11. Bourrillon and Got, *Biochim. Biophys. Acta,* 58, 63 (1962).
12. Berman, *Biochim. Biophys. Acta,* 83, 27 (1964).
13. Kono, Kakuma, Homma, and Fukuda, *Biochim. Biophys. Acta,* 88, 155 (1964).
14. Ludewig and·Glover, *Arch. Biochem. Biophys.,* 113, 654 (1966).
15. Fishkin and Berenson, *Arch. Biochem. Biophys.,* 95, 130 (1961).
16. Bezkorovainy and Rafelson, *J. Lab. Clin. Med.,* 64, 212 (1964).
17. Sinohara and Asano, *J. Biochem.* (Tokyo), 62, 129 (1967).
18. Ito, Okabe, and Namba, *Endocrinol. Jap.,* 12, 249 (1965).
19. Williams, Ellenbogen, and Esposito, *Proc. Soc. Exp. Biol. Med.,* 87, 400 (1954).
20. Eylar, *J. Theor. Biol.,* 10, 89 (1965).
21. Maser and Rice, *Biochim. Biophys. Acta,* 63, 255 (1962).
22. Gross, Dumsha, and Glazer, *Biochim. Biophys. Acta,* 30, 293 (1958).
23. Kefalides, *Biochem. Biophys. Res. Commun.,* 22, 26 (1966).
24. Bosmann and Jackson, *Biochim. Biophys. Acta,* 170, 6 (1968).
25. Aketa, Tsuzuki, and Onitake, *Exp. Cell Res.,* 50, 676 (1968).
26. Mullinger and Manley, *Biochim. Biophys. Acta,* 170, 282 (1968).
27. Wagh and Roberts, *Biochemistry,* 11, 4222 (1972).
28. van Nieuw Amerongen, van den Eijnden, Heijlman, and Roukema, *J. Neurochem.,* 19, 2195 (1972).
29. van Nieuw Amerongen and Roukema, *J. Neurochem.,* 23, 85 (1947).
30. Bogoch, in *Protein Metabolism of the Nervous System,* Plenum Press, New York, 1970, 555.

Table 4 (continued)
GLYCOPROTEINS, TISSUES

31. **Arima, Muramatsu, Saigo, and Egami,** *Jap. J. Exp. Med.,* 39, 301 (1969).
32. **Margolis, Margolis, and Shelanski,** *Biochem. Biophys. Res. Commun.,* 47, 432 (1972).
33. **Feit and Shelanski,** *Biochem. Biophys. Res. Commun.,* 66, 920 (1975).
34. **Heller,** *Biochemistry,* 7, 2907 (1968).
35. **Fukushi and Spiro,** *J. Biol. Chem.,* 244, 2041 (1969).
36. **Kefalides and Denduchis,** *Biochemistry,* 8, 4613 (1969).
37. **Spiro,** *J. Biol. Chem.,* 242, 1915 (1967).
38. **Beisswenger and Spiro,** *Science,* 168, 596 (1970).
39. **Ohno, Riquetti, and Hudson,** *J. Biol. Chem.,* 250, 7780 (1975).
40. **Spiro,** *J. Biol. Chem.,* 244, 602 (1969).
41. **Sastry and Ramachandran,** *Biochim. Biophys. Acta,* 97, 281 (1965).
42. **Katzman and Jeanloz,** *Fed. Proc.,* 29, 599 (1970).
43. **Spiro and Bhoyroo,** in *Glycoproteins, Part B,* Gottschalk, Ed., Elsevier, 1972, 986.

Table 4A
GLYCOPROTEINS, CELL MEMBRANES

Protein	Source	Carbohydrate content g/100 g				Reference
		Neutral sugar	Acetyl hexosamine	Sialic acid	Fucose	
1. Glycophorin	Human erythrocytes	14.9	17.7	25.4	1.0	1, 2
2. Anion exchange protein	Human erythrocytes	3.8	1.7	0.6	1.4	3
3. GP-II	Human erythrocytes	4.9	5.4	7.8	–	4
4. GP-III		7.6	7.2	8.1	–	4
5. Major glycoprotein	Horse erythrocytes	9.4	12.4	33.7	0.6	5
6. Major glycoprotein	Swine erythrocytes	18.0	23.4	27.2	1.7	5
7. Major glycoprotein	Swine erythrocytes	6.4	8.6	36.3	–	5
8. Lipopolysaccharide receptor	Human erythrocytes	7.8	12.7	16.3	0.9	6
9. H-2^b Alloantigen fragment	Mouse spleen cells	5.4	4.3	1.3	–	7, 8
10. H-2^d Alloantigen fragment	Mouse spleen cells	3.5	3.5	1.0	–	7, 8
11. Glycopeptide of (Na$^+$ + K$^+$) adenosine-triphosphatase	Canine renal medulla	2.2 – 3.7	4.8	1.7	–	9

Compiled by Morris Soodak.

REFERENCES

1. **Javaid and Winzler,** *Biochemistry,* 13, 3635 (1974).
2. **Tomita and Marchesi,** *Proc. Natl. Acad. Sci. U.S.A.,* 72, 2964 (1975).
3. **Ho and Guidotti,** *J. Biol. Chem.,* 250, 675 (1975).
4. **Fujita and Cleve,** *Biochim. Biophys. Acta,* 382, 172 (1975).
5. **Fujita and Cleve,** *Biochim. Biophys. Acta,* 406, 206 (1975).
6. **Springer, Adye, Bezkorovainy, and Jirgensons,** *Biochemistry,* 13, 1379 (1974).
7. **Shimada and Nathanson,** *Biochemistry,* 8, 4048 (1969).
8. **Schwartz, Kato, Cullen, and Nathanson,** *Biochemistry,* 12, 2157 (1973).
9. **Kyte,** *J. Biol. Chem.,* 247, 7642 (1972).

Table 4B
GLYCOPROTEINS, ENVELOPED VIRUSES

Protein	Source	Carbohydrate content g/100 g				
		Neutral sugar	Acetyl hexosamine	Sialic acid	Fucose	Reference
1. Membrane protein	Sindbis virus	5.1	8.7	0.7–1.2	0.5	1
2. Membrane protein E_1	Semliki Forest virus	2.8	3.4	1.7	0.2	2
3. Membrane protein E_2	Semliki Forest virus	5.8	3.8	2.8	0.3	2
4. Membrane protein E_3	Semliki Forest virus	24.0	34.0	17.0	7.0	2
5. Membrane protein glycoprotein	Vesicular stomatitis virus	4.6	3.7	2.3	0.3	3, 4

Compiled by Morris Soodak.

REFERENCES

1. Strauss, Burge, and Darnell, *J. Mol. Biol.,* 47, 437 (1970).
2. Garoff, Simons, and Renkonen, *Virology,* 61, 493 (1974).
3. Ethchison and Holland, *Virology,* 60, 217 (1974).
4. Ethchison and Holland, *Proc. Natl. Acad. Sci. U.S.A.,* 71, 4011 (1974).

Table 5
GLYCOPROTEINS, BODY FLUIDS

Glycoprotein	Source	Carbohydrate content, g/100 g				
		Neutral sugar	Acetyl hexosamine	Sialic acid	Fucose	Reference
1. Tamm and Horsfall glycoprotein	Human urine	8.1	8.8	9.1	1.1	1–4
2. Tamm and Horsfall glycoprotein	Sheep urine	30.6	6.2	4.0	–	5
3. L-type Bence Jones protein	Human urine	6.1	4.4	3.1	1.6	6
4. Glycoprotein	Human plasmacytoma urine	16.6	12.5	6.8	4.5	7
5. B_{12}-binding proteins	Leukemic human urine	14–28	15–25	3.7–8.0	–	8
6. Pleural fluid glycoprotein	Human	14.2	13.9	11.9	–	9, 10
7. Synovial fluid glycoprotein	Human	11.5	11.6	4.9	1.0	11, 12
8. α-Globulin	Human amniotic fluid	47.0	24.5	9.0	–	13
9. Glycoprotein	Human cancer ascites fluid	19.8	12.3	30.5	5.0	14
10. Ascites fluid	Yoshida rat tumor	3.25	2.2	12.4	11.0	15, 16
11. Soluble glycoprotein	Boar seminal plasma	10.8	5.7	0.5	–	17
12. Insoluble glycoprotein	Boar seminal plasma	10.0	7.6	0.4	–	18
13. Tamm and Horsfall glycoprotein	Human urine	11.7	11.2	4.4	0.8	19
14. N_0 glycoprotein antigen	Normal human gastric juice	28.5	32.6	11.4	16.1	20
15. I Glycoprotein antigen	Normal human gastric juice	28.8	23.9	14.8	17.3	20
16. Sulfated glycoprotein, allantoic antigen	Chick allantoic fluid	26.0	32.1	1.0	4.3	21

Compiled by Morris Soodak.

Table 5 (continued)
GLYCOPROTEINS, BODY FLUIDS

REFERENCES

1. Eylar, *J. Theor. Biol.,* 10, 89 (1965).
2. Gottschalk, *Nature,* 170, 662 (1952).
3. Maxfield, in *Glycoproteins,* Gottschalk, Ed., Elsevier, Amsterdam, 1966, 446.
4. Rosenfeld and Yusipova, *Biokhimiya,* 32, 111 (1967).
5. Cornelius, Bishop, Berger, and Pangborn, *Am. J. Vet. Res.,* 22, 1000 (1961).
6. Edmundson, Sheber, Ely, Simonds, Hutson, and Rossiter, *Arch. Biochem. Biophys.,* 127, 725 (1968).
7. Weicker, Huhnstock, and Grässlin, *Clin. Chim. Acta,* 9, 19 (1964).
8. Kallee, Debiasi, Karypidis, Heide, and Schwick, *Acta Isot.,* 4, 103 (1964).
9. Bourrillon, Michon, and Got, *Biochim. Biophys. Acta,* 47, 243 (1961).
10. Bourrillon, Got, and Meyer, *Biochim. Biophys. Acta,* 74, 255 (1963).
11. Miki and Noma, in *Biochemistry and Medicine of Mucopolysaccharides,* Egami and Oshima, Eds., Maruzen, Tokyo, 1962, 191.
12. Buddecke, in *Glycoproteins,* Gottschalk, Ed., Elsevier, Amsterdam, 1966, 558.
13. Lambotte and Gosselin-Rey, *Arch. Int. Physiol. Biochem.,* 75, 109 (1967).
14. Turumi, Takahashi, and Saito, *Fukushima J. Med. Sci.,* 3, 31 (1956).
15. Marcante, *Clin. Chim. Acta,* 8, 799 (1963).
16. Caputo, Marcante, and Zito, *Br. J. Exp. Pathol.,* 47, 599 (1966).
17. McIntosh and Boursnell, *Biochim. Biophys. Acta,* 130, 252 (1966).
18. Rottenberg and Boursnell, *Biochim. Biophys. Acta,* 117, 157 (1966).
19. Fletcher, Neuberger, and Ratcliffe, *Biochem. J.,* 120, 417 (1970).
20. Häkkinen, *Transplant. Rev.,* 20, 61 (1974).
21. How and Higginbothan, *Carbohydr. Res.,* 12, 355 (1970).

Table 6
GLYCOPROTEINS, EGGS, MILK

Eggs	Source	Carbohydrate content, g/100 g			Reference
		Neutral sugar	Acetyl hexosamine	Sialic acid	
1. Avidin	Hen egg	5.8	14.7	0	1, 2
2. Ovalbumin	Hen egg	2.0	1.2	0	1, 3, 4
3. Ovoglycoprotein	Hen egg	13.6	17.0	3.0	5
4. Ovomucin	Hen egg	15.4	14.7	5.8	1, 6
5. Ovomucoid	Hen egg	5.7–10.5	11.3–21.0	0.4–2.0	1, 7–9
6. Conalbumin	Hen egg	0.8	1.7	0	10
7. Transferrin	Hen egg	0.8–1.4	1.5–1.7	0.4	11
8. Vitello mucoid	Hen egg	23.7	26.4	–	12
9. Ovoinhibitor of chymotrypsin	Hen egg	3.9	6.9	0.1	13
10. κ Casein	Cow milk	1.4	1.5	2.4	1, 14
11. Casein	Cow milk	0.3	0.22	0.3	15, 16
12. Casein	Human milk	2.5	2.7	0.8	16
13. Casein	Polar bear milk	2.8	1.3	1.9	15
14. Casein	Sheep milk	0.23	0.15	0.11	16
15. Casein	Goat milk	0.22	0.16	0.30	16
16. Casein	Whale milk	0.59	0.42	0.37	16
17. Casein	Horse milk	0.55	0.44	0.56	16

Table 6 (continued)
GLYCOPROTEINS, EGGS, MILK

Eggs	Source	Carbohydrate content, g/100 g			Reference
		Neutral sugar	Acetyl hexosamine	Sialic acid	
18. Casein	Reindeer milk	0.44	0.23	0.46	16
19. M_1-Glycoprotein	Cow colostrum	10.8	10.2	11.1	17
20. M_1-Glycoprotein	Cow milk	8.4	8.2	5.9	18
21. M_2-Glycoprotein	Cow colostrum	7.2	6.5	6.4	17
22. M_2-Glycoprotein	Cow milk	4.9	4.8	5.3	17
23. Phosphoglycoprotein	Cow milk	3.1	2.4	4.0	17
24. Lactotransferrin	Human milk	3.9	3.0	0.87	19
25. Glycoprotein a	Cow milk	3.1	–	–	20
26. Interfacial glycoprotein	Cow milk	5.6	4.8	4.7	21

Compiled by Richard J. Winzler.

REFERENCES

1.   Eylar, *J. Theor. Biol.,* 10, 89 (1965).
2.   Melamed and Green, *Biochem. J.,* 89, 591 (1963).
3.   Fletcher, Marshall, and Neuberger, *Biochim. Biophys. Acta,* 74, 311 (1963).
4.   Neuberger and Marshall, in *Glycoproteins,* Gottschalk, Ed., Elsevier, Amsterdam, 1966, 299.
5.   Ketterer, *Biochem. J.,* 96, 372 (1965).
6.   Gottschalk and Lind, *Br. J. Exp. Pathol.,* 30, 85 (1949).
7.   Chatterjee and Montgomery, *Arch. Biochem. Biophys.,* 99, 426 (1962).
8.   Bragg and Hough, *Biochem. J.,* 78, 11 (1961).
9.   Montreuil, Castiglioni, Adam-Chosson, Caner, and Queval, *J. Biochem.,* 57, 514 (1965).
10.   Williams, *Biochem. J.,* 83, 355 (1962).
11.   Williams, *Biochem. J.,* 108, 57 (1968).
12.   Osaki and Yosizawa, *Tohoku J. Exp. Med.,* 51, 62 (1949).
13.   Davis, Zahnley, and Donovan, *Biochemistry,* 8, 2044 (1969).
14.   Jolles, Alais, and Jolles, *Arch. Biochem. Biophys.,* 98, 56 (1962).
15.   Baker, Huang, and Harington, *Biochem. Biophys. Res. Commun.,* 13, 227 (1963).
16.   Johansson and Svennerholm, *Acta Physiol. Scand.,* 37, 324 (1956).
17.   Bezkorovainy, *Arch. Biochem. Biophys.,* 110, 558 (1965).
18.   Bezkorovainy, *J. Dairy Sci.,* 50, 1368 (1967).
19.   Montreuil, Tonnelat, and Mullet, *Biochim. Biophys. Acta,* 45, 413 (1960).
20.   Groves and Gordon, *Biochemistry,* 6, 2388 (1967).
21.   Jackson, Coulson, and Clark, *Arch. Biochem. Biophys.,* 97, 373 (1962).

Table 7
GLYCOPROTEINS, PLANTS

Protein	Source	Carbohydrate content, g/100 g		Reference
		Neutral sugar	Acetyl hexosamine	
1. Hemagglutinin	Soybean	1.2–9.8	0	1–4
2. Hemagglutinin	Phaseolus vulgaris	6.0–11.8	3.4	5, 6
3. Taste modifying glycoprotein	Miracle fruit	6.7	–	7
4. Globulin	Soybeans	3.8	1.5	8
5. Stellacyanin	Japanese lac tree	20.0	25.0	9
6. Glycopeptides from extensin	Various plant cell walls	70.5	–	10
7. Agglutinin	Soybean (Glycine max)	4.5	1.2	11–13
8. Agglutinin	Horse gram (Dolichos biflorus)	2.5	1.25	14–16
9. Lectin A	Lotus tetragonolobulus	8	1.4	17, 18
10. Lectin B	Lotus tetragonolobulus	4	0.8	17, 18
11. Lectin C	Lotus tetragonolobulus	8	1.2	17, 18
12. Agglutinin A	Red kidney bean (Phaseolus vulgaris)	10.0	2.9	19
13. Agglutinin B	Red kidney bean (P. vulgaris)	10.5	2.4	19
14. Agglutinin	Robina pseudoacacia	6.7	4.0	20
15. Agglutinin	Lima bean (Phaseolus lunatus)	4.3	1.4	21, 22
16. Agglutinin	Castor bean (Ricinus communis)	3.7	–	23
17. Agglutinin	Wax bean (Phaseolus vulgaris)	9.5	1.1	24–26
18. Agglutinin I	Goose (Ulex europeus)	3.8	1.7	27–29
19. Agglutinin II	Goose (U. europeus)	19.9	2.2	27–29
20. Lectin	Bandeiraea simplicifolia	7.8	2.8	30
21. Agglutinin	Sophora japonica	5.9	2.2	31
22. Mitogen	Wistaria floribunda	7.9	4.3	32, 33
23. Agglutinin	Bauhinia purpurea alba	7.7	4.1	34
24. Mitogen	Pokeweed	2.7	1.7	35, 36

Compiled by Morris Soodak.

REFERENCES

1. Lis, Sharon, and Katchalski, *J. Biol. Chem.,* 241, 684 (1966).
2. Lis, Fridman, Sharon, and Katchalski, *Arch. Biochem. Biophys.,* 117, 301 (1966).
3. Liener and Pallansch, *J. Biol. Chem.,* 197, 29 (1952).
4. Wada, Pallansch, and Liener, *J. Biol. Chem.,* 233, 395 (1958).
5. Borjeson, Bouveng, Gardell, and Thunell, *Biochim. Biophys. Acta,* 82, 158 (1964).
6. Pusztai, *Biochem. J.,* 101, 379 (1966).
7. Kurihara and Beidler, *Science,* 161, 1241 (1968).
8. Koshiyama, *Agric. Biol. Chem.,* 30, 646 (1966).
9. Peisach, Levine, and Blumberg, *J. Biol. Chem.,* 242, 2847 (1967).
10. Lamport, *Biochemistry,* 8, 1155 (1969).
11. Lis, Sharon, and Katchalski, *J. Biol. Chem.,* 241, 684 (1966).

Table 7 (continued)
GLYCOPROTEINS, PLANTS

12. Sharon and Lis, *Science,* 177, 949 (1972).
13. Lis and Sharon, *Annu. Rev. Biochem.,* 42, 541 (1973).
14. Font, Leseney, and Bourrillon, *Biochim. Biophys. Acta,* 243, 434 (1971).
15. Etzler, in *Methods in Enzymology,* Vol. 28B, Ginsburg, Ed., 1972, 340.
16. Etzler and Kabat, *Biochemistry,* 9, 899 (1970).
17. Kalb, *Biochim. Biophys. Acta,* 168, 532 (1968).
18. Yariv, Kalb, and Blumberg, in *Methods in Enzymology,* Vol. 28B, Ginsburg, Ed., 1972, 356.
19. Dahlgren, Porath, and Lindahl-Kiessling, *Arch. Biochem. Biophys.,* 137, 306 (1970).
20. Font and Bourrin, *Biochim. Biophys. Acta,* 243, 111 (1971).
21. Galbraith and Goldstein, in *Methods in Enzymology,* Vol. 28B, Ginsburg, Ed., 1972, 318.
22. Gould and Scheinberg, *Arch. Biochem. Biophys.,* 141, 607 (1970).
23. Waldschmidt-Leitz and Keller, *Z. Physiol. Chem.,* 351, 990 (1970).
24. Takahashi, Ramachandramurthy, and Liener, *Biochim. Biophys. Acta,* 133, 123 (1967).
25. Takahashi and Liener, *Biochim. Biophys. Acta,* 154, 560 (1968).
26. Sela, Lis, and Sharon, *Biochim. Biophys. Acta,* 310, 273 (1973).
27. Matsumoto and Osawa, *Biochim. Biophys. Acta,* 194, 180 (1969).
28. Matsumoto and Osawa, *Arch. Biochem. Biophys.,* 140, 484 (1970).
29. Osawa and Matsumoto, in *Methods in Enzymology,* Vol. 28B, Ginsburg, Ed., 1972, 323.
30. Hayes and Goldstein, *J. Biol. Chem.,* 249, 1904 (1974).
31. Poretz, Riss, Timberlake, and Chien, *Biochemistry,* 13, 250 (1974).
32. Osawa and Toyoshima, in *Methods in Enzymology,* Vol. 28B, Ginsburg, Ed., 1972, 328.
33. Toyoshima, Akiyama, Nakano, Tonomura, and Osawa, *Biochemistry,* 10, 4457 (1971).
34. Irimura and Osawa, *Arch. Biochem. Biophys.,* 151, 475 (1972).
35. Reisfield, Börjeson, Chessin, and Small, *Proc. Natl. Acad. Sci. U.S.A.,* 58, 2020 (1967).
36. Börjeson, Reisfeld, Chessin, Welsh, and Douglas, *J. Exp. Med.,* 124, 859 (1966).

Table 8
GLYCOPROTEINS, HORMONES

| Hormone | Source | Carbohydrate content, g/100 g | | | | |
		Neutral sugar	Acetyl hexosamine	Sialic acid	Fucose	Reference
1. Chorionic gonadotropin	Human urine	11.0–11.2	10.8–11.1	8.5–9.0	1.2	1–4
2. Serum gonadotropin	Pregnant mare serum	18.6	17.5	10.4	1.4	1, 3, 6
3. Interstitial cell stimulating hormone	Ovine pituitary	5.0	9.6	0.5	1.1	3, 7, 8
4. Interstitial cell stimulating hormone	Human pituitary	1.1	2.5	1.0	0.4	3, 9
5. Follicle stimulating hormone	Ovine pituitary	4.0–5.7	8.6–5.8	12.9	–	3, 10
6. Follicle stimulating hormone	Porcine pituitary	2.6	4.6	–	1.1	3, 9
7. Follicle stimulating hormone	Human pituitary	3.9–12.2	2.9–9.0	1.4–5.1	–	11, 12
8. Lutinizing hormone	Human pituitary	11.3	4.9	2.0	–	11
9. Lutinizing hormone	Ovine pituitary	6.1–11.8	7.4–9.5	0.3–0.5	–	11, 13
10. Lutinizing hormone	Bovine pituitary	11.9	5.9	0.3	–	11
11. Thyrotropin	Bovine pituitary	9.4	13.2	–	1.3	3
12. Thyrotropin	Human pituitary	5.9	4.1	–	0.5	14
13. Erythropoietin	Ovine anemic plasma	29.2	21.5	13.0	–	3, 15
14. Gonadotropin transportation protein	Equine urine	14.0	13.3	12.0	–	1, 16
15. Thyroglobulin	Human thyroid	4.8	4.2	1.1	0.5	17, 18

Table 8 (continued)
GLYCOPROTEINS, HORMONES

Hormone	Source	Carbohydrate content, g/100 g				Reference
		Neutral sugar	Acetyl hexosamine	Sialic acid	Fucose	
16. Thyroglobulin	Porcine thyroid	4.0	3.4	1.2	0.5	17, 18
17. Thyroglobulin	Ovine thyroid	4.0	3.2	1.5	0.4	17, 18
18. Thyroglobulin	Bovine thyroid	3.7	3.2	1.4	0.4	17, 18
19. Chorionic gonadotropin (native)	Human urine	12.5	11.8	8.7	0.9	19
20. Chorionic gonadotropin α subunit	Human urine	13.8	11.6	8.8	0.06	19
21. Chorionic gonadotropin β subunit	Human urine	12.5	14.3	10.2	1.3	19
22. Follicle stimulating hormone (native)	Ovine pituitary	7.4	10.2	5.6	0.7	20
23. Follicle stimulating hormone α subunit	Ovine pituitary	7.4	10.3	3.6	0.4	20
24. Follicle stimulating hormone β subunit	Ovine pituitary	7.3	10.7	7.6	1.3	20
25. Follicle stimulating hormone (native)	Equine pituitary	8.7	9.2	6.8	1.0	21
26. Follicle stimulating hormone α subunit	Equine pituitary	7.0	8.1	5.6	0.5	21
27. Follicle stimulating hormone β subunit	Equine pituitary	8.4	9.4	7.4	1.4	21
28. Luteinizing hormone native	Human pituitary	7.3	7.2	2.8	0.9	22
29. Luteinizing hormone α subunit	Human pituitary	8.2	10.8	4.3	0.7	22
30. Luteinizing hormone β subunit	Human pituitary	4.0	2.7	1.8	1.6	22
31. Luteinizing hormone native	Ovine pituitary	6.5	9.4	–	1.4	23
32. Luteinizing hormone α subunit	Ovine pituitary	8.3	10.3	–	1.3	23
33. Luteinizing hormone β subunit	Ovine pituitary	4.3	7.0	–	1.0	23
34. Thyrotropin native	Human pituitary	4.4	6.6	1.8	0.5	24
35. Thyrotropin α subunit	Human pituitary	6.2	9.7	2.9	0.2	24
36. Thyrotropin β subunit	Human pituitary	2.8	4.1	0.4	0.5	24
37. Thyrotropin native	Bovine pituitary	5.6	11.0	–	0.76	25
38. Thyrotropin α subunit	Bovine pituitary	7.5	14.0	–	0.35	25
39. Thyrotropin β subunit	Bovine pituitary	3.2	7.3	–	1.1	25
40. Erythropoieten	Ovine anemic plasma	9.0	9.2	10.8	–	26, 27

Compiled by Morris Soodak.

Table 8 (continued)
GLYCOPROTEINS, HORMONES

REFERENCES

1.   **Eylar,** *J. Theor. Biol.,* 10, 89 (1965).
2.   **Got, Bourrillon, and Michon,** *Bull. Soc. Chim. Biol.,* 42, 41 (1960).
3.   **Papkoff,** in *Glycoproteins,* Gottschalk, Ed., Elsevier, Amsterdam, 1966, 532.
4.   **Bahl,** *J. Biol. Chem.,* 244, 567 (1969).
5.   **Got and Bourrillon,** *Biochim. Biophys. Acta,* 42, 505 (1960).
6.   **Bourrillon, Michon, and Got,** *Bull. Soc. Chim. Biol.,* 41, 493 (1959).
7.   **Li and Starman,** *Nature,* 202, 291 (1964).
8.   **Li,** *J. Natl. Cancer Inst. Monogr.,* 12, 181 (1963).
9.   **Steelman and Segeloff,** *Recent Prog. Horm. Res.,* 15, 115 (1959).
10. **Papkoff, Gospodarowicz, and Li,** *Arch. Biochem. Biophys.,* 120, 434 (1967).
11. **Kathan, Reichert, and Ryan,** *Endocrinology,* 81, 45 (1967).
12. **Papkoff, Mahlmann, and Li,** *Biochemistry,* 6, 3976 (1967).
13. **Walborg and Ward,** *Biochim. Biophys. Acta,* 78, 304 (1963).
14. **Kim, Shome, Liao, and Pierce,** *Anal. Biochem.,* 20, 258 (1967).
15. **Goldwasser, White, and Taylor,** *Biochim. Biophys. Acta,* 64, 487 (1962).
16. **Bourrillon, Got, and Marcy,** *Bull. Soc. Chim. Biol.,* 40, 87 (1958).
17. **McQuillan and Trikojus,** in *Glycoproteins,* Gottschalk, Ed., Elsevier, Amsterdam, 1966, 516.
18. **Spiro,** *Fed. Proc.,* 22, 538 (1963).
19. **Bahl,** in *Hormonal Proteins and Peptides,* Vol. 1, Li, Ed., Academic Press, New York, 1973, 171.
20. **Grimek and McShan,** *J. Biol. Chem.,* 249, 5725 (1974).
21. **Landefield and McShan,** *J. Biol. Chem.,* 249, 3527 (1974).
22. **Closset, Vandalem, Hennen, and Lequin,** *Eur. J. Biochem.,* 57, 325 (1975).
23. **Papkoff,** in *Hormonal Proteins and Peptides,* Vol. 1, Li, Ed., Academic Press, New York, 1973, 68.
24. **Cornell and Pierce,** *J. Biol. Chem.,* 248, 4327 (1973).
25. **Pierce, Liao, and Carlsen,** in *Hormonal Proteins and Peptides,* Vol. 1, Li, Ed., Academic Press, New York, 1973, 28.
26. **Goldwasser and Kung,** *Fed. Proc.,* 30, 1128 (1971).
27. **Winzler,** in *Hormonal Proteins and Peptides,* Vol. 1, Li, Ed., Academic Press, New York, 1973, 2.

Table 9
GLYCOPROTEINS, ENZYMES

Enzyme	Source	Carbohydrate content, g/100 g				
		Neutral sugar	Acetyl hexosamine	Sialic acid	Fucose	Reference
1. Cholinesterase	Horse serum	–	–	3.2		1
2. Cholinesterase	Human serum	3.6–9.3	2.9–8.4	1.8–6.0		2–4
3. Non-specific esterase	Rat tissues	6.2	4.6	6.5		5
4. Glucose oxidase	*Aspergillus niger*	14.2	2.0	–		6
5. Peroxidase	Horse radish	16.2	1.9	–		7
6. Chloroperoxidase	Caldariomyces fumago	25–30	–	–		8
7. Monamine oxidase	Bovine plasma	4.6	–	–		9, 10
8. Ribonuclease B	Bovine pancreas	6.3	5.0	–		9, 11
9. Ribonuclease B	Bovine pancreatic juice	7.4	3.1	0.2		12
10. Ribonuclease C	Bovine pancreatic juice	6.4	5.5	2.5		12
11. Ribonuclease D	Bovine pancreatic juice	6.4	4.6	4.8		12
12. Ribonuclease	Porcine pancreatic juice	11.1–24.6	7.4–16.8	0–2.8		13
13. Deoxyribonuclease	Bovine pancreas	1.1–2.9	1.3–1.4	0		14, 15
14. Deoxyribonuclease	Hog spleen	–	4.1	–		16
15. DPN ase	Neurospora	82.2	2.0	–		17

Table 9 (continued)
GLYCOPROTEINS, ENZYMES

Enzyme	Source	Carbohydrate content, g/100 g				Reference
		Neutral sugar	Acetyl hexosamine	Sialic acid	Fucose	
16. DPN ase	*B. subtilis*	52.8	17.9	–		17
17. DPN ase inhibitor	*B. subtilis*	72.7	0	–		17
18. Alkaline phosphatase	Human placenta	9–11	10–11	4–6		18
19. Taka-amylase A	*Aspergillus oryzae*	2.74	0.91	–		19
20. α-Amylase	*Aspergillus oryzae*	3.0–7.5	0.9–1.4	0		20, 21
21. α-Galactosidase I	Vicia faba	25.0	–	–		22
22. α-Galactosidase II	Vicia faba	2.8	–	–		22
23. Invertase	Yeast	47.0	3.1	–		23
24. β-Glucuronidase	Bovine liver	2.8–3.8	–1%	–		24
25. Bromelain	Pineapple stem	1.5–2.1	2.7–3.6	0		25
26. Pepsinogen	Hog stomach	1.5	–	–		27
27. γ-Glutamyl transferase	–	21.0	9.5–12.0	3.5–5.3		9, 28, 29
28. Carboxylesterase	Porcine liver	1.8	0.2	–	0.2	30
29. Carboxylesterase	Porcine kidney	1.4	0.6	–	0.2	30
30. Carboxylesterase	Bovine liver	0.8	0.2	–	0.2	30
31. Carboxylesterase	Human liver microsomes	2.5	0.5	–	–	31
32. Desoxyribonuclease A	Bovine pancreas	3.1	1.2	–	–	32
33. Desoxyribonuclease B	Bovine pancreas	2.9	2.2	0.9	–	32
34. Desoxyribonuclease C	Bovine pancreas	2.5	1.2	–	–	32
35. Desoxyribonuclease D	Bovine pancreas	2.3	2.1	0.8	–	32
36. α-L-Fucosidase	Rat epididymus	0.3	0.1	–	0.02	33
37. Lactoperoxidase	Bovine milk	1.5	6.8	–	–	34
38. Lipase	Porcine pancreas	1.3	1.2	–	–	35
39. Monoamine oxidase	Bovine plasma	2.4	1.3	0.8	–	36
40. N-Acetyl-β-D-glucosaminidase	*Aspergillus oryzae*	3.8	1.1	–	–	37
41. β-Lactamase II	*Bacillus cereus*	10.4	8.0	–	1.4	38
42. Ribonuclease R₁	*Rhizopus oligosporus*	7.6	3.4	–	0.5	39
43. Ribonuclease R₂	*Rhizopus oligosporus*	7.0	2.7	–	0.5	39

Compiled by Morris Soodak.

REFERENCES

1. Heilbronn, *Biochim. Biophys. Acta,* 58, 222 (1962).
2. Schultze and Heremans, in *Molecular Biology of Human Proteins,* Vol. 1, Elsevier, 1965, 204.
3. Yamashina, *Arch. Kemi.,* 9, 225 (1956).
4. Haupt, Heide, Zwisler, and Schwick, *Blut,* 14, 65 (1966).
5. Dugan, Radhakrishnamurthy, and Berenson, *Enzymologia,* 33, 215 (1967).
6. Pazur, Kleppe, and Cepure, *Arch. Biochem. Biophys.,* 111, 351 (1965).
7. Shannon, Kay, and Lew, *J. Biol. Chem.,* 241, 2166 (1966).
8. Morris and Hager, *J. Biol. Chem.,* 241, 1763 (1966).
9. Eylar, *J. Theor. Biol.,* 10, 89 (1965).
10. Yamada, Gee, Ebata, and Yasunobu, *Biochim. Biophys. Acta,* 81, 165 (1964).
11. Plummer and Hirs, *J. Biol. Chem.,* 239, 2530 (1964).
12. Plummer, *J. Biol. Chem,* 243, 5961 (1968).
13. Reinhold, Dunne, Wriston, Schwarz, Sarda, and Hirs, *J. Biol. Chem.,* 243, 6482 (1968).

Table 9 (continued)
GLYCOPROTEINS, ENZYMES

14. Catley, Moore, and Stein, *J. Biol. Chem.*, 244, 933 (1969).
15. Price, Liu, Stein, and Moore, *J. Biol. Chem.*, 244, 917 (1969).
16. Bernardi, Appella, and Zito, *Biochemistry*, 4, 1725 (1965).
17. Everse and Kaplan, *J. Biol. Chem.*, 243, 6072 (1968).
18. Usategui-Gomez, *Proc. Soc. Exp. Biol. Med.*, 120, 385 (1965).
19. Hanafusa, Ikenaka, and Akabori, *J. Biochem.* (Tokyo), 42, 55 (1955).
20. McKelvy and Lee, *Arch. Biochem. Biophys.*, 132, 99 (1969).
21. Arai, Minoda, and Yamada, *Agric. Biol. Chem.*, 33, 922 (1969).
22. Dey and Pridham, *Biochem. J.*, 113, 49 (1969).
23. Neumann and Lampen, *Biochemistry*, 6, 468 (1967).
24. Plapp and Cole, *Biochemistry*, 6, 3676 (1967).
25. Ota, Moore, and Stein, *Biochemistry*, 3, 180 (1964).
26. Murachi, Suzuki, and Takahashi, *Biochemistry*, 6, 3730 (1967).
27. Neumann, Zehavi, and Tanksley, *Biochem. Biophys. Res. Commun.*, 36, 151 (1969).
28. Szewczuk and Connell, *Biochim. Biophys. Acta*, 83, 218 (1964).
29. Szewczuk and Baranowski, *Biochem. Z.*, 338, 317 (1963).
30. Klapp, Kirsch, and Borner, *Z. Physiol. Chem.*, 351, 81 (1970).
31. Junge, Heymann, Krisch, and Hollandt, *Arch. Biochem. Biophys.*, 165, 749 (1974).
32. Liao, *J. Biol. Chem.*, 249, 2354 (1974).
33. Carlsen and Pierce, *J. Biol. Chem.*, 247, 23 (1972).
34. Rombauts, Schroeder, and Morrison, *Biochemistry*, 6, 2965 (1967).
35. Garner and Smith, *J. Biol. Chem.*, 247, 561 (1972).
36. Watanabe and Yasunobu, *J. Biol. Chem.*, 245, 4612 (1970).
37. Mega, Ikenaka, and Matsushima, *J. Biochem.* (Tokyo), 68, 109 (1970).
38. Kuwahara, Adams, and Abraham, *Biochem. J.*, 118, 475 (1970).
39. Woodroof and Glitz, *Biochemistry*, 10, 1532 (1971).

CHARACTERIZATION OF HISTONES

Robert J. DeLange

The second edition of the *Handbook of Biochemistry and Molecular Biology* contained an extensive "Table of Histone Fractions" (C-56 to C-61) in which histone fractions prepared by various procedures were characterized. It is now generally accepted that almost all eukaryotic tissues contain only five major histone fractions (e.g., see the five calf thymus fractions below). In some tissue (e.g., nucleated erythrocytes) or organisms (e.g., trout) special histone fractions (5 and 6 below) are found. Each major histone fraction can often by separated into subfractions which originate through sidechain modifications (acetylation, methylation, phosphorylation, etc.) or through limited sequence heterogeneity.

Wherever possible the items in this table have been derived from sequence data (see Table 2: Histone Sequences). Other items have been selected and updated from the reviews by DeLange and Smith.[1-4]

Table 1
CHARACTERIZATION OF HISTONES

Histone[a]	$\dfrac{\text{Lys}}{\text{Arg}}$ ratio	Total residues	Molecular weight	NH$_2$-Terminal	COOH-Terminal	Evolutionary conservation of sequence	Eukaryotes in General — Sequence heterogeneity (same tissue)	Eukaryotes in General — Sidechain modifications
Calf Thymus								
1 (F1, I)	~21.0	~223	~22,130	Ac-Ser	Lys	Most variable	14% (Residue 1–73, rabbit thymus)	Phosphorylated, ADP-ribosylated in some species
2A (F2A2, IIb1)	1.17	129	14,004	Ac-Ser	Lys	Intermediate	Residue 16 in rat	Appear to be acetylated, methylated and phosphorylated in some species
2B (F2B, IIb2)	2.50	125	13,774	Pro	Lys	Intermediate	Not demonstrated	methylated and phosphorylated in some species
3 (F3, III)	0.72	135	15,324	Ala	Ala	Highly conserved (3% difference, pea and calf)	Residue 96 in pea and calf	Acetylated and methylated in most tissues, phosphorylated in some species
4 (F2A1, IV)	0.79	102	11,282	Ac-Ser	Gly	Most conserved (2% difference, pea and calf)	Not demonstrated	
Nucleated Erythrocytes								
5 (F2C, V)	~2.2	~197	~21,450	Thr	Lys	Insufficient data	Residue 15 in chicken	Phosphorylated
Trout Tissues								
6 (T)	~3.0	~122	~14,500	Pro	?	Found only in trout thus far	Not demonstrated	?

Compiled by Robert J. DeLange.

[a]The histone nomenclature is that which was recently approved by a number of investigators in this field at the Ciba Foundation Symposium on "The Structure and Function of Chromatin," April 2 to 5, 1974, London. The two older systems of nomenclature, which were most commonly used, are shown in parentheses. Histone T has been given the designation 6 in this table, although this was not discussed at the symposium.

REFERENCES (see also Table 2, Histone Sequences)

1. DeLange and Smith, *Ann. Rev. Biochem.*, 40, 279 (1971).
2. DeLange and Smith, *Acc. Chem. Res.*, 5, 368 (1972).
3. DeLange and Smith, in *The Structure and Function of Chromatin*, Ciba Symp., April 1974, in press.
4. DeLange and Smith, in *The Proteins*, Vol. 4, 3rd ed., Neurath and Hill, Eds., Academic Press, New York, in press.

Table 2
HISTONE SEQUENCES[a]

Histone 1[b](I, F1)

					10							
Calf thymus-a[b](1)	Ac- Ser- Glu- Ala- Pro- Ala- Glu- Thr- Ala- Ala- Pro- Ala- Pro- Ala- Pro-
Rabbit thymus-d(1)	Ac- Ser- Glu- Ala- Pro- Ala- Glu- Thr- Ala- Ala- Pro- Ala- Pro- Ala-
Rabbit thymus-c(1)	Ac- Ser- Glu- Ala- Pro- Ala- Glu- Thr- Ala- Ala- Pro- Ala- Pro- Ala- Glu-
Trout testis(2)	————————————————Unknown————————

	20										30		
	Lys- Ser- Pro- Ala- Lys- Thr- Pro- Val- Lys- Ala- Ala- Lys- Lys- Lys- Lys- Pro- Ala- Gly- Ala- Arg-
	Lys- Ser- Pro- Ala- Lys- Thr- Pro- Val- Lys- Ala- Arg- Lys- Lys- Lys- Ser- Ala- Gly- Ala- Ala- Lys-
	Lys- Ser- Pro- Ala- Lys- Lys- Lys- -Lys- Ala- Ala- Lys- Lys- Pro- Gly- Ala- Gly- Ala- Ala- Lys-
	————————————————Unknown————————

	40									50			
	Arg- Lys- Ala- Ser- Gly- Pro- Pro- Val- Ser- Glu- Leu- Ile- Thr- Lys- Ala- Val- Ala- Ala- Ser- Lys-
	Arg- Lys- Ala- Ser- Gly- Pro- Pro- Val- Ser- Glu- Leu- Ile- Thr- Lys- Ala- Val- Ala- Ala- Ser- Lys-
	Arg- Lys- Ala- Ala- Gly- Pro- Pro- Val- Ser- Glu- Leu- Ile- Thr- Lys- Ala- Val- Ala- Ala- Ser- Lys-
	————————————————Unknown————————

	60									70		
	Glu- Arg- Ser- Gly- Val- Ser- Leu- Ala- Ala- Leu- Lys- Lys- Ala- Leu- Ala- Ala- Ala- Gly- Tyr
	Glu- Arg- Ser- Gly- Val- Ser- Leu- Ala- Ala- Leu- Lys- Lys- Ala- Leu- Ala- Ala- Ala- Gly- Tyr
	Glu- Arg- Asn- Gly- Leu- Ser- Leu- Ala- Ala- Leu- Lys- Lys- Ala- Leu- Ala- Ala- Gly- Gly- Tyr- Asp-
	——Arg- Ser- Gly- Val- Ser- Leu- Ala- Ala- Leu- Lys- Lys- Ser- Leu- Ala- Ala- Gly- Gly- Tyr- Asp-

	80									90		
	Val- Glu- Lys- Asn- Asn- Ser- Arg- Ile- Lys- Leu- Gly- Leu- Lys- Ser- Leu- Val- Ser- Lys- Gly- Thr-
	Val- Glu- Lys- Asn- Asn- Ser- Arg- Val- Lys- Ile- Ala- Val- Lys- Ser- Leu- Val- Thr- Lys- Gly- Thr-

	100								110			
	Leu- Val- Glu- Thr- Lys- Gly- Thr- Gly- Ala- Ser- Gly- Ser- Phe- Lys- Leu- Asp- Lys- Lys- Ala- Ala-
	Leu- Val- Glu- Thr- Lys- Gly- Thr- Gly- Ala- Ser- Gly- Ser- Phe- Lys- Leu- Asn- Lys- Lys- Ala-

	120							130			
	Ser- Gly- Glu- Ala- Lys- Pro- Lys- Pro- -Lys- Lys- Ala- Gly- Ala- Ala- Lys- Pro- Lys- Lys- Pro-
	-Val- Glu- Ala- Lys- -Lys- Pro- Ala- Lys- Lys- Ala- - Ala- Ala- - Pro- Lys- Ala- Lys-

	140								150			
	Ala- Gly- Ala- Ala- Lys- Lys- Pro- Ala- Gly- Ala, Ala, Lys, Ala, Pro, Thr, Pro, Lys* Val- Ala-
	Lys- Val- Ala- Ala- Lys- Lys- Pro- Ala- - Ala- Ala- Lys- Ala- Pro- Lys- - Lys- Val- Ala- Ala-

	160								170			
	Lys* Lys- Ala- Val- Lys* Ala- Lys- Lys* Ser- Pro- Lys* Lys- Ala- Lys* Lys- Pro- Lys* Ala- Pro- Lys*
	Lys- Lys- Ala- Val- Ala- Ala- Lys- Lys- Ser- Pro- Lys- Lys- Ala- Lys- Lys- Pro- Ala- Thr- Pro- Lys-

	180							190			
	Ser- Ala- Ala- Lys* Ser- Pro- Ala- Lys- Pro- -Lys* Ala- Ala- Lys- Pro- Lys- Ala- Pro- Lys- Pro-
	Lys- Ala- Ala- Lys- Ser- Pro- Lys- Lys- Ala- Thr- Lys- Ala- Ala- Lys- Pro- Lys- Ala- Ala- Lys- Pro-

[a]See the previous table for nomenclature and characterization of histones.

[b]Subfractions of histone 1 have been isolated and studied. These have usually been designated by numbers (1, 2, 3, etc.) but since these might be confused with the designations in the new system of nomenclature, letters (a for 1, b for 2, etc.) have been substituted here. The numbers above residues indicate relative positions only, but are not residue numbers (due to deletions, etc.). * indicates start and end of peptides.

Table 2 (continued)
HISTONE SEQUENCES[a]

Histone 1[b](I, F1)

```
              200                                        210
Lys* Ala-  Ala-Lys*Lys* Ala- Ala-Lys* Ser-Pro-  Ala-Lys* Ala-  Val-Lys- Pro-Lys* Ala- Ala-  Ala-
Lys- Ala-  Ala- Lys- Lys- Ala-Ala- Lys- Ser-Pro- Lys- Lys-    - Val-Lys-    - Lys- Ala- Ala-  Ala-
```

```
              220
Lys-Pro-Lys* Ala-  Ala-Gly- Ala-Lys*Lys-Lys-COOH
Lys-      - Lys- Ala- Pro-     - Ala- Lys-Lys-COOH
```

The positioning of peptides from calf thymus-a and rabbit thymus-d is by analogy with rabbit thymus-c.

The positioning of peptides in rabbit thymus-c beyond position 108 is based on analogy with trout testis histone and partially on overlapping thermolysin peptides. There is apparently serine-threonine heterogeneity at positions 163, 175, 203, and probably 179 in rabbit thymus-c.[3]

HISTONE 2A(IIb1, F2a2); CALF THYMUS SEQUENCE SHOWN[4,5]

```
                         10
Ac- Ser-Gly-Arg-Gly-Lys-Gln-Gly- Gly- Lys- Ala-Arg- Ala- Lys- Ala-Lys-Thr-Arg- Ser- Ser-
```

```
20                       30
Arg- Ala-Gly-Leu-Gln-Phe- Pro- Val- Gly- Arg- Val- His- Arg- Leu-Leu- Arg-Lys-Gly-Asn-Tyr-
```

```
40                       50
Ala-Glu-Arg- Val-Gly- Ala-Gly- Ala- Pro- Val- Tyr-Leu- Ala- Ala- Val-Leu-Glu-Tyr-Leu-Thr-
```

```
60                       70
Ala-Glu-  Ile-Leu-Glu-Leu- Ala-Gly-Asn- Ala- Ala-Arg-Asp-Asn- Lys-Lys-Thr-Arg-  Ile-  Ile-
```

```
80                       90
Pro- Arg- His-Leu-Gln-Leu- Ala-  Ile- Arg-Asn-Asp-Glu- Glu-Leu-Asn-Lys-Leu-Leu- Gly-Lys-
```

```
100                      110
Val-Thr-  Ile- Ala-Gln-Gly-Gly- Val- Leu- Pro-Asn-  Ile- Gln- Ala- Val-Leu-Leu- Pro-Lys-Lys-
```

```
120                      129
Thr- Glu- Ser- His- His-Lys- Ala-Lys- Gly- Lys-COOH
```

In the rat histone, Residue 16 is Thr (some molecules) or Ser (other molecules) and Residue 99 is Arg.[6] Residue 6 is Thr in trout testis histone 2A;[7] only the first 11 residues are known.

Table 2 (continued)
HISTONE SEQUENCES[a]

HISTONE 2B(IIb2, F2B); CALF THYMUS SEQUENCE SHOWN[8]

10
HN- Pro- Glu- Pro- Ala- Lys- Ser- Ala- Pro- Ala- Pro- Lys- Lys- Gly- Ser- Lys- Lys- Ala- Val- Thr-

20 30
Lys- Ala- Gln- Lys- Lys- Asp- Gly- Lys- Lys- Arg- Lys- Arg- Ser- Arg- Lys- Glu- Ser- Tyr- Ser- Val-

40 50
Tyr- Val- Tyr- Lys- Val- Leu- Lys- Gln- Val- His- Pro- Asp- Thr- Gly- Ile- Ser- Ser- Lys- Ala- Met-

60 70
Gly- Ile- Met- Asn- Ser- Phe- Val- Asn- Asp- Ile- Phe- Glu- Arg- Ile- Ala- Gly- Glu- Ala- Ser- Arg-

80 90
Leu- Ala- His- Tyr- Asn- Lys- Arg- Ser- Thr- Ile- Thr- Ser- Arg- Glu- Ile- Gln- Thr- Ala- Val- Arg-

100 110
Leu- Leu- Leu- Pro- Gly- Glu- Leu- Ala- Lys- His- Ala- Val- Ser- Glu- Gly- Thr- Lys- Ala- Val- Thr-

120 125
Lys- Tyr- Thr- Ser- Ser- Lys- COOH

Residues 9 and 10 are not present and the sequence of Residues 21—23 is Ser (or Thr)-Ala-Gly in the trout testis histone;[9] only the first 22 residues are known.

Table 2 (continued)
HISTONE SEQUENCES[a]

HISTONE 3(III, F3); CALF THYMUS SEQUENCE SHOWN[10]

10
H$_2$N- Ala- Arg- Thr- Lys- Gln- Thr- Ala- Arg- Lys(CH$_3$)$_{0-3}$- Ser- Thr- Gly- Gly- Lys(Ac)$_{0,1}$ Ala-Pro-

20 30
Arg- Lys- Gln- Leu- Ala- Thr- Lys(Ac)$_{0,1}$ Ala- Ala- Arg-Lys (CH$_3$)$_{0-3}$- Ser-Ala-Pro-Ala-Thr-Gly-

40 50
Gly- Val- Lys- Lys- Pro- His- Arg- Tyr- Arg- Pro- Gly- Thr- Val- Ala-Leu- Arg- Glu- Ile- Arg- Arg-

60 70
Tyr- Gln- Lys- Ser- Thr- Glu- Leu- Leu- Ile- Arg- Lys- Leu- Pro- Phe- Gln- Arg- Leu- Val- Arg-Glu-

80 90
Ile- Ala- Gln- Asp- Phe- Lys- Thr- Asp-Leu- Arg- Phe- Gln- Ser- Ser- Ala- Val-Met- Ala-Leu-Gln-

100 110
Glu- Ala- Cys- Glu- Ala- Tyr- Leu- Val-Gly- Leu- Phe- Glu- Asp- Thr-Asn-Leu- Cys-Ala- Ile- His-

120 130
Ala- Lys- Arg- Val- Thr- Ile-Met- Pro- Lys- Asp- Ile- Gln- Leu- Ala- Arg- Arg- Ile- Arg-Gly-Glu-

135
Arg- Ala-COOH

In the pea seedling histone, Residue 41 is Phe, 53 is Lys, 90 is Ser, and 96 is Ala (60%) or Ser (40%). ϵ-N-Acetyllysine and ϵ-N-trimethyllysine are absent.[11]

In the fish testis histone, Residue 96 is Ser and Residues 14 and 23 are not acetylated.[12]

In the chicken erythrocyte histone, Residue 96 is Ser, and evidence was obtained for methylation of Residue 36 in addition to Residues 9 and 27.[13] The sites of ϵ-N-acetylation and several amides were not determined.

Some molecules of calf thymus histone contain Ser at Residue 96.[14]

Table 2 (continued)
HISTONE SEQUENCES[a]

HISTONE 4(IV, F2a1); CALF THYMUS SEQUENCE SHOWN[15,16]

10
Ac- Ser- Gly- Arg-Gly-Lys-Gly- Gly- Lys-Gly-Leu- Gly- Lys-Gly-Gly- Ala- Lys(Ac)$_{0,1}$ Arg-His-

20   30
Arg-Lys(CH$_3$)$_{1,2}$ - Val-Leu-Arg-Asp-Asn- Ile-Gln-Gly- Ile-Thr-Lys-Pro- Ala- Ile-Arg-Arg-

40   50
Leu- Ala- Arg-Arg-Gly-Gly- Val- Lys- Arg- Ile- Ser- Gly-Leu- Ile-Tyr-Glu- Glu- Thr-Arg-Gly-

60   70
Val- Leu- Lys- Val-Phe-Leu-Glu-Asn- Val- Ile-Arg-Asp- Ala- Val-Thr-Tyr- Thr- Glu- His- Ala-

80   90
Lys- Arg- Lys- Thr- Val-Thr- Ala-Met-Asp- Val- Val- Tyr- Ala-Leu-Lys-Arg- Gln- Gly-Arg-Thr-

100
Leu- Tyr- Gly- Phe-Gly-Gly-COOH

In pea seedling histone 4,[15] Residue 60 is Ile, and Residue 77 is Arg; Residue 20 is not
  methylated and at least one other lysyl residue (Residue 8?) is ε-N-acetylated.
Other lysyl residues in calf thymus histone 4 can also be ε-N-acetylated to a small extent.[17]
The sequences of rat[18] and pig[19] histone 4 are identical to the calf thymus sequence.
Sea urchin histone 4 apparently contains a cysteine residue.[20]

HISTONE 5(V, F2c) (INCOMPLETE); CHICKEN ERYTHROCYTE SEQUENCE SHOWN[21]

1   10   15   20
Thr- Glu- Ser- Leu- Val- Leu- Ser- Pro- Ala- Pro- Ala- Lys- Pro- Lys- Gln- Val- Lys- Ala- Ser- Arg- Arg-Ser- Ala- Ser- His-
                                                                          Arg

                 30   40   50
Pro- Thr- Tyr- Ser- Glu- Met- Ile- Ala- Ala- Ala- Ile- Arg- Ala- Glu- Lys- Ser- Arg- Gly- Gly- Ser- Ser- Arg-Gln- Ser- Ile-

                                                     60   70
Gln- Lys- Tyr- Ile- Lys- Ser- His- Tyr-Lys- Val- Gly- His- Asn- Ala- Asp- Leu- Gln- Ile- Lys- Leu-

HISTONE 6(T) (INCOMPLETE); TROUT TESTIS[22]

10
HN- Pro- Lys- Arg- Lys-Ser- Ala-Thr- Lys-Gly- Asp- Glu-Pro- Ala- Arg- Arg-Ser- Ala- Arg-Leu-

20
Ser- Gly- Arg- Pro- Val-Pro- Lys- Pro- Ala-Ala - - - - - -

Compiled by Robert J. Delange.

Table 2 (continued)
HISTONE SEQUENCES[a]

REFERENCES

1. Rall and Cole, *J. Biol. Chem.,* 246, 7175 (1971); Cole, personal communication.
2. Personal communicaiton from Professor G. H. Dixon; McLeod and Dixon, unpublished results.
3. Jones, Rall, and Cole, *J. Biol. Chem.,* 249, 2548 (1974).
4. Yeoman, Olson, Sugano, Jordan, Taylor, Starbuck, and Busch, *J. Biol. Chem.,* 247, 6018 (1972).
5. Sautiere, Tyrou, Laine, Mizon, Ruffin, and Biserte, *Eur. J. Biochem.,* 41, 563 (1974).
6. Sautiere, personal communication.
7. Candido and Dixon, *J. Biol. Chem.,* 247, 3868 (1972).
8. Iwai, Hayashi, and Ishikawa, *J. Biochem.,* 72, 357 (1972).
9. Candido and Dixon, *Proc. Natl. Acad. Sci. U.S.A.,* 69, 2015 (1972).
10. DeLange, Hooper, and Smith, *Proc. Natl. Acad. Sci. U.S.A.,* 69, 882 (1972); *J. Biol. Chem.,* 248, 3261 (1973).
11. Patthy and Smith, *J. Biol. Chem.,* 248, 6834 (1973).
12. Hooper and Smith, *J. Biol. Chem.,* 248, 3275 (1973).
13. Brandt and von Holt, *Eur. J. Biochem.,* 46, 419 (1974).
14. Patthy and Smith, *J. Biol. Chem.,* 250 (1975), in press.
15. DeLange, Fambrough, Smith, and Bonner, *Proc. Natl. Acad. Sci. U.S.A.,* 61, 1145 (1968); *J. Biol. Chem.,* 244, 319 (1969); *J. Biol. Chem.,* 244, 5669 (1969).
16. Ogawa, Quagliarotti, Jordan, Taylor, Starbuck, and Busch, *J. Biol. Chem.,* 244, 4387 (1969).
17. Wangh, Ruiz-Carrillo, and Allfrey, *Arch. Biochem. Biophys.,* 150, 44 (1972).
18. Sautiere, Tyrou, Moschetto, and Biserte, *Biochimie,* 53, 479 (1971).
19. Sautiere, Lambelin-Breynaert, Moschetto, and Biserte, *Biochimie,* 53, 711 (1971).
20. Subirana, *FEBS Lett.,* 16, 133 (1971); Sautiere, personal communication.
21. Garel, Mazen, Champagne, Sautiere, Kmiecik, Loy, and Biserte, *FEBS Lett.,* 50, 195 (1975).
22. Huntley and Dixon, *J. Biol. Chem.,* 247, 4916 (1972).

MOLECULAR PARAMETERS OF THE CONTRACTILE PROTEINS[a]

Protein	Localization in myofibril	% Total protein	Intrinsic sedimentation coefficient (Svedberg)	Intrinsic viscosity (ml/g)	Molecular weight	Chain weight	% α-helix	References
Myosin	Thick filament	55	6.4	210	470,000	200,000 20,700[b] 18,000 16,500	57	1−8
C-protein	Thick filament	2	4.6	14	140,000	140,000	<10	9
Paramyosin	Thick filament	5−50	3.1	190	200,000	100,000	>90	10−12
M-proteins	M-line	<2	5.0	−	160,000	160,000	−	13, 14
			5.4	4.5	86,000	43,000	26	15, 16
G-actin	Thin filament	25	3.3	4	41,780	41,780[b]	26	17−20
Tropomyosin	Thin filament	5	2.6	34	65,000	32,760[b]	90	21−23
Troponin	Thin filament	5	4.0	4	80,000	37,000 24,000 17,850[b]	35	24−27
α-Actinin	Z-line	<2	6.2	9	180,000	90,000	60	28, 29

Compiled by S. Lowey.

[a]Parameters determined for proteins from vertebrate fast skeletal muscles, with the exception of paramyosin.
[b]Molecular weights based on sequence data.

REFERENCES

1. **Lowey, Slayter, Weeds, and Baker,** *J. Mol. Biol.,* 42, 1 (1969).
2. **Gershman, Stracher, and Dreizen,** *J. Biol. Chem.,* 244, 2726 (1969).
3. **Gazith, Himmelfarb, and Harrington,** *J. Biol. Chem.,* 245, 15 (1970).
4. **Godfrey and Harrington,** *Biochemistry,* 9, 894 (1970).
5. **Frank and Weeds,** *Eur. J. Biochem.,* 44, 317 (1974).
6. **Lowey and Risby,** *Nature,* 234, 81 (1971).
7. **Sarkar, Sreter, and Gergely,** *Proc. Natl. Acad. Sci. U.S.A.,* 68, 946 (1971).
8. **Weeds and Lowey,** *J. Mol. Biol.,* 61, 701 (1971).
9. **Offer, Moos, and Starr,** *J. Mol. Biol.,* 74, 653 (1973).
10. **Bullard, Luke, and Winkelman,** *J. Mol. Biol.,* 75, 359 (1973).
11. **Stafford and Yphantis,** *Biochem. Biophys. Res. Commun.,* 49, 848 (1972).
12. **Lowey, Holtzer, and Kucera,** *J. Mol. Biol.,* 7, 234 (1963).
13. **Trinick,** *Fed. Proc.,* in press.
14. **Masaki and Takaiti,** *J. Biochem.* (Tokyo), 75, 367 (1974).
15. **Eaton and Pepe,** *J. Cell Biol.,* 55, 681 (1972).
16. **Morimoto and Harrington,** *J. Biol. Chem.,* 247, 3052 (1972).
17. **Elzinga, Collins, Kuehl, and Adelstein,** *Proc. Natl. Acad. Sci. U.S.A.,* 70, 2687 (1973).
18. **Cohen,** *Arch. Biochem. Biophys.,* 117, 289 (1966).
19. **Rees and Young,** *J. Biol. Chem.,* 242, 4449 (1967).
20. **Nagy,** *Biochim. Biophys. Acta,* 115, 498 (1966).
21. **Holtzer, Clark and Lowey,** *Biochemistry,* 4, 2401 (1965).
22. **Woods,** *J. Biol. Chem.,* 242, 2859 (1967).
23. **Stone, Sodek, Johnson, and Smillie,** *Proc. IX FEBS Meet.,* Budapest, 1974.
24. **Collins, Potter, Horn, Wilshire, and Jackman,** in *Calcium Binding Proteins,* Drabikowski, Strzelecka-Golaszewska, and Carafoli, Eds., Elsevier, Amsterdam, 1974.
25. **Greaser and Gergely,** *J. Biol. Chem.,* 246, 4226 (1971).
26. **Perry, Cole, Head, and Wilson,** *Cold Spring Harbor Sym. Quant. Biol.,* 37, 251 (1972).
27. **Ebashi, Ohtsuki, and Mihaski,** *Cold Spring Harbor Sym. Quant. Biol.,* 37, 215 (1972).
28. **Goll, Suzuki, and Singh,** *Biophys. J.,* 11, 107a (1971).
29. **Suzuki, Goll, Stromer, Singh, and Temple,** *Biochim. Biophys. Acta,* 295, 188 (1973).

SUBUNIT CONSTITUTION OF PROTEINS

Dennis W. Darnall and Irving M. Klotz

The wide response from readers to our previously published protein subunit tables[1-4] indicates the usefulness of such compilations for teaching and research purposes. We have therefore prepared a new updated listing of proteins with subunits held together by noncovalent bonds. Individual polypeptide chains held together by disulfide bridges have not been individually classified as subunits; insulin, for example, is listed as having a subunit molecular weight of 5733 even though this is the combined weight of the disulfide-linked A and B chains. Kleine[4a] has compiled a list of proteins whose subunits are associated through both noncovalent and covalent (disulfide) bonds.

In many instances, the subunit listed may not be the minimal subunit obtainable, but instead the minimal subunit that has been unequivocally obtained under conditions that eliminate cleavage of peptide or disulfide bonds. For some proteins, two or more stages of dissociation can be clearly recognized; in such instances two or more entries specifying the relations between the different aggregates are given. Parentheses around molecular weights or subunit numbers indicate uncertainty in the value.

The most accessible references are given for each entry; they do not necessarily indicate the source most deserving of credit for establishing the subunit stoichiometry. These sources are mentioned in the cited works.

SUBUNIT CONSTITUTION OF PROTEINS

Protein	Source	Organ	Molecular weight	Subunits		References
				No.	Molecular weight	
Insulin	Bovine		11,466	2	5,733	5
S-100 protein	Bovine	Brain	19,500	(4–3)	(4,100–7,000)	6
Mercaptopyruvate sulfur transferase (EC 2.8.1.2)	*Escherichia coli*		23,800	2	12,000	7
$\Delta_{5\rightarrow4}$-3-Oxosteroid isomerase (EC 5.3.3.1)	*Pseudomonas*		26,300	2	13,000	8
Nerve growth factor	Mouse	Submaxillary gland	26,518	2	13,259	9
Leuteinizing hormone	Ovine		27,322	1	12,500	10
				1	14,830	
Cytochrome (CC')	*Pseudomonas*		28,000	2	14,000	11
Leuteinizing hormone	Human	Pituitary gland	28,260	1	13,853	12
				1	14,407	
Ribonuclease	Bull	Semen	29,000	2	14,000	13
Interstitial cell-stimulating hormone	Ovine	Pituitary gland	30,000	1	13,700	14
				1	16,300	
Phospholipase A$_2$	*Crotalus*	Venom	30,000	2	15,000	14a
Leuteinizing hormone	Rat	Pituitary gland	31,000	2	15,500	15
Superoxide dismutase	*Neurospora*		31,000	2	16,800	16
Superoxide dismutase	Human	Erythrocyte	32,000	2	16,000	17
Thyrotropin	Bovine		32,000	1	15,000	17a
				1	15,000	
Follicle stimulating hormone	Ovine		33,000	1	18,500	17b
				1	18,500	
Lactose specific factor III	*E. coli* and *Staphylococcus*		33,000	4	8,000	18
Follicle stimulating hormone	Equine •	Pituitary gland	33,800	1	16,500	19
				1	16,000	
Adenine phosphoribosyltransferase (EC 2.4.2.7)	Human	Erythrocyte	34,000	3	11,000	20
Hemoglobin I	Blood clam	Erythrocyte	34,000	2	17,500	21
Follicle stimulating hormone	Human	Pituitary gland	35,000	2	17,500	22

SUBUNIT CONSTITUTION OF PROTEINS (continued)

Protein	Source	Organ	Molecular weight	Subunit No.	Subunit Molecular weight	References
β-Lactoglobulin	Bovine	Milk	35,000	2	17,500	23
Lactose-specific phosphocarrier protein	*Staphylococcus*		35,000	3	12,000	24
Agglutinin	Wheat	Germ	35,000	2	17,000	25
DNA dependent RNA polymerase	*Halobacterium*		36,000	1	18,000	26
				1	18,000	
β-Hydroxydecanoyl thioester dehydrase	*E. coli*		36,000	2	18,000	27
Rhodanese (EC 2.8.1.1)	Bovine	Liver	37,000	2	18,500	28
Chorionic gonadotropin	Human		37,900	1	14,900	29
				1	23,000	
Chymotrypsin inhibitor I	Potato		39,000	4	9,800	30
Superoxide dismutase	*E. coli*		39,500	2	21,600	31
Hemerythrin	*Phascolosoma*	Coelomic fluid	40,600	3	12,700	31a
Proteinase inhibitor I	Potato		42,000	2	19,300	32
			19,300	2	9,400	
Dethiobiotin synthetase	*E. coli*		42,000	2	24,500	32a
Dihydropteridine reductase	Sheep	Liver	42,000	2	21,000	33
Nucleoside phosphotransferase	Carrot		44,000	1	22,000	34
				1	22,000	
Biotin carboxyl carrier protein	*E. coli*		45,000	2	22,500	35
Catabolite gene-activator protein	*E. coli*		45,000	2	22,000	36
Cytoplasmic protein	*Neurospora*		45,000	3	15,000	37
Growth hormone	Bovine	Pituitary gland	48,000	2	25,000	38
Factor X	Bovine	Plasma	48,000	1	20,000	39
				1	30,000	
Hemagglutinin LcH	*Lens*		49,000	2	24,500	39a
Phycocyanin	*Chroomonas*		50,000	2	16,000	40
				2	10,000	
Phenylalanine hydroxylase-stimulating protein	Rat	Liver	51,500	4	12,500	41
Galactokinase (EC 2.7.1.6)	Human	Erythrocyte	53,000	2	27,000	42
Triosephosphate isomerase (EC 5.3.1.1)	Rabbit	Muscle	53,000	2	26,500	43

SUBUNIT CONSTITUTION OF PROTEINS (continued)

Protein	Source	Organ	Molecular weight	Subunit No.	Subunit Molecular weight	References
Malate dehydrogenase (EC 1.1.1.37)	*Neurospora*		54,000	4	13,500	44
Transglutaminase	Guinea pig	Hair follicle	54,000	2	27,000	45
Azoferredoxin	*Clostridium*		55,000	2	27,500	46
Hyaluronidase (EC 3.2.1.35)	Bovine	Testicle	55,000	4	14,000	47
5-10-Methylenetetrahydrofolate dehydrogenase (EC 1.5.1.5)	*Clostridium*		55,000	2	30,000	48
ω-Amidase	Rat	Liver	58,000	2	27,000	49
Alcohol dehydrogenase (EC 1.1.1.1)	*Drosophila*		60,000	8	7,400	49a
NADP-linked isocitrate dehydrogenase (EC 1.1.1.42)	Porcine	Heart	60,000	2	32,000	50
Deoxycytidylate deaminase	*Staphylococcus*		60,000	2	29,000	51
Hydrogenase (EC 1.12.1.1)	*Clostridium*		60,000	2	30,000	52
Lactose synthetase	Bovine	Milk	60,000	1	22,300	53
				1	36,600	
Transcobalamin II	Human	Plasma	60,000	1	38,000	54
				1	25,000	
Nuclear DNA polymerase (EC 2.7.7.7)	Rat	Liver	60,000	2	29,000	55
Aldose reductase (EC 1.1.1.21)	*Rhodotorula*		61,000	1	22,300	56
				1	36,600	
Prealbumin	Human	Plasma	62,000	4	15,500	57
Phosphoglucomutase (EC 2.7.5.a)	Rabbit	Muscle	62,000	1	31,000	58
				1	31,000	
Erythrocuprein	Bovine	Blood	64,000	4	16,000	59
Serine dehydratase (EC 4.2.1.13)	Rat	Liver	64,000	2	34,000	60
D-Galactose dehydrogenase (EC 1.1.1.48)	*Pseudomonas*		64,000	2	32,000	60a
T2 Phage induced thymidylate synthetase	*Lactobacillus*		64,400	2	31,500	61
Hemoglobin	Mammalian	Erythrocytes	64,500	4	16,000	62
Tu-Ts Complex	*E. coli*		65,000	1	41,500	63
				1	28,500	
L-3-Hydroxyacyl-CoA dehydrogenase (EC 1.1.1.35)	Pig	Heart	65,000	2	31,000	64
Inorganic pyrophosphatase (EC 3.6.1.1)	Yeast		65,000	2	32,000	65

SUBUNIT CONSTITUTION OF PROTEINS (continued)

Protein	Source	Organ	Molecular weight	Subunit		References
				No.	Molecular weight	
Thiogalactoside transacetylase (EC 2.3.1.18)	E. coli		65,300	2	29,700	66
Phosphoglycerate mutase (EC 2.7.5.3)	Pig	Muscle	66,000	2	33,000	67
Thioredoxin reductase	E. coli		66,000	2	32,000	67a
Malate dehydrogenase (EC 1.1.1.37)	Rat	Liver	66,300	2	37,500	68
Malate dehydrogenase (EC 1.1.1.37)	Pig	Heart	67,000	2	35,000	69
Glucokinase (EC 2.7.1.12)	Bacillus		67,000	2	34,500	70
O-Acetylserine sulfhydrylase A	Salmonella		68,000	2	34,000	71
Tropomyosin B	Rabbit	Muscle	68,000	2	33,500	72
Transaldolase III (EC 2.2.1.2)	Candida		68,000	2	34,000	73
Adenylate kinase (EC 2.7.4.3)	Rat	Liver	68,000	3	23,000	74
Glycerol-3-phosphate dehydrogenase (EC 1.1.1.8)	Chicken, rabbit, honeybee	Muscle, thorax	68,000	2	34,000	75
17β-Estradiol dehydrogenase	Human	Placenta	68,000	2	33,000	76
Avidin	Chicken	Egg white	68,300	4	18,000	77
Hemoglobin III	Blood clam	Erythrocyte	69,000	4	17,500	78
D-Glycerate dehydrogenase	Beef	Liver	70,000	2	34,000	78a
D-Lactate dehydrogenase (EC 1.1.1.28)	Limulus	Muscle	70,000	2	35,000	79
Fructose diphosphate aldolase (EC 4.1.2.13)	E. coli		70,000	2	35,000	80
Malate-lactate transhydrogenase (EC 1.1.99.7)	Viellonella		70,000	2	35,000	81
Thymidylate synthetase (EC 2.1.1.6)	Lactobacillus		70,000	2	35,000	82
Hydroxypyruvate reductase (EC 1.1.1.29)	Pseudomonas		70,000	2	35,000	83
Nucleoside diphosphate kinase (EC 2.7.4.6)	Pea	Seed	70,000	4	17,000	84
NAD-Glycohydrolase (EC 3.2.2.5)	Mouse, rat, rabbit	Liver	70,000	2	38,000	85
Protein P11	Bacteriophage T4D		70,000	3	24,000	85a
2-Keto-3-deoxy-6-phosphogluconate aldolase (EC 4.1.2.12)	Pseudomonas		72,000	3	24,000	86

SUBUNIT CONSTITUTION OF PROTEINS (continued)

Protein	Source	Organ	Molecular weight	Subunit No.	Subunit Molecular weight	References
Malate dehydrogenase (EC 1.1.1.37)	Bovine	Heart	72,000	2	37,000	87
2,4-Diaminopentanoic acid C_4 dehydrogenase	Clostridium		72,000	2	40,000	88
Propylamine transferase	E. coli		73,000	2	37,000	89
L-Histidinol phosphate aminotransferase (EC 2.6.1.9)	Salmonella		74,000	2	37,000	90
Tryptophanyl tRNA synthetase (EC 6.1.1.2)	E. coli		74,000	2	37,000	91
Sedoheptulose 1,7-diphosphatase	Candida		75,000	2	35,000	92
Electron-transferring flavoprotein	Peptostreptococcus		75,000	1 / 1	41,000 / 33,000	93
Glutathione peroxidase (EC 1.11.1.9)	Rat	Liver	76,000	4	19,000	93a
Chorismate mutase-prephenate dehydrogenase	Aerobacter		76,000	2	40,000	94
Diacetyl reductase (EC 1.1.1.15)	Beef	Liver	76,000	3	26,000	95
Hypoxanthine-guanine phosphoribosyltransferase (EC 2.4.2.8)	Chinese hamster, human	Brain, erythrocyte	78,000	3	25,000	96
Histidyl tRNA synthetase	Salmonella		78,000	2	40,000	97
Cyclic AMP dependent protein kinase	Bovine	Sperm	78,000	1 / 1	35,000 / 40,000	98
Glycerol 1-phosphate dehydrogenase (EC 1.1.1.6)	Yeast, rabbit	Muscle	78,000	2	40,000	99
Hydroxyindole-O-methyl transferase (EC 2.1.1.4)	Bovine	Pineal gland	78,000	2	39,000	100
Uridine diphosphogalactose 4-epimerase (EC 5.1.3.2)	Yeast		79,000	2	39,000	101
Luciferase	Photobacterium		79,000	1 / 1	42,000 / 37,000	102
Phosphotransacetylase (EC 2.3.1.8)	Viellonella		80,000	2	(40,000)	103
Creatine kinase (EC 2.7.3.2)	Chicken	Muscle	80,000	2	40,000	104
Alcohol dehydrogenase (EC 1.1.1.1)	Horse	Liver	80,000	2	41,000	105

SUBUNIT CONSTITUTION OF PROTEINS (continued)

Protein	Source	Organ	Molecular weight	Subunit No.	Subunit Molecular weight	References
Aldolase (EC 4.1.2.13)	Yeast		80,000	2	40,000	106
Lombricine kinase (EC 2.7.3.5)	Cancer and Homarus		80,000	2	40,000	107
Taurocyamine kinase (EC 2.7.3.4)	Lumbricus		80,000	2	40,000	108
Prephenoloxidase	Silkworm	Hemolymph	80,000	2	40,000	109
Troponin	Rabbit	Muscle	80,000	1	37,000	110
				1	24,000	
				1	20,000	
L-Erythro-3,5-diaminohexanoate dehydrogenase	Clostridium		80,000	2	39,300	111
Histidinol dehydrogenase (EC 1.1.1.23)	Salmonella		80,000	2	40,000	112
Galactose 1-phosphate uridyltransferase (EC 2.7.7.12)	E. coli		80,000	2	40,000	113
Anthranilate synthase	Pseudomonas		81,000	1	63,000	114
				1	18,000	
Enolase (EC 4.2.1.11)	Rabbit	Muscle	82,000	2	42,000	115
ATP-Creatine transphosphorylase	Rabbit	Muscle	82,600	2	41,300	116
NADP-Specific isocitrate dehydrogenase	E. coli		83,000	2	43,000	116a
Glutathione peroxidase	Bovine	Blood	83,800	4	21,000	117
Purine nucleoside phosphorylase (EC 2.4.1.2)	Calf	Spleen	84,600	3	28,000	118
Histidyl tRNA synthetase	E. coli		85,000	2	42,500	119
Succinate dehydrogenase (EC 1.3.99.1)	Rhodospirillum		85,000	1	60,000	120
				1	25,000	
Glutamate decarboxylase (EC 4.1.1.15)	Mouse	Brain	85,000	2	44,000	121
Haptoglobin 1-1	Human	Serum	85,000	2	40,000	122
Acid phosphatase (EC 3.1.3.2)	Neurospora		85,000	2	42,000	123
Alkaline phosphatase (EC 3.1.3.1)	E. coli		86,000	2	43,000	124
Anthranilate synthetase	Acinobacter		86,000	1	70,000	125
				1	14,000	
Histidinol dehydrogenase (EC 1.1.1.23)	E. coli		87,000	2	40,000	126

SUBUNIT CONSTITUTION OF PROTEINS (continued)

Protein	Source	Organ	Molecular weight	Subunit No.	Subunit Molecular weight	References
Enolase (EC 4.2.1.11)	Yeast		88,000	2	44,000	127
Procarboxypeptidase A	Bovine	Pancreas	88,000	1	40,000	128
High density lipoprotein	Human	Serum	88,000	2	23,000 27,000	129
M-line Protein	Chicken	Muscle	88,000	2	17,000	130
Putrecine oxidase	Micrococcus		88,000	2	43,000	131
Enolase (EC 4.2.1.11)	E. coli		90,000	2	46,000	131a
Galactose 1-phosphate uridyl-transferase	Human	Liver	90,000	(3—4)	46,000 30,000	132
D-Erythulose reductase	Beef	Liver	90,000	4	22,000	133
2-Deoxycitrate synthase	Penicillium		90,000	2	45,000	134
Purine nucleoside phosphorylase	Bacillus		90,000	2	47,000	135
			47,000	2	24,000	
Pyruvate dehydrogenase (EC 1.2.4.1)	E. coli		90,000	2	45,000	136
Carboxypeptidase G_1	Pseudomonas		92,000	2	46,000	136a
Salycilate hydroxylase (EC 1.14.1.a)	Pseudomonas		92,000	2	52,000	137
Luciferase	Firefly	Lanterns	92,000	2	52,000	138
Isocitrate dehydrogenase	Bacillus		92,500	2	45,000	138a
Adenylate kinase (EC 2.7.4.3)	Brevibacterium		92,400	2	46,000	139
L-6-Hydroxynicotine oxidase	Arthrobacter		93,000	2	47,000	139a
6-Phosphogluconate dehydrogenase (EC 1.1.1.44)	Sheep	Liver	94,000	2	47,000	140
Dipeptidase M	E. coli		94,000	2	47,000	141
Prolyl tRNA synthetase	E. coli		94,000	2	47,000	141a
Tyrosyl tRNA synthetase	Bacillus		95,000	2	45,000	142
Ceramide trihexosidase	Human	Plasma	95,000	4	22,000	143
α-Amylase	Bacillus		96,000	2	48,000	144
			48,000		24,000	
Allophycocyanin	Synechococcus		96,000	(?)	17,250	145
				(?)	15,200	
Glyoxylic acid reductase	Spinach	Leaf	97,500	2	47,000	145a

SUBUNIT CONSTITUTION OF PROTEINS (continued)

Protein	Source	Organ	Molecular weight	Subunit No.	Subunit Molecular weight	References
L-Ribulokinase (EC 2.7.1.16)	E. coli		98,000	2	50,000	146
N-Formimino-L-glutamate imino-hydrolase	Pseudomonas		100,000	2	50,000	147
D-Amino acid oxidase (EC 1.4.3.3)	Pig	Kidney	100,000	2	50,000	148
Diacetyl (acetoin) reductase (EC 1.1.1.5)	Aerobacter		100,000	4	25,000	149
Cysteamine oxygenase (EC 1.13.1.22)	Horse	Kidney	100,000	2	50,000	150
Galactokinase (EC 2.7.1.6)	Yeast		100,000	4	23,000	151
Citrate synthase (EC 4.1.3.7)	Pig	Heart	100,000	2	50,000	152
Aspartate aminotransferase (EC 2.6.1.1)	Chicken	Heart	100,000	2	50,000	153
Seryl tRNA synthetase (EC 6.1.1.11)	E. coli		100,000	2	50,000	154
β-D-N-Acetylhexose amidase	Human	Placenta	100,000	6	17,000	155
3-α-Hydroxysteroid dehydrogenase (EC 1.1.1.50)	Pseudomonas		100,000	2	50,000	155a
6-Phosphogluconate dehydrogenase (EC 1.1.1.44)	Bacillus, rat	Liver	101,000	2	51,000	156
Glutamyl tRNA synthetase	E. coli		102,000	1	56,000	157
				1	46,000	
D-Galactose dehydrogenase	Pseudomonas		102,000	4	25,000	158
Glutathione reductase	Sea urchin	Egg	102,000	2	52,000	159
Acid phosphomonoesterase I (EC 3.1.3.2)	Human	Prostate	102,000	2	50,000	160
Hexokinase (EC 2.7.1.1)	Yeast		102,000	2	51,000	161
Nucleoside diphosphokinase (EC 2.7.4.6)	Yeast		102,000	6	17,000	162
Delta hemolysin	Staphylococcus		103,000	5	21,000	163
			21,000	4	5,000	
Uridine phosphorylase	Rat	Liver	103,000	4	26,000	164
6-Phosphogluconate dehydrogenase (EC 1.1.1.44)	Human	Erythrocyte	104,000	2	52,000	165
Aspartokinase III (EC 2.7.2.4)	E. coli		105,000	2	50,000	166
Ribosephosphate isomerase (EC 5.3.1.6)	Candida		105,000	4	26,000	167
Isocitrate dehydrogenase (TPN)	Rhodopseudomonas		105,000	2	50,000	168

SUBUNIT CONSTITUTION OF PROTEINS (continued)

Protein	Source	Organ	Molecular weight	Subunit No.	Subunit Molecular weight	References
Lipoamide dehydrogenase (EC 1.6.4.3)	E. coli		106,000	2	53,000	169
20-β-Hydroxysteroid dehydrogenase (EC 1.1.1.53)	Streptomyces		106,000	4	27,000	170
Glycyl-L-leucine hydrolase	Monkey	Small intestine	107,000	2	54,000	171
Sulphatase A (EC 3.1.6.1)	Ox	Liver	107,000	2	50,000	171a
Ornithine transcarbamylase (EC 2.1.3.3)	Streptococcus, bovine	Liver	108,000	3	36,000	172
Concanavalin A	Jack bean		108,000 / 54,000	2 / 2	54,000 / 27,000	173
Lipoxygenase (EC 1.99.2.1)	Soybean		108,000	2	54,000	174
Hemerythrin	Golfingia	Erythrocyte	108,000	8	13,500	175
Glutamine transaminase	Rat	Liver	110,000	2	54,000	176
Tubulin	Pig	Brain	110,000	1 / 1	56,000 / 53,000	177
Ribitol dehydrogenase (EC 1.1.1.56)	Klebsiella		110,000	4	27,000	177a
Tryptophanyl tRNA synthetase (EC 6.1.1.2)	Bovine	Pancreas	110,000	2	58,000	178
Phosphoglycerate mutase (EC 2.7.5.3)	Yeast		110,000	4	27,000	179
Urocanase	Pseudomonas		110,000	2	54,000	180
Histidine decarboxylase (EC 4.1.1.22)	Micrococcus		110,000	3 / 3	29,000 / 7,000	181
Phosphatase (EC 3.1.3.2)	Sweet potato	Tuber	110,000	2	55,000	182
Dihydrolipoyl dehydrogenase	E. coli		112,000	2	56,000	183
Canavalin			113,000	6	19,500	184
Adenosylmethionine decarboxylase	E. coli		113,000	(8)	(15,000)	184a
D-Ribulose 1,5-diphosphate carboxylase	Rhodospirillum		114,000	2	56,000	185
Lectin	Navy bean		114,000	4	30,000	186
DNA Modification methylase	E. coli		115,000	1 / 1	60,000 / 55,000	187

SUBUNIT CONSTITUTION OF PROTEINS (continued)

Protein	Source	Organ	Molecular weight	Subunit No.	Subunit Molecular weight	References
Monoamine oxidase	Pig	Liver	115,000	2	60,000	188
Sulfite oxidase (EC 1.8.3.1)	Bovine	Liver	115,000	2	55,000	188a
DNA Modification enzyme	Bacteriophage P1		115,000	1 1	70,000 45,000	189
Tyrosine aminotransferase (EC 2.6.1.5)	Rat	Liver	115,000	4	32,000	190
Tyrosine tRNA synthetase (EC 6.1.1.1)	*Saccharomyces*	Liver	116,000	4	31,500	191
Arginase (EC 3.5.3.1)	Human	Liver	118,000	4	30,000	192
Phosphoglucose isomerase	Yeast		119,400	4	30,000	193
α-Ketoglutaric semialdehyde dehydrogenase	*Pseudomonas*		120,000	2	60,000	194
Agglutinin	Soybean		120,000	4	30,000	195
Aspartokinase (EC 2.7.2.4)	*Bacillus*		120,000	2 2	43,000 17,000	196
L-Asparaginase (EC 3.5.1.1)	*Proteus*		120,000	4	30,000	197
Leucine tRNA synthetase (EC 6.1.1.4)	Yeast		120,000	2	60,000	198
Cyclic AMP-dependent protein kinase	Bovine	Sperm	120,000	1 1	78,000 35,000	199
Protein toxin B	*Pasteurella*		120,000 24,000	(5—6) 2	24,000 12,000	200
Pyrophosphatase (EC 3.6.1.1)	*E. coli*		120,000	6	20,000	201
Seryl tRNA synthetase (EC 6.1.1.11)	Yeast		120,000	2	60,000	201a
Anti-A1 lectin	*Dolichos*		120,000	4	30,000	201b
Aldolase (EC 4.1.2.13)	Spinach	Leaf	120,000	4	30,000	202
Oestradiol-receptor	Calf	Uterine	120,000	2	55,000	202a
DDT-Dehydrochlorinase (EC 4.5.1.1)	Housefly		120,000	4	30,000	203
Alkaline phosphatase (EC 3.1.3.1)	*Bacillus*		121,000	2	55,000	204
Tryptophan oxygenase (EC 1.13.1.12)	*Pseudomonas*		122,000	4	31,000	205
Fatty acylthiokinase I	*E. coli*		122,000	4	30,000	206
Cyclic AMP-dependent protein kinase	Rabbit	Muscle	123,000	1 1	82,000 49,000	207

SUBUNIT CONSTITUTION OF PROTEINS (continued)

Protein	Source	Organ	Molecular weight	Subunit No.	Subunit Molecular weight	References
Ceruloplasmin	Human	Serum	124,000	2	53,000	208
				2	16,000	
Methylmalonyl-CoA mutase (EC 5.4.99.2)	*Propionibacterium*		124,000	1	66,000	209
Glutathione reductase (EC 1.6.4.2)	*E. coli*		124,000	1	61,000	210
Deoxycytidylate deaminase (T-2 bacteriophage induced)	*E. coli*		124,000	2	56,000	211
				6	20,200	
Hybridase	Rat	Liver	125,000	1	85,000	212
				1	43,000	
Uricase (EC 1.7.3.3)	Pig	Liver	125,000	4	32,000	213
Xanthosine 5'-phosphate aminase (EC 6.3.4.1)	*E. coli*		126,000	2	63,000	214
Tyrosinase (EC 1.10.3.1)	Mushroom		128,000	4	32,000	215
Fructose diphosphatase (EC 3.1.3.11)	Swine	Kidney	130,000	4	34,000	216
Glucose 6-phosphate dehydrogenase (EC 1.1.1.49)	Rat	Mammary gland	130,000	2	63,000	217
D-Gluconate dehydratase (EC 4.2.1)	*Clostridium*		131,000	2	64,000	218
Ornithine aminotransferase (EC 2.6.1.13)	Rat	Liver	132,000	4	33,000	219
Methylmalonate semialdehyde dehydrogenase	*Pseudomonas*		132,000	2	59,000	219a
L-Asparaginase (EC 3.5.1.1)	*E. coli*		133,000	4	33,000	220
Aspartokinase (EC 2.7.2.4)	*Pseudomonas*		133,000	3	43,000	221
Phosphoglucose isomerase (EC 5.3.1.9)	Human, rabbit	Muscle	134,000	2	61,000	222
L-Amino acid oxidase (EC 1.4.3.2)	Rattlesnake	Venom	135,000	2	70,000	223
Myrokinase (EC 3.2.3.1)	Rape seed		135,000	2	65,000	224
L-Asparaginase (EC 3.5.1.1)	*Erwinia*		135,000	4	32,500	225
Aminotripeptidase (EC 3.4.1.3)	Swine	Kidney	137,200	2	71,100	226
Aspartate carbamoyltransferase (EC 2.1.3.2)	Yeast		138,000	(6)	21,000	227
Lysine tRNA synthetase (EC 6.1.1.6)	Yeast		138,000	2	72,000	228
Nucleoside diphosphokinase	Pig	Kidney	138,000	6	21,000	229
L-Asparaginase (EC 3.5.1.1)	*Achrombacteraceae*		138,000	4	35,000	230

SUBUNIT CONSTITUTION OF PROTEINS (continued)

Protein	Source	Organ	Molecular weight	Subunit No.	Subunit Molecular weight	References
Transketolase (EC 2.2.1.1)	Yeast		140,000	2	70,000	231
Protein phosphokinase (EC 2.7.1.37)	Bovine	Brain	140,000	1 / 1	80,000 / 60,000	232
Succinyl-CoA synthetase (EC 6.2.1.5)	E. coli		140,000	2 / 2	38,500 / 29,500	233
C-Reactive protein	Rabbit	Blood	140,000	6	23,000	234
Lactate dehydrogenase (EC 1.1.1.27)	Porcine	Heart	140,000	4	35,000	235
L-Rhamnulose 1-phosphate aldolase (EC 1.4.2.b)	E. coli		140,000	4	35,000	236
Ascorbate oxidase (EC 1.10.3.3)	Zucchini squash		140,000	2	65,000	237
Cyclic GMP protein kinase	Lobster	Muscle	140,000	1	100,000	238
Cyclic AMP-dependent protein kinase I	Beef	Brain	140,000	1 / 1	40,000 / 100,000	239
Exonuclease I	E. coli		140,000	1	40,000	239a
Phytohemagglutinin	Phaseolus		140,000	2 / 2	70,000 / 35,000	240
Fructose 1,6-diphosphatase (EC 3.1.3.11)	Rabbit	Liver, kidney, muscle	140,000	4	36,000	241
Esterase	Rat	Liver	140,000	2	70,000	242
3,4-Dihydroxyphenylacetate-2,3-dioxygenase (EC 1.13.1)	Pseudomonas		140,000	4	35,000	243
Aldolase (EC 4.1.2.13)	E. coli		140,000	4	35,000	244
L-Erythro-3,5-diaminohexanoate dehydrogenase	Clostridium		140,000 / 68,000	2 / 2	68,000 / 37,000	245
Alkaline phosphatase (EC 3.1.3.1)	Calf	Intestine	140,000	2	69,000	246
Alcohol dehydrogenase (EC 1.1.1.1)	Yeast		141,000	4	35,000	246a
Anthranilate synthetase	Serratia		141,000	2 / 2	60,000 / 21,000	247
Lectin	Navy bean		143,000	4	37,000	248
Tryptophan synthetase (EC 4.2.1.20)	Yeast		143,000	4	37,000	248a

SUBUNIT CONSTITUTION OF PROTEINS (continued)

Protein	Source	Organ	Molecular weight	Subunit No.	Subunit Molecular weight	References
Glyceraldehyde 3-phosphate dehydrogenase (EC 1.2.1.12)	Rabbit, lobster, pig	Muscle	144,000	2	72,000	249
			72,000	2	37,000	
Tartaric acid dehydrase (EC 4.2.1.c)	Pseudomonas		145,000	4	39,000	250
Phosphofructokinase (EC 2.7.1.11)	Clostridium		145,000	4	35,000	251
Platelet factor XIII	Human	Platelet	146,000	2	75,000	252
L-Asparaginase (EC 3.5.1.1)	Serratia		147,000	4	37,000	252a
Malate dehydrogenase (EC 1.1.1.37)	Bacillus		148,000	4	37,000	253
Tryptophan synthetase (EC 4.2.1.20)	E. coli		148,000	2	45,000	254
				2	28,700	
Aldolase C	Rabbit	Brain	148,000	4	37,000	255
5'-Nucleotidase (EC 3.1.3.5)	Mouse	Liver	150,000	2	75,000	256
Pyridoxamine pyruvate transaminase (EC 2.6.1.a)	Pseudomonas		150,000	4	38,000	257
Alcohol dehydrogenase (EC 1.1.1.1)	Yeast		150,000	4	37,000	258
Pyruvate dehydrogenase	Bovine	Heart, kidney	154,000	2	41,000	259
				2	36,000	
α-Acetylgalactosaminidase	Beef	Liver	155,000	4	42,000	260
Aspartic β-semialdehyde dehydrogenase (EC 1.2.1.11)	Yeast		156,000	4	41,000	261
Alkaline phosphate (EC 3.1.3.1)	Pig	Kidney	156,000	4	39,000	261a
D-Xylose isomerase (EC 5.3.1.5)	Streptomyces		157,000	4	40,000	262
Crotonase (EC 4.2.1.17)	Clostridium		158,000	4	43,000	263
Aldolase	Rabbit	Muscle	160,000	4	40,000	264
Lac repressor	E. coli		160,000	4	40,000	265
Cystathionine γ-synthetase (EC 4.2.1.21)	Salmonella		160,000	4	40,000	266
3-Deoxy-D-arabinoheptulosonate 7-Phosphate synthetase-chorismate mutase	Bacillus		160,000	4	38,500	266a
Cystathionase (EC 4.2.1.15)	Rat	Liver	160,000	8	20,000	267
Trimethylamine dehydrogenase	Bacterium 4B 6		160,000	2	80,000	268
Threonine deaminase (EC 4.2.1.16)	Clostridium		160,000	4	40,000	269
Phosphoglycerate dehydrogenase	E. coli		163,000	4	41,000	270

SUBUNIT CONSTITUTION OF PROTEINS (continued)

Protein	Source	Organ	Molecular weight	Subunit No.	Subunit Molecular weight	References
Succinic semialdehyde dehydrogenase (EC 1.2.1.b)	Pseudomonas		164,000	3	54,500	271
Crotonase (EC 4.2.1.17)	Beef	Liver	164,000	6	28,000	272
Quinolate phosphoribosyltransferase	Pseudomonas		165,000	3	54,000	273
Carboxyl esterase (EC 3.1.1.1)	Beef	Liver	167,000	2	85,000	274
Tryptophan oxygenase (EC 1.13.1.12)	Rat	Liver	167,000	2	43,000	275
				2	44,000	
Palmitoyl-CoA synthetase (EC 6.2.1.3)	Rat	Liver	168,000	6	27,000	276
Molybdoferedoxin	Clostridium		168,000	2	59,000	277
				1	50,700	
Aspartokinase II-homoserine dehydrogenase II	E. coli		169,000	4	43,000	278
Acetoacetyl CoA thiolase (EC 2.3.1.9)	Avian	Liver	169,000	4	41,000	279
Thiolase (EC 2.3.1.9)	Pig	Heart	169,000	4	41,000	279a
Methionine tRNA synthetase (EC 6.1.1.10)	E. coli		170,000	2	85,000	280
β-Lysine mutase	Clostridium		170,000	2	52,000	281
				2	32,000	
Carbamylphosphate synthetase (EC 2.7.2.5)	E. coli		170,000	1	130,000	282
				1	42,000	
Aspartase (EC 4.3.1.1)	E. coli		170,000	4	45,000	283
Cyclic AMP-dependent protein kinase	Bovine	Heart	174,000	2	49,000	284
				2	38,000	
Carboxylesterase (EC 3.1.1.1)	Pig	Liver	180,000	3	60,000	285
D-α-Ornithine 5,4-aminomutase	Clostridium		180,000	2	95,000	286
Thetin homocysteine methylpherase (EC 2.1.1.10)	Horse	Liver	180,000	(3–4)	50,000	287
Carbonic anhydrase	Parsley	Leaf	180,000	6	29,000	287a
L-Leucine:2-oxoglutarate aminotransferase (EC 2.6.1.6)	Salmonella		183,000	6	81,500	288
L-Threonine dehydratase	Clostridium		184,000	4	46,000	289
Aminoacyl transferase I	Rabbit	Reticulocytes	186,000	3	62,000	290

SUBUNIT CONSTITUTION OF PROTEINS (continued)

Protein	Source	Organ	Molecular weight	Subunit No.	Subunit Molecular weight	References
α-Dialkyl amino acid transaminase	*Pseudomonas*		188,000	4	47,000	291
Histidine decarboxylase (EC 4.1.1.22)	*Lactobacillus*		190,000	5	9,000	292
				5	29,700	
Prohistidine decarboxylase	*Lactobacillus*		190,000	5	37,000	293
Fumarase (EC 4.2.1.2)	Swine	Heart	194,000	4	48,500	294
Pyruvate kinase (EC 2.7.1.40)	*Saccharomyces*		195,000	4	49,000	294a
Threonine deaminase (EC 4.2.1.16)	*Salmonella*		195,000	4	48,500	295
Dipeptidyl transferase (EC 3.4.4.9)	Beef	Spleen	197,000	2	100,000	296
				4	24,500	
Phosphoenolpyruvate carboxylase (EC 4.1.1.31)	*Salmonella*		198,000	4	49,200	297
Glutamine phosphoribosylpyrophosphate amidotransferase (EC 2.4.2.14)	Pigeon	Liver	200,000	2	100,000	298
				2	50,000	
Polynucleotide phosphorylase	*E. coli*		200,000	2	95,000	299
α-Isopropylmalate synthase	*Salmonella*		200,000	4	50,000	300
Aspartylkinase (lysine sensitive) (EC 2.7.2.4)	*E. coli*		200,000	2	100,000	301
				2	48,000	
Succinic dehydrogenase (EC 1.3.99.1)	Beef	Heart	200,000	2	100,000	302
				1	70,000	
				1	30,000	
Neuraminidase (EC 3.2.1.18)	*Influenza*		200,000	4	50,000	303
Tyrosinase (EC 1.10.3.1)	Frog	Epidermis	200,000	4	50,000	304
Mo-Fe Protein	Soybean	Nodule	200,000	4	50,000	305
Peroxidase	Pig	Thyroid	200,000	3	70,000	305a
Cytochrome oxidase	Bovine	Heart	200,000	2	100,000	305b
3-Hydroxy-3-methylglutaryl CoA reductase (EC 1.1.1.34)	Rat	Liver	200,000	3	65,000	305c
Aliphatic amidase (EC 3.5.1.4)	*Pseudomonas*		200,000	6	33,000	306
β-Amylase (EC 3.2.1.2)	Sweet potato	Tuber	201,000	4	50,000	307
Argininosuccinase (EC 4.3.2.1)	Steer	Liver	202,000	2	100,000	308
				2	50,000	

SUBUNIT CONSTITUTION OF PROTEINS (continued)

Protein	Source	Organ	Molecular weight	Subunit No.	Subunit Molecular weight	References
NAD-Linked malic enzyme (EC 1.1.1.38)	E. coli		200,000	4	52,500	309
Threonine deaminase (EC 4.2.1.16)	E. coli		204,000	4	51,000	310
Glucose 6-phosphate dehydrogenase (EC 1.1.1.49)	Human	Erythrocyte	204,800 101,400	2 2	101,400 51,300	311
Qβ Replicase	E. coli		205,000	1 1 1 1	70,000 65,000 45,000 35,000	312
Isocitrate lyase	Pseudomonas		206,000	4	48,200	313
Glucose 6-phosphate dehydrogenase (EC 1.1.1.49)	Neurospora		206,000 104,000	2 2	104,000 57,000	314
Tryptophanase (EC 4.2.1.e)	Bacillus		208,000	4	50,500	315
Pyruvate decarboxylase (EC 4.1.1.1)	Yeast		209,000	2	108,000	316
Invertase (EC 3.2.1.26)	Neurospora		210,000	4	51,500	317
Phenylalanine hydroxylase	Rat	Liver	210,000	4	51,000	317a
Cytidine triphosphate synthetase (EC 6.3.4.2)	E. coli		210,000 105,000	2 2	105,000 50,000	318
High density protein	Porcine	Plasma	210,000	4	28,000	319
Phosphoribosyl ATP pyrophosphate phosphoribosyl transferase	Salmonella		215,000	6	36,000	320
Serine transhydroxymethylase (EC 2.1.2.1)	Rabbit	Liver	215,000	4	47,000	321
Acetyl-CoA carboxylase (EC 6.4.1.2)	Rat	Liver	215,000	1 1	118,000 125,000	322
Pyruvate kinase (EC 2.7.1.40)	Bovine	Liver	215,000	4	54,000	323
α-Glucan phosphorylase (EC 2.4.1.1)	Potato		215,000	2	108,000	324
2-Oxoglutarate dehydrogenase	Pig	Heart	216,000	2	105,000	325
Tryptophanase (EC 4.2.1.e)	Aeromonas		216,000	4	54,000	326
α-L-Fucosidase	Rat	Epididymes	216,000	2 2	47,000 59,300	327
Glycerol kinase (EC 2.7.1.30)	E. coli		217,000	4	55,000	328

SUBUNIT CONSTITUTION OF PROTEINS (continued)

Protein	Source	Organ	Molecular weight	Subunit No.	Subunit Molecular weight	References
Adenosine deaminase (EC 3.5.4.4)	*Aspergillus*		217,000	2	105,000	328a
Acetol acetate-forming enzyme	*Aerobacter*		220,000	4	58,000	329
3-Aminopropanal dehydrogenase	*Pseudomonas*		220,000	3	74,000	329a
Tryptophanase (EC 4.2.1.e)	*Bacillus, E. coli*		220,000	2	110,000	330
			110,000	2	55,000	
Paramyosin	*Venus*	Muscle	220,000	2	110,000	331
Chorismate mutase prephenate dehydratase	*Salmonella*		220,000	2	109,000	332
Sucrase-isomaltase complex (EC 3.2.1)	Rabbit	Small intestine	220,000	1	110,000	333
				1	120,000	
ATPase (Na and K dependent)	Canine	Renal medulla	(220,000)	1	(135,000)	334
				2	(35,000)	
α-Acetohydroxyacid isomeroreductase	*Salmonella*		220,000	4	55,000	335
L(+) Hydroxybutyryl-CoA dehydrogenase (EC 1.1.1.35)	*Clostridium*		220,000	8	26,000	336
Pyruvate kinase (EC 2.7.1.40)	Frog	Muscle	220,000	4	55,000	337
Arylamidase	Human	Liver	223,500	6	38,100	338
Ornithine transcarbamylase (EC 2.1.3.3)	*Streptococcus*, bovine	Liver	223,000	3	74,000	339
			74,000	2	38,000	
Ribonucleoside diphosphate reductase	Bacteriophage T4		225,000	2	85,000	340
				2	35,000	
Glycyl tRNA synthetase (EC 6.1.1.e)			227,000	2	33,000	341
				2	80,000	
4-Aminobutanal dehydrogenase (EC 1.2.1.e)	*Pseudomonas*		228,000	3	75,000	342
Cysteine desulfhydrase	*Salmonella*		229,000	6	37,000	343
Pyruvate kinase (EC 2.7.1.40)	Bovine	Muscle	230,000	4	57,000	344
Glyoxylate carboligase (EC 4.1.1.b)	*Pseudomonas*		230,000	2	115,000	345
			115,000	2	61,000	
Collagen	Chicken	Leg tendon	231,000	12	18,500	345a
C-Phyocyanin	*Synechococcus*		232,000	6	19,000	346
				6	17,700	

SUBUNIT CONSTITUTION OF PROTEINS (continued)

Protein	Source	Organ	Molecular weight	Subunit No.	Subunit Molecular weight	References
Catalase (EC 1.11.1.6)	Bovine	Liver	232,000	4	57,500	347
Cytochrome b$_2$	Yeast		235,000	4	57,000	348
Pyruvate kinase (EC 2.7.1.40)	Rabbit	Muscle	237,000	4	57,200	349
Formyltetrahydrofolate synthetase (EC 6.3.4.3)	Clostridium		240,000	4	60,000	350
δ-Aminolevulinate dehydratase (EC 4.2.1.24)	Rhodopseudomonas		240,000	2	120,000	351
			120,000	3	40,000	
Anthranilate synthetase complex	Neurospora		240,000	6	40,000	352
Protein toxin A	Pasteurella		240,000	10–12	24,000	353
			24,000	2	12,000	
Protocollagen proline hydroxylase	Chick	Embryo	240,000	2	60,000	354
				2	65,000	
Glucocerebrosidase	Human	Placenta	240,000	4	60,000	355
Pyruvate kinase (EC 2.7.1.40)	E. coli		240,000	4	60,000	356
2,5-Dihydroxypyridine oxygenase	Pseudomonas		242,000	6	39,500	357
Citrate synthase (EC 4.1.3.7)	Acinetobacter		242,000	(4)	59,000	357a
Aldehyde dehydrogenase (EC 1.2.1.3)	Horse	Liver	245,000	4	57,000	358
Ribonucleotide diphosphate reductase	E. coli		245,000	1	78,000	359
				1	80,000	
				1	80,000	
Melilotate hydroxylase	Bacterial		250,000	4	64,000	360
Glycogen synthase (EC 2.1.1.11)	Rabbit	Muscle	(250,000)	(3)	90,000	361
DNA Restriction endonuclease	E. coli		250,000	1	135,000	362
				1	60,000	
				1	55,000	
Cystathionine synthetase (EC 4.2.1.21)	Rat	Liver	250,000	(2)	51,000	363
				(2)	73,000	
Uridine diphosphate galactose 4-epimerase (EC 5.1.3.2)	Yeast		250,000	2	125,000	364
			125,000	2	60,000	
Malic enzyme (EC 1.1.1.39)	Ascaris	Muscle	250,000	4	64,000	365
Citrate synthase (EC 4.1.3.7)	Azotobacter		250,000	(4)	59,000	365a

SUBUNIT CONSTITUTION OF PROTEINS (continued)

Protein	Source	Organ	Molecular weight	Subunit No.	Subunit Molecular weight	References
Phytochrome	Rye, oat	Shoots	252,000	6	42,000	366
Leucine aminopeptidase (EC 3.4.1.1)	Swine	Kidney	255,000	4	63,500	367
Malic enzyme (EC 1.1.1.40)	Pigeon	Liver	260,000	4	65,000	368
Butyrylcholinesterase (EC 3.1.1.8)	Horse	Serum	260,000	4–6	42,200–75,000	369
L-Phenylalanine tRNA synthetase (EC 6.1.1.b)	E. coli		267,000	2	94,000	370
				2	39,000	
Glutamine phosphoribuosylpyrophosphate amidotransferase (EC 2.4.2.14)	Human	Placenta	270,000	2	133,000	371
Glycollate oxidase (EC 1.1.3.1)	Spinach	Leaves	270,000	2	140,000	372
Phenylalanine tRNA synthetase (EC 6.1.1.b)	Yeast		276,000	2	75,000	373
				2	63,000	
AMP Deaminase (EC 3.5.4.6)	Rabbit, chicken	Muscle	278,000	4	69,000	374
Mandelate racemase (EC 5.1.2.2)	Pseudomonas		278,000	4	69,500	375
Acetylcholinesterase (EC 3.1.1.7)	Electrophorus	Tissue	280,000	4	70,000	376
Protein kinase	Bovine	Heart	280,000	(?)	42,000	377
				(?)	55,000	
β-Glucuronidase (EC 3.2.1.31)	Rat	Liver	280,000	4	75,000	378
Anthranilate synthetase complex	Salmonella		280,000	2	62,000	379
				2	62,000	
δ-Aminolevulinic acid dehydratase (EC 4.2.1.24)	Bovine	Liver	282,000	2	140,000	380
Glucose-6-phosphate dehydrogenase (EC 1.1.1.49)	Bovine	Adrenal gland	284,000	4	64,600	381
Lysine 2,3-amino mutase	Clostridium		285,000	6	48,000	382
Glutamate dehydrogenase (EC 1.4.1.3)	Neurospora		288,400	6	48,800	383
Dopamine-β-hydroxylase (EC 1.14.2.1)	Bovine	Adrenal gland	290,000	4	75,000	384
Nitrogenase	Klebsiella		295,000	1	229,000	385
				1	66,700	
			229,000	2	51,300	
				2	59,600	
			66,700	2	34,000	

SUBUNIT CONSTITUTION OF PROTEINS (continued)

Protein	Source	Organ	Molecular weight	Subunit No.	Subunit Molecular weight	References
Arginine decarboxylase (EC 4.1.1.19)	E. coli		296,000	4	75,000	386
Uridine diphosphate glucose dehydrogenase (EC 1.1.1.22)	Bovine	Liver	300,000	6	52,000	387
Glycogen synthetase (EC 2.4.1.11)	Yeast		300,000	4	77,000	388
Edestin	Hemp	Seed	300,000	6	50,000	389
Excelsin	Brazil nut		300,000	6	50,000	390
Isocitrate dehydrogenase (EC 1.1.1.41)	Yeast		300,000	8	39,000	391
Monoamine oxidase	Rat	Liver	300,000	4	75,000	391a
Cysteine synthetase (EC 4.2.1.22)	Salmonella		309,000	1	160,000	392
				2	68,000	
Aspartyl transcarbamylase (EC 2.1.3.2)	E. coli		310,000	2	100,000	393
			100,000	3	34,000	
			34,000	3	33,000	
				2	17,000	
Glutamate decarboxylase (EC 4.1.1.15)	E. coli		310,000	6	50,000	394
Cholinesterase (EC 3.1.1.8)	Horse	Serum	315,000	4	77,300	395
Carbamoylphosphate synthase (EC 2.7.2.5)	Rat	Liver	316,000	2	160,000	396
Glutamate dehydrogenase (EC 1.4.1.3)	Beef	Liver	320,000	6	57,000	397
Plasma factor VII	Human	Plasma	320,000	2	75,000	398
				2	88,000	
L-Phenylalanine ammonia lyase (EC 4.1.3.5)	Maize, potato, wheat		320,000	4	83,000	399
Chloroplast-coupling factor	Spinach	Chloroplasts	325,000	6	62,000	400
Leucine aminopeptidase (EC 3.4.1.1)	Bovine	Lens	327,000	6	54,000	401
Nitrogenase	Clostridium		330,000	1	220,000	402
			220,000	2	55,000	
			55,000	2	50,700	
				2	59,500	
				2	27,500	
Glutamate dehydrogenase (EC 1.4.1.3)	Neurospora		330,000	6	51,500	403
Glutamine synthetase (EC 6.3.1.2)	Hamster	Liver	335,000	8	42,000	404

SUBUNIT CONSTITUTION OF PROTEINS (continued)

Protein	Source	Organ	Molecular weight	Subunit No.	Subunit Molecular weight	References
Acetoacetate decarboxylase (EC 4.1.1.4)	Clostridium		340,000	6	62,000	405
			62,000	2	29,000	
Aspartokinase l-homoserine dehydrogenase I			340,000	4	85,000	406
Phosphoenolpyruvate carboxylase (EC 4.1.1.31)	Maize		340,000	2	160,000	407
Peptidase	Sheep	Erythrocyte	340,000	6	60,000	408
Glutamine synthetase (EC 6.3.1.2)	Chicken	Neural retina	340,000	8	42,000	409
Aminopeptidase	Clostridium		340,000	6	60,000	410
Arachin	Arachis		345,000	2	180,000	411
			180,000	6	30,000	
N-Methylglutamate synthetase	Pseudomonas		350,000	12	(30,000)	412
ATPase (EC 3.6.1.3)	Micrococcus		350,000	3	52,500	413
				3	47,000	
				1	41,000	
Glutamine synthetase (EC 6.3.1.2)	Rat	Liver	352,000	8	44,000	414
Enolase (EC 4.2.1.11)	Thermus		355,000	8	44,000	415
Phycocyanin	Anacystis, Nostic		360,000	12—24	15,000—30,000	416
Phosphofructokinase (EC 2.7.1.11)	Rabbit	Muscle	360,000	4	80,000	417
L-Arabinose isomerase (EC 5.3.1.4)	E. coli		360,000	6	60,000	418
Glutamine synthetase (EC 6.3.1.2)	Neurospora		360,000	4	90,000	419
Aspartate transcarbamylase (EC 2.1.3.2)	Pseudomonas		360,000	2	180,000	420
ATPase (EC 3.6.1.3)	Rat	Liver	360,000	6	53,000	421
				1	28,000	
				1	12,500	
				1	9,000	
Phosphoenolpyruvate carboxylase (EC 4.1.1.31)	E. coli W		361,000	4	88,200	422
D-Ribulose-1,5-biphosphate carboxylase (EC 4.1.1.39)	Chlorobium		361,000	6	53,000	423
Glycogen synthetase (EC 2.4.1.11)	Swine	Kidney	370,000	4	92,000	424

SUBUNIT CONSTITUTION OF PROTEINS (continued)

Protein	Source	Organ	Molecular weight	Subunit No.	Subunit Molecular weight	References
Phosphorylase A (EC 2.4.1.1)	Rabbit	Muscle	370,000	4	92,500	425
High density lipoprotein	Bovine	Serum	376,000	4	28,000	426
Polysaccharide depolymerase	Aerobacter		379,000	4	63,200	426a
				(3—4)	36,400	
ATPase (EC 3.6.1.3)	Streptococcus		385,000	12	33,000	427
Pyruvate phosphate dikinase	Maize		387,000	2	195,000	428
			195,000	2	94,000	
Phosphoenolpyruvate carboxylase (EC 4.1.1.31)	Salmonella		400,000	4	100,000	429
Aminopeptidase IA (EC 3.4.11.1)	Bacillus		100,000	2	50,000	430
Aminopeptidase IB (EC 3.4.11.1)	Bacillus		400,000	10	36,000	431
				2	36,000	
				8	36,000	
				4	36,000	
Aminopeptidase IC (EC 3.4.11.1)	Bacillus		400,000	6	36,000	432
				6	36,000	
RNA Polymerase (EC 2.7.7.6)	E. coli		400,000	2	39,000	433
				1	155,000	
				1	165,000	
RNA Polymerase (EC 2.7.7.6)	Rat	Liver	400,000	1	190,000	434
				1	150,000	
				1	35,000	
				1	25,000	
Cholesterol esterase (EC 3.1.1.13)	Rat	Pancreas	400,000	6	65,000	435
Lipovitellin	Chicken	Egg yolk	400,000	2	200,000	436
Glutamine synthetase	Ovine	Brain	400,000	8	49,000	437
Phosphoenolpyruvate carboxylase (EC 4.1.1.31)	E. coli B		402,000	4	99,600	438
Sucrose synthetase (EC 2.1.1.14)	Phaeolus		405,000	4	94,000	439
Phosphoenolpyruvate carboxytrans-phosphorylase (EC 4.1.1.38)	Entamoeba		408,000	2	200,000	440
			200,000	2	100,000	
Ribonucleotide reductase	Euglena		440,000	4	100,000	440a

SUBUNIT CONSTITUTION OF PROTEINS (continued)

Protein	Source	Organ	Molecular weight	Subunit No.	Subunit Molecular weight	References
Adenosine triphosphate sulfurylase (EC 2.7.7.4)	*Penicillium*		440,000	8	56,000	441
Apoferritin	Horse	Spleen	443,000	24	18,500	442
RNA Polymerase	*Physarum*		460,000	1 1 1 2 1	200,000 135,000 45,000 24,000 17,000	443
RNA Polymerase	*Bacillus*		466,000	1 1 1 2	160,000 155,000 63,000 44,000	444
Myosin	Rabbit	Muscle	468,000	2 1 2 1	212,000 21,000 19,000 17,000	445
Urease (EC 3.5.1.5)	Jack bean		480,000	2	240,000	446
Uridine diphosphate glucose pyrophosphorylase	Calf	Liver	240,000	3	83,000	447
Isocitrate lyase (EC 4.1.3.1)	*Turbatrix*	Liver	480,000	8	60,000	448
Fatty acid synthetase	Pigeon	Liver	480,000	4	123,000	449
Fatty acid synthetase	Chicken	Liver	500,000	2	240,000	450
Ribulose 1,5-diphosphate carboxylase (EC 4.1.1.39)	*Hydrogenomonas*		515,000	12–14	40,700	451
Pyruvate carboxylase	Pig	Liver	520,000	4	130,000	452
β-Galactosidase (EC 3.2.1.23)	*E. coli*		540,000	4	135,000	453
Ribulose diphosphate carboxylase (EC 4.1.1.39)	Spinach	Leaves	550,000	8 6	52,000 24,500	454
L-Malic enzyme (EC 1.1.1.40)	*E. coli*		550,000	2 8	12,000 67,000	455
Phosphoenolpyruvate carboxylase (EC 4.1.1.31)	Spinach	Leaf	560,000	4	130,000	455a

SUBUNIT CONSTITUTION OF PROTEINS (continued)

Protein	Source	Organ	Molecular weight	Subunit No.	Subunit Molecular weight	References
Citrate lyase (EC 4.1.3.6)	Aerobacter		575,000	2	290,000	456
			290,000	2	137,000	
			137,000	2	74,000	
Ribulose 1,5-diphosphate carboxylase (EC 4.1.1.39)	Chlorella		588,000	8	58,200	457
				8	15,300	
Phosphofructokinase (EC 2.7.1.11)	Yeast		590,000	6	100,000	458
Glutamine synthetase (EC 6.3.1.2)	E. coli		592,000	12	48,500	459
Glyceraldehyde 3-phosphate dehydrogenase	Spinach	Chloroplasts	600,000	4	145,000	460
Acetolactate synthase (EC 4.1.3.18)	Pseudomonas		600,000	(8)	60,000	461
				(8)	60,000	
Glutamine synthetase (EC 6.3.1.2)	Bacillus		600,000	12	50,000	462
Ovomacroglobulin	Chicken	Egg white	650,000	2	325,000	463
Globulin	French bean		654,000	4	163,000	464
			163,000	1	43,000	
				1	47,000	
				1	53,000	
Pyruvate carboxylase (EC 6.4.1.1)	Chicken	Liver	660,000	4	165,000	465
			165,000	4	45,000	
Thyroglobulin	Bovine	Thyroid	669,000	2	335,000	466
Isocitrate dehydrogenase (EC 1.1.1.41)	Bovine	Heart	670,000	2	330,000	467
			330,000	8	41,000	
L-Aspartate β-decarboxylase (EC 4.1.1.12)	Alcaligines		675,000	6	112,000	468
			112,000	2	57,000	
Sulfite reductase (EC 1.8.1.2)	E. coli, Salmonella		700,000	4	53,000	469
				8	58,000	
Nitrate reductase	E. coli		773,000	4	142,000	469a
				4	58,000	
Propionylcarboxylase (EC 6.4.1.3)	Pig	Heart	700,000	4	175,000	470
Lysine decarboxylase (EC 4.1.1.18)	E. coli		780,000	10	80,000	471
RNA Polymerase (EC 2.7.7.6)	Azotobacter		782,000	2	391,000	472

SUBUNIT CONSTITUTION OF PROTEINS (continued)

Protein	Source	Organ	Molecular weight	Subunit		References
				No.	Molecular weight	
Transcarboxylase (EC 2.1.3.1)	*Propionibacterium*		792,000	1	360,000	473
				3	144,000	
			360,000	3	120,000	
			120,000	2	60,000	
			144,000	2	60,000	
				2	12,000	
Phosphofructokinase (EC 2.7.1.11)	Chicken	Liver	800,000	2	400,000	474
			400,000	2	210,000	
			210,000	2	100,000	
			100,000	2	60,000	
L-Aspartate β-decarboxylase (EC 4.1.1.12)			800,000	2	400,000	475
			400,000	4	100,000	
α-Crystallin			810,000	(30)	26,000	476
Arginine decarboxylase (EC 4.1.1.19)	*E. coli*		820,000	5	160,000	477
			160,000	2	82,000	
Cytochrome P-450	Bovine	Adrenocortical mitochondria	850,000	2	470,000	478
			470,000	8	53,000	
RNA Polymerase (EC 2.7.7.6)	*E. coli* B		880,000	2	440,000	479
Hemocyanin	*Loglio, Cancer,* ghost shrimp		300,000–9,000,000		760,000	480
					430,000	
					380,000	
					74,000	
					70,000	
Lipoate succinyltransferase	Pig	Heart	1,000,000	(24)	41,000	481
Monoamine oxidase	Pig	Liver	1,200,000	(8)	146,000	482
Phosphorylase kinase (EC 2.7.1.38)	Rabbit	Muscle	1,330,000	4	118,000	483
				4	108,000	
				8	41,000	
Fatty acid synthetase	*Mycobacterium*		1,390,000	(?)	250,000	484
Dihydrolipoyl transacetylase (EC 2.3.1.12)	*E. coli*		1,700,000	24	65,000	485
Chlorocruorin	*Spirographis*	Blood	2,750,000	12	250,000	486

SUBUNIT CONSTITUTION OF PROTEINS (continued)

Protein	Source	Organ	Molecular weight	Subunit No.	Subunit Molecular weight	Reference
α-Ketoglutarate dehydrogenase complex	E. coli		2,784,000	12	94,000	487
				12	54,000	
				24	42,000	
Hemoglobin	Arenicola	Blood	2,850,000	12	230,000	488
			230,000	4	54,000	
Erythrocruorin	Cirraformia	Plasma	3,000,000	162	18,500	489
Dihydrolipoyl transacetylase	Bovine	Kidney, heart	3,120,000	60	52,000	490
Phage fll			3,620,000	180	13,750	491
Pyruvate dehydrogenase complex	E. coli		4,600,000	24	96,000	492
				24	65,000	
				12	56,000	
Cowpea chlorotic mottle virus			4,608,000	180	19,600	492
Bromgrass mosaic virus			4,600,000	180	20,000	494
Broad bean mottle virus			4,800,000	180	20,900	495
Turnip yellow mosaic virus			5,000,000	180	20,133	496
Poliomyelitis virus			5,500,000	130	27,000	497
Cucumber mosaic virus			5,500,000	180	25,000	498
Alfalfa mosaic virus			7,400,000	(140)	51,600	499
			51,600	2	24,500	
Acetyl CoA carboxylase	Chicken	Liver	4–10,000,000	(?)	470,000	500
			470,000	2	117,000	
				1	129,000	
				1	139,000	
Rhinovirus 1A			8,400,000	60	96,000	501
Bushy stunt virus			9,000,000	180	42,000	502
Polyoma virus			24,000,000	420	50,200	503
Potato virus X			35,000,000	650	52,000	504
Tobacco mosaic virus			40,000,000	2130	17,500	505

Compiled by Dennis W. Darnall and Irving M. Klotz.

REFERENCES

1. Darnall and Klotz, *Arch. Biochem. Biophys.,* 149, 1 (1972).
2. Klotz, Langerman, and Darnall, *Annu. Rev. Biochem.,* 39, 25 (1970).
3. Klotz and Darnall, *Science,* 166, 126 (1969).
4. Klotz, *Science,* 155, 697 (1967).
4a. Kleine, *Fortschr. Arzneimittelforsch.,* 16, 365 (1972).
5. Crowfoot, *Proc. R. Soc. London, Ser. A,* 164, 580 (1938); Moody, Dissertation, University of Wisconsin, 1944; Waugh, *Adv. Protein Chem.,* 9, 325 (1954).
6. Stewart, *Biochem. Biophys. Res. Commun.,* 46, 1405 (1972); Dannies and Levine, *Biochem. Biophys. Res. Commun.,* 37, 587 (1969).
7. Vachek and Wood, *Biochim. Biophys. Acta,* 258, 133 (1972).
8. Weintraub, Vincent, Baulieu, and Alfsen, *FEBS Lett.,* 37, 82 (1973).
9. Angeletti and Bradshaw, *Proc. Natl. Acad. Sci. U.S.A.,* 68, 2417 (1971).
10. Liu, Nahm, Sweeney, Holcomb, and Ward, *J. Biol. Chem.,* 247, 4365 (1972); Liu, Nahm, Sweeney, Lamkin, Baker, and Ward, *J. Biol. Chem.,* 247, 4351 (1972).
11. Cusanovich, Tedro, and Kamen, *Arch. Biochem. Biophys.,* 141, 557 (1970).
12. Bishop and Ryan, *Biochemistry,* 12, 3077 (1973).
13. D'Alessio, Parente, Guida, and Leone, *FEBS Lett.,* 27, 285 (1972).
14. Sairam, Papkoff, and Li, *Arch. Biochem. Biophys.,* 153, 554 (1972); Sairam, Samy, Papkoff, and Li, *Arch. Biochem. Biophys.,* 153, 572 (1972).
14a. Wells, *Biochemistry,* 10, 4074 (1971).
15. Ward, Reichart, Fitak, Nahm, Sweeney, and Neill, *Biochemistry,* 10, 1796 (1971).
16. Misra and Fridovich, *J. Biol. Chem.,* 247, 3410 (1972).
17. Hartz and Deutsch, *J. Biol. Chem.,* 247, 7043 (1972).
17a. Hennen, Maghuin-Rogister, and Mamoir, *FEBS Lett.,* 9, 20 (1970).
17b. Grimek and McShan, *J. Biol. Chem.,* 249, 5725 (1974).
18. Schrecker and Hengstenberg, *FEBS Lett.,* 13, 209 (1971).
19. Landefeld and McShan, *J. Biol. Chem.,* 249, 3527 (1974).
20. Thomas, Arnold, and Kelley, *J. Biol. Chem.,* 248, 2529 (1973).
21. Ohnoki, Mitomi, Hata, and Satake, *J. Biochem.,* 73, 717 (1973).
22. Ryan, Jiang, and Hanlon, *Biochemistry,* 10, 1321 (1971); Saxena and Rathnam, *J. Biol. Chem.,* 246, 3549 (1971).
23. Bull, *J. Am. Chem. Soc.,* 68, 745 (1946); Townend and Timasheff, *J. Am. Chem. Soc.,* 79, 3613 (1957).
24. Hays, Simoni, and Roseman, *J. Biol. Chem.,* 248, 941 (1973).
25. Nagata and Burger, *J. Biol. Chem.,* 249, 3116 (1974).
26. Louis and Fitt, *Biochem. J.,* 127, 69 (1972).
27. Helmkamp and Bloch, *J. Biol. Chem.,* 244, 6014 (1969).
28. Volini, DeToma, and Westley, *J. Biol. Chem.,* 242, 5220 (1967).
29. Bellisario, Carlsen, and Bahl, *J. Biol. Chem.,* 248, 6796 (1973); Carlsen, Bahl, and Swaminathan, *J. Biol. Chem.,* 248, 6810 (1973).
30. Melville and Ryan, *Arch. Biochem. Biophys.,* 138, 700 (1970).
31. Keele, McCord, and Fridovich, *J. Biol. Chem.,* 245, 6176 (1970).
31a. Liberatore, Truby, and Klippenstein, *Arch. Biochem. Biophys.,* 160, 223 (1974).
32. Kiyohara, Iwasaki, and Yoshikawa, *J. Biochem.,* 73, 89 (1973).
32a. Krell and Eisenberg, *J. Biol. Chem.,* 245, 6558 (1970).
33. Craine, Hall, and Kaufman, *J. Biol. Chem.,* 247, 6082 (1972).
34. Rodgers and Chargaff, *J. Biol. Chem.,* 247, 5448 (1972).
35. Fall and Vagelos, *J. Biol. Chem.,* 247, 8005 (1972).
36. Riggs, Reiness, and Zubay, *Proc. Natl. Acad. Sci. U.S.A.,* 68, 1222 (1971).
37. Shannon and Hill, *Biochemistry,* 10, 3021 (1971).
38. Edelhoch, Condliffe, Lippoldt, and Burger, *J. Biol. Chem.,* 241, 5205 (1966).
39. Radcliffe and Barton, *J. Biol. Chem.,* 247, 7735 (1972).
39a. Howard, Sage, Stein, Young, Leon, and Dykes, *J. Biol. Chem.,* 246, 1590 (1971).
40. MacColl, Habig, and Berns, *J. Biol. Chem.,* 248, 7080 (1973).
41. Huang, Max, and Kaufman, *J. Biol. Chem.,* 248, 4235 (1973).
42. Blume and Buetler, *J. Biol. Chem.,* 246, 6507 (1971).
43. Hartman, *Biochemistry,* 10, 146 (1971).
44. Munkres, *Biochemistry,* 4, 2180, 2186 (1965).
45. Chung and Folk, *Proc. Natl. Acad. Sci. U.S.A.,* 69, 303 (1972).
46. Nakos and Mortenson, *Biochemistry,* 10, 455 (1971).

47. Khorlin, Vikha, and Milishnikov, *FEBS Lett.*, 31, 107 (1973).
48. O'Brien, Brewer, and Ljungdahl, *J. Biol. Chem.*, 248, 403 (1973).
49. Hersh, *Biochemistry*, 10, 2881 (1971).
49a. Jacobson and Pfuderer, *J. Biol. Chem.*, 245, 3938 (1970).
50. Magar and Robbins, *Biochim. Biophys. Acta*, 191, 173 (1969).
51. Bessman, Diamond, Debeer, and Duncan, *Fed. Proc., Fed. Am. Soc. Exp. Biol.*, 30, 1121 (1971); Duncan, Diamond, and Bessman, *J. Biol. Chem.*, 247, 8136 (1972).
52. Nakos and Mortenson, *Biochemistry*, 10, 2442 (1971).
53. Trayer and Hill, *J. Biol. Chem.*, 246, 6666 (1971); Magee, Mawal, and Ebner, *Fed. Proc., Fed. Am. Soc. Exp. Biol.*, 31, 499 (1972).
54. Allen and Majerus, *J. Biol. Chem.*, 247, 7709 (1972).
55. Hains, Wickremasinghe, and Johnston, *Eur. J. Biochem.*, 31, 119 (1972).
56. Sheys and Doughty, *Biochim. Biophys. Acta*, 235. 414 (1971).
57. Rask, Peterson, and Nilson, *J. Biol. Chem.*, 246, 6087 (1971); Gonzalez and Offord, *Biochem. J.*, 125, 309 (1971).
58. Duckworth and Sanwal, *Biochemistry*, 11, 3182 (1972).
59. Weser, Bunnenberg, Cammack, Djerassi, Flohé, Thomas, and Voelter, *Biochim. Biophys. Acta*, 243, 203 (1971).
60. Inoue, Kasper, and Pitol, *J. Biol. Chem.*, 246, 2626 (1971).
60a. Blachnitzky, Wengenmayer, and Kurz, *Eur. J. Biochem.*, 47, 235 (1974).
61. Galivan, Maley, and Maley, *Biochemistry*, 13, 2282 (1974).
62. Braunitzer, Hilse, Rudloff, and Hilschmann, *Adv. Protein Chem.*, 19, 1 (1964).
63. Hachmann, Miller, and Weissbach, *Arch. Biochem. Biophys.*, 147, 457 (1971).
64. Noyes and Bradshaw, *J. Biol. Chem.*, 248, 3060 (1973); Noyes, Glatthaar, Garavelli, and Bradshaw, *Proc. Natl. Acad. Sci. U.S.A.*, 71, 1334 (1974).
65. Heinrikson, Sterner, Noyes, Cooperman, and Bruckmann, *J. Biol. Chem.*, 248, 2521 (1973); Avaeva, Libedeva, Biesembaeva, and Egorov, *FEBS Lett.*, 24, 169 (1972); Ridling, Yang, and Butler, *Arch. Biochem. Biophys.*, 153, 714 (1972).
66. Brown, Brown, and Zabin, *J. Biol. Chem.*, 242, 4254 (1967).
67. Scopes and Penny, *Biochim. Biophys. Acta*, 236, 409 (1971).
67a. Thelander, *Eur. J. Biochem.*, 4, 407 (1968).
68. Mann and Vestling, *Biochemistry*, 8, 1105 (1969).
69. Noyes, Glatthaar, Garavelli, and Bradshaw, *Proc. Natl. Acad. Sci. U.S.A.*, 71, 1334 (1974).
70. Hengartner and Zuber, *FEBS Lett.*, 37, 212 (1973).
71. Becker, Kredich, and Tomkins, *J. Biol. Chem.*, 244, 2418 (1969).
72. Holtzer, Clark, and Lowey, *Biochemistry*, 4, 2401 (1965); Olander, Emerson, and Holtzer, *J. Am. Chem. Soc.*, 89, 3058 (1967); Woods, *J. Biol. Chem.*, 242, 2859 (1967).
73. Tsolas and Horecker, *Arch. Biochem. Biophys.*, 136, 303 (1970).
74. Criss, Sapico, and Litwack, *J. Biol. Chem.*, 245, 6346 (1970).
75. White, *Arch. Biochem. Biophys.*, 147, 123 (1971).
76. Burns, Engel, and Bethune, *Biochemistry*, 11, 2699 (1972); Jarabak and Street, *Biochemistry*, 10, 3831 (1971); Burns, Engel, and Bethune, *Biochem. Biophys. Res. Commun.*, 44, 786 (1971).
77. Green, *Biochem. J.*, 92, 16c (1964); Green and Ross, *Biochem. J.*, 110, 59 (1968).
78. Ohnoki, Mitomi, Hata, and Satake, *J. Biochem.*, 73, 717 (1973).
78a. Rosenblum, Antkowiak, Sallach, Flanders, and Fahien, *Arch. Biochem. Biophys.*, 144, 375 (1971).
79. Long and Kaplan, *Arch. Biochem. Biophys.*, 154, 696 (1973).
80. Stribling and Perham, *Biochem. J.*, 131, 833 (1973).
81. Allen, *Eur. J. Biochem.*, 35, 338 (1973).
82. Dunlap, Harding, and Huennekens, *Biochemistry*, 10, 88 (1971); Loeble and Dunlap, *Biochem. Biophys. Res. Commun.*, 49, 1671 (1972).
83. Utting and Kohn, *Fed. Proc., Fed. Am. Soc. Exp. Biol.*, 30, 1057 (1971).
84. Edlund, *FEBS Lett.*, 13, 56 (1971).
85. Green and Dobrjansky, *Biochemistry*, 10, 4533 (1971).
85a. Terzaghi and Terzaghi, *J. Biol. Chem.*, 249, 5119 (1974).
86. Hammerstedt, Mohler, Decker, and Wood, *J. Biol. Chem.*, 246, 2069, 2075 (1971); Vandlen, Ersfeld, Tulinsky, and Wood, *J. Biol. Chem.*, 248, 2251 (1973).
87. Wolfenstein, England, and Listowsky, *J. Biol. Chem.*, 244, 6415 (1969).
88. Somack and Costilow, *J. Biol. Chem.*, 248, 385 (1973).
89. Bowman, Tabor, and Tabor, *J. Biol. Chem.*, 248, 2480 (1973).
90. Henderson and Snell, *J. Biol. Chem.*, 248, 1906 (1973).
91. Joseph and Muench, *J. Biol. Chem.*, 246, 7610 (1971).
92. Traniello, Calcagno, and Pontremoli, *Arch. Biochem. Biophys.*, 146, 603 (1971).
93. Whitefield and Mayhew, *J. Biol. Chem.*, 249, 2801 (1974).
93a. Nakamura, Hosoda, and Hayashi, *Biochim. Biophys. Acta*, 358, 251 (1974).

94. Koch, Shaw, and Gibson, *Biochim. Biophys. Acta,* 212, 387 (1970); *Biochim. Biophys. Acta,* 229, 805 (1971).
95. Burgos and Martin, *Biochim. Biophys. Acta,* 268, 261 (1972).
96. Olsen and Milman, *J. Biol. Chem.,* 249, 4030, 4038 (1974); Arnold and Kelley, *J. Biol. Chem.,* 246, 7398 (1971).
97. DeLorenzo, Di Natale, and Schecter, *J. Biol. Chem.,* 249, 908 (1974).
98. Garbers, First, and Lardy, *J. Biol. Chem.,* 248, 875 (1973).
99. Pfeiderer and Auricchio, *Biochem. Biophys. Res. Commun.,* 16, 53 (1964); Deal and Holleman, *Fed. Proc., Fed. Am. Soc. Exp. Biol.,* 23, 264 (1964).
100. Jackson and Lovenberg, *J. Biol. Chem.,* 246, 4280 (1971).
101. Wilson and Hogness, *J. Biol. Chem.,* 244, 2132 (1969).
102. Meighen, Nicoli, and Hastings, *Biochemistry,* 10, 4062 (1971); Gunsalus-Miguel, Meighen, Nicoli, Nealson, and Hastings, *J. Biol. Chem.,* 247, 398 (1972).
103. Whiteley and Pelroy, *J. Biol. Chem.,* 247, 1911 (1972).
104. Dawson, Eppenberger, and Kaplan, *J. Biol. Chem.,* 242, 211 (1967); Bayley and Thomson, *Biochem. J.,* 104, 33c (1967).
105. Theorell and Winer, *Arch. Biochem. Biophys.,* 83, 291 (1959); Li and Vallee, *Biochemistry,* 3, 869 (1964); Drum, Harrison, Li, Bethune, and Vallee, *Proc. Natl. Acad. Sci. U.S.A.,* 57, 1434 (1967); Castellino and Barker, *Biochemistry,* 7, 2207 (1968); Weber and Osborn, *J. Biol. Chem.,* 244, 4406 (1969); Green and McKay, *J. Biol. Chem.,* 244, 5034 (1969).
106. Harris, Kobes, Teller, and Rutter, *Biochemistry,* 8, 2442 (1969).
107. Oriol, Landon, and Thoai, *Biochim. Biophys. Acta,* 207, 514 (1970).
108. Oriol, Landon, and Thoai, *Biochim. Biophys. Acta,* 207, 514 (1970).
109. Ashida, *Arch. Biochem. Biophys.,* 144, 749 (1971).
110. Ebashi, Wakabayashi, and Ebashi, *J. Biochem.,* 69, 441 (1971); Greaser and Gergely, *J. Biol. Chem.,* 248, 2124 (1973); Dabrowska, Barlyko, Nowak, and Drabikowski, *FEBS Lett.,* 29, 239 (1973).
111. Baker and van der Drift, *Biochemistry,* 13, 292 (1974).
112. Yang, Lee, and Haslam, *J. Molec. Biol.,* 81, 517 (1973).
113. Saito, Ozutsumi, and Kurahashi, *J. Biol. Chem.,* 242, 2362 (1967).
114. Queener, Queener, Meeks, and Gunsalus, *J. Biol. Chem.,* 248, 151 (1973).
115. Winstead and Wold, *Biochemistry,* 3, 791 (1964); Winstead and Wold, *Biochemistry,* 4, 2145 (1965); Cardenas and Wold, *Biochemistry,* 7, 2736 (1968).
116. Yue, Palmieri, Olsen, and Kuby, *Biochemistry,* 6, 3205 (1967).
116a. Burke, Johanson, and Reeves, *Biochim. Biophys. Acta,* 351, 333 (1974).
117. Flohé, Eisele, and Wendel, *Hoppe-Seyler's Z. Physiol. Chem.,* 352, 151 (1971).
118. Edwards, Edwards, and Hopkinson, *FEBS Lett.,* 32, 235 (1973).
119. Kalousek and Konigsberg, *Biochemistry,* 13, 999 (1974).
120. Hatefi, Davis, Baltscheffsky, Baltscheffsky, and Johansson, *Arch. Biochem. Biophys.,* 152, 613 (1972).
121. Wu, Matsuda, and Roberts, *J. Biol. Chem.,* 248, 3029 (1973).
122. Waks and Alfsen, *Arch. Biochem. Biophys.,* 123, 133 (1968).
123. Jacobs, Nye, and Brown, *J. Biol. Chem.,* 246, 1419 (1971).
124. Reynolds and Schlesinger, *Biochemistry,* 8, 588 (1969); Schlesinger and Barrett, *J. Biol. Chem.,* 240, 4284 (1965).
125. Sawula and Crawford, *J. Biol. Chem.,* 248, 3573 (1973).
126. Loper, *J. Biol. Chem.,* 243, 3264 (1968); Yourno, *J. Biol. Chem.,* 243, 3277 (1968); Lew and Roth, *Biochemistry,* 10, 204 (1971).
127. Mann, Castellino, and Hargrave, *Biochemistry,* 9, 4002 (1970).
128. Brown, Greenshields, Yamasaki, and Neurath, *Biochemistry,* 2, 867 (1963); Teller, *Biochemistry,* 9, 4201 (1970).
129. Scanu, Edelstein, and Lim, *Fed. Proc., Fed. Am. Soc. Exp. Biol.,* 31, 829 (1972).
130. Morimoto and Harrington, *J. Biol. Chem.,* 247, 3052 (1972).
131. DeSa, *J. Biol. Chem.,* 247, 5527 (1972).
131a. Spring and Wold, *J. Biol. Chem.,* 246, 6797 (1971).
132. Tedesco, *J. Biol. Chem.,* 247, 6631 (1972).
133. Uehara, Tanimoto, and Soto, *J. Biochem.,* 75, 333 (1974).
134. Måhlén, *Eur. J. Biochem.,* 22, 104 (1971).
135. Gilpin and Sadoff, *J. Biol. Chem.,* 246, 1475 (1971).
136. Reed, *Curr. Top. Cell. Regul.,* 1, 233 (1969).
136a. McCullough, Chabner, and Bertino, *J. Biol. Chem.,* 246, 7207 (1971).
137. White-Stevens and Kamin, *J. Biol. Chem.,* 247, 2358 (1972).
138. Travis and McElroy, *Biochemistry,* 5, 2170 (1966).
138a. Howard and Becker, *J. Biol. Chem.,* 245, 3186 (1971).
139. Takai, Kurashina, Suzuki-Hori, Okamoto, and Hayaishi, *J. Biol. Chem.,* 249, 1965 (1974).
139a. Dai, Decker, and Sund, *Eur. J. Biochem.,* 4, 95 (1968).
140. Silverberg and Dalziel, *Eur. J. Biochem.,* 38, 229 (1973).
141. Brown, *J. Biol. Chem.,* 248, 409 (1973).
141a. Lee and Muench, *J. Biol. Chem.,* 244, 223 (1969).

142. Koch, *Biochemistry*. 13, 2307 (1974).
143. Mapes, Suelter, and Sweeley, *J. Biol. Chem.*, 248, 2471 (1973).
144. Robyt and Ackerman, *Arch. Biochem. Biophys.*, 154, 445 (1973); Mitchell, Riquetti, Loring, and Carraway, *Biochim. Biophys. Acta*, 295, 314 (1973); Connellan and Shaw, *J. Biol. Chem.*, 245, 2845 (1970); Robyt, Chittenden, and Lee, *Arch. Biochem. Biophys.*, 144, 160 (1971).
145. Glazer and Cohen-Bazire, *Proc. Natl. Acad. Sci. U.S.A.*, 68, 1398 (1971).
145a. Kohn, Warren, and Carroll, *J. Biol. Chem.*, 245, 3821 (1970).
146. Lee, Patrick, and Barnes, *J. Biol. Chem.*, 245, 1357 (1970).
147. Wickner and Tabor, *J. Biol. Chem.*, 247, 1605 (1972).
148. Fonda and Anderson, *J. Biol. Chem.*, 243, 5635 (1968).
149. Hetland, Olsen, Christensen, and Størmer, *Eur. J. Biochem.*, 20, 200 (1971).
150. Federici, Barra, Fiori, and Costa, *Physiol. Chem. Phys.*, 3, 448 (1971); Gavallini, Cannella, Federici, Dupre, Fiori, and DelGrosso, *Eur. J. Biochem.*, 16, 537 (1970).
151. Rustum and Barnard, *Fed. Proc., Fed. Am. Soc. Exp. Biol.*, 30, 1122 (1971).
152. Wu and Yang, *J. Biol. Chem.*, 245, 212 (1970); Singh, Books, and Srere, *J. Biol. Chem.*, 245, 4636 (1970); Moriyama and Srere, *J. Biol. Chem.*, 246, 3217 (1971).
153. Bertland and Kaplan, *Biochemistry*, 7, 134 (1968).
154. Katze and Konigsberg, *J. Biol. Chem.*, 245, 923 (1970); Boeker, Hays, and Cantorie, *Biochemistry*, 12, 2379 (1973).
155. Srivastava, Yoshida, Awasthi, and Beutler, *J. Biol. Chem.*, 249, 2049 (1974).
155a. Skålhegg, *Eur. J. Biochem.*, 46, 117 (1974).
156. Veronese, Boccu, Fontana, Benassi, and Scoffone, *Biochim. Biophys. Acta*, 334, 31 (1974); Proscal and Holten, *Biochemistry*, 11, 1310 (1972).
157. Lapointe and Söll, *J. Biol. Chem.*, 247, 4966 (1972).
158. Wengenmayer, Ueberschär, and Kurz, *Eur. J. Biochem.*, 43, 49 (1974).
159. Ii and Sakai, *Biochim. Biophys. Acta*, 350, 141 (1974).
160. Derechin, Ostrowski, Galka, and Barnard, *Biochim. Biophys. Acta*, 250, 143 (1971).
161. Pringle, *Biochem. Biophys. Res. Commun.*, 39, 46 (1970); Schmidt and Colowick, *Fed. Proc., Fed. Am. Soc. Exp. Biol.*, 29, 334 (1970).
162. Palmieri, Yue, Jacobs, Maland, Wu, and Kuby, *Fed. Proc., Fed. Am. Soc. Exp. Biol.*, 29, 914 (1970); Palmieri, Yue, Jacobs, Maland, Wu, and Kuby, *J. Biol. Chem.*, 248, 4486 (1973).
163. Kantor, Temples, and Shaw, *Arch. Biochem. Biophys.*, 151, 142 (1972).
164. Bose and Yamada, *Biochemistry*, 13, 2051 (1974).
165. Pearse and Rosemeyer, *Eur. J. Biochem.*, 42, 225 (1974).
166. Richard, Mazat, Gros, and Patte, *Eur. J. Biochem.*, 40, 619 (1973).
167. Domagk, Doering, and Chilla, *Eur. J. Biochem.*, 38, 259 (1973).
168. Chung and Braginski, *Arch. Biochem. Biophys.*, 153, 357 (1973).
169. Burleigh and Williams, *J. Biol. Chem.*, 247, 2077 (1972).
170. Blomquist, *Arch. Biochem. Biophys.*, 159, 590 (1973).
171. Das and Radhakrishnan, *Biochem. J.*, 135, 609 (1973).
171a. Roy and Jerfy, *Biochim. Biophys. Acta*, 207, 156 (1970).
172. Marshall and Cohen, *J. Biol. Chem.*, 247, 1641 (1972).
173. Wang, Cunningham, and Edleman, *Proc. Natl. Acad. Sci. U.S.A.*, 68, 1130 (1971); Hardman, Wood, Schiffey, Edmundson, and Ainsworth, *Proc. Natl. Acad. Sci. U.S.A.*, 68, 1393 (1971).
174. Stevens, Brown, and Smith, *Arch. Biochem. Biophys*, 136, 413 (1970).
175. Klotz and Keresztes-Nagy, *Biochemistry*, 2, 445, 923 (1963).
176. Cooper and Meister, *Biochemistry*, 11, 661 (1972).
177. Feit, Slusarek, and Shelanski, *Proc. Natl. Acad. Sci. U.S.A.*, 68, 2028 (1971).
177a. Taylor, Rigby, and Hartley, *Biochem. J.*, 141, 693 (1974).
178. Preddie, *J. Biol. Chem.*, 244, 3958 (1969); Gros, Lemaire, Rapenbusch, and Labouesse, *J. Biol. Chem.*, 247, 2931 (1972); Penneys and Muench, *Biochemistry*, 13, 560 (1974).
179. Sasaki, Sugimoto, and Chiba, *Agric. Biol. Chem.*, 34, 135 (1970); Campbell, Hodgson, Watson, and Scopes, *J. Mol. Biol.*, 61, 257 (1971).
180. Lynch and Phillips, *J. Biol. Chem.*, 247, 7799 (1972).
181. Prozorovski and Jörnvall, *Eur. J. Biochem.*, 42, 405 (1974).
182. Uehara, Fujimoto, and Taniguchi, *J. Biochem.*, 75, 627 (1974).
183. Reed, *Curr. Top. Cell. Regul.*, 1, 233 (1969).
184. McPherson and Rich, *J. Biochem.*, 74, 155 (1973).
184a. Wickner, Tabor, and Tabor, *J. Biol. Chem.*, 245, 2132 (1970).
185. Tabita and McFadden, *J. Biol. Chem.*, 249, 3459 (1974).
186. Andrews, *Biochem. J.*, 139, 421 (1974).
187. Lautenberger and Linn, *J. Biol. Chem.*, 247, 6176 (1972).
188. Oreland, Kinemuchi, and Stigbrand, *Arch. Biochem. Biophys.*, 159, 854 (1973).
188a. Cohen and Fridovich, *J. Biol. Chem.*, 246, 367 (1971).

189. Evans and Gurd, *Biochem. J.*, 133, 189 (1973).
190. Auricchio, Valeriote, Tomkins, and Riley, *Biochim. Biophys. Acta*, 221, 307 (1970).
191. Kucan and Chambers, *J. Biochem.*, 73, 811 (1973).
192. Carvajal, Venegas, Oestreicher, and Plaza, *Biochim. Biophys. Acta*, 250, 437 (1971).
193. Low and Reithel, *Fed. Proc., Fed. Am. Soc. Exp. Biol.*, 33, 1478 (1974).
194. Koo and Adams, *J. Biol. Chem.*, 249, 1704 (1974).
195. Lotan, Siegelman, Lis, and Sharon, *J. Biol. Chem.*, 249, 1219 (1974).
196. Biswas and Paulus, *J. Biol. Chem.*, 248, 2894 (1973).
197. Tosa, Sano, Yamamoto, Nakamura, and Chibata, *Biochemistry*, 12, 1075 (1973).
198. Chirikjian, Wright, and Fresco, *Proc. Natl. Acad. Sci. U.S.A.*, 69, 1638 (1972).
199. Garbers, First, and Lardy, *J. Biol. Chem.*, 248, 875 (1973).
200. Montie and Montie, *Biochemistry*, 10, 2094 (1971).
201. Wong, Hall, and Josse, *J. Biol. Chem.*, 245, 4335 (1970).
201a. Heider, Gottschalk, and Cramer, *Eur. J. Biochem.*, 20, 144 (1971).
201b. Pere, Font, and Bourrillon, *Biochim. Biophys. Acta*, 365, 40 (1974).
202. Rapoport, Davis, and Horecker, *Arch. Biochem. Biophys.*, 132, 286, (1969).
202a. Erdose and Fries, *Biochem. Biophys. Res. Commun.*, 58, 932 (1974).
203. Dinamarca, Levenbook, and Valdes, *Arch. Biochem. Biophys.*, 147, 374 (1971).
204. Hulett-Cowling and Campbell, *Biochemistry*, 10, 1371 (1971).
205. Poillon, Maeno, Koike, and Feigelson, *J. Biol. Chem.*, 244, 3447 (1969).
206. Bonner and Bloch, *J. Biol. Chem.*, 247, 3123 (1972).
207. Corbin, Bronstrom, King, and Krebs, *J. Biol. Chem.*, 247, 7791 (1972).
208. Freeman and Daniel, *Biochemistry*, 12, 4806 (1973).
209. Zagalak and Rétey, *Eur. J. Biochem.*, 44, 529 (1974).
210. Mavis and Stellwagen, *J. Biol. Chem.*, 243, 809 (1968).
211. Maley and Maley, *Fed. Proc., Fed. Am. Soc. Exp. Biol.*, 30, 1113 (1971).
212. Roewekamp and Sekeris, *Eur. J. Biochem.*, 43, 405 (1974).
213. Pitts, Priest, and Fish, *Biochemistry*, 13, 889 (1974).
214. Sakamoto, Hatfield, and Moyed, *J. Biol. Chem.*, 247, 5880 (1972).
215. Zito and Kertesz, in *Biological and Chemical Aspects of Oxygenases,* Bloch and Hayaishi, Eds., Maruzen, Tokyo, 1966, 290; Bouchilloux, McMahill, and Mason, *J. Biol. Chem.*, 238, 1699 (1963).
216. Mendicino, Kratowich, and Oliver, *J. Biol. Chem.*, 247, 6643 (1972).
217. Levy, Raineri, and Nevaldine, *J. Biol. Chem.*, 241, 2181 (1966).
218. Bender and Gottschalk, *Eur. J. Biochem.*, 40, 309 (1973).
219. Peraino, Bunville, and Tahmisian, *J. Biol. Chem.*, 244, 2241 (1969).
219a. Bannerjee, Sanders, and Sokatch, *J. Biol. Chem.*, 245, 1828 (1970).
220. Frank, Pekar, Veros, and Ho, *J. Biol. Chem.*, 245, 3716 (1970).
221. Dungan and Datta, *J. Biol. Chem.*, 248, 8534 (1973).
222. Carter and Yoshida, *Biochim. Biophys. Acta*, 181, 12 (1969); Yoshida and Carter, *Biochim. Biophys. Acta*, 194, 151 (1969); Blackburn and Noltman, *J. Biol. Chem.*, 247, 5668 (1972).
223. deKok and Rawitch, *Biochemistry*, 8, 1405 (1969).
224. Lönnerdal and Janson, *Biochim. Biophys. Acta*, 315, 421 (1973).
225. Cammack, Marlborough, and Miller, *Biochem. J.*, 126, 361 (1972).
226. Chenoweth, Brown, Valenzuela, and Smith, *J. Biol. Chem.*, 248, 1684 (1973).
227. Aitken, Bhatti, and Kaplan, *Biochim. Biophys. Acta*, 309, 50 (1973).
228. Rymo, Lundvik, and Lagerkvist, *J. Biol. Chem.*, 247, 3888 (1972); Lagerkvist, Rymo, Lindqvist, and Anderson, *J. Biol. Chem.*, 247, 3897 (1972).
229. Hossler and Rendi, *Biochem. Biophys. Res. Commun.*, 43, 530 (1971).
230. Roberts, Holcenberg, and Dolowy, *J. Biol. Chem.*, 247, 84 (1972).
231. Heinrich and Wiss, *FEBS Lett.*, 14, 251 (1971).
232. Tao, Salas, and Lipmann, *Proc. Natl. Acad. Sci. U.S.A.*, 67, 408 (1970); Miyamoto, Petzold, Harris, and Greengard, *Biochem. Biophys. Res. Commun.*, 44, 305 (1971).
233. Bridger, *Biochem. Biophys. Res. Commun.*, 42, 948 (1971); Leitzmann, Wu, and Boyer, *Biochemistry*, 9, 2338 (1970).
234. Kushner and Somerville, *Biochim. Biophys. Acta*, 207, 105 (1970); Gotschlich and Edelman, *Proc. Natl. Acad. Sci. U.S.A.*, 54, 558 (1965).
235. Castellino and Barker, *Biochemistry*, 7, 2207 (1968); Heck, *J. Biol. Chem.*, 244, 4375 (1969); Schwert, Miller, and Peanasky, *J. Biol. Chem.*, 242, 3245 (1967); Adams, Ford, Koekoek, Lentz, McPherson, Rossman, Smiley, Schevitz, and Wonacott, *Nature*, 227, 1098 (1970).
236. Vance and Feingold, *Fed. Proc., Fed. Am. Soc. Exp. Biol.*, 30, 1057 (1971).
237. Strothkamp and Dawson, *Biochemistry*, 13, 434 (1974).

238. Miyamoto, Petzold, Kuo, and Greengard, *J. Biol. Chem.*, 248, 179 (1973).
239. Miyamoto, Petzold, Kuo, and Greengard, *J. Biol. Chem.*, 248, 179 (1973).
239a. Ray, Reuben, Molineux, and Gefter, *J. Biol. Chem.*, 249, 5379 (1974).
240. Oh and Conrad, *Arch. Biochem. Biophys.*, 152, 631 (1972).
241. Traniello, Melloni, Pontremoli, Sia, and Horecker, *Arch. Biochem. Biophys.*, 149, 222 (1972); Tashima, Tholey, Drummond, Bertrand, Rosenberg, and Horecker, *Arch. Biochem. Biophys.*, 149, 118 (1972); Black, Tol, Fernando, and Horecker, *Arch. Biochem. Biophys.*, 151, 576 (1972).
242. Haugen and Suttie, *J. Biol. Chem.*, 249, 2717 (1974).
243. Ono-Kamimoto and Senoh, *J. Biochem.*, 75, 321 (1974).
244. Stribling and Perham, *Biochem. J.*, 131, 833 (1973).
245. Baker, Jeng, and Barker, *J. Biol. Chem.*, 247, 7724 (1972).
246. Fosset, Chappelet-Tordo, and Lazdunski, *Biochemistry*, 13, 1783, (1974).
246a. Buhner and Sund, *Eur. J. Biochem.*, 11, 73 (1969).
247. Zalkin and Hwang, *J. Biol. Chem.*, 246, 6899 (1971).
248. Andrews, *Biochem. J.*, 139, 421 (1974).
248a. Wolf and Hoffmann, *Eur. J. Biochem.*, 45, 269 (1974).
249. Deal and Holleman, *Fed. Proc., Fed. Am. Soc. Exp. Biol.*, 23, 264 (1964); Harris and Perham, *J. Mol. Biol.*, 13, 876 (1965); Harrington and Karr, *J. Mol. Biol.*, 13, 885; Jaenicke, Schmid, and Knof, *Biochemistry*, 7, 919 (1968); Hoagland and Teller, *Biochemistry*, 8, 594 (1969).
250. Hurlbert and Jakoby, *J. Biol. Chem.*, 240, 2772 (1965).
251. Uyeda and Kurooka, *Fed. Proc., Fed. Am. Soc. Exp. Biol.*, 29, 399 (1970); Uyeda and Kurooka, *J. Biol. Chem.*, 245, 3315 (1970).
252. Schwartz, Pizzo, Hill, and McKee, *J. Biol. Chem.*, 248, 1395 (1973).
252a. Whelan and Wriston, *Biochim. Biophys. Acta*, 365, 212 (1974).
253. Yoshida, *J. Biol. Chem.*, 240, 1113 (1965).
254. Henning, Helinski, Chao, and Yanofsky, *J. Biol. Chem.*, 237, 1523 (1962); Carlton and Yanofsky, *J. Biol. Chem.*, 237, 1531 (1962); Wilson and Crawford, *Bacteriol. Proc.*, 1964, 92 (1964); Goldberg, Creighton, Baldwin, and Yanofsky, *J. Mol. Biol.*, 21, 71 (1966); Yanofsky, Drapeau, Guest, and Carlton, *Proc. Natl. Acad. Sci. U.S.A.*, 57, 296 (1967); Hathaway and Crawford, *Biochemistry*, 9, 1801 (1970).
255. Lee and Horecker, *Arch. Biochem. Biophys.*, 162, 401 (1974).
256. Evans and Gurd, *Biochem. J.*, 133, 189 (1973).
257. Kolb, Cole, and Snell, *Biochemistry*, 7, 2946 (1968).
258. Pfleiderer and Auricchio, *Biochem. Biophys. Res. Commun.*, 16, 53 (1964); Harris, *Nature*, 203, 30 (1964).
259. Barrera, Namihira, Hamilton, Munk, Eley, Linn, and Reed, *Arch. Biochem. Biophys.*, 148, 343 (1972).
260. Wang and Weissman, *Fed. Proc., Fed. Am. Soc. Exp. Biol.*, 29, 333 (1970); Wang and Weissman, *Biochemistry*, 10, 1067 (1971).
261. Holland and Westhead, *Biochemistry*, 12, 2264 (1974).
261a. Wachsmuth and Hiwada, *Biochem. J.*, 141, 273 (1974).
262. Berman, Rubin, Carrell, and Glusker, *J. Biol. Chem.*, 249, 3983 (1974).
263. Waterson, Castellino, Hass, and Hill, *J. Biol. Chem.*, 247, 5266 (1972).
264. Stellwagen and Schachman, *Biochemistry*, 1, 1056 (1962); Deal, Rutter, and Van Holde, *Biochemistry*, 2, 246 (1963); Schachman and Edelstein, *Biochemistry*, 5, 2681 (1966); Penhoet, Kochman, Valentine, and Rutter, *Biochemistry*, 6, 2940 (1967); Sia and Horecker, *Arch. Biochem. Biophys.*, 123, 186 (1968); Kawahara and Tanford, *Biochemistry*, 5, 1578 (1966).
265. Adler, Beyreuther, Fanning, Geisler, Cronenborn, Klemp, Muller-Hill, Pfahl, and Schmitz, *Nature*, 237, 322 (1972).
266. Kaplan and Flavin, *J. Biol. Chem.*, 241, 5781 (1966).
266a. Huang, Nakatsukasa, and Nester, *J. Biol. Chem.*, 249, 4467 (1974).
267. Churchich and Dupourque, *Biochem. Biophys. Res. Commun.*, 46, 524 (1972).
268. Colby and Zatman, *Biochem. J.*, 121, 9P (1971).
269. Whiteley, *J. Biol. Chem.*, 241, 4890 (1966).
270. Winicov and Pizer, *J. Biol. Chem.*, 249, 1348 (1974).
271. Rosemblatt, Callewaert, and Chen, *J. Biol. Chem.*, 248, 6014 (1973).
272. Hass and Hill, *J. Biol. Chem.*, 244, 6080 (1969).
273. Packman and Jakoby, *J. Biol. Chem.*, 242, 2075 (1967).
274. Benöhr, and Krisch, *Z. Physiol. Chem.*, 348, 1115 (1967).
275. Schutz and Feigelson, *J. Biol. Chem.*, 247, 5327 (1972).
276. Bar-Tana and Rose, *Biochem. J.*, 131, 443 (1973).
277. Dalton, Morris, Ward, and Mortenson, *Biochemistry*, 10, 2066 (1971).
278. Cohen, *Curr. Top. Cell. Regul.*, 1, 183 (1969).
279. Clinkenbeard, Sugiyama, Moss, Reed, and Lane, *J. Biol. Chem.*, 248, 2275 (1973).
279a. Gehring and Riepertinger, *Eur. J. Biochem.*, 6, 281 (1968).
280. Koch and Bruton, *FEBS Lett.*, 40, 180 (1974).

281. Baker, van der Drift, and Stadtman, *Biochemistry,* 12, 1054 (1973).
282. Trotta, Burt, Haschmeyer, and Meister, *Proc. Natl. Acad. Sci. U.S.A.,* 68, 2599 (1971); Matthews and Anderson, *Biochemistry,* 11, 1176 (1972).
283. Suzuki, Yamaguchi, and Tokushige, *Biochem. Biophys. Acta,* 321, 369 (1973); Rudolph and Fromm, *Arch. Biochem. Biophys.,* 147, 92 (1971).
284. Erlichman, Rubin, and Rosen, *J. Biol. Chem.,* 248, 7607 (1973).
285. Junge, Krisch, and Hollandt. *Eur. J. Biochem.,* 43, 379 (1974).
286. Somack and Costilow, *Biochemistry,* 12, 2597 (1973).
287. Durell and Cantoni, *Biochim. Biophys. Acta,* 35, 515 (1959); Klee, *Biochim. Biophys. Acta,* 59, 562 (1962).
287a. Tobin, *J. Biol. Chem.,* 245, 2656 (1970).
288. Lipscomb, Horton, and Armstrong, *Biochemistry,* 13, 2070 (1974).
289. Simon, Schorr, and Phillips, *J. Biol. Chem.,* 249, 1993 (1974).
290. McKeehan and Hardesty, *J. Biol. Chem.,* 244, 4330 (1969).
291. Lamartiniere and Dempey, *Fed. Proc., Fed. Am. Soc. Exp. Biol.,* 30, 1121 (1971).
292. Riley and Snell, *Biochemistry,* 9, 1485 (1970).
293. Recsei and Snell, *Biochemistry,* 12, 365 (1973).
294. Kanarek, Marler, Bradshaw, Fellows, and Hill, *J. Biol. Chem.,* 239, 4207 (1964); Penner and Cohen, *J. Biol. Chem.,* 246, 4261 (1971).
294a. Bornmann and Hess, *Eur. J. Biochem.,* 47, 1 (1974); Fell, Liddle, Peacocke, and Dwek, *Biochem. J.,* 139, 665 (1974).
295. Zarlengo, Robinson, and Burns., *J. Biol. Chem.,* 243, 186 (1968).
296. Metrione, Okuda, and Fairclough, *Biochemistry,* 9, 2427 (1970).
297. Maeba and Sanwal, *J. Biol. Chem.,* 244, 2549 (1969).
298. Rowe and Wyngaarden, *J. Biol. Chem.,* 243, 6373 (1968).
299. Lehrach, Schafer, and Scheit, *FEBS Lett.,* 14, 343 (1971).
300. Leary and Kohlhaw, *J. Biol. Chem.,* 247, 1089 (1972); Bartholomew and Calvo, *Biochim. Biophys. Acta,* 250, 568, 577 (1971).
301. Niles and Westhead, *Biochemistry,* 12, 1715 (1973).
302. Coles, Tisdale, Kenney, and Singer, *Biochem. Biophys. Res. Commun.,* 46, 1843 (1972).
303. Kendal and Eckert, *Biochim. Biophys. Acta,* 258, 484 (1972).
304. Barisas and McGuire, *J. Biol. Chem.,* 249, 3151 (1974).
305. Israel, Howard, Evans, and Russell, *J. Biol. Chem.,* 249, 500 (1974).
305a. Danner and Morrison, *Biochim. Biophys. Acta,* 235, 44 (1971).
305b. Love, Chan, and Stotz, *J. Biol. Chem.,* 245, 6664 (1970).
305c. Higgens, Brady, and Rudney, *Arch. Biochem. Biophys.,* 163, 271 (1974).
306. Brown, Symth, Clarke, and Rosemyer, *Eur. J. Biochem.,* 34, 177 (1973).
307. Colman and Matthews, *J. Mol. Biol.,* 60, 163 (1971); Spradlin and Thoma, *J. Biol. Chem.,* 245, 117 (1970).
308. Schulze, Lusty, and Ratner, *J. Biol. Chem.,* 245, 4534 (1970).
309. Yamaguchi, Tokushige, and Katsuki, *J. Biochem.,* 73, 169 (1973).
310. Calhoun, Rimerman, and Hatfield, *J. Biol. Chem.,* 248, 3511 (1973).
311. Bonsignore, Cancedda, Lorenzoni, Cosulich, and DeFlora, *Biochem. Biophys. Res. Commun.,* 43, 94 (1971).
312. Kamen, *Nature,* 228, 527 (1970); Kondo, Gallerani, and Weissman, *Nature,* 228, 525 (1970).
313. McFadden, Rao, Cohen, and Roche, *Biochemistry,* 7, 3574 (1968).
314. Scott, *J. Biol. Chem.,* 246, 6353 (1971).
315. Hoch and DeMoss, *J. Biol. Chem.,* 247, 1750 (1972).
316. Gounaris, Turkenkopf, Buckwald, and Young, *J. Biol. Chem.,* 246, 1302 (1971).
317. Meachum, Calvin, and Braymer, *Biochemistry,* 10, 326 (1971).
317a. Kaufman and Fisher, *J. Biol. Chem.,* 245, 4745 (1970).
318. Long, Levitzki, and Koshland, *J. Biol. Chem.,* 245, 80, (1970).
319. Cox and Tanford, *J. Biol. Chem.,* 243, 3083 (1968); Scanu, Reader, and Edelstein, *Fed. Proc., Fed. Am. Soc. Exp. Biol.,* 26, 435 (1967).
320. Voll, Appella, and Martin, *J. Biol. Chem.,* 242, 1760 (1967).
321. Martinez-Carrion, Critz, and Quashnock, *Biochemistry,* 11, 1613 (1972); Schirch, Edmiston, Chen, Barra, Bossa, Hinds, and Fassella, *J. Biol. Chem.,* 248, 6456 (1973).
322. Inoue and Lowenstein, *J. Biol. Chem.,* 247, 4825 (1972).
323. Cardenas and Dyson, *J. Biol. Chem.,* 248, 6938 (1973).
324. Iwata and Fukui, *FEBS Lett.,* 36, 322 (1973).
325. Koike, Hamada, Tanaka, Otsuka, Ogashara, and Koike, *J. Biol. Chem.,* 249, 3836 (1974).
326. Cowell and DeMoss, *J. Biol. Chem.,* 248, 6262 (1973).
327. Carlson and Pierce, *J. Biol. Chem.,* 247, 23 (1972).
328. Thorner and Paulus, *J. Biol. Chem.,* 246, 3885 (1971).
328a. Wolfenden, Tomozawa, and Bamman, *Biochemistry,* 7, 3965 (1968).

329. Huseby, Christensen, Olsen, and Størmer, *Eur. J. Biochem.,* 20, 209 (1971).

329a. Callewaert, Rosemblatt, and Chen, *Biochemistry,* 13, 4181 (1974).

330. Hoch and DeMoss, *Biochemistry,* 5, 3137 (1966); Morino and Snell, *J. Biol. Chem.,* 242, 5591 (1967).

331. Lowey, Kucera, and Holtzer, *J. Mol. Biol.,* 7, 234 (1963); Olander, Emerson, and Holtzer, *J. Am. Chem. Soc.,* 89, 3058 (1967); McCubbin and Kay, *Biochim. Biophys. Acta,* 154, 239 (1968).

332. Schmidt and Zalkin, *J. Biol. Chem.,* 246, 6002 (1971).

333. Cogoli, Eberle, Sigrist, Joss, Robinson, Mosimann, and Semenza, *Eur. J. Biochem.,* 33, 40 (1973); Mosimann, Semenza, and Sund, *Eur. J. Biochem.,* 36, 489 (1973).

334. Kyte, *J. Biol. Chem.,* 247, 7642 (1972).

335. Shematek, Divin, and Arfin, *Arch. Biochem. Biophys.,* 158, 126 (1973).

336. Madon, Hillmer, and Gottschalk, *Eur. J. Biochem.,* 32, 51 (1973).

337. Flanders, Bamburg, and Sallach, *Biochim. Biophys. Acta,* 242, 566 (1971).

338. Little, Riley, and Behal, *Fed. Proc., Fed. Am. Soc. Exp. Biol.,* 30, 1121 (1971); Little and Behal, *Biochim. Biophys. Acta,* 243, 312 (1971).

339. Marshall and Cohen, *J. Biol. Chem.,* 247, 1641 (1972).

340. Berglund, *J. Biol. Chem.,* 247, 7270 (1972).

341. Ostrem and Berg, *Proc. Natl. Acad. Soc. U.S.A.,* 67, 1967 (1970).

342. Callewaert, Rosenblatt, and Tchen, *J. Biol. Chem.,* 249, 1737 (1974).

343. Kredich, Keenan, and Foote, *J. Biol. Chem.,* 247, 7157 (1972).

344. Cardenas, Dyson, and Standholm, *J. Biol. Chem.,* 248, 6931 (1973).

345. Chung, Tan, and Suzuki, *Biochemistry,* 10, 1205 (1971).

345a. Kakiuchi and Kobayashi, *J. Biochem.,* 69, 43 (1971).

346. Glazer, Fang, and Brown, *J. Biol. Chem.,* 248, 5679 (1973).

347. Tanford and Lovrien, *J. Am. Chem. Soc.,* 84, 1892 (1962); Schroeder, Shelton, Shelton, and Olson, *Biochim. Biophys. Acta,* 89, 47 (1964); Weber and Sund, *Angew. Chem.,* 77, 621 (1965); Schroeder, Shelton, Shelton, Roberson, and Apell, *Arch. Biochem. Biophys.,* 131, 653 (1969).

348. Monteilhet and Risler, *Eur. J. Biochem.,* 12, 165 (1970); Lederer and Simon, *Eur. J. Biochem.,* 20, 469 (1971).

349. Morawiecki, *Biochim. Biophys. Acta,* 44, 604 (1960); Steinmetz and Deal, *Biochemistry,* 5, 1399 (1966).

350. Curthoys, Straus, and Rabinowitz, *Biochemistry,* 11, 345 (1972).

351. Heyningen and Shemin, *Biochemistry,* 10, 4676 (1971).

352. Gaertner and DeMoss, *J. Biol. Chem.,* 244, 2716 (1969).

353. Montie and Montie, *Biochemistry,* 10, 2094 (1971).

354. Berg, Olsen, and Kivirikko, *Fed. Proc., Fed. Am. Soc. Exp. Biol.,* 31, 479 (1972).

355. Pentchev, Brady, Hibbert, Gal, and Shapiro, *J. Biol. Chem.,* 248, 5256 (1973).

356. Waygood and Sanwal, *J. Biol. Chem.,* 249, 265 (1974).

357. Gauthier and Rittenberg, *J. Biol. Chem.,* 246, 3737 (1971).

357a. Johnson and Hanson, *Biochim. Biophys. Acta,* 350, 336 (1974).

358. Feldman and Weiner, *J. Biol. Chem.,* 247, 260 (1972).

359. Thelander, *J. Biol. Chem.,* 248, 4591 (1973).

360. Strickland and Massey, *J. Biol. Chem.,* 248, 2944 (1973).

361. Smith, Brown, and Larner, *Biochim. Biophys. Acta,* 242, 81 (1971).

362. Eskin and Linn, *J. Biol. Chem.,* 247, 6183 (1972).

363. Kashiwamata, Kotake, and Greenburg, *Biochim. Biophys. Acta,* 212, 501 (1970).

364. Darrow and Rodstrom, *J. Biol. Chem.,* 245, 2036 (1970).

365. Fodge, Gracy, and Harris, *Biochim. Biophys. Acta,* 268, 271 (1972).

365a. Johnson and Hanson, *Biochim. Biophys. Acta,* 350, 336 (1974).

366. Correll, Steers, Towe, and Shropshire, *Biochim. Biophys. Acta,* 168, 46 (1968).

367. Melius, Moseley, and Brown, *Biochim. Biophys. Acta,* 221, 62 (1970).

368. Nevaldine, Bassel, and Hsu, *Biochim. Biophys. Acta,* 336, 283 (1974).

369. Berman, *Biochemistry,* 12, 1710 (1973).

370. Hanke, Bartmann, Hennecke, Kosakowski, Jaenicke, Holler, and Böck, *Eur. J. Biochem.,* 43, 601 (1974).

371. Holmes, Wyngaarden, and Kelley, *J. Biol. Chem.,* 248, 6035 (1973).

372. Frigerio and Harbury, *J. Biol. Chem.,* 231, 135 (1958).

373. Schmidt, Wang, Stanfield, and Reid, *Biochemistry,* 10, 3264 (1971).

374. Boosman, Sammons, and Chilson, *Biochem. Biophys. Res. Commun.,* 45, 1025 (1971).

375. Fee, Hegeman, and Kenyon, *Biochemistry,* 13, 2529 (1974).

376. Rosenberry, Chen, and Bock, *Biochemistry,* 13, 3068 (1974).

377. Rubin, Erlichman, and Rosen, *J. Biol. Chem.,* 247, 36 (1972).

378. Stahl and Touster, *Fed. Proc., Fed. Am. Soc. Exp. Biol.,* 30, 1121 (1971).

379. Henderson and Zalkin, *J. Biol. Chem.,* 246, 6891 (1971).

380. Wu, Shemin, Richards, and Williams, *Proc. Natl. Acad. Sci. U.S.A.,* 71, 1767 (1974).

381. Singh and Squire, *Biochemistry,* 13, 1819 (1974).

382. Zappia and Barker, *Biochim. Biophys. Acta,* 207, 505 (1970).
383. Blumenthal and Smith, *J. Biol. Chem.,* 248, 6002 (1973).
384. Craine, Daniels, and Kaufman, *J. Biol. Chem.,* 248, 7838 (1973); Wallace, Krantz, and Lovenberg, *Proc. Natl. Acad. Sci. U.S.A.,* 70, 2253 (1973).
385. Eady and Postgate, *Nature,* 249, 805 (1974).
386. Wu and Morris, *J. Biol. Chem.,* 248, 1687 (1973).
387. Gainey, Pestell, and Phelps, *Biochem. J.,* 129, 821 (1972).
388. Huang and Cabib, *J. Biol. Chem.,* 249, 3851 (1974).
389. Schepman, Wichertjes, and Van Bruggen, *Biochim. Biophys. Acta,* 271, 279 (1972).
390. Schepman, Wichertjes, and Van Bruggen, *Biochim. Biophys. Acta,* 271, 279 (1972).
391. Barnes, Kuehn, and Atkinson, *Biochemistry,* 10, 3939 (1971).
391a. Youdim and Collins, *Eur. J. Biochem.,* 18, 73 (1971).
392. Kredich, Becker, and Tomkins, *J. Biol. Chem.,* 244, 2428 (1969).
393. Gerhart and Schachman, *Biochemistry,* 4, 1054 (1965); Schachman and Edelstein, *Biochemistry,* 5, 2681 (1966); Changeux, Gerhart, and Schachman, *Biochemistry,* 7, 531 (1968); Weber, *Nature,* 218, 116 (1968); Wiley and Lipscomb, *Nature,* 218, 1119 (1968).
394. Strausbauch and Fischer, *Biochemistry,* 9, 226 (1970).
395. Main, Tarkan, Aull, and Soucie, *J. Biol. Chem.,* 247, 566 (1972).
396. Virden, *Biochem. J.,* 127, 503 (1972).
397. Eisenberg and Tomkins, *J. Mol. Biol.,* 31, 37 (1968); Cassman and Schachman, *Biochemistry,* 10, 1015 (1971); Reisler and Eisenberg, *Biochemistry,* 10, 2659 (1971); Josephs, *J. Mol. Biol.,* 55, 147 (1971).
398. Schwartz, Pizzo, Hill, and McKee, *J. Biol. Chem.,* 248, 1395 (1973).
399. Nari, Mouttet, Fouchier, and Ricard, *Eur. J. Biochem.,* 41, 499 (1974); Havir and Hanson, *Biochemistry,* 12, 1583 (1973).
400. Farron, *Biochemistry,* 9, 3823 (1970).
401. Melbye and Carpenter, *J. Biol. Chem.,* 246, 2459 (1971).
402. Eady and Postgate, *Nature,* 249, 805 (1974).
403. Strickland, Jacobson, and Strickland, *Biochim. Biophys. Acta,* 251, 21 (1971).
404. Tiemeier and Milman, *J. Biol. Chem.,* 247, 2272 (1972).
405. Tagaki and Westheimer, *Biochemistry,* 7, 891 (1968); *Biochemistry,* 7, 895 (1968).
406. Starnes, Munk, Maul, Cunningham, Cox, and Shive, *Biochemistry,* 11, 677 (1972); Clark and Ogilvie, *Biochemistry,* 11, 1278 (1972).
407. Kerr and Robertson, *Biochem. J.,* 125, 34P (1971).
408. Witheiler and Wilson, *J. Biol. Chem.,* 247, 2217 (1972).
409. Sarkar, Fischman, Goldwasser, and Moscona, *J. Biol. Chem.,* 247, 7743 (1972).
410. Kessler and Yaron, *Biochem. Biophys. Res. Commun.,* 50, 405 (1973).
411. Tombs and Lowe, *Biochem. J.,* 105, 181 (1967).
412. Pollock and Hersch, *J. Biol. Chem.,* 246, 4737 (1971).
413. Andreu, Albendea, and Muñoz, *Eur. J. Biochem.,* 37, 505 (1973).
414. Tate and Meister, *Proc. Natl. Acad. Sci. U.S.A.,* 68, 781 (1971).
415. Stellwagen, Cronlund, and Barnes, *Biochemistry,* 12, 1552 (1973).
416. O'Carra and Killilea, *Biochem. Biophys. Res. Commun.,* 45, 1192 (1971); Hattori, Crespi, and Katz, *Biochemistry,* 4, 1225 (1965); Bloomfield, Van Holde, and Dalton, *Biopolymers,* 5, 149 (1967); Jennings, *Biopolymers,* 6, 1177 (1968); Glazer and Fang, *J. Biol. Chem.,* 248, 663 (1973); Glazer and Cohen-Bazire, *Proc. Natl. Acad. Sci. U.S.A.,* 68, 1398 (1973); Bennett and Bogorad, *Biochemistry,* 10, 3625 (1971); MacColl, Lee, and Berns, *Biochem. J.,* 122, 421 (1971); Berns, *Biochem. Biophys. Res. Commun.,* 38, 65 (1970); Neufeld and Riggs, *Biochim. Biophys. Acta,* 181, 234 (1969).
417. Aaronson and Frieden, *J. Biol. Chem.,* 247, 7502 (1972); Coffee, Aaronson, and Frieden, *J. Biol. Chem.,* 248, 1381 (1973); Uyeda, *Biochemistry,* 8, 2366 (1969); Paetkau, Younathan, and Lardy, *J. Mol. Biol.,* 33, 721 (1968).
418. Patrick and Lee, *J. Biol. Chem.,* 244, 4277 (1969).
419. Kapoor, Bray, and Ward, *Arch. Biochem. Biophys.,* 134, 423 (1969).
420. Adair and Jones, *J. Biol. Chem.,* 247, 2308 (1972).
421. Lambeth and Lardy, *Eur. J. Biochem.,* 22, 355 (1971); Catterall and Pedersen, *J. Biol. Chem.,* 246, 4987 (1971); Senior and Brooks, *Arch. Biochem. Biophys.,* 140, 257 (1970); Senior and Brooks, *FEBS Lett.,* 17, 327 (1971); Brooks and Senior, *Biochemistry,* 11, 4675 (1972).
422. Yoshinaga, Teraoka, Izui, and Katsuki, *J. Biochem.,* 75, 913 (1974).
423. Tabita, McFadden, and Pfennig, *Biochim. Biophys. Acta,* 341, 187 (1974).
424. Issa and Medicino, *J. Biol. Chem.,* 248, 685 (1973).
425. Seery, Fischer, and Teller, *Biochemistry,* 9, 3591 (1970); Madsen and Cori, *J. Biol. Chem.,* 223, 1055 (1956); Seery, Fischer, and Teller, *Biochemistry,* 6, 3315 (1967); DeVincenzi and Hedrick, *Biochemistry,* 6, 3489 (1967).
426. Jonas, *J. Biol. Chem.,* 247, 7767 (1972).
426a. Yurwicz, Ghalambor, Duckworth, and Heath, *J. Biol. Chem.,* 246, 5607 (1971).

427. Schnebli, Vatter, and Abrams, *J. Biol. Chem.*, 245, 1122 (1970).

428. Sugiyama, *Biochemistry*, 12, 2862 (1973).

429. Smando, Waygood, and Sanwal, *J. Biol. Chem.*, 249, 182 (1974).

430. Stoll, Ericsson, and Zuber, *Proc. Natl. Acad. Sci. U.S.A.*, 70, 3781 (1973).

431. Stoll, Ericsson, and Zuber, *Proc. Natl. Acad. Sci. U.S.A.*, 70, 3781 (1973).

432. Stoll, Ericsson, and Zuber, *Proc. Natl. Acad. Sci. U.S.A.*, 70, 3781 (1973).

433. Burgess, *J. Biol. Chem.*, 244, 6168 (1969); Johnson, DeBacker, and Boezi, *J. Biol. Chem.*, 246, 1222 (1971).

434. Weaver, Blath, and Rutler, *Proc. Natl. Acad. Sci. U.S.A.*, 68, 2994 (1971).

435. Hyun, Steinberg, Treadwell, and Vahouny, *Biochem. Biophys. Res. Commun.*, 44, 819 (1971).

436. Bernardi and Cook, *Biochim. Biophys. Acta*, 44, 96, 105 (1960); Burley and Cook, *Can. J. Biochem. Physiol.*, 40, 363 (1962).

437. Tate and Meister, *Proc. Natl. Acad. Sci. U.S.A.*, 68, 781 (1971).

438. Smith, *J. Biol. Chem.*, 246, 4234 (1971).

439. Delmer, *J. Biol. Chem.*, 247, 3822 (1972).

440. Haberland, Willard, and Wood, *Biochemistry*, 11, 712 (1972).

440a. Hamilton, *J. Biol. Chem.*, 249, 4428 (1974).

441. Tweedie and Segel, *J. Biol. Chem.*, 246, 2438 (1971).

442. Crichton, Eason, Barclay, and Bryce, *Biochem. J.*, 131, 855 (1973); Hoy, Harrison, and Hoare, *J. Mol. Biol.*, 84, 515 (1974); Bryce and Crichton, *J. Biol. Chem.*, 246, 4198 (1971); Bjork and Fish, *Biochemistry*, 10, 2844 (1971).

443. Gornick, Vuturo, West, and Weaver, *J. Biol. Chem.*, 249, 1792 (1973).

444. Spiegelman and Whiteley, *J. Biol. Chem.*, 249, 1476 (1974).

445. Holtzer and Lowey, *J. Am. Chem. Soc.*, 81, 1370 (1959); Mueller, *J. Biol. Chem.*, 239, 797 (1964); Tonomura, Appel, and Morales, *Biochemistry*, 5, 515 (1966); Richards, Chung, Menzel, and Olcott, *Biochemistry*, 6, 528 (1967); Gershman, Stracher, and Dreizen, *J. Biol. Chem.*, 244, 2726 (1969); Kominz, Carroll, Smith, and Mitchell, *Arch. Biochem. Biophys.*, 79, 191 (1959); Frederiksen and Holtzer, *Biochemistry*, 7, 3935 (1968); Weeds and Lowey, *J. Mol. Biol.*, 61, 701 (1971).

446. Contaxis and Reithel, *J. Biol. Chem.*, 246, 677 (1971); Bailey and Boulter, *Biochem. J.*, 113, 669 (1969); Creeth and Nichol, *Biochem. J.*, 77, 230 (1960); Reithel, Robbins, and Gorin, *Arch. Biochem. Biophys.*, 108, 409 (1964).

447. Levine, Gillett, Turnquist, and Hansen, *Fed. Proc., Fed. Am. Soc. Exp. Biol.*, 30, 1121 (1971); Levine, Gillett, Hageman, and Hansen, *J. Biol. Chem.*, 244, 5729 (1969).

448. Reiss and Rothstein, *Biochemistry*, 13, 1796 (1974).

449. Lornitzo, Qureshi, and Porter, *J. Biol. Chem.*, 249, 1654 (1974).

450. Yun and Hsu, *J. Biol. Chem.*, 247, 2689 (1973).

451. Kuehn and McFadden, *Biochemistry*, 8, 2403 (1969).

452. Warren and Tipton, *Biochem. J.*, 139, 297 (1974).

453. Craven, Steers, and Anfinsen, *J. Biol. Chem.*, 240, 2468 (1965); Fowler and Zabin, *J. Biol. Chem.*, 245, 5032 (1970).

454. Rutner, *Biochem. Biophys. Res. Commun.*, 39, 923 (1970); Kawashima and Wildman, *Biochem. Biophys. Res. Commun.*, 41, 1463 (1970); Trown, *Biochemistry*, 4, 908 (1965); Haselkorn, Fernández-Morán, Kieras, and van Bruggen, *Science*, 150, 1598 (1965).

455. Spina, Bright, and Rosenbloom, *Biochemistry*, 9, 3794 (1970).

455a. Miziorko, Nowak, and Mildvan, *Arch. Biochem. Biophys.*, 163, 378 (1974).

456. Bowen and Mortimer, *Biochem. J.*, 117, 71P (1970); Mahadik and SivaRaman, *Biochem. Biophys. Res. Commun.*, 32, 167 (1968).

457. Sugiyama, Ito, and Akazawa, *Biochemistry*, 10, 3406 (1971).

458. Wilgus, Pringle, and Stellwagen, *Biochem. Biophys. Res. Commun.*, 44, 89 (1971).

459. Woolfolk and Stadtman, *Arch. Biochem. Biophys.*, 122, 174 (1967); Valentine, Shapiro, and Stadtman, *Biochemistry*, 7, 2143 (1968).

460. Pupillo and Piccari, *Arch. Biochem. Biophys.* 154, 324 (1973).

461. Arfin and Koziell, *Biochim. Biophys. Acta*, 321, 356 (1973).

462. Tate and Meister, *Proc. Natl. Acad. Sci. U.S.A.*, 68, 781 (1971).

463. Donovan, Mapes, Davis, and Hamburg, *Biochemistry*, 8, 4190 (1969).

464. Sun, McLeester, Bliss, and Hall, *J. Biol. Chem.*, 249, 2118 (1974).

465. Valentine, Wrigley, Scrutton, Irias, and Utter, *Biochemistry*, 5, 3111 (1966).

466. Steiner and Edelhoch, *J. Am. Chem. Soc.*, 83, 1435 (1961); Edelhoch and de Crombrugghe, *J. Biol. Chem.*, 241, 4357 (1966).

467. Giorgio, Yip, Fleming, and Plaut, *J. Biol. Chem.*, 245, 5469 (1970); Harvey, Giorgio, and Plaut, *Fed. Proc., Fed. Am. Soc. Exp. Biol.*, 29, 532 (1970).

468. Bowers, Czubaroff, and Haschemyer, *Biochemistry*, 9, 2620 (1970); Tate and Meister, *Biochemistry*, 9, 2626 (1970).

469. Siegal and Kamin, *Fed. Proc., Fed. Am. Soc. Exp. Biol.*, 30, 1261 (1971); Siegel and Davis, *J. Biol. Chem.*, 249, 1587 (1974).

469a. MacGregor, Schnaitman, Normansell, and Hodgins, *J. Biol. Chem.*, 249, 5321 (1974).

470. Kaziro, Ochoa, Warner, and Chen, *J. Biol. Chem.,* 236, 1917 (1961).

471. Sabo, Boeker, Byers, Waron, and Fischer, *Biochemistry,* 13, 662 (1974).

472. Lee-Huang and Warner, *J. Biol. Chem.,* 244, 3793 (1969).

473. Green, Valentine, Wrigley, Ahmad, Jacobson, and Wood, *J. Biol. Chem.,* 247, 6284 (1972).

474. Kono, Uyeda, and Oliver, *J. Biol. Chem.,* 248, 8592 (1973).

475. Kakimoto, Kato, Shibatani, Nishimura, and Chibata, *J. Biol. Chem.,* 245, 3369 (1970).

476. Bloemendal, Bont, Jongkind, and Wisse, *Exp. Eye Res.,* 1, 300 (1962); Bloemendal, Bont, Benedett, and Wisse, *Exp. Eye Res.,* 4, 319 (1965).

477. Boeker, Fischer, and Snell, *J. Biol. Chem.,* 244, 5239 (1969); Boeker and Snell, *J. Biol. Chem.,* 243, 1678 (1968).

478. Shikita and Hall, *J. Biol. Chem.,* 248, 5605 (1973).

479. Stevens, Emery, and Sternberger, *Biochem. Biophys. Res. Commun.,* 24, 929 (1966); Richardson, *Proc. Natl. Acad. Sci. U.S.A.,* 55, 1616 (1966).

480. Carpenter and Van Holde, *Biochemistry,* 12, 2231 (1973); Loehr and Mason, *Biochem. Biophys. Res. Commun.,* 51, 741 (1973); Roxby, Miller, Blair, and Van Holde, *Biochemistry,* 13, 1662 (1974); DePhillips, Nickerson, Johnson, and Van Holde, *Biochemistry,* 8, 3665 (1969); Pickett, Riggs, and Larimer, *Science,* 151, 1005 (1966); Van Holde and Cohen, *Biochemistry,* 3, 1803 (1964); Fernández-Morán, van Bruggen, and Ohtsuki *J. Mol. Biol.,* 16, 191 (1966); Lontie and Witters, in *The Biochemistry of Copper,* Peisach, Aisen, and Blumberg, Eds., Academic Press, New York, 1966, 455.

481. Tanaka, Koike, Otsuka, Hamada, Ogasahara, and Koike, *J. Biol. Chem.,* 249, 191 (1974).

482. Carper, Stoddard, and Martin, *Biochim. Biophys. Acta,* 334, 287 (1974); Carper, Stoddard, and Martin, *Biochem. Biophys. Res. Commun.,* 54, 721 (1973).

483. Hayakawa, Perkins, Walsh, and Krebs, *Biochemistry,* 12, 567, 574 (1973); Cohen, *Eur. J. Biochem.,* 34, 1 (1973).

484. Vance, Mitsuhashi, and Block, *J. Biol. Chem.,* 248, 2303 (1973).

485. Eley, Namihira, Hamilton, Munk, and Reed, *Arch. Biochem. Biophys.,* 152, 655 (1972).

486. Guerritore, Bonacci, Brunori, Antonini, Wyman, and Rossi-Fanelli, *J. Mol. Biol.,* 13, 234 (1965).

487. Pettit, Hamilton, Munk, Namihira, Eley, Williams, and Reed, *J. Biol. Chem.,* 248, 5282 (1973).

488. Waxman, *J. Biol. Chem.,* 246, 7318 (1971).

489. Swaney and Klotz, *Arch. Biochem. Biophys.,* 147, 475 (1971).

490. Barrera, Namihira, Hamilton, Munk, Eley, Linn, and Reed, *Arch. Biochem. Biophys.,* 148, 343 (1972).

491. Hohn and Hohn, *Adv. Virus Res.,* 16, 43 (1970).

492. Eley, Namihira, Hamilton, Munk, and Reed, *Arch. Biochem. Biophys.,* 152, 655 (1972); Vogel, Hoehn, and Henning, *Proc. Natl. Acad. Sci. U.S.A.,* 69, 1615 (1972).

493. Bancroft, Hiebert, Rees, and Markham, *Virology,* 34, 224 (1968).

494. Bockstahler and Kaesberg, *Biophys. J.,* 2, 1 (1962).

495. Miki and Knight, *Virology,* 25, 478 (1965).

496. Markham, *Discuss. Faraday Soc.,* 11, 221 (1951); Harris and Hindley, *J. Mol. Biol.,* 3, 117 (1961); Finch and Klug, *J. Mol. Biol.,* 15, 344 (1966); Peter, Stehdin, Reinbolt, Collet, and Duranton, *Virology,* 49, 615 (1972).

497. Anderer and Restle, *Z. Naturforsch.,* 19b, 1026 (1964).

498. Yamazaki and Kaesberg, *Biochim. Biophys. Acta,* 53, 173 (1961); Van Regenmortel, Hendry, and Baltz, *Virology,* 49, 647 (1972).

499. Kelley and Kaesberg, *Biochim. Biophys. Acta,* 55, 236 (1962); Kelley and Kaesberg, *Biochim. Biophys. Acta,* 61, 865 (1962); Kruseman, Kraal, Jaspers, Bol, Brederode, and Veldstra, *Biochemistry,* 10, 447 (1971).

500. Guchhart, Zwergel, and Lane, *J. Biol. Chem.,* 249, 4776 (1974).

501. Medappa, McLean, and Rueckert, *Virology,* 44, 259 (1971).

502. Hersh and Schachman, *Virology,* 6, 234 (1958); Michelin-Lausarot, Ambrosino, Steere, and Reichmann, *Virology,* 41, 160 (1970); Weber and Rosenbusch, *Virology,* 41, 763 (1970).

503. Fine, Mass, and Murakami, *J. Mol. Biol.,* 36, 167 (1968).

504. Reichmann, *J. Biol. Chem.,* 235, 2959 (1960); Reichmann and Hatt, *Biochim. Biophys. Acta,* 49, 153 (1961).

505. Anderer, *Adv. Protein Chem.,* 18, 1 (1963); Caspar, *Adv. Protein Chem.,* 18, 37 (1963).

MOLAR ABSORPTIVITY AND $A_{1cm}^{1\%}$ VALUES FOR PROTEINS AT SELECTED WAVELENGTHS OF THE ULTRAVIOLET AND VISIBLE REGION

Protein	$\epsilon^a(\times\ 10^{-4})$	$A_{1cm}^{1\%}$ [b]	nm[c]	Ref.	Comments[d]
Acetoacetate decarboxylase (EC 4.1.1.4) C. acetobutylicum	3.05	10.5	280	567	MW = 29,000 (567) subunit
Acetolactate synthase (EC 4.1.3.18) Aerobacter aerogenes	–	8.3	280	568	pH 6, 50 mM P$_i$; data from Figure 1 (568)
Acetyl coenzyme A carboxylase (EC 6.4.1.2) Chicken liver	–	11.6	280	569	Calc. using OD$_{280}$ × 0.86 = mg/ml (569)
Acetylcholinesterase (EC 3.1.1.7)					
Electrophorus electricus	–	21.4	280	1322	Kjeldahl or Dumas[e]
	–	21.8	280	1322	Microninhydrin[e]
	–	18.8	280	1322	Nitrogen from amino acid analysis
	–	17.6	280	1322	DR
	–	18.2	280	1322	Dry wt
	52.7	22.9	280	1	pH 7.0, 0.1 M NaCl, 0.03 M NaP$_i$, MW = 230,000 (1)
	–	16.1	280	2	0.02 M AcONH$_4$
	–	19.0	280	877	
β-*N*-Acetyl-D-glucosaminidase (EC 3.2.1.30)					
Aspergillus oryzae Beef spleen	29.3	20.9	280	777	MW = 140,000 (777)
Enzyme A	–	12.8	278	1323	–
Enzyme B	–	12.7	278	1323	–
O-Acetylserine sulfhydrase A (cysteine synthase) (EC 4.2.99.8)					
Salmonella typhimurium	8.2	12.1	280	634	MW = 68,000 (634)
	0.76	1.12	412	634	MW = 68,000 (634)
Acetylserotonin methyltransferase (EC 2.1.1.4) (see hydroxyindole-*O*-methyltransferase)					

[a] ϵ is the molar absorption coefficient with units of M^{-1} cm^{-1} and is either the value reported in the reference cited or calculated from the $A_{1cm}^{1\%}$ value and the molecular weight.

[b] $A_{1cm}^{1\%}$ is the absorbance for a 1% solution in a 1-cm cuvette and is either the value reported in the reference cited or calculated from the ϵ and the molecular weight. The relationship between ϵ, $A_{1cm}^{1\%}$ and molecular weight, MW, is $10\epsilon = (A_{1cm}^{1\%})$ (MW).

[c] Refers to the wavelength cited and may not be the peak of the absorption band.

[d] Abbreviations used: SC, corrected for light scattering; P$_i$, phosphate; GdmCl, guanidinium chloride; PP$_i$, pyrophosphate; Gro-P, glyerophosphate; S$_2$ threitol, dithiothreitol; NaDodSO$_4$, sodium dodecyl sulfate; HSEtOH, 2-mercaptoethanol; Gly$_2$, glycyl-glycine; ImzAc, imidazoleacetate; Tes, *N*-Tris(hydroxymethyl)methyl-2-aminoethanesulfonic acid; SucNBr, *N*-bromosuccinimide; albumin, bovine serum albumin.

Methods of protein determination: Dry wt, dry weight; AA, amino acid analysis; Refr., R, refractometry; Biuret, colorimetric method; Folin, colorimetric method; N, nitrogen determination; UC, ultracentrifuge; FC, fringe counting (interferometry); DR, differential refractometry; Kjedahl: Lowry.

[e] Methods for determining nitrogen concentration in order to determine protein concentration.

MOLAR ABSORPTIVITY AND $A_{1cm}^{1\%}$ VALUES FOR PROTEINS AT SELECTED WAVELENGTHS OF THE ULTRAVIOLET AND VISIBLE REGION (continued)

Protein	$\epsilon^a(\times\ 10^{-4})$	$A_{1cm}^{1\%}$ [b]	nm[c]	Ref.	Comments[d]
Acid deoxyribonuclease (deoxyribonuclease II) (EC 3.1.4.6) Pig spleen	4.6	12.1	280	635	MW = 38,000 (635)
Aconitase (aconitate hydrase) (EC 4.2.1.3) Pig heart	–	13.7	280	778	–
Actin Muscle	–	11.08	280	1324	–
F-Actin Rabbit	–	11.08	280	1156	Kjeldahl
Rabbit muscle	–	9.65	280	3	0.1 M KCl
	–	11.49	280	4	–
	–	11.5	280	636	–
G-Actin Rabbit muscle	5.05	10.97	280	5	MW = 46,000 (5)
β-Actinin Rabbit muscle	–	9.8	278	878	
Acyl phosphatase (EC 3.6.1.7) Horse muscle	1.09	11.58	280	639	pH 5.3, 0.05 M; AcO$^-$, MW = 9,400 (639)
Acyl-carrier-protein *Escherichia coli*	0.27	3.0[g]	275	637	MW = 9,100 (637); pH 7.0, 0.01 M KP$_i$, data from Figure 2 (638)
	0.18	–	275	638	–
Acyl-CoA dehydrogenase (see fatty acyl-CoA dehydrogenase)					
Adenosine deaminase (EC 3.5.4.4) Calf spleen	2.7	8.15	278	6	Est. from Figure 1 (6) MW = 33,120 (6)
Aspergillus oryzae	27.8	13.0	280	879	MW = 214,000 (879)
Adenosine 5'-phosphate deaminase (EC 3.5.4.17) Rabbit muscle	–	9.13	280	880	Dry wt
	–	9.3	280	881	–
Adenosine 5'-phosphate nucleosidase (EC 3.2.2.4) *Azotobacter vinelandii*	5.58	9.73	280	1325	pH 8, 0.05 M triethanolamine · HCl containing 0.1 mM S$_2$ threitol and 1 mM EDTA; MW = 57,300 (1325)

[g]Optical density used for calculation corrected for light scattering by extrapolation from 350 nm.

MOLAR ABSORPTIVITY AND $A_{1cm}^{1\%}$ VALUES FOR PROTEINS AT SELECTED WAVELENGTHS OF THE ULTRAVIOLET AND VISIBLE REGION (continued)

Protein	$\epsilon^a (\times 10^{-4})$	$A_{1cm}^{1\%}$ [b]	nm[c]	Ref.	Comments[d]
Adenosine triphosphate sulfurylase (sulfate adenylyltransferase) (EC 2.7.7.4) see also ATP-sulfurylase.					
Penicillium chrysogenum	—	8.71	278	882	—
Adenovirus					
Hexon	—	14.6	279	1326	pH 7, 0.01 M NaP$_i$
Adenylic acid deaminase (EC 3.5.4.6)					
Rat muscle	28.5	9.84	280	1327	Dry wt, MW = 290,000 (1327)
Adrenodoxin					
Beef	1.14	—	276	1157	—
Beef adrenals	1.3	—	276	779	—
	1.26	—	325	779	—
	1.26	—	340	779	—
	0.98	—	414	779	—
	0.84	—	455	779	—
Beef adrenal cortex	0.579	—	276	1328	Values cited are per mole of Fe
	0.641	—	320	1328	
	0.496	—	414	1328	
	0.421	—	455	1328	
	—	6.75	276	1328	
	—	5.78	414	1328	
Apo-	0.76	—	276	779	
	0.35	—	276	1157	
Aequorin					
Aequorea	8.65	27.0[h]	280	780	MW = 32,000 (780); protein det. by dry wt
Apoaequorin-SH	—	18.2	280	780	
Apoaequorin-SO	—	18.2	280	780	
Agglutinin					
Wheat germ	10.9		280	1329	pH 7.0, 0.01 M NaP$_i$
	12		272	1329	
Alanine dehydrogenase (EC 1.4.1.1)					
Bacillus subtilis	51.2	22.3	280	883	pH 8, TrisCl, 0.05 M, MW = 230,000 (883)
Alanine racemase (EC 5.1.1.1)					
Pseudomonas putida	—	10.8	275	781	pH 7.4, 0.005 M KP$_i$, data from Figure 2 of Reference 781
Albocuprein					
Human, brain					
I	8.4	11.65	280	1158	pH 6, 0.05 M NaCl/0.05 M AcONa, dry wt, MW = 72,000 (1158)

[h] 1 mg of aequorin in 1 ml added to 10^{-4} EDTA gives an OD of 2.52 at 280 nm. After freeze-drying and redissolving, the OD is now 2.25 at 280 nm. (780)

MOLAR ABSORPTIVITY AND $A_{1cm}^{1\%}$ VALUES FOR PROTEINS AT SELECTED WAVELENGTHS OF THE ULTRAVIOLET AND VISIBLE REGION (continued)

Protein	$\epsilon^a(\times 10^{-4})$	$A_{1cm}^{1\%}$ [b]	nm[c]	Ref.	Comments[d]
Albocuprein (continued)					
II	1.2	8.63	280	1158	pH 6, 0.05 M NaCl/0.05 M AcONa, dry wt, MW = 14,000 (1158)
Albumin	–	10.6	278	1159	
Beef serum	–	6.49	280	1160	6 M GdmCl
	–	6.62	278	1161	pH 2, 0.01 N HCl
	–	3.58	255	1162 ⎫	
	–	6.14	280	1162 ⎬	pH 7.0, 0.01 M P$_i$
	–	0.50	310	1162 ⎭	
	–	6.8	280	12	–
	–	6.67	279	13	–
	–	6.6	280	14	–
	–	6.6	279.5	15	–
	–	6.6	279	16	–
	· 3.96	–	280	17	pH 7
	4.36	6.61	280	11	MW = 66,000 (11)
	–	6.3	280	640	Water
	–	270	210	640	Water
	–	840	191	640	Water
	4.69	6.9	279	641	MW = 68,000 (641)
	–	6.7	278	642	–
	2.77	–	288	643	–
	4.24	–	279	643	–
	19.43	–	234	643	–
	–	6.75	278	644	–
	–	6.2	280	645	–
	–	3.7	253.7	645	–
Beef, mercapto- (see also mercaptoalbumin, beef)	–	6.82	279	894	–
	–	3.03	253	1330 ⎫	pH 6.2
	–	6.54	278	1330 ⎭	
	–	6.67	277.5	884	–
	4.37	–	280	885	2°C, alcohol-water mixtures
	4.6	–	280	886	–
	–	8.2	280	887	–
	–	650	191.4	887	–
	4.2	6.2	280	888	pH 7.96, TrisCl, MW = 68,000 (888), Water (889)
Fragment F$_2$	–	5.51	278	890	Dry wt
	1.71	5.51	278	647	MW = 31,000 (647)
Fragment F$_3$	2.74	7.55	278	647	MW = 36,300 (647)
	–	7.55	278	890	Dry wt
S-Carboxymethyl-	–	5.96	278	891	pH 8, 6 M GdmCl– 0.02 M EDTA
S-Cysteinyl-	–	5.96	278	891	pH 8, 6 M GdmCl– 0.02 M EDTA
	–	6.14	278	892	6 M GdmCl

MOLAR ABSORPTIVITY AND $A_{1cm}^{1\%}$ VALUES FOR PROTEINS AT SELECTED WAVELENGTHS OF THE ULTRAVIOLET AND VISIBLE REGION (continued)

Protein	$\epsilon^a(\times 10^{-4})$	$A_{1cm}^{1\%}$ [b]	nm[c]	Ref.	Comments[d]
Albumin (continued)					
Polypeptidyl derivatives					
Gly-261[i]	4.9	5.8	278	893	—
L-Phe-31	4.8	6.4	278	893	—
L-Phe-36	4.7	6.3	278	893	—
DL-Phe-48	4.9	6.3	278	893	—
L-Glu-13	4.8	6.8	278	893	—
L-Glu-41	4.8	6.4	278	893	—
L-Glu-73	4.7	5.9	278	893	—
L-Glu-218	4.6	4.7	278	893	—
L-Glu-275	4.9	4.7	278	893	—
L-Lys-2	4.9	7.0	278	893	—
L-Lys-14	5.0	7.1	278	893	—
Methylated	—	6.5	280	645	—
	—	3.7	253.7	645	—
Acetylated	—	6.9	280	645	80% acetylated
	—	3.7	253.7	645	on amino groups
Diazotized	—	7.5	280	645	—
	—	7.3	253.7	645	—
Guanidinated	—	9.2	280	645	—
	—	11.0	253.7	645	45 groups
Iodinated	—	12.0	312	645 ⎫	
	—	7.5	280	645 ⎬	32 mol I/mol
	—	18.6	253.7	645 ⎭	
Glutaraldehyde modified	—	26.2	280	646	pH 8, borate
S-β-Pyridylethyl-	—	12.0	274	1161	pH 2, 0.01 N HCl
Cow's milk	4.55	6.6	280	901	MW = 69,000 (901)
Human serum	—	193	210	648	—
	—	143	215	648	—
	—	4.3	254	648	—
	—	7.15	280	648	—
	—	3.92	255	1162 ⎫	
	—	5.94	280	1162 ⎬	pH 7.0, 0.01 M P_i
	—	0.90	310	1162 ⎭	
	—	6	280	7	
	4.0	5.8	280	8	MW = 69,000 (8)
	3.6	5.31	280	9	MW = 68,000 (10)
	3.5	5.3	280	11	MW = 66,000 (11)
	3.52	5.03	277.5	895	pH 2 MW = 70,000 (895)
	—	7.15	280	896	—
	—	193	210	896	—
	—	143	215	896	—
	—	4.3	254	896	—
N$_2$ph-[q]	—	30.8	290	897	0.1 M NaOH
	—	11.7	360	897	0.1 M NaOH
Human mercapto-					
Fraction I	—	5.7	280	898	—
Fraction II	—	5.61	280	898	—
Fraction III	—	5.31	280	898	—
Fraction IV	—	5.8	280	898	—
Pig serum	—	6.72	280	899	pH 8.6, 0.2 M TrisCl, 22°C, dry wt
Rabbit serum	—	6.6	280	900	—

[i]Gly-261 means that 261 moles of glycine have been attached to albumin. Other derivatives have been prepared similarly using other amino acids and are so indicated.

[q]N$_2$ph- = 2,4-dinitrophenyl.

MOLAR ABSORPTIVITY AND $A_{1cm}^{1\%}$ VALUES FOR PROTEINS AT SELECTED WAVELENGTHS OF THE ULTRAVIOLET AND VISIBLE REGION (continued)

Protein	$e^a (\times 10^{-4})$	$A_{1cm}^{1\%}$ [b]	nm[c]	Ref.	Comments[d]
Albuminoid (insoluble protein)					
Young rat, lens	–	18.4	280	782 ⎫	
Young rat X-rayed, lens	–	18.6	280	782 ⎪	
Old rat, lens	–	17.7	280	782 ⎬ pH 9.8, 0.05 M borate/8 M
Medium-aged dogfish, lens	–	22.4	280	782 ⎪ urea
Old dogfish, lens	–	21.2	280	782 ⎭	
Albuminoid sulfonated					
Rat lens	–	17.3	280	902	–
	–	18.0	280	903	–
Dogfish lens	–	25.1	280	903	–
Beef lens	–	9.8	280	903	–
Human lens					
0–10 years old	–	11.1	280	903	–
11–20 years old	–	15.0	280	903	–
40–49 years old	–	15.9	280	903	–
50–59 years old	–	15.0	280	903	–
60–69 years old	–	16.5	280	903	–
70–79 years old	–	17.1	280	903	–
80– years old	–	18.0	280	903	–
Alcohol dehydrogenase					
(EC 1.1.1.1)					
Horse liver	–	4.55	280	904	–
	–	4.2	280	905	Dilute neutral buffer
	–	4.26	280	906	pH 7.2, 3 M GdmCl
	3.59	–	280	907	MW = 79,000 (907)
	3.82	4.55	280	18	–
	–	4.5	280	19	–
	3.83	4.6	280	20	MW = 83,300 (20)
	3.54	4.2	280	22	MW = 84,000 (21)
Zn-	3.44	4.3	280	908 ⎫	
Co-	3.92	4.9	280	910 ⎬ MW 80,000 (909)
Cd-	4.56	5.7	280	910 ⎪	
	1.02	1.25	245	910 ⎭	
Human liver	–	4.6	280	911	pH 7.0, 0.03 M P$_i$, 0.07 M NaCl
	5.3	6.1	280	23	pH 7.0, 0.1 μM NaP$_i$, MW = 87,000 (23)
Yeast	–	12.1	280	912	MW = 140,000 (913)
	20.78	14.8	278	913	MW = 141,000 (914)
	–	14.6	280	914	
	18.9	–	280	24	pH 8.1, 0.08 M glycine
	–	12.6	280	25	–
Arachis hypogea					
(peanuts)	7.2	6.4	278	915	MW = 112,000 (915)
Drosophila melanogaster	3.96	9.0	280	916	MW = 44,000 (916)
Aldehyde dehydrogenase					
(EC 1.2.1.3-.5)					
Horse liver	–	20.8	280	1163	pH 7, AcONH$_4$, dry wt
Pseudomonas aeruginosa	19.5	10.4	280	783	pH 7.0, 1 mM KP$_i$, protein con. det. by dry wt, MW = 187,000 (783)

MOLAR ABSORPTIVITY AND $A_{1cm}^{1\%}$ VALUES FOR PROTEINS AT SELECTED WAVELENGTHS OF THE ULTRAVIOLET AND VISIBLE REGION (continued)

Protein	$\epsilon^a(\times 10^{-4})$	$A_{1cm}^{1\%}$ [b]	nm[c]	Ref.	Comments[d]
Aldehyde dehydrogenase (continued)					
Yeast	13.4	6.7	280	26	Reference states: "1 mg enzyme . . . OD equals 0.67," no volume given, assumed 1 ml, MW = 200,000 (26)
Aldehyde oxidase (EC 1.2.3.1)					
Rabbit liver	6.3	—	450	27	—
	2.2	—	550	27	—
Aldolase (EC 4.1.2.13)					
Rabbit muscle	—	9.1	280	28	—
	—	12.1	280	29	—
	11.8	8.32	277	30	pH 2
	13.3	9.38	280	30	pH 5.7, MW = 142,000 (31)
	—	7.8	276	32	pH 2
	11.8	7.4	280	917	pH 7.5–13, MW = 160,000 (918)
	—	8.4	280	919	pH 12.5, 0.1 M borate
	—	9.6	289.5	919	
	—	8.16[j]	276	920	3 M GdmCl
	—	8.20[j]	276	920	5 M GdmCl
	—	8.21[j]	276	920	6 M GdmCl
	—	8.23[j]	276	920	7 M GdmCl
Succinyl-	—	8.2	276.5	921	Dry wt/KN[k]
Rabbit liver	13.09	8.5	280	33	MW = 154,000 (33)
	13.3	8.40	280	922	MW = 158,000 (922)
Gallus domesticus, muscle					
(chicken)	16	10.3	280	923	pH 6.5, MW = 158,000 (923)
Liver	13.7	8.6	280	924	pH 7.5, 2 mM Tris, 0.2 mM EDTA MW = 160,000 (925)
Rat muscle	15.0	9.39	280	926	MW = 160,000 (926)
Gradus morhua					
(codfish) muscle	15.2	9.5	280	927	pH 7.5, MW = 160,000 (927)
Spinach	20.76	17.3	280	34	pH 7.4, 0.05 M P$_i$, MW = 120,000 (34)
	—	13.3	280	930	pH 7.5, 0.1 M P$_i$, Lowry[l]
	—	11.0	280	930	Dry wt
Yeast	7.95	10.6	280	35	MW = 75,000 (35)
	8	10	280	928	MW = 80,000 (928), dry wt
	8.15	10.2	280	929	MW = 80,000 (929)
	8.0	10.1	280	929	Dry wt
	8.5	10.6	280	929	DR[m]
	7.9	9.9	280	929	FD[n]

[j]Calculated from an equation in Reference 920.
[k]KN, protein concentration determined by Kjeldahl nitrogen.
[l]Lowry, protein concentration determined by Lowry method using bovine serum albumin as standard.
[m]DR, protein concentration determined by differential refractometry.
[n]FD, protein concentration determined by fringe displacement method.

MOLAR ABSORPTIVITY AND $A_{1cm}^{1\%}$ VALUES FOR PROTEINS AT SELECTED WAVELENGTHS OF THE ULTRAVIOLET AND VISIBLE REGION (continued)

Protein	$\epsilon^a (\times\ 10^{-4})$	$A_{1cm}^{1\%}$ [b]	nm[c]	Ref.	Comments[d]
Aldolase (continued)					
Lobster muscle					
(*Homarus americanus*)	17.9	11.2	280	1164	MW = 160,000 (1164)
Shark muscle					
(*Mustelus canis*)	–	8.64	278.4	1165⎱	Dry wt, Kjeldahl
	–	8.60	280	1165⎰	
Rabbit liver	–	8.9	280	1166	Dry wt
Rabbit brain	–	8.8	280	1166	Refract
Aldolase, L-Rhamnulose 1-phosphate (EC 4.1.2.19)					
Escherichia coli	23.4	17.3	280	1167	MW = 135,000 (1167)
Aldolase, 3-Deoxy-2-keto-6-phosphogluconate (EC 4.1.2.14)					
Pseudomonas putida	–	8.63	280	1168	0.1 N NaOH
Aldose 1-epimerase (EC 5.1.3.3)					
Escherichia coli K12	–	10.8	280	1331	–
Allantoicase (EC 3.5.3.4), 0.9S					
Pseudomonas aeruginosa	–	27.3	280	1169	Calcd from data in Figure 4 (1169), pH 7.7
	–	26.0	280	1169	Calcd from data in Figure 4 (1169), pH 4.6
	–	24.3	280	1169	Calcd from data in Figure 4 (1169), pH 4.6, in the presence of 0.1 M glycolate
	–	31.7	280	1169	Lowry
Allergen					
Short ragweed pollen					
Antigen E	–	11.3	280	1170	pH 7.15
Antigen Ra.3	16.4	10.9	280	1171	pH 7.3, 0.005 M NH$_4$ HCO$_3$, MW = 15,000 (1171)
Atopic					
Rye grass pollen					
I-B	–	15.0	280	1332⎱	pH 7, MW =
	–	2.18	305	1332⎰	34,000 (1332)
II-B	–	10.3	280	1332⎱	pH 7, MW =
	–	0.88	305	1332⎰	11,000 (1332)
B	–	14.1	280	1332⎫	
	–	3.10	305	1332⎬ pH 7	
D[IEP]	–	14.7	280	1332⎪	
	–	4.75	305	1332⎭	
K	–	14.8	280	1332	pH 7, MW = 38,200 (1332)
Pool Cc	–	7.63	280	1332	
	–	1.05	305	1332	
Trifidin A	–	4.1	280	1332	
	–	1.20	305	1332	
Ipecac IPC-D	–	10.5	280	1332	pH 7
	–	4.10	305	1332	
Liquorice SL-F	–	11.0	305	1332	
Pyrethrum					
Whole dialysate	–	76.8	280	1332	
	–	69.8	305	1332	

MOLAR ABSORPTIVITY AND $A_{1cm}^{1\%}$ VALUES FOR PROTEINS AT SELECTED WAVELENGTHS OF THE ULTRAVIOLET AND VISIBLE REGION (continued)

Protein	$\epsilon^a(\times 10^{-4})$	$A_{1cm}^{1\%}$ [b]	nm[c]	Ref.	Comments[d]
Allergen (continued)					
Kapok KP-E	—	76.2	280	1332 ⎫	
	—	64.4	305	1332 ⎬ pH 7	
Cotton CL-E	—	20.6	280	1332 ⎪	
	—	13.0	305	1332 ⎭	
Cotton seed CS 60C	—	6.58	255	1162	—
	—	6.85	280	1162	—
	—	4.89	310	1162	—
Castor bean [CB-1A] SRI	—	3.38	280	1332 ⎫ pH 7	
	—	5.21	305	1332 ⎭	
Human dandruff HD-E	—	9.32	255	1162	—
	—	10.02	280	1162	—
	—	4.36	310	1162	—
Horse dandruff	—	8.28	255	1162	—
	—	9.62	280	1162	—
	—	3.44	310	1162	—
Whole dialysate	—	6.40	280	1332 ⎫	
	—	1.20	305	1332 ⎪	
Feathers FE-B	—	58.0	280	1332 ⎪	
	—	43.3	305	1332 ⎪	
Caddis fly Pool 2	—	35.2	280	1332 ⎬ pH 7	
	—	15.5	305	1332 ⎪	
Alternaria	—	8.00	280	1332 ⎪	
	—	4.14	305	1332 ⎪	
Trichophytin	—	5.10	305	1332 ⎭	
	—	7.96	255	1162	—
	—	7.94	280	1162	—
	—	4.80	310	1162	—
House dust HE-E	—	8.1	305	1332	pH 7
	—	13.80	255	1162	—
	—	13.64	280	1162	—
	—	8.36	310	1162	—
Tomato TO-G	—	21.0	280	1332 ⎫ pH 7	
	—	12.5	305	1332 ⎭	
	—	14.56	255	1162	—
	—	13.80	280	1162	—
	—	9.40	310	1162	—
Cow's milk VM-5	—	9.20	280	1332 ⎫ pH 7, MW =	
	—	1.95	305	1332 ⎭ 36,000 (1332)	
Egg white VE₉	—	4.44	280	1332 ⎫ pH 7, MW =	
	—	0.40	305	1332 ⎭ 31,500 (1332)	
Hay HH-C	—	84.2	280	1332 ⎫ pH 7	
	—	67.0	305	1332 ⎭	
Succus liquiritiae	—	15.80	255	1162	—
	—	14.02	280	1162	—
	—	10.58	310	1162	—
Radix ipecacuanhae	—	9.66	255	1162	—
	—	11.02	280	1162	—
	—	3.84	310	1162	—
Alliin lyase (EC 4.4.1.4) Garlic (*Allium sativum*)	—	16.6	280	1172	Calcd from data in Figure 4 (1172), pH 7.5, 10% glycerol-0.02 M P_i

MOLAR ABSORPTIVITY AND $A_{1cm}^{1\%}$ VALUES FOR PROTEINS AT SELECTED WAVELENGTHS OF THE ULTRAVIOLET AND VISIBLE REGION (continued)

Protein	$\epsilon^a(\times 10^{-4})$	$A_{1cm}^{1\%}$ [b]	nm[c]	Ref.	Comments[d]
Amandin					
Almonds	–	7	280	1173	From Figure 1 (1173), pH 5.7
Amidophosphoribosyltransferase (EC 2.4.2.14) (see glutamine phosphoribosylpyrophosphate-amidotransferase and phosphoribosyldiphosphate amido-transferase)					
Amine dehydrogenase (amine oxidase)					
Pseudomonas AM 1	11.3	8.46	280	1174	pH 7.5, 0.05 M P$_i$, Lowry, MW = 133,000 (1174)
Amine oxidase (EC 1.4.3.4)					
Aspergillus niger	–	11.8	280	1179	–
Beef plasma	–	9.8	280	1180	–
Amine oxidase (EC 1.4.3.6) (see diamine oxidase and monoamine oxidase)					
D-Amino-acid oxidase (EC 1.4.3.3)					
Pig kidney	–	15.6	277	1175	–
	–	126	220	1175	–
	7.31	16.0	280	36	MW = 182,000 (36)
Batch I enzyme	–	23.0	274	37	–
Apo-	–	15.4	278	37	pH 8.5, 0.1 M PP$_i$
Batch II enzyme	–	19.8	274	37	pH 8.5, 0.1 M PP$_i$
Apo-	15.1	15.1	280	38	pH 8.3, M/60 PP$_i$, MW = 100,000 (39)
Apo-	17.5	14.0	280	40	pH 8.3, 0.1 M PP$_i$, MW = 125,000 (40)
Apo-	–	14	280	1176	pH 8.3, 0.1 M PP$_i$
L-Amino-acid oxidase (EC 1.4.3.2)					
Crotalus adamanteus	23.6	17.9	275	41	0.1 M KCl, MW = 132,000 (41)
	2.35	1.78	390	41	0.1 M KCl, MW = 132,000 (41)
	2.26	1.71	462	41	0.1 M KCl, MW = 132,000 (41)
Rat kidney	8.5	9.55	275	42	MW = 89,000 (42)
	1.07	1.20	358	42	MW = 89,000 (42)
	1.27	1.43	455	42	MW = 89,000 (42)
Amino-acid racemase (EC 5.1.1.10)					
Pseudomonas striata	–	8.3	280	1177	Data from Figure 3 (1177), pH 7.0, 0.01 M KP$_i$
Aminoacyl-tRNA: ribosome binding enzyme					
Rabbit reticulocytes	17.9	9.6	280	1178	Calc. from data in Figure 3 (1178), pH 7.5, 0.01 M KP$_i$, MW = 186,000 (1178)

MOLAR ABSORPTIVITY AND $A_{1cm}^{1\%}$ VALUES FOR PROTEINS AT SELECTED WAVELENGTHS OF THE ULTRAVIOLET AND VISIBLE REGION (continued)

Protein	$\epsilon^a (\times\ 10^{-4})$	$A_{1cm}^{1\%}$ [b]	nm[c]	Ref.	Comments[d]
Aminopeptidase (EC 3.4.11.1-.2) (see also leucine amino- peptidase)					
Pig kidney	–	16.3	280	43	pH 7.2, 0.06 M P_i
	–	12.28	266	43	pH 7.2, 0.06 M P_i
	–	125	225	43	pH 7.2, 0.06 M P_i
	–	168	215	43	pH 7.2, 0.06 M P_i
Pig kidney, particulate (EC 3.4.11.2)	–	16	280	1182	Estimated from figure in Reference 1182
	47.3	16.9	280	1183	MW = 280,000 (1183), refr. and Lowry
Rat kidney	–	16.1	280	44	–
Aeromonas proteolytica	4.18	14.4	278.5	1181	MW = 29,000 (1181)
B. stearothermophilus	41	10.2	280	1184	pH 7.2, 0.05 M TrisCl, 0.001 M Co$^+$, MW = 400,000 (1184)
Aminopeptidase (microsomal) (EC 3.4.11.2)					
Pig kidney	45.5	16.2	280	1186	MW = 280,000 (1186)
Aminopeptidase P (aminoacyl- proline aminopeptidase, EC 3.4.11.9)					
Escherichia coli B.	–	10.3	280	1187	Kjeldahl
Aminotransferase alanine[o] (EC 2.6.1.2)					
Rat liver	–	6.85	278.7	1333	DR, pH 7.0, 50 mM KP_i containing 0.5 mM S_2 threitol
Amylase (EC 3.2.1.1-.3)					
Human plasma	–	9	280	1188	Water, calcd from data in Figure 4 (1188)
Rat pancreas	–	20	280	1189	–
B. subtulis Takamine	–	25.2	280	1190	–
B. subtilis Kalle	–	25.2	280	1190	–
α-Amylase (EC 3.2.1.1)	–	24.4	290	1191	0.1 N NaOH
Human saliva	–	26	280	45	–
Pig pancreas	–	26	280	45	–
	12.8	25	280	46	Water, MW = 51,300 (47)
B. subtilis	–	25.3	280	45	–
A. oryzae	–	19.7	280	45	–
B. macerans	9.9	7.11	280	48	pH 6.2, 0.01 M P_i, MW = 139,000 (48)
Pirkka barley	8.55	15.0	280	1334	Dry wt, pH 7.0, 0.05 M NaP_i, MW = 57,200 (1334)
B. subtilis	9.35	19.8	280	1192	MW = 47,300 (1193)
	12.5	25.6	280	1195	MW = 49,000 (1195)

[o]Pyridoxal or pyridoxamine form.

MOLAR ABSORPTIVITY AND $A_{1cm}^{1\%}$ VALUES FOR PROTEINS AT SELECTED WAVELENGTHS OF THE ULTRAVIOLET AND VISIBLE REGION (continued)

Protein	$\epsilon^a (\times 10^{-4})$	$A_{1cm}^{1\%\,b}$	nm[c]	Ref.	Comments[d]
α-Amylase (continued)					
B. stearothermophilus	13.8	28.7	280	1194	MW = 48,000 (1194)
Rat pancreas	–	16.4	–[p]	1196	
Pig pancreas	12.0	24	280	1197	MW = 50,000 (1197), pH 7.4, Tris
B. subtilis	8.3	19.8	280	1198	MW = 41,900 (1198)
B. subtilis var. *saccharitikus*					
Fukumoto	8.2	20.0	280	1199	MW = 41,000 (1199) pH 6.8, 0.01 M AcO⁻
	7.9	19.3	280	1199	0.1 N NaOH
β-Amylase (EC 3.2.1.2)					
Sweet potato	–	17.7	280	1201	–
	26	17.1	280	49	pH 4.8, 0.016 M AcO⁻
Iodosobenzoate oxidized		17.05	–	1202	
α-Amylase inhibitor					
Wheat					
I	2.7	15.0	280	1200	MW = 18,215 (1200)
II	2.6	10.0	280	1200	MW = 26,200 (1200)
Anaphylatoxin					
Rat serum		4.1	280	1335	pH 7.2
Anthranilate synthase (EC 4.1.3.27)					
Salmonella typhimurium					
Component I	3.3	5.2	278	1202	pH 7.4, 0.05 M TrisCl, MW = 64,000 (1202)
Escherichia coli	2.18	3.64	280	1204	pH 7.0, 0.05 M KP$_i$, MW = 60,000 (1204), calcd from data in Figure 5 (1204)
	2.94	4.91	295	1204	0.1 N NaOH, MW = 60,000 (1204), calcd from data in Figure 5 (1204)
Anthranilate synthase: anthranilate phosphoribosyl-transferase complex (EC 4.1.3.27: 2.4.2.18)					
Salmonella typhimurium	16.1	5.75	280	1205	pH 7.4, 0.05 M KP$_i$, MW = 280,000 (1205)
Antibody					
Rabbit					
Anti-N$_2$ph[q]	–	15.7	279	1206	–
Anguilla rostrata (eel)					
Anti-human blood group					
H[O]	–	12.696	278	1207	Water

[p] Wavelength not cited.

[q] N$_2$ph, 2,4-dinitrophenyl-.

MOLAR ABSORPTIVITY AND $A_{1cm}^{1\%}$ VALUES FOR PROTEINS AT SELECTED WAVELENGTHS OF THE ULTRAVIOLET AND VISIBLE REGION (continued)

Protein	$\epsilon^a (\times 10^{-4})$	$A_{1cm}^{1\%}$ [b]	nm[c]	Ref.	Comments[d]
Antigen					
Human					
Hepatitis associated (Australia)	—	9.42	260	1208	—
Blood group N active					
Erythrocyte, NN	65	10.90	274	1209	MW = 595,000 (1209)
Meconium	23.2	4.49	274	1209	MW = 520,000 (1209)
Paramecium aurelia					
Immobilization	—	11.9	277	1210	From Figure 1 (1210)
Apocytochrome *c*					
Horse heart	—	9.2	277	1336	—
Apolipoprotein Glu-II					
Human plasma	1.91	10.97	276	1337 ⎱	pH 8.0, 0.01% EDTA,
	1.80	10.35	280	1337 ⎰	MW = 17,380 (1337)
α-L-Arabinofuranosidase (EC 3.2.1.55)					
Aspergillus niger	12.7	23.1	280	1211	pH 7.0, 0.02 *M* NaP$_i$, MW = 53,000 (1211), data from Figure 8 (1211)
Arachin	—	7.98	278	50	8 *M* urea, 0.1 *M* sulfite
	—	7.85	278	50	6 *M* GdmCl, 0.1 *M* sulfite
Arachis hypogaea (peanut)	—	8.8	281.5	1338	pH 10.5, 0.1 *M* P$_i$
Arginase (EC 3.5.3.1)					
Rat liver	—	10.9	280	51	
Beef liver	—	9.6	278	1212	Dry wt
Pig liver	—	13.0	280	1213 ⎱	
Chicken liver	—	260	210	1339 ⎱	pH 7.5, 0.05 *M*
	—	22	340	1339 ⎰	TrisCl
Arginine decarboxylase (EC 4.1.1.19)					
Escherichia coli	133	15.7	280	1214	MW = 850,000 dry wt and refr.
Arginine kinase (EC 2.7.3.3)					
Sipunculus nudus (Marine worm)	8.4	9.8	280	1215	MW = 86,000 (1215)
Cancer pagurus (Crab)	2.9	7.35	278	1216	MW = 39,500 (1216)
Homarus vulgaris (Lobster)	2.9	7.35	278	1216	MW = 39,500 (1216)
	—	7.8	275	1217	
	—	8.1	280	1218	pH 8.0 0.01 *M* P$_i$
	—	6.1	271	1219	Alkaline sol.
Arginine racemase (EC 5.1.1.9)					
Pseudomonas graveolens	15.5	9.3	280	1220	Dry wt, MW = 167,000 (1220)

MOLAR ABSORPTIVITY AND $A_{1cm}^{1\%}$ VALUES FOR PROTEINS AT SELECTED WAVELENGTHS OF THE ULTRAVIOLET AND VISIBLE REGION (continued)

Protein	$\epsilon^a (\times 10^{-4})$	$A_{1cm}^{1\%}$ [b]	nm[c]	Ref.	Comments[d]
Argininosuccinase (EC 4.3.2.1)					
Beef liver	25.8	13.0[r]	280	1221	
	–	7.1[r]	260	1221	pH 7.5, 0.05 M KP$_i$,
Beef kidney	25.0	12.5[r]	280	1221	MW = 202,000 (1221)
	–	6.8[r]	260	1221	
Aromatic α-ketoacid reductase (diiodophenyl pyruvate reductase, EC 1.1.1.96)					
Rat kidney	–	10	280	1222	pH 6.5, 0.005 M NaP$_i$
Ascorbate oxidase (EC 1.10.3.3)					
Cucumis sativas	1480[s]	1120[s]	280	1340	pH 7.0, 0.1 M P$_i$,
	61.6[s]	46.8[s]	607	1340	MW = 132,000 (1340)
Cucumber	0.53	–	330	1223	–
	0.97	–	607	1223	–
	0.36	–	760	1223	–
Corcubita pepo condensa (yellow crookneck squash)	28.5	–	280	1224	Dry wt
Corcubita pepo medullosa (green zucchini)	28.5	–	280	1224	Dry wt
Asparaginase (EC 3.5.1.1)					
Proteus vulgaris	–	6.6	280	1225	Dry wt, pH 7.0, 0.05 M NaP$_i$
Escherichia coli	9.9	7.46	280	1226	MW = 133,000 (1226)
	–	7.2	278	1342	pH 7
	–	9.9	292	1342	pH 13
E. coli HAP	8.83	6.26	278	1227	MW = 141,000 (1227)
E. coli B	9.2	7.1	278	649	MW = 130,000 (649) pH 7.3, P$_i$
	–	7.1	278	650	pH 5, 0.05 M AcONa
	–	7.1	278	650	pH 8.5, 0.05 M Tris
	–	6.5	276	650	7 M urea
	–	6.5	276	650	5 M guanidine
Succinylated monomer	–	6.7	276	1228	–
Erwinia carotovora	8.2	6.1	280	1341	pH 7.4
Asparaginase A	–	7.5	278	652	pH 7, P$_i$, calcd from opt. factor of 1.325
	–	9.5	290.5	652	0.1 N NaOH, calcd from opt. factor of 1.059
Escherichia coli ATCC 9637	–	7.9	277	651	pH 7, M/15 P$_i$, data from Figure 2 (651)
Deaminated	–	7.9	277	651	pH 7, M/15 P$_i$, data from Figure 2 (651) 651

[r]Amino acid analysis used for determining protein concentration.
[s]Values calculated from data and are for 8 Cu atoms per molecule.

MOLAR ABSORPTIVITY AND $A_{1cm}^{1\%}$ VALUES FOR PROTEINS AT SELECTED WAVELENGTHS OF THE ULTRAVIOLET AND VISIBLE REGION (continued)

Protein	$\epsilon^a (\times 10^{-4})$	$A_{1cm}^{1\%}$ [b]	nm[c]	Ref.	Comments[d]
Aspartate aminotransferase					
(EC 2.6.1.1)					
Pig heart muscle	11.1	14.1	280	1230	MW = 78,600 (1230)
Apo-	10.6	13.5	280	1231	pH 7.4, 0.1 M P$_i$, MW = 78,600 (1230)
Pig heart	13.5	15	280	1232 }	MW = 90,000 (1232)
Apo-	12.8	14.2	280	1232 }	
Pig heart	–	12.6	280	1233	pH 5.3, 0.1 M AcO$^-$ data from Figure 1 (1233)
α form	–	14.8	278	1234	–
Chicken heart	–	14.0	280	1235	pH 7.4, 0.1 M TrisCl, pH 7.5, 0.05 M KP$_i$
Soluble mixture	–	14.2	280	1235	
		15	280	53	pH 7.5
α soluble	–	13.7	280	1235	–
β soluble	–	14.5	280	1235	–
γ soluble	–	14.0	280	1235	–
Mitochondrial	–	13.2	280	1235	–
Apo-	–	14.5	280	53	pH 7.5
Rat brain					
Cytoplasmic	10.8	13.5	280	1236 }	pH 7.5, MW = 80,000 (1236)
Mitochondrial	8.65	10.8	280	1236 }	
Ox heart	–	14.40	278	1237	–
Apo-	–	14.14	278	1238	–
Beef kidney	12.5	13.4	280	1343	pH 8, 10 mM P$_i$
L-Aspartate β-decarboxylase					
(EC 4.1.1.12)					
A. faecalis	88	11	278	52	pH 6.8, 0.1 M P$_i$, MW = 800,000 (52)
Pseudomonas dacunhae	–	10.0	280	1229	pH 6.8, 0.1 M KP$_i$
D-Aspartate oxidase					
(EC 1.4.3.1)					
Octopus vulgaris	–	18.25	275	1344	Dry wt
Aspartate kinase					
(see aspartokinase)					
Aspartate transcarbamoylase					
(EC 2.1.3.2)					
Escherichia coli	18.2	5.9	280	54	MW = 310,000 (54)
	18.2	5.9	279	55	MW = 310,000 (54)
	18.4	5.9	284	56	MW = 310,000 (54)
Mercury derivative	–	7.6	280	1239	–
Regulatory subunit	–	8.0	280	1240	–
	–	7.2	280	1239	–
	–	12	284	57	In presence of HSEtOH MW = 30,000 (56)
Mercury derivative	–	8.3	280	1239	–
Zinc derivative	–	3.2	280	1239	–
Apo-	–	3.2	280	1239	–
Catalytic subunit	–	7.0	280	1240	–
	7.2	7.2	284	56	MW = 100,000 (56)
Permanganate modified	–	7.0	280	1241	–
5-Thio-2-nitrobenzoate					
derivative	–	9.2	280	1242	–

MOLAR ABSORPTIVITY AND $A_{1cm}^{1\%}$ VALUES FOR PROTEINS AT SELECTED WAVELENGTHS OF THE ULTRAVIOLET AND VISIBLE REGION (continued)

Protein	$\epsilon^a (\times 10^{-4})$	$A_{1cm}^{1\%}$ [b]	nm[c]	Ref.	Comments[d]
Aspartokinase (EC 2.7.2.4)					
Bacillus polymyxa	7.74	6.7	280	1243	pH 6.5, 6 M GdmCl/0.02 M KP_i MW = 116,000 (1243)
Escherichia coli					
Lysine sensitive	4.7	3.6	276	1244	MW = 130,000 (1244)
Aspartokinase I: homoserine dehydrogenase I (EC 2.7.2.4:1.1.1.3)					
Escherichia coli	5.4	6.3	278	1345	AA, pH 7.2, 20 mM KP_i containing 0.15 M KCl, 2 mM Mg titriplex, 1 mM L-threonine, and 1 mM S_2 threitol, MW = 86,000 (1345)
Threonine-sensitive	—	4.6	280	1245	—
	—	4.7	280	1246	6 M GdmCl
Aspartokinase II: homoserine dehydrogenase II (EC 2.7.2.4:1.1.1.3)					—
Escherichia coli K12	—	8.7	280	1247	
Aspergillopeptidase A (EC 3.4.23.6)					
A. saitoi	4.5	13.15	280	58	MW = 34,500 (58)
Aspergillopeptidase B (EC 3.4.21.15)					
A. oryzae	1.73	9.08	278	59	pH 5, 0.1 M AcO⁻ MW = 18,000 (59)
	1.62	9.00	280	59	pH 5, 0.1 M AcO⁻ MW = 18,000 (59)
ATP:AMP Phosphotransferase (EC 2.7.4.3)					
Human	1.43	6.67	279	1346	Dry wt, MW = 22,000 (1346)
ATP citrate-lyase (EC 4.1.3.8)					
Rat liver	—	11.8	280	1248	Calcd from data in Figure 5 (1248)
ATP Phosphoribosyltransferase (see phosphoribosyltransferase and phosphoribosy-ATP)	—	7.1	280	1248	Calcd from data in Figure 5 (1248), corrected for light scattering
ATP-Sulfurylase (sulfate adenylyltransferase) (EC 2.7.7.4)					
Penicillium chrysogenum	—	8.71	278	1347	Kjeldahl[e]
Avidin					
Egg white	8.3	15.7	280	1249	MW = 53,000 (1249)
		53	233	1250	—

MOLAR ABSORPTIVITY AND $A_{1cm}^{1\%}$ VALUES FOR PROTEINS AT SELECTED WAVELENGTHS OF THE ULTRAVIOLET AND VISIBLE REGION (continued)

Protein	$e^a(\times 10^{-4})$	$A_{1cm}^{1\%}$ [b]	nm[c]	Ref.	Comments[d]
Avimanganin					
Chicken liver					
Mitochondria	0.0508	0.057	480	1251 }	MW = 89,000 (1251)
	0.0250	0.028	600	1251 }	
Azobacterflavoprotein					
Azobacter vinelandii,					
oxidized	1.06	–	452	60	pH 7.0, 25 mM KP$_i$
	–	16.5	274	60	pH 7.0, 25 mM KP$_i$
Azurin					
Pseudomonas fluorescens	0.0285	–	459	1252	–
	0.350	–	625	1252	–
	0.032	–	781	1252	–
Pseudomonas aeruginosa	0.027	–	467	1252	–
	0.350	–	625	1252	–
	0.039	–	820	1252	–
Bacteriochlorophyll-protein complex					
Chloropseudomonas ethylicum	–	90	371	61	MW = 37,940/subunit (61)
Bacteriocin DF 13					
Enterobacter cloacae DF 13	5.5	9.87	280	931	pH 7.0, 0.06 M P$_i$, MW = 56,000 (931)
Biotin carboxylase (EC 6.3.4.14)					
Escherichia coli	–	6.25	280	932	–[t]
	–	6.25[u]	280	785	–
Brain					
Pig, basic	–	4	280	439	pH 6.8, 0.05 N P$_i$, est. from Figure 7 (439) no value cited for wavelength
Bromelain (EC 3.4.22.4-.5)	6.33	19.0	280	62	MW = 33,315 (62)
	6.68	20.1	280	63	MW = 33,000 (64)
C1 Inactivator					
Human plasma	–	4.5	280	1253	–
C1q component of complement					
Human serum	26.4	6.8	278	1254	1% NaDodSO$_4$ MW = 388,000 (1254)
C2 component of complement					
Guinea pig	–	13.9	280	786	–
Caerulein					
Hyla caerulae	0.725	–	280	787	80% ethanol
Calcitonin					
Salmon	0.15	4.5	280	788	–

[t]OD_{280} = (mg protein/ml)/I.6.
[u]Calculated from equation: $\frac{mg\ protein}{ml}$ = 1.6 × OD_{280}. (785)

MOLAR ABSORPTIVITY AND $A_{1cm}^{1\%}$ VALUES FOR PROTEINS AT SELECTED WAVELENGTHS OF THE ULTRAVIOLET AND VISIBLE REGION (continued)

Protein	$\epsilon^a (\times 10^{-4})$	$A_{1cm}^{1\%}$ [b]	nm[c]	Ref.	Comments[d]
Calcium-binding phosphoprotein					
Pig brain	0.36	3.1	Not cited	1348	Kjeldahl[e], MW = 11,500 (1348)
Chicken intestinal mucosa	–	21.9	280	451	pH 8.1, 0.2% TrisCl, 0.77% glycine, 0.1 mM glutathione, est. from Figure 8 (451)
Carbamate kinase (EC 2.7.2.2)					
S. faecalis	1.798	5.8	280	789	pH 7.5, 50 mM Na P$_i$, MW = 31,000 (789)
Carbamoylphosphate synthase (EC 2.7.2.5-.9)					
Rat liver	--	8.4	280	1349	DR
Carbonic anhydrase (carbonate dehydratase, EC 4.2.1.1)					
Beef					
B	5.7	19.0	280	1255	MW = 30,000 (1255)
	5.6	18.0	280	66	MW = 31,000 (66)
Beef erythrocyte					
B	5.2	16.8	280	656	MW = 31,000 (656), pH 7.4
Ox, rumen					
Isozyme a	–	17	280	1256	–
Isozyme b	–	17	280	1256	–
Gallus domesticus (Chicken)	5.6	20	280	1257	MW = 28,000 (1258)
Carcharhinus leucas (Bull shark)	7.5	20.9	280	1259	MW = 36,000 (1259)
Galeocerdo cuvieri (Tiger shark)	6.3	16.0	280	1259	MW = 39,500 (1259)
Guinea pig					
Blood/mucosa					
G.I. tract					
High activity	–	17.1	280	1260	Dry wt
	–	16.9[r]	280	1260	–
	–	16.7[v]	280	1260	--
	–	17.0	280	1260	Value used
Low activity	–	16.4	280	1260	Dry wt
	–	16.5[r]	280	1260	–
	–	16.1[v]	280	1260	–
	–	16.5	280	1260	Value used
Human erythrocytes					
A	4.67	16.3	280	653	MW = 28,600 (653)
A, generated from B in vitro	4.29	16.0	280	653	MW = 26,800 (653)
B	4.56	16.3	280	653	MW = 28,000 (653)
	4.9	16.3	280	65	pH 7.0, 0.1 ionic strength Na P$_i$, MW = 30,000 (65)
	4.89	16.3[w]	280	1262	MW = 30,000 (1262)

[v]Calculated according to Wetlaufer, *Adv. Protein Chem.*, 17, 362 (1962).
[w]The value of $A_{1cm}^{1\%}$ of 16.3 and ϵ = 4.89 × 10^4 taken from Rickli, Ghazanfar, Gibbons, and Edsall, *J. Biol. Chem.*, 239, 1065 (1964), and a MW of 30,000 assumed (Reference 1262).

MOLAR ABSORPTIVITY AND $A_{1\,cm}^{1\%}$ VALUES FOR PROTEINS AT SELECTED WAVELENGTHS OF THE ULTRAVIOLET AND VISIBLE REGION (continued)

Protein	$\epsilon^a(\times 10^{-4})$	$A_{1cm}^{1\%}$ [b]	nm[c]	Ref.	Comments[d]
Carbonic anhydrase (continued)	2.05	–	250	1262	
B	5.74	–	240	1262	
	12.8	–	235	1262	
	24.5	–	230	1262	MW = 30,000 (1262)
	36.0	–	225	1262	
	45.5	–	220	1262	
	54.5	–	215	1262	
B, nitrated	–	17.5	280	654	pH 9
C	5.34	17.8	280	65	pH 7.0, 0.1 ionic strength Na P_i, MW = 30,000 (68)
	5.05	18.7	280	653	MW = 27,000 (653)
	–	17.8	280	655	–
	5.34	17.8[x]	280	1262	
	2.01	–	250	1262	
	5.47	–	240	1262	
	12.9	–	235	1262	MW = 30,000 (1262)
	25.0	–	230	1262	
	37.0	–	225	1262	
	48.0	–	220	1262	
	60.0	–	215	1262	
D	4.52	16.0	280	653	MW = 28,200 (653)
D, generated from B in vitro	4.14	16.1	280	653	MW = 25,700 (653)
G	4.9	18.5	280	653	MW = 26,500 (653)
H	5.14	18.5	280	653	MW = 27,800 (653)
O	3.98	15.4	280	653	MW = 25,800 (653)
P	4.21	15.4	280	653	MW = 27,300 (653)
Pmut	4.47	16	280	1261	MW = 28,000 (1261)
Horse erythrocyte	4.38	15.9	280	653	MW = 27,500 (653)
B	3.95	13.6	280	657	MW = 29,007 (657)
C	3.73	13.4	280	657	MW = 27,918 (657)
Rat erythrocyte					
1a	5.22	18	280	658	MW = 29,000 (658)
2	5.22	18	280	658	MW = 29,000 (658)
3	4.93	17	280	658	MW = 29,000 (658)
Prostate					
1b	5.22	18	280	658	MW = 29,000 (658)
Parsley (*Petroselinum crispum* var. *latifolium*)	3.18	11.3	280	659	MW = 28,150 (659)
Apocarbonic anhydrase	5.7	–	280	660	–
Precursor of above	5.7	–	280	660	–
Pig erythrocytes					
B	5.3	17.4	280	1263	MW = 30,375 (1263)
	5.6	18.5	280	1264	MW = 30,000 (1264)
C	4.6	15.2	280	1264	MW = 30,000 (1264)
Spinach	–	8.6	280	1265	pH 7.2, 0.03 M P_i
Monkey					
B	4.88	–	280	67	–
C, *M. mulata*	5.35	17.8	280	67	MW = 30,000 (67)
Neisseria sicca	3.575	12.5	280	1350	Kjeldahl,[e] MW = 28,600 (1350)

[x]The value of $A_{1cm}^{1\%}$ of 17.8 and an ϵ of 5.34 × 10⁴ taken from Nyman and Lindskog, *Biochim. Biophys. Acta,* 85, 141 (1964).

MOLAR ABSORPTIVITY AND $A_{1cm}^{1\%}$ VALUES FOR PROTEINS AT SELECTED WAVELENGTHS OF THE ULTRAVIOLET AND VISIBLE REGION (continued)

Protein	$\epsilon^a (\times 10^{-4})$	$A_{1cm}^{1\%\ b}$	nm[c]	Ref.	Comments[d]
Carboxylesterase (EC 3.1.1.1)					
Ox liver	20.1	13.40	280	1266	pH 7.92, 0.15 M Tris, MW = 150,000 (1267)
Pig kidney					
Microsomes	15.3	9.4	280	1268	MW = 163,000 (1268)
Liver	20.27	13.05	280	69	pH 8.16, 0.15 M Tris, MW = 172,000 (69)
S-Carboxymethylalbumin, beef (see Albumin, beef)					
Carboxypeptidase (EC 3.4.12.1-.3)					
Beef pancreas	–	19.4	278	70	10% LiCl
	6.32	18.1	278	71	–
	6.42	–	278	72	–
	6.45	–	278	73	–
	6.49	–	278	74	–
	6.67	–	278	75	–
	–	23	280	76	–
	7.9	–	280	77	–
	8.6	25	278	78	MW = 34,300 (78)
Carboxypeptidase A					
Ox					
Squalis acanthias	6.48	18.5	280	790	MW = 35,000 (790)
Co II	–	0.0205	530	1351	–
	–	0.0195	572	1351	–
Co III	–	0.0500	503	1351	–
A_1, Pig	6.72	19.6	278	79	MW = 34,800 (79)
A_2, Pig	6.72	19.6	278	79	–
A_α, Beef	6.49	18.8	278	80	pH 7.0, 0.5 M NaCl, 0.01 M Tris, MW = 34,600 (80)
A, Acetyl-, beef	6.17	–	280	72	In the absence of β-phenylpropionate
	6.01	–	278	73	In the absence of β-phenylpropionate
	5.9	–	280	72	In the presence of β-phenylpropionate
	5.78	–	278	73	In the presence of β-phenylpropionate
A, Arsonilazo-, beef	7.32	–	278	1353	–
A, Succinyl-, beef	–	18.3	280	81	–
	6.47	–	278	71	–
Carboxypeptidase B					
Pig	7.34	21.4	278	82	pH 8.0, 0.005 M Tris
Beef	7.35	21	280	83	MW = 34,600 (83)
Barley	–	16.5	280	791	Water
P. omnivorum	5.53	17.6	278	792	pH 6.5, MW = 31,400 (792)
Gossypium hirsutum	17.3	20.5	280	1352	MW = 84,500 (1352)
Carnitine acetyltransferase (EC 2.3.1.7)					
Pigeon breast muscle	4.8	8.25	280	1269	

MOLAR ABSORPTIVITY AND $A_{1\,cm}^{1\%}$ VALUES FOR PROTEINS AT SELECTED WAVELENGTHS OF THE ULTRAVIOLET AND VISIBLE REGION (continued)

Protein	$\epsilon^a (\times 10^{-4})$	$A_{1cm}^{1\%\,b}$	nm^c	Ref.	Comments[d]
Carotenoid-protein complex					
Pecten maximus					
ovary	—	9.7	280	1270	pH 7, 0.2 M P$_i$
Plesionika edwardsi	—	13	280	1270	pH 7, 0.2 M P$_i$, data from Figure 3 (1270)
Carotenoproteins[y]					
Aristeus antennatus					
Carapace					
α (+ salt)	12.3	—	593	1271	—
Stomach					
α (+ salt)	12.4	—	588	1271	—
Scyllarus arctus					
Carapace					
α (+ salt)	12.0	—	616	1271	—
Clibanarius erythropus					
Exoskeleton					
α (+ salt)	12.6	—	620	1271	—
Labidocera acutifrons	—	152	640	1354	pH 7, 0.02 M P$_i$
Casein	—	10.0	280	84	—
French Friesian cows	—	6.7	278	1272	—
Bovine					
βA	—	4.6	280	1273	—
βB	—	4.7	280	1273	—
βC	—	4.6	280	1273	—
30% acid prep.	—	8.4	278	85	—
2% acid prep.	—	8.6	278	85	—
3% Am. sulfate prep.	—	8.7	278	85	—
20% Am. sulfate prep.	—	8.7	278	85	—
Calcium gel prep.	—	8.1	278	85	—
α$_s$	2.73	10.1	280	86	MW = 27,000 (85)
α$_{s1}$	2.73	10.1	280	87	MW = 27,000 (85)
β	1.15	4.6	280	88	MW = 25,000 (85)
k	2.44	12.2	280	89	MW = 29,000 (85)
Catalase (EC 1.11.16)					
Human erythrocyte	28	12.5	280	90	MW = 225,000 (90), $A_{1cm}^{1\%}$ = 16.9–18.7 at 405 nm (90)
	—	14.4	280	91	—
	—	17.8	405	91	—
	30.8	13.2	280	1355 ⎫	MW = 232,400 (1355)
	39.7	17.1	405	1355 ⎭	
Beef erythrocyte	43.1	16.8	405	661	MW = 257,000 (661)
Liver	82.2	36.5	276	662	MW = 225,000 (662), pH 7.3, 0.05 M P$_i$
	63.6	28.2	404	662	MW = 225,000 (662), pH 7.3, 0.05 M P$_i$
	31	—	405	92	pH 7.4
	38	—	280	92	pH 7.4
	32.4	13.5	405	93	MW = 240,000 (93)
	31.4	12.9	276	1274 ⎫	MW = 240,000 (1274)
	16.6	6.9	405	1274 ⎭	

[y] All species are decapods.

MOLAR ABSORPTIVITY AND $A^{1\%}_{1cm}$ VALUES FOR PROTEINS AT SELECTED WAVELENGTHS OF THE ULTRAVIOLET AND VISIBLE REGION (continued)

Protein	$\epsilon^a (\times 10^{-4})$	$A^{1\%\ b}_{1cm}$	nm^c	Ref.	Comments^d
Catalase (continued)					
Horse liver	67.5	27	275	663	pH 8, 0.004 M P$_i$, data from Figure 1 (663)
	8.3	–	623	664	
	9.4	–	536	664	–
	78.5	–	400	664	–
	85.0	–	280	664	–
	65	–	280	665	pH 7.15, P$_i$, MW = 225,000, data from Figure 4 (665)
Horse blood	65	–	280		pH 7.15, P$_i$, MW = 225,000 data from Figure 4 (665)
Rat liver	–	15.8	276	1356	From Figure 1 (1356)
	–	16	407	666	–
	39.7	15.5	276	667	MW = 256,000 (667)
	43.0	16.8	407	667	MW = 256,000 (667)
Spinach	–	1.97	502	668⎱	pH 7.4, 0.1 M Tricene-NaOH
	–	1.38	620	668⎰	
	–	14.8	278	94	pH 7.4, 0.1 M Tricene-NaOH
	–	14.9	404	94	pH 7.4, 0.1 M Tricene-NaOH
Commercial					
Crystalline	41	16.4	276	669⎫	
	30	12.0	406	669	
Lyophilized-A	32	12.8	276	669⎬ MW = 250,000 (669)
	20	8.0	406	669	
Lyophilized-B	34	13.6	276	669	
	7.3	2.9	406	669⎭	
Cathepsin B (EC 3.4.22.1)					
Beef spleen	–	–^{z}	280	1357	pH 7.6, 50 mM P$_i$
Cellulase (EC 3.2.1.4)					
Trichoderma koningi					
I	5.73	22	280	95	MW = 26,000 (95)
II	12.5	25	280	95	MW = 50,000 (95)
Stereum sanguinolentum					
P 1 Fraction	8	38	280	670	pH 5.4, 0.1 M, AmAc, MW = 21,000 (670)
Penicillium notatum	–	26	280	671	–
	9.1	–	280	672	pH 5, 0.05 M AcO$^-$ MW = 35,500 (672)
Cerebrocuprein					
Human brain	–	7.35	265	1275	–
	–	0.075	675	1275	–
Beef brain	–	0.18	660	1276	–
Ceruloplasmin					
Human	–	0.684	610	1277	–
	1.13	–	610	1278	–
	–	15.03	280	1277	–
	0.40	–	332	1278	–

^{z}14–15.

MOLAR ABSORPTIVITY AND $A_{1cm}^{1\%}$ VALUES FOR PROTEINS AT SELECTED WAVELENGTHS OF THE ULTRAVIOLET AND VISIBLE REGION (continued)

Protein	$\epsilon^a (\times 10^{-4})$	$A_{1cm}^{1\%}$ [b]	nm[c]	Ref.	Comments[d]
Ceruloplasmin (continued)					
Human					
(continued)					
	0.12	–	459	1278	–
	0.22	–	794	1278	–
	–	16.3	280	8	–
	23.7	14.9	280	97	MW = 159,000 (97)
	8.15	–	280	98	–
	10	–	605	98	–
	–	14.6	280	99	MW = 160,000 (99)
	–	14.4	279	100	–
	–	0.68	610	100	–
Form I	–	15.5	280	1279	–
IIa/IIb	–	16.2	280	1279	–
IIIb	–	16.0	280	1279	–
Asialoceruloplasmin	–	15.0	280	1277	–
	–	0.64	610	1277	–
Pig	1.01	0.63	610	1280	MW = 160,000 (1281)
2 days old	–	0.52	610	1282	–
	–	11.5	280	1282	–
10 weeks old	–	0.61	610	1282	–
	–	12.8	280	1282	–
Rat, copper deficient	–	0.64	610	1283	–
Rabbit	–	13.1	280	101	–
	–	0.618	610	101	–
Chitinase (EC 3.2.1.14)					
Streptomyces	–	15.0	280	1358	–
Chloride peroxidase					
(EC 1.11.1.10)					
Caldariomyces fumago	7.53	–	403	102	–
	1.15	–	515	102	–
	1.08	–	542	102	–
	0.42	–	650	102	–
Choleragen					
(Cholera enterotoxin)	9.6	11.41	280	1284	pH 7.5, 0.2 M Tris, MW = 84,000 (1284)
	8.75	10.39	280	1284	pH 8.0, 5 M GdmCl, MW = 84,000 (1284), dry wt
Choleragenoid					
(Toxoid)	1.43	9.56	280	1284	pH 7.5, 0.2 M Tris MW = 15,000 (1284), dry wt
	1.36	9.09	280	1284	pH 8.0, 5 M GdmCl, MW = 15,000 (1284), dry wt
Cholinesterase (EC 3.1.1.8)					
Amiarus nebulosus					
(a fish)	–	3.5	275	1285	pH 7, 0.02 M KP$_i$/0.13 M KCl, data from Figure 4 (1285)

MOLAR ABSORPTIVITY AND $A_{1cm}^{1\%}$ VALUES FOR PROTEINS AT SELECTED WAVELENGTHS OF THE ULTRAVIOLET AND VISIBLE REGION (continued)

Protein	$\epsilon^a(\times 10^{-4})$	$A_{1cm}^{1\%}$ [b]	nm[c]	Ref.	Comments[d]
Chorionic gonadotropin					
Human	1.4	5.2	280	1286	MW = 27,000 (1286)
	–	3.88	278	1287	–
Chorismate mutase					
(EC 5.4.99.5): Prephenate					
dehydrogenase (EC 1.3.1.12)					
E. coli	7.8	9.5	280	1288	MW = 82,000 (1289)
Aerobacter aerogenes	7.2[aa]	9.5[aa]	278	1290	pH 8.0, 0.1 M TrisCl, MW = 76,000 (1290)
Chymopapain (EC 3.4.22.6)					
I	6.16	18.40	280	104	pH 5.0, AcO⁻, MW = 33,500 (104)
II	6.18	18.45	280	104	pH 5.0, AcO⁻, MW = 33,500 (104)
III	6.23	18.60	280	104	pH 5.0, AcO⁻, MW = 33,500 (104)
IV	6.16	18.40	280	104	pH 5.0, AcO⁻, MW = 33,500 (104)
B	–	19.6	280	103	pH 7.2, 0.1 M cacodylate, est. from Figure 3 (103)
Chymosin					
(see rennin)					
Chymotrypsin (EC 3.4.21.1)					
Purified, commercial	5.205	–	281	1292	–
Commercial					
Diazoacetyl derivative	2.77	–	250	1293	–
Above photolytically modified	1.75	–	250	1293	–
Chymotrypsin II					
Human	4.75	19.0	280	1299	MW=25,000(1299)
Chymotrypsin A (EC 3.4.21.1)					
Beef pancreas	–	20.2	280	110	0.001 M HCl
Chymotrypsin C (EC 3.4.21.2)					
Pig pancreas	5.95	25	278	111	MW=23,800 (111)
Chymotrypsin, iPr$_2$P-[bb]					
Beef pancreas	5.0	–	280	112	–
α-Chymotrypsin					
Commercial	–	20.8	280	1294	–
	–	300	210	1294	–
	–	920	191	1294	–
	–	19.95	280	1295 ⎫	
Polyvalyl-	–	18.51	280	1295 ⎬ pH 6.2, P$_i$	
Trans-cinnamoyl-	1.78	–	292	1296	–
Denatured derivative	1.78	–	281	1296	–
Furylacryloyl-	1.98	–	320	1296	–
Denatured derivative	2.00	–	310	1296	–
Indolacryloyl-	1.78	–	360	1296	–
Denatured derivative	1.90	–	335	1296	–
Beef	–	18.7	280	1297	–
Pancreas	–	21.5	280	1298	–

[aa]The same values used at 280 nm in Reference 1291.
[bb]iPr$_2$P, -diisopropylphospho.

MOLAR ABSORPTIVITY AND $A_{1cm}^{1\%}$ VALUES FOR PROTEINS AT SELECTED WAVELENGTHS OF THE ULTRAVIOLET AND VISIBLE REGION (continued)

Protein	$\epsilon^a(\times 10^{-4})$	$A_{1cm}^{1\%}$ [b]	nm[c]	Ref.	Comments[d]
α-Chymotrypsin (continued)	4.46	20.75	282	105	MW=21,500 (105)
	–	20.7	282	106	–
	5.0	20	280	107	MW=25,000 (107)
	–	20.4	282	108	–
	–	18.9	280	109	–
Chymotrypsinogen					
Commercial	–	19.7	282	1300	Dry wt
Carbon disulfide derivative	6.0	–	285	1301	Data from Figure 4 (301)
Alkaline denatured	3.66	–	292.8	1302	–
	4.56	–	285.5	1302	–
	23.07	–	230	1302	–
Acid denatured	3.66	–	293	1302	–
	4.56	–	285.5	1302	–
	4.75	–	276	1302	–
	23.06	–	230.5	1302	–
Rat pancreas	4.0	16	280	1303	MW=25,000 (1303)
Chymotrypsinogen A					
Commercial	–	20.3	282	1304	pH 2.0
Beef	–	20.3	282	1305	pH 9.3, glycine buffer, 0.1 μM, dry wt
	5.15	20	280	115	MW=25,761 (115)
Pig	4.43	18	280	116	MW=24,600 (116)
Spiny Pacific dogfish	–	21.4[r]	280	1306	–
(*Squalus acanthias*)	–	21.7	280	1306	Refractometry
Chymotrypsinogen B					
Beef	–	18.7	280	117	0.001 M HCl
	–	18.4	280	115	–
Pig, pancreas	6.0	23.8	278	1307	MW=26,000 (1307)
Chymotrypsinogen C					
Pig	7.6	23.8	278	118	MW=31,800 (118)
Chymotrypsinogen, *S*-sulfo					
Beef pancreas	–	17.9	280	119	–
α-Chymotrypsinogen					
Beef pancreas	–	20.6	282	113	–
	5.02	20.0	282	114	MW=25,100 (114)
Citramalate hydrolase (citramalate lyase) (EC 4.1.3.22)					
Clostridium tetanomorphum					
Component I	–	9.0	280	673	–
Component II	–	9.5	280	673	–
Citrate condensing enzyme [citrate (*re*)-synthase] (EC 4.1.3.28)					
Pig heart	13.15	15.5	280	120	MW=85,000 (120)
Citrate synthase (EC 4.1.3.7)					
Pig heart	17.1	17.8	280	1308	MW=96,000 (1308)
	15	15	280	674	MW=100,000 (674)

MOLAR ABSORPTIVITY AND $A_{1cm}^{1\%}$ VALUES FOR PROTEINS AT SELECTED WAVELENGTHS OF THE ULTRAVIOLET AND VISIBLE REGION (continued)

Protein	$\epsilon^a (\times 10^{-4})$	$A_{1cm}^{1\%}$ [b]	nm[c]	Ref.	Comments[d]
Cobratoxin					
Naja naja atra	1.45	13.2	280	675	MW=11,000 (675)
	–	12.1	280	676	Data from Figure 1 (676)
Cocoonase (EC 3.4.21.4)					
Silkmoth					
Antherea polyphemus	–	11	280	1309	1 *N* NaOH
	–	13	280	1310	0.1 *N* NaOH
Bombyx mori	–	9.8	280	1310	0.1 *N* NaOH
Cocosin					
Coconuts	–	7.0	280	677	–
	–	2.7	255	677	–
Cocytotaxin					
Rat serum	–	2.90	280	1363	pH 7.2, 0.05 *M* NaKP$_i$
Colicin					
D, *Escherichia coli* K12	–	7.26	280	1359	Neutral sol
E$_1$, *E. coli*	–	7.36	280	1311	pH 7.0, 0.01 *M* KP$_i$
E$_2$, *E. coli* W 3110	5.84	9.73	280	121	MW=60,000
E$_3$, *E. coli* W 3110	7.45	12.42	280	121	MW=60,000
1$_a$, *E. coli* W 3110-r	–	10.3	280	678	–
1$_b$, *E. coli* W 3110-r	–	10.9	280	678	–
Colipase					
Pig pancreas	–	4.0	280	1360	–
Coliphage N4					
Phenolic subunits	1.29	7.0	280	679	pH 7.2, 0.1 *M* Tris, MW=18,500 (680)
Collagen proline hydroxylase					
Rat skin, newborn	–	12.3	280	1313	pH 7.0
Collagenase (EC 3.4.24.3)					
Clostridium histolyticum					
A	–	14.7	280	1312	–
B	–	13.8	280	1312	–
C	–	16.8	280	1312	–
Complement system (see also C1, C1q, and C2)					
Guinea pig serum	–	13.2	280	1314	–
Conalbumin					
Hen	8.5	11.1	280	122	0.02 *M* HCl, MW=76,600
Egg	–	12.0	280	681	–
Egg white	8.8	–	280	1315	–
Iron	12.2	–	280	1315	–
	–	0.62	470	1316	pH5–10
Copper	12.2	–	280	1315	–

MOLAR ABSORPTIVITY AND $A_{1cm}^{1\%}$ VALUES FOR PROTEINS AT SELECTED WAVELENGTHS OF THE ULTRAVIOLET AND VISIBLE REGION (continued)

Protein	$\epsilon^a(\times 10^{-4})$	$A_{1cm}^{1\%}$ [b]	nm[c]	Ref.	Comments[d]
Concanavalin A					
Canavalia ensiformis	7.98	11.4	280	933	pH 7.0, 1 N NaCl, MW=70,000 (933)
	7.75	11.4	280	934	1 M NaCl, MW = 68,000 (935)
	—	13.7	280	936	pH 6.8, 0.05 M P_i, 0.2 M NaCl
	—	12.4	280	936	pH 5.2, 0.05 M AcONa, 0.2 M NaCl
Copper, blue					
Pseudomonas	—	6.7	280	444	—
C-Reactive protein					
Human	—	20	280	123	—
	—	18	280	124	—
Creatine kinase (EC 2.7.3.2)					
Rabbit muscle	3.28	—	250	1317	0.1 M Tris, 0.02 M EDTA
	3.7	8.88	280	793	pH 7.0, 0.05 M P_i, MW=41,300 (793)
	7.3	8.9	280	125	pH 7, MW=81,000 (125)
	7.1	8.7	280	125	pH 9.8, MW=81,000 (125)
Cyrrinus carpio L.	8.0	9.38	280	794	MW=85,100 (794)
Calf brain	—	8.24	280	795	pH 7.0, 0.05 M NaP$_i$
Human muscle	7.1	8.8	280	796	pH 8.0, MW=81,000 (796)
Crotonase (Enoyl-CoA hydratase, EC 4.2.1.17)					
Beef liver	—	5.76	280	797	—
	—	5.56	280	797	pH 7.4, 6 M GdmCl or 8 M urea
Clostridium acetobutylicum	—	8.90	280	1362	pH 8, 0.05 M TrisCl, 0.1 M KCl
Crotoxin					
Crotalus terrificus terrificus	—	19	280	798	Neutral sol.
Crustacyanin					
Homarus grammarus	—	11.5	278	937	pH 7
	—	37.2	633	937	pH 7
	—	35.6[cc]	600	937	Water
	—	5.4[cc]	320	937	pH 7
	—	5.3[cc]	360	937	Water
	—	5.7[cc]	370	937	pH 7
Cryoglobulin, 6.6S					
Human blood	—	13.3	280	938	Dry wt

[cc]Data from Figure 1 of Reference 937.

MOLAR ABSORPTIVITY AND $A_{1cm}^{1\%}$ VALUES FOR PROTEINS AT SELECTED WAVELENGTHS OF THE ULTRAVIOLET AND VISIBLE REGION (continued)

Protein	$\epsilon^a(\times 10^{-4})$	$A_{1cm}^{1\%}$ [b]	nm[c]	Ref.	Comments[d]
Cryptocytochrome c					
Pseudomonas					
denitrificans					
Aerobic cells					
Ferri form	16.0	–	–[dd]	1371	
	1.97	–	500	1371	
	0.65	–	642	1371	
Ferro form	18.3	–	426	1371	
	17.3	–	438	1371	
	1.97	–	–[ee]	1371	
CO-ferro form	44.1	–	419	1371	
	2.2	–	540	1371	
	1.86	–	570	1371	
NO-ferro form	17.2	–	396	1371	pH 7, dry wt
	2.1	–	490	1371	
	2.0	–	–[ff]	1371	
Anaerobic cells	1.9	–	570	1371	
Ferri form	16.0	–	–[dd]	1371	
	1.8	–	500	1371	
	0.6	–	642	1371	
Ferro form	17.8	–	426	1371	
	17.1	–	438	1371	
	1.8	–	–[ee]	1371	
Crystallin					
Beef					
α	–	9.6	280	1318	–
	–	8.85	280	126	Value drops to 8.0 after aging 7 months
	73.1	8.7	280	127	MW=840,000 (127)
α'	–	8.0	280	129	–
α''	–	8.3	280	129	–
γ	–	21.0	280	1318	–
	–	21.0	280	130	–
B_s	5.28	18.6	278	131	pH 8.2, 0.0005 M P_i, MW=28,402 (131)
	5.25	18.5	280	131	pH 8.2, 0.0005 M P_i, MW=28,402 (131)
Calf					
α	–	9.9	280	1319	7 M urea
Dogfish					
α	–	8.5	280	1318	–
Medium age	–	8.80	280	939	pH 9.8, 0.05 M borate–8 M urea
Old	–	7.88	280	939	
β					
Medium age	–	15.9	280	939	pH 9.8, 0.05 M borate–8 M urea
Old	–	15.9	280	939	
γ		22.4	280	1318	–
Medium age	–	23.1	280	939	pH 9.8, 0.05 M borate–8 M urea
Old	–	22.2	280	939	
Fox					
α	55.2	8.9	280	127	MW=620,000 (127)

[dd]400–402 nm.

[ee]550–560 nm.

[ff]530–540 nm.

MOLAR ABSORPTIVITY AND $A_{1cm}^{1\%}$ VALUES FOR PROTEINS AT SELECTED WAVELENGTHS OF THE ULTRAVIOLET AND VISIBLE REGION (continued)

Protein	$\epsilon^a(\times 10^{-4})$	$A_{1cm}^{1\%}$ [b]	nm[c]	Ref.	Comments[d]
Crystallin (continued)					
Horse					
α	103.8	9.6	280	127	MW=1,050,000 (127)
Human					
α, 0–10 years in age	–	10.2	280	1318	–
β, 0–10 years in age	–	16.7	280	1318	–
γ, 0–10 years in age	–	15.7	280	1318	–
α, 11–20 years in age	–	11.6	280	1318	–
β, 11–20 years in age	–	16.5	280	1318	–
γ, 11–20 years in age	–	15.6	280	1318	–
α, 40–49 years in age	–	16.2	280	1318	–
β, 40–49 years in age	–	16.0	280	1318	–
γ, 40–49 years in age	–	15.4	280	1318	–
α, 50–59 years in age	–	16.0	280	1318	–
β, 50–59 years in age	–	16.8	280	1318	–
γ, 50–59 years in age	–	15.2	280	1318	–
α, 60–69 years in age	–	17.3	280	1318	–
β, 60–69 years in age	–	15.7	280	1318	–
γ, 60–69 years in age	–	15.9	280	1318	–
α, 70–79 years in age	–	15.7	280	1318	–
β, 70–79 years in age	–	16.3	280	1318	–
γ, 70–79 years in age	–	15.9	280	1318	–
α, 80–89 years in age	–	15.6	280	1318	–
β, 80–89 years in age	–	14.8	280	1318	–
γ, 80–89 years in age	–	17.3	280	1318	–
Mink					
α	55.4	8.8	280	127	MW = 630,000 (127)
Pig					
α	70.6	8.5	280	127	MW = 830,000 (127)
Rabbit					
α	43.8	8.4	280	127	MW - 575,000 (127)
	–	8.3	280	128	–
β	–	21.5	280	128	–
γ	–	17.6	280	128	–
Rat					
α	–	8.0	280	1318	–
Young	–	7.73	280	939 ⎫	
X-Rayed	–	7.49	280	939 ⎬ pH 9.8, 0.05 M borate–8 M urea
Old	–	7.88	280	939 ⎭	
β					
Young	–	18.6	280	939 ⎫	
X-Rayed	–	16.8	280	939 ⎬ pH 9.8, 0.05 M borate–8 M urea
Old	–	16.5	280	939 ⎭	
γ	–	18.0	280	1318	–
Young	–	19.9	280	939 ⎫	
Young, X-rayed	–	19.8	280	939 ⎬ pH 9.8, 0.05 M borate–8 M urea
Old	–	16.6	280	939 ⎭	
Sulfonated		17.1	280	132	–
Sheep					
α	100.1	8.4	280	127	MW = 840,000 (127)
Cyanocobalamin-protein					
Pig pyloric mucosa	–	9	278	1320	–
Cyanocobalamin-binding factor					
Pig pyloris	–	8	278	1321	Water, dry wt

MOLAR ABSORPTIVITY AND $A_{1cm}^{1\%}$ VALUES FOR PROTEINS AT SELECTED WAVELENGTHS OF THE ULTRAVIOLET AND VISIBLE REGION (continued)

Protein	$\epsilon^a (\times 10^{-4})$	$A_{1cm}^{1\%}$ [b]	nm[c]	Ref.	Comments[d]
Cystathione γ-lyase (EC 4.4.1.1) (see homoserine deaminase)					
Cysteamine dioxygenase (EC 1.13.11.19)					
Horse kidney	11.2	13.5	280	940	–
Cysteine desulfyhydrase (cystathionine γ-lyase, EC 4.4.1.1)					
S. typhimurium	7.85	21.2	280	1503a	MW = 37,000 (1503a)
Cysteine synthase (EC 4.2.99.8)					
S. typhimurium	28.4	9.2	280	575	MW = 309,000 (575)
S-Cysteinylalbumin (see albumin, beef)					
Cytochrome b					
B. anitratum	–	9.5	280	570	pH 7, 0.06 M P_i, data from Figure 5 (570)
Beef heart	11.4[gg]	–	429	1364	pH 7.4, 0.01 M P_i, 0.001 M
	2.07[gg]	–	_[hh]	1364	NaDodSO$_4$, values based on
	1.32[jj]	–	562.5	1364	pyridine hemechromogen value
	–	36.6	418	1364	From Figure 2 (1364)
	–	7.2	562	1364	
	–	3.7	532	1364	
Cytochrome b_2					
Yeast, oxidized	0.92	–	560	571	
	1.13	–	530	571	
	12.95	–	413	571	
	3.44	–	_[kk]	571	
	8.35	–	280	571	pH 7.0, 0.2 M P_i, 0.2 mM
Reduced with 19.5 mM lactate	3.09	–	557	571	EDTA
	1.56	–	528	571	
	18.3	–	424	571	
	3.9	–	328	571	
	8.8	–	269	571	
Yeast					
Fraction B	–	11.0	275	1365	10% AcOH
Fraction C	–	6.7	275	1365	
Core, ferri-form	12.0	–	413	1366	–
Apo-	4.81	2.05	278	1367	MW = 235,000 (1368)
Polypeptide, oxidized	2.3[ll]	–	260	572	
	2.2[ll]	–	280	572	
	11.2[ll]	–	413	572	MW = 11,000 (572)
Reduced	15.8[ll]	–	423	572	
	13.4[ll]	–	528	572	
	2.68[ll]	–	557	572	

[gg] Absolute reduced.
[hh] 562.5–600 nm.
[jj] Reduced-oxidized.
[kk] 360–365 nm.
[ll] Per heme.

MOLAR ABSORPTIVITY AND $A_{1cm}^{1\%}$ VALUES FOR PROTEINS AT SELECTED WAVELENGTHS OF THE ULTRAVIOLET AND VISIBLE REGION (continued)

Protein	$\epsilon^a(\times 10^{-4})$	$A_{1cm}^{1\%}$ [b]	nm[c]	Ref.	Comments[d]
Cytochrome b_5					
Rat liver					
Oxidized	11.7	—	413	1369	—
Reduced	17.1	—	423	1369	—
	1.34	—	526	1369	—
	2.56	—	556	1369	—
Cytochrome b_{562}					
E. coli B					
Reduced	3.16	—	562	1370	—
	1.74	—	531.5	1370	—
	18.01	—	427	1370	—
Oxidized	0.97	—	564	1370	—
	1.06	—	530	1370	—
	11.74	—	418	1370	—
	2.1	—	280	1370	From OD_{562}/OD_{280} = 1.5 (1370)
Cytochrome *c*	—	19.5	280	1372 ⎫	
	—	290	210	1372 ⎬ Dry wt	
	—	870	191	1372 ⎭	
Beef heart	2.77	—	550	137	—
	2.42	—	280	138	Calcd from OD_{550}/OD_{280} = 1.26
	2.9	—	550	139	Reduced with dithionite
	3.053	23.94	550	138	MW = 12,750 (138)
Dithionite reduced	—	23.94	550	1373	pH 6.8, 40 mM NH$_4$ P$_i$
Horse heart	—	17.1	280	1374	—
	—	650	—mm	1374	—
	1.12	—	528	1375	—
	—	9.05	528	1376	—
	2.77	—	550	140	pH 6.8, 0.1 M P$_i$, reduced
	3.18	—	—	140	pH 6.8, 0.1 M P$_i$, reduced, wavelength 268–272 nm
	—	1.12	528	140	pH 6.8, 0.1 M P$_i$, oxidized
	—	2.32	280	140	pH 6.8, 0.1 M P$_i$, oxidized
NO-ferro form	0.54	—	570	1377 ⎫	
	0.65	—	540	1377 ⎪	
NO-ferri form	0.56	—	571	1377 ⎬ Wavelength determined
	0.69	—	561	1377 ⎪ from frequency
	0.67	—	540	1377 ⎪	
	0.67	—	527	1377 ⎭	
Human heart	—	23.1	550	135	—
	2.77	21.9	550	136	Reduced, MW = 12,600 (136)
	2.33	18.5	280	136	Oxidized, MW = 12,600 (136)
Micrococcus dentrificans					
NaBH$_4$ reduced	2.68	—	550	1378	—
	2.23	—	280	1378	From OD_{550}/OD_{280} = 1.2 (1378)
Camelus dromedarius (camel)					
Reduced	2.4	—	550	1379 ⎫	
	0.646	—	535	1379 ⎬ pH 7.5
	1.382	—	520	1379 ⎪	
	10.95	—	417	1379 ⎭	

mm191–194 nm.

MOLAR ABSORPTIVITY AND $A_{1cm}^{1\%}$ VALUES FOR PROTEINS AT SELECTED WAVELENGTHS OF THE ULTRAVIOLET AND VISIBLE REGION (continued)

Protein	$\epsilon^a (\times 10^{-4})$	$A_{1cm}^{1\%}$ [b]	nm[c]	Ref.	Comments[d]
Cytochrome c (continued)					
Oxidized	0.056	—	695	1379 ⎫	
	0.946	—	530	1379 ⎪	
	9.197	—	409	1379 ⎬ pH 7.5	
	2.440	—	360	1379 ⎪	
	2.033	—	280	1379 ⎭	
Cytochrome c_1					
Beef heart	—	14	276	1380	—
Cytochrome c_2					
Rhodospirillum rubrum					
Reduced	2.81[nn]	—	550	1381 ⎫	
	1.7[nn]	—	521	1381 ⎪	
	14.3[nn]	—	415	1381 ⎪	
	3.71[nn]	—	316	1381 ⎬	
	3.38[nn]	—	272	1381 ⎪ pH 6.90, 0.1 M NaP$_i$	
Oxidized	1.05[nn]	—	525	1381 ⎪	
	11.5[nn]	—	410	1381 ⎪	
	2.95[nn]	—	357	1381 ⎪	
	2.47[nn]	—	275	1381 ⎭	
Cytochrome c_{550}					
Bacillus subtilis					
Reduced	2.55	—	550	1384	—
	1.31	—	520	1384	—
	12.75	—	414	1384	—
	3.14	—	316	1384	—
	3.94	—	279	1384	—
Oxidized	1.18	—	528	1384	—
	9.25	—	407	1384	—
	3.06	—	279	1384	—
Spirillum itersonii					
Reduced[oo]	2.76	26.4	550	1385 ⎫	
	1.61	15.5	522	1385 ⎬ MW = 10,411 (1385)	
	14.6	14.0	416	1385 ⎪	
Oxidized[oo]	11.9	11.4	412	1385 ⎭	
Thiobacillus novellus					
Reduced	2.58	—	550	1386 ⎫	
	13.4	—	414.5	1386 ⎬ At 77°K	
Oxidized	2.92	—	280	1386 ⎭	
Cytochrome c_{551}					
Thiobacillus novellus					
Reduced	1.96	—	551	1386	—
	13.9	—	416	1386	—
Oxidized	15.1	—	280	1386	—
Cytochrome $c_{551.5}$					
Chloropseudomonas					
ethylica					
Reduced[pp]	19.91	—	418	1387 ⎫ pH 7.0, 0.05 M	
	1.58	—	523	1387 ⎬ TrisCl	
	3.08	—	551.5	1387 ⎭	

[nn]The value given here is 1/1000 of the value cited in Reference 1364. "I believe that a concentration of moles/1000 ml was used rather than mole/liter, which resulted in a very large value." [Kirschenbaum, *Anal. Biochem.*, 55, 166 (1973)].
[oo]Based on one atom of Fe per molecule.
[pp]All values per heme.

MOLAR ABSORPTIVITY AND $A_{1cm}^{1\%}$ VALUES FOR PROTEINS AT SELECTED WAVELENGTHS OF THE ULTRAVIOLET AND VISIBLE REGION (continued)

Protein	ϵ^a ($\times 10^{-4}$)	$A_{1cm}^{1\%}$ [b]	nm[c]	Ref.	Comments[d]
Cytochrome $c_{551.5}$ (continued)					
Oxidized[pp]	2.75	–	351	1387	
	12.44	–	408	1387	pH 7.0, 0.05 M
	1.03	–	528	1387	TrisCl
Cytochrome c_{552}					
Chromatium strain D					
Reduced α peak	3.12	4.35	550	1388	MW = 72,000 (1388)
Cytochrome c_{552} [I]					
P. stutzeri, reduced	3.1[qq]	–	552	573	–
	1.63[qq]	–	523	573	–
	15.08[qq]	–	418	573	–
Oxidized	1.05[qq]	–	530	573	–
	10.04[qq]	–	410	573	–
	2.51[qq]	–	284	573	–
Cytochrome c_{552} [II]					
P. stutzeri, reduced	1.95[qq]	–	552	573	–
	1.66[qq]	–	523	573	–
	15.35[qq]	–	417	573	–
Oxidized	0.99[qq]	–	525	573	–
	12.21[qq]	–	410	573	–
	1.69[qq]	–	277	573	–
Cytochrome c_{553}					
Petalonia fascia (an alga)					
Ferro form	2.68	25.5	273	1389	
	2.19	20.9	293	1389	
	4.40	41.9	317.5	1389	
	19.71	187.5	415.5	1389	
	0.38	3.6	471	1389	
	1.86	17.8	521.5	1389	MW = 10,500 (1389)
	2.85	27.1	553	1389	
Ferri form	2.68	25.5	269	1389	
	2.21	21.0	292	1389	
	3.62	34.5	360	1389	
	13.53	129.0	409	1389	
	1.29	12.3	528	1389	
Monochrysis lutheri (an alga)					
Reduced	2.59		553	1390	pH 7, P_i value
	15.65		416	1390	based on iron determination
Cytochrome c_{554}					
Bacillus subtilis					
Reduced	3.39	–	279	1384	–
	2.0	–	554	1384	–
	1.45	–	550	1384	–
	1.48	–	521	1384	–
	14.36	–	417	1384	–
	4.0	–	316	1384	–
Oxidized	2.4	–	280	1384	–
	1.02	–	523	1384	–
	10.67	–	409	1384	–

[qq]All ϵ values are based on heme content.

MOLAR ABSORPTIVITY AND $A_{1cm}^{1\%}$ VALUES FOR PROTEINS AT SELECTED WAVELENGTHS OF THE ULTRAVIOLET AND VISIBLE REGION (continued)

Protein	$\epsilon^a(\times 10^{-4})$	$A_{1cm}^{1\%}$ [b]	nm[c]	Ref.	Comments[d]
Cytochrome c_{555}					
Crithidia fasciculata					
Reduced[pp]	2.97	24.8	555.5	1391	
	1.68	14.0	525	1391	
	15.4	128	420	1391	
Oxidized[pp]	11.2	93	413	1391	MW = 12,051 (1391)
	1.22	10.2	533	1391	
	0.9	7.5	555.5	1391	
	0.84	7.0	565	1391	
Cytochrome $c_{555(550)}$					
Chloropseudomonas ethylica					
Reduced[pp]	15.34	—	417.5	1387	
	1.71	—	523	1387	
	2.04	—	555	1387	
Oxidized[pp]	4.02	—	275	1387	pH 7.0, 0.05 *M* TrisCl
	3.01	—	358	1387	
	13.20	—	412	1387	
	!.13	—	525	1387	
Cytochrome $c_{557(551)}$					
Alcaligenes faecalis					
Reduced[rr]	4.46	—	557	1383	Dry wt
	3.72	5.7	557	1383	
	2.89	4.44	525	1383	
	28.3	43	420	1383	
Reduced-CO[rr]	3.72	5.7	557	1383	pH 7, 0.05 *M* P$_i$,
	2.89	4.44	525	1383	MW = 65,000 (1383)
	40.0	61.4	416	1383	
Oxidized[rr]	2.16	3.32	530	1383	
	26.2	40.2	408	1383	
Cytochrome *cc'*					
Pseudomonas denitrificans					
Ferri form[pp]	0.37	—	635	1382	
	1.02	—	495	1382	
	8.0	—	400	1382	
	3.08	—	280	1382	
Ferro form[pp]	0.71	—	550	1382	pH 7.3, 0.02 *M*
	8.75	—	434	1382	Tris, 0.5 *M* NaCl
	9.70	—	426	1382	
CO-ferro form[pp]	1.05	—	564	1382	
	1.18	—	534	1382	
	21.0	—	418	1382	
Ferri form[pp]	0.63	—	575	1382	
	0.94	—	538	1382	
	9.90	—	413	1382	
	2.51	—	348	1382	3 *N* NaOH
Ferro form[pp]	2.49	—	550	1382	
	1.32	—	522	1382	
	15.62	—	416	1382	
Ferri form[pp]	0.245	—	635	1382	pH 5
	0.41	—	635	1382	pH 10

[rr]As diheme derivative.

MOLAR ABSORPTIVITY AND $A_{1cm}^{1\%}$ VALUES FOR PROTEINS AT SELECTED WAVELENGTHS OF THE ULTRAVIOLET AND VISIBLE REGION (continued)

Protein	$\epsilon^a(\times 10^{-4})$	$A_{1cm}^{1\% \, b}$	nm^c	Ref.	Comments^d
Cytochrome *cd*					
Alcaligenes faecalis					
Reduced^{pp}	18.9	21.0	418	1383	
	4.45	4.95	460	1383	
	2.87	3.18	556	1383	pH 7, 0.05 *M* P$_i$,
	2.65	2.94	525	1383	MW = 90,000 (1383)
	1.85	2.06	625	1383	
Oxidized^{pp}	15.1	16.8	412	1383	
	2.12	2.35	640	1383	
	13.8	15.3	280	1383	pH 7, 0.05 *M* P$_i$, MW = 90,000 (1383), value calcd from A$_{412}$/A$_{280}$ = 1.2 (1383)
Cytochrome P-450					
Rat liver microsomes					
Males	7.96	–	450	1392	Biuret
Females	9.41	--	450	1392	Biuret
Cytochrome oxidase					
(EC 1.9.3.1-.2)					
Pseudomonas (EC 1.9.3.2)	17,600	–	280	574	Data from Figure 2 (574)
Reduced	18,200	–	418	574	
	3,600	–	625	574	pH 6.0, 0.1 *M* NaP$_i$
Oxidized	14,900	–	408	574	
	3,020	–	630	574	
Pseudomonas aeruginosa	22	18.5	280	1393	Dry wt
Beef heart	–	17.4	280	133	–
Cytochrome *c* peroxidase (EC 1.11.1.5)					
Pseudomonas	6.48	12.1	280	1394	Dry wt, MW = 53,500 (1394)
Yeast	8.1	–	408	941	Dimethyl protoheme
	9.3	23.2	408	134	pH 6, MW = 40,000 (134)
	7.4	18.5	282	134	pH 6, MW = 40,000 (134)
Apo-	5.5	13.75	282	134	pH 6, MW = 40,000 (134)
Cytocuprein					
Human	–	210	210	576	
	–	5.8	268	576	
	–	5.5	280	576	
	–	0.08	675	576	Data from Figure 5 (576)
Apocytocuprein					
Human	–	210	210	576	
	–	4.3	268	576	
	–	4.2	280	576	
3-Deoxy-2-keto-6-phosphogluconate aldolase (see aldolase)					
Deoxyribonuclease I					
(EC 3.1.4.5) (DNase)					
Beef pancreas	–	11.5	280	141	–
	–	12.8	280	942	–

MOLAR ABSORPTIVITY AND $A_{1cm}^{1\%}$ VALUES FOR PROTEINS AT SELECTED WAVELENGTHS OF THE ULTRAVIOLET AND VISIBLE REGION (continued)

Protein	$\epsilon^a(\times 10^{-4})$	$A_{1cm}^{1\%}$ [b]	nm[c]	Ref.	Comments[d]
Deoxyribonuclease I (EC 3.1.4.5) (DNase) Beef pancreas (continued)	–	11.1	280	942	Nbs inactivated
	–	11.8	280	799	2.5 mM HCl
	3.72	12.0	–	577	pH 4.7, 0.2 M AcONa, MW = 31,000 (577)
	–	12.3	280	578	pH 7.6, 0.1 M KP$_i$
	–	13.9	280	578	pH 13, 0.1 N NaOH
	–	12.3	280	142	pH 7.6, 0.1 M KP$_i$
	–	13.9	280	142	pH 13, 0.1 N NaOH
Deoxyribonuclease inhibitor Calf spleen	6.05	10.2	280	579	pH 7.6, 0.1 M KP$_i$, MW = 59,400 (579)
	6.2	10.4	280	579	0.1 N NaOH, MW = 59,400 (579)
Deoxyribonucleic acid polymerase (DNA polymerase, EC 2.7.7.7) E. coli	9.26	8.5	280	580	10 mM NH$_4$HCO$_3$, MW = 109,000 (580)
	11.5	10.5	290	580	0.1 M NaOH–5 mM NH$_4$HCO$_3$, data from Figure 3 (580)
Large fragment Ehrlich ascites tumor	–	9.3	278	1401	
	–	23.9	280	581	pH 7.0, 0.2 M KP$_i$–0.001 M EDTA, 0.01 M 2–HSEtOH, data from Figure 4 (581)
	–	19.2	290	581	
	–	20.4	280	581	pH 10, data from Figure 4 (581)
	–	17.5	290	581	
Dextranase (EC 3.2.1.11) *Aspergillus carneus*	–	17.8	280	1395	Dry wt
Diamine oxidase (amine oxidase, EC 1.4.3.6) Pig kidney	10.6	12.8	280	143	MW = 87,000 (143)
Dihydrofolate reductase (tetrahydrofolate dehydrogenase) (EC 1.5.1.3) *Streptomyces faecium*	4.47	22.0	280	1396	DR, MW = 20,300 (1396)
Lactobacillus casei	2.15	–	278	1397	pH 7
	2.76	–	268	1397	+ NADPH[ss]
	0.72	–	340	1397	
E. coli, Methotrexate resistant	4	23.8	280	1398	Microbiuret and dry wt, MW = 16,810 (1398)
E. coli	–	19.1	280	1399	pH 7.0, 0.04 M
T4 phage	–	12.0	280	1399	KP$_i$
Chicken liver	3.4	15.5	278	682	MW = 22,000 682
Streptococcus faecium	–	20	280	683	

[ss]NADPH, nicotinamide adenine dinucleotide phosphate, reduced.

MOLAR ABSORPTIVITY AND $A_{1cm}^{1\%}$ VALUES FOR PROTEINS AT SELECTED WAVELENGTHS OF THE ULTRAVIOLET AND VISIBLE REGION (continued)

Protein	$e^a(\times 10^{-4})$	$A_{1cm}^{1\%}$ [b]	nm[c]	Ref.	Comments[d]
Dihydrolipoyl transacetylase (lipoate acetyltransferase, EC 2.3.1.12)					
E. coli	–	4.5	280	684	pH 7.0, P_i
Dihydroorotic dehydrogenase (orotate reductase, EC 1.3.1.14)					
Zymobacterium oroticum	3.5	–	450	27	–
	0.59	–	550	27	–
3,4-Dihydroxy-9,10-secoandrosta-1,3,5[10]-triene-9,17-dione 4,5-dioxygenase (steroid 4,5-dioxygenase) (EC 1.13.11.25)					
Nocardia restrictus	26.8	9.3	280	800	MW = 286,000 (800)
Dimethylglycine dehydrogenase (EC 1.5.99.2)					
Rat liver, mitochondria	–	17	280	685	pH 7.5, 0.0075 M KP$_i$, data from Figure 4 (685)
Dipeptidase (EC 3.4.13.11)					
Pig kidney	4.2	8.96	280	686	pH 8.0, 0.002 M Tris, MW = 47,200 (686)
p-Diphenol oxidase (monophenol monooxygenase) (EC 1.14.18.1)					
Polyporus versicolor	–	11.5	280	1400	–
Diphtheria toxin	–	12.7	278	145	pH 6.8, $M/15$ P$_i$, 0.175 M NaCl
	–	12.9	278	145	0.1 N HCl
	–	14.1	293	145	0.1 N NaOH
DPNase (NAD nucleosidase, EC 3.2.2.5)					
Pig brain	8.0	3.1	280	144	MW = 26,000 (144)
Edeine A					
B. brevis Vm 4	0.131	–	270	582	Water
Edeine B					
B. brevis Vm 4	0.131	–	270	582	Water
Elastase (EC 3.4.21.11)	–	11.0	280	943	–
Pig	5.74	22.2	280	801	0.1 M NaOH
	5.23	20.2	280	801	pH 5.0, 0.05 M AcONa
	–	18.5	280	583	–
	4.85	–	280	944	ϵ = [M/2.06] $\times$ 10^5
	–	22.0	280	945	–
Elastase-like enzyme					
Streptomyces griseus					
I	2.94	10.5	280	946	MW = 28,000 (946)
II	0.85	12.1	280	946	MW = 7,000 (946)
III	1.58	11.3	280	946	MW = 14,000 (946)

MOLAR ABSORPTIVITY AND $A_{1cm}^{1\%}$ VALUES FOR PROTEINS AT SELECTED WAVELENGTHS OF THE ULTRAVIOLET AND VISIBLE REGION (continued)

Protein	$\epsilon^a (\times 10^{-4})$	$A_{1cm}^{1\%}$ [b]	nm[c]	Ref.	Comments[d]
Elastoidin, soluble					
Prionace glauca, pectoral fins					
(Great blue shark)	–	1.86	277	947	0.5 M AcOH,
					data from Figure 2 (947)
Elinin					
Human erythrocyte	4420	22.1	274	948	MW = 20 × 10⁶ (948)[tt]
Elongation factor 2					
ADP ribosylated					
Rat liver					
Aminoethylated	9.7	–	276	959 ⎰	pH 7.8, 5 M GdmCl,
	9.4	–	280	959 ⎱	50 mM Tris
	9.9	–	273	959 ⎰	
	8.9	–	280	959 ⎱	pH 3.3, 50 mM
Endopolygalacturonase					
Verticillium albo-atrum	–	12.7[uu]	280	802	–
	–	11.2[uu]	280	802	–
Aspergillus niger	–	1.29	280	1404	–
Enolase (2-phosphoglycerate					
hydro-lyase) (EC 4.2.1.11)					
Oncorhynchus kisutch					
(Coho salmon)	7.8	7.4	280	949	MW = 105,560 (949)
Oncorhynchus keta					
(Chun salmon)	8.6	8.7	280	949	MW = 99,260 (949)
Yeast	7.8	8.9	280	949	MW = 88,000 (949)
	–	9.0	280	584	–
Rabbit	–	9.0	280	949	–
Rabbit muscle	–	8.95	280	141	–
	7.65	9	280	146	MW = 85,000 (146)
	6.1	9	280	147	MW = 67,200 (148)
	–	8.85	280	1407	Kjeldahl[e]
E. coli	–	5.7	280	1405 ⎰	DR
	–	6.1	277	1405 ⎱	
Rhesus monkey	7.2	8.8	280	1406	MW = 82,000 (1406)
Trout	–	7.9	280	949	–
Enterokinase (Enteropeptidase)					
(EC 3.4.21.9)					
Pig duodena	–	17.8	280	1408	–
Enterotoxin A					
Staphylococcus aureus	4.09	14.6	277	1409	MW = 28,000 (1409)
Enterotoxin B					
Staphylococci	–	12.1	277	149	–
	–	15	277	149	N = 16.1% (149)
	–	14	277	150	–

[tt]In Figure 3 of Reference 948: OD = 0.375 for a 0.4% solution. This is equivalent to an $A_{1cm}^{1\%}$ of 9.35. The value cited in the paper was "$E_{0.1cm}^{0.1\%}$ = 0.221" from which an $A_{1cm}^{1\%}$ was obtained.
[uu]Two different preparations.

MOLAR ABSORPTIVITY AND $A_{1cm}^{1\%}$ VALUES FOR PROTEINS AT SELECTED WAVELENGTHS OF THE ULTRAVIOLET AND VISIBLE REGION (continued)

Protein	$\epsilon^a (\times 10^{-4})$	$A_{1cm}^{1\%}$ [b]	nm[c]	Ref.	Comments[d]
Enterotoxin B, nitrated					
Staphylococcal	6.29	20.6	277	687 ⎱	pH 8, MW = 30,500 (687)
	2.42	7.9	428	687 ⎰	
	2.16	7.1	360	687	pH 6.2, MW = 30,500 (687)
Enterotoxin C					
Staphylococcus aureus	–	12.1	277	688	–
Enterotoxin E					
Staphylococcus aureus	–	11.9	280	1410	–
	–	12.5	277	1410	
Enzyme, thrombin-like					
Crotalus adamanteus	4.84	14.8	280	1411	Dry wt, MW =
trans-Epoxysuccinate hydratase					32,700 (1411)
(see tartrate epoxidase)					
Erabutoxin a					
Laticauda semifasciata	0.7	–	280	803	Water
Erythrocruorin					
Lumbricus terrestris	1.18	5.13	504	804 ⎫	
	11.36	49.4	430	804	
Oxygenated	1.37	5.95	542	804	
	11.27	49.0	417	804	
	5.22	22.7	283	804 ⎬	pH 7.0, 0.1 M P$_i$,
Plus CO	1.374	5.97	538	804	MW = 23,230 (804)
	18.83	81.9	420	804	
Ferric derivative	1.165	5.07	500	804	
	10.0	43.7	395	804 ⎭	
Erythrocruorin, oxy					
Cirraformia grandis	4.09	22.1	280	1412 ⎫	
(annelid worm)					
	2.58	13.9	345	1412	Dry wt, MW =
	9.68	52.2	415	1412 ⎬	18,500 (1412)
	1.21	6.59	539	1412	
	1.18	6.38	574	1412 ⎭	
Erythrocuprein (superoxide dismutase, EC 1.15.1.1)					
Beef	0.0313	–	680	1413	–
	0.984	–	259	1414	–
Apo-	0.241	–	252	1415 ⎫	
	0.367	–	259	1415	
	0.330	–	262	1415	
	0.419	–	264	1415 ⎬	pH 7.2, 10 mM P$_i$
	0.413	–	268	1415	
	0.383	–	275	1415 ⎭	
	0.330	–	252	1415 ⎫	
	0.420	–	259	1415	
	0.420	–	261	1415	
	0.540	–	269	1415 ⎬	GdmCl, pH 5.9
	0.540	–	275	1415	
	0.460	–	281	1415 ⎭	
	0.790	–	295	1415 ⎱	pH 11.7, GdmCl
	3.81	–	246	1415 ⎰	

MOLAR ABSORPTIVITY AND $A_{1cm}^{1\%}$ VALUES FOR PROTEINS AT SELECTED WAVELENGTHS OF THE ULTRAVIOLET AND VISIBLE REGION (continued)

Protein	$\epsilon^a(\times 10^{-4})$	$A_{1cm}^{1\%}$ [b]	nm[c]	Ref.	Comments[d]
Erythrocuprein (continued)					
Human	–	5.06	265	805 ⎫	
	–	0.075	655	805 ⎬	pH 6.5, 0.15 M NaCl
	–	0.077	675	805 ⎭	
Human blood	0.0284	0.085	655	98	MW = 33,200 (432)
	1.84	5.5	265	98	MW = 33,200 (432)
	0.0350	0.104	655	151	MW = 33,600 (151)
	1.87	5.58	265	151	MW = 33,600 (151)
Erythropoietin					
Human urine	–	9.26	279	1416	–
Esterase (carboxylesterase, EC 3.1.1.1)					
Pig liver	–	13.8	280	585	pH 8.0
Goat intestine	–	14.73	275	586	–
	–	13.75	280	586	–
Excelsin					
Brazil nuts	–	9	279	806	pH 5.5
	–	14	279	806	pH 12.2
Exo-1,3-3-glucosidase (EC 3.2.1.58) (see glucanase)					
Factor					
Antihemorrhagic					
Trimeresurus flavoviridis (snake) serum	–	8.8	280	1417	–
Direct lytic (acetate)					
Haemachatus haemachates (snake) venom	0.29	4.2	278	1418	MW = 7,000 (1418)
Epidermal growth					
Mice (adult male, albino), submaxillary gland	1.81	30.9	280	1419	pH 5.6, 0.1 M AcONa or water DR, MW = 6,045 (1419)
X					
Beef plasma	–	12.4	280	1420	–
X, activated					
Beef plasma	4.1	8.6	280	1421	Dry wt, MW = 48,000 (1421)
X, thrombokinase, activated Stuart factor					
Human	–	5.8	280	587	–
XIII					
Human plasma	–	13.8	280	1503b	Dry wt
Fatty acid synthetase (EC 2.3.1.41?) (see also 3-oxoacyl-[ACP] reductase)					
Chicken liver	48.3	9.65	279	588	MW = 500,000 (588)
Pigeon liver	46.4	8.35	279	589	MW = 545,000 (589)
	38.7	8.6	280	153	MW = 450,000 (153)

MOLAR ABSORPTIVITY AND $A_{1cm}^{1\%}$ VALUES FOR PROTEINS AT SELECTED WAVELENGTHS OF THE ULTRAVIOLET AND VISIBLE REGION (continued)

Protein	$\epsilon^a(\times 10^{-4})$	$A_{1cm}^{1\%}$ [b]	nm[c]	Ref.	Comments[d]
Fatty acyl-CoA dehydrogenase (acyl-CoA dehydrogenase) (EC 1.3.1.8 or .9 or EC 1.3.99.3)					
Pig liver	—	13.8	275	152	pH 7.5, 0.036 M P$_i$, est. from Figure 3 (152)
Ferredoxin					
Alfalfa	0.95	—	277	154	Per mole iron
	1.83	16.6	277	154	MW = 11,000 (154)
	—	7.2	465	154	—
	—	7.9	422	154	—
	—	10.6	331	154	—
Scenedesmus	1.5	13.1	276	155	MW = 11,500 (155)
	1.33	—	330	155	—
	0.98	—	421	155	—
Apo-	0.855	15.8	280	156	pH 5.4, est. from Figure 1 (156), apoprotein made with α,α'-dipyridyl
	0.765	14.2	280	156	pH 7.4, est. from Figure 1 (156), apoprotein made with mersalyl
Azotobacter vinelandii					
Oxidized, FdI	2.7	19.2	400	1425	MW = 14,140 (1425)
Bacillus polymyxa					
Oxidized	—	21	279	950	
	—	12.3	325	950	
	—	11.1	400	950	pH 7.3, 25 mM TrisCl Data from Figure 2 (950)
	—	5.1	500	950	
Reduced with Na$_2$S$_2$O$_4$	—	6.9	400	950	
	—	3.1	500	950	
Chlorobium thiosulfatophilum	2.04	34	280	951	pH 7.3, 0.3 M Tris, 0.54 M NaCl, Data from Figure 1 (951), MW = 6,000 (951)
E. coli	0.96	7.6	416	952	MW = 12,600 (952)
	1.78	13.3	277	952	
Clostridium pasteurianum	2.16	41.5	285	952	
	2.16	41.5	300	952	MW = 5,200
	1.73	33.2	390	952	
	3.0	—	390	952	
	—	34.0	390	811	—
	2.1	35.0	390	812	MW = 6,000 (812)
Monomer	2.600	—	390	1428	Dry wt
	3.126	—	390	1428	Kjeldahl[e]
Dimer	1.600	—	390	1428	Dry wt
	1.540	—	390	1428	Kjeldahl[e]
lyophilized	1.617	—	390	1428	Dry wt
	1.260	—	390	1428	Kjeldahl[e]
C. tartarivorum	3.106	—	280	810	—
	2.422	—	290	810	—
C. thermosaccharolyticum	3.09	—	280	810	—
	2.413	—	390	810	—

MOLAR ABSORPTIVITY AND $A_{1cm}^{1\%}$ VALUES FOR PROTEINS AT SELECTED WAVELENGTHS OF THE ULTRAVIOLET AND VISIBLE REGION (continued)

Protein	$\epsilon^a(\times 10^{-4})$	$A_{1cm}^{1\%}$ b	nm^c	Ref.	Comments^d
Ferredoxin (continued)					
C. acidi urici	–	37.0	390	811	–
	3.06	–	390	813	–
	3.15	58.4	280	156	pH 7.4, 0.1 *M* Tris, est. from Figure 1 of Reference 156, MW = 5,400 (156)
	3.06^{vv}	–	390	955	–
	2.98	–	390	955	Dry wt^{ww}
	3.36^{xx}	–	390	955	–
	3.02^{yy}	–	390	955	–
C. tetanomorphum	–	35.0	390	811	–
C. butyricun	–	31.0	390	811	–
C. cylindrosporum	–	29.6	390	811	–
Cyperus rotundus L.	0.88	–	465	807	–
	0.98	–	420	807	–
	1.45	–	330	807	–
	2.42	–	275	807	–
Cotton, type I	0.655	–	460	808	–
	0.758	–	419	808	–
	1.082	–	325	808	–
Bacterial	2.45	–	390	809	–
Horsetail leaves	0.88	–	421	956	–
	1.17	–	276	956	From A_{421}/A_{276} = 0.75
Methanobacterium omelianskii	–	39	280	957	pH 7.3, 0.07 *M* TrisCl, data from Figure 1 (957)
Pseudomonas oleovorans	–	3.36	497	958	pH 7.3, 0.1 *M* Tris, one atom Fe
	–	20.2	280	958	From A_{280}/A_{497} = 6.3
	–	5.74	495	958	pH 7.3, 0.1 *M* Tris,
	–	21.2	280	958	two atoms Fe
Apo-	–	18	277	958	From A_{280}/A_{495} = 3.7, 50% AcOH
Rhodospirillum rubrum					
Type I	2.43	27.9	385	1426	MW = 8,700 (1426)
Type II	0.88	11.7	385	1426	MW = 7,500 (1426)
Parsley	1.24	11.47	255	1427	
	1.27	11.75	260	1427	
	1.5	13.9	277	1427	
	0.98	9.09	294	1427	
	1.22	11.34	330	1427	MW = 10,800 (1427)
	0.74	6.92	390	1427	
	0.92^{zz}	8.65	422	1427	
	0.82	7.59	448	1427	
	0.84	7.79	463	1427	
	0.97	0.90	690	1427	
	1.01	–	420	814	–
Spinach	0.88	–	420	812	–
	0.94	–	420	814	–
Urea-oxygen denat.	2.0	–	275	815	Contains Fe
	1.3	–	275	815	Lacks Fe

^{vv}Protein concentration determined by release of C-terminal amino acid by carboxypeptidase A.
^{ww}MW = 6232, 8 eq. Fe and 8 eq. S.
^{xx}Protein concentration determined by aspartic acid analysis.
^{yy}Protein concentration determined by glutamic acid analysis.
^{zz}This value cited in Reference 1427. "I calculated 9350 as ϵ." [Kirschenbaum, *Anal. Biochem.*, 55, 166 (1973)].

MOLAR ABSORPTIVITY AND $A_{1cm}^{1\%}$ VALUES FOR PROTEINS AT SELECTED WAVELENGTHS OF THE ULTRAVIOLET AND VISIBLE REGION (continued)

Protein	$\epsilon^a (\times 10^{-4})$	$A_{1cm}^{1\%}$ [b]	nm[c]	Ref.	Comments[d]
Ferredoxin: NADP reductase (EC 1.6.7.1)					
Spinach	1.074	–	456	1429	–
Ferritin					
Apo-, horse spleen	–	–	280	157	$A_{1cm}^{1\%} = 8.6-9.7$
Rat liver	–	2	320	1430	–
	–	302	260	1430	–
Fetuin					
Commercial	–	4.5	278	1431	This value good for native, reduced and carboxymethylated, oxidized, and neuraminidase-treated fetuin
Calf serum	–	4.1	278	1432	Water
	–	5.3	278	1432	–
Calf spleen	–	4.5	278	158	Australian sample
	–	4.8	278	158	Colorado sample
Fibrin					
Beef	–	16.16	283	816 ⎱	6 M alkaline urea
	–	16.19	290	816 ⎰	
	–	16.84	282	1433	40% urea, 0.2 N NaOH
Human	–	15	280	1434	–
Fibrin stabilizing factor					
Beef plasma	–	14.15	283	1422	8 M urea, 0.2 M NaOH
Fibrinogen					
Beef					
α chain	–	11.8	280	1435 ⎫	
β chain	–	17.4	280	1435 ⎬	0.1 M NaOH
γ chain	–	20.4	280	1435 ⎭	
Beef	–	16.51	282	1433	8 M urea, 0.2 N NaOH
	–	15.4	279	1433	pH 2, 0.3 M NaCl
	–	14.0	279	1433	pH 6, 0.3 M NaCl
	–	15.6	282	1433	pH 11, 0.3 M NaCl
	–	15.04[a*]	280	1436	0.3 M NaCl
	–	15.49[a*]	280	1436	pH 6.9, 0.05 M P_i
	–	15.16[a*]	280	1436	2% AcOH
	–	15.29[a*]	280	1436	pH 5.3, 1 M NaBr
	–	15.45[a*]	280	1436	30% urea
	–	15.60[a*]	280	1436	pH 5.45, 6 N GdmCl
	–	15.94[a*]	280	1436	pH 5.8, 2 M KCNS

[a*] Read against value at 320 nm.

MOLAR ABSORPTIVITY AND $A_{1cm}^{1\%}$ VALUES FOR PROTEINS AT SELECTED WAVELENGTHS OF THE ULTRAVIOLET AND VISIBLE REGION (continued)

Protein	$\epsilon^a(\times 10^{-4})$	$A_{1cm}^{1\%}$ [b]	nm[c]	Ref.	Comments[d]
Fibrinogen (continued)					
Beef (continued)	–	15.06	280	161	pH 7.1, ionic strength = 0.3
	–	16.01	282.5	161	pH 12.8, 0.1 M KOH
	–	15.92	289.5	161	pH 12.8, 0.1 M KOH
	–	15.17	278.5	161	pH 5.8, 5 M GdmCl
	–	15.06	279	161	pH 7.4, 5 M urea
	–	15.00	278	161	pH 7.6
	–	15.1	280	162	–
	–	15	280	689	–
	–	15.87	283	816 ⎫	6 M alkaline urea
	–	15.88	290	816 ⎭	
	–	15.04	280	816	3 M NaCl
	–	15.5	280	816	–
Fragment D	–	20.7	280	816	–
Fragment E	–	13.0	280	816	–
Calf	–	15.9	280	1439 ⎫	
Dog	–	15.8	280	1439 ⎪	0.2 M KCl, corrected for
Elephant	–	15.7	280	1439 ⎬	scattering
Goat	–	15.6	280	1439 ⎪	
Human	–	13.9	280	1437 ⎭	pH 7.1, 0.055 M Na$_3$Cit·2H$_2$O
	–	17.65	282	1437	Alkaline urea
	46.4	13.6	280	96	MW = 341,000 (159)
	45.3	15.1	280	690	MW = 300,000 (690)
Fraction I-4	–	14.5	280	819	
Fraction I-8	–	15.6	280	819	
Fragment X	34.1	14.2	280	690	MW = 240,000 (690)
Fragment Y	27.3	17.6	280	690	MW = 155,000 (690)
Fragment D	16.7	20.8	280	690	MW = 83,000 (690)
	–	20.8	280	1438	
Fragment E	5.1	10.2	280	690	MW = 50,000 (690)
	–	10.2	280	1438	
Knot	–	11.8	280	1438	
Human, high solubility	–	16.8	282	160	0.1 N NaOH−5 M urea
Low solubility	–	16.7	280	160	0.1 N NaOH−5 M urea
Panulirus interruptus	52.5	12.5	280	820	MW = 420,000 (820)
Sheep	–	15.5	280	1439	0.2 M KCl, corrected for scattering
Fibroin					
Bombyx mori L	–	11.3	276	691	pH 4.3−7.3, 0.2 M NaCl
Ficin (EC 3.4.22.3)	5.8	22.4	280	163	MW = 26,000 (163)
	5.9	22.6	280	164	MW = 26,000 (163)
	4.6	–	280	821	
Fraction III	5.1	21	280	165	0.05 M NH$_4$HCO$_3$, MW = 24,500 (165)
Ficus glabrata					
Component G	5.4	–	280	1440	–
F. carica var. *kodata*	–	20.2	280	822	–
Flavodoxin					
Desulfovibrio gigas					
Oxidized	4.7	–	273	959	–
	0.82	–	374	959	–
	1.02	–	456.5	959	–

MOLAR ABSORPTIVITY AND $A_{1cm}^{1\%}$ VALUES FOR PROTEINS AT SELECTED WAVELENGTHS OF THE ULTRAVIOLET AND VISIBLE REGION (continued)

Protein	$e^a(\times 10^{-4})$	$A_{1cm}^{1\%}$ [b]	nm[c]	Ref.	Comments[d]
Flavodoxin (continued)					
Semiquinone	0.87	—	349	959	—
	0.41	—	580	959	—
Desulfovibrio vulgaris					
Oxidized	4.8	—	273	959	—
	0.87	—	375.5	959	—
	1.07	—	456–7	959	—
Semiquinone	0.9	—	349	959	—
	0.47	—	580	959	—
Apo-	2.00	—	278	1442	—
E. coli	0.825	—	467	960	—
	0.382	—	580	960	—
	5.0	—	274	960	From $\epsilon_{274}/\epsilon_{467} = 6.67$ (960), MW = 14,500 (960)
Clostridium MP					
Oxidized	4.68	—	272	961	—
	0.91	—	376	961	—
	1.04	—	445	961	—
Semiquinone	0.84	—	350	961	—
	0.485	—	376	961	—
	0.24	—	445	961	—
	0.462	—	575	961	—
Reduced	0.175	—	445	961	—
Clostridium pasteurianum	4	27.4	272	166	MW = 14,600 (166)
	0.79	5.3	372	166	MW = 14,600 (166)
	0.91	6.2	443	166	MW = 14,600 (166)
Oxidized	4.58	—	272	961	—
	0.847	—	374	961	—
	1.04	—	443	961	—
Semiquinone	0.766	—	350	961	—
	0.465	—	374	961	—
	0.208	—	443	961	—
	0.455	—	575	961	—
Reduced	0.16	—	443	961	—
	—	36.8	274	167	pH 7.3, 0.02 *M* TrisCl
Apo-	2.5	—	278	962	—
	2.52	—	282	962	—
Peptostreptococcus elsdenii					
Oxidized	4.76	—	272	963	pH 7.8, 0.002 *M* NaP$_i$
	0.63	—	350	963	—
	0.875	—	377	963	—
	1.02	—	445	963	—
Semiquinone	0.765	—	350	963	—
	0.5	—	377	963	—
	0.21	—	445	963	—
	0.45	—	580	963	—
Reduced	0.16–0.18	—	445	963	—
Apo-	2.67	—	278	962	—
Rhodospirillum rubrum					
Oxidized	1.12	—	460	1441	—
	1.13	—	376	1441	—
	5.42	—	272	1441	—
Semiquinone	0.5	—	627	1441	—
	0.45	—	588	1441	—
	1.09	—	353	1441	—
	6.07	—	273	1441	—

MOLAR ABSORPTIVITY AND $A_{1cm}^{1\%}$ VALUES FOR PROTEINS AT SELECTED WAVELENGTHS OF THE ULTRAVIOLET AND VISIBLE REGION (continued)

Protein	$\epsilon^a (\times 10^{-4})$	$A_{1cm}^{1\%}$ [b]	nm[c]	Ref.	Comments[d]
Flavodoxin (continued)					
Apo-	3.50	–	276	1442	–
Chlorella fuscea	1.0	–	464	1443	–
	5.46	–	275	1443	–
	0.905	--	379	1443	–
Flavoprotein	7.65	–	276	964 ⎫	
Egg yolk	1.0	–	375	964 ⎬	pH 7.2, 0.1 M NaP$_i$
	1.32	–	458	964 ⎭	
Azobacter vinelandii	–	13.4	280	445	Biuret
Apo-	–	13.4	280	445	Lowry
Follicle stimulating hormone (FSH)					
Sheep	1.23	4.9	275	168	Neutral and acidic solutions, MW = 25,000 (168)
Human	2.3	9.2	276	169	MW = 25,000 (169)
Formiminoglutamase (EC 3.5.3.8)					
Pseudomonas ATCC 11,299b	–	14.7	280	1444	Dry wt, pH 7.4, 1 mM KP$_i$, 20 mM NaCl, 1 mM HSEtOH
Formylglycinamide ribonucleotide amidotransferase (phosphoribosylformyl-glycinamidine synthetase, EC 6.3.5.3)					
Chicken liver	18.6	–	280	170	pH 6.5, 0.1 M P$_i$
Formyltetrahydrofolate synthetase (EC 6.3.4.3)					
Clostridium acidi-urici	–	3.70	280	1445	Dry wt and DR
C. cylindrosporum	12.7	5.3	280	1446	–
	12.2	5.3	280	171	MW = 230,000 (171)
C. thermoaceticum	–	7.37	280	823	pH 8.1, TrisCl
β-Fructofuranosidase (see invertase)					
Fructokinase					
Rat liver	–	20	280	1447	pH 7, 0.12 M P$_i$, data from Figure 9 (1447)
Fructose-1,6-bisphosphatase (hexose bisphosphatase) (EC 3.1.3.11); (see also aldolase)					
Chicken					
Muscle	–	7.4	280	1448	–
Liver	–	7.4	280	1448	–
Pig kidney	–	8.9	280	1448	pH 8.0, 0.05 M TrisCl
	–	7.55	280	172	–
Rabbit					
Liver	–	8.9	280	824	–
	–	5.3	260	824	–
	–	8.3	280	174	–

MOLAR ABSORPTIVITY AND $A_{1cm}^{1\%}$ VALUES FOR PROTEINS AT SELECTED WAVELENGTHS OF THE ULTRAVIOLET AND VISIBLE REGION (continued)

Protein	$\epsilon^a(\times 10^{-4})$	$A_{1cm}^{1\%}$ [b]	nm[c]	Ref.	Comments[d]
Fructose-1,6-bisphosphatase (continued)					
Liver (neutral)	5.18	3.70	280	1450	Dry wt, MW = 140,000 (1450)
Liver (alkaline)	11.6	8.9	280	1451	MW = 130,000 (1451)
Muscle	–	6.1	260	1451	–
	–	9.4	280	173	
Kidney	9.7	6.9	280	1452	Dry wt, MW = 140,000 (1452)
L-Fucose binding					
Lotus tetragonolobus					
A	21.4	17.8	280	452	MW = 120,000 (452)
B	12.2	20.9	280	452	MW = 58,000 (452)
C	20.4	17.4	280	452	MW = 117,000 (452)
Fumarase (fumarate hydratase) (EC 4.2.1.2)					
Pig heart	9.9	5.1	280	175	pH 7.3, MW = 194,000 (175)
Pig heart muscle	11.65	5.3	280	1334	pH 7.2, 0.005 M P_i, MW = 220,000 (1453)
α-Galactosidase (EC 3.2.1.22)					
Sweet almonds	8.32	25.5	276	176	MW = 33,000 (176)
	7.99	24.3	280	176	MW = 33,000 (176)
Vicia sativa	2.7	9	280	591	pH 7.2, TrisCl 10 mM MW = 30,000 (591), data from Figure 4 (591)
α-Galactosidase I					
Vicia faba	37.6	18	280	590	MW = 209,000 (590)
α-Galactosidase II					
Vicia faba	7.6	20	280	590	MW = 38,000 (590)
	7.2	19	278	590	MW = 38,000 (590)
β-Galactosidase (EC 3.2.1.23)					
E. coli ML 309	143.3	19.1	280	177	MW = 750,000 (178)
E. coli K12		19.1	280	1454	–
Galactosyl transferase A protein (lactose synthase, EC 2.4.1.22)					
Milk	5.3	12	279	1455	MW = 44,000 (1455)
Galactothermin		7.4	280	1456	pH 6.8, 0.01 M P_i
Human milk	1.04	7.4	277	965	0.1 M HCl, MW = 14,000 (965)
	1.20	8.55	290	965 ⎱	0.1 M NaOH,
	1.13	8.15	283	965 ⎰	MW = 14,000 (965)
Gastricsin					
Human gastric juice	–	4.56	247	1457	–
(pepsin C, EC 3.4.23.3)	–	12.83	278	1457	–
Gastrin					
Human	4.83	15.32	278	179	MW=31,500 (179)
Pig	5.02	15.3	280	180	MW=32,500 (180)

MOLAR ABSORPTIVITY AND $A_{1cm}^{1\%}$ VALUES FOR PROTEINS AT SELECTED WAVELENGTHS OF THE ULTRAVIOLET AND VISIBLE REGION (continued)

Protein	$\epsilon^a (\times 10^{-4})$	$A_{1cm}^{1\%}$ [b]	nm[c]	Ref.	Comments[d]
Gliadin					
α-, Wheat	2.9	5.8	276	181	pH 0.7, 6 M HCl, or pH 3.0, 0.001 M HCl, MW=50,000 (181)
	4.45	8.9	290	181	0.1 N NaOH, MW=50,000 (181)
Hard, red winter wheat	–	5.7	276	1458	Dry wt, corrected for light scattering
α_{1b}-	–	5.6	276	1458	–
α_{1c}-	–	5.6	276	1458	–
α_2-	–	5.6	276	1458	–
Globin					
Human	–	8.75	280	182	–
	–	8.5	280	183	
	–	8.74	280	184	pH 4.8, ionic strength 0.05
	3.36	8.0	280	185	MW=42,000 (185)
Horse	–	10.62	280	1459	Dry wt
Beef	–	8.5	280	1460	–
Dog	–	7.9	280	1460	–
Rabbit	–	8.9	280	1460	–
Rat	–	8.5	280	1460	–
Chironomus thummi thummi	–	8.7	280	1460	–
	–	7.23	282	1461	Dry wt
Globulin					
α-, Human hepatoma	–	5.26	278	1462	Dry wt
α-, Human	1.0		280	1463	–
β-, Human	1.54		280	1463	–
β_{1c}-, Human	–	10.0	280	186	–
β_2-, micro-, Human	1.99	16.8	280	187	pH 7.0, 0.1 M P$_i$, MW=11,815 (187)
γ-, low MW Human	1.72	10.1	280	188	MW=17,000 (188)
Corticosteroid binding					
Human	3.78	7.4	279	189	MW=51,700 (189)
Rabbit	3.43	8.4	279	190	MW-40,700 (190)
Rat	3.84	6.2	279	191	MW=61,000 (191)
Thyroxine binding					
Human	5.1	8.9	280	192	MW=58,000 (192)
Rabbit lens					
1	–	8.5	–[k†]	440	–
2	–	4.5	–[k†]	440	–
3	–	4.5	–[k†]	440	–
4	–	4.4	–[k†]	440	–
5	–	4.2	–[k†]	440	–
Globulin, 75, soybean	–	6.9	280	1464	Dry wt
Globulin, 11 S (Glycinin)	9.85	5.47	280	446	MW=180,000 (447)
Soybean seed (Glycine max)	–	8.04	280	1465	–
Globulin, cold-insoluble					
Human plasma	–	12.8	280	825	pH 7.0, 0.05 M P$_i$, 0.15 M NaCl
	–	14.8	282	825	0.1 N NaOH–5 M urea

MOLAR ABSORPTIVITY AND $A_{1cm}^{1\%}$ VALUES FOR PROTEINS AT SELECTED WAVELENGTHS OF THE ULTRAVIOLET AND VISIBLE REGION (continued)

Protein	$\epsilon^a(\times 10^{-4})$	$A_{1cm}^{1\%}$ [b]	nm[c]	Ref.	Comments[d]
Glucagon					
Commercial	–	23.0	280	1466 ⎫	
	0.72	–	260	1466 ⎬ pH 10.0, 0.2 M P$_i$	
	1.1	–	250	1466 ⎭	
Pig	–	23.0	279	1467	0.1 M glycine/0.1 M NaCl/0.1 N NaOH, pH 10.4
Beef	0.83	23.8	278	193	pH 2, MW=3647 (193)
Glucanase, exo-β-D-[1 →, 3] (EXO-1-β- glucosidase, EC 3.2.1.58)					
Basidomycete QM 806	9.2	18.1	280	194	MW=51,000 (194)
Gluconolactonase (see lactonase)					
Glucose dehydrogenase (EC 1.1.1.47 or .118 or .119)					
Bacterium anitratum	–	9.1	280	966	pH 7.0, 0.06 M P$_i$, data from Figure 1(966)
Soluble	1.56	–	350	195	Oxidized state
	3.89	–	339	195	Reduced state
Glucose isomerase (EC 5.3.1.18)					
Streptomyces	15.7	10	280	692	MW=157,000 (692)
Bacillus coagulans	17.0	10.6	280	826	pH 6.0, 0.1 M AcONa, MW=160,000 (826)
	17.0	10.6	280	826 ⎫	0.1 M NaOH, data from Figure 7 (826),
	23.8	14.9	290	826 ⎭	MW=160,000.
Glucose oxidase (EC. 1.1.3.4)					
A. niger	31.1	16.7	280	196	MW=186,000 (196)
Penicillium amagasukiense	18.8	11.9	278	967	MW=158,000 (967)
Apo-	19.2	–	278	1468	–
Glucose-6-phosphate dehydrogenase (EC 1.1.1.49)					
Human erythrocyte	–	6.15	280	693	–
Leuconostoc mesenteroides	11.9	11.5	280.5	968	MW=103,700 (968)
D-Glucose-6-phosphate ketol isomerase (EC 5.2.1.9)					
Pea	9.6	8.75	280	197	MW=110,000 (197)
α-Glucosidase (EC 3.2.1.20)					
Beef liver	14.35	13.4	280	694 ⎫	MW=107,000 (694)
	11.2	10.5	288	694 ⎭	
β-Glucosidase (EC 3.2.1.21)					
Aspergillus wentii	33.8	19.1	278	1469	MW=170,000 (1470)
Sweet almonds	–	21.8	278	696	–
A	–	18.8	278	695	–
B	–	18.2	278	695	–

MOLAR ABSORPTIVITY AND $A_{1cm}^{1\%}$ VALUES FOR PROTEINS AT SELECTED WAVELENGTHS OF THE ULTRAVIOLET AND VISIBLE REGION (continued)

Protein	$\epsilon^a (\times 10^{-4})$	$A_{1cm}^{1\%}$ [b]	nm[c]	Ref.	Comments[d]
β-Glucuronidase (EC 3.2.1.31)					
Beef liver	47.6	17	280	198	pH 5, MW=280,000 (198)
Glutamate decarboxylase (EC 4.1.1.15)					
E. coli	51.0	17	280	827	pH 7.0, MW=300,000 (827)
Glutamate dehydrogenase (EC 1.4.1.2-4)					
Commercial	46.5	_[b*]	279	1471	MW=56,100 (1472)
Clostridium SB4	−	10.7	280	1473	DR
Beef liver	−	9.3	280	1474	Dry wt, pH 7.0, 0.11 M P$_i$
	−	8.9	280		Dry wt, pH 7.0, 0.11 M P$_i$,
	−	9.5	280	1474 ⎱	corrected for light
				1475 ⎰	scattering, pH 7.6, M/15 KNaP$_i$
Frog liver	−	9.5	280	828	−
Pig liver	−	9.7	279	829	−
Glutamate dehydrogenase, (NAD dependent) (EC 1.4.1.2)					
Cl. SB$_4$	29.4	10.7	280	830	pH 7, 50 mM KP$_i$ or pH 7.4, 50 mM TrisCl, MW=275,000 (830)
Glutamate mutase (methylaspartate mutase, EC 5.4.99.1)					
C. tetanomorphum H 1					
Component E	7.94	6.2	280	831	MW=128,000 (831)
Component S	1.1	6.44	280	199	MW=17,000 (199)
Glutamic dehydrogenase (EC 1.4.1.4)					
E. coli	−	12.9	278	1476	Dry wt, Kjeldahl, and ash[e]
Beef liver	−	10.0	280	200	−
	−	9.73	279	201	pH 7, 0.2 M P$_i$
	−	8.55	276	202	5.1 M GdmCl
	−	9.71	279	203	−
	−	8.20	279	203	6 M GdmCl
Frog, liver	23.8	9.5	280	204	pH 8.0, 0.1 M Tris- AcO⁻, 0.0001 M EDTA, MW=250,000 (204)
Glutamin-(asparagin-)ase (EC 3.5.1.38)					
Achromobacteracae	−	10.2	280	1477	pH 7.2, 0.01 M NaP$_i$
Glutaminyl-peptide glutaminase (see peptidoglutaminase II)					

[b*]Duplicate analyses: 8.31 and 8.36.

MOLAR ABSORPTIVITY AND $A_{1cm}^{1\%}$ VALUES FOR PROTEINS AT SELECTED WAVELENGTHS OF THE ULTRAVIOLET AND VISIBLE REGION (continued)

Protein	$\epsilon^a (\times 10^{-4})$	$A_{1cm}^{1\%\ b}$	nm[c]	Ref.	Comments[d]
Glutamine cyclotransferase (glutaminyl-tRNA cyclotransferase, EC 2.3.2.5)	3.6	14.3	280	205	MW=25,000 (205)
Glutamine phosphoribosylpyrophosphateamidotransferase (amidophosphoribosyl transferase, EC 2.4.2.14)					
Pigeon liver	1.02	–	415	206	–
	8.18	8.18	279	207	MW=100,000 (207)
Glutamine synthetase (EC 6.3.1.2)					
Pig brain	51	13.8	279	1478	pH 7.2, MW= 370,000 (1478)
Sheep brain	–	10.0	280	208	–
	–	13.5	280	832	–
E. coli	52.4	7.7	280	209	pH 7.0, 0.01 M imidazole HCl, MW=680,000 (210)
Glutamine transaminase (EC 2.6.1.15)					
Rat liver	–	6.5	280	1479 ⎫	Lowry[e], pH 7.2,
	–	3.26	260	1479 ⎬	0.005 M KP$_i$
	–	0.78	415	1479 ⎭	
γ-Glutamylcysteine synthetase (EC 6.3.2.2)					
Rat kidney	10.6	11.5	280	1480	Lowry[e], MW= 92,000 (1480)
Glutathione peroxidase (EC 1.11.1.9)					
Beef blood	6.3	7.5	280	969	MW=84,000 (969)
Glutathione reductase (EC 1.6.4.2)					
Yeast	12.0	10.5	280	211	MW = 118,000 (211), protein det. by dry wt
	–	18.6	280	212	Used biuret method for protein det.
	–	14.5	280	212	Protein det. by dry wt
	18.6	15.4	280	213	MW=121,000 (213)
Human red blood cell	19.5	16.3	275	214	Estimated from Figure 8 (214), Reference 214, MW=120,000 (214)
Penicillium chrysogenum	–	18.6	280	833	
Rice embryos	9.51	9.1	275	1481 ⎫	
	18.3	17.6	280	1481 ⎪	
	1.09	10.5	370	1481 ⎬	MW=104,000 (1481)
	1.10	10.6	379	1481 ⎪	
	1.16	11.1	463	1481 ⎭	
Gluten					
Wheat	–	6.20	276	1482 ⎫	
S-β-(1-Pyridylethyl)-	–	7.07	275	1482 ⎬	pH 2, 0.01 N HCl
Acrylonitrile derivative	–	6.58	276	1482 ⎭	

MOLAR ABSORPTIVITY AND $A_{1cm}^{1\%}$ VALUES FOR PROTEINS AT SELECTED WAVELENGTHS OF THE ULTRAVIOLET AND VISIBLE REGION (continued)

Protein	$\epsilon^a (\times 10^{-4})$	$A_{1cm}^{1\%}$ [b]	nm[c]	Ref.	Comments[d]
Glyceraldehyde-3-phosphate dehydrogenase (EC 1.2.1.12)					
Pig	–	9.6	280	970	5 M GdmCl
	14	10	280	971	0.1 N NaOH, MW=140,000 (971)
	–	9.6	280	972	–
Apo-	12.7	9.1	280	971	0.1 N NaOH, MW=140,000 (971)
Pig, skeletal	–	9	280	221	0.1 NaOH, coenzyme free
Rabbit	17.5	12.7	280	973	MW=145,000 (973)
	–	10.6	276	974	–
	–	8.15	280	974	Charcoal treated enzyme
Rabbit muscle	–	10.3	280	1483	–
Denatured	3.0	–	337	1484	–
Furylacryloyl-	3.0	–	344	1484	–
Chicken heart	14	10.2	280	1485	MW=137,600 (1485)
Lobster	–	5.55	290	975	–
	14	10	280	976	MW=140,000 (977)
	–	9.6	280	978	With 4 NAD$^+$
	–	8.0	280	979	Charcoal treated
Lobster, muscle	14.8	10.1	276	218	MW=146,650 (219)
	–	9.6	280	220	pH 8.5, 0.05 M, Na PP$_i$
Yeast	–	8.94	280	215	–
	10.9	9.08	280	216	MW=120,000 (216)
	12.4	8.6	280	217	MW=144,700 (217)
	–	9.4	280	980	–
	–	8.6	280	981	–
Apo-	–	8.94	280	982	–
	–	8.2	280	1485	–
Human erythrocyte	13.6	9.9	280	983	pH 8.0, 0.05 M TrisCl, MW=137,000 (983)
Beef liver	13.0	9.13	280	984	MW=142,000 (984)
E. coli	14.4	10	280	1486	pH 8, 20 mM TrisCl, 2 mM EDTA
Rabbit muscle	11.5	9.8	280	222	MW=118,000 (222)
	11.4	8.29	280	223	MW=138,000 (223) Charcoal treated
	14.9	10.3	280	217	pH 8, PP$_i$, MW=144,000 (217)
Glycerol dehydrogenase (EC 1.1.1.6)					
Aspergillus niger	–	17	280	985	pH 7.2, P$_i$
Glycerol kinase (EC 2.7.1.30)					
E. coli	–	15.6	280	986	pH 6.5, 6 M GdmCl, data from Figure 7 (986)
	–	14.1	280	986 }	pH 12.5, data from Figure 7 (986)
	–	14.7	290	986 }	
Glycerol-3-phosphate dehydrogenase (EC 1.1.1.8, EC 1.1.99.5)					
Apis mellifera, thorax	–	3.3	280	226	pH 7.8
	2.5	–	280	227	pH 7.5

MOLAR ABSORPTIVITY AND $A_{1cm}^{1\%}$ VALUES FOR PROTEINS AT SELECTED WAVELENGTHS OF THE ULTRAVIOLET AND VISIBLE REGION (continued)

Protein	$\epsilon^a(\times 10^{-4})$	$A_{1cm}^{1\%}$ [b]	nm[c]	Ref.	Comments[d]
Glycerol-3-phosphate dehydrogenase (continued)					
Chicken					
Liver	3.59[d*]	—	280	1488	—
Muscle	2.83[d*]	—	280	1488	—
Rabbit muscle	4.2	7.0	280	987 ⎫	
	4.4	7.3	280	988 ⎬	MW = 60,000 (987)
	3.8	6.3	280	989 ⎭	
	—	6.3	280	224	—
	—	5.3	280	224	Charcoal treated
	—	7.5	280	141	—
	3.8	6.5	280	225	pH 6.6, 0.02 M P_i, MW = 58,300 (225)
	3.52	6.0	280	225	pH 7.2, 0.1 M P_i, MW = 58,300 (225)
Rat liver					
Fraction 1 enzyme	11.0	18.3	280	1487 ⎫	pH 7.2, 0.1 M P_i,
Fraction 2 enzyme	4.0	6.7	280	1487 ⎭	MW = 60,000 (1487)
Glycinin					
Soybean flakes	32.2	9.2	280	990	MW = 350,000 (991)
Glycocyaminekinase (guanidinoacetate kinase, EC 2.7.3.1)					
Nepthys coeca	7.5	8.6	278	228	pH 8.1, Tris-AcO⁻ 0.1 M 10^{-4} M EDTA, MW = 87,500 (228)
Glyco-α-lactalbumin					
Cow's milk	—	17.7	280	1489	—
Glycollate oxidase (EC 1.1.3.1)					
Spinach	—	7.7	280	1490	Calcd from Figure 3 (1490), pH 8.3
Glycoprotein					
Envelope-specific, *E. coli*	—	11	278	229	—
Tamm-Horsfall					
Rabbit urine	—	6.7	277	1503e	Water and 6 M GdmCl
Human urine	—	13	277	1491	Calcd from Figure 1 (1491)
	—	10.8	277	1492	—
	—	9.5	277	1493	6 M GdmCl
	—	9.4	277	1493	Water
Acceptor of glycosyl transferase					
Rat intestinal mucosa	—	4.8	278	1494	—
Glycoprotein (α-globulin)					
Mouse plasma	—	8.8	278	1495	—
Tumor	—	11.6	278	1495	—
α_1-Glycoprotein					
Human					
Easily precipitable	4.5	9.0	280	230	MW = 50,000 (230)
Tryptophan poor	—	6.0	280	231	—

[d*] Average of values obtained for two different preparations.

MOLAR ABSORPTIVITY AND $A_{1cm}^{1\%}$ VALUES FOR PROTEINS AT SELECTED WAVELENGTHS OF THE ULTRAVIOLET AND VISIBLE REGION (continued)

Protein	$\epsilon^{a}(\times\,10^{-4})$	$A_{1cm}^{1\%}$ [b]	nm[c]	Ref.	Comments[d]
α_1-Glycoprotein (continued)					
Acid					
(orosomucoid)	3.9	8.9	280	232	MW = 44,100 (233)
Liver	4.0	9.16	279	1497	MW = 43,600 (1497)
Blood					
Variant pI 3.0	–	9.38	278	1498	–
Variant pI 3.2	–	9.31	278	1498	–
Variant pI 3.4	–	9.32	278	1498	–
Pool	–	9.33	278	1498	–
Serum	–	18.2	280	1503c	pH 7.0, 1/15 $M\,P_i$
Anti-trypsin	2.3	5.3	280		MW = 45,000 (234)
Chimpanzee plasma	–	8.52	278	1496	–
α_2-Glycoprotein					
Neuramino	–	5.0	280	235	–
Zinc	–	18	280	236	–
Histidine-rich 3.8S					
Human serum	–	5.85	280	1499	–
Macroglobulin					
Heat labile	68.9	8.4	280	8	MW = 820,000 (238)
α_{2HS}-Glycoprotein	2.7	5.6	280	230	MW = 49,000 (237)
α_2-β_1-Glycoprotein					
Human plasma	–	11.5	278	1500	pH 6, 0.1 M NaCl
α_3-Glycoprotein, 8S					
Human serum	–	10.0	280	1501	–
α_{IX}-Glycoprotein	–	6.0	280	230	–
β-Glycoprotein					
Glycine-rich					
Human serum	–	6.2	280	1502	–
β_1-Glycoprotein					
Sialic acid free					
Human plasma	–	11.2	278	1503	pH 7
γ-Glycoprotein					
Glycine-rich					
Human serum	–	10.0	280	1503d	–
Glyoxylate reductase					
(EC 1.1.1.26)					
Spinach leaves	–	9.76	280	834	Protein det. by Lowry method
	–	10.7	280	834	Protein det. by dry wt and N
Isozyme R_f 0.22	–	10.8	280	834	–
Isozyme R_f 0.19	–	14.5	280	834	–
Isozyme R_f 0.17	–	13.8	280	834	–
Gramicidin A, HO-NBzl[e]*	7.25[f]*	–	410	1503f	
	7.27[g]*	–	410	1503f	0.1 N Na$_2$CO$_3$
	3.58[f]*	–	270	1503f	

[e]*HO-NBzl, 2-hydroxy-5-nitrobenzyl group.
[f]*Same fraction of two fractions.
[g]*Same fraction of two fractions.

MOLAR ABSORPTIVITY AND $A_{1cm}^{1\%}$ VALUES FOR PROTEINS AT SELECTED WAVELENGTHS OF THE ULTRAVIOLET AND VISIBLE REGION (continued)

Protein	$\epsilon^a (\times 10^{-4})$	$A_{1cm}^{1\%}$ [b]	nm[c]	Ref.	Comments[d]
Gramicidin A, HO-NBzl[e]* (continued)	3.73[g]*	–	270	1503f ⎫	
Gramicidin D, HO-NBzl[e]*	7.18	–	410	1503f ⎬ 0.1 N Na$_2$CO$_3$	
	4.17	–	270	1503f ⎭	
Growth hormone					
Beef	–	7.1	280	239	–
	–	6.5	280	835	–
Haptoglobin					
Human, type I					
Plasma	–	11.0	278	697	0.1 N HCl, pH 1
	–	11.9	278	697	0.02 M HCO$_3^-$, pH 8.6
	–	11.1	278	697	0.1 N NaOH, pH 12
Urine	–	12.7	278	697	0.02 M HCO$_3^-$, pH 8.6
Type II	–	15.6	278	698	–
	–	11.2	280	697	0.1 N HCl, pH 1
	–	12.1	280	697	0.02 M HCO$_3^-$, pH 8.6
	–	11.5	280	697	0.1 N NaOH, pH 12
Type II complex with hemoglobin	–	14.5	280	698	–
	–	29	408	698	–
Canine	9.4	11.6	–	699	MW = 81,000 (699)
Haptoglobin 1-1, human	10.5	–	280	1503g	–
Haptoglobulin					
Human	12.1	12.1	280	240	MW = 100,000 (240)
	–	11.6	278	241	–
Hemagglutinin					
P. lunatus	–	12.3	280	836	–
Agaricus campestris	10.5	16.4	280	837	MW = 64,000 (837)
Pisum sp. (peas)		10	280	1504	From Figure 1 (1504)
Phaseolus vulgaris	9.6	10.5[h]*	280	1505	pH 2, 0.01 M HCl
Lens culinaris	5.6	12.5	280	1506	MW = 44,050 (1506)
L. culinaris A	6.2	12.6	280	1507 ⎫ MW = 49,000 (1507)	
L. culinaris B	6.2	12.6	280	1507 ⎭	
Hemerythrin					
Phascolosoma gouldii					
Coelomic fluid					
Methemerythrin					
Azide	0.019	–	680	1508 ⎫	
	0.370	–	446	1508	
	0.675	–	326	1508	
Bromide	0.0165	–	677	1508	
	0.540	–	387	1508 ⎬ ϵ/dimeric iron unit	
	0.650	–	331	1508	
Chloride	0.018	–	656	1508	
	0.600	–	380	1508	
	0.660	–	329	1508 ⎭	

[h]*Calculated from the amino acid analysis.

MOLAR ABSORPTIVITY AND $A_{1cm}^{1\%}$ VALUES FOR PROTEINS AT SELECTED WAVELENGTHS OF THE ULTRAVIOLET AND VISIBLE REGION (continued)

Protein	$\epsilon^a(\times\ 10^{-4})$	$A_{1cm}^{1\%\ b}$	nm^c	Ref.	Comments^d
Hemerythrin (continued)					
Cyanate	0.0166	–	650	1508	
	0.650	–	377	1508	
	0.655	–	334	1508	
Cyanide	0.014	–	695	1508	
	0.077	–	493	1508	
	0.530	–	374	1508	
	0.640	–	330	1508	
Fluoride	0.500	–	362	1508	ϵ/dimeric iron unit
	0.560	–	317	1508	
Hydroxide	0.016	–	597	1508	
	0.590	–	362	1508	
	0.680	–	320	1508	
Thiocyanate	0.020	–	674	1508	
	0.510	–	452	1508	
	0.720	–	327	1508	
Water	0.640	–	355	1508	
	0.630	–	340	1508	ϵ/dimeric iron unit, plus iodide
Oxyhemerythrin	0.220	–	500	1508	ϵ/dimeric iron unit
	0.680	–	330	1508	
Dendrostomum pyroides	–	30.3	280	838	Protein det. by Biuret method
	–	31.0	280	838	Protein det. by Lowry method
Golfingia gouldii	–	25.8	280	839	–
Hemocyanin					
Busycon canaliculatum	–	15.1	280	1512	
	–	3.28	345	1512	FC, pH 8.2 Tris cont
Cancer borealis	–	14.0	280	1512	0.01 M MgCl$_2$;^{9} SC
	–	2.29	336	1512	
Cancer magister	–	15.0	280	1512	
25S particle	–	14.7	279	843	pH 7.0, Tris
5S particle	–	14.1	–	843	pH 10, Bic
Carcinus meanas	–	14.2	280	1512	FC, pH 8.2 Tris cont
	–	2.33	335	1512	0.01 M MgCl$_2$, SC
C. sapidus	–	12.4	278	243	–
Dolabella auricularia (gastropod)	9.84	17.6	278	1509	MW = 55,800 (1509)
Eriphia spinifrons (arthropod)	–	12.07	278	1510	pH 7.2, 0.01 M P$_i$
	–	12.7	278	1510	pH 9.7, 0.05 M CO$_3^{2-}$
	–	16.1	278	243	
E. moschata	–	14.9	278	243	–
Homarus americanus	–	14.3	280	1512	FC, pH 8.2 Tris cont
	–	2.69	335	1512	0.01 M MgCl$_2$, SC
	–	13.4	280	840	pH 9.6, 0.014 M Ca^{++}, glycine buffer, protein det. by dry wt
H. vulgaris	–	14.4	278	243	–
Levantina hierosolima	–	4.0	345	846	pH 6.6, oxygenated
	–	3.4	348	846	pH 12
Limulus polyphemus	–	13.9	280	1512	FC, pH 8.2 Tris cont
	–	2.23	340	1512	0.01 M MgCl$_2$, SC
	–	11.2	278	243	–
Loligo pealii	–	2.79	345	1512	FC, pH 8.2 Tris cont 0.01 M MgCl$_2$, SC
	–	15.8	278	243	–

MOLAR ABSORPTIVITY AND $A_{1cm}^{1\%}$ VALUES FOR PROTEINS AT SELECTED WAVELENGTHS OF THE ULTRAVIOLET AND VISIBLE REGION (continued)

Protein	$e^a (\times 10^{-4})$	$A_{1cm}^{1\%}$ [b]	nm[c]	Ref.	Comments[d]
Hemocyanin (continued)					
Helix pomatia	—	16.1	278	841	pH 9.2, borate
α-	—	13.8	278	242	—
β-	—	14.1	278	242	—
Succinylated	—	14.16	278	842	pH 9.2, 0.1 M $Na_2B_4O_7$
Megathura crenulata	—	17.7	280	844	Protein det. by dry wt
	—	15.5	280	845	
Murex trunculus (whelk)	—	13.9	280	1511	pH 9.2
	—	18.9	278	243	—
Apo-	—	13.9	280	1511	pH 9.2
M. brandacis	—	18.1	278	243	—
O. vulgaris	—	13.5	278	243	—
O. macropus	—	16.6	278	243	—
Pagarus pollicarus	—	15.6	280	1512 }	FC, pH 8.2 Tris cont
	—	2.58	335	1512 }	0.01 M $MgCl_2$, SC
P. vulgaris	—	13.8	278	243	—
Hemoglobin					
Aphrodite aculeata (polychaet annelid)[i*]					
nerves/ganglia	12.6	—	425	1513	—
	1.41	—	549	1513	—
	1.48	—	566	1513	—
HbO_2	14.7	—	414	1513	—
	1.70	—	541	1513	—
	1.84	—	577	1513	—
HbCO	22.0	—	419	1513	—
	1.82	—	537	1513	—
	1.84	—	567	1513	—
HbCN	14.6	—	434	1513	—
	1.5	—	536	1513	—
	1.9	—	564	1513	—
Aplysia californica (mollusc)[i*] nerves	12.0	—	435	1513	—
	1.3	—	560	1513	—
HbO_2	13.0	—	416	1513	—
	1.5	—	543	1513	—
	1.4	—	578	1513	—
HbCO	17.0	—	423	1513	—
	1.5	—	541	1513	—
	1.4	—	571	1513	—
HbCN	15.0	—	437	1513	—
	1.6	—	539	1513	—
	1.8	—	568	1513	—
Ascaris, bodywall					
Oxygenated	0.22		621	700	
	1.11		578	700	
	1.25		543	700	
	11.7		412	700	
Deoxygenated	0.3		615	700	0.1 M P_i, pH 7.0, ϵ_M values expressed per mole heme
	1.32		556	700	
	11.1		429	700	
Carbon monoxide	0.31		614	700	
	1.32		566	700	
	1.33		538	700	
	18.2		419	700	

[i*] c/mole heme.

MOLAR ABSORPTIVITY AND $A_{1cm}^{1\%}$ VALUES FOR PROTEINS AT SELECTED WAVELENGTHS OF THE ULTRAVIOLET AND VISIBLE REGION (continued)

Protein	$\epsilon^a (\times 10^{-4})$	$A_{1cm}^{1\%}$ [b]	nm[c]	Ref.	Comments[d]
Hemoglobin (continued)					
Cyanide	0.31		614	700	
	2.1		562	700	
	1.47		532	700	
	14.7		429	700	
Methemoglobin, acid	0.23		635	700	0.1 M P$_i$, pH 7.0, ϵ_M values
	1.02		499	700	expressed per mole heme
	14.1		405	700	
Methemoglobin, cyanide	1.2		542	700	
	10.9		417	700	
Ascaris lumbricoides,					
perienteric fluid[i]*	1.22	–	553.5	1515	
	10.9	–	429.5	1515	
HbO$_2$	1.04	–	576.5	1515	pH 7.0, 0.1 M P$_i$
	1.23	–	542	1515	
	10.95	–	412	1515	
HbO$_2$[j]*	11.0	–	412	1514	
	1.23	–	541	1514	
	1.04	–	575	1514	At 273°K
Hb IV[j]*	9.75	–	411	1514	
	1.06	–	542	1514	
	0.87	–	576	1514	
HbCO	1.28	–	568	1515	
	1.27	–	538	1515	
	16.8	–	417.5	1515	
HbCN	2.02	–	564	1515	
	1.48	–	534	1515	
	14.9	–	431	1515	
MetHb					
Acid	0.36	–	632	1515	pH 7.0, 0.1 M P$_i$
	1.06	–	501	1515	
	15.5	–	404	1515	
CN$^-$	1.22	–	540	1515	
	11.4	–	417	1515	
N$_3^-$	0.31	–	630	1515	
	1.08	–	540	1515	
	11.7	–	414	1515	
Biomphalaria glabrata					
(mollusc)	–	23.8	280	1516	Lowry, dry wt
Glycera dibranchiata					
(common bloodworm)					
Fe^{2+}-O$_2$ (P)[k]*	1.51	–	576	1517	
	1.47	–	540	1517	
	14.2	–	414	1517	
Fe^{2+}-O$_2$ (M)[l]*	1.71	–	575	1517	
	1.65	–	540	1517	
	14.2	–	420	1517	pH 7, 0.1 M P$_i$, ϵ/mole
Fe^{2+}-CO (P)	1.46	–	569	1517	hematin
	1.51	–	538	1517	
	19.8	–	419	1517	
Fe^{2+}-CO (M)	1.62	–	569	1517	
	1.63	–	538	1517	
	23.2	–	422	1517	

[j]*Prepared from deoxyhemoglobin.

[k]*(P) = polymer.

[l]*(M) = monomer.

MOLAR ABSORPTIVITY AND $A_{1cm}^{1\%}$ VALUES FOR PROTEINS AT SELECTED WAVELENGTHS OF THE ULTRAVIOLET AND VISIBLE REGION (continued)

Protein	$\epsilon^a (\times 10^{-4})$	$A_{1cm}^{1\% \, b}$	nm^c	Ref.	Comments[d]
Hemoglobin (continued)					
Fe^{2+}-H_2O (P)	1.34	–	555	1517	
	13.5	–	428	1517	
Fe^{2+}-H_2O (M)	1.47	–	564	1517	
	11.7	–	429	1517	
Fe^{3+}-H_2O (P)	0.32	–	632	1517	
	0.95	–	503	1517	
	15.0	–	408	1517	
Fe^{3+}-H_2O (M)	0.32	–	637	1517	
	1.48	–	505	1517	
	12.9	–	391	1517	pH 7, 0.1 M P_i,
Fe^{3+}-CN^- (P)	1.11	–	536	1517	ϵ/mole hematin
	13.9	–	418	1517	
Fe^{3+}-CN^- (M)	1.20	–	545	1517	
	14.3	–	420	1517	
Fe^{3+}-N_3^- (P)	0.89	–	573	1517	
	1.13	–	539	1517	
	12.5	–	416	1517	
Fe^{3+}-N_3^- (M)	1.14	–	575	1517	
	1.20	–	542	1517	
	13.1	–	419	1517	
Fe^{3+}-OH^- (P)	0.87	–	574	1517	
	0.97	–	537	1517	pH 10.5 glycine buffer,
	11.4	–	411	1517	ϵ/hematin
Fe^{3+}-OH^- (M)	0.76	–	576	1517	
	0.99	–	534	1517	pH 9.9, methylamine
	9.5	–	398	1517	buffer, ϵ/hematin
Fe^{3+}-F^- (M)	1.07	–	593	1517	pH 7, 0.1 M P_i,
	12.3	–	395	1517	ϵ/hematin
Cucumaria miniata Brandt (echinoderm)					
HbCO	1.33	–	570	1518	
	1.31	–	539	1518	
	12.33	–	416	1518	
Cucumaria piperata Stimpson (echinoderm)					
HbCO	1.66	–	570	1518	
	1.65	–	539	1518	AA
	14.89	–	417	1518	
Molpadia intermedia Ludwig (echinoderm)					
HbCO[m*]	1.31	–	570	1518	
	1.46	–	540	1518	
	14.69	–	418	1518	
Thunnus orientalis (tuna)[n*]					
	12.4	–	428–9	1519	–
	1.37	–	555	1519	–
HbO_2	13.3	–	411	1519	–
	1.49	–	540	1519	–
	1.53	–	575	1519	–
HbCO	20.3	–	418	1519	–
	1.51	–	538	1519	–
	1.44	–	568	1519	–

[m*]Half-molecule of Hb-2.

[n*]Hb concentration determined using the value of $\epsilon = 1.15 \times 10^4$ at 540 nm for MetHbCN.

MOLAR ABSORPTIVITY AND $A_{1cm}^{1\%}$ VALUES FOR PROTEINS AT SELECTED WAVELENGTHS OF THE ULTRAVIOLET AND VISIBLE REGION (continued)

Protein	$\epsilon^a (\times 10^{-4})$	$A_{1cm}^{1\%}$ [b]	nm[c]	Ref.	Comments[d]
Hemoglobin (continued)					
MetHb	14.4	—	405	1519	—
	0.98	—	500	1519	—
	0.42	—	630	1519	—
MetHbCN	11.5	—	418	1519	—
Anguilla japonica (eel)[n]*					
Component F					
	12.6	—	430–1	1519	—
	1.33	—	555	1519	—
HbO$_2$	11.5	—	415	1519	—
	1.39	—	540	1519	—
	1.42	—	575	1519	—
HbCO	17.8	—	420–1	1519	—
	1.46	—	537–9	1519	—
	1.34	—	568–70	1519	—
MetHb	9.3	—	408–9	1519	—
	0.9	—	520–2	1519	—
	0.3	—	630–2	1519	—
MetHbCN	8.5	—	421	1519	—
Component S					
	13.5	—	430–1	1519	—
	1.35	—	555	1519	—
HbO$_2$	13.2	—	410–2	1519	—
	1.39	—	540	1519	—
	1.43	—	575	1519	—
HbCO	17.5	—	418–9	1519	—
	1.43	—	538–9	1519	—
	1.38	—	567–9	1519	—
MetHb	15.3	—	404–5	1519	—
	0.99	—	500	1519	—
	0.44	—	630	1519	—
MetHbCN	10.1	—	419	1519	—
Misgurnus anguillicaudalus (loach)[n]*					
Component F					
	12.4	—	430	1520	—
	1.20	—	555	1520	—
HbO$_2$	13.2	—	413	1520	—
	1.48	—	540–1	1520	—
	1.53	—	576	1520	—
HbCO	20.6	—	418	1520	—
	1.53	—	538	1520	—
	1.46	—	568	1520	—
MetHb	19.4	—	404–5	1520	—
	0.43	—	630	1520	—
	0.96	—	—[o]*	1520	—
MetHbCN	12.8	—	419	1520	—
Component S					
	12.2	—	430	1520	—
	1.18	—	555	1520	—
HbO$_2$	12.4	—	413	1520	—
	1.47	—	541	1520	—
	1.53	—	576–7	1520	—
HbCO	20.4	—	419	1520	—
	1.5	—	538	1520	—
	1.43	—	568	1520	—

[o]*499–502 nm.

MOLAR ABSORPTIVITY AND $A_{1cm}^{1\%}$ VALUES FOR PROTEINS AT SELECTED WAVELENGTHS OF THE ULTRAVIOLET AND VISIBLE REGION (continued)

Protein	$\epsilon^a (\times 10^{-4})$	$A_{1cm}^{1\%}$ [b]	nm[c]	Ref.	Comments[d]
Hemaglobin (continued)					
MetHb	18.2	—	405	1520	—
	0.96	—	—P*	1520	—
	0.43	—	630	1520	—
MetHbCN	11.7	—	419	1520	—
Oncorhynchus tshawytscha					
(chinook salmon)					
HbCS-1					
MetHb	0.44	—	628	1521	
	0.88	—	498	1521	
	13.2	—	405	1521	
	4.6	—	276	1521	
CarboxyHb	1.34	—	566	1521	
	1.38	—	537	1521	
	16.4	—	419	1521	
HbCS-2					Dry wt
MetHb	0.41	—	625	1521	
	0.89	—	499	1521	
	12.3	—	404	1521	
	3.8	—	276	1521	
CarboxyHb	1.29	—	564	1521	
	1.36	—	534	1521	
	16.9	—	417	1521	
Salmo gairdneri					
(rainbow trout)					
Hemoglobin RT-1					
MetHb	0.39	—	628	1521	
	0.71	—	497	1521	
	13.7	—	406	1521	
	4.5	—	276	1521	
CarboxyHb	1.32	—	567	1521	
	1.39	—	536	1521	
	17.5	—	419	1521	
Hemoglobin RT-3					Dry wt
MetHb	0.35	—	628	1521	
	0.79	—	500	1521	
	13.6	—	405	1521	
	4.2	—	277	1521	
CarboxyHb	1.28	—	566	1521	
	1.34	—	536	1521	
	16.2	—	419	1521	
Chironomus plumosus,					
larvae haemolymph					
MetHbCN	1.28	4.0	540	1522	MW = 32,000 (1522)
Sheep					
Hb A, β chain	1.53	9.5	280	1523	MW = 16,134 (1523)
Hb B, β chain	1.46	9.0	280	1523	MW = 16,245 (1523)
Hb C, β chain	1.56	9.9	280	1523	MW = 15,788 (1523)
Beef					
MetHb	—	5.32	550	1524	FC
Horse					
Hb	13.5[n]*	—	430–1	1519	--
	1.37[n]*	—	555	1519	—
	59.6		406	1526	--
	--	16.8	280	1527	—
	—	630	191–4	1527	—

P*499–503 nm.

MOLAR ABSORPTIVITY AND $A_{1cm}^{1\%}$ VALUES FOR PROTEINS AT SELECTED WAVELENGTHS OF THE ULTRAVIOLET AND VISIBLE REGION (continued)

Protein	$\epsilon^a (\times 10^{-4})$	$A_{1cm}^{1\%}$ [b]	nm[c]	Ref.	Comments[d]
Hemoglobin (continued)					
Apo-	–	8.5	280	1528	–
	4.95	–	280	1529	Corrected for scattering
HbO$_2$	12.4[n*]	–	412–3	1519	–
	1.52[n*]	–	541	1519	–
	1.59[n*]	–	576	1519	–
HbCO	20.6[n*]	–	419–20	1519	See footnote q*
	1.49[n*]	–	538	1519	–
	1.50[n*]	–	568	1519	–
MetHb	66	–	406	1525	–
	13.9[n*]	–	405–6	1519	–
	0.97[n*]	–	500	1519	–
	0.43[n*]	–	630	1519	–
MetHbCN	9.5[n*]	–	420	1519	–
Mouse					
Hb	–	17.5	280	1530	–
Human					
Hb A	12.7	–	430	709	pH 6.4, 0.1 M P$_i$
Oxy	0.43	–	523	710	–
	1.57	–	542	710	–
	1.64	–	578	710	–
Deoxy	1.36	–	558	710	–
	0.27	–	583	710	–
Hb	3.14	–	275	1531	–
	11.8	–	430	1531	–
	1.29	–	552.5	1531	–
	0.0396	–	755	1531	–
Apo-	–	8.5	280	702	0.1 N NaOH
HbO$_2$	3.60	–	275	1531	–
	2.88	–	350	1531	–
	12.85	–	415	1531	–
	1.42	–	541.5	1531	–
	1.54	–	576	1531	–
	–	8.5	541	703	–
	5.6	–	540	706	–
	5.9	–	576	706	–
Hb FII	–	22.0	280	1532 ⎫	
Hb AII	–	21.8	280	1532 ⎪	
Hb β_4	–	22.2	280	1532 ⎬ From Figure 5 (1532)	
Hb γ_4	–	23.9	280	1532 ⎭	
HbCO	6.63	–	429	1533	–
	20.8	–	419	1533	–
	17.4	–	419	1534	–
	1.43	–	540	702	–
	–	8.4	540	703	–
Hb-BuNC[r*]	18.75	–	429	1533	–
	9.65	–	419	1533	–
MesoHb	11.5	–	421	1535	–
	1.2	–	550	1535	–
MesoHbO$_2$	1.28	–	543	1535	–
	1.06	–	568	1535	–
MesoHbCO	21.0	–	410	1535	–
	1.39	–	532	1535	–
	1.26	–	560	1535	–

[q*] $\epsilon = 18.5{-}19.0 \times 10^4$ at 419 nm; $1.25{-}1.43 \times 10^4$ at 538 nm; $1.22{-}1.37 \times 10^4$ at 569 nm. See Reference 1534.
[r*] BuNC, butylisocyanide.

MOLAR ABSORPTIVITY AND $A^{1\%}_{1cm}$ VALUES FOR PROTEINS AT SELECTED WAVELENGTHS OF THE ULTRAVIOLET AND VISIBLE REGION (continued)

Protein	$\epsilon^a (\times 10^{-4})$	$A^{1\%}_{1cm}{}^{b}$	nm^c	Ref.	Comments^d
Hemoglobin (continued)					
MesoHb$^+$	14.4	–	396	1535 ⎱	
	0.89	–	495	1535 ⎬ pH 7	
	0.38	–	620	1535 ⎰	
DeuteroHb	11.5	–	421	1535	–
	1.13	–	544	1535	–
DeuteroHbO$_2$	11.6	–	403	1535	–
	1.21	–	532	1535	–
	0.91	–	565	1535	–
DeuteroHbCO	20.0	–	408	1535	–
	1.22	–	528	1535	–
	0.93	–	556	1535	–
DeuteroHb$^+$	11.6	–	394	1535 ⎱	
	0.71	–	500	1535 ⎬ pH 7	
	0.28	–	620	1535 ⎰	
ChloroHb	13.0	–	443	1535	–
	1.85	–	567	1535	–
ChloroHbCo	14.0	–	432	1535	–
	1.70	–	550	1535	–
	2.00	–	590	1535	–
ChloroHb$^+$	13.0	–	421	1535 ⎱	
	1.45	–	550	1535 ⎬ pH 7	
	1.30	–	598	1535 ⎰	
HematoHb	11.3	–	423	1535	–
	1.33	–	552	1535	–
HematoHbCO	18.3	–	412	1535	–
	1.33	–	536	1535	–
	1.10	–	564	1535	–
HematoHb$^+$	15.0	–	398	1535 ⎱	
	1.00	–	501	1535 ⎬ pH 7	
	0.41	–	628	1535 ⎰	
Fetal (7% adult)	–	9.87	576	1536 ⎱	
	–	10.0	541	1536 ⎫	
	–	18.8	290	1536 ⎬ pH 7, 0.02 M P$_i$	
	–	24.0	280	1536 ⎭	
	–	24.8	270	1536 ⎰	
Adult	–	10.2	576	1536 ⎱	
	–	9.6	541	1536 ⎫	
	–	18.8	290	1536 ⎬ pH 7, 0.02 M P$_i$	
	–	23.3	280	1536 ⎭	
	–	25.8	270	1536 ⎰	
MetHbCn	4.350	–	540	1537 ⎱ pH 7.4	
	10.800	–	281	1537 ⎰	
	4.388	–	540	1538	Fe det
	4.284	–	540	1539	Fe det
	4.360	–	540	1540	Nitrogen det
Cyanmethemoglobin	1.1	–	540	702	–
	–	20	280	704	–
	–	80	416	704	–
	4.59	–	541	705	–
	12.02	–	280	705	–
MetHb	–	5.97	540	708	–
No source cited					
60% MetHb	–	15.6	280	701	–
	–	330	210	701	–
	–	910	191	701	–

MOLAR ABSORPTIVITY AND $A_{1cm}^{1\%}$ VALUES FOR PROTEINS AT SELECTED WAVELENGTHS OF THE ULTRAVIOLET AND VISIBLE REGION (continued)

Protein	$\epsilon^a (\times 10^{-4})$	$A_{1cm}^{1\%}$ [b]	nm[c]	Ref.	Comments[d]
Hemoglobin chains					
α-HgBzO^{-S}*	11.1	–	428	709	
β-HgBzO$^-$	11.2	–	428	709	
α-HgBzO$^-$ + β-HgBzO$^-$	12.6	–	430	709	pH 6.4, 0.1 M P$_i$
α-SH^t*	11.1	–	429	709	
β-SH^t*	11.0	–	428	709	
α-SH + β-SH	12.4	–	430	709	
Hemoglobins, synthetic					
ProtoHb	14.0	–	430	1541	
	1.34	–	555	1541	
ProtoHbO$_2$	13.5	–	414	1541	pH 7.0, 0.1 M P$_i$
	1.48	–	541	1541	
	1.57	–	577	1541	
ProtometHb	15.4	–	406	1541	
	0.95	–	500	1541	pH 6.5, 0.1 M P$_i$
	0.38	–	630	1541	
DimethylprotoHb	14.0	–	430	1541	
	1.36	–	555	1541	
DimethylprotoHbO$_2$	13.3	–	414	1541	pH 7.0, 0.1 M P$_i$
	1.46	–	540	1541	
	1.53	–	577	1541	
DimethylprotoMetHb	15.1	–	405	1541	
	0.94	–	500	1541	pH 6.5, 0.1 M P$_i$
	0.39	–	630	1541	
MesoHb	13.5	–	421	1541	
	1.33	–	546	1541	
MesoHbO$_2$	13.9	–	404	1541	pH 7.0, 0.1 M P$_i$
	1.38	–	534	1541	
	1.24	–	567	1541	
MesometHb	18.5	–	395	1541	pH 6.5, 0.1 M P$_i$
	0.73	–	620	1541	
Deoxy Hb	–	4.25	524.5	707	–
DimethylmesoHb	13.5	–	421	1541	
	1.37	–	545	1541	
DimethylmesoHbO$_2$	13.8	–	404	1541	
	1.40	–	534	1541	
	1.30	–	568	1541	pH 7.0, 0.1 M P$_i$
EtioHb	13.4	–	420	1541	
	1.20	–	540	1541	
EtioHbO$_2$	13.8	–	401	1541	
	1.33	–	532	1541	
	1.15	–	566	1541	
EtiometHb	18.1	–	395	1541	
	1.32	–	490	1541	pH 6.5, 0.1 M P$_i$
	0.65	–	620	1541	
HematoHb	13.7	–	426	1541	
	1.42	–	551	1541	
HematoHbO$_2$	13.4	–	409	1541	
	1.47	–	537	1541	
	1.28	–	572	1541	pH 7.0, 0.1 M P$_i$
DeuteroHb	11.9	–	420	1541	
	1.29	–	542	1541	
DeuteroHbO$_2$	11.5	–	402	1541	
	1.35	–	531	1541	
	0.94	–	564	1541	

^{S}*-HgBzO$^-$, paramercuribenzoate.
^{t}*SH, free sulfhydryl group.

MOLAR ABSORPTIVITY AND $A_{1cm}^{1\%}$ VALUES FOR PROTEINS AT SELECTED WAVELENGTHS OF THE ULTRAVIOLET AND VISIBLE REGION (continued)

Protein	$\epsilon^a(\times 10^{-4})$	$A_{1cm}^{1\%}$ [b]	nm[c]	Ref.	Comments[d]
Human Hb, subunits					
αCO	—	8.4	540	1542	—
α-HgBzO$^-$-CO[u*]	1.4	—	540	1543	—
Ferric α-HgBzO^{-}[u*]	11.65	—	410	1544	ϵ/mole heme
	1.04	—	533	1544	
α(Mn^{2+})	1.19	—	585	1545	—
	1.23	—	570	1545	
	2.03	—	555	1545	
	1.09	—	535	1545	
	15.7	—	430	1545	
	10.0	—	420	1545	
Apo-α	1.62	—	280	1545	
βCO	—	8.4	540	1542	—
β-Fe^{2+}-CO	0.55	—	585	1545	—
	1.52	—	570	1545	—
	1.30	—	555	1545	—
	1.54	—	535	1545	—
	6.78	—	430	1545	—
	20.9	—	420	1545	—
Apoβ	1.62	—	280	1545	—
Hemoglobin-reductase complex, Candida-					
Mycoderma (yeast)	12	—	278	1546	pH 6.0, 0.1 M K P$_i$
	14	—	415	1546	
	17	—	420	1546	+ Na$_2$S$_2$O$_4$ + CO
	11	—	423	1546	+ Na$_2$S$_2$O$_4$
Hemopexin	13.5	16.9	280	246	MW = 80,000 (246)
Rabbit	—	21.8	280	1547	pH 7.1, 0.05 M KP$_i$
	—	19.2	414	1547	0.05 M KCl, dry wt
Rabbit blood	—	23.9	280	992	
	—	26.4	280	992	Contains 1 heme
	—	23.2	413.5	992	
Apo-	—	19.7	280	1547	pH 7.1, 0.05 M KP$_i$ 0.05 M KCl, dry wt
Human blood	—	23.8	280	992	—
	—	26.4	280	992	Contains 1 heme
	—	23.0	414	992	
Hemoprotein					
Chironomus thummi	—	71.5	415	244	Ferric form, Calcd from equation in Reference 244, mg = absorbance × f where f = 0.14
559, Beef heart	6.95	—	400	245	pH 9.5, oxidized
	8.84	—	266	245	pH 9.5, oxidized
	6.94	—	394	245	pH 12, oxidized
	1.37	—	559	245	pH 9.5, reduced
	0.958	—	530	245	pH 9.5, reduced
	8.95	—	423	245	pH 9.5, reduced
	2.5	—	557	245	pH 12, reduced
	1.48	—	528	245	pH 12, reduced
	1.33	—	422	245	pH 12, reduced, also data for oxidized and KCN and reduced and CO in Reference 245

[u*]-HgBzO$^-$, *p*-mercuribenzoate.

MOLAR ABSORPTIVITY AND $A_{1cm}^{1\%}$ VALUES FOR PROTEINS AT SELECTED WAVELENGTHS OF THE ULTRAVIOLET AND VISIBLE REGION (continued)

Protein	$\epsilon^a(\times 10^{-4})$	$A_{1cm}^{1\%}$ [b]	nm[c]	Ref.	Comments[d]
Hepatocuprein					
Human liver	–	0.075	675	592	–
	–	5.87	265	592	–
	–	7.00	265	592	–
	–	5.6	278	593	pH 8.6
Hexokinase					
B, Yeast	–	9.16	280	247	pH 5.0, 5 mM, sodium succinate
	–	9.20	278	247	pH 5.0, 5 mM sodium succinate
Yeast	12.5	13	278	248	MW = 96,000 (248)
	10	9	280	1549	MW = 106,000 (1549) refractometry
Rat brain	5.1	–	280	1548	–
Hexosebisphosphatase (see fructose-1,6-bisphosphatase)					
High potential iron protein					
Chromatium, strain D	1.61	16.1	388	993	MW = 10,074 (993), reduced with HSEtOH
	1.86	18.6	450	993 ⎫	MW = 10,074 (993),
	2.0	20.0	375	993 ⎬	oxidized with ferricyanide,
	2.18	21.8	325	993 ⎭	pH 7.0, 0.05 M P$_i$
	41.3	4.1	283	443	Reduced, MW = 10,074 (443)
	39.3	3.9	283	443	Oxidized, MW = 10,074 (443)
Rhodopseudomonas gelatinosa	1.53	16	388	993	MW = 9,579, pH 7.0, 0.05 M P$_i$ reduced with HSEtOH
	1.69	17.6	450	993 ⎫	MW = 9,579
	1.88	19.6	375	993 ⎬	pH 7.0, 0.05 M P$_i$,
	2.11	22.2	325	993 ⎭	oxidized with ferricyanide
	35.4	3.7	283	443	Reduced, MW = 9,579 (443)
	33.8	3.5	283	443	Oxidized, MW = 9,579 (443)
Histaminase					
Pig plasma	–	9.2	278	994 ⎫	Data from Figure 4 (994)
	–	0.169	470	994 ⎭	
Histidine ammonia lyase					
Pseudomonas ATCC 11299b	–	5.6	280	995	Data from Figure 7 (995)
	–	4.0	280	995	Data from Figure 7 (995), NaBH$_4$ added
	–	3.9	280	995	Data from Figure 7 (995), cysteine added
Pseudomonas	–	5.0	280	996	pH 7.2, 0.1 M KP$_i$
	10.3	4.80	279	1550	MW = 215,000 (1550), dry wt
Histidine decarboxylase (EC 4.1.1.22)					
Lactobacillus 30a	–	16.1	280	249	pH 4.8, 0.2 M AcONH$_4$
	–	16.2	280	249	pH 8.0, 0.05 M NH$_4$HCO$_3$

MOLAR ABSORPTIVITY AND $A_{1cm}^{1\%}$ VALUES FOR PROTEINS AT SELECTED WAVELENGTHS OF THE ULTRAVIOLET AND VISIBLE REGION (continued)

Protein	$\epsilon^a (\times 10^{-4})$	$A_{1cm}^{1\%}$ [b]	nm[c]	Ref.	Comments[d]
Histidine decarboxylase (EC 4.1.1.22)					
Lactobacillus 30a (continued)	--	17.3	280	250	pH 4.8, 0.2 M AcONH$_4$
	6.3	16.1	280	997	MW = 38,800 (997)
Chain I	1.1	12.1	280	997	MW = 9,000 (997)
Chain II	5.1	17.2	280	997	MW = 29,700 (997)
Histidinol dehydrogenase (EC 1.1.1.23)					
S. typhimurium	3.6	4.8	280	251	pH 6, 0.05 M P$_i$, MW = 75,000 (251), data from Figure 3 (251)
S. typhimurium LT-2	3.8	4.78	280	252	pH 7.5, MW = 80,000 (253)
LT-7	–	4.63	280	254	pH 8.0, 0.01 M NH$_4$HCO$_3$
N. crassa	–	12.11	280	255	Water
Histidinol dehydrogenase – Histidinolphosphate amino transferase (EC 1.1.1.23, EC 2.6.1.9)					
Salmonella typhimurium LT 2 Strain TM 220	–	15.3	279	998	pH 7.5, 0.1 M triethanol-amine ·HCl Data from Figure 9 (998)
Histone					
Chicken erythocyte	2.9	18.6	275	999	MW = 15,714 HCl der. used
Calf thymus F-8$_{2a}$	–	1.5	276	1000	–
F-8$_{b[z]}$	–	4.0	276	1000	–
F-6$_{3bb}$	–	3.8	276	1000	–
P-4$_4$C$_b$	–	4.5	276	1000	–
P-8$_{a[z]}$	–	4.6	276	1000	–
Histone f2 Calf thymus	–	12	276	1552	pH 6.5
Histone IV Calf thymus	0.0047	–	230	1551	ϵ/mole of residue/l
Homoserine deaminase (cystathione γ-lyase, EC 4.4.1.1)					
Rat liver	–	6.64	280	1001	pH 7.5, 0.2 M KP$_i$
Homoserine dehydrogenase (EC 1.1.1.3)					
Rhodospirillum rubrum	–	3.85[l]	280	1002	pH 7.5, 0.05 M KP$_i$ 0.05 M KCl, 0.001 M EDTA

MOLAR ABSORPTIVITY AND $A_{1cm}^{1\%}$ VALUES FOR PROTEINS AT SELECTED WAVELENGTHS OF THE ULTRAVIOLET AND VISIBLE REGION (continued)

Protein	$\epsilon^a (\times 10^{-4})$	$A_{1cm}^{1\%}$ [b]	nm[c]	Ref.	Comments[d]
Homoserine dehydrogenase – Aspartokinase (EC 1.1.1.3, EC 2.7.2.4)					
E. coli K-12	15.8	4.4	278	256	pH 7.2, 20 mM KP$_i$, 0.15 M KCl, 2 mM Mg titriplex and 4 mM DL-threonine
	16.5	4.6	278	1003	MW = 360,000 (1003)
Hormone					
Chorionic gonadotropin, human	–	5.47	276	1553	–
	–	5.72	276	1553	Asialo der.
Chorionic somatomam-motropin, human	–	8.22	277	1554	pH 8.2, 0.1 M Tris
Follicle-stimulating, human	–	4.40	250	1555	–
	–	5.09	260	1555	–
	–	6.54	270	1555	–
	–	7.17	277	1555	–
	–	7.10	280	1555	–
	–	4.96	290	1555	–
Growth					
Beef pituitary	–	7.30	277	1556	0.1 N AcOH
Human pituitary	2	9.31	277	1557	pH 2.0–8.5, MW = 21,500 (1557)
Sheep pituitary	–	7.30	277	1556	0.1 N AcOH
Lactogen (MPL-2), monkey placenta	1.92	9.12	277	1558	MW = 21,000 (1558)
Lactogenic, sheep pituitary	–	9.09	277	1559	pH 8.0–8.5, dil NH$_4$OH
	–	8.94	278	1560	–
Interstitial cell stimu-lating, sheep	–	4.39	276	1561	–
α subunit	–	5.86	276	1560	–
β subunit	–	3.01	276	1561	–
Parathyroid					
Pig	0.53	5.6	280	1562	MW = 9423 (1562)
Beef	0.63	6.6	280	1562	MW = 9563 data from Figure 5 (562) 0.1 N AcOH
Prolactin					
Sheep	–	9.71	278	1563	–
Hyaluronidase inhibitor					
Human blood	–	8.5	280	1005	–
Hyaluronoglucuronidase (EC 3.2.1.36)					
Beef, testicular	–	9.6	280	1004	–
L-α-Hydroxyacid oxidase (EC 1.1.3.15)					
Rat liver	–	2.3	330	257	pH 7.9, 0.005 M NaP$_i$
	–	14.7	280	257	pH 7.9, 0.005 M NaP$_i$

MOLAR ABSORPTIVITY AND $A_{1cm}^{1\%}$ VALUES FOR PROTEINS AT SELECTED WAVELENGTHS OF THE ULTRAVIOLET AND VISIBLE REGION (continued)

Protein	$\epsilon^a (\times 10^{-4})$	$A_{1cm}^{1\%}$ [b]	nm[c]	Ref.	Comments[d]
p-Hydroxybenzoate hydroxylase (EC 1.14.99.13)					
Pseudomonas desmolytica	–	10.8	280	1008	Data from Figure on p. 331, Reference 1008
Pseudomonas putida	10.4	–	278	1009	pH 7.5, 0.05 M KP$_i$,, data from Figure (1009), 2° C
Hydrogenase (EC 1.12.2.1)					
Desulfovibrio vulgaris	4.1	9.1	277	1006	
Clostridium pasteurianum	0.82	–	400	1007	MW = 45,000 (1006)
(EC 1.12.7.1)	2.45	–	280	1007	–
Hydroxyindole-*O*-methyltransferase (acetylserotonin methyltransferase, EC 2.1.1.4)					
Beef pineal gland					
Fraction A	6.2	7.68	280	1564	pH 7.7, 0.05 M TrisCl, data from Figure 6 (1564), MW = 81,000 (1564), SC
Fraction B	5.8	7.68	280	1564	pH 7.7, 0.05 M TrisCl, data from Figure 6 (1564), MW = 76,000 (1564), SC
L-6-Hydroxynicotine oxidase (EC 1.5.3.5)					
Arthrobacter oxidans	19.3	20.7	274	258	pH 7.5, 0.1 M P$_i$, wavelength est. from Figure 3 (258), MW = 93,000 (258)
Hydroxynitrilase (oxynitrilase, mandelonitrite lyase, EC 4.1.2.10)					
Prunaceae	11.2	15	275	259	pH 7.5, 0.1 M PP$_i$
Prunus communis Stokes	11.4	14.3	275	260	MW = 80,000 (260)
Prunoideae amygdalus	–	15.8	275	1565 ⎱	pH 7.5, 0.1 M P$_i$, data
Maloideae communis	–	18.3	275	1565 ⎰	from Figure 1 (1565)
Almonds	–	1.500	460	1010	–
	–	1.654	390	1010	–
Isoenzyme I	–	14.14	275	1010	–
Isoenzyme II	–	13.93	275	1010	–
Isoenzyme III	–	14.22	275	1010	–
Hydroxyproline 2-epimerase (EC 5.1.1.8)					
Pseudomonas putida	–	11.89	280	1011	Dry wt
Hydroxypyruvate reductase (EC 5.1.1.1.81)					
Pseudomonas acidovorans	6.6	8.8	280	261	pH 7.5, MW = 75,000 (261)

MOLAR ABSORPTIVITY AND $A_{1cm}^{1\%}$ VALUES FOR PROTEINS AT SELECTED WAVELENGTHS OF THE ULTRAVIOLET AND VISIBLE REGION (continued)

Protein	$\epsilon^a (\times\ 10^{-4})$	$A_{1cm}^{1\%}$ [b]	nm[c]	Ref.	Comments[d]
Imidazole acetate monooxygenase (EC 1.14.13.5)					
Pseudomonas ATCC 11299B	9.61	10.7	270	847	
	1.08	1.2	383	847	MW = 90,000 (847)
	1.07	1.19	442	847	
Imidazolylacetolphosphate: L-glutamate aminotransferase					
Salmonella typhimurium		9.54	279	1012	pH 6, 0.1 *M* TEA, pyridoxamine enzyme, data from Figure 3 (1012)
	6.49	11.8	295	262	0.1 *N* NaOH, est. from Figure 9 (262), MW = 59,000 (262)
	5.78	9.86	280	262	pH 7.5, 0.01 *M* Tris, MW = 59,000 (262)
Immunoglobulins					
Human					
Bence-Jones Protein	6.4	14.2	280	1013	MW = 45,000 (1013)
	2.7	12.2	280	1014	MW = 22,000 (1014)
Hac	–	10.5	280	1567	Neutral pH
Sch	–	10.5	280	1567	Neutral pH
Nu	–	13.0	280	1567	Neutral pH
MIg	–	14.5	280	1567	Neutral pH
λ chain	–	14.6	280	1568	–
Normal light chain	–	14.3	280	1568	–
½ normal light chain	–	14.5	280	1568	–
Constant half of λ chain	–	14.6	280	1568	–
Variable half of λ chain	–	14.6	280	1568	–
Variable fragment	1.3	11.6	280	1014	MW = 11,000 (1014)
L chain	–	9.8	280	1569	pH 6.8, 0.01 *M* P_i
J chain	–	6.5	280	1569	pH 6.8, 0.01 *M* P_i
H chain	–	10.7	280	1569	pH 6.8, 0.01 *M* P_i
IgG	–	13.3	278	1570	pH 7.4, 0.01 *M* NaP_i – 0.15 *M* NaCl
γ chain	–	13.8	282	1570	0.1 *N* NaOH
Light chain	–	11.0	282	1570	0.1 *N* NaOH
Heavy chain	7	14	280	1014	MW = 50,000 (1014)
Doty, type γ_1, K	–	14.10	280	1019	Dry wt
Sackfield	–	10.69	280	1019	Dry wt
Atypical	–	10.52	280	1018	–
Serum	21.6	13.3	277.5	1017	pH 2, MW = 162,000 (1017)
	–	13.42	280	595	–
IgG	21.2	13.8	280	1571	MW = 153,000 (1571)
IgA	21.7	13.4	280	1571	MW = 162,000 (1571)
Serum	–	13.4	280	1015	–
Secretory	–	13.9	280	1016	–
Colostrum	48.3	12.37	280	595	0.01 *N* HCl, MW = 390,000 (595)
IgE(PS)	–	12.5	280	1572	–
IgE(ND)	–	15.33	280	1572	–

MOLAR ABSORPTIVITY AND $A_{1cm}^{1\%}$ VALUES FOR PROTEINS AT SELECTED WAVELENGTHS OF THE ULTRAVIOLET AND VISIBLE REGION (continued)

Protein	$\epsilon^a (\times 10^{-4})$	$A_{1cm}^{1\%}$ [b]	nm[c]	Ref.	Comments[d]
Immunoglobulins (continued)					
IgM	–	14.5	280	1573	–
Serum	–	13.3	280	1020	–
Waldenstrom	–	13.5	280	600	0.25 M AcOH
IgG	–	13.5	280	1574	–
γ chain	–	14.0	280	1574	–
Light chain	–	12.0	280	1574	–
IgND, Myeloma	–	14	280	1021	–
Kappa chain					
urine	1.8	10.7	280	594	MW = 17,000 (594)
Chicken					
Anti-N$_2$ ph[v*]	–	17.7	280	1022	Neutral buffer
	–	24.6	290	1022	0.1 N NaOH
	–	13.0	290	1022	0.1 N NaOH, another sample
Pig					
Anti-N$_2$ ph-BGG[w*] λ chain	–	11.3	280	1023	–
π chain	–	9.8	280	1023	–
ρ chain	–	12.6	280	1023	From antibody
ρ chain	–	12.2	280	1023	Non-specific
γ chain	–	14.2	280	1023	–
Light chain fraction	–	11.3	280	1023	From antibody
	–	11.0	280	1023	Non-specific
Epinephelus itaiva					
(Giant grouper)					
IgG 6.4S	–	16.57	280	1024	0.3 M KCl
	–	17.82	280	1024	0.1 M NaOH
	–	16.50	280	1024	5.0 M GdmCl
IgG 16S	–	13.78	280	1024	0.3 M KCl
	–	15.03	280	1024	0.1 N NaOH
	–	13.53	280	1024	5.0 M GdmCl
Lepisosteus osseus [Gar]					
IgG 14S	–	15	280	1025	0.85% NaCl
Rabbit	21.9	14.6	280	1026 ⎰	pH 8, Tris,
	8.98	6.0	250	1026 ⎱	MW = 150,000 (1026)
IgG	–	14.0	280	1027	–
IgM	–	13.0	280	1029	–
Anti-N$_2$ ph[v*]	23.4	14.6	279	1030	MW = 160,000 (1030)
Anti azobenzenearsonate					
IgG	–	14.6	280	1031	–
IgM	–	13.4	280	1031	–
Polyalanylated Ab	–	14.6	280	1032	–
Goat					
IgG	18.7	13.0	280	1033	pH 7.0, MW = 144,000 (1033)
	–	14	280	1034	
	20.2	13.1	280	1579	pH 7.2, 5 mM NaP$_i$, 0.2 M NaCl, MW = 146,000 (1579)
Heavy chain	6.4	12.0	280	1033	pH 3.0 0.1 M AcOH MW = 53,600 (1033)
Light chain	2.9	12.7	280	1033	MW = 23,000 (1033)
IgM	–	13	280	1034	–

[v*] N$_2$ ph, dinitrophenyl.
[w*] N$_2$ ph-BGG, dinitrophenyl bovine γ-globulin.

MOLAR ABSORPTIVITY AND $A_{1cm}^{1\%}$ VALUES FOR PROTEINS AT SELECTED WAVELENGTHS OF THE ULTRAVIOLET AND VISIBLE REGION (continued)

Protein	$\epsilon^a (\times 10^{-4})$	$A_{1cm}^{1\%}$ [b]	nm[c]	Ref.	Comments[d]
Immunoglobulins (continued)					
Horse	–	14.4	277	1580	–
IgG [Anti-tick borne	–	15.06	278	1034	–
encephalitis]	–	1.37	330	1035	–
	–	1.27	333	1035	–
	–	1.22	335	1035	–
	–	1.20	338	1035	–
	–	1.18	340	1035	–
Heavy chain, reduced					
and alkylated	–	14.3	280	1036	–
Light chain, reduced					
and alkylated	–	12.7	280	1036	–
Anti-polysaccharide	–	15.0	287	1566	0.1 N NaOH
Type II	–	14.6	280	1566	pH 8.0, borate-NaCl
Mouse, myeloma protein with					
antibody activity					
IgA, MOPC-315 tumor	–	13.5	275	1578	pH 7.4, 0.01 M KP$_i$ –
					0.15 – NaCl
Fraction III	–	14	278	1037	–
7S Monomer	15	12.5	278	1037	MW = 120,000 (1037)
Fab fragment	7.7	14.0	278	1037	MW = 55,000 (1037)
Rabbit					
IgG	–	13.5	280	1574	–
γ chain	–	14.0	280	1574	–
Light chain	–	12.0	280	1574	–
IgA	–	13.5	280	1575	0.1 N NaOH
Anti-benzylpenicilloyl-	–	13.8	280	1576	Saline + 0.25 N AcOH
	–	15.4	294	1576	0.1 N NaOH
Anti-N$_2$ ph^v*	–	15.5	280	1577	Dry wt
Heavy chain	–	15.4	280	1577 ⎫	
Light chain	–	13.2	280	1577 ⎭	Refractometry
Zebra	–	15.2	277	1580	–
Donkey	–	13.2	277	1580	–
Mule	–	14.7	277	1580	–
Hinny	–	14.1	277	1580	–
Dog					
Colostral IgA	–	11.80	280	1581	pH 7.4, phosphate
					buffered saline, FC
α chain	–	10.07	280	1581	6 M GdmCl, FC
L chain	–	7.32	280	1581 ⎫	pH 7.4, phosphate
Serum IgA	–	14.08	280	1581 ⎭	buffered saline, FC
α chain	–	10.63	280	1581 ⎫	
L chain	–	8.37	280	1581 ⎭	6 M GdmCl, FC
Beef					
IgG	–	13.7	280	1582	
Sheep					
IgA	–	12	280	596	
Rat					
IgG	–	14.6	280	599	
IgM	–	12.5	280	599	
Guinea pig					
γM	–	10.5	280	848 ⎫	
γ_2 G	–	12.3	280	848 ⎬	0.5% (NH$_4$)$_2$ CO$_3$
γ_1 G	–	13.0	280	848 ⎭	
γ_2 chain	–	13.2	280	848 ⎫	
γ_1 chain	–	13.5	280	848 ⎭	0.1 M AcOH

MOLAR ABSORPTIVITY AND $A_{1cm}^{1\%}$ VALUES FOR PROTEINS AT SELECTED WAVELENGTHS OF THE ULTRAVIOLET AND VISIBLE REGION (continued)

Protein	$e^a (\times 10^{-4})$	$A_{1cm}^{1\%}$ [b]	nm[c]	Ref.	Comments[d]
Immunoglobulins (continued)					
μ chain	—	9.2	280	848 ⎱	0.1 M AcOH
Light chain	—	12.0	280	848 ⎰	
F ab [trypsin]	—	15	280	597	
F c [trypsin]	—	15	280	597	
IgGs, colostrum	—	13.68	278	598	P_i buffered saline
	—	13.70	278	598	0.1 N HCl
	—	14.77	290	598 ⎱	0.1 N NaOH, after 10–20
	—	14.73	283	598 ⎰	min
IgG$_1$, serum	—	13.57	278	598	P_i buffered saline
	—	13.44	278	598	0.1 N HCl
	—	14.77	290	598 ⎱	0.1 N NaOH, after 10–20
	—	14.65	283	598 ⎰	min
IgG$_2$, serum	—	13.52	278	598	P_i buffered saline
	—	13.32	278	598	0.1 N HCl
	—	14.87	290	598 ⎱	0.1 N NaOH, after 10–20
	—	14.73	283	598 ⎰	min
Immunoglobulins, specific					
Rabbit anti-HGG[x]*	—	14.4	279	263	0.25 M acetic acid
Anti-BSA[y]*	—	15.0	279	263	0.25 M acetic acid
Anti-N$_2$ph[v]*	—	15.8	278	263	pH 7.4, 0.15 M NaCl, 0.02 M P$_i$
Anti-N$_2$ph[v]* pepsin frag.	—	16.9	278	263	pH 7.4, 0.15 M NaCl, 0.02 M KP$_i$
Anti-N$_2$ph[v]*	—	16.8	278	263	pH 7.4, 0.15 M NaCl, 0.02 M KP$_i$
Anti-N$_2$ph[v]* Fab	—	16.5	278	263	pH 7.4, 0.15 M NaCl, 0.02 M KP$_i$
Anti-N$_2$ph,[v]* Fc	—	13.5	278	263	pH 7.4, 0.15 M NaCl, 0.02 M KP$_i$
Anti-N$_2$ph,[v]* pepsin frag.	—	18.1	278	263	pH 7.4, 0.15 M NaCl, 0.02 M KP$_i$
Anti-N$_2$ph,[v]*	—	15.4	278	263	pH 7.4, 0.04 M KP$_i$
	—	15.7	278	263	pH 7.4, 0.04 M KP$_i$
	—	15.5	278	263	pH 7.4, 0.04 M KP$_i$
	—	15.6	278	263	pH 7.4, 0.04 M KP$_i$
	—	16.4	278	263	pH 7.4, 0.04 M KP$_i$
	—	13.6	278	263	pH 7.4, 0.15 M NaCl, 0.01 M P$_i$
	—	5.3	251	263	pH 7.4, 0.15 M NaCl, 0.01 M P$_i$
γG-Anti-N$_2$ph[v]*	—	16.2	278	263	pH 7.4, 0.04 M KP$_i$
	—	16.0	278	263	pH 7.4, 0.04 M KP$_i$
	—	15.9	278	263	pH 7.4, 0.02 M KP$_i$
γG-Anti-N$_3$ph[z]*	—	15.8	278	263	pH 7.4, 0.15 M NaCl 0.02 M KP$_i$
	—	14.9	278	263	pH 7.4, 0.15 M NaCl, 0.02 M KP$_i$
	—	15.3	278	263	pH 7.4, 0.15 M NaCl, 0.02 M KP$_i$
Anti-p-azobenzenearsonate	—	14.8	278	263	pH 7.4, 0.04 M KP$_i$
Light chains	—	12.8	278	263	pH 8.0, 0.05 M, Na dodecyl sulfate

[x]*HGG, human gamma globulin.
[y]*Albumin, bovine serum albumin.
[z]*N$_3$ph, trinitrophenyl.

MOLAR ABSORPTIVITY AND $A_{1cm}^{1\%}$ VALUES FOR PROTEINS AT SELECTED WAVELENGTHS OF THE ULTRAVIOLET AND VISIBLE REGION (continued)

Protein	$\epsilon^a (\times 10^{-4})$	$A_{1cm}^{1\%}$ [b]	nm[c]	Ref.	Comments[d]
Immunoglobulins, specific (continued)					
γG-Anti-phenyl[*p*-aminobenzoylamino]acetate	–	13.9	279	263	pH 7.4, 0.15 *M* NaCl, 0.02 *M* P$_i$
Horse, γG-Anti-lac	–	14.7	280	263	Neutral solvent
γA-Anti-lac	--	14.7	280	263	Neutral solvent
Pepsin fragment	–	14.6	280	263	Neutral solvent
γG, Human	–	14.3	280	263	pH 7.5, 0.2 *M* NaCl
γ$_1$-globulin, Human	–	14.7	280	263	pH 6.0, 0.1 *M* NaCl
γM, Human	–	11.85	280	263	pH 7.5, 0.2 *M* NaCl
γM, Human, subunit	–	12.0	280	263	pH 7.5, 0.2 *M* NaCl
γG, Horse	–	13.8	280	263	pH 6.5, 0.0175 *M* NaP$_i$
	–	13.8	280	263	8 *M* urea, neutral solution
Heavy chains	–	15.4	280	263	0.04 *M* NaDodSO$_4$
	–	15.2	280	263	1 *N* propionic acid
Light chains	–	14.0	280	263	0.04 *M* NaDodSO$_4$
	–	13.6	280	263	1 *N* propionic acid
γG, Rabbit	–	14.6	280	263	–
	–	15.0	278	263	pH 7.4, 0.04 *M* KP$_i$
	–	15.4	278	263	pH 7.4, 0.04 *M* KP$_i$
	–	14.5	278	263	pH 7.4, 0.04 *M* KP$_i$
	–	15.1	278	263	pH 7.4, 0.04 *M* KP$_i$
	–	14.7	278	263	pH 7.4, 0.04 *M* KP$_i$
	–	14.9	278	263	pH 7.4, 0.04 *M* KP$_i$
	–	14.6	278	263	pH 7.4, 0.15 *M* NaCl, 0.02 *M* P$_i$
	–	13.5	280	263	0.01 *N* HCl
Heavy chains	–	13.7	280	263	0.01 *N* HCl
Light chains	–	11.8	280	263	0.01 *N* HCl
Fd fragment	–	14.4	280	263	0.01 *N* HCl
Heavy chains, mildly reduced and alkylated		14.5	280	263	pH 7.2, 0.04 *M* DodSO$_4$ 0.01 *M* P$_i$
Light chains mildly reduced and alkylated	–	13.2	280	263	pH 7.2, 0.04 *M* DodSO$_4$ 0.01 *M* P$_i$
γG, Rabbit	–	13.6	280	263	pH 7, 5 *M* GdmCl
Heavy chains	–	13.7	280	263	pH 7, 5 *M* GdmCl
Light chains	–	11.4	280	263	pH 7, 5 *M* GdmCl
Fab fragment	7.5	15.0	278	263	MW=50,000 (263)
γG, Rabbit	–	13.8	278	263	–
Fab fragment	–	15.3	278	263	–
Fc fragment	–	12.2	278	263	
Fab fragment	–	15.0	280	263	pH 7.5, 0.1 *M* P$_i$
5S pepsin fragment	–	14.8	280	263	
γA, Rabbit, colostrum	–	13.5	280	263	0.1 *N* NaOH
	–	12.8	280	263	5 *M* GdmCl
α-chains	–	10.6	280	263	5 *M* GdmCl
γ$_1$, Guinea pig	–	15	278	263	–
γ$_2$, Guinea pig	–	13.2	278	263	–
γ-globulin fraction, chicken	–	13.5	280	263	–
γG, Lemon shark	–	13.85	280	263	0.3 *M* KCl
	–	14.04	280	263	0.1 *N* NaOH
	–	12.82	280	263	5 *M* GdmCl

MOLAR ABSORPTIVITY AND $A_{1cm}^{1\%}$ VALUES FOR PROTEINS AT SELECTED WAVELENGTHS OF THE ULTRAVIOLET AND VISIBLE REGION (continued)

Protein	$\epsilon^a (\times 10^{-4})$	$A_{1cm}^{1\%}$ [b]	nm[c]	Ref.	Comments[d]
Immunoglobulins, specific (continued)					
γM, Lemon shark	–	13.39	280	263	0.3 M KCl
	–	13.75	280	263	0.1 N NaOH
	–	12.79	280	263	5 M GdmCl
γG Heavy chains	–	11.74	280	263	5 M GdmCl
γG Light chains	–	13.1	280	263	5 M GdmCl
Indole-3-glycerophosphate synthase (EC 4.1.1.48)	3.6	8	280	264	pH 7.0, 5 mM P$_i$
	4.3	9.5	280	264	0.1 N NaOH, est. from Figure 8 (264)
Inhibitor					
Amylase, *Colocasia esculenta*	–	10.7	280	1583	pH7
Phospholipase A, *Bothrops neuwiedii* (snake) venom	–	9.09	280	1584	–
	–	16.36	260	1584	–
	–	201.60	230	1584	–
Proteinase, potato	–	9.18	280	1585	–
Proteinase IIa, potato	–	10.03	278	1586	pH 6, 0.1 M NaCl,
Proteinase IIb, potato	–	10.06	278	1586	dry wt
Trypsin, *Phaseolus aureus* Roxb. (mungbean)					
Type A	–	3.7	280	1587	–
Type B	–	3.7	280	1587	–
Trypsin-chymotrypsin *Arachis hypogaea* (groundnut)	0.1958	2.5	280	1038	MW=7832 (1038)
Protease					
Barley	–	8.82	280	1039	–
Insulin	–	10.4	276	711	–
	0.553		277.5	712	–
	–	10.3	275	713	0.01 N HCl
	–	10.52	276	714	pH 7.0, 0.03 M P$_i$
	0.608	–	277	849	
Beef	0.57	10	280	265	MW=5733 (265)
	–	10.4	277	11	–
	0.61	10.6	278	266	pH 7.0, 0.025 M P$_i$, MW=5734 (266)
	0.52	–	280	267	–
	–	9.91	276	1588	pH 7.2, 0.01 M NaP$_i$– 0.1 N NaCl
Crystalline	0.6740	–	276	1589	–[a†]
N$^{\alpha A1}$-Acetyl-	0.6110	–	276	1589	–[a†]
	0.5790	–	276	1589	–[b†]
I$_{A-a}$	0.6480	–	276	1589	–[a†]
Amorphous	0.6870	–	276	1589	–[a†]
	0.6230	–	276	1589	–[b†]
N$^{\epsilon B29}$-Acetyl-	0.6630	–	276	1589	–[a†]
	0.6180	–	276	1589	–[b†]
N$^{\alpha A1}$, N$^{\epsilon B29}$-Diacetyl-	0.6440	–	275	1589	–[a†]
	0.6020	–	275	1589	–[b†]

[a†]Concentration calculated from the optical density at 210 nm.

[b†]Weighed sample used.

MOLAR ABSORPTIVITY AND $A_{1cm}^{1\%}$ VALUES FOR PROTEINS AT SELECTED WAVELENGTHS OF THE ULTRAVIOLET AND VISIBLE REGION (continued)

Protein	$\epsilon^a(\times\ 10^{-4})$	$A_{1cm}^{1\%}$ [b]	nm[c]	Ref.	Comments[d]
Insulin (continued)					
$N^{\alpha A1}, N^{\alpha B1}, N^{\epsilon B29}$- Triacetyl-	0.6390	–	275	1589	_a†
	0.6270	–	275	1589	_b†
Diacetyl-	0.3420	–	264	1589	_a†
	0.3020	–	264	1589	_b†
Triacetyl-	0.3460	–	264	1589	_a†
	0.3480	–	264	1589	_b†
β-chain	0.31	–	276	268	–
Interferon					
Chick embryo	–	8.6	280	269	–
Intrinsic factor, pig pylorus	–	9.2	278	1590	Dry wt
Invertase (β-fructofuranosidase, EC 3.2.1.26)					
Neurospora crassa	–	18.6	280	850	–
	–	18.6	280	270	–
Yeast	62.1	23	280	271	MW=270,000 (271)
Isoamylase (EC 3.2.1.68)					
Pseudomonas sp. str Sb-15	–	22.6	280	851	pH 4.5, 0.01 M AcO⁻
Isocitrate dehydrogenase (EC 1.1.1.41-.42)					
Saccharomyces cerevisiae (baker's yeast)	–	6.9	280	1591	–
	–	3.5	260	1591	–
Pig heart	5.3	9.1	280	272	MW=58,000 (272)
Pig liver					
Cytoplasm	4.73	12.6	280	716	MW=37,500 (716)
Acobacter vinelandii	7.12	8.9	280	715	MW=80,000 (715)
Isocitrate lyase (EC 4.1.3.1)					
P. indigofera	38	17.1	280	852	MW=222,000 (852)
Isomerase [*N*-(5-phospho-D-ribosylformimino)-5-amino-(5″-phosphoribosyl)-4-imidazolecarboxamide isomerase, EC 5.3.1.16)					
S. typhimurium	3.3	11.4	280	273	pH 8, 0.05 M TrisCl
C_{55}-Isoprenoid alcohol phosphokinase	3.188	18.7	280	1592	MW=17,000 (1592)
Staphylococcus aureus	2.200	12.9	288	1592	
β-Isopropylmalate dehydrogenase (EC 1.1.1.85)					
Salmonella typhimurium	5.34	7.63	278	853	MW=70,000 (853)
Kallikrein (kininogenin, EC 3.4.21.8)	–	16.6	280	1593	pH 7 P_i buffer, dry wt
Pig pancreas	4.92	20.5	280	1594	MW=24,000 (98)
Kallikrein A,					
Cinnamoyl-	1.80	–	298	1594	pH 8.8
Indoleacrylyl-	2.24	–	353	1594	

MOLAR ABSORPTIVITY AND $A_{1cm}^{1\%}$ VALUES FOR PROTEINS AT SELECTED WAVELENGTHS OF THE ULTRAVIOLET AND VISIBLE REGION (continued)

Protein	$\epsilon^a(\times 10^{-4})$	$A_{1cm}^{1\%}$ [b]	nm[c]	Ref.	Comments[d]
Kallikrein B					
Cinnamoyl-	1.80	--	298	1594 ⎱ pH 8.8	
Indoleacrylyl-	2.17	--	353	1594 ⎰	
Kallikrein inactivator					
Lung	5.5	8.4	276	274	MW=6511 (274)
Kerateine fractions					
Wool, *S*-carboxymethyl-					
SCMKA2	—	8.6	276	1595	—
SCMKB1	—	5.5	276	1595	—
SCMKB2	—	5.9	276	1595	—
Feather rachis,					
S-carboxymethyl-					
SCMK	—	7.0	276	1595	—
Keratinase	—	8.4	280	275	pH 8, 0.1 *M* Tris
S. fradiae (EC 3.4.99.8)	—	10.42	280	276	pH 5.0, AcO⁻
Trichophyton mentagrophytes (EC 3.4.99.12)	—	13.2	280	854	
Keratinase (EC 3.4.99.11) conjugate					
S. fradiae	—	58.1	280	277	—
Keratins, zinc precipitatable fractions					
Hair					
Lincoln sheep	—	6.6	277	1596	—
Pig	—	6.5	277	1596	—
Cattle	—	8.3	277	1596	—
Macaca irus (monkey)	—	8.7	277	1596	—
Lama glauca (llama)	—	9.8	277	1596	—
Guinea pig	—	9.9	277	1596	—
Merino sheep	—	10.2	277	1596	—
Rat	—	11.7	277	1596	—
Mouse	—	14.2	277	1596	—
Horny keratin					
Diceros bicornis (rhinocerous) horn	—	6.4	277	1596	—
Fingernail	—	7.0	277	1596	—
Sheep horn	—	7.6	277	1596	—
Cattle horn	—	8.0	277	1596	—
Sheep hoof	—	8.8	277	1596	—
Erethizon dorsatum (porcupine) quill	—	9.3	277	1596	—
Balaenoptera muscelus (whale) baleen	—	12.1	277	1596	—
Erinaceus europaeus (hedgehog) quill	—	12.3	277	1596	—
Histrix cristata (porcupine) quill	—	15.3	277	1596	—
Tachyglossus aculeatus aculeatus (echidna) quill	—	20.4	277	1596	—

MOLAR ABSORPTIVITY AND $A_{1cm}^{1\%}$ VALUES FOR PROTEINS AT SELECTED WAVELENGTHS OF THE ULTRAVIOLET AND VISIBLE REGION (continued)

Protein	$\epsilon^a(\times\,10^{-4})$	$A_{1cm}^{1\%}$ [b]	nm[c]	Ref.	Comments[d]
α-Keto acid dehydrogenase complex					
Pig heart	–	5.1	280	278	pH 7, 0.05 M KP$_i$, est. from Figure 6 in Reference 278,
3-Keto-Δ5-steroid isomerase					
P. testosteroni	–	4.13	280	279	–
Kininogenin (see kallikrein)					
Lac repressor, *E. coli*	–	6.9	280	1569	pH 6.8, 0.01 M P$_i$
Laccase (monophenol monooxygenase, EC 1.14.18.1)					
Rhus vernicfera	0.26	–	330	856	–
	0.52	–	614	856	–
	0.09	–	788	856	–
Oxidized	9.35	–	280	717	–
Reduced	0.57	–	614	717	–
	0.28	–	333	717	–
	0.55	–	614	717	–
Polyporus versicolor	8.4	13.7	280	718	MW=61,000 (718)
	0.41	–	610	718	–
	0.33	–	330	858	–
	0.08	–	440	858	–
	0.46	–	610	858	–
	0.2	–	720	858	–
Rhus succedanea	0.32	–	325	857	–
	0.45	–	610	857	–
	0.12	–	770	857	–
Fungal	7.4	11.6	280	859	MW=64,000 (859)
	7.44	–	280	1040	–
α-Lactalbumin	–	20.9	280	280	–
	–	20.1	280	281	--
Human	2.04	–	280	1041	–
	–	16.2	280	1597	–
	–	19.0	–[c†]	1598	–
American Indian	–	14.1	280	1042	–
Caucasian	–	15.3	280	1042	–
Negro	–	15.2	280	1042	–
Japanese	–	15.0	280	1042	–
Beef	–	20.1	280	1042	–
	–	20.5	280	601	–
A, Droughtmaster	–	20.2	281.5	1043	–
B, Droughtmaster	–	20.9	281.5	1043	–
Cow's milk	2.9	–	280	719	–
	–	20.9	280	720	20 mM Tris, pH 7.4
Carboxyl modified with glucineamide	3.0	–	280	719	–
Goat	–	17.3	280	1042	–
Sheep	–	16.7	280	1042	–
Pig	–	18.1	280	1042	–
Guinea pig	–	16.7	–[c†]	1598	–
Camel	–	19.0	–[c†]	1598	–

[c†]No wavelength cited.

MOLAR ABSORPTIVITY AND $A_{1cm}^{1\%}$ VALUES FOR PROTEINS AT SELECTED WAVELENGTHS OF THE ULTRAVIOLET AND VISIBLE REGION (continued)

Protein	$\epsilon^a (\times 10^{-4})$	$A_{1cm}^{1\%}$ [b]	nm[c]	Ref.	Comments[d]
β-Lactamase II (cephalosporinase, EC 3.5.2.8)					
Bacillus cereus 569H	–	8.7	277	1044	MW = 22,500
Hp	2.41	10.7	–[c†]	1599	MW = 22,500 refractometry
β-Lactamase I *B. cereus*					
569/H	2.44	8.76	–[c†]	1599	MW = 27,800 (1599), refractometry
Lactate dehydrogenase (EC 1.1.1.27 and/or .28)					
Beef					
Heart	–	14.55	280	1600	5 *M* GdmCl, Kj
	21	15	280	284	–
	20	–	280	285	–
	–	15	280	1050	–[e†]
	–	13.8	280	1051	–
	–	14.2	280	1052	–
	–	14.5	280	1053	–
	–	14.9	280	1054	–
H_1 [92%]	–	14.6	280	1048	–
$H_2 M_2$ [80%]	–	14.4	280	1048	–
Skeletal muscle	20	–	280	285	–
Muscle	18.1	12.9	280	284	–
Isozyme A	21	15.6	280	286	MW = 134,000 (286)
Isozyme B	19.6	14.5	280	286	MW = 134,000 (286)
Duck, M_4	18.9	13.5	280	287	MW = 140,000 (287)
Turkey, M_4	20.4	14.6	280	287	MW = 140,000 (287)
Pheasant, M_4	19.1	13.7	280	287	MW = 140,000 (287)
Ostrich, M_4	18.6	13.3	280	287	MW = 140,000 (287)
Rhea, M_4	18.1	13	280	287	MW = 140,000 (287)
Halibut, M_4	18.9	13.5	280	287	MW = 140,000 (287)
Skeletal muscle	20	–	280	285	–
Bullfrog, M_4	19.3	13.8	280	287	MW = 140,000 (287)
Tuna, M_4	17.5	12.5	280	287	MW = 140,000 (287)
Dogfish, M_4	20.8	14.8	280	287	MW = 140,000 (287)
	19.7	–	280	287	Using Kjeldahl N
Chicken heart muscle	18	–	280	285	–
	19	13.6	280	284	–
Skeletal muscle	22	–	280	285	–
	21.8	15.6	280	284	–
Rabbit	20	–	280	285	–
M_4	–	14.4	280	286	–
	20.1	–	280	287	–
Muscle	16.2	12.3	280	288	pH 7.6, 0.1 *M* NaP$_i$, MW = 132,000 (288)
	16.7	12.6	280	289	0.2 *N* NaOH, MW = 132,000 (289)
	–	14.0	280	141	–
Skeletal muscle					
V[96%]	20.68	14.6	279	1049	IV[2%], II/III [0.5%], I [1.5%], MW = 142,000 (1049)
	–	8.9	280	1050	–[d†]

[d†] mg protein/ml = (1.13) (OD$_{280}$).
[e†] mg/ml = (0.67) (OD$_{280}$).

MOLAR ABSORPTIVITY AND $A_{1cm}^{1\%}$ VALUES FOR PROTEINS AT SELECTED WAVELENGTHS OF THE ULTRAVIOLET AND VISIBLE REGION (continued)

Protein	$\epsilon^a (\times 10^{-4})$	$A_{1cm}^{1\%}$ [b]	nm[c]	Ref.	Comments[d]
Lactate dehydrogenase (continued)					
Pig heart	–	12.9	280	1045	0.1 NaOH, coenzyme free
	–	14.5	280	1046	–
	–	151	215	1047	0.9% NaCl
M_4 [96%]	–	14.0	280	1048	–
I [97%]	19.59	13.8	279	1049	II [2.5%], III [0.5%], MW = 142,000 (1049)
Heart [H_4]	–	13.7	280	286	–
Human	–	14.6	280	290	pH 7, P_i
Heart	–	16.4	280	1055	–
Uterus	18.7	12.3	280	295	MW = 152,000 (295)
Uterine myoma	17.7	12.4	280	295	MW = 143,000 (295)
Rat liver	–	12.6	280	291	MW = 132,000 (292)
	–	12.8	280	1058	–
Jensen sarcoma	–	11.7	280	1059	pH 7.4
Yeast	23.2	–	424	293	–
	–	29	423	1056	–
E. coli B	–	0.727[ft]	280	294	–
Homarus americanus	21.8	15.6	280	1057	MW = 140,000 (1057)
Lactate-malate dehydrogenase (see malate-lactate transhydrogenase)					
L-Lactate oxidase (lactate 2-monooxygenase, EC 1.13.12.4)					
M. smegmatis	–	21.9	280	282	pH 7.0, NaP$_i$
	–	2.62	452	282	–
M. phlei	81.5	20.4	280	283	pH 7.0, 0.1 M P_i, MW = 399,000 (283)
	–	2.26	454	283	H 7.0, 0.1 M P_i
Lactoferrin					
Beef	–	14.5	280	1060	–
	–	11.3	280	1060	6 M GdmCl
Cow's milk	–	15.1	280	306	–
	–	0.547	450	307	–
	–	0.460	470	1603	pH 7
Apo-	–	12.7	280	1603	–
Human milk	–	0.540	465	602	pH 8.2
	11.096	–	280	1601	–
	–	14.6	280	1602	–
	–	0.510	470	1603	pH 7
Apo-	8.512	–	280	1601	–
	–	11.2	280	1602	–
	–	10.9	280	1603	–
β-Lactoglobulin					
Beef	–	9.5	280	296	–
	–	9.66	278.5	297	–
	–	9.7	280	298	–
	3.66	–	280	299	–
	–	9.6	278	300	pH 5.3, ionic strength 0.1, AcO⁻

[ft]This may be in error. Value should be 7.27.

MOLAR ABSORPTIVITY AND $A_{1cm}^{1\%}$ VALUES FOR PROTEINS AT SELECTED WAVELENGTHS OF THE ULTRAVIOLET AND VISIBLE REGION (continued)

Protein	$\epsilon^a(\times 10^{-4})$	$A_{1cm}^{1\%}$ [b]	nm[c]	Ref.	Comments[d]
β-Lactoglobulin (continued)					
β-A, β-B	–	9.6	278	301	–
β-C	–	9.5	278	301	–
β-A, β-B	–	9.4	278	302	–
β-1	–	9.5	290	303	0.1 N NaOH
	–	9.5	282	303	0.1 N NaOH
β-2	–	9.9	288	303	0.1 N NaOH
Sheep, β-A	–	9.2	278	304	0.1 M NaCl
β-B	–	8.35	278	304	0.1 M NaCl
Goat	–	9.4	278	305	–
Pig's milk	1.05	5.65	280	603	0.1 M NaOH
Buffalo	–	9.4	279	1061	–
Cow's milk	–	5.12	255	1604 ⎫	
	–	10.80	280	1604 ⎬	pH 7.0, 0.01 M P$_i$
β-Lactoglobulin-B	–	1.00	310	1604 ⎭	
Cow's milk	3.5	10.0	280	1062	MW = 35,000 (1062)
MalNEt[g†]	3.5	10.0	280	1062	–
ClHgBzO$^{-h†}$	3.7	10.56	280	1062	–
Lactollin					
Cow's milk	7.1	16.5	280	1063	MW = 43,000 (1063)
Lactonase (gluconolactonase) (EC 3.1.1.17)					
Actinoplanes missouriensis	–	4	280	1064	pH 7.0, 0.07 M KP$_i$, data from Figure 8 (1064)
Lactoperoxidase (peroxidase, EC 1.11.1.7)					
Cow's milk					
B-1	–	14.9	280	1065	–
B-2$_I$	–	15.0	280	1065	–
B-2$_{II}$	–	14.9	280	1065	–
B-3	–	14.9	280	1065	–
A	–	15.5	280	1065	–
Cow's milk	–	15.2	280	1066	–
	–	15.41	280	1067	–
	–	1.37	497	1067	–
Cow's milk	11.6	–	412	604 ⎫	
+ hydrogen peroxide	8.89	–	425	604 ⎬	Data from Figure 2 (604)
+ Na$_2$S$_2$O$_4$	7.26	–	437	604 ⎭	
Cow's milk	16.1	–	280	604 ⎫	
+ hydrogen peroxide	12.3	–	280	604 ⎬	Calculated from
+ Na$_2$S$_2$O$_4$	10.1	–	280	604 ⎭	$\epsilon_{412}/\epsilon_{280}$ = 0.72
Lactose synthetase, A protein, (see also galactosyl transferase)					
Cow's milk	–	16.1	280	1605	Refractometry
Lactosiderophilin lactotransferrin					
Human milk	10.9	11.7	280	308	pH 7, sat. with iron, MW = 93,000 (308)
	–	0.500	452.5	308	–

[g†] MalNEt, N-ethylmaleimide.

[h†] ClHgBzO$^-$, p-chloromercuribenzoate.

MOLAR ABSORPTIVITY AND $A_{1cm}^{1\%}$ VALUES FOR PROTEINS AT SELECTED WAVELENGTHS OF THE ULTRAVIOLET AND VISIBLE REGION (continued)

Protein	$\epsilon^a (\times 10^{-4})$	$A_{1cm}^{1\%}$ [b]	nm[c]	Ref.	Comments[d]
Leghemoglobin					
Soybean	1.51	–	574	1068	
	1.5	–	541	1068	pH 6.4, 0.01 M P_i
	13.9	–	411	1068	
Lupinus luteus (yellow lupin),					
Root nodules	3.8	–	275	1606	–
$Lb^{3+}H_2O/OH^-$	16.0	–	403.5	1607	
	1.03	–	498	1607	pH 6.0
	0.44	–	624	1607	
	14.3	–	404	1607	
	1.05	–	495	1607	
	0.98	–	535	1607	pH 8.5
	0.7	–	574	1607	
	0.35	–	622	1607	
	12.3	–	411	1607	
	1.35	–	544	1607	pH 10.5
	1.12	–	574	1607	
$LL^{3+}F^-$	15.1	–	402.5	1607	
	0.99	–	495	1607	
	0.47	–	615	1607	pH 6.0
	14.5	–	403	1607	
	1.4	–	490	1607	
	0.84	–	538	1607	pH 8.5
	0.75	–	575	1607	
$Lb^{2+}NCS^-$	12.4	–	410	1607	pH 6.0
	1.33	–	534	1607	
	12.3	–	410	1607	pH 8.5
	1.14	–	534	1607	
	12.3	–	410	1607	pH 10.5
	1.63	–	534	1607	
$Lb^{2+}N_2^-$	13.0	–	414	1607	pH 6.0
	1.28	–	545	1607	
	13.3	–	414	1607	pH 8.5
	1.37	–	545	1607	
$Lb^{2+}CN^-$	11.9	–	416	1607	pH 6.0
	1.38	–	545	1607	
	12.0	–	416	1607	pH 8.5
	1.56	–	544	1607	
	12.0	–	416	1607	pH 10.5
	1.65	–	543	1607	
Lb^{2+} imidazole	13.2	–	407.5	1607	pH 6.0
	1.32	–	533	1607	
	13.0	–	408	1607	pH 8.5
	1.38	–	535	1607	
	12.6	–	410	1607	pH 10.5
	1.4	–	533	1607	
Lb^{2+} pyridine	13.5	–	406	1607	
	1.27	–	530	1607	pH 6.0
	1.1	–	562	1607	
	13.3	–	406.5	1607	
	1.26	–	537	1607	pH 8.5
	1.08	–	574	1607	
	12.9	–	408	1607	
	1.52	–	530	1607	pH 10.5
	1.22	–	558	1607	

MOLAR ABSORPTIVITY AND $A_{1cm}^{1\%}$ VALUES FOR PROTEINS AT SELECTED WAVELENGTHS OF THE ULTRAVIOLET AND VISIBLE REGION (continued)

Protein	ϵ^a ($\times 10^{-4}$)	$A_{1cm}^{1\%}$ [b]	nm[c]	Ref.	Comments[d]
Leghemoglobin (continued)					
Lb^{2+}(desoxy)	10.7	–	421	1607	
	10.5	–	428	1607	pH 6.0
	1.38	–	555	1607	
	10.5	–	421	1607	
	10.5	–	428	1607	pH 8.5
	1.4	–	555	1607	
	10.3	–	421	1607	
	10.3	–	428	1607	pH 10.5
	1.3	–	555	1607	
Lb^{2+}NCS$^-$	11.2	–	421	1607	
	10.7	–	427	1607	pH 6.0
	1.43	–	555	1607	
	10.8	–	421.5	1607	
	10.0	–	427	1607	pH 8.5
	1.23	–	555	1607	
	11.3	–	421	1607	
	11.3	–	428	1607	pH 10.5
	1.43	–	555	1607	
Lb^{2+}N$_3^-$	13.2	–	416.5	1607	
	1.52	–	525	1607	pH 6.0
	1.47	–	561.5	1607	
	11.1	–	417.5	1607	
	1.39	–	550	1607	pH 8.5
	13.4	–	416.5	1607	
	1.35	–	525	1607	pH 10.5
	1.74	–	553	1607	
Lb^{2+}CN$^-$	11.3	–	421	1607	
	9.4	–	425	1607	
	1.18	–	535	1607	pH 6.0
	1.29	–	560	1607	
	1.13	–	561	1607	
	13.1	–	430	1607	
	1.47	–	535	1607	pH 8.5
	2.0	–	562.5	1607	
	16.6	–	431.5	1607	pH 10.5
	2.0	–	534	1607	pH 10.5
	2.43	–	562.5	1607	
Lb^{2+} imidazole	12.4	–	422	1607	
	1.09	–	527	1607	pH 6.0
	1.6	–	556.5	1607	
	16.0	–	422	1607	
	1.43	–	527	1607	pH 8.5
	2.62	–	556.5	1607	
	15.1	–	422	1607	
	1.46	–	527	1607	pH 10.5
	2.90	–	556.5	1607	
Lb^{2+} pyridine	19.6	–	419	1607	
	1.26	–	467	1607	pH 6.0
	2.2	–	523.5	1607	
	3.72	–	555	1607	
	18.5	–	420	1607	
	2.26	–	523.5	1607	pH 8.5
	3.68	–	555	1607	
	17.7	–	419.5	1607	
	1.98	–	523.5	1607	pH 10.5
	3.46	–	555	1607	

MOLAR ABSORPTIVITY AND $A^{1\%}_{1cm}$ VALUES FOR PROTEINS AT SELECTED WAVELENGTHS OF THE ULTRAVIOLET AND VISIBLE REGION (continued)

Protein	$e^a (\times 10^{-4})$	$A^{1\%}_{1cm}$ [b]	nm[c]	Ref.	Comments[d]
Leghemoglobin (continued)					
Lb²⁺ nicotinic acid	11.6	–	418	1607 ⎫	
	1.46	–	524	1607 ⎬ pH 6.0	
	2.53	–	554.5	1607 ⎭	
	9.4	–	420	1607 ⎫	
	1.36	–	524	1607 ⎬ pH 8.5	
	1.7	–	554.5	1607 ⎭	
	10.0	–	420	1607 ⎫	
	1.1	–	524	1607 ⎬ pH 10.5	
	1.5	–	554.5	1607 ⎭	
Lb²⁺ ethyl isocyanide	15.7	–	426	1607 ⎫	
	1.78	–	526	1607 ⎬ pH 6.0	
	1.98	–	556	1607 ⎭	
	14.0	–	426	1607 ⎫	
	1.68	–	526	1607 ⎬ pH 8.5	
	1.98	–	556	1607 ⎭	
	14.8	–	426	1607 ⎫	
	1.5	–	526	1607 ⎬ pH 10.5	
	1.98	–	556	1607 ⎭	
Lb²⁺CO	17.1	–	417.5	1607 ⎫	
	1.34	–	540	1607 ⎬ pH 6.0	
	1.33	–	562	1607 ⎭	
	19.0	–	417	1607 ⎫	
	1.29	–	536	1607 ⎬ pH 8.5	
	1.24	–	562	1607 ⎭	
	20.6	–	417	1607 ⎫	
	1.7	–	537.5	1607 ⎬ pH 10.5	
	1.7	–	562	1607 ⎭	
Legumin					
Vicia sativa	–	7.5	280	1069	pH 6.5 and pH 12.4
Leucine aminopeptidase (amino peptidase, EC 3.4.11.1-.2)					
Pig kidney	–	8.4	280	309	pH 8.0, 0.005 M Tris, 0.005 M MgCl$_2$, est. from Figure 5 (309)
Beef lens	–	9.2	280	605	–
	–	10	280	606	pH 8.0, 0.1 M Tris
Leucine binding					
E. coli	–	6.5	280	453	pH 6.9
Leucine dehydrogenase (EC 1.4.1.9)					
B. subtilis SJ-2	103.5	45	280	1070	MW = 230,000 (1070)
Bacillus sphaericus		6.44	280	1608	–
Lipase (triacylglycerol lipase, EC 3.1.1.3)					
Pig pancreas	6.65	13.3	280	1071	MW = 50,000 (1071)
N$_3$ ph-[i†]	–	14.2	280	1071	–
		11	280	1609	–
Rat pancreas	–	12	280	1072	–
Lipoamide dehydrogenase					
C. krusei	6.2	11.76	273	310	pH 7.0, MW = 53,000 (310)

[i†] N$_3$ ph, trinitrophenyl.

MOLAR ABSORPTIVITY AND $A_{1cm}^{1\%}$ VALUES FOR PROTEINS AT SELECTED WAVELENGTHS OF THE ULTRAVIOLET AND VISIBLE REGION (continued)

Protein	$\epsilon^a (\times 10^{-4})$	$A_{1cm}^{1\%}$ [b]	nm[c]	Ref.	Comments[d]
Lipoate acetyltransferase (EC 2.3.1.12) (see dihydrolipoyl transacetylase)					
Lipoprotein, HDL$_2$, human					
Apo-	—	18.2	280	1610	—
Fraction III	—	12.2	280	1610	—
Fraction IV	—	9.2	280	1611	—
Carboxymethyl- low density	—	9.2	280	1611	—
Apo-	—	8.0	—[c†]	1612 ⎱	Dry wt, 7.5 M GdmCl
	—	7.7	—[c†]	1612 ⎰	
Lipoprotein, very low density					
Human, ApoLP-Val	—	9.1	280	1073 ⎱	pH 7.5, 0.02 M KP$_i$
ApoLP-Ala	—	16.1	280	1073 ⎰	
Low density Rat serum	—	11.7	280	1074 ⎱	pH 8.6, data from
	—	8.0	290	1074 ⎰	Figure 6 (1074)
	—	11.3	280	1074 ⎱	pH 11.6, data from
	—	9.7	290	1074 ⎰	Figure 6 (1074)
Apo-	—	10.0	280	1074 ⎱	pH 8.6, data from
	—	7.3	290	1074 ⎰	Figure 6 (1074)
	—	10.4	280	1074 ⎱	pH 11.6, data from
	—	10.0	290	1074 ⎰	Figure 6 (1074)
High density Rat serum	—	12.5	280	1074 ⎱	pH 8.6, data from
	—	9.6	290	1074 ⎰	Figure 5 (1074)
	—	12.3	280	1074 ⎱	pH 11.6, data from
	—	11.0	290	1074 ⎰	Figure 5 (1074)
Apo-	—	11.3	280	1074 ⎱	pH 8.6, data from
	—	8.8	290	1074 ⎰	Figure 6 (1074)
	—	11.7	280	1074 ⎱	pH 11.6, data from
	—	11.2	290	1074 ⎰	Figure 6 (1074)
Human skin, high density	—	10.8	278	1074	
Lipovitellin					
Leucophaea maderae (cockroach)	—	8.3	280	1613	—
Lipoxygenase (EC 1.13.11.12)					
Soybean	18.8	17.4	280	721	MW = 108,000 (721)
Isoenzyme		14	280	1614	
Peas (*Pisum sativum L*)	9.5	13.2	278	722	0.05 M TrisCl, pH 7.2, MW = 72,000 (722)
Lombricine kinase (EC 2.7.3.5)					
Lumbricus terrestris	—	11.4	280	607	—
Photinus pyralis	7.5	7.5	278	608	MW = 100,000 (608)
Renilla reniformis	—	10.4	280	860 ⎱	pH 7.5, 100 mM KP$_i$ plus 1.4 mM HSEtOH, plus 2.2 mM EDTA,
	—	0.38	500	860 ⎰	data from Figure 4 (860)
Diplocardia longa (earthworm)	54	18	278	1615	Biuret, MW = 300,000 (1615)
Bacterial	6.3	8.3	280	311	MW = 76,000 (311)

MOLAR ABSORPTIVITY AND $A\,^{1\%}_{1cm}$ VALUES FOR PROTEINS AT SELECTED WAVELENGTHS OF THE ULTRAVIOLET AND VISIBLE REGION (continued)

Protein	$e^a(\times 10^{-4})$	$A\,^{1\%}_{1cm}$ b	nm^c	Ref.	Comments^d
Lutenizing hormone- releasing factor					
Synthetic					
[Gly^2]LRF	1.22	−	244	1423	
	0.574	−	280	1423	
	0.561	−	288	1423	
	0.410	−	294	1423	0.1 N NaOH
des-His^2-LRF	1.0465	−	244	1423	
	0.5310	−	280	1423	
	0.5177	−	288	1423	
	0.3706	−	294	1423	
Lysin					
Tegula pfeifferi					
Egg, membrane	2.1	23.8	280	861	pH 6.0, 0.03 M P$_i$
ε-Lysine acylase (lysine acetyltransferase, EC 2.3.1.32)					
A. pestifer	−	12.0	280	862	−
Lysine decarboxylase (EC 4.1.1.18)					
E. coli B	−	11.3	280	724	Data from Figure 2 (724), pH 6
	−	15.2	290	724	Data from Figure 2 (724), 2 N NaOH
Bacterium cadaveris	−	10.1	280	725	pH 6.2, 0.01 M KP$_i$
L-Lysine 6-aminotransferase (EC 2.6.1.36)					
Achromobacter liquidum	8.5	7.35	280	723	pH 7.4, MW = 116,000 (723)
L-Lysine 2-monooxygenase (EC 1.13.12.2)					
Pseudomonas fluorescens	34.2	17.9	280	726	MW = 191,000 (726)
Lysozyme (EC 3.2.1.17)	−	22.8	280	1616	−
	−	320	210	1616	−
	−	910	191	1616	−
Egg white	3.79	−^{c†}		1617	−
	−	26.04	280	1618	−^{j†}
	−	12.60	255	1604	−
	−	24.80	280	1604	−
	−	0.72	310	1604	−
	3.60	24.7	280	1619	MW = 14,600 (1620)
	−	25.5	277.5	1621	Neutral pH
	−	20.2	290	1621	
	−	27.4	281	1622	pH 5–6, 0.1 M KCl, dry wt
	−	27.2	280	1623	0.02 N HCl
	3.9		280	1624	0.2 M AcO$^-$, pH 4.75
	3.8		280	1624	Carboxyl modified with aminomethyl-sulfonic acid
	−	26	280	11	−
	−	25.32	280	312	−
	−	26.9	280	313	−
	3.88	−	281	314	0.1 N HCl, MW = 14,700 (314)
	−	23.05	281	315	−
	−	27.3	282	316	−
	−	26.35	280	317	pH 5.4
	3.65	25.5	280	318	MW = 14,307 (318)
	−	26.5	280	319	pH 7.0, 0.2 M NaP$_i$
	390	−	281	320	pH 3.9
	−	23.3	290	321	4.8 M guanidine-0.01 M HCl

^{j†}Values given for photo-oxidation products in Reference 1618.

MOLAR ABSORPTIVITY AND $A_{1cm}^{1\%}$ VALUES FOR PROTEINS AT SELECTED WAVELENGTHS OF THE ULTRAVIOLET AND VISIBLE REGION (continued)

Protein	$\epsilon^a(\times\ 10^{-4})$	$A_{1cm}^{1\%\ b}$	nm^c	Ref.	Comments^d
Lysozyme (continued)					
Oxidized, tryptophan-108 oxidized to oxindole	–	22.7	280	1625	–
N-Bromosuccinimide Modified	3.37	–	280	1626	–
Reduced, carboxy-methylated and N-bromosuccimide modified	3.37	–	280	1626	See Figure 1 of Reference 1626
Azophenyl-p-sulfonic acid modified					
S-1	1.85	12.6	500	1627 ⎱	0.1 N NaOH
S-2	3.60	24.5	330	1627 ⎰	
Azophenyl-p-carboxylic acid modified					
B-1	8.58	58.4	340	1627	pH 7
B-1	2.28	15.5	500	1627	0.1 N NaOH
B-2	8.54	58.1	330	1627	pH 6
	2.84	19.3	478	1627	0.1 N NaOH
Azotetrazole modified					
T-1-2	5.58	38.0	330	1627	pH 7
	2.87	19.6	510	1627 ⎱	0.1 N NaOH
T-2	1.96	13.3	478	1627 ⎰	
Azophenyldiethyl-methylammonium chloride modified					
A-1	4.09	27.8	330	1627	pH 7
A-2	1.36	9.25	330	1627	pH 6
Mouse	–	21.7	280	1628	pH 6.5, 0.2 M P$_i$, dry wt
Human					
Milk	–	25.1	280	1629	–
	–	25.65	280	1633	pH 6.0, 0.1 M AcONa – 0.1 M NaCl
	3.1	20.7	280	322	Est. from Figure 3 (322), MW = 15,00● pH 5.5, est. from Figure (323), MW =
Tear	3.6	24.2	280	323	14,900,
Urine	–	25.1	280	1629	–
	3.51	24.7	280	1630	MW = 14,200 (1631)
	–	24.6	281	1632	pH 5.8
Duck	–	26.6	–^{c†}	1634	–
Goose	–	14.8	–^{c†}	1634	–
Chalaropis sp.	–	24.8	280	1634	Dry wt
Bacteriophage T4	2.4	12.8	280	324	MW = 19,000 (324)
Papaya	5.95	23.8	280	325	MW = 25,000 (325)
Bacteriophage λ	1.9	10.7	280	326	MW = 17,900 (326)
Lysyl-tRNA synthetase (EC 6.1.1.6)					
Yeast	–	6.4	280	1075	pH 7, 0.1 M P$_i$, 0.1 M EDTA, data from Figure 4 (1075)
α$_2$-Macroglobulin					
Beef plasma	80	10	280	863	pH 8.0 MW = 800,000 (863)

MOLAR ABSORPTIVITY AND $A_{1cm}^{1\%}$ VALUES FOR PROTEINS AT SELECTED WAVELENGTHS OF THE ULTRAVIOLET AND VISIBLE REGION (continued)

Protein	$\epsilon^a (\times 10^{-4})$	$A_{1cm}^{1\%}$ [b]	nm[c]	Ref.	Comments[d]
α_2-Macroglobulin (continued)					
Human	66.5	8.1	280	864	MW = 820,000 (864)
Pig	98	10.2	280	865	MW = 960,000 (865)
Mouse	–	7.42	280	1635	Dry wt
Rabbit	–	9	277	1028	–
Malate dehydrogenase (EC 1.1.1.37)					
Pig heart	1.78	–	280	867	–
	–	4.6	280	327	–
	–	2.8	280	328	–
	–	3.8	280	329	–
Mitochondria	1.98	3.05	280	337	pH 8.0, 0.05 μM Tris-AcO⁻, MW = 65,000 (337)
	–	2.5	280	1636	–
5S protein	1.97	2.9	280	868	MW = 68,000 (868)
9S protein	7.74	5.6	280	868	MW = 138,000 (868)
Supernatant	6.9	9.3	280	869	MW = 74,000 (869)
Chicken					
Intramitochondrial	–	2.9	280	870	–
Extramitochondrial	–	13.1	280	870	–
Rat liver					
Mitochondrial	3.4	5.08	280	871	MW = 66,300 (871)
Horse heart	–	2.8	280	327	–
Ox heart	–	8.5	280	330	–
Beef heart	3.25	5.0	280	331	MW = 65,000 (331)
B. subtilis	2.44	6.6	280	332	pH 7.7, 0.05 M P_i, MW = 37,000 (332)
	7.8	6.67	280	333	MW = 117,000 (333)
B. stearothermophilus	–	5.82	280	333	–
E. coli	2.03	3.39	280	333	MW = 60,000 (333)
Ostrich heart	8.96	12.8	280	334	pH 7.5, 0.1 M P_i, MW = 70,000 (334)
Tuna heart	20.8	31	280	335	pH 7.5, 0.1 M P_i, MW = 67,000 (335)
P. acidovorans	3.4	8.0	–[k†]	336	MW = 43,000 (336)
Malate-lactate transhydrogenase (EC 1.1.99.7) (lactate-malate dehydrogenase)					
Veillonelia alcalescens	–	12.7	280	1637	–
Malic enzyme [malate dehydrogenase (decarboxylating)] (EC 1.1.1.38-.40)					
Pigeon liver	25.8	9.2	278	338	pH 7.0, 0.042 M Tris, MW = 280,000 (338), protein contains NADP[l†]
	24.1	8.6	278	338	pH 7.0, 0.042 M Tris, MW = 280,000 (338)

[k†]No value cited for wavelength.

[l†]NADP, nicotinamide adenine dinucleotide phosphate.

MOLAR ABSORPTIVITY AND $A_{1cm}^{1\%}$ VALUES FOR PROTEINS AT SELECTED WAVELENGTHS OF THE ULTRAVIOLET AND VISIBLE REGION (continued)

Protein	$\epsilon^a(\times 10^{-4})$	$A_{1cm}^{1\%}$ [b]	nm[c]	Ref.	Comments[d]
Malic enzyme (continued)					
E. coli K 10 [HfrC]	26.5	4.8	279	872	MW = 550,000 (872), protein con. det. from tyr/trp content
	28.1	5.1	279	872	MW = 550,000 (872), protein con. det. by Lowry
Malic enzyme, NAD-linked [malate dehydrogenase (decarboxylating)] (EC 1.1.1.38-.39)					
E. coli	–	10.2	278	1638	Dry wt
Mandelonitrite lyase (see hydroxynitrilase)					
α-Mannosidase (EC 3.2.1.24)					
Vicia sativa	–	13.3	280	727	Data from Figure 3 (727), pH 7.2, Tris-0.01 *N* HCl
Soybeans	–	20	280	1639	
α-Melanotropin; 5-glutamine, nitrophenyl sulfenyl					
Synthetic	1.65	–	282	873 }	0.001 *N* HCl
	0.4	–	365	873 }	
Melilotate hydroxylase (melilotate 3-monooxygenase, EC 1.14.13.4)					
Arthrobacter, apoenzyme	--	11.1	277	339	pH 7.3, 0.15 *M* P$_i$, 0.1 *M* KCl, 1 m*M* cysteine, est. from Figure 6 (339)
Holoenzyme	–	18.9	280	339	pH 7.3, 0.15 *M* P$_i$, 0.1 *M* KCl, 1 m*M* cysteine, est. from Figure 6 (339)
Mercaptoalbumin, human see albumin, human					
Mercaptopyruvate sulfurtransferase (EC 2.8.1.2)					
E. coli	2.23	9.3	280	1640	pH 7.5, 0.05 *M* Tris, 0.8 *M* KCl, MW = 24,000 (1640)
Meromyosin, heavy	–	6	280	1641	Kjeldahl, absorption corrected for light scattering
Rabbit	–	–	–	1642	See footnote[m†]
Metallothionein					
Chicken liver	8.06	–	205	1643	pH 6.6
Rat liver	8.06	--	205	1643	pH 6.6

[m†] $\epsilon_{211} - \epsilon_{350} = 776$ cm^2/g in 0.5 *N* NaOH.

MOLAR ABSORPTIVITY AND $A_{1cm}^{1\%}$ VALUES FOR PROTEINS AT SELECTED WAVELENGTHS OF THE ULTRAVIOLET AND VISIBLE REGION (continued)

Protein	$\epsilon^a (\times 10^{-4})$	$A_{1cm}^{1\%}$ [b]	nm[c]	Ref.	Comments[d]
Metapyrocatechase (catechol 2,3-dioxygenase, (EC 1.1.3.11.2)					
Pseudomonas arvilla	18.9	13.5	280	1076	MW = 140,000 (1076)
	–	13.2	280	1644	See footnote[n†]
Methemoglobin (see hemoglobin)					
Methemoglobin, N_3^-	0.34	–	630	700	0.1 M P_i, pH 7.0,
	1.07	–	542	700	ϵ_M values expressed per mole
	11.7	–	414	700	heme
Methemoglobin reductase (EC 1.6.2.1.-2?)					
Human, erythrocytes	1.13	–	462	1645	Oxidized enzyme
Form I	–	7.45	278	728	–
Form II	–	17.9	268	728	–
Methionyl-tRNA synthetase (EC 6.1.1.10)					
E. coli K 12	–	13.9	283	1077	Native enzyme
	–	16.2	283	1077	Trypsin modified
	–	20	280	874	pH 7.4, 0.02 M KP$_i$
β-Methylaspartase (methylaspartate ammonia-lyase, EC 4.3.1.2)					
C. tetranomorphum	5.63	5.63	279	340	pH 7.0, 0.5 M Me$_4$NCl, MW = 100,000 (340)
	–	6.60	280	341	pH 6.5, 0.005 M P$_i$
Methylaspartate mutase (EC 5.4.99.1) (see glutamate mutase)					
Metmyoglobins (see myoglobin)					
β$_2$-Microglobulin					
Human urine	1.985	17	280	866	pH 7.0, MW = 11,600 (866)
Mitogenic components *Phaseolus vulgaris*					
A	–	6	280	1646	pH 7.0, 5 mM NaP$_i$,
B	–	3.6	280	1646	0.1 M NaCl
Molybdoferredoxin *Clostridium pasteurianum*	22.5	13.4	280	1078	pH 7.0, 0.1 M Tes buffer, data from Figure 7 (1078), MW = 168,000 (1078)
Monellin *Dioscoreophyllum cumminsil*	1.47	13.7	277	1647	pH 7.2
	1.83	17	290	1647	pH 12.8

[n†]Absorption at 280 nm taking the value of 1.32 ml/cm-mg.

MOLAR ABSORPTIVITY AND $A_{1cm}^{1\%}$ VALUES FOR PROTEINS AT SELECTED WAVELENGTHS OF THE ULTRAVIOLET AND VISIBLE REGION (continued)

Protein	$\epsilon^a(\times 10^{-4})$	$A_{1cm}^{1\%}$ [b]	nm[c]	Ref.	Comments[d]
Monoamine oxidase (amine oxidase, EC 1.4.3.4)					
Beef kidney	53.9	18.6	280	342	pH 7.6, 0.05 M P$_i$, MW = 290,000 (342)
	4.7	1.62	455	342	pH 7.6, 0.05 M P$_i$, MW = 290,000 (342)
Plasma	25.0	9.8	280	343	MW = 255,000 (343)
Myeloperoxidase (peroxidase, EC 1.11.1.7?)					
Canine uterine pus	22	—	280	875	Data from Figure 3 (875)
Human leukocyte	—	24	280	344	pH 7.0, 0.2 M P$_i$
Myoglobin	17.1	—	409	609	
	0.97	—	500	609	
Synthetic					
Proto-Mb$^+$	18.8	—	409	1648	
	1.16	—	502	1648	
	0.47	—	630	1648	
+CN$^-$	12.6	—	422	1648	
	1.30	—	540	1648	
+N$_3$$^-$	12.5	—	419	1648	pH 6, 0.1 M P$_i$
	1.22	—	540	1648	
	0.97	—	574	1648	
+F$^-$	16.0	—	406	1648	
	1.09	—	490	1648	
	0.98	—	605	1648	
+H$_2$O$_2$	12.3	—	421	1648	—
	1.25	—	546	1648	—
+OH$^-$	11.7	—	413	1648	—
	1.16	—	542	1648	—
	1.10	—	583	1648	—
+Na$_2$S$_2$O$_4$	13.5	—	434	1648	
	1.54	—	556	1648	
+Na$_2$S$_2$O$_4$ + CO	20.1	—	422	1648	pH 6, 0.1 M P$_i$
	1.78	—	541	1648	
	1.56	—	578	1648	
Meso-Mb$^+$	17.2	—	395	1648	
	0.88	—	495	1648	
	0.41	—	622	1648	
+CN$^-$	12.4	—	411	1648	
	0.992	—	531	1648	
+N$_3$	12.0	—	409	1648	pH 6, 0.1 M P$_i$
	0.9	—	530	1648	
	0.68	—	564	1648	
+F$^-$	15.4	—	394	1648	
	0.83	—	486	1648	
	0.69	—	598	1648	
+H$_2$O$_2$	11.6	—	408	1648	—
	1.02	—	536	1648	—
+OH$^-$	11.4	—	398	1648	—
	0.91	—	531	1648	—
	0.82	—	569	1648	—
	0.85	—	587	1648	—

MOLAR ABSORPTIVITY AND $A_{1cm}^{1\%}$ VALUES FOR PROTEINS AT SELECTED WAVELENGTHS OF THE ULTRAVIOLET AND VISIBLE REGION (continued)

Protein	$\epsilon^a (\times 10^{-4})$	$A_{1cm}^{1\%}$ [b]	nm[c]	Ref.	Comments[d]
Myoglobin (continued)					
+Na$_2$S$_2$O$_4$	11.1	—	421	1648	
	1.27	—	544	1648	
+Na$_2$S$_2$O$_4$ + CO	20.6	—	409	1648	
	1.46	—	530	1648	
	1.16	—	558	1648	
Deutero-Mb$^+$	12.8	—	392	1648	
	0.67	—	495	1648	
	0.29	—	620	1648	
+CN$^-$	10.8	—	409	1648	
	0.83	—	532	1648	
+N$_3$$^-$	10.5	—	408	1648	pH 6, 0.1 M P$_i$
	0.84	—	530	1648	
	0.61	—	560	1648	
+I$^-$	10.5	—	392	1648	
	0.69	—	483	1648	
	0.54	—	595	1648	
+Na$_2$S$_2$O$_2$	9.76	—	419	1648	
	1.0	—	542	1648	
+Na$_2$S$_2$O$_4$ + CO	17.2	—	408	1648	
	1.21	—	528	1648	
	0.81	—	556	1648	
+H$_2$O$_2$	9.64	—	409	1648	—
	0.72	—	534	1648	—
+OH$^-$	8.87	—	400	1648	—
	0.67	—	532	1648	—
	0.53	—	568	1648	—
	0.53	—	587	1648	—
Hemato-Mb$^+$	14.5	—	400	1648	
	0.94	—	499	1648	
	0.39	—	627	1648	
+CN$^-$	10.8	—	415	1648	
	1.05	—	537	1648	
+N$_3$$^-$	10.3	—	412	1648	
	0.96	—	535	1648	
	0.74	—	566	1648	pH 6, 0.1 M P$_i$
+I$^-$	13.2	—	400	1648	
	0.88	—	489	1648	
	0.66	—	600	1648	
+Na$_2$S$_2$O$_4$	11.0	—	426	1648	
	1.30	—	550	1648	
+Na$_2$S$_2$O$_4$ + CO	16.3	—	413	1648	
	1.50	—	534	1648	
+H$_2$O$_2$	10.2	—	413	1648	—
	1.07	—	540	1648	—
+OH$^-$	10.6	—	402	1648	—
	0.92	—	536	1648	—
	0.83	—	589	1648	—
Proto monomethyl Mb$^+$	16.6	—	406	1648	
	1.05	—	501	1648	
	0.45	—	630	1648	
+CN$^-$	11.7	—	418	1648	pH 6, 0.1 M P$_i$
	1.18	—	540	1648	
+N$_3$$^-$	11.5	—	418	1648	
	1.11	—	540	1648	
	0.90	—	572	1648	

MOLAR ABSORPTIVITY AND $A_{1cm}^{1\%}$ VALUES FOR PROTEINS AT SELECTED WAVELENGTHS OF THE ULTRAVIOLET AND VISIBLE REGION (continued)

Protein	$e^a(\times 10^{-4})$	$A_{1cm}^{1\%}$ [b]	nm[c]	Ref.	Comments[d]
Myoglobin (continued)					
+F[-]	13.9	--	404	1648	
	1.00	–	488	1648	
	0.95	–	604	1648	
+Na$_2$S$_2$O$_4$	12.7	–	431	1648	
	1.49	–	556	1648	pH 6, 0.1 M P$_i$
+Na$_2$S$_2$O$_4$ + CO	17.5	–	419	1648	
	1.64	–	539	1648	
	1.37	–	573	1648	
+H$_2$O$_2$	11.1	–	418	1648	
	1.08	–	544	1648	
OH[-]	11.1	–	410	1648	
	1.03	--	540	1648	
	0.96	–	580	1648	
Proto dimethyl					
Mb[+]	14.5	–	407	1648	
	0.88	–	502	1648	
	0.40	–	630	1648	
+CN[-]	10.5	–	420	1648	
	1.00	–	541	1648	
+N$_3$[-]	10.5	–	418	1648	pH 6, 0.1 M P$_i$
	1.01	–	541	1648	
	0.83	–	572	1648	
+F[-]	11.7	–	405	1648	
	0.86	–	486	1648	
	0.83	–	605	1648	
+Na$_2$S$_2$O$_4$	10.5		432	1648	
	1.30		556	1648	
+Na$_2$S$_2$O$_4$ + CO	15.9		421	1648	pH 6, 0.1 M P$_i$
	1.41		540	1648	
	1.23		570	1648	
+H$_2$O$_2$	9.15		418	1648	
	0.94		545	1648	
+OH[-]	7.88		410	1648	
	0.96		538	1648	
	0.86		578	1648	
Oxyforms					
Proto	14.5		417	1649	
	1.73		543	1649	
	1.79		581	1649	
Meso	13.9		404	1649	
	1.50		533	1649	
	1.32		568	1649	
Deutero	11.4		402	1649	
	1.27		532	1649	
	0.89		565	1649	
Hemato	11.9		410	1649	
	1.40		538	1649	
	1.16		574	1649	
Protomonoester	11.4		415	1649	
	1.40		541	1649	
	1.35		579	1649	
Protodiester	9.65		416	1649	
	1.64		542	1649	
	1.68		579	1649	

MOLAR ABSORPTIVITY AND $A^{1\%}_{1cm}$ VALUES FOR PROTEINS AT SELECTED WAVELENGTHS OF THE ULTRAVIOLET AND VISIBLE REGION (continued)

Protein	$\epsilon^a (\times 10^{-4})$	$A^{1\%}_{1cm}$ [b]	nm[c]	Ref.	Comments[d]
Myoglobin (continued)					
Albacore tuna					
(*Thunnus germo*)	–	6.4	555	1081	–
	–	61	430	1081	–
MetMb	–	2.1	630	1081	–
	–	1.7	580	1081	–
	–	4.8	502	1081	–
	–	81	408	1081	–
Myoglobin, carboxy-	–	15.5	280	1081	–
	–	6.7	570	1081	–
	–	7.5	538	1081	–
Beef, MetMbCN	–	95	421	1081	–
	2.96	–	280	1086	} pH 7.2, 0.01 M P$_i$ containing
	0.945	–	340	1086	} 0.01% KCN
Bluefin tuna					
(*Thunnus thynnus*)	–	7.1	558	1081	–
	–	56	431	1081	–
MetMb	–	2.1	634	1081	–
	–	1.7	580	1081	–
	–	4.8	504	1081	–
	–	86	407	1081	–
Myoglobin, carboxy-	–	13.3	275	1081	–
	–	7.6	570	1081	–
	–	8.4	540	1081	–
Camel	–	104	420	1081	–
	3.13	–	280	615	}
	17.2	–	409	615	}
	0.796	–	470	615	} Ferri form, acidic
	0.987	–	503	615	}
	0.360	–	580	615	}
	0.366	–	630	615	}
	3.43	–	280	615	}
	10.4	–	414	615	} Basic
	0.948	–	542	615	}
	0.93	–	587	615	}
	3.33	–	280	615	}
	2.87	–	360	615	} Cyanide form
	11.5	–	423	615	}
MetMbCN	1.132	–	542	615	}
	2.90	–	280	1086	} pH 7.2, 0.01 M P$_i$
Chicken	0.915	–	340	1086	} containing 0.01% KCN
(*Gallus gallus*)					
Muscle	3.1	–	280	1651	–
MetMbCN					
	0.96	–	534	1650	–
MetMb	10.2	–	423	1650	–
	0.32	–	632	1650	–
	0.84	–	504	1650	–
Muscle, distrophic					
MetMbCN					
	0.95	–	543	1650	–
	10.0	–	423	1650	–
	0.32	–	632	1650	–
	0.84	–	504	1650	–
Chinook salmon					
(*Oncorhynchus tschawtscha*)	–	7.0	558	1081	–
	–	50	428	1081	–

MOLAR ABSORPTIVITY AND $A_{1cm}^{1\%}$ VALUES FOR PROTEINS AT SELECTED WAVELENGTHS OF THE ULTRAVIOLET AND VISIBLE REGION (continued)

Protein	$\epsilon^a (\times 10^{-4})$	$A_{1cm}^{1\%}$ [b]	nm[c]	Ref.	Comments[d]
Myoglobin (continued)					
MetMb	—	2.2	632	1081	—
	—	1.9	580	1081	—
	—	5.1	502	1081	—
	—	85	404	1081	—
	—	20.4	280	1081	—
Myoglobin, carboxy-	—	7.4	569	1081	—
	—	7.7	539	1081	—
	—	95	420	1081	—
Cormorant					
(*Phalacrocorax*)	—	6.4	558	1081	—
	—	64	435	1081	—
MetMb	—	1.8	634	1081	—
	—	1.4	580	1081	—
	—	4.8	504	1081	—
	—	90	409	1081	—
	—	17.1	280	1081	—
Myoglobin, carboxy-	—	6.5	579	1081	—
	—	7.7	542	1081	—
	—	99	432	1081	—
Fin whale, component VII	3.2	18.28	280	1083	MW = 17,504 (1083)
Goat					
MetMbCN	2.95	—	280	1086 ⎫	
	1.01	—	340	1086 ⎬ pH 7.2, 0.01 M P$_i$ containing
Hamster					0.01% KCN
MetMbCN	—	53.9	420	1087 ⎭	
Habor seal	2.99	—	280	614 ⎱ Ferri form, pH 6.2
	16.2	—	409	614 ⎰
Horse	1.33	—	555	1082	—
	11.3	—	434	1082	—
MetMbCN	2.95	—	280	1086 ⎱ pH 7.2, 0.01 M P$_i$
	0.95	—	340	1086 ⎰ containing 0.01% KCN
MetMb	—	17.9	280	1088	—
Myoglobin, carboxy-	1.18	—	578	1082	—
	1.4	—	540	1082	—
	17.8	—	423	1082	—
Mb$^+$	16.0	—	408	1082	—
	1.02	—	505	1082	—
	0.42	—	630	1082	—
Mb-OH	9.0	—	414	1082	—
	0.91	—	540	1082	—
	0.86	—	580	1082	—
Human	3.05	17.4	280	610	MW = 17,510 (610)
Apo-	1.55	9.2	280	610	MW = 16,900 (610)
MetMbCN	3.07	—	280	1086 ⎱ Ph 7.2, 0.01 M P$_i$
	1.04	—	340	1086 ⎰ containing 0.01% KCN
Humpback whale					
(*Megaptera nodosa*)	—	6.5	558	1081	—
	—	61	434	1081	—
MetMb	—	1.9	634	1081	—
	—	1.5	580	1081	—
	—	5.0	504	1081	—
	—	85	409	1081	—
	—	15.9	281	1081	—
Myoglobin, carboxy-	—	6.1	577	1081	—
	—	7.2	543	1081	—
	—	96	423	1081	—

MOLAR ABSORPTIVITY AND $A_{1cm}^{1\%}$ VALUES FOR PROTEINS AT SELECTED WAVELENGTHS OF THE ULTRAVIOLET AND VISIBLE REGION (continued)

Protein	$\epsilon^a (\times 10^{-4})$	$A_{1cm}^{1\%}$ [b]	nm[c]	Ref.	Comments[d]
Myoglobin (continued)					
Lamb					
MetMbCN	3.09	–	280	1086 ⎱	pH 7.2, 0.01 M P$_i$
	1.08	–	340	1086 ⎰	containing 0.01% KCN
Molluscs					
(*Aplysia depilans* and					
Aplysia limacina)	1.3	–	555	1082	–
	11.3	–	438	1082	–
(*Acanthopleura granulata*)					
Myoglobin, carboxy-					
Type 1'	1.22	–	570	1094	–
	1.38	–	538	1094	–
	14.94	–	419	1094	–
Type 2'	1.24	–	572	1094	–
	1.32	–	538	1094	–
	17.90	–	419	1094	–
Type 3'	1.30	–	572	1094	–
	1.33	–	538	1094	–
	17.93	–	419	1094	–
(*Buccinum undatum L.*)					
Myoglobin, carboxy-	1.35	–	570	1095	–
	1.42	–	538	1095	–
	18.44	–	418	1095	–
	3.4	–	280	1095	–
(*Aplysia depilans* and					
Aplysia limacina)					
Myoglobin, carboxy-	1.37	–	571	1082	–
	1.42	–	541	1082	–
	17.6	–	424	1082	–
Mb$^+$	9.9	–	400	1082	–
	1.31	–	505	1082	–
	0.38	–	640	1082	–
Mb-OH	9.1	–	412	1082	–
	0.9	–	543	1082	–
	0.87	–	580	1082	–
	0.87	–	600	1082	–
Mb-O$_2$	10.8	–	416	1082	–
	1.32	–	542	1082	–
	1.33	–	578	1082	–
Monkey					
MetMbCn	3.01	–	280	1086 ⎱	pH 7.2, 0.01 M P$_i$
	1.02	–	340	1086 ⎰	containing 0.01% KCN
Pelican					
(*Pelecanus occidentalis*)	–	6.2	556	1081	–
	–	57	433	1081	–
MetMb	–	1.6	632	1081	–
	–	1.3	581	1081	–
	–	4.8	504	1081	–
	–	78	409	1081	–
	–	14.6	280	1081	–
Myoglobin, carboxy-	–	6.2	577	1081	–
	–	7.3	540	1081	–
	–	87	423	1081	–
Penguin					
(*Aptenodytes forsteri*)	3.2	–	280	1651	–
Porpoise	2.98	–	280	614 ⎱	Ferri form, pH 6.2
	16.2	–	409	614 ⎰	

MOLAR ABSORPTIVITY AND $A_{1cm}^{1\%}$ VALUES FOR PROTEINS AT SELECTED WAVELENGTHS OF THE ULTRAVIOLET AND VISIBLE REGION (continued)

Protein	$\epsilon^a(\times 10^{-4})$	$A_{1cm}^{1\%}$ [b]	nm^c	Ref.	Comments[d]
Myoglobin (continued)					
Skip jack tuna					
(*Katsuwonus pelamis*)	–	6.4	556	1081	–
	–	62	431	1081	–
MetMb	–	2.0	632	1081	–
	–	1.5	578	1081	–
	–	4.7	502	1081	–
	–	89	406	1081	–
	–	12.2	275	1081	–
Myoglobin, carboxy-	–	6.7	569	1081	–
	–	7.3	539	1081	–
	–	102	421	1081	–
Sperm whale	10.5	–	434	1079 ⎫	pH 9.1, 0.1 *M* borax, data from
	1.08	–	558	1079 ⎭	
	17.9	100	408	1080 ⎫	Figure 14 (1079), MW = 17,800
	3.45	19	280	1080 ⎭	(1080)
	3.79	–	280	611	Fe 0.31% (611)
	3.34	–	280	612 ⎫	
	16.8	–	409	612 ⎬	Ferri form
	0.367	–	634	612 ⎭	
	3.06	–	280	614 ⎫	Ferri form, pH 6.2
	16.4	–	409	614 ⎭	
	30.6	–	589	1652	–
	1.15	–	557	1652	–
	0.64	–	521	1652	–
Apo-	1.59	8.9	280	1080	MW = 17,800 (1080)
	1.54	–	280	1084	Dry wt, corrected for light scattering
	–	9.3	280	1085	Neutral
	–	9.2	280	1085	0.1 *N* NaOH
	1.58	–	280	613	pH 6.8, 0.1 *M* P_i
MbNO	0.7	–	583	1652	
	1.0	–	546	1652	–
MetMbCN	0.2	–	583	1652	–
	0.95	–	539	1652	–
MbCO	0.65	–	582	1652	–
	0.78	–	572	1652	–
	1.36	–	544	1652	–
	0.4	–	522	1652	–
MbO$_2$	1.08	–	582	1652	–
	0.41	–	571	1652	–
	1.06	–	550	1652	–
	0.79	–	533	1652	–
MetMbNO	0.77	–	575	1652	–
	0.215	–	562	1652	–
MetMbN$_3^-$	0.36	–	587	1652	–
	0.38	–	570	1652	–
	0.62	–	546	1652	–
	0.38	–	514	1652	–
MetMb	–	93.4	410	1088	–
	14.4	–	407	1079 ⎫	pH 6.5, P_i buffer containing 25
	0.82	–	502	1079 ⎬	m*M* Na$_2$S$_2$O$_4$, data from
	0.34	–	633	1079 ⎭	Figure 14 (1079) Water/0.1 *N* Na
	–	18.0	280	1089	Water/0.1 *N* NaOH
	0.35	–	630	1090	–
	0.93	–	505	1090	–
	16.0	–	409	1090	–

MOLAR ABSORPTIVITY AND $A_{1cm}^{1\%}$ VALUES FOR PROTEINS AT SELECTED WAVELENGTHS OF THE ULTRAVIOLET AND VISIBLE REGION (continued)

Protein	ϵ^a ($\times$ 10^{-4})	$A_{1cm}^{1\%}$ [b]	nm[c]	Ref.	Comments[d]
Myoglobin (continued)					
MetMb (continued)	3.1	—	280	1090	—
	0.35	—	630	1090	
	0.91	—	505	1090	Regenerated
	15.9	—	409	1090	
	3.2	—	280	1090	
	—	17.9	280	1085	0.1 N NaOH
Myoglobin, carboxy-	3.24	18.2	280	1091	pH 8.8, dry wt, MW = 17,800 (1092), data from Figure 2 (1091)
	1.21	6.8	578	1093	
	1.38	7.2	540	1093	MW = 17,816 (1093)
	18.6	104	423	1093	
Mb-Fe^{3+}	0.9	5.05	503	1093	—
	16.8	94	409	1093	—
	3.42	19.3	280	1093	—
MbO$_2$	1.45	8.13	581	1093	—
	1.36	7.6	543	1093	—
	12.5	70	418	1093	—
Mb-Fe^{2+}	1.17	6.6	555	1093	—
	11.6	65	434	1093	—
Mb-Fe^{3+}-N$_3$	11.1	62.1	421	1093	—
Yellowfin tuna					
(*Neothunnis macropterus*)	—	6.6	556	1081	—
	—	60	431	1081	—
MetMb	—	2.1	631	1081	—
	—	1.6	578	1081	—
	—	4.8	501	1081	—
	—	85	406	1081	—
	—	13.9	275	1081	—
Myoglobin, carboxy-	—	7.0	568	1081	—
	—	7.7	538	1081	—
	—	107	420	1081	—
Myokinase	1.15	5.38	277	876	pH 7.0, 0.15 M KCl–0.01 M KP$_i$, MW = 21,400 (876) Protein con. det. by dry wt
Rabbit muscle	—	5.2	280	141	—
	1.1	5.3	279	345	pH 6.9, MW = 21,000 (345)
	—	11.8	279	346	pH 7.0, 0.01 M P$_i$
Myosin					
Rabbit muscle	—	5.2	280	347	pH 7.0, 0.5 M KCl, 20 mM Tris
	—	2.5	250	347	pH 7.0, 0.5 M KCl, 20 mM Tris
	—	6.47	280	348	—
	—	5.07	276	348	pH 7.5, 5 M GdmCl
	—	5.88	280	348	0.5 M KCl[o†]
	—	4.83	280	348	5 M GdmCl[o†]
	—	5.43	280	349	—
	—	5.9	280	350	pH 7, 0.3 M KCl, 0.01 M Tris

[o†]These values have been corrected for light scattering.

MOLAR ABSORPTIVITY AND $A_{1cm}^{1\%}$ VALUES FOR PROTEINS AT SELECTED WAVELENGTHS OF THE ULTRAVIOLET AND VISIBLE REGION (continued)

Protein	$\epsilon^a(\times 10^{-4})$	$A_{1cm}^{1\%}$ [b]	nm[c]	Ref.	Comments[d]
Myosin (continued)					
Rabbit muscle (continued)	–	5.87	277	351	Neutral KCl
	–	5.50	280	616 ⎫	pH 7.3, 0.5 M KCl–0.01 M
	...	5.52	280	616 ⎟	EDTA + 0.05 M P$_i$
	–	5.55	280	616 ⎬	+ 0.2 M P$_i$
	–	5.58	280	616 ⎭	+ 0.5 M P$_i$
	–	5.35	280	1097	Free of low molecular weight protein
Light chains	0.7	3.5	280	1098	MW = 20,200 (1098)
Meromyosin, heavy	--	6.35	280	1099	–
Meromyosin, light	--	3.0	280	1100	--
Rabbit filament	–	2.00	278	1653	–
Skeletal muscle	–	5.6	280	1654	–
Heavy meromyosin	--	6.25	280	352	–
	–	6.47	279	353	–[c]†
		6.47	280	1654	–
Beef heart	–	5.7	280	618	–
Light meromyosin, fraction I	–	3.69	279	353	Trypsin digestion for 25 sec
	–	3.29	279	353	Trypsin digestion for 25 min
Light component		3.5	280	1654	–
Light subfragment I					
Beef heart	–	6.4	280	618	–
Meromyosin, light					
Rabbit	–	2.97	278	619	pH 3, 0.1 M NaCl–HCl
Light chain	–	4.03	280	1655	N
Slime mold	–	5.2	278	617	pH 7.4, 0.5 M KCl–0.01 M Tris, data from Figure 3 (617)
Human heart	–	5.1	280	1096	pH 6.8, 0.4 M KCl-borate, corrected for scattering
Paramyosin	–	3.04	277	354	–
Heavy alkali subunit	–	5.77	277	351	Neutral KCl
	–	5.19	277	351	6 M GdmCl
	–	5.29	277	351	5 M GdmCl
Light alkali subunit	–	4.34	277	351	Neutral KCl
	–	4.79	277	351	5 M GdmCl
S-1 fragment	–	7.9	280	355	–
S-n fragment	–	8.0	280	355	–
Chicken					
Red/slow muscle	–	5.33	280	356	–
White/fast muscle	–	5.05	280	356	–
Pectoral muscle	–	5.25	280	1655	–
Pig muscle					
Chesterwhite	–	4.9	280	347	pH 7.0, 0.5 M KCl, 20 mM Tris
	–	2.4	250	347	pH 7.0, 0.5 M KCl, 20 mM Tris
Poland China	–	4.9	280	347	pH 7.0, 0.5 M KCl, 20 mM Tris
	–	2.5	250	347	pH 7.0, 0.5 M KCl, 20 mM Tris
Poland China [PSE]	–	5.2	280	347	pH 7.0, 0.5 M KCl, 20 mM Tris
Myrosinase (thioglucosidase)					
(EC 3.2.3.1)					
Sinapsis alba					
(white mustard seed)	–	15	278	1656	–

MOLAR ABSORPTIVITY AND $A_{1cm}^{1\%}$ VALUES FOR PROTEINS AT SELECTED WAVELENGTHS OF THE ULTRAVIOLET AND VISIBLE REGION (continued)

Protein	$\epsilon^a(\times 10^{-4})$	$A_{1cm}^{1\%}$ [b]	nm[c]	Ref.	Comments[d]
NAD nucleosidase (EC 3.2.2.5) (see DPNase)					
NADH-cytochrome b_5 reductase					
Rat liver, microsomes	--	23	280	729	pH 7.5, 0.01 M KP$_i$
NADH:FAD oxidoreductase					
E. coli B	16.4	13	272	1101	MW = 126,000 (1101) pH 7.0, 0.05 M KP$_i$, data from Figure 1 (1101)
	2.33	1.85	380	1101	_p†
	2.24	1.78	448	1101	_p†
NADPH-adrenodoxin reductase					
Beef adrenocortical mitochondria	–	18	272	1657	From Figure 2 (1657)
	–	2.0	378	1657	
NADPH oxidase					
Rabbit liver, microsomes	–	19.1	280	730	pH 7.4, 0.34 M P$_i$
NADPH-sulfite reductase (EC 3.2.2.5)					
Bull semen	–	10.9	278	1658	pH 7.4, 0.01 M NaP$_i$, dry wt
NADPH-sulfite reductase (EC 1.8.1.2)					
E. coli B	–	4.60	386	1659	Dry wt
Nagarse [BPN']	–	8.8	280	357	–
Nerve growth factor					
Mouse, 2.5S	2.2	16.4	280	1660	AA, MW = 13,259
Salivary gland	–	14.7	280	1661	pH 4, 0.1 M AcO$^-$ or pH 4, 0.1 M AcO$^-$ 8 M urea, from Figure 1 (1661)
Submaxillary gland	4.15	13.86	280	1102	pH 5.0, 0.05 M AcO$^-$, MW = 30,000 (1102)
Naja naja	–	11.8	280	1661	pH 4, 0.1 M AcO$^-$ from Figure 1 (1661)
	–	12.9	280	1661	pH 4, 0.1 M AcO$^-$ 8 M urea, from Figure 1 (1661)
Vipera russelli venom	–	9.9	282	1662	
Cobra venom	3.22	12.7	280	620	Water MW = 25,300 (620)
Neurophysin-II					
Beef pituitary	0.395	–	260	621	–
Neurotoxin					
Naja naja siamensis (Thiland cobra)	0.83	10.6	279	1103	Neutral sol MW = 7820 (1130)

p‡Calculated from OD$_{272}$/OD$_{448}$ = 7.33, and OD$_{380}$/OD$_{448}$ = 1.04.

MOLAR ABSORPTIVITY AND $A_{1cm}^{1\%}$ VALUES FOR PROTEINS AT SELECTED WAVELENGTHS OF THE ULTRAVIOLET AND VISIBLE REGION (continued)

Protein	$\epsilon^a(\times 10^{-4})$	$A_{1cm}^{1\%}$ [b]	nm[c]	Ref.	Comments[d]
Neurotoxin (continued)					
Reduced and carboxy-methylated	0.68	–	279	1103	10% AcOH
Toxin 3	–	10.6	279	1664	
Toxin 3C	0.89	12.9	279	1103	MW = 6793 (1103)
Toxin 5	1.3	19.1	280	1103	MW = 6875 (1103)
Toxin 7C	0.9	12.9	279	1103	MW = 6985 (1103)
Naja naja naja (Indian cobra)					
Toxin 3	0.85	10.9	279	1103	MW = 7834 (1103)
	–	10.9	279	1664	–
Reduced and carboxy-methylated	0.66	–	279	1103	–
Toxin 4	0.85	10.9	279	1103	MW = 7807 (1103)
Reduced and carboxy-methylated	0.65	–	279	1103	–
Naja haje (cobra)					
Toxin I[q†]	0.917	13.28	280	1104	MW = 6843 (1104)
Toxin I[r†]	0.923	13.49	279	1104	MW = 6843 (1104)
Toxin II[q†]	0.828	12.07	280	1104	MW = 6857 (1104)
Toxin II[r†]	1.123	16.31	278	1104	MW = 6887 (1104)
Toxin III[r†]	0.898	11.48	280	1104	MW = 7806 (1104)
Naja nigricollis	0.87	12.8	279	731	pH 7.3, 0.03 M NaP$_i$, MW = 6787 (731)
	0.84	12.4	279	731	pH 2.3, MW = 6787 (731)
Carboxymethyl-	0.66	–	279	731	–
I	0.901	13.26	280	1663	MW = 6794
II	0.886	13.03	280	1663	MW = 6796
Notechis scutatus scutatus (Australian tiger snake)					
Venom[s†]	2.8	20.6	278	1665	Acid pH
Androctonus australis Hector (scorpion)					
Toxin I	1.071	15.75	275	1105 ⎱	MW = 6808 (1105)
	1.03	15.12	280	1105 ⎰	
Toxin II	1.801	24.80	276	1105 ⎱	MW = 7249 (1105)
	1.67	23.08	280	1105 ⎰	
Toxin III	1.191	17.45	277.5	1105 ⎱	MW = 6826 (1105)
	1.169	17.13	280	1105 ⎰	
Buthus occitanus tunetanus (scorpion)					
Toxin I	1.888	–	278	1105	MW[t†]
	–	26.75	280	1105	–
Toxin II	2.126	28.20	278	1105 ⎱	MW = 7539 (1105)
	2.079	27.58	280	1105 ⎰	
Toxin III	1.915	26.34	278	1105 ⎱	MW = 7270 (1105)
	1.857	25.55	280	1105 ⎰	
Leirus quinquestriatus quinquestriatus (scorpion)					
Toxin I	1.511	21.81	275	1105 ⎱	MW = 6928 (1105)
	1.467	21.17	280	1105 ⎰	
Toxin II	1.340	–	277	1105	MW[u†]
	–	19.86	280	1105	–

[q†]From Miami Serpentarium.
[r†]From Institute Pasteur.
[s†]Protein called "Notexin," Reference 1665.
[t†]MW = 6919–6933.
[u†]MW = 6511–6545.

MOLAR ABSORPTIVITY AND $A_{1cm}^{1\%}$ VALUES FOR PROTEINS AT SELECTED WAVELENGTHS OF THE ULTRAVIOLET AND VISIBLE REGION (continued)

Protein	$\epsilon^a(\times 10^{-4})$	$A_{1cm}^{1\%}$ [b]	nm[c]	Ref.	Comments[d]
Neurotoxin (continued)					
Toxin III	2.153	—	276.5	1105	MW[v†]
	—	30.53	280	1105	—
Toxin IV	2.002	27.38	278	1105	
	1.966	26.89	280	1105	MW = 7313 (1105)
Toxin V	2.135	28.61	275	1105	
	2.082	27.90	280	1105	MW = 7462 (1105)
Nicotin oxidase (nicotine dehydrogenase, EC 1.5.99.4)					
A. oxydans	—	23.6	280	358	pH 7.9, 0.1 M PyP$_i$, mM EDTA
Nicotinic acid hydroxylase (nicotinate dehydrogenase, EC 1.5.1.13)					
Clostridium	33	11	275	1106	MW = 300,000 (1106) data from Figure 6 (1106)
Nitrate reductase					
Micrococcus dentrificans	41.9	26	280	1107	pH 7.4, 0.2 M P$_i$, MW = 161,129 (1107) data from Figure 3 (1107)
Nitrite reductase (EC 1.7.99.3)					
Achromobacter cycloclastes	0.16	—	400	1666	
	0.40	—	464	1666	Dry wt
	0.20	—	590	1666	
	0.17	—	700	1666	
Achromobacter fisheri	1.49	—	525	1667	
	16.6	—	409	1667	pH 7.0, oxidized
	11.8	—	278–80	1667	
	4.26	—	551	1667	
	2.6	—	523	1667	pH 7.0, reduced
	21.9	—	420	1667	
Chlorella fusca	2.2	—	384	1668	
Nitrogenase					
Azotobacter vinelandii	47.0	17.4	280	1108	
	8.5	3.15	412	1108	MW = 270,000 (1108)
Mo-Fe protein	47	17.4	280	1669	pH 7.4, 0.25 M NaCl 0.01 M TrisCl under N_2 MW =
	8.5	3.2	418	1669	270,000 (1669)
Klebsiella pneumonia					
Kp 1	26	—	258.5	1670	
	30	—	269	1670	
	33	—	277.5	1670	Reduced and oxidized
	32	—	282	1670	
	25	—	289	1670	
	35	—	430	1670	
	50	—	430	1670	Oxidized
Kp 2	11	—	258	1670	
	11.2	—	268	1670	Reduced and oxidized
	0.4–0.5	—	460	1670	Reduced
	1.0	—	460	1670	Oxygen inactivated

[v†]MW = 6764–6792.

MOLAR ABSORPTIVITY AND $A_{1cm}^{1\%}$ VALUES FOR PROTEINS AT SELECTED WAVELENGTHS OF THE ULTRAVIOLET AND VISIBLE REGION (continued)

Protein	$\epsilon^a (\times 10^{-4})$	$A_{1cm}^{1\%}$ [b]	nm[c]	Ref.	Comments[d]
Non-histone					
Rat liver	1.8	12.3	275	442	pH 8, 0.01 M Tris, 0.01% NaDodSO$_4$, est. from Figure 1 (442)
Nuclease					
Bacteriophage T4	—	14.8	280	1109	pH 7.5[w†]
Staphylococcal	—	9.7	277	1110	
Performic acid oxidized	—	7.0	274	1110	—
Lightly acetylated	—	9.7	277	1110	—
Heavily acetylated	—	9.7	277	1110	—
Trifluoroacetylated	—	9.0	277	1110	—
S. aureus	—	11.6	280	359	pH 7.2, 0.15 N NaP$_i$
3-Nucleotidase (see phosphodiesterase, 2':3'-cyclic)					
Nucleoside triphosphate– adenylate kinase (EC 2.7.4.10)					
Beef heart mitochondria	4.55	8.75	280	1111	pH 8.5, 0.05 M N(EtOH)$_3$HCl MW = 52,000 (1111), data from Figure 3 (1111)
Octopine dehydrogenase (EC 1.5.1.11)	4.33	—	280	1671	—
Pecten maximus	—	11.4	280	732	—
Ornithine-oxoacid aminotransferase (EC 2.6.1.13)					
Rat liver	—	10.6	280	736 ⎫	Data from Figure 3 (736),
With added ornithine	—	11.0	280	736 ⎬	ph 8, 0.2 M TrisCl
Apo-	—	10.8	280	736 ⎭	
Ornithine transcarbamylase (EC 2.1.3.3)					
S. faecalis	—	8.32	280	1672 ⎱	pH 7.0, 50 mM NaP$_i$, dry
Beef liver		12.3	280	1672 ⎰	wt and nitrogen
Orosomucoid					
Rabbit serum	1.93	6.05	276	1673 ⎫	
	1.94	6.06	277	1673 ⎬	MW = 32,000
	1.93	6.04	278	1673	
	1.92	6.00	280	1673 ⎭	
Orotate reductase (EC 1.3.1.14) (see dihydroorotic dehydrogenase)					
Ovalbumin	2.85	—	280	17	pH 4.9
	—	7.5	280	16	—
	—	7.35	280	361	—
	—	7.5	280	733 ⎫	
	—	260	210	733 ⎬	0.1 M NaOH
	—	750	191	733 ⎭	

[w†]14 mM Tris-HCl, 140 mM KCl, 1.4 mM β-mercaptoethanol (HSEtOH), and 7.1% glycerol. Protein determined by method and absorption corrected for scattering.

MOLAR ABSORPTIVITY AND $A_{1\,cm}^{1\%}$ VALUES FOR PROTEINS AT SELECTED WAVELENGTHS OF THE ULTRAVIOLET AND VISIBLE REGION (continued)

Protein	$\epsilon^a (\times 10^{-4})$	$A_{1cm}^{1\%}$ [b]	nm[c]	Ref.	Comments[d]
Ovalbumin (continued)	–	7.15	280	734	
	2.94	7.35	280	1674	MW = 45,000
	--	3.40	255	1675 ⎫	
	–	7.60	280	1675 ⎬	pH 7.0, 0.01 M P_i
	–	0.48	310	1675 ⎭	
	--	7.37	280	1676	Native and SDS denatured. Refr.
	–	7.54	280	1676	6 M GdmCl Refr.
	1.579	–	293	1112	--
	2.792	--	287	1112	–
	21.42	–	232	1112	–
Chicken	3.15	7.01	280	1113	MW = 45,000 (1114)
Chicken egg	3.218	–	280	1677	–
	–	6.9	280	1678	–
	–	7.14	280	1679	
Turkey	3.74	8.32	280	1113	MW = 45,000 (1114)
Duck	3.72	8.28	280	1113	MW = 45,000 (1114)
Ovoglycoprotein					
Chicken egg	0.93	3.8	280	1680	MW = 24,440 (1680)
Ovoinhibitor					
Chicken	–	7.1	280	1681	pH 6.8, 0.01 M P_i
Chicken egg white	–	6.5–6.9	278	1682	
	–	7.4	278	1115	
Quail egg white	–	7.1	278	1116 ⎱	
	–	7.1	278	1116 ⎰	pH 8, 0.1 M Tris
Ovomacroglobulin					
Chicken egg white	–	8.6	278	1117	Dry wt
Ovomucin					
Chicken egg white	–	10.3	290	1118	pH 13
	–	9.3	277.5	1118 ⎱	
	–	5.6	290	1118 ⎰	Neutral sol.
Ovomucoid	–	4.55	280	362	
	1.19	4.13	280	363	MW = 28,800 (363)
Chicken		4.1	280	1681	pH 6.8, 0.01 M P_i
	1.17	4.3	277	1683	pH 7.5, water: glycerol, 1:1
Chicken egg white	1.1	4.10	278	1119	MW = 27,300 (1119)
Turkey	–	4.15	280	1684	pH 7.8, Tris
Ovorubin					
Pomacea canaliculata australis	–	10.5	280	735	Data from Figure 3 (735)
Ovotransferrin (conalbumin)					
Chicken	0.475	0.620	470	1685	pH 6–9 MW = 76,600
	8.5	11.1	280	1685	0.02 M HCl, MW = 76,600 (1685)
	–	11.6	280	1686	–
	7.96	–	280	1687	pH 6.5, 6 M Gdn 0.02 M P_i

MOLAR ABSORPTIVITY AND $A_{1cm}^{1\%}$ VALUES FOR PROTEINS AT SELECTED WAVELENGTHS OF THE ULTRAVIOLET AND VISIBLE REGION (continued)

Protein	$\epsilon^a (\times 10^{-4})$	$A_{1cm}^{1\%}$ [b]	nm[c]	Ref.	Comments[d]
Ovotransferrin (continued)					
Chicken (continued)	9.03	--	280	1687	pH 10.5, 0.1 M glycine-NaOH
	--	11.2	280	1688	--
Iodate oxidized, pH 8.5	9.2	--	280	1687	
Iodate oxidized at pH 5.0	9.5	--	280	1687	pH 6.5, 6 M Gdn 0.02 M P$_i$
Iron free, oxidized	7.96	--	280	1687	
Iodate oxidized at pH 8.5	9.33	--	280	1687	
					pH 10.5, 0.1 M glycine-NaOH
Iodate oxidized at pH 5.0	9.80	--	280	1687	
Iron free, oxidized	9.33	--	280	1687	
Oxaloglycolate reductase (decarboxylating) (EC 1.1.1.92)					
P. putida	7.2	11.8	280	364	MW = 61,000 (364)
3-Oxoacyl-[acyl-carrier-protein] reductase (EC 1.1.1.100) (see also fatty acid synthetase)					
Pig liver	46	--	279	1424	--
3-Oxosteroid $\Delta^{4,5}$- isomerase (steroid Δ-isomerase) (EC 5.3.3.1) (see also 3-keto-Δ^5- steroid isomerase)		3.28	280	1689	Refr.
Beef adrenals	11.3	10.1	--	855	MW = 112,000 (855)
Oxynitrilase (see hydroxynitrilase)	--	3.72	277	1689	Refr.
Oxytyrosinase (monophenol monoxygenase or tyrosinase, EC 1.14.18.1, or tyrosine 3-monooxygenase, EC 1.14.16.2)					
Mushroom	0.9	--	345	1690	Value per mole Cu
	0.6	--	600	1690	
Papain (EC 3.4.22.2)	--	25.0	278	365	--
	4.9	--	280	367	MW = 20,700 (366)
	5.1	--	280	368	--
	1.2	--	295	365	pH 5--8
Papaya	--	21.5	280	1691	0.01 N HCl, from Figure 1 (1691)
	--	27	278	1692	
Papaya latex					
Trans-cinnamoyl-	2.6	--	326	1120	--
Furylacryloyl-	3.0	--	360	1121	--
	3.0	--	337	1121	Denatured
Indolacryloyl-	4.3	--	398	1121	--
	4.3	--	373	1121	Denatured
Mercury derivative	5.1	--	280	1122	--

MOLAR ABSORPTIVITY AND $A_{1cm}^{1\%}$ VALUES FOR PROTEINS AT SELECTED WAVELENGTHS OF THE ULTRAVIOLET AND VISIBLE REGION (continued)

Protein	$e^a(\times 10^{-4})$	$A_{1cm}^{1\%}$ [b]	nm[c]	Ref.	Comments[d]
Paramyosin					
Venus mercenaria	–	$3.24^{x\dagger}$	277	1123	pH 7.4, 1 M KCl, 0.1 M KP$_i$
Reduced and carboxy-					
methylated	–	$3.39^{y\dagger}$	277	1123	–
Crassostrea commercialis					
(oyster)	5.95	2.86	276	1124	pH 7.0, 1.1 M NaCl –
					25 mM NaP$_i$
					MW = 208,000 (1124)
Aulacomya magellanica					
(mollusc)	8.76	3.4	280	1125 ⎱	pH 7.5, 1 M KCl
	9.55	3.7	276	1125 ⎰	MW ≈ 258,000 (1125)
Parathyroid hormone					
Beef	0.72	7.6	280	369	pH 4.7, AcONH$_4$
					MW = 9,500 (369)
Parvalbumin					
Chondrostoma nasus	–	1.84	260	1693	Water, from Figure 7 (1693)
	–	1.57	260	1693	0.1 N NaOH,
					from Figure 7 (1693)
Merluccius merluccius					
(hake)	0.202	1.8	259	1694	MW = 11,500 (1694)
					AA
	0.215	–	259	1695	–
Raja clavata					
(thornback ray)	0.142	–	275	1695	–
Penicillinase (EC 3.5.2.6)					
Bacillus cereus 569	–	$6.0^{z\dagger}$	280	1126	–
	3.2	10.5	280	1127	Biuret
Bacillus cereus 569/H	–	$6.35^{z\dagger}$	280	1126	–
Bacillus cereus 5/B	–	$7.35^{z\dagger}$	280	1126	–
B. licheniformis 6346/C	–	$5.65^{z\dagger}$	280	1126	–
B. licheniformis 749/C[c]	–	$5.45^{z\dagger}$	280	1126	–
B. licheniformis 749/C	–	$4.75^{z\dagger}$	280	1126	–
Staphylococcus aureus PC 1	–	$7.38^{z\dagger}$	280	1126	–
E. coli K 12	6.1	21.0	280	1128	pH 6.8, 0.01 M KP$_i$
					MW = 29,000 (1128)
Penicillocarboxypeptidase-S					
(peptidase B)					
Penicillium janthinelium	–	26	280	1696	–
Penicillopepsin (EC 3.4.23.7)					
Penicillium janthinellum	4.32	13.5	280	1129	MW = 32,000 (1129)
Pepsin	–	5.38	255	1675 ⎱	
	–	12.20	280	1675 ⎬	pH 7.0, 0.01 M P$_i$
	–	0.28	310	1675 ⎰	
	1.94	–	292.8	1697 ⎱	
	3.68	–	286	1697 ⎬	Denatured at pH
	5.02	–	279	1697 ⎰	1.45, 3 hr, 39°C
	23.87	–	230	1697	

^{x†}In GdmCl: $\epsilon_{277} = 0.035$ (M GdmCl) + 3.24.

^{y†}In GdmCl: $\epsilon_{277} = 0.34$ (M GdmCl) + 3.39.

$^{z\dagger}A_{1cm}^{1\%}$/mg N.

MOLAR ABSORPTIVITY AND $A_{1cm}^{1\%}$ VALUES FOR PROTEINS AT SELECTED WAVELENGTHS OF THE ULTRAVIOLET AND VISIBLE REGION (continued)

Protein	$\epsilon^a (\times 10^{-4})$	$A_{1cm}^{1\%}$ [b]	nm[c]	Ref.	Comments[d]
Pepsin (continued)	--	13.1	280	1698 ⎫	
	–	290	210	1698 ⎬ Water, dry wt	
	–	920	191	1698 ⎭	
	5.247	–	278	1699	–
Beef	4.94	14.81	280	1700	MW = 33,367 (1700)
	5.17	14.3	280	370	
	5.09	–	278	371	–
Pig	–	14.1	280	1701	–
	–	780	191–4	1701	–
Chicken	5.34	15.2	276	1702 ⎱	pH 7.5, 0.1 M P_i,
	–	150	220	1702 ⎰	MW = 35,000 (1702)
Human	6.2	17.3	278	179	pH 5, MW = 34,000 (179)
Pepsinogen					
Beef	5.6	12.5	278	373	MW = 41,000 (373)
	5.1	13.05	280	374	MW = 39,000 (374)
Succinyl-	5.5	–	278	375	pH 7.7, 0.032 M P_i
Bovine 1	–	13.45	280	1703	
2	–	13.05	280	1703	–
4	–	13.25	280	1703	–
Dog	–	12.79	280	1704	–
Chicken	5.59	13.0	278	1702	pH 1.83, 0.02 N HCl, MW = 43,000 (1702)
	–	12.66	280	1702	
	–	139.9	220	1702	–
Peptidoglutaminase I peptidyl-glutaminase, EC 3.5.1.43)	–	7.27	280	1705	–
Peptidoglutaminase II (glutaminyl-peptide glutaminase) (EC 3.5.1.44)	–	11.62	280	1705	–
Peroxidase (EC 1.11.1.7) (see also lactoperoxidase, myeloperoxidase)					
Raphanus sativus (Japanese radish)	1.175	–	500	739	–
Isoenzyme 3	2.72	–	280	1130	–
	11.14	–	404	1130	–
	1.17	–	502	1130	–
	0.323	–	644	1130	–
Isoenzyme 5	3.82	–	280	1130	–
	11.66	–	405	1130	–
	1.156	–	502	1130	--
	0.338	–	644	1130	--
Isoenzyme 16	2.97	–	280	1130	–
	10.42	–	403	1130	--
	1.186	–	504	1130	–
	0.339	–	645	1130	–
Apo-	1.72	–	--	739	–
Brassica napus L					
P_1	3.8	–	276	740	–
P_2	3.4	–	276	740	–

MOLAR ABSORPTIVITY AND $A_{1\,cm}^{1\%}$ VALUES FOR PROTEINS AT SELECTED WAVELENGTHS OF THE ULTRAVIOLET AND VISIBLE REGION (continued)

Protein	$\epsilon^a (\times 10^{-4})$	$A_{1\,cm}^{1\%\ b}$	nm^c	Ref.	Comments^d
Peroxidase (continued)					
P$_3$	3.0	–	277	740	–
P$_6$	3.0	–	278	740	–
P$_7$	2.9	–	278	740	–
Japanese radish, a					
(see also peroxidase a, apo-)					
Oxidized	3.38	–	276	379	–
	11.10	–	405	379	–
	1.18	–	500	379	–
	0.33	–	645	379	–
Japanese radish, c					
Oxidized	3.0	–	280	379	–
	10.64	–	420	379	–
	1.17	–	540	379	–
Reduced	10.38	–	425	379	–
	1.26	–	560	379	–
Horseradish	–	22	403	1706	–
	9.1	–	403	1707	–
	10.2	–	403	737	–
	9.1	–	403	376	–
	–	13.4	275	377	pH 7.0
	10.4	–	498	378	Ferric derivative
Nitroso-	11.0	–	420	378	–
Carbon monoxide	15.3	–	422	378	–
Apoenzyme-A1	0.92	–	276	738 ⎫	pH 6.8, 50 mM NaP$_i$
Apoenzyme-C	1.3	–	277	738 ⎬	
A$_1$	10.2	–	401	1708 ⎫	
A$_2$	10.2	–	401	1708 ⎪	
A$_3$	9.7	–	401	1708 ⎬	MW = 40,000 (1708)
B	9.5	–	401	1708 ⎪	
C	9.5	–	401	1708 ⎭	
Component I	1.31^{a‡}	–	440	1709	–
	1.74^{a‡}	–	432	1709	–
	1.86^{a‡}	–	430	1709	–
	2.83^{a‡}	–	420	1709	–
	4.12^{a‡}	–	410	1709	–
	4.80^{a‡}	–	400	1709	–
	4.62^{a‡}	–	390	1709	–
	4.30^{a‡}	–	380	1709	–
	4.00^{a‡}	–	370	1709	–
	3.60^{a‡}	–	360	1709	–
	3.28^{a‡}	–	350	1709	–
Fraction Ib	2.78^{b‡}	–	280	1710	–
	11.5^{b‡}	–	403	1710	–
Fraction IIIb	2.87^{b‡}	–	280	1710	–
	9.98^{b‡}	–	403	1710	–
Fraction Vb	3.68^{b‡}	–	280	1710	–
	9.50^{b‡}	–	280	1710	–
Fraction VI	4.08^{b‡}	–	403	1710	–
	12.29^{b‡}	–	403	1710	–
Native, oxidized	0.323	–	641	1711 ⎫	pH 7.0, 10 mM
	1.095	–	498	1711 ⎬	NaP$_i$
	10.00	–	402	1711 ⎭	

^{a‡}All values based on $\epsilon_m = 9.14 \times 10^4$ at 430 nm.
^{b‡}Based on hemin content.

MOLAR ABSORPTIVITY AND $A_{1cm}^{1\%}$ VALUES FOR PROTEINS AT SELECTED WAVELENGTHS OF THE ULTRAVIOLET AND VISIBLE REGION (continued)

Protein	$\epsilon^a (\times 10^{-4})$	$A_{1cm}^{1\%\ b}$	nm^c	Ref.	Comments[d]
Peroxidase (continued)					
Protoperoxidase	0.289	—	641	1711	pH 7.0, 10 mM NaP$_i$, oxidized
	1.046	—	499	1711	
	9.43	—	402	1711	
Dipropenyldeutero-hematinperoxidase	0.287	—	635	1711	
	1.122	—	500	1711	
	10.22	—	402	1711	
Dibutenyldeutero-hematinperoxidase	0.298	—	636	1711	
	1.123	—	502	1711	
	11.18	—	404	1711	
Deuterohematinperoxidase	0.272	—	627	1711	pH 7.0, 10 mM NaP$_i$
	0.708	—	497	1711	
	11.26	—	393	1711	
Mesoperoxidase	0.275	—	634	1711	
	0.942	—	492	1711	
	8.80	—	395	1711	
Diacetyldeutero-hematinperoxidase	0.226	—	642	1711	
	0.732	—	508	1711	
	8.01	—	412	1711	
Hematoperoxidase	0.299	—	633	1711	
	0.850	—	497	1711	
	10.27	—	397	1711	
Native, reduced	1.260	—	556	1711	
	8.86	—	436	1711	
Protoperoxidase	1.255	—	555	1711	
	7.53	—	433	1711	
Dipropenyldeutero-hematinperoxidase	1.213	—	555	1711	
	8.45	—	433	1711	
Dibutenyldeutero-hematinperoxidase	1.251	—	556	1711	pH 7.0, 10 mM NaP$_i$, reduction by dithionite
	8.49	—	433	1711	
Deuterohematinperoxidase	1.000	—	546	1711	
	7.90	—	427	1711	
Mesoperoxidase	1.156	—	549	1711	
	8.80	—	428	1711	
Diacetyldeutero-hematinperoxidase	1.037	—	562	1711	
	7.82	—	447	1711	
Hematoperoxidase	1.093	—	549	1711	
	8.91	—	431	1711	
Fig latex (*Ficus carica*)	10.1	—	403	1713	—
	1.16	—	500	1713	—
	0.33	—	640	1713	—
Reduced	9.21	—	438	1713	—
	1.26	—	556	1713	—
Reduced + CO	15.52	—	423	1713	—
	1.31	—	543	1713	—
	1.38	—	573	1713	—
+OH⁻	10.6	—	417	1713	—
	1.02	—	544	1713	—
	0.79	—	574	1713	—
+NaN$_3$	11.9	—	415	1713	—
	0.93	—	535	1713	—
	0.23	—	637	1713	—

MOLAR ABSORPTIVITY AND $A_{1cm}^{1\%}$ VALUES FOR PROTEINS AT SELECTED WAVELENGTHS OF THE ULTRAVIOLET AND VISIBLE REGION (continued)

Protein	$\epsilon^a (\times 10^{-4})$	$A_{1cm}^{1\%}$ [b]	nm[c]	Ref.	Comments[d]
Peroxidase (continued)					
+NaF	15.55	–	404	1713	–
	0.93	–	490	1713	–
	0.57	–	561	1713	–
	0.78	–	613	1713	–
+NaCN	10.25	–	421	1713	–
	1.17	–	540	1713	–
Complex I	5.8	–	402	1713	–
	0.76	–	562	1713	–
	0.62	–	657	1713	–
Complex II	9.6	–	419	1713	–
	0.9	–	527	1713	–
	0.95	–	558	1713	–
Complex III	10.45	–	418	1713	–
	1.17	–	546	1713	–
	1.01	–	583	1713	–
FPO-A	3.78	–	280	1713 ⎫	
FPO-B	3.31	–	280	1713 ⎬ Calculated	
FPO-C	3.45	–	280	1713 ⎭	
Pig intestinal mucosa	–	15.5	280	1714	–
	–	14.07	417	1714	–
	–	1.54	490	1714	–
	–	1.27	543	1714	–
	–	1.59	596	1714	–
	–	0.89	642	1714	–
Peroxidase II					
Horseradish, oxidized	3.27	–	270	279	–
	10.77	–	403	279	–
	1.19	–	497	379	–
	0.34	–	641.5	379	–
Reduced	9.17	–	437	379	–
	1.33	–	556	379	–
Peroxidase *a*, apo- Japanese radish roots *(Raphanus sativus)*	1.22	2.7	280		MW = 45,300 (77)
Phenolase, mushroom (monophenol monooxygenase, EC 1.14.18.1)	–	26.92	282	380	–
Phenolhydroxylase (phenol 2-monoxygenase) (EC 1.14.13.7) *Trichosporon cutaneum*	–	9.87	276	1715	pH 7.6, 0.1 M KP$_i$, from Figure 5 (1715)
Phenoloxidase, pre- (ct. pre-phenoloxidase)					
L-Phenylalanine ammonia lyase (EC 4.3.1.5) Maize (*Zea mays*)	–	8.9	280	1716	AA

MOLAR ABSORPTIVITY AND $A^{1\%}_{1\,cm}$ VALUES FOR PROTEINS AT SELECTED WAVELENGTHS OF THE ULTRAVIOLET AND VISIBLE REGION (continued)

Protein	$\epsilon^a (\times 10^{-4})$	$A^{1\%}_{1\,cm}$ [b]	nm[c]	Ref.	Comments[d]
Phenylalanine 4-monooxygenase (EC 1.14.16.1)					
Rat liver	–	5.6	280	1717	pH 8.6, 0.01 M TrisCl, from Figure 3 (1717)
Phenylalanine monooxygenase-stimulating protein					
Rat liver	4.67	9.06	280	1718	Biuret MW = 51,500 (1718)
Phenylpyruvate tautomerase (EC 5.3.2.1)					
Pig thyroid	5.95	13.5	280	741	pH 6.2, 0.1 M P_i, MW = 44,000 (741)
Phosphatase, acid (EC 3.1.3.2)					
Human prostate	–	24	280	1746	–
Phaseolus mungo	8.75	16	278	1747	pH 5.6, MW = 55,000 (1747)
Rat, liver	4.7	4.7	278	387	MW = 100,000 (387)
	6.18	6.18	278	1724	MW = 100,000 (1725)
Sweet potato	–	9.1	280	1726 ⎱	pH 6.0, 0.01 M KP_i,
	–	0.21	555	1726 ⎰	from Figure 2 (1726)
N. crassa	9.18	10.8	280	1727	Water, MW = 85,000 (1727)
Phosphatase, alkaline (EC 3.1.3.1)					
Escherichia coli	6.4	7.2	278	742	MW = 89,000 (742)
	0.0260	–	640	1722	–
	0.0220	–	605	1722	–
	0.0378	–	555	1722	–
	0.0335	–	510	1722	–
	–	7.2	278	1721	–
E. coli C90	6.2	7.7	280	381	MW = 80,000 (381)
E. coli K 12	5.6	7.0	278	382	MW = 80,000 (382)
	6.6	7.7	280	743 ⎫	
Azatryptophan substituted for tryptophan	11	12.8	280	743 ⎬	MW = 86,000 (744)
Tryptazan substituted for tryptophan	9.6	11.1	280	743 ⎭	
B. licheniformis	7.5[c‡]	6.2	278	1719	MW = 121,000 (1719)
B. subtilis	0.0305	–	620	1720	–
	0.0335	–	596	1720	–
	0.0500	–	567	1720	–
	0.0382	–	517	1720	–
Micrococcus sodonensis	13.7	17.3	280	1723	–
N. crassa	17.4	11.3	280	383	pH 8.3, 0.01 M Tris, HCl, MW = 154,000 (383)
A. nidulans	–	10.2	278	384	pH 7.0
Intestinal	7.0	–	280	385	–
Placental, human	9.75	7.8	278	386	pH 7.0, 0.05 M P_i, MW = 125,000 (386)

[c‡] On p. 1369 of Reference 1719, value is given as 7.25×10^4.

MOLAR ABSORPTIVITY AND $A_{1cm}^{1\%}$ VALUES FOR PROTEINS AT SELECTED WAVELENGTHS OF THE ULTRAVIOLET AND VISIBLE REGION (continued)

Protein	$\epsilon^a(\times\,10^{-4})$	$A_{1cm}^{1\%}$ [b]	nm[c]	Ref.	Comments[d]
Phosphatase, alkaline (continued)					
Calf intestine	–	10	280	745	pH 7.7
Milk	–	11.5	280	745	
Pig kidney	–	13.9	280	746	–
	–	12.0	260	746	–
Phosphodiesterase 2':3'-cyclic (2':3'-cyclic-nucleoside monophosphate phosphodiesterase, EC 3.1.4.16)					
B. subtilis	–	11	280	1728	pH 7.5, 0.05 M TrisCl
Phosphoenolpyruvate carboxykinase (EC 4.1.1.32,.38, .49) Baker's yeast	32	12.7	280	1729	pH 7.01 M EDTA – 0.025 M Na-Borate MW = 252,000 (1724)
Phosphoenolpyruvate carboxylase (EC 4.1.1.31)					
E. coli	63.4	15.5	280	1730	MW = 402,000 (1730) dry wt
E. coli, str. B	–	10.9	280	1731	pH 8.5, 0.1 M TrisCl, 10 mM MgCl$_2$, 10 mM KHCO$_3$ 0.01 N NaOH
S. typhimurium	–	15.5	280	1732	–
Phosphofructokinase Rabbit muscle	–	10.2	279	388	–
	–	9.4	279	389	pH 7
	–	8.7	283	389	0.1 N NaOH
	–	8.7	290	389	0.1 N NaOH
	–	10.9	290	390	0.1 N NaOH
Sheep heart	–	10.0	280	391	–
Yeast	72.4	12.4	279	392	MW = 584,000 (392)
	54.2	9.5	279	1733	pH 7.0, 0.1 M P$_i$, MW = 570,000 (1733)
Phosphoglucoisomerase Rabbit muscle	17.2	13.2	280	393	pH 7.0, 0.01 M P$_i$, MW = 130,000 (393)
Phosphoglucomutase (EC 2.7.5.1) Rabbit muscle	–	7.7	278	394	–
	5.98	7.8	278	395	MW = 77,000 (395)
	4.77	7.7	278	1734	MW = 62,000 (1735)
Yeast	8.2	11.8	280	1736	MW = 69,500 (1736), turbidimetric
6-Phosphogluconate dehydrogenase (EC 1.1.1.44)					
C. utilis, Type I	12.8	12.7	280	396	MW = 101,000 (396)
Type II	14.1	12.7	280	396	MW = 111,000 (396)
Sheep liver	--	10.3	280	622	–
	10.7	11.4	280	1737	MW = 94,000 (1737), dry wt

MOLAR ABSORPTIVITY AND $A_{1cm}^{1\%}$ VALUES FOR PROTEINS AT SELECTED WAVELENGTHS OF THE ULTRAVIOLET AND VISIBLE REGION (continued)

Protein	$\epsilon^a(\times 10^{-4})$	$A_{1cm}^{1\%}{}^b$	nm^c	Ref.	Comments^d
Phosphoglucose isomerase (glucosephosphate isomerase, EC 5.3.19)					
Rabbit muscle	17.4	13.2	280	623	pH 7.0, 0.01 M P$_i$, MW = 132,000 (623)
Human muscle	–	12	280	624	pH 7.2
Human erythrocytes	–	13.1	280	1738	Neutral pH, refr.
Phosphoglycerate dehydrogenase					
E. coli	11.2	6.7	280	397	MW = 165,000 (397)
3-Phosphoglycerate kinase (EC 2.7.2.3)					
Rabbit muscle	3.12	6.9	280	1739	MW = 45,200 (1739)
Yeast	2.24	4.9	280	1739	MW = 45,800 (1739)
	–	5.0	280	141	–
Rabbit muscle	2.16	5.7	280	1740	MW = 38,000 (1740)
Yeast	–	5.35^{d‡}	278	1741	Dry wt
Phosphoglycerate phosphomutase (EC 5.4.2.1)					
Yeast, Component I	–	14.2	280	1742	–
Component II	–	14.9	280	1742	–
Phosphoglyceromutase (EC 2.7.5.3)					
Rabbit muscle	–	12.5	280	398	–
Yeast	–	14.2	280	399	–
Sheep muscle	–	7.1	280	1743	pH 7.0, 0.1 μm P$_i$
Phospholipase A$_2$ (EC 3.1.1.4) *Agkistrodon halys blomhoffi* (snake)					
A-I	2.06^{e‡}	14.9	278.5	1744	pH 7.2, 0.1 M NaP$_i$, MW = 13,800 (1744)
A-II	2.06^{f‡}	15.1	278.5	1744	pH 7.2, 0.1 M NaP$_i$, MW = 13,700 (1744)
Pig pancreas	–	14.2	280	1745	pH 8. AA
Crotalus adamanteus	6.76	22.7	280	401	MW = 29,864 (401)
Phosphomonoesterase, acid (see phosphatase, acid)					

^{d‡}Value may be in error by 10%. (1741).

^{e‡}From Figure 12 of Reference 1744 the following values were estimated: ϵ_M = 7500 (277 nm, pH 1.2, 0.1 M HCl); ϵ_M = 9500 (278.5 nm, pH 7.2, 0.1 M NaP$_i$); ϵ_M = 9800 (290 nm, pH 12.5, 0.1 M NaOH). The value at pH 7.2 does not agree with the value calculated from the molecular weight, 13,800, and the $A_{1cm}^{1\%}$ value of 14.9.

^{f‡}From Figure 13 of Reference 1744, the following values were estimated: ϵ_M = 8800 (277 nm, pH 1.2, 0.1 M HCl); ϵ_M = 10,000 (278.5 nm, pH 7.2, 0.1 M NaP$_i$); ϵ_M = 11,200 (290 nm, pH 12.5, 0.1 N NaOH). The value at pH 7.2 does not agree with the value calculated from the molecular weight, 13,700, and the $A_{1cm}^{1\%}$ of 15.

MOLAR ABSORPTIVITY AND $A_{1cm}^{1\%}$ VALUES FOR PROTEINS AT SELECTED WAVELENGTHS OF THE ULTRAVIOLET AND VISIBLE REGION (continued)

Protein	$\epsilon^a (\times 10^{-4})$	$A_{1cm}^{1\%}$ [b]	nm[c]	Ref.	Comments[d]
5-Phosphoriboisomerase (ribosephosphate isomerase, EC 5.3.1.6)					
Spinach	—	4.35	280	1748	Ref./Biuret
Phosphoribosyl-ATP pyrophosphorylase (ATP phosphoribosyl transferase, EC 2.4.2.17)					
S. typhimurium	16.0	7.45	280	400	MW = 215,000 (400)
Phosphoribosyldiphosphate amidotransferase					
Chicken liver	—	12.5	280	1749	pH 8.0, 0.05 *M* TrisCl from Figure 1 (1749)
N-(5'-Phospho-D-ribosyl formimino)-5-amino-(5''-phosphoribosyl)-4-imidazolecarboxamide isomerase (see isomerase)					
Phosphoribosyl formylglycin-amidine synthetase (see formylglycinamide ribonucleotide amidotransferase)					
Phosphoribosyltransferase (ATP phosphoribosyltransferase, EC 2.4.2.17)					
S. typhimurium	—	10.7	280	1750	pH 7.5, 6 *M* GdmCl– 0.025 *M* TrisCl
Phosphorylase					
Rabbit muscle	—	11.8	277	405	pH 8.0, 0.1 *M* HCO$_3^-$
Pig liver	—	11.9	277	406	—
Frog muscle	—	12.8	288	407	—
Rabbit muscle	—	13.2	280	407	—
	—	12.3[g‡]	280	408	pH 6.8, 5 m*M* P$_i$ dry wt
	—	13.1	279	409	pH 7, 0.01 *M* NaP$_i$
	—	11.5	278	410	—
	—	11.9	278	411	—
	—	13.2[h‡]	280	404	—
Apo-	—	12.0	278	411	—
Human	28.8	11.9	278	410	MW = 242,000 (410)
Potato	—	11.7	278	1757	pH 7.5, 0.005 *M* TrisCl
Rat muscle	—	12.5	280	1751	—
Beef spleen	21.8	11.5	278	1752 ⎰	pH 7.4, 0.01 *M*
	1.1	0.565	333	1752 ⎱	TrisCl, 2 m*M* HSEtOH, MW = 190,000 (1752), dry wt

[g‡] $A_{1cm}^{1\%}$ by amino acid anaylsis, 13.2.

[h‡] $A_{1cm}^{1\%}$ by Biuret and amino acid analysis, 13.5; by refractive index increment, 13.2.

MOLAR ABSORPTIVITY AND $A_{1cm}^{1\%}$ VALUES FOR PROTEINS AT SELECTED WAVELENGTHS OF THE ULTRAVIOLET AND VISIBLE REGION (continued)

Protein	$\epsilon^a (\times 10^{-4})$	$A_{1cm}^{1\%}$ [b]	nm[c]	Ref.	Comments[d]
Phosphorylase (continued)					
Rabbit muscle	–	13.1	280	1753 ⎫	pH 7, 50 mM P$_i$,
Pacific dogfish				⎬	1.0 mM EDTA,
(*Squalus sucklii*)	25.8	12.9	280	1753 ⎭	1.5 mM HSEtOH
Baker's yeast					MW = 200,000 (1753)
(*Saccharomyces cerevisiae*)					
a	15.3	14.9	280	1754	Refr. MW = 103,000 (1754)
b	15.3	14.9	280	1754	Refr. MW = 103,000 (1754)
Silky shark					
(*Carcharhinus falciformis*)					
b	–	13.0	280	1755	pH 7.5, 0.15 M Tris, 0.01 M EDTA, N
Lobster					
(*Homarus americanus*)					
b	–	13.5	280	1756	pH 6.9, 0.04 M Tris, 0.01 M EDTA
Phosphorylase kinase					
(EC 2.7.1.38)					
Rabbit muscle	–	12.4	280	1758	Refr.
	–	11.8	280	1759	pH 7.0, 5 mM P$_i$ – 0.2 mM EDTA, Biuret
Phosphorylase, purine-nucleoside					
Beef liver	–	16	280	402	–
Human erythrocyte	8.91	11	280	403	pH 7.5, MW = 81,000 (403)
Phosphotransacetylase (phosphate acetyltransferase, EC 2.3.1.8)					
C. kluyveri	–	4.2	280	415	Calculated from data
Phosphovitin					
Salmon	1.8	9.4[i‡]	280	1760	pH 2, from Figure 2 (1760), MW = 19,000 (1760)
Trout	1.8	9.3[i‡]	280	1760	pH 2, from Figure 2 (1760), MW = 19,350 (1760)
Phosphovitin kinase					
Calf brain	–	15.7	280	1761	Lowry
Phycocyanin					
Chroomonas sp.	–	114	645	1763	pH 6.0, dry wt
Synechococcus sp.	10.40	–	352	1764 ⎫	MW = 36,700 (1764)
	10.65	–	662.5	1764 ⎭	
α-Subunit	3.26	–	352	1764 ⎫	MW = 17,700 (1764)
	3.32	–	662.5	1764 ⎭	
β-Subunit	6.63	–	352	1764 ⎫	MW = 19,000 (1764)
	6.95	–	662.5	1764 ⎭	
Allophycocyanin	3.17	–	352	1764 ⎫	MW = 16,500 (1764)
	3.22	–	662.5	1764 ⎭	

[i‡] Same values at pH 7.

MOLAR ABSORPTIVITY AND $A_{1cm}^{1\%}$ VALUES FOR PROTEINS AT SELECTED WAVELENGTHS OF THE ULTRAVIOLET AND VISIBLE REGION (continued)

Protein	$\epsilon^a (\times 10^{-4})$	$A_{1cm}^{1\%}$ [b]	nm[c]	Ref.	Comments[d]
Phycocycenin (continued)		72	610	1765	
Synechcococcus lividus	--	72	610	1765	pH 7.0, 0.01 M P_i,
Plectonema calothricoides	--	72	620	1765	from Figure 1 (1765)
Phormidium luridum	--	19	620	1765	pH 6.2 M urea, from Figure 2 (1765)
	--	9.2	620	1765	pH 6, 4 M urea
	--	10	620	1765	pH 6, 6 M and 8 M urea
Coccochloris elabens	--	60	620	1766	pH 6.0
Phycocyanin, anacystis	9.9	7.9	615	412	MW = 12,500 (412)
C-Phycocyanin					
Alga (*Anacystis nidulans*)					
Monomer	23	--	615	1762	pH 7.0, 0.05 M P_i
Hexamer	33	--	621	1762	pH 5.5, 0.2 M AcOH
α-Subunit	9.8	--	620	1762	pH 7.0, 0.05 M P_i
β-Subunit	14.3	--	608	1762	
Phycoerythrin					
Porphyridium cruentin	2.4	2.73	565	412	MW = 87,000 (412)
Phytochrome					
Oat seedlings	--	12.5	280	1767	--
	--	12.5	660	1767	--
Winter rye					
(*Secale cereale*)	9	--	280	1768	Red-absorbing
	9	--	280	1768	Far-red-absorbing
	7	--	665	1768	Red-absorbing
	4	--	730	1768	Far-red-absorbing
Phytohemagglutinin					
Robinia pseudoacacia	--	9.65	278	1769	--
Pigeon droppings					
Old	--	27.5	280	1770	pH 7, dialyzed and
Fresh	--	30.5	280	1770	lyophilized extracts
Pigment					
Serum, eel					
(*Anguilla japonica*)	4.41	--	279	1771	
	0.948	--	383–4	1771	MW = 89,100 (1771)
	0.222	--	704–5	1771	
Pinguinain (EC 3.4.99.18)	--	22.0	280	413	--
	4.72	24.6	280	414	pH 7.3 and pH 4.6, MW = 19,200 (414)
Plasma					
Human, basic B_2	0.299	3.32	278	438	pH 6.0, 0.1 M NaCl, MW = 9,000 (438)
Plasmin (EC 3.4.21.7)					
Human, urokinase activated	--	16.7	280	418	0.1 N NaOH
Streptokinase activated	--	17.3	280	418	0.1 N NaOH
Human	--	19.4	280	419	0.01 N HCl
	--	20.0	280	419	0.01 N HCl
Plasmin, iPr_2 P-human	--	16.8	280	1772	AA

MOLAR ABSORPTIVITY AND $A_{1cm}^{1\%}$ VALUES FOR PROTEINS AT SELECTED WAVELENGTHS OF THE ULTRAVIOLET AND VISIBLE REGION (continued)

Protein	$e^a(\times 10^{-4})$	$A_{1cm}^{1\%}$ [b]	nm[c]	Ref.	Comments[d]
Plasminogen					
Human	—	17.1	280	418	0.1 N NaOH
	—	18.4	280	419	0.01 N HCl
	—	21.7	280	419	0.01 N HCl, ascending portion of peak on DEAE-cellulose
	—	19.8	280	419	0.01 N HCl, descending portion of peak on DEAE-cellulose
	—	16.1	280	1773	—
Human, A	—	16.8	280	1772	AA
B	—	16.8	280	1772	AA
Plastocyanin					
Comfrey (*Symphytum officinale*)	0.45	—	597	1774	⎫
Elder (*Sambucus nigra*)	0.45	—	597	1774	
Nettle (*Urtica dioica*)	0.45	—	597	1774	Extinction coefficient of copper chromophore
Dog's mercury (*Mercurialis perennis*)	0.45	—	597	1774	
Goose grass (*Galium aparine*)	0.45	—	597	1774	
Lettuce (*Lactuca sativa*)	4.5 [j‡]	—	597	1774	⎭
Spinach (*Spinacea oleracea*)	0.118	—	460	1775	—
	0.980	—	597	1775	—
	0.330	—	770	1775	—
French bean (*Phaseolus vulgaris*)	0.45	4.2	597	1776	⎱ MW = 10,690 (1776)
	0.45	4.2	278	1776	⎰
Pokeweed mitogen *Phytolacca americana*	5.9	18.5	280	416	pH 6 and water, MW = 32,000 (416)
Polymerase, DNA (EC 2.7.7.7)	92.7	8.5	280	417	10 mM NH₄HCO₃, MW = 109,000 (417)
Polynucleotide phosphorylase (EC 2.7.7.8)					
M. luteus	—	4.30	280	1777	N: assumed 16.5%
Form I	—	5.30	280	1777	Lowry
Form T	—	4.40	280	1777	N: assumed 16.5%
	—	4.40	280	1777	Lowry
Polyphenol oxidase (monophenol monooxygenase, EC 1.14.18.1)					
Mushroom	—	26.92	280	420	—
Camellia sinensis L.	—	13.5	279	421	pH 6.8, 0.3 M NaPᵢ
	—	0.84	611	421	—

j‡This is given as 45 × 10³ M (1774). It may be a typographical error.

**MOLAR ABSORPTIVITY AND $A_{1cm}^{1\%}$ VALUES FOR PROTEINS AT SELECTED WAVELENGTHS
OF THE ULTRAVIOLET AND VISIBLE REGION (continued)**

Protein	$\epsilon^a(\times 10^{-4})$	$A_{1cm}^{1\%}$ [b]	nm[c]	Ref.	Comments[d]
Polysaccharide depolymerase					
Aerobacter aerogenes	41	10.8	280	1778	MW = 379,000 (1778)
Porphyrenglobin					
Human, fast moving	13.3	–	403	1779	–
	1.01	–	506	1779	–
	0.91	–	542	1779	–
	0.57	–	568	1779	–
	0.37	–	621.5	1779	–
Slow moving	13.3	–	403	1779	–
	0.99	–	506	1779	–
	0.92	–	542	1779	–
	0.58	–	568	1779	–
	0.38	–	621.5	1779	–
Postalbumin, 4.6S					
Human serum	–	8.0	280	1781	–
Post-γ-globulin					
Human urine	–	9.1	280	1780	pH 7
Prealbumin					
Human serum	8.5	13.3	280	1782	MW = 64,000 (1782)
	–	14.4	280	1783	pH 7.0
	–	14.1	280	1784	Dry wt
	–	12.2	280	1785	–
Tryptophan-rich	–	13.2	280	1786	–
Thyroxine-binding	–	12.3	280	1787	–
	9.93	13.6	280	1788	MW = 73,000 (1788)
Cynomolgus monkey					
(*Macaca irus*)	8.71	15	280	1789	MW = 58,000 (1789)
Rhesus monkey					
Pt-1-1	–	14.4	280	1790	–
Pt-2-2	–	14.4	280	1790	–
Chicken egg yolk	–	18.5	280	1791	pH 7.0, 0.05 M NaP$_i$
Mouse					
Urinary	–	6.0	280	422	–
Serum	–	8.0	280	422	–
Pre-phenoloxidase					
Bombyx mori	–	13.0	280	1792	–
	0.029	–	650	1792	Per atom copper
Principle, sweet					
Dioscoreophyllum cumminsii	–	16.2	278	1793	pH 5.6
	–	17.6	288	1793	pH 13
Procarboxypeptidase					
A					
Cobalt derivative	0.0110	–	500	1794	–
	0.0140	–	555	1794	–
	0.0140	–	570	1794	–
Shrimp					
(*Penaeus setiferus*)	–	25.8	280	1795	AA
Spiny Pacific dogfish					
(*Squalus acanthias*)	–	16.5	280	1796	pH 8.0, 0.1 M Tris, 0.01 M CoCl$_2$, Refr.

MOLAR ABSORPTIVITY AND $A_{1cm}^{1\%}$ VALUES FOR PROTEINS AT SELECTED WAVELENGTHS OF THE ULTRAVIOLET AND VISIBLE REGION (continued)

Protein	$\epsilon^a(\times 10^{-4})$	$A_{1cm}^{1\%}$ [b]	nm[c]	Ref.	Comments[d]
Procarboxypeptidase (continued)					
Beef pancreas	18.2	19	280	77	MW = 96,000 (77)
S_5	11.5	17.7	280	1797	Refr. MW = 63,000 (1791)
S_6	16.5	19	280	77	MW = 87,000 (77)
B					
Shrimp					
(*Penaeus setiferus*)	–	27.8	280	1795	UC
African lungfish					
(*Protopterus aethiopicus*)	7.3	16.2	280	1798	Refr. MW = 45,000 (1798)
Beef	9.2	16	280	83	MW = 57,400 (83)
Proelastase					
Pig pancreas	4.4	17.0	280	1799	MW = 25,840 (1799)
	–	15.8	280	1800	–
Progesterone-binding globulin					
Pregnant guinea pig serum	–	7.3	280	1801	–
Proinsulin					
Beef	–	7.0	276	1802	–
	–	5.9	276	1803	pH 7.2, Ph, 0.1 N NaCl, 10^{-5} M EDTA
Pig	–	6.67	276	423	pH 7.0, 0.03 M P$_i$
Prolactin					
Sheep	2.05	9.09[k‡]	280	1804	–
Proline hydroxylase (EC 1.14.11.2), collagen (see collagen)					
Protease (proteinase)					
Chinese gooseberry					
(*Actinidia chinensis*)		21.2	280	1809	–
Acremonium Kiliense	2.73	10.1	280	1826	1mM HCl, dry wt, MW = 27,000
Agkistrodonhalys blomhoffi					
a	4.54	9.08	280	424	Water, MW = 50,000 (424)
b	7.0	7.4	280	425	MW = 95,000 (425)
c	7.7	10.98	280	424	Water, MW = 70,000 (424)
Alternaria tenuissima	3.2	13	277	426	0.1 M AcONa, MW = 24,750 (426)
Aspergillus candidus		7.1	280	1827	Dry wt
Aspergillus flavus	1.61	9.04	280	1828	pH 5, MW = 17,800 (1828)
Aspergillus sojae	2.03	8.98	280	1829	MW = 22,600 (1829)
I	6.96	16.7	280	1810	pH 7.3, 50 mM TrisCl, dry wt, MW = 41,700 (1810)
II	1.78	9.0	280	1810	pH 7.3, 50 mM TrisCl, dry wt, MW = 19,800 (1810)

[k‡]Personal communication from Dr. Li. The value cited in Reference 1804 is in error. The correct value is 9.09.

MOLAR ABSORPTIVITY AND $A_{1cm}^{1\%}$ VALUES FOR PROTEINS AT SELECTED WAVELENGTHS OF THE ULTRAVIOLET AND VISIBLE REGION (continued)

Protein	$\epsilon^a (\times 10^{-4})$	$A_{1cm}^{1\%}$ [b]	nm[c]	Ref.	Comments[d]
Protease (continued)					
Bacillus natto	2.38	8.8	280	1811	—
B. subtilis	2.7	10	278	427	MW = 27,000 (427)
Neutral	—	13.6	280	428	—
B. subtilis NRRL B3411					
A	—	14.8	280	1825	—
B	—	14.7	280	1825	—
B. subtilis var. *amylosac-chariticus*	—	13.6–8	280	1812	—
B. thermoproteolyticus	6.63	17.65	280	429	pH 7.0, 0.05 M Tris, MW = 37,500 (429)
	7.34	19.6	280	429	0.1 N NaOH
Lotus seed (*Nelumbo nucifera Gaertn*)	—	10	278	1813	pH 4.0, 50 mM AcO$^-$
Mouse, submaxillary					
A	7.5	24.9	280	1814	MW = 30,000 (1814)
B	7.25	25.9	280	1814	MW = 28,000 (1814)
Myxobacter, α-lytic	1.94		280	1815	
Myxobacter, AL-1	2.2	15.8	280	430	0.1 M P_i, MW = 14,000 (430)
Rhizopus chinensis	—	12.6	280	431	
S. fradiae	—	8.4	280	275	pH 8, 0.1 M Tris
Streptomyces griseus str. K		8.1	280	1820	Folin
Streptomyces naraensis	4.16	11.22	280	1824	pH 7.5, 0.005 M TrisCl, MW = 37,000 (1824)
Streptomyces rectus var. *proteolyticus*	3.92	18.2	280	1830	MW = 21,500 (1830)
S. griseus K1					
I	—	16.2	278	433	—
III	—	11.5	278	433	—
IV	—	10.2	278	433	—
S. maraensis, neutral	—	9.15	280	434	pH 7.5, 0.005 M Tris, HCl, 0.005 M Ca(OH)$_2$
Tricophyton granulosum extracellular	7.6	22.15	274	435	pH 6.0, 0.1 M AcO$^-$, MW = 34,300 (435)
Sorangium, α-lytic	1.94	—	280	1816	
	—	9.7	280	1817	—
Worm (*Schistosoma mansoni*)	2.42[l‡]	—	280	1818	pH 3.95
	3.75[l‡]	—	280	1818	pH 13
Vibrio B-30	—	10	280	1819	—
Streptococcal	—	16.4	280	1821	—
zymogen	—	13.7	280	1821	—
Yeast, A	—	11.9	280	1822	—
	—	11.9	280	436	pH 6.2, 0.01 M NaP$_i$
C	—	16.6	280	1822	
Yeast	—	16.6	280	436	pH 6.2, 0.01 M NaP$_i$
(*Saccharomyces cerevisiae*)	—	14.82	280	1823	pH 7.0, 0.01 M NaP$_i$, 0.1 M KCl, dry wt
Protease, acid					
Cladosporium sp. 45-2	—	10.7	280	1805	Dry wt
Rhizopus chimensis	—	12.1	280	1806	From Figure 1 (1806)
Rhodotorula glutinis K-24	—	12.9	280	1807	—
	—	14.0	280	1808	pH 4, 0.01 M citrate

[l‡]MW = 28,000.

MOLAR ABSORPTIVITY AND $A_{1cm}^{1\%}$ VALUES FOR PROTEINS AT SELECTED WAVELENGTHS OF THE ULTRAVIOLET AND VISIBLE REGION (continued)

Protein	$\epsilon^a(\times 10^{-4})$	$A_{1cm}^{1\%}$ [b]	nm[c]	Ref.	Comments[d]
Protease inhibitor, soybean	0.35	4.4	280	537	MW = 7975 (437)
Proteinase (see protease)					
Proteolipid Beef brain white matter					
Crude	–	7.6	278	1831	–
Purified	–	14	278	1831	–
Prothrombin Beef	–	13.4	280	454	pH 6. Est. from Figure 1 (454)
	–	10.8	280	455	0.1 *N* NaOH
	–	15.3	280	456	pH 6, P$_i$NaCl
	–	14.8	280	456	0.1 *N* NaOH
Prep. I	–	13.2	280	1832	–
Prep. II	–	14.4	280	1832	–
Human	–	15.2	280	457	–
	–	13.6	280	458	pH 6 and 7, 0.1 *M* P$_i$
A-DEAE purified	–	11.7	280	459	pH 7.4, 0.02 *M* Tris, 0.1 *M* NaCl
B-DEAE purified	–	12.6	280	459	pH 7.4, 0.02 *M* Tris, 0.1 *M* NaCl
C-Disc electrophoresis prep.	–	13.8	280	459	pH 7.4, 0.02 *M* Tris, 0.1 *M* NaCl
D,E-NIH-DEAE purified	–	13.3	280	459	pH 7.4, 0.02 *M* Tris, 0.1 *M* NaCl
F-NIH-DEAE purified	–	11.6	280	459	pH 7.4, 0.02 *M* Tris, 0.1 *M* NaCl
G-NIH=Disc electrophoresis prep.	–	12.9	280	459	pH 7.4, 0.02 *M* Tris, 0.1 *M* NaCl
Protocatechuate 3,4-dioxygenase (EC 1.13.11.3) *Pseudomonas aeruginosa*	–	13.2	280	1833	pH 8.5
	92.4	13.2	280	460	pH 8.5, 50 m*M* TrisCl, MW = 700,000 (460)
Protocatechuate 4,5-dioxygenase (EC 1.13.11.8) Pseudomonad	11.15	7.45	280	1834	pH 75., 0.05 *M* KP$_i$ + 10% EtOH, MW = 150,000 (1834), from Figure 3 (1834)
Protocollagen hydroxylase (proline, 2-oxoglutarate dioxygenase, EC 1.14.11.2) Chicken embryo	–	49.5	230	1835	–
Protoheme, P-450 particle Rabbit liver	9.6	–	414	441	Reduced
	2.15	–	543	441	Reduced
	13.8	–	415–8	441	Oxidized
	1.9	–	532	441	Oxidized
	1.85	–	567	441	Oxidized

MOLAR ABSORPTIVITY AND $A_{1cm}^{1\%}$ VALUES FOR PROTEINS AT SELECTED WAVELENGTHS OF THE ULTRAVIOLET AND VISIBLE REGION (continued)

Protein	$e^a(\times 10^{-4})$	$A_{1cm}^{1\%}$ [b]	nm^c	Ref.	Comments[d]
Putrescine oxidase (EC 1.4.3.10)					
Micrococcus rubens	1.08	–	458	1836	–
	11.45	–	275	1836	–
Apo-		10.0	280	1836	–
Pyocin R					
P. aeruginosa R	–	16.9	280	461	–
Sheath	–	18.1	280	1837	0.1 N NaOH – 0.1 M NaCl
S-β-Pyridylethylalbumin (see albumin, beef)					
Pyridoxamine-pyruvate transaminase (EC 2.6.1.30)					
Pseudomonas MA-1	–	9.75	280	462	–
Pyrocatechase (catechol 1,2-dioxygenase, EC 1.13.11.1)					
P. arvilla	0.47	0.52	440	463	pH 8.0, 0.05 M Tris, MW = 90,000 (463)
	8.04	8.93	280	463	pH 8.0, 0.05 M Tris, MW = 90,000 (463)
Pyrophosphatase, inorganic					
Yeast	–	14.5	280	464	0.1 M HCl or water
Pyruvate dehydrogenase (EC 1.2.4.1)					
E. coli K-12	30	10	276	747	pH 7, 0.05 M KP$_i$ MW 3 × 10⁶ (747), data from Figure 3A (747)
	1.07	0.40	355	465	MW = 265,000 (465)
	1.13	0.43	370	465	MW = 265,000 (465)
	1.05	0.40	415	465	MW = 265,000 (465)
	1.46	0.55	438	465	MW = 265,000 (465)
	1.27	0.47	460	465	MW = 265,000 (465)
Pyruvate kinase					
Rabbit muscle	–	5.4	280	466	–
Human	–	5.4	280	466	–
Rat	–	5.4	280	466	–
Yeast	10.8	6.53	280	748	MW = 166,000 (748)
Quinolinate phosphoribosyltransferase (nicotinatemononucleotide pyrophosphorylase, EC 2.4.2.19)					
Pseudomonad	6.04	3.4	278	467	pH 7, 0.05 M KP$_i$, MW = 178,000 (467)
RNA polymerase (EC 2.7.7.6)					
E. coli	24.0	5.41	280	1135	MW = 440,000 (1135)
E. coli B	–	5.9	280	749	Con. by Biuret and Lowry (749)
	–	6.7	280	749	Con. det by refract. incr. with BSA as standard (749)
	–	6.5	280	749	–
	–	11.8	278	750	Calc. from data in Figure 1 (750)

MOLAR ABSORPTIVITY AND $A_{1cm}^{1\%}$ VALUES FOR PROTEINS AT SELECTED WAVELENGTHS OF THE ULTRAVIOLET AND VISIBLE REGION (continued)

Protein	$\epsilon^a (\times 10^{-4})$	$A_{1cm}^{1\%}$ [b]	nm[c]	Ref.	Comments[d]
Relaxing protein					
Rabbit muscle	–	3.32	278	1131	–
Rennin (chymosin, EC 3.4.23.4)	–	14.3	278	468	–
Retinol-binding protein					
Human serum	3.91	18.5	280	1132	MW = 21,000 (1132)
Retinol-transporting protein					
Human urine	–	18.7	280	1133	–
Rhodanese, (thiosulfate sulfur transferase, EC 2.8.1.1)					
Beef liver	–	17.5	280	469	–
Rhodopsin	4.2	–	498	470	pH 6.5 $M/15$ P_i, 1% Emulphogene BC720
	1.1	–	350	470	pH 6.5 $M/15$ P_i, 1% Emulphogene BC720
	7.4	–	280	470	pH 6.5 $M/15$ P_i, 1% Emulphogene BC720
	4.06	–	498	471	–
	–	9.8	278	471	–
Cattle	4.06	10.1	500	625	MW = 40,000 (626)
	8.12	20.2	278	625	Calc. from $\epsilon_{278}/\epsilon_{500} = 2.0$
	7.17	–	279	627 ⎫	In cetyltrimethylammonium bromide
	1.06	–	345	627 ⎬	
	3.97	–	498	627 ⎭	
Ribitol dehydrogenase					
Aerobacter aerogenes	–	11.1	280	628	pH 7.4, data from Figure (628)
Ribonuclease	1.19	–	278	472	0.2 M NaCl, 99.85% D_2O
Beef pancreatic	0.88	6.95	280	473	–
	–	6.9	280	474	–
	0.98	7.2	227.5	475	–
	–	7.2	280	476	–
	0.98	7.2	278	477	pH 6.5, MW = 13,700 (477)
	1.13	8.3	278	478	MW = 13,683 (478)
	1.06	–	279.5	479	Ethylene glycol
	1.14	–	278	480	Ethylene glycol
Corn	4.32	18.8	280	485	MW = 23,000 (485)
Ribonuclease I (EC 3.1.4.22)					
Beef pancreatic	0.91	7.14	277.5	481	–
	–	6.95	280	482	–
41-N$_2$ ph	–	11.2	280	488	–
	–	6.7	280	484	–
Sheep pancreas	–	7.1	280	483	–

MOLAR ABSORPTIVITY AND $A_{1cm}^{1\%}$ VALUES FOR PROTEINS AT SELECTED WAVELENGTHS OF THE ULTRAVIOLET AND VISIBLE REGION (continued)

Protein	$\epsilon^a(\times 10^{-4})$	$A_{1cm}^{1\%}$ [b]	nm[c]	Ref.	Comments[d]
Ribonuclease II (T$_2$) (EC 3.1.4.23)					
A. oryzae	7.16	19.9	281	487	MW = 36,000 (487)
Ribonuclease P					
Pig pancreas	0.41	3	280	486	MW = 13,500 (486)
Ribonuclease S	–	6.95	280	482	–
S Protein	–	7.84	280	482	–
Ribonucleotide-diphosphate reductase (EC 1.17.4.1) *E. coli B*					
Protein B-1	8.2	10.5	280	751 $\rbrace$	MW = 78,000 (751)
Protein B-2	11.5	14.8	280	751	
	0.4	–	410	752 $\rbrace$	G-200 prep.
	0.6	–	360	752	
	0.33	–	410	752 $\rbrace$	Gel electrophoresis prep.
	0.56	–	360	752	
Ribosephosphate isomerase (see 5-phosphoriboisomerase)					
L-Ribulokinase, *E. coli*	–	15.1	280	489	–
E. coli B/r	15.2	15.5	280	753	pH 7.6, MW = 98,000 (753)
Ribulosebisphosphate carboxylase (EC 4.1.1.39)					
Spinacea oleracea	–	16	280	490	–
Hydrogenomonas eutropha	80.0	15.51	280	1134	pH 8.0, 0.02 TrisSO$_4$ 0.1 M MgCl$_2$, MW = 515,000 (1134)
Hydrogenomonas facilis	62.6	12.28	280	1134	pH 8.0, 0.02 TrisSO$_4$, 0.01 M MgCl$_2$, MW = 551,000 (1134)
L-Ribulose phosphate 4-epimerase (EC 5.1.3.4)					
E. coli	16.2	15.7	280	491	pH 7.75, 10 mM NH$_4$HCO$_3$ MW = 103,000 (491)
L-Rhamnulose-1-phosphate (see aldolase)					
Rubredoxin					
Desulfovibrio desulfurican	1.342	–	280	492	Oxidized
	0.596	–	380	492	Oxidized
	0.512	–	490	492	Oxidized
	1.413	–	277	492	Reduced
	0.543	–	312	492	Reduced
	0.219	–	335	492	Reduced
Peptostreptococcus elsdenii	1.83	–	280	493	–
	0.819	–	350	493	–
	0.94	–	378	493	–
	0.763	–	390	493	–
	0.38	–	566	493	–

MOLAR ABSORPTIVITY AND $A_{1cm}^{1\%}$ VALUES FOR PROTEINS AT SELECTED WAVELENGTHS OF THE ULTRAVIOLET AND VISIBLE REGION (continued)

Protein	$\epsilon^a (\times 10^{-4})$	$A_{1cm}^{1\%}$ [b]	nm[c]	Ref.	Comments[d]
Rubredoxin (continued)					
M. lactilyticus	2.2	36.7	280	494	MW = 6,000 (494)
	0.91	15.3	490	494	--
M. aerogenes	1.83	30.5	280	495	MW = 6,000 (495)
	0.84	14	350	495	MW = 6,000 (495)
	0.92	15.3	378	495	MW = 6,000 (495)
	0.765	12.8	490	495	MW = 6,000 (495)
	0.35	5.8	570	495	MW = 6,000 (495)
P. oleovarans	1.11	8.7	495	496	MW = 12,800 (496)
	1.08	8.4	280	496	MW = 12,800 (496)
Rubredoxin, S-aminoethyl					
M. aerogenes	0.85	14.1	280	495	MW = 6,000 (495)
Rubredoxin, Apo					
C. pasteurianum	1.85	30.8	280	497	MW = 6,000 (497)
Rubredoxin reductase (EC 1.6.7.2)					
Pseudomonas oleovorans	7.3	—	272	1402	—
	1.0	—	378	1402	—
	1.1	—	450	1402	—
Serine dehydratase (EC 4.2.1.13, .14, .16)					
Rat liver	6.66	10.4	280	629 ⎫	pH 7.2, 0.025 M KP$_i$,
					0.001 M EDTA,
					0.001 S$_2$ threitol,
Aposerine dehydratase	6.02	9.4	280	629 ⎭	data from Figure 14 (629)
D-Serine dehydratase (EC 4.2.1.14)					
E. coli K 12	5.33	14.2	280	1136⎱	
	0.533	1.42	415	1136⎰	MW = 37,300 (1136)
E. coli	4.78	10.5	280	1137	pH 6.5–7.8, 0.1 M KP$_i$,
					dry wt/R,
					MW = 45,500 (1137)
	4.11	9.02	280	1137	pH 6.5, 0.4 M imidazole-
					citrate,
					MW = 45,500 (1137)
	4.35	9.56	280	1137	0.1 N NaOH,
					MW = 45,500 (1137)
Apo-	—	9.78	280	1137	pH 7.8, 0.1 M KP$_i$
	—	9.56	280	1137	0.1 N NaOH
Serine hydroxymethylase					
Rabbit liver, soluble	—	9	278	630⎱	pH 7.1, 0.05 M KP$_i$,
Mitochondrial	—	7	278	630⎰	data from Figure 4 (630)
Siderophilin					
Human	—	14	280	7	See transferrin
	—	10.9	280	8	—
	—	11.2	280	308	—
Pig	—	11.0	278	498	0.1 N HCl
	—	13.8	280	498	pH 7
	—	13.3	290	498	0.1 N NaOH

MOLAR ABSORPTIVITY AND $A_{1cm}^{1\%}$ VALUES FOR PROTEINS AT SELECTED WAVELENGTHS OF THE ULTRAVIOLET AND VISIBLE REGION (continued)

Protein	$\epsilon^a(\times 10^{-4})$	$A_{1cm}^{1\%}$ [b]	nm[c]	Ref.	Comments[d]
Spectrin					
Human erythrocytes	–	8.8	280	631	–
Stellacyanin					
Rhus vernicifera	2.6	–	280	499	0.1 N NaOH
	2.8	–	290	499	0.1 N NaOH
Oxidized	2.32	–	280	717	–
	0.096	–	450	717	–
	0.48	–	604	717	–
	0.079	–	850	717	–
Reduced	0.088	–	450	717	–
	0.408	–	604	717	–
	0.079	–	850	717	–
Steroid 4,5-dioxygenase (EC 1.13.11.25) (see 3,4-dihydroxy-9,10-secoandrosta-1,3,5[10]-triene-9,17-dione 4,5-dioxygenase)					
Streptavidin					
S. avidinii	–	34	280	500	–
	–	31	282	501	–
Streptokinase, Streptococcus	–	9.49	280	502	–
Subtilisin (EC 3.4.21.14)	–	8.6	280	357	–
Subtilisin, Thiol-	3.31	–	278	506	–
Subtilisin BPN'	–	8.8	280	754	–
Subtilisin BPN', iPr$_2 P$[bb]	3.23	11.7	278	504	pH 7.0, 0.05 M AcONa, MW = 27,600 (504)
Subtilisin Carlsberg, iPr$_2 P$[bb]	–	8.6	278.1	505	–
Subtilisin, Nova	3.11	–	278	506	–
Subtilopeptidase (subtilisin, EC 3.4.21.14)	–	10.7	280	507	–
Succinyl-Co A synthetase (EC 6.2.1.4, .5)					
E. coli	7.2	5.11	280	508	pH 7.2, 1 mM KP$_i$, MW = 141,000 (508)
Sulfate adenylyltransferase (see ATP-sulfurylase)					
Sulphatase, (aryl sulfatase, EC 3.1.6.1)					
Beef liver					
A	7.5	7.0	280	509	pH 7.5, MW = 107,000 (509)
B	3.5	14.0	280	509	pH 7.5, MW = 25,000 (509)

MOLAR ABSORPTIVITY AND $A_{1cm}^{1\%}$ VALUES FOR PROTEINS AT SELECTED WAVELENGTHS OF THE ULTRAVIOLET AND VISIBLE REGION (continued)

Protein	$\epsilon^a (\times 10^{-4})$	$A_{1cm}^{1\%}$ [b]	nm[c]	Ref.	Comments[d]
Sulphatase (continued)					
$B_{\alpha 2}$	—	20	280	509	Protein det. by Folin Method
	—	13.3	280	509	Refractometric method used to det. protein
B_β	—	19.9	280	509	Protein det. by Folin method
	—	13.8	280	509	Refractometric method used to det. protein
Tartrate dehydrogenase (EC 1.1.1.93)					
Pseudomonas putida	20.9	14.4	280	510	MW = 145,000 (510)
Tartrate epoxidase (*trans*-epoxysuccinate hydratase, EC 4.2.1.37)					
P. putida	—	14.0	280	632	—
Tartronic semialdehyde reductase (EC 1.1:1.60)					
P. putida	7.1	6.83	280	511	MW = 104,000 (511)
Taurocyamine kinase					
Arenicola marinae	—	9.7	280	607	—
Tetanus toxin					
C. tetani	—	7.8	280	512	—
Thermolysin (EC 3.4.24.4)					
B. thermoproteolyticus Rokko	6.63	17.6	280	1138	MW = 37,500 (1138)
Thioredoxin					
Bacteriophage T4	6.3	6.06[m‡]	280	1139	MW = 104,000 (1139)
E. coli	1.37	11.4	280	513	MW = 12,000 (513)
Thioredoxin reductase (EC 1.6.4.5)					
E. coli	0.33	14.0	280	514	MW = 66,000 (515)
Threonine aldolase (EC 4.1.2.5)					
Candida humicola	11.6	4.17	280	755	pH 6.4, 0.03 M KP$_i$, 0.005 M HSEtOH, 0.001 M EDTA, MW = 277,000 (755), calc. from Figure 2 (755)
Threonine deaminase (threonine dehydratase, EC 4.2.1.1.6)					
S. typhimurium	18.0	9.3	278	516	pH 7.4, 0.05 M KPh, 0.8 mM L-isoleucine, 0.5 mM EDTA 0.5 mM S$_2$ threitol
E. coli	2.6	1.75	415	517	MW = 147,000 (517)
	8.1	5.5	277	517	MW = 147,000 (517)
Rhodospirillum rubrum	—	3.82	278	1140	pH 6.8, 25 mM KP$_i$,
	—	1.31[n‡]	412	1140	data from Figure 7 (1140)

[m‡] OD = 1 when con. = 1.65 mg/ml.

[n‡] OD$_{412}$ = OD$_{278}$/2.9.

MOLAR ABSORPTIVITY AND $A_{1cm}^{1\%}$ VALUES FOR PROTEINS AT SELECTED WAVELENGTHS OF THE ULTRAVIOLET AND VISIBLE REGION (continued)

Protein	$\epsilon^a (\times 10^{-4})$	$A_{1cm}^{1\%}$ [b]	nm[c]	Ref.	Comments[d]
Thrombin (EC 3.4.21.5)					
Human	–	16.2	280	518	–
Beef	–	19.5	280	519	–
Thyrocalcitonin	0.76	21	280	520	MW = 3,604 (520)
Pig	0.757	21	280	756	0.1 M AcOH
Thyroid stimulating hormone					
Beef	2.6	10.5	292	521	0.1 N NaOH, est. from Figure 1 (521), MW = 25,000 (521)
Human	2.5	9.9	292	521	0.1 N NaOH, est. from Figure 1 (521), MW = 25,000 (521)
Thyroglobulin, beef					
19S	–	10.0	280	503	pH 7.4, KCl-P$_i$
19S	–	10.5	280	96	–
19S	65	10	280	757 }	MW = 650,000
	1310	201	210	757 }	
27S	132	10.8	280	757 }	MW = 1,220,000
	2700	221	210	757 }	
27S Iodoprotein	–	10.8	280	503	pH 7.4, KCl-P$_i$
Human	–	10.5	280	360	–
Lamprey, 12S	–	8.8	280	372	–
Transaldolase (EC 2.2.1.2), *C. utilis*	–	11	280	522	0.1 N NaOH
Transcortin, Human plasma	–	7.4	280	523	–
Transferrin, Human plasma	–	14.1	280	524	Iron saturated
	–	11.2	280	525	–
Human apoenzyme	–	11.4	280	524	–
Pig	–	14.1	280	526	Iron saturated
	–	0.6	470	526	–
Apoenzyme	–	11.4	280	526	–
Transglutaminase					
Pig heart	–	15.8	280	527	–
Guinea pig liver	14.2	15.8	280	528	MW = 90,000 (528)
Tripeptide synthetase (glutathione synthetase, EC 6.3.2.3)					
Yeast	18.5	15	280	529	MW = 123,000 (529)
Tropomyosin B					
Rabbit muscle	–	3.3	277	530	pH 7.0, ionic strength 1.1
	–	3.1	276	530	8 M urea
Troponin					
Rabbit muscle	–	42	260	758	pH 7.5, 2 mM TrisCl, data from Figure 1 (758)
Trypsin (see also cocoonase) (EC 3.4.21.4)					
Beef pancreas	0.154	–	280	531	–
	–	16.6	280	532	...

MOLAR ABSORPTIVITY AND $A_{1cm}^{1\%}$ VALUES FOR PROTEINS AT SELECTED WAVELENGTHS OF THE ULTRAVIOLET AND VISIBLE REGION (continued)

Protein	$\epsilon^a(\times 10^{-4})$	$A_{1cm}^{1\%}$ [b]	nm[c]	Ref.	Comments[d]
Trypsin (continued)					
Beef pancreas (continued)	--	15.6	280	533	–
	--	15.0	280	534	–
	--	17.1	280	535	Acid sol.
	–	14.4	280	536	–
	–	16	–[k†]	537	1 mM HCl
	–	12.9	280	538	–
	–	17.24	280	539	–
	–	15.5	280	540	–
Acetyltrypsin B	–	14.4	280	541	pH 7.5
N_2ph-Trypsin	–	14.9	280	535	Acid sol.
Sheep pancreas	–	17.4	280	540	–
Pig pancreas	–	15.0	280	540	–
Trypsin inhibitor					
Inter-α, human	–	7.1	280	542	–
Beef pancreas	–	7.9	280	535	Acid sol.
	–	8.2	280	535	Neutral sol.
	–	8.25	280	543	Acid sol.
	–	8.35	280	543	Neutral sol.
	0.38	6.2	276.1	544	MW = 6,155 (544)
	0.36	5.9	280	544	MW = 6,155 (544)
Kazals	–	6.5	280	535	Neutral sol.
Pig pancreas					
I	–	5.18	280	545	pH 7.8
II	–	6.06	280	545	pH 7.8
Soybean	–	9.1	280	535	Acid sol.
	–	9.54	280	535	Neutral sol.
	–	10.5	280	546	–
	–	4.8	280	547	–
A_2	–	9.94	280	548	–
I	–	9.44	280	549	–
F_1	–	7.16	280	550	–
F_2	–	10.4	280	550	–
F_3	–	6.34	280	550	–
Colostrum, beef	–	5.0	280	535	Acid sol.
Barley	1.78	12.7	280	759	MW = 14,000 (759)
Trypsin-Trypsin inhibitor complex					
Pancreatic inhibitor, beef	–	12.3	280	535	Acid sol.
Soybean inhibitor	–	13.1	280	535	–
Colostrum inhibitor, beef	–	11.9	280	535	Acid sol.
Trypsinogen					
Beef pancreas	–	13.9	280	536	–
Pig pancreas	–	13.9	280	536	–
Sheep pancreas	–	14.1	280	551	–
Beef pancreas, S-Sulfo-	–	14.2	280	552	–
Tryptophan oxygenase (EC 1.13.11.11)					
Pseudomonas acidovorans	14.6	12.0	280	760 }	MW = 121,000 (760)
	22.9	18.8	405	760 }	

MOLAR ABSORPTIVITY AND $A_{1cm}^{1\%}$ VALUES FOR PROTEINS AT SELECTED WAVELENGTHS OF THE ULTRAVIOLET AND VISIBLE REGION (continued)

Protein	$\epsilon^a(\times 10^{-4})$	$A_{1cm}^{1\%}$ [b]	nm[c]	Ref.	Comments[d]
Tryptophan synthase (EC 4.2.1.20) *E. coli*					
A protein	1.37	4.05	278	761	pH 7.2, 0.01 M KP$_i$ MW = 29,500 (761)
	1.4	4.75	293	761	0.1 M NaOH, MW = 29,500 (761)
B protein	–	6.5	280	762	pH 7.3, 0.05 M KPh conc. 1γ/ml, pyridoxal-P and 0.001 M HSEtOH
	–	1.11	300	762 ⎫	
	–	0.71	335	762 ⎬ Calc. from ratios of ODs	
	–	1.14	414	762 ⎬	
	–	3.2	290	762 ⎭	
	6.2	–	278	763	pH 7.5, 0.01 M KP$_i$, 0.01 M HSEtOH
Apo B protein	–	5.8	280	762	
B component	–	5.7	278	554	
Tryptophanase (EC 4.1.99.1) *E. coli*					
Apo-	1.74	7.95	278	553	pH 7.5, 0.1 M KP$_i$ 2 mM, EDTA, 2 mM mercapto-ethanol, MW = 22,000 (553)
Reduced holoenzyme	0.34	1.5	336	553	MW = 22,000 (553)
	1.79	8.14	277	553	MW = 22,000 (553)
Tryptophanyl-tRNA synthetase (EC 6.1.1.2)					
Beef pancreas	9	8.4	280	764	pH 7.5, 0.2 M KCl, MW = 108,000 (764)
	–	8.0	280	764	8 M urea, pH 8
Tyramine oxidase (amine oxidase, EC 1.4.3.4) *Sarcina lutea*	34.8	27	280	555	MW = 129,000 (555)
β-Tyrosinase (tyrosine phenol-lyase) (EC 4.1.99.2) *E. intermedia*	–	83.7	280	556	–
Tyrosine aminotransferase (EC 2.6.1.5) Rat liver	9.1	10	277	557	pH 7.0, 0.1 M P$_i$, MW = 91,000 (557) est. from Figure 6 (557)
UDPG dehydrogenase (EC 1.1.1.22) Beef liver	–	9.8	277	633	–
Umecyanin *Armoacia lapathiofolia* (Horseradish root)	0.34	2.3	610	765 ⎫	
	0.012	0.083	400	765 ⎬ pH 5.75, 30 mM AcONa	
	0.021	0.144	330	765 ⎬ MW = 14,600 (765)	
	1.27	8.7	280	765 ⎭	

MOLAR ABSORPTIVITY AND $A_{1cm}^{1\%}$ VALUES FOR PROTEINS AT SELECTED WAVELENGTHS OF THE ULTRAVIOLET AND VISIBLE REGION (continued)

Protein	$\epsilon^a(\times 10^{-4})$	$A_{1cm}^{1\%}$ [b]	nm[c]	Ref.	Comments[d]
Urease (EC 3.5.1.5)					
Beans	–	5.5	280	766	
Cajanus indicus	–	20.1	280	767 }	pH 7.0, 0.05 M Tris-AcO$^-$
	–	20.6	277	767 }	
Jack bean	28.4	5.89	280	768	MW = 483,000 (768)
Carboxymethyl-	–	7.2	–[o‡]	769	–
Jack bean meal	37	7.7	272	558	MW = 480,000 (558)
	–	7.71	272	559	–
	–	7.54	278	560	pH 7, 0.02 M P_i
	–	6.4	278	560	pH 7, 0.02 M P_i, HSEtOH added
Uricase, pig liver	–	11.3	276	561	1% sodium carbonate
Urocanase (EC 4.2.1.99)					
Pseudomonas putida	–	8.3	280	1141	pH 7.5, 0.2 M KP_i data from Figure 6 (1141) and corrected for scattering
	–	8.03	280	770	Calc. from data in Figure 4 (770), corrected for scattering of light
Urokinase (EC 3.4.99.26)					
Human urine, S_1	4.23	13.2	280	1142	pH 6.5, MW=32,000 (1142)
S_2	7.48	13.6	280	1142	pH 6.5, MW=55,000 (1142)
Human placenta	–	10.7	280	771	–
Virus					
Barley stripe mosaic	–	34	240	562	–
	–	26	260	562	–
	–	25	280	562	–
	–	17	280	449	–
	–	28	260	449	–
	–	17	280	1143	Dry wt, Lowry, and Biuret
Broad bean mottle	–	3.2	276.5	1144	Water
	–	5.5	292	1144	0.1 N NaOH, data from Figure 1 (1144)
Brome mosaic	–	36	240	562	–
	–	48	260	562	–
	–	31	280	562	–
Bromegrass mosaic	–	4.6[p‡]	260	1145	–
	–	7.6[p‡]	280	1145	–
	2.0	5.0	260	1145 }	pH 6.0, 1 M CaCl$_2$ –0.05 M Nacacodylate MW=40,000 (1145),
	3.52	8.8	280	1145 }	data from Figure 1 (1145)
Foot-and-mouth disease	–	11.1	276	1146	–
Mouse-Elberfeld (ME)	–	14.9	280	1147	0.002 M AcOH
Southern bean mosaic	–	48	240	562	–
	–	58	260	562	–
	–	37	280	562	–
	–	5.85	260	1148 }	0.1 N NaOH
	–	12[p‡]	260	1148 }	

[o‡] Wavelength not cited; may be 277 nm.

[p‡] Calculated from the content of tyrosine, tryptophan, and phenylalanine.

MOLAR ABSORPTIVITY AND $A_{1cm}^{1\%}$ VALUES FOR PROTEINS AT SELECTED WAVELENGTHS OF THE ULTRAVIOLET AND VISIBLE REGION (continued)

Protein	$\epsilon^a(\times 10^{-4})$	$A_{1cm}^{1\%}$ [b]	nm[c]	Ref.	Comments[d]
Virus (continued)					
Tobacco etch	–	9.5	280	1149	pH 6.5, 0.02 M P$_i$, 6 M GdmCl
Tobacco mosaic	–	57	240	562	–
	–	32.4	260	562	–
	–	27	280	562	–
	–	27	260	563	pH 7.5, 0.033 M P$_i$
	–	26	263	774	
	–	27	260	1150	
	–	13	281	1150	
	–	13	281	774	pH 7.1, 0.033 M NaP$_i$
	–	13	281	563	pH 7.5, 0.033 M P$_i$
PM 2					
Non-functioning	–	13.7	280	450	–
Tomato ringspot	–	10.3	260	1151	–
Turnip yellow mosaic					
Artificial top component	–	11.8	275	1152	pH 7.0, 0.01 M NaP$_i$
White clover mosaic, str.					
WCD-17	1.88	13.2	280	1153	Water, MW=14,300 (1153)
Mengo					
L-	–	17.1	280	772	–
M-	–	16.8	280	772	–
S-	–	17.0	280	772	–
Alfalfa mosaic	–	52	260	773	–
Potato X	–	12.3	280	775	pH 7.5, 0.05 M P$_i$
Visual pigment, beef	3.7	13.7	280	564	MW=27,000 (564)
	2.3	8.5	500	564	MW=27,000 (564)
Vitellogenin					
Xenopus laevis	–	7.5	280	1154	Dry wt
Wool, helix-rich fraction	–	5.9–6.5	277	776	–
Xanthine oxidase (EC 1.2.3.2)					
milk	–	2.3	450	565	pH 7.8, 0.05 M P$_i$
	–	11.5	280	565	–
	20.4	11.26	280	566	pH 8.0, 0.02 M P$_i$, MW=181,000 (566)
	2.2	–	550	27	–
	–	2.41	450	1155	–
High mol. wt. fract.	–	0.87	450	1155	–

Compiled with the assistance of Waldo E. Cohn, and material supplied by Donald M. Kirschenbaum.

REFERENCES

1. Kremzner and Wilson, *Biochemistry*, 3, 1902 (1964).
2. Leuzinger, Baker, and Cauvin, *Proc. Natl. Acad. Sci. U.S.A.*, 59, 620 (1968).
3. Nanninga, *Biochim. Biophys. Acta*, 82, 507 (1964).
4. Eisenberg and Moos, *J. Biol. Chem.*, 242, 2945 (1967).
5. Rees and Young, *J. Biol. Chem.*, 242, 4449 (1967).
6. Pfrogner, *Arch. Biochem. Biophys.*, 119, 147 (1967).
7. Tombs, Souter, and Maclagan, *Biochem. J.*, 73, 167 (1959).
8. Schonenberger, *Z. Naturforsch.*, 10b, 474 (1955).
9. Hunter and McDuffie, *J. Am. Chem. Soc.*, 81, 1400 (1959).
10. Phelps and Putnam, in *The Plasma Proteins*, Putnam, Ed., Academic Press, New York, 1960, 143.
11. Wetlaufer, *Adv. Protein Chem.*, 17, 378 (1962).
12. Van Kley and Stahmann, *J. Am. Chem. Soc.*, 81, 4374 (1959).
13. Everett, *J. Biol. Chem.*, 238, 2676 (1963).
14. Tanford and Roberts, *J. Am. Chem. Soc.*, 74, 2509 (1952).
15. Kolthoff, Shore, Tan, and Matsuoka, *Anal. Biochem.*, 12, 497 (1965).
16. Foster and Yang, *J. Am. Chem. Soc.*, 76, 1015 (1954).
17. Weber, in *The Biochemists Handbook*, Long, Ed., E. and F. N. Spon, Ltd., 1961, 82.
18. Bonnichsen, *Acta Chem. Scand.*, 4, 715 (1950).
19. Bonnichsen and Brink, *Methods Enzymol.*, 1, 495 (1955).
20. Theorell, Taniguchi, Akeson, and Skursky, *Biochem. Biophys. Res. Commun.*, 24, 603 (1966).
21. Sund and Theorell, *Enzymology*, 7, 25 (1963).
22. Rosenberg, Theorell, and Yonetani, *Arch. Biochem. Biophys.*, 110, 413 (1965).
23. Mourad and Woronick, *Arch. Biochem. Biophys.*, 121, 431 (1967).
24. Ohta and Ogura, *J. Biochem.* (Tokyo), 58, 73 (1965).
25. Hayes and Velick, *J. Biol. Chem.*, 207, 225 (1954).
26. Steinman and Jakoby, *J. Biol. Chem.*, 242, 5019 (1967).
27. Rajagopalan and Handler, *J. Biol. Chem.*, 239, 1509 (1964).
28. Baranowski and Niederland, *J. Biol. Chem.*, 180, 543 (1949).
29. Taylor, Green, and Cori, *J. Biol. Chem.*, 173, 591 (1948).
30. Donovan, *Biochemistry*, 3, 67 (1964).
31. Stellwagen and Schachman, *Biochemistry*, 1, 1056 (1962).
32. Sia and Horecker, *Arch. Biochem. Biophys.*, 123, 186 (1968).
33. Rajkumar, Woodfin, and Rutter, *Methods Enzymol.*, 9, 491 (1966).
34. Fluri, Ramasarma, and Horecker, *Eur. J. Biochem.*, 1, 117 (1967).
35. Rutter, Hunsley, Groves, Calder, Rajkumar, and Woodfin, *Methods Enzymol.*, 9, 479 (1966).
36. Massey, Palmer, and Bennett, *Biochim. Biophys. Acta*, 48, 1 (1961).
37. Antonini, Brunori, Bruzzesi, Chiancone, and Massey, *J. Biol. Chem.*, 241, 2358 (1960).
38. Yagi, Naoi, Harada, Okamura, Hidaka, Ozawa, and Kotaki, *J. Biochem.* (Tokyo), 61, 580 (1967).
39. Kotaki, Harada, and Yagi, *J. Biochem.* (Tokyo), 61, 598 (1967).
40. Miyake, Aki, Hashimoto, and Yamano, *Biochim. Biophys. Acta*, 105, 86 (1965).
41. Wellner and Meister, *J. Biol. Chem.*, 235, 2013 (1960).
42. Nakano and Danowski, *J. Biol. Chem.*, 241, 2075 (1966).
43. Wachsmuth, Fritze, and Pfleiderer, *Biochemistry*, 5, 169 (1966).
44. Hanson, Hutter, Mannsfeldt, Kretschmer, and Sohr, *Hoppe-Seyler's Z. Physiol. Chem.*, 348, 680 (1967).
45. Fischer and Stein, in *The Enzymes*, Vol. 4, 2nd ed., Bover, Lardy, and Myrbäck, Eds., Academic Press, New York, 1960, 319.
46. Caldwell, Adams, Kung, and Toralballa, *J. Am. Chem. Soc.*, 74, 4033 (1952).
47. Caldwell, Dickey, Hanrahan, Kung, Kung, and Misko, *J. Am. Chem. Soc.*, 76, 143 (1954).
48. DePinto and Campbell, *Biochemistry*, 7, 114 (1968).
49. Englard and Singer, *J. Biol. Chem.*, 187, 213 (1950).
50. Tombs and Lowe, *Biochem. J.*, 105, 181 (1967).
51. Schmike, *J. Biol. Chem.*, 239, 3808 (1964).
52. Wilson and Meister, *Biochemistry*, 5, 1166 (1966).
53. Bertland and Kaplan, *Biochemistry*, 7, 134 (1968).
54. Gerhart and Schachman, *Biochemistry*, 4, 1054 (1965).
55. Dratz and Calvin, *Nature*, 211, 497 (1966).
56. Gerhart and Holoubek, *J. Biol. Chem.*, 242, 2886 (1967).
57. Gerhart and Schachman, *Biochemistry*, 7, 538 (1968).
58. Ichishima and Yoshida, *Biochim. Biophys. Acta*, 110, 155 (1965).
59. Subramanian and Kalnitsky, *Biochemistry*, 3, 1868 (1964).

60. Hinkson and Bulen, *J. Biol. Chem.,* 242, 3345 (1967).
61. Thornber and Molson, *Biochemistry,* 7, 2242 (1968).
62. Murachi and Yasui, *Biochemistry,* 4, 2275 (1965).
63. Murachi, Inagami, and Yasui, *Biochemistry,* 4, 2815 (1965).
64. Murachi, Yasui, and Yasuda, *Biochemistry,* 3, 48 (1964).
65. Rickli, Ghazanfar, Gibbons, and Edsall, *J. Biol. Chem.,* 239, 1065 (1964).
66. Lindskog, *Biochim. Biophys. Acta,* 39, 218 (1960).
67. Duff and Coleman, *Biochemistry,* 5, 2009 (1966).
68. Edsall, Mehta, Meyers, and Armstrong, *Biochem. Z.,* 345, 9 (1966).
69. Horgan, Webb, and Zerner, *Biochem. Biophys. Res. Commun.,* 23, 23, (1966).
70. Vallee, Rupley, Coombs, and Neurath, *J. Biol. Chem.,* 235, 64 (1960).
71. Bethune, Ulmer, and Vallee, *Biochemistry,* 6, 1955 (1967).
72. Simpson, Riordan, and Vallee, *Biochemistry,* 2, 616 (1963).
73. Riordan and Vallee, *Biochemistry,* 2, 1460 (1963).
74. McClure, Neurath, and Walsh, *Biochemistry,* 3, 1897 (1964).
75. Smith and Stockell, *J. Biol. Chem.,* 207, 501 (1954).
76. Blostein and Rutter, *J. Biol. Chem.,* 238, 3280 (1963).
77. Keller, Cohen, and Neurath, *J. Biol. Chem.,* 223, 457 (1956).
78. Neurath, *Methods Enzymol.,* 1, 77 (1955).
79. Folk and Schirmer, *J. Biol. Chem.,* 238, 3884 (1963).
80. Bargetzi, Sampathkumar, Cox, Walsh, and Neurath, *Biochemistry,* 2, 1468 (1963).
81. Freisheim, Walsh, and Neurath, *Biochemistry,* 6, 3010 (1967).
82. Folk, Piez, Carroll, and Gladner, *J. Biol. Chem.,* 235, 2272 (1960).
83. Cox, Wintersberger, and Neurath, *Biochemistry,* 1, 1078 (1962).
84. Herskovits, *Biochemistry,* 5, 1018 (1966).
85. McKenzie, *Adv. Protein Chem.,* 22, 55 (1967).
86. Herskovits, *J. Biol. Chem.,* 240, 628 (1965).
87. Thompson and Kiddy, *J. Dairy Sci.,* 47, 626 (1964).
88. Thompson and Pepper, *J. Dairy Sci.,* 47, 633 (1964).
89. Zittle and Custer, *J. Dairy Sci.,* 46, 1183 (1963).
90. Bonnichsen, *Methods Enzymol.,* 2, 781 (1955).
91. Stansell and Deutsch, *J. Biol. Chem.,* 240, 4299 (1965).
92. Hiraga, Anan, and Abe, *J. Biochem.* (Tokyo), 56, 416 (1964).
93. Samejima and Yang, *J. Biol. Chem.,* 238, 3256 (1963).
94. Gregory, *Biochim. Biophys. Acta,* 159, 429 (1968).
95. Iwasaki, Hayashi, and Funatsu, *J. Biochem.* (Tokyo), 55, 209 (1964).
96. Edelhoch, *J. Biol. Chem.,* 235, 1326 (1960).
97. Kasper and Deutsch, *J. Biol. Chem.,* 238, 2325 (1963).
98. Markowitz, Cartwright, and Wintrobe, *J. Biol. Chem.,* 234, 40 (1959).
99. Schwick and Heide, in *Protides of the Biological Fluids,* Vol. 14, Peeters, Ed., Elsevier, Amsterdam, 1967, 55.
100. Sgouris, Coryell, Gallick, Storey, McCall, and Anderson, *Vox Sang.,* 7, 394 (1962).
101. Morell, Irvine, Sternlieb, Scheinberg, and Ashwell, *J. Biol. Chem.,* 243, 155 (1968).
102. Morris and Hager, *J. Biol. Chem.,* 241, 1763 (1966).
103. Tsunoda and Yasunobu, *J. Biol. Chem.,* 241, 4610 (1966).
104. Kunimitsu and Yasunobu, *Biochim. Biophys. Acta,* 139, 405 (1967).
105. Schwert and Kaufman, *J. Biol. Chem.,* 190, 807 (1951).
106. Narasinga, Rao, and Kegeles, *J. Am. Chem. Soc.,* 80, 5724 (1958).
107. Dixon and Neurath, *J. Biol. Chem.,* 225, 1049 (1957).
108. Morimoto and Kegeles, *Biochemistry,* 6, 3007 (1967).
109. Moon, Sturtevant, and Hess, *J. Biol. Chem.,* 240, 4204 (1965).
110. Laskowski, *Methods Enzymol.,* 2, 8 (1955).
111. Folk and Cole, *J. Biol. Chem.,* 240, 193 (1965).
112. Wootton and Hess, *J. Am. Chem. Soc.,* 84, 440 (1962).
113. Schwert, *J. Biol. Chem.,* 190, 799 (1951).
114. Wilcox, Cohen, and Tan, *J. Biol. Chem.,* 228, 999 (1957).
115. Guy, Gratecos, Rovery, and Desnuelle, *Biochim. Biophys. Acta,* 115, 404 (1966).
116. Charles, Gratecos, Rovery, and Desnuelle, *Biochim. Biophys. Acta,* 140, 395 (1967).
117. Smillie, Enenkel, and Kay, *J. Biol. Chem.,* 241, 2097 (1966).
118. Folk and Schirmer, *J. Biol. Chem.,* 240, 181 (1965).
119. Pechere, Dixon, Maybury, and Neurath, *J. Biol. Chem.,* 233, 1364 (1958).
120. Srere, *J. Biol. Chem.,* 241, 2157 (1966).
121. Herschman and Helinski, *J. Biol. Chem.,* 242, 5360 (1967).
122. Warner and Weber, *J. Am. Chem. Soc.,* 75, 5094 (1953).

123. Gotschlich and Edelman, *Proc. Natl. Acad. Sci. U.S.A.*, 54, 558 (1965).
124. Wood and McCarty, *J. Clin. Invest.*, 30, 616 (1951).
125. Noda, Kuby, and Lardy, *J. Biol. Chem.*, 209, 203 (1954).
126. Wisse, Zweers, Jongkind, Bont, and Bloemendal, *Biochem. J.*, 99, 179 (1966).
127. Bjork, *Exp. Eye Res.*, 7, 129 (1968).
128. Mason and Hines, *Invest. Ophthalmol.*, 5, 601 (1966).
129. Bjork, *Exp. Eye Res.*, 2, 339 (1963).
130. Bjork, *Exp. Eye Res.*, 3, 254 (1964).
131. Van Dam, *Exp. Eye Res.*, 5, 255 (1966).
132. Zigman and Lerman, *Biochim. Biophys. Acta*, 154, 423 (1968).
133. Yonetani, *J. Biol. Chem.*, 236, 1680 (1961).
134. Yonetani, *J. Biol. Chem.*, 242, 5008 (1967).
135. Paleus, *Arch. Biochem. Biophys.*, 96, 60 (1962).
136. Matsubara, Chu, and Yasunobu, *Arch. Biochem. Biophys.*, 101, 209 (1962).
137. Paul, *Acta Chem. Scand.*, 5, 389 (1951).
138. Flatmark, *Acta Chem. Scand.*, 18, 1517 (1964).
139. Flatmark, *Acta Chem. Scand.*, 20, 1476 (1966).
140. Margoliash and Frohwirt, *Biochem. J.*, 71, 570 (1959).
141. Jirgensons, *J. Biol. Chem.*, 240, 1064 (1965).
142. Linberg, *Biochemistry*, 6, 335 (1967).
143. Mondovi, Rotilio, Costa, Finazzi-Agro, Chiancone, Hansen, and Beinert, *J. Biol. Chem.*, 242, 1160 (1967).
144. Swislocki and Kaplan, *J. Biol. Chem.*, 242, 1083 (1967).
145. Raynaud, *Proc. 2nd Meet. Fed. Eur. Biochem. Soc.*, Vienna 1965, 1, 199 (1967).
146. Holt and Wold, *J. Biol. Chem.*, 236, 3227 (1961).
147. Bucher, *Methods Enzymol.*, 1, 427 (1955).
148. Malmstrom, Kimmel, and Smith, *J. Biol. Chem.*, 234, 1108 (1959).
149. Bergdoll, Chu, Huang, Rowe, and Shih, *Arch. Biochem. Biophys.*, 112, 104 (1965).
150. Schantz, Roessler, Wagman, Spero, Dunnery, and Bergdoll, *Biochemistry*, 4, 1011 (1965).
151. Stansell and Deutsch, *J. Biol. Chem.*, 240, 4306 (1965).
152. Crane, Mii, Hauge, Green, and Beinert, *J. Biol. Chem.*, 218, 701 (1956).
153. Yang, Butterworth, Bock, and Potter, *J. Biol. Chem.*, 242, 3501 (1967).
154. Keresztes-Nagy and Margoliash, *J. Biol. Chem.*, 241, 5955 (1966).
155. Matsubara, *J. Biol. Chem.*, 243, 370 (1968).
156. Malkin and Rabinowitz, *Biochemistry*, 5, 1262 (1966).
157. Hofmann and Harrison, *J. Mol. Biol.*, 6, 256 (1963).
158. Verpoorte, Green, and Kay, *J. Biol. Chem.*, 240, 1156 (1965).
159. Caspary and Kekwick, *Biochem. J.*, 56, 35 (1954).
160. Mosesson, Alkjaersig, Sweet, and Sherry, *Biochemistry*, 6, 3279 (1967).
161. Mihalyi, *Biochemistry*, 7, 208 (1968).
162. Mihalyi and Godfrey, *Biochim. Biophys. Acta*, 67, 73 (1963).
163. Gould and Liener, *Biochemistry*, 4, 90 (1965).
164. Hornby, Lilly, and Crook, *Biochem. J.*, 98, 420 (1966).
165. Englund, King, Craig, and Walti, *Biochemistry*, 7, 163 (1968).
166. Knight and Hardy, *J. Biol. Chem.*, 241, 2752 (1966).
167. Knight, D'Eustachio, and Hardy, *Biochim. Biophys. Acta*, 113, 626 (1966).
168. Papkoff, Gospodarowicz, and Li, *Arch. Biochem. Biophys.*, 120, 434 (1967).
169. Papkoff, Mahlmann, and Li, *Biochemistry*, 6, 3976 (1967).
170. Mizobuchi and Buchanan, *J. Biol. Chem.*, 243, 4842 (1968).
171. Himes and Cohn, *J. Biol. Chem.*, 242, 3628 (1967).
172. Marcus and Hubert, *J. Biol. Chem.*, 243, 4923 (1968).
173. Fernando, Enser, Pontremoli, and Horecker, *Arch. Biochem. Biophys.*, 126, 599 (1968).
174. Pontremoli, Grazi, and Accorsi, *Biochemistry*, 7, 3628 (1968).
175. Kanarek and Hill, *J. Biol. Chem.*, 239, 4202 (1964).
176. Malhotra and Dey, *Biochem. J.*, 103, 508 (1967).
177. Wallenfells and Golker, *Biochem. Z.*, 346, 1 (1966).
178. Wallenfells and Malhotra, in *The Enzymes*, Vol. 4, 2nd ed., Boyer, Lardy, and Myrback, Eds., Academic Press, New York, 1960, 413.
179. Mills and Tang, *J. Biol. Chem.*, 242, 3093 (1967).
180. Chiang, Sanchez-Chiang, Mills, and Tang, *J. Biol. Chem.*, 242, 3098 (1967).
181. Bernardin, Kasarda, and Mecham, *J. Biol. Chem.*, 242, 445 (1967).
182. Vodrazka, Hrkal, Cejka, and Sipalova, *Collect. Czech. Chem. Commun.*, 32, 3250 (1967).
183. Gibson and Antonini, *J. Biol. Chem.*, 238, 1384 (1963).

184. Hrkal and Vodrazka, *Biochim. Biophys. Acta,* 133, 527 (1967).
185. Rossi-Fanelli, Antonini, and Caputo, *J. Biol. Chem.,* 234, 2906 (1959).
186. Schultze, Heide, and Haupt, *Klin. Wochenschr.,* 40, 729 (1962).
187. Berggard and Bearn, *J. Biol. Chem.,* 243, 4095 (1968).
188. Deutsch, *Science,* 141, 435 (1963).
189. Muldoon and Westphal, *J. Biol. Chem.,* 242, 5636 (1967).
190. Chader and Westphal, *J. Biol. Chem.,* 243, 928 (1968).
191. Chader and Westphal, *Biochemistry,* 7, 4272 (1968).
192. Giorgio and Tabachnick, *J. Biol. Chem.,* 243, 2247 (1968).
193. Gratzer, Bailey, and Beaven, *Biochem. Biophys. Res. Commun.,* 28, 914 (1967).
194. Huotari, Nelson, Smith, and Kirkwood, *J. Biol. Chem.,* 243, 952 (1968).
195. Hauge, *Methods Enzymol.,* 9, 107 (1966).
196. Swoboda and Massey, *J. Biol. Chem.,* 240, 2209 (1965).
197. Takeda, Hizukuri, and Nikuni, *Biochim. Biophys. Acta,* 146, 568 (1967).
198. Plapp and Cole, *Arch. Biochem. Biophys.,* 116, 193 (1966).
199. Switzer and Barker, *J. Biol. Chem.,* 242, 2658 (1967).
200. Tomkins, Yielding, Curran, Summers, and Bitensky, *J. Biol. Chem.,* 240, 3793 (1965).
201. Eisenberg and Tomkins, *J. Mol. Biol.,* 31, 37 (1968).
202. Olson and Anfinsen, *J. Biol. Chem.,* 197, 67 (1952).
203. Reithel and Sakura, *J. Phys. Chem.,* 67, 2497 (1963).
204. Fahien, Wiggert, and Cohen, *J. Biol. Chem.,* 240, 1083 (1965).
205. Messer and Ottesen, *C. R. Trav. Lab. Carlsberg Ser. Chim.,* 35, 1 (1965).
206. Rowe and Wyngaarden, *Fed. Proc.,* 27, No. 2, 340 (1968).
207. Rowe and Wyngaarden, *J. Biol. Chem.,* 243, 6373 (1968).
208. Pamiljans, Krishnaswamy, Dumville, and Meister, *Biochemistry,* 1, 153 (1962).
209. Shapiro and Stadtman, *J. Biol. Chem.,* 242, 5069 (1967).
210. Woolfolk, Shapiro, and Stadtman, *Arch. Biochem. Biophys.,* 116, 177 (1966).
211. Colman and Black, *J. Biol. Chem.,* 240, 1796 (1965).
212. Massey and Williams, *J. Biol. Chem.,* 240, 4470 (1965).
213. Mavis and Stellwagen, *J. Biol. Chem.,* 243, 809 (1968).
214. Icen, *Scand. J. Clin. Lab. Invest., Suppl.,* 20, 96 (1967).
215. Kirschner and Voigt, *Hoppe-Seyler's Z. Physiol. Chem.,* 349, 632 (1968).
216. Krebs, *Methods Enzymol.,* 1, 407 (1955).
217. Jaenicke, Schmid, and Knof, *Biochemistry,* 7, 919 (1968).
218. Trentham, *Biochem. J.,* 109, 603 (1968).
219. Davidson, Sajgo, Noller, and Harris, *Nature,* 216, 1181 (1967).
220. Allison, *Methods Enzymol.,* 9, 212 (1966).
221. Mora and Elodi, *Eur. J. Biochem.,* 5, 574 (1968).
222. Velick, Hayes, and Harting, *J. Biol. Chem.,* 203, 527 (1953).
223. Fox and Dandliker, *J. Biol. Chem.,* 221, 1005 (1958).
224. Ankel, Bucher, and Czok, *Biochem. Z.,* 332, 315 (1960).
225. Fondy, Levin, Sollohub, and Ross, *J. Biol. Chem.,* 243, 3148 (1968).
226. Marquardt and Brosemer, *Biochim. Biophys. Acta,* 128, 454 (1966).
227. Brosemer and Marquardt, *Biochim. Biophys. Acta,* 128, 464 (1966).
228. Pradel, Kassab, Conlay, and Thoai, *Biochim. Biophys. Acta,* 154, 305 (1968).
229. Okuda and Weinbaum, *Biochemistry,* 7, 2819 (1968).
230. Schultze and Heremans, in *Molecular Biology of Human Proteins,* Vol. 1, Elsevier, New York, 1966, 176.
231. Haupt and Heide, *Clin. Chim. Acta,* 10, 555 (1964).
232. Schmidt, *J. Am. Chem. Soc.,* 72, 2816 (1950).
233. Smith, Brown, Weimer, and Winzler, *J. Biol. Chem.,* 185, 569 (1950).
234. Bundy and Mehl, *J. Biol. Chem.,* 234, 1124 (1959).
235. Schultze, Heide, and Haupt, *Naturwissenschaften,* 49, 133 (1962).
236. Burgi and Schmid, *J. Biol. Chem.,* 236, 1067 (1961).
237. Schmid and Burgi, *Biochim. Biophys. Acta,* 47, 440 (1961).
238. Schonenberger, Schmidtberger, and Schultze, *Z. Naturforsch.,* 13b, 761 (1958).
239. Edelhoch, Condliffe, Lippoldt, and Burger, *J. Biol. Chem.,* 241, 5205 (1966).
240. Polonovski and Sayle, *Bull. Soc. Chim. Biol.,* 21, 661 (1939).
241. Lisowska and Dobryszycka, *Biochim. Biophys. Acta,* 133, 338 (1967).
242. Heirwegh, Borginon, and Lontie, *Biochim. Biophys. Acta,* 48, 517 (1961).
243. Ghiretti-Magaldi, Nuzzolo, and Ghiretti, *Biochemistry,* 5, 1943 (1966).
244. Formanek and Engel, *Biochim. Biophys. Acta,* 160, 151 (1968).
245. Schichi and Kuroda, *Arch. Biochem. Biophys.,* 118, 682 (1967).

246. Schultze, Heide, and Haupt, *Naturwissenschaften.*, 48, 696 (1961).
247. Derechin, Ramel, Lazarus, and Barnard, *Biochemistry*, 5, 4017 (1966).
248. McDonald, *Methods Enzymol.*, 1, 269 (1955).
249. Riley and Snell, *Biochemistry*, 7, 3520 (1968).
250. Chang and Snell, *Biochemistry*, 7, 2005 (1968).
251. Loper and Adams, *J. Biol. Chem.*, 240, 788 (1965).
252. Yourno and Ino, *J. Biol. Chem.*, 243, 3273 (1968).
253. Yourno, *J. Biol. Chem.*, 243, 3277 (1968).
254. Loper, *J. Biol. Chem.*, 243, 3264 (1968).
255. Bennett, Creaser, and McDonald, *Biochem. J.*, 109, 307 (1968).
256. Truffa-Bachi, Van Rapenbusch, Jannin, Gros, and Cohen, *Eur. J. Biochem.*, 5, 73 (1968).
257. Nakano, Ushijima, Saga, Tsutsumi, and Asami, *Biochim. Biophys. Acta*, 167, 9 (1968).
258. Dai, Decker, and Sund, *Eur. J. Biochem.*, 4, 95 (1968).
259. Becker and Pfeil, *Biochem. Z.*, 346, 301 (1966).
260. Becker, Benthin, Eschenhof, and Pfeil, *Biochem. Z.*, 337, 156 (1963).
261. Kohn and Jakoby, *J. Biol. Chem.*, 243, 2494 (1968).
262. Martin and Goldberger, *J. Biol. Chem.*, 242, 1168 (1967).
263. Little and Donahue, *Meth. Immunol. Immunochem.*, 2, 343 (1968).
264. Creighton and Yanofsky, *J. Biol. Chem.*, 241, 4616 (1966).
265. Porter, *Biochem. J.*, 53, 320 (1953).
266. Weil, Seibles, and Herskovits, *Arch. Biochem. Biophys.*, 111, 308 (1965).
267. Praissman and Rupley, *Biochemistry*, 7, 2431 (1968).
268. Nakaya, Horinishi, and Shibata, *J. Biochem. (Tokyo)*, 61, 345 (1967).
269. Lampson, Tytell, Nemes, and Hilleman, *Proc. Soc. Exp. Biol. Med.*, 112, 468 (1963).
270. Metzenberg, *Arch. Biochem. Biophys.*, 100, 503 (1963).
271. Neumann and Lampen, *Biochemistry*, 6, 468 (1967).
272. Colman, *J. Biol. Chem.*, 243, 2454 (1968).
273. Margolies and Goldberger, *J. Biol. Chem.*, 242, 256 (1967).
274. Anderer and Hornle, *J. Biol. Chem.*, 241, 1568 (1966).
275. Morihara, Oka, and Tsuzuki, *Biochim. Biophys. Acta*, 139, 382 (1967).
276. Nickerson and Durand, *Biochim. Biophys. Acta*, 77, 87 (1963).
277. Nickerson, Noval, and Robison, *Biochim. Biophys. Acta*, 77, 73 (1963).
278. Hirashima, Hayakawa, and Koike, *J. Biol. Chem.*, 242, 902 (1967).
279. Kawahara, Wang, and Talalay, *J. Biol. Chem.*, 237, 1500 (1962).
280. Wetlaufer, *C. R. Trav. Lab. Carlsberg Ser. Chim.*, 32, 125 (1960).
281. Kronman and Andreotti, *Biochemistry*, 3, 1145 (1964).
282. Sullivan, *Biochem. J.*, 110, 363 (1968).
283. Takemori, Nakazawa, Nakai, Suzuki, and Katagiri, *J. Biol. Chem.*, 243, 313 (1968).
284. Pesce, McKay, Stolzenbach, Cahn, and Kaplan, *J. Biol. Chem.*, 239, 1753 (1964).
285. DiSabato, Pesce, and Kaplan, *Biochim. Biophys. Acta*, 77, 135 (1963).
286. Markert and Appella, *Ann. N. Y. Acad. Sci.*, 94, 678 (1961).
287. Pesce, Fondy, Stolzenbach, Castillo, and Kaplan, *J. Biol. Chem.*, 242, 2151 (1967).
288. Fromm, *J. Biol. Chem.*, 238, 2938 (1963).
289. Schellenberg, *J. Biol. Chem.*, 242, 1815 (1967).
290. Jaenicka, *Biochem. Z.*, 338, 614 (1963).
291. Gibson, Davisson, Bachhawat, Ray, and Vestling, *J. Biol. Chem.*, 203, 397 (1953).
292. Wieland and Pfleiderer, *Ann. N.Y. Acad. Sci.*, 94, 691 (1961).
293. Appleby and Morton, *Biochem. J.*, 73, 539 (1969).
294. Tarmy and Kaplan, *J. Biol. Chem.*, 243, 2579 (1968).
295. Okabe, Hayakawa, Hamada, and Koike, *Biochemistry*, 7, 79 (1968).
296. Polis, Schmuckler, Custer, and McMeekin, *J. Am. Chem. Soc.*, 72, 4965 (1950).
297. Baker and Saroff, *Biochemistry*, 4, 1670 (1965).
298. Wetlaufer and Lovrien, *J. Biol. Chem.*, 239, 596 (1964).
299. Gordon, Basch, and Kalan, *J. Biol. Chem.*, 236, 2908 (1961).
300. Townend, Winterbottom, and Timasheff, *J. Am. Chem. Soc.*, 82, 3161 (1960).
301. Townend, Herskovits, Swaisgood, and Timasheff, *J. Biol. Chem.*, 239, 4196 (1964).
302. Tanford and Nozaki, *J. Biol. Chem.*, 234, 2874 (1959).
303. Ogston and Tombs, *Biochem. J.*, 66, 399 (1957).
304. Bell and McKenzie, *Biochim. Biophys. Acta*, 147, 123 (1967).
305. Ghose, Chaudhuri, and Sen, *Arch. Biochem. Biophys.*, 126, 232 (1968).
306. Groves, *J. Am. Chem. Soc.*, 82, 3345 (1960).
307. Masson and Heremans, in *Protides of the Biological Fluids*, Peeters, Ed., Vol. 14, Elsevier, Amsterdam, 1967, 115.

308. Montreuil, Tonnelat, and Mullet, *Biochim. Biophys. Acta,* 45, 413 (1960).
309. Spackman, Smith, and Brown, *J. Biol. Chem.,* 212, 255 (1955).
310. Kawahara, Misaka, and Nakanishi, *J. Biochem.* (Tokyo), 63, 77 (1968).
311. Hastings, Riley, and Massa, *J. Biol. Chem.,* 240, 1473 (1965).
312. Bruzzesi, Chiancone, and Antonini, *Biochemistry,* 4, 1796 (1965).
313. Hamaguchi and Kurono, *J. Biochem.* (Tokyo), 54, 111 (1963).
314. Fromageot and Schnek, *Biochim. Biophys. Acta,* 6, 113 (1950).
315. Chandan, Parry, and Shahani, *Biochim. Biophys. Acta,* 110, 389 (1965).
316. Glazer, *Aust. J. Chem.,* 12, 304 (1959).
317. Sophianopoulos, Rhodes, Holcomb, and Van Holde, *J. Biol. Chem.,* 237, 1107 (1962).
318. Praissman and Rupley, *Biochemistry,* 7, 2446 (1968).
319. Canfield, *J. Biol. Chem.,* 238, 2691 (1963).
320. Wetlaufer and Stahmann, *J. Am. Chem. Soc.,* 80, 1493 (1958).
321. Yutani, Yutani, Imanishi, and Isemura, *J. Biochem.* (Tokyo), 64, 449 (1968).
322. Jolles and Jolles, *Biochemistry,* 6, 411 (1967).
323. Bonavida, Sapse, and Sercarz, *J. Lab. Clin. Med.,* 70, 951 (1967).
324. Tsugita, Inouye, Terzaghi, and Streisinger, *J. Biol. Chem.,* 243, 391 (1968).
325. Howard and Glazer, *J. Biol. Chem.,* 242, 5715 (1967).
326. Black, Ph.D. thesis, Stanford University, 1967.
327. Wolfe and Neilands, *J. Biol. Chem.,* 221, 61 (1956).
328. Thorne, *Biochim. Biophys. Acta,* 59, 624 (1962).
329. Pfliederer and Hohnholz, *Biochem. Z.,* 331, 245 (1959).
330. Davies and Kun, *Biochem. J.,* 66, 307 (1957).
331. Grimm and Doherty, *J. Biol. Chem.,* 236, 1980 (1961).
332. Yoshida, *J. Biol. Chem.,* 240, 1113 (1965).
333. Murphey, Barnaby, Lin, and Kaplan, *J. Biol. Chem.,* 242, 1548 (1967).
334. Kitto, *Biochim. Biophys. Acta,* 139, 16 (1967).
335. Kitto and Lewis, *Biochim. Biophys. Acta,* 139, 1 (1967).
336. Kohn and Jakoby, *J. Biol. Chem.,* 243, 2472 (1968).
337. Harada and Wolfe, *J. Biol. Chem.,* 243, 4123 (1968).
338. Hsu and Lardy, *J. Biol. Chem.,* 242, 520 (1967).
339. Levy, *J. Biol. Chem.,* 242, 747 (1967).
340. Hsiang and Bright, *J. Biol. Chem.,* 242, 3079 (1967).
341. Barker, Smyth, Wilson, and Weissbach, *J. Biol. Chem.,* 234, 320 (1959).
342. Erwin and Hellerman, *J. Biol. Chem.,* 242, 4230 (1967).
343. Yamada and Yasunobu, *J. Biol. Chem.,* 237, 1511 (1962).
344. Rohrer, van Wartburg, and Aebi, *Biochem. Z.,* 344, 478 (1966).
345. Noda and Kuby, *J. Biol. Chem.,* 226, 551 (1957).
346. Callaghan and Weber, *Biochem. J.,* 73, 473 (1959).
347. Quass and Briskey, *J. Food Sci.,* 33, 180 (1968).
348. Kielley and Harrington, *Biochim. Biophys. Acta,* 41, 401 (1960).
349. Gellert and Englander, *Biochemistry,* 2, 39 (1963).
350. Nanninga, *Biochim. Biophys. Acta,* 82, 507 (1964).
351. Frederiksen and Holtzer, *Biochemistry,* 7, 3935 (1968).
352. Morita and Yagi, *Biochem. Biophys. Res. Commun.,* 22, 297 (1966).
353. Young, Himmelfarb, and Harrington, *J. Biol. Chem.,* 239, 2822 (1964).
354. Riddiford, *J. Biol. Chem.,* 241, 2792 (1966).
355. Yagi, Yazawa, and Yasui, *Biochem. Biophys. Res. Commun.,* 29, 331 (1967).
356. Wu, *Biochemistry,* 8, 39 (1969).
357. Hagihara, in *The Enzymes,* Vol. 4, 2nd ed., Boyer, Lardy, and Myrbäck, Eds., Academic Press, New York, 1960, 193.
358. Hochstein and Dalton, *Biochim. Biophys. Acta,* 139, 56 (1967).
359. Taniuchi and Anfinsen, *J. Biol. Chem.,* 241, 4366 (1966).
360. Edelhoch and Lippoldt, *J. Biol. Chem.,* 237, 2788 (1962).
361. Cunningham and Nuenke, *J. Biol. Chem.,* 234, 1447 (1959).
362. Edelhoch and Steiner, *J. Biol. Chem.,* 240, 2877 (1965).
363. Chatterjee and Montgomery, *Arch. Biochem. Biophys.,* 99, 426 (1962).
364. Kohn and Jakoby, *J. Biol. Chem.,* 243, 2486 (1968).
365. Glazer and Smith, *J. Biol. Chem.,* 236, 2948 (1961).
366. Smith, Light, and Kimmel, *Biochem. Soc. Symp.,* 21, 88 (1962).
367. Finkle and Smith, *J. Biol. Chem.,* 230, 669 (1958).
368. Whitaker and Bender, *J. Am. Chem. Soc.,* 87, 2728 (1965).
369. Potts, Aurbach, and Sherwood, *Recent Prog. Horm. Res.,* 22, 114 (1966).

370. Edelhoch, *J. Am. Chem. Soc.*, 79, 6100 (1957).
371. Perlman, *J. Biol. Chem.*, 241, 153 (1966).
372. Aloj, Salvatore, and Roche, *J. Biol. Chem.*, 242, 3810 (1967).
373. Perlman, Oplatka, and Katchalsky, *J. Biol. Chem.*, 242, 5163 (1967).
374. Chow and Kassel, *J. Biol. Chem.*, 243, 1718 (1968).
375. Gounaris and Perlman, *J. Biol. Chem.*, 242, 2739 (1967).
376. Keilin and Hartree, *Biochem. J.*, 49, 88 (1951).
377. Maehly, *Methods Enzymol.*, 2, 801 (1955).
378. Wittenberg, Antonini, Brunori, Noble, Wittenberg, and Wyman, *Biochemistry*, 6, 1970 (1967).
379. Paul, in *The Enzymes*, Vol. 8, 2nd ed., Boyer, Lardy, and Myrbäck, Eds., Academic Press, New York, 1963, 233.
380. Kertesz and Zito, in *Oxygenases*, Hayashi, Ed., Academic Press, New York, 1962, 307.
381. Rothman and Byrne, *J. Mol. Biol.*, 6, 330 (1963).
382. Plocke, Levinthal, and Vallee, *Biochemistry*, 1, 373 (1962).
383. Kadner, Nye, and Brown, *J. Biol. Chem.*, 243, 3076 (1968).
384. Dorn, *J. Biol. Chem.*, 243, 3500 (1968).
385. Neumann, *J. Biol. Chem.*, 243, 4671 (1968).
386. Harkness, *Arch. Biochem. Biophys.*, 126, 503 (1968).
387. Igarashi and Hollander, *J. Biol. Chem.*, 243, 6084 (1968).
388. Parmeggiani, Luft, Love, and Krebs, *J. Biol. Chem.*, 241, 4625 (1966).
389. Paetkau and Lardy, *J. Biol. Chem.*, 242, 2035 (1967).
390. Younathan, Paetkau, and Lardy, *J. Biol. Chem.*, 243, 1603 (1968).
391. Froede, Geraci, and Mansour, *J. Biol. Chem.*, 243, 6021 (1968).
392. Lindell and Stellwagen, *J. Biol. Chem.*, 243, 907 (1968).
393. Chatterjee and Noltman, *Eur. J. Biochem.*, 2, 9 (1967).
394. Najjar, *J. Biol. Chem.*, 175, 281 (1948).
395. Najjar, *Methods Enzymol.*, 1, 294 (1955).
396. Rippa, Signorini, and Pontremoli, *Eur. J. Biochem.*, 1, 170 (1967).
397. Sugimoto and Pizer, *J. Biol. Chem.*, 243, 2090 (1968).
398. Zwaig and Milstein, *Biochem. J.*, 98, 360 (1966).
399. Sugimoto, Sasaki, and Chiba, *Arch. Biochem. Biophys.*, 113, 444 (1966).
400. Voll, Appella, and Martin, *J. Biol. Chem.*, 242, 1760 (1967).
401. Wells and Hanahan, *Biochemistry*, 8, 414 (1969).
402. Korn and Buchanan, *J. Biol. Chem.*, 217, 183 (1955).
403. Agarwal and Parks, *Fed. Proc.*, 27, 585, Abstr. No. 2072 (1968).
404. Buc and Buc, in *Symposium on Regulation of Enzyme Activity and Allosteric Interactions*, 4th Federation of European Biochemical Societies, Oslo, 1967. Academic Press, 1967, 109.
405. Velick and Wicks, *J. Biol. Chem.*, 190, 741 (1951).
406. Appleman, Krebs, and Fischer, *Biochemistry*, 5, 2101 (1966).
407. Metzger, Glaser, and Helmreich, *Biochemistry*, 7, 2021 (1968).
408. Kastenschmidt, Kastenschmidt, and Helmreich, *Biochemistry*, 7, 3590 (1968).
409. Gold, *Biochemistry*, 7, 2106 (1968).
410. Appleman, Yunis, Krebs, and Fischer, *J. Biol. Chem.*, 238, 1358 (1963).
411. Shaltiel, Hedrick, and Fischer, *Methods Enzymol.*, 11, 675 (1967).
412. Brody and Brody, *Biochim. Biophys. Acta*, 50, 348 (1961).
413. Toro-Goyco and Matos, *Nature*, 210, 527 (1966).
414. Toro-Goyco, Maretzki, and Matos, *Arch. Biochem. Biophys.*, 126, 91 (1968).
415. Bergmeyer, Holz, Klotzsch, and Lang, *Biochem. Z.*, 338, 114 (1963).
416. Reisfeld, Borjeson, Chessin, and Small, *Proc. Natl. Acad. Sci. U.S.A.*, 58, 2020 (1967).
417. Jovim, Englund, and Bertsch, *J. Biol. Chem.* (1969).
418. Robbins, Summaria, Elwyn, and Barlow, *J. Biol. Chem.*, 240, 541 (1965).
419. Robbins and Summaria, *J. Biol. Chem.*, 238, 952 (1963).
420. Kertesz and Zito, *Biochim. Biophys. Acta*, 96, 447 (1965).
421. Gregory and Bendall, *Biochem. J.*, 101, 569 (1966).
422. Reuter, Hamoir, Marchand, and Kennes, *Eur. J. Biochem.*, 5, 233 (1968).
423. Frank and Veros, *Biochem. Biophys. Res. Commun.*, 32, 155 (1968).
424. Oshima, Matsuo, Iwanaga, and Suzuki, *J. Biochem.* (Tokyo), 64, 227 (1968).
425. Oshima, Iwanaga, and Suzuki, *J. Biochem.* (Tokyo), 54, 215 (1968).
426. Jonsson, *Arch. Biochem. Biophys.*, 129, 62 (1969).
427. Ganno, *J. Biochem.* (Tokyo), 58, 556 (1965).
428. McConn, Tsuru, and Yasunobu, *J. Biol. Chem.*, 239, 3706 (1964).
429. Ohta, Ogura, and Wada, *J. Biol. Chem.*, 241, 5919 (1966).
430. Jackson and Wolfe, *J. Biol. Chem.*, 243, 879 (1968).
431. Fukumoto, Tsuru, and Yamamoto, *Agric. Biol. Chem.*, 31, 710 (1967).

432. Kimmel, Markowitz, and Brown, *J. Biol. Chem.*, 234, 46 (1959).
433. Narahashi, Shibuya, and Yanagita, *J. Biochem.* (Tokyo), 64, 427 (1968).
434. Hiramatsu, *J. Biochem.* (Tokyo), 62, 353 (1967).
435. Day, Toncic, Stratman, Leeman, and Harmon, *Biochim. Biophys. Acta*, 167, 597 (1968).
436. Hata, Hayashi, and Doi, *Agric. Biol. Chem.*, 31, 357 (1967).
437. Frattali, *J. Biol. Chem.*, 244, 274 (1969).
438. Iwasaki and Schmid, *J. Biol. Chem.*, 242, 5247 (1967).
439. Tomasi and Kornguth, *J. Biol. Chem.*, 242, 4933 (1967).
440. Wood, Massi, and Solomon, *J. Biol. Chem.*, 234, 329 (1959).
441. Miyake, Gaylor, and Mason, *J. Biol. Chem.*, 243, 5788 (1968).
442. Marushige, Britlag, and Bonner, *Biochemistry*, 7, 3149 (1968).
443. Dus, De Klerk, Sletten, and Bartsch, *Biochim. Biophys. Acta*, 140, 291 (1967).
444. Tang and Coleman, *J. Biol. Chem.*, 243, 4286 (1968).
445. Hinkson, *Biochemistry*, 7, 2666 (1968).
446. Koshiyama, *Cereal Chem.*, 45, 405 (1968).
447. Koshiyama, *Cereal Chem.*, 45, 394 (1968).
448. Shore and Shore, *Biochemistry*, 6, 1962 (1967).
449. Gumpf and Hamilton, *Virology*, 35, 87 (1968).
450. Zaitlin and McCaughey, *Virology*, 26, 500 (1965).
451. Wasserman, Corradino, and Taylor, *J. Biol. Chem.*, 243, 3978 (1968).
452. Kalb, *Biochim. Biophys. Acta*, 168, 532 (1968).
453. Penrose, Nichoalds, Piperno, and Oxender, *J. Biol. Chem.*, 243, 5921 (1968).
454. Lamy and Waugh, *J. Biol. Chem.*, 203, 489 (1953).
455. Shulman and Hearon, *J. Biol. Chem.*, 238, 155 (1963).
456. Tishkoff, Williams, and Brown, *J. Biol. Chem.*, 243, 4151 (1968).
457. Aronson, *Thromb. Diath. Haemorrh.*, 16, 491 (1966).
458. Shapiro and Waugh, *Thromb. Diath. Haemorrh.*, 16, 469 (1966).
459. Lanchantin, Hart, Friedman, Saavedra, and Mehl, *J. Biol. Chem.*, 243, 5479 (1968).
460. Fujisawa and Hayashi, *J. Biol. Chem.*, 243, 2673 (1968).
461. Yui, Ishii, and Egami, *J. Biochem.* (Tokyo), 65, 37 (1969).
462. Ayling and Snell, *Biochemistry*, 7, 1616 (1968).
463. Kojima, Fujisawa, Nakazawa, Nakazawa, Kanetsuna, Taniuchi, Nozaki, and Hayaishi, *J. Biol. Chem.*, 242, 3270 (1967).
464. Kunitz, *J. Gen. Physiol.*, 35, 423 (1952).
465. Williams and Hager, *Methods Enzymol.*, 9, 265 (1966).
466. Bucher and Pfleiderer, *Methods Enzymol.*, 1, 435 (1955).
467. Packman and Jakoby, *J. Biol. Chem.*, 240, PC4107 (1965).
468. Foltman, *C. R. Trav. Lab. Carlsberg Ser. Chim.*, 34, 319 (1964).
469. Wang and Volini, *J. Biol. Chem.*, 243, 5465 (1968).
470. Shichi, Lewis, Irreverre, and Stone, *J. Biol. Chem.*, 244, 529 (1969).
471. Shields, Dinovo, Henriksen, Kimbel, and Millar, *Biochim. Biophys. Acta*, 147, 238 (1967).
472. Meadows and Jardetzky, *Proc. Natl. Acad. Sci. U.S.A.*, 61, 406 (1968).
473. Hill and Schmidt, *J. Biol. Chem.*, 237, 389 (1962).
474. Taborsky, *J. Biol. Chem.*, 234, 2915 (1959).
475. Sela, Anfinsen, and Harrington, *Biochim. Biophys. Acta*, 26, 502 (1957).
476. Jaenicke, Schmid, and Knof, *Biochemistry*, 7, 919 (1968).
477. Tanford, Hauenstein, and Rands, *J. Am. Chem. Soc.*, 77, 6409 (1955).
478. Blumenfeld and Levy, *Arch. Biochem. Biophys.*, 76, 97 (1958).
479. Sage and Singer, *Biochim. Biophys. Acta*, 29, 663 (1958).
480. Sage and Singer, *Biochemistry*, 1, 305 (1962).
481. Bigelow, *J. Biol. Chem.*, 236, 1706 (1961).
482. Sherwood and Potts, *J. Biol. Chem.*, 240, 3799 (1965).
483. Keller, Cohen, and Neurath, *J. Biol. Chem.*, 233, 344 (1958).
484. Aqvist and Anfinsen, *J. Biol. Chem.*, 234, 1112 (1959).
485. Wilson, *J. Biol. Chem.*, 242, 2260 (1967).
486. Yamasaki, Murakami, Irie, and Ukita, *J. Biochem.* (Tokyo), 63, 25 (1968).
487. Uchida, *J. Biochem.* (Tokyo), 60, 115 (1966).
488. Ettinger and Hirs, *Biochemistry*, 7, 3374 (1968).
489. Lee and Bendet, *J. Biol. Chem.*, 242, 2043 (1967).
490. Akoyunoglou, Argyroudi-Akoyunoglou, and Methenitou, *Biochim. Biophys. Acta*, 132, 481 (1967).
491. Lee, Patrick, and Masson, *J. Biol. Chem.*, 243, 4700 (1968).
492. Newman and Postgate, *Eur. J. Biochem.*, 7, 45 (1968).

493. Bachmayer, Yasunobu, Peel, and Mayhew, *J. Biol. Chem.*, 243, 1022 (1968).
494. Lovenberg, in *Protides of the Biological Fluids,* Vol. 14, Peeters, Ed., Elsevier, Amsterdam 1967, 165.
495. Bachmayer, Benson, Yasunobu, Garrard, and Whiteley, *Biochemistry*, 7, 986 (1968).
496. Peterson and Coon, *J. Biol. Chem.*, 243, 329 (1968).
497. Lovenberg and Williams, *Biochemistry*, 8, 141 (1969).
498. Laurell, *Acta Chem. Scand.*, 7, 1407 (1953).
499. Peisach, Levine, and Blumberg, *J. Biol. Chem.*, 242, 2841 (1967).
500. Green and Malamed, *Biochem. J.*, 100, 614 (1966).
501. Chaiet and Wolf, *Arch. Biochem. Biophys.*, 106, 1 (1964).
502. Taylor and Botts , *Biochemistry*, 7, 232 (1968).
503. Salvatore, Vecchio, Salvatore, Cahnman, and Robbins, *J. Biol. Chem.*, 240, 2935 (1965).
504. Matsubara, Kaspar, Brown, and Smith, *J. Biol. Chem.*, 240, 1125 (1965).
505. Landon, Evans, and Smith, *J. Biol. Chem.*, 243, 2165 (1968).
506. Neet, Nanci, and Koshland, *J. Biol. Chem.*, 243, 6392 (1968).
507. Gounaris and Ottesen, *C. R. Trav. Lab. Carlsberg Ser. Chim.*, 35, 37 (1965).
508. Ramaley, Bridger, Moyer, and Boyer, *J. Biol. Chem.*, 242, 4287 (1967).
509. Allen and Roy, *Biochim. Biophys. Acta*, 168, 243 (1968).
510. Kohn, Packman, Allen, and Jakoby, *J. Biol. Chem.*, 243, 2479 (1968).
511. Kohn, *J. Biol. Chem.*, 243, 4426 (1968).
512. Murphy, Plummer, and Miller, *Fed. Proc.*, 27, 268 Abstr. No. 298 (1968).
513. Holmgren and Reichard, *Eur. J. Biochem.*, 2, 187 (1967).
514. Thelander, *J. Biol. Chem.*, 242, 852 (1967).
515. Thelander and Baldesten, *Eur. J. Biochem.*, 4, 420 (1968).
516. Burns and Zarlengo, *J. Biol. Chem.*, 243, 178 (1968).
517. Shizuta, Nakazawa, Tokushige, and Hayaishi, *J. Biol. Chem.*, 244, 1883 (1969).
518. Kezdy, Lorand, and Miller, *Biochemistry*, 2302 (1965).
519. Winzor and Scheraga, *Arch. Biochem. Biophys.*, 104, 202 (1964).
520. Brewer, Keutmann, Potts, Reisfeld, Schlueter, and Munson, *J. Biol. Chem.*, 243, 5739 (1968).
521. Shome, Brown, Howard, and Pierce, *Arch. Biochem. Biophys.*, 126, 456 (1968).
522. Lai, Chen, and Tsolas, *Arch. Biochem. Biophys.*, 121, 790 (1967).
523. Seal and Doe, *J. Biol. Chem.*, 237, 3136 (1962).
524. Aisen, Aasa, Malstrom, and Vanngard, *J. Biol. Chem.*, 242, 2484 (1967).
525. Koechlin, *J. Am. Chem. Soc.*, 74, 2649 (1952).
526. Leibman and Aisen, *Arch. Biochem. Biophys.*, 121, 717 (1967).
527. Folk, Mullooly, and Cole, *J. Biol. Chem.*, 242, 1838 (1967).
528. Folk and Cole, *J. Biol. Chem.*, 241, 5518 (1966).
529. Mooz and Meister, *Biochemistry*, 6, 1722 (1967).
530. Woods, *J. Biol. Chem.*, 242, 2859 (1967).
531. Mee, Navon, and Stein, *Biochim. Biophys. Acta*, 104, 151 (1965).
532. Mihalyi and Harrington, *Biochim. Biophys. Acta*, 36, 447 (1959).
533. Green, *Biochem. J.*, 66, 407 (1957).
534. Kassell, Radicevic, Berlow, Peanasky, and Laskowski, *J. Biol. Chem.*, 238, 3274 (1963).
535. Laskowski and Laskowski, *Adv. Protein Chem.*, 9, 203 (1954).
536. Davie and Neurath, *J. Biol. Chem.*, 212, 515 (1955).
537. Labeyrie, Groudinsky, Jacquot-Armand, and Naslin, *Biochim. Biophys. Acta*, 128, 492 (1966).
538. Shaw, Mares-Guia, and Cohen, *Biochemistry*, 4, 2219 (1965).
539. Meloun, Fric, and Sorm, *Eur. J. Biochem.*, 4, 112 (1968).
540. Buck, Vithayathil, Bier, and Nord, *Arch. Biochem. Biophys.*, 97, 417 (1962).
541. Labouesse and Gervais, *Eur. J. Biochem.*, 2, 215 (1967).
542. Heide, Heilburger, and Haupt, *Clin. Chim. Acta*, 11, 82 (1956).
543. Pharo, Sordahl, Edelhoch, and Sanadi, *Arch. Biochem. Biophys.*, 125, 416 (1968).
544. Greene, Rigbi, and Fackre, *J. Biol. Chem.*, 241, 5610 (1966).
545. Greene, DiCarlo, Sussman, Bartelt, and Roark, *J. Biol. Chem.*, 243, 1804 (1968).
546. Kunitz, *J. Gen. Physiol.*, 30, 291 (1947).
547. Birk, Gertler, and Khalef, *Biochem. J.*, 87, 281 (1963).
548. Rackis, Sasame, Mann, Anderson, and Smith, *Arch. Biochem. Biophys.*, 98, 471 (1962).
549. Wu and Scheraga, *Biochemistry*, 1, 698 (1962).
550. Frattali and Steiner, *Biochemistry*, 7, 521 (1968).
551. Schyns, Bricteux-Gregoire, and Florkin, *Biochim. Biophys. Acta*, 175, 97 (1969).
552. Pechere, Dixon, Maybury, and Neurath, *J. Biol. Chem.*, 233, 1364 (1958).
553. Morino and Snell, *J. Biol. Chem.*, 242, 2800 (1967).
554. Hathaway, Kida, and Crawford, *Biochemistry*, 8, 989 (1969).

555. Kumagai, Matsui, Ogata, and Yamada, *Biochim. Biophys. Acta,* 171, 1 (1968).
556. Yamada, Kumagi, Matsui, Ohgishi, and Ogata, *Biochem. Biophys. Res. Commun.,* 33, 10 (1968).
557. Hayashi, Granner, and Tomkins, *J. Biol. Chem.,* 242, 3998 (1967).
558. Gorin and Chin, *Biochim. Biophys. Acta,* 99, 418 (1965).
559. Gorin, Fuchs, Butler, Chopra, and Hersh, *Biochemistry,* 1, 911 (1962).
560. Gorin and Chin, *Anal. Biochem.,* 17, 49 (1966).
561. Mahler, Hubscher, and Baum, *J. Biol. Chem.,* 216, 625 (1955).
562. *Iscotables,* Instrumentation Specialities Co., Inc., Lincoln, Nebraska, 1967, 9.
563. Shalaby, Banerjee, and Lauffer, *Biochemistry,* 7, 955 (1968).
564. Heller, *Biochemistry,* 7, 2906 (1968).
565. Avis, Bergel, and Bray, *J. Chem. Soc.* (Lond.), 1219 (1956).
566. Massey, Brumby, Komai, and Palmer, *J. Biol. Chem.,* 224, 1682 (1969).
567. O'Leary and Westheimer, *Biochemistry,* 7, 913 (1968).
568. Stormer, *J. Biol. Chem.,* 243, 3740 (1968).
569. Gregolin, Ryder, and Lane, *J. Biol. Chem.,* 243, 4227 (1968).
570. Hauge, *Arch. Biochem. Biophys.,* 94, 308 (1961).
571. Pajot and Groudinsky, *Eur. J. Biochem.,* 12, 158 (1970).
572. Labeyrie, Groudinsky, Jacquot-Armand, and Naslin, *Biochim. Biophys. Acta,* 128, 492 (1966).
573. Kodama and Shidara, *J. Biochem.* (Tokyo), 65, 356 (1969).
574. Horio, Higashi, Yamanaka, Matabara, and Okunuki, *J. Biol. Chem.,* 23, 944 (1961).
575. Kredich, Becker, and Tomkin, *J. Biol. Chem.,* 244, 2428 (1969).
576. Carrico and Deutsch, *J. Biol. Chem.,* 245, 723 (1970).
577. Price, Liu, Stein, and Moore, *J. Biol. Chem.,* 244, 917 (1969).
578. Lindberg, *Biochemistry,* 6, 335 (1967).
579. Lindberg, *Biochemistry,* 6, 323 (1967).
580. Jovin, Englund, and Bertsch, *J. Biol. Chem.,* 244, 2996 (1969).
581. Roychoudhury and Bloch, *J. Biol. Chem.,* 244, 3359 (1969).
582. Roncari, Kurylo-Borowska, and Craig, *Biochemistry,* 5, 2153 (1966).
583. Bender, Begue-Canton, Blakeley, Brubacher, Feder, Gunter, Kesdy, Killhefer, Jr., Marshall, Miller, Roeske, and Stoops, *J. Am. Chem. Soc.,* 88, 5890 (1966).
584. Warburg and Christian, *Biochem. Z.,* 310, 384 (1941).
585. Barker and Jencks, *Biochemistry,* 8, 3879 (1969).
586. Malhotra and Philip, *Indian J. Biochem.,* 3, 7 (1966).
587. Lanchantin, Friedman, and Hart, *J. Biol. Chem.,* 244, 865 (1969).
588. Hsu and Yun, *Biochemistry,* 9, 239 (1970).
589. Hsu, Wasson, and Porter, *J. Biol. Chem.,* 240, 3736 (1965).
590. Dey and Pridham, *Biochem. J.,* 113, 49 (1969).
591. Petek, Villarroya, and Courtois, *Eur. J. Biochem.,* 8, 395 (1969).
592. Carrico and Deutsch, *J. Biol. Chem.,* 244, 6087 (1969).
593. Porter, Sweeney, and Porter, *Arch. Biochem. Biophys.,* 105, 319 (1964).
594. Deutsch, *Immunochemistry,* 2, 207 (1965).
595. Newcomb, Normansell, and Stanworth, *J. Immunol.,* 101, 905 (1968).
596. Heimer, Jones, and Maurer, *Biochemistry,* 8, 3937 (1969).
597. Doi and Jirgensons, *Biochemistry,* 9, 1066 (1970).
598. Kickhofen, Hammer, and Scheel, *Hoppe-Seylers Z. Physiol. Chem.,* 349, 1755 (1968).
599. Binaghi and Oriol, *Bull. Soc. Chim. Biol.,* 50, 1035 (1968).
600. Mihaesco and Mihaesco, *Biochem. Biophys. Res. Commun.,* 33, 869 (1968).
601. Krigbaum and Kugler, *Biochemistry,* 9, 1216 (1970).
602. Johansson, *Acta Chem. Scand.,* 23, 683 (1969).
603. Kessler and Brew, *Biochim. Biophys. Acta,* 200, 449 (1970).
604. Morrison, Hamilton, and Stotz, *J. Biol. Chem.,* 228, 767 (1957).
605. Frohne and Hanson, *Hoppe-Seylers Z. Physiol. Chem.,* 350, 207 (1969).
606. Kretschmer, *Hoppe-Seylers Z. Physiol. Chem.,* 349, 846 (1968).
607. Oriol-Audit, Landon, Robin, and van Thoai, *Biochim. Biophys. Acta,* 188 132 (1969).
608. Lee and McElroy, *Biochemistry,* 8, 130 (1969).
609. Antonini, *Physiol. Rev.,* 45, 123 (1965).
610. Harris and Hill, *J. Biol. Chem.,* 244, 2195 (1969).
611. Straus, Gordon, and Wallach, *Eur. J. Biochem.,* 11, 201 (1969).
612. Willick, Schonbaum, and Kay, *Biochemistry,* 8, 3729 (1969).
613. Stryer, *J. Mol. Biol.,* 13, 482 (1965).
614. Hapner, Bradshaw, Hartzell, and Gurd, *J. Biol. Chem.,* 243, 683 (1968).
615. Awad and Kotite, *Biochem. J.,* 98, 909 (1966).

616.   Godfrey and Harrington, *Biochemistry,* 9, 886 (1970).
617.   Adelman and Taylor, *Biochemistry,* 8, 4976 (1969).
618.   Tada, Bailin, Barany, and Barany, *Biochemistry,* 8, 4842 (1969).
619.   Woods, *Int. J. Protein Res.,* 1, 29 (1969).
620.   Angeletti, *Proc. Natl. Acad. Sci., U.S.A.,* 65, 668 (1970).
621.   Furth and Hope, *Biochem. J.,* 116, 545 (1970).
622.   Villet and Dalziel, *Biochem. J.,* 115, 639 (1969).
623.   Pon, Schnackerz, Blackburn, Chatterjee, and Noltmann, *Biochemistry,* 9, 1506 (1970).
624.   Carter and Yoshida, *Biochim. Biophys. Acta,* 181, 12 (1969).
625.   Wald and Brown, *J. Gen. Physiol.,* 37, 189 (1953–54).
626.   Hubbard, *J. Gen. Physiol.,* 37, 381 (1953–54).
627.   Schichi, *Biochemistry,* 9, 1973 (1970).
628.   Nordlie and Fromm, *J. Biol. Chem.,* 234, 2522 (1959).
629.   Nakagawa and Kinnura, *J. Biochem.* (Tokyo), 66, 669 (1969).
630.   Fojioka, *Biochim. Biophys. Acta,* 185, 338 (1969).
631.   Marchesi, Steers, Marchesi, and Tillack, *Biochemistry,* 9, 50 (1970).
632.   Allen and Jakoby, *J. Biol. Chem.,* 244, 2078 (1968).
633.   Zalitis and Feingold, *Arch. Biochem. Biophys.,* 132, 457 (1969).
634.   Becker, Kredich, and Tomkins, *J. Biol. Chem.,* 244, 2418 (1969).
635.   Bernardi, *Adv. Enzymol.,* 31, 1 (1968).
636.   Eisenberg, Zobel, and Moos, *Biochemistry,* 7, 3186 (1968).
637.   Pugh and Wakil, *J. Biol. Chem.,* 240, 4727 (1965).
638.   Sauer, Pugh, Wakil, Delany, and Hill, *Proc. Natl. Acad. Sci., U.S.A.,* 52, 1360 (1964).
639.   Ramponi, Guerritore, Treves, Nassi, and Baccari, *Arch. Biochem. Biophys.,* 130, 362 (1969).
640.   Webster, *Biochim. Biophys. Acta,* 207, 371 (1970).
641.   Koberstein, Weber, and Jaenicke, *Z. Naturforsch. B,* 23, 474 (1968).
642.   Andersson, *Int. J. Protein Res.,* 1, 151 (1969).
643.   Glazer and Smith, *J. Biol. Chem.,* 235, PC43 (1960).
644.   Bonewell and Rossini, *Ital. J. Biochem.,* 18, 457 (1969).
645.   Kaldor, Saifer, and Westley, *Arch. Biochem. Biophys.,* 99, 275 (1962).
646.   Habeeb and Hiramoto, *Arch. Biochem. Biophys.,* 126, 16 (1968).
647.   Pederson and Foster, *Biochemistry,* 8, 2357 (1969).
648.   Groulade, Chicault, and Waltzinger, *Bull. Soc. Chim. Biol.,* 48, 1609 (1967).
649.   Frank and Veros, *Fed. Proc.,* 28, 728 (1969).
650.   Frank, Peker, Veros, and Ho, *J. Biol. Chem.,* 245, 3716 (1970).
651.   Wagner, Irion, Arens, and Bauer, *Biochem. Biophys. Res. Commun.,* 37, 383 (1969).
652.   Arens, Rauenbusch, Irion, Wagner, Bauer, and Kaufmann, *Hoppe-Seylers Z. Physiol Chem.,* 351, 199 (1970).
653.   Funakoshi and Deutsch, *J. Biol. Chem.,* 244, 3438 (1969).
654.   Verpoorte and Linnblow, *J. Biol. Chem.,* 243, 5993 (1968).
655.   Armstrong, Myers, Verpoorte, and Edsall, *J. Biol. Chem.,* 241, 5137 (1966).
656.   Carpy, *Biochem. Biophys. Acta,* 151, 245 (1968).
657.   Furth, *J. Biol. Chem.,* 243, 4832 (1968).
658.   McIntosh, *Biochem. J.,* 114, 463 (1969).
659.   Tobin, *J. Biol. Chem.,* 245, 2656 (1970).
660.   Emery, *Biochemistry,* 8, 877 (1969).
661.   Deisseroth and Dounce, *Arch. Biochem. Biophys.,* 131, 18 (1969).
662.   Petit and Tauber, *J. Biol. Chem.,* 195, 703 (1952).
663.   Stern and Lavin, *Science,* 88, 263 (1938).
664.   Agner, *Biochem. J.,* 32, 1702 (1938).
665.   Bonnichsen, *Arch. Biochem.,* 12, 83 (1947).
666.   Ushijima and Nakano, *Biochim. Biophys. Acta,* 178, 429 (1969).
667.   Greenfield and Price, *J. Biol. Chem.,* 220, 607 (1956).
668.   Gregory, *Biochim. Biophys. Acta,* 159, 429 (1968).
669.   Tanford and Lovrien, *J. Am. Chem. Soc.,* 84, 1892 (1962).
670.   Bjorndal and Eriksson, *Arch. Biochem. Biophys.,* 124, 149 (1968).
671.   Eriksson and Pettersson, *Arch. Biochem. Biophys.,* 124, 160 (1968).
672.   Pettersson and Eaker, *Arch. Biochem. Biophys.,* 124, 154 (1968).
673.   Wang and Barker, *Methods Enzymol.,* 13, 331 (1969).
674.   Wu and Yang, *J. Biol. Chem.,* 245, 212 (1970).
675.   Yang, *J. Biol. Chem.,* 240, 1616 (1965).
676.   Chang and Hayashi, *Biochem. Biophys. Res. Commun.,* 37, 841 (1969).
677.   Sjogren and Spychalski, *J. Am. Chem. Soc.,* 52, 4400 (1930).

678. Konisky and Richards, *J. Biol. Chem.*, 245, 2973 (1970).
679. Schito, *G. Microbiol.*, 14, 77 (1966).
680. Schito, Molina, and Pesce, *Biochem. Biophys. Res. Commun.*, 28, 611 (1967).
681. Tan and Woodworth, *Biochemistry*, 8, 3711 (1969).
682. Freisheim and Huennekens, *Biochemistry*, 8, 2271 (1969).
683. Nixon and Blakley, *J. Biol. Chem.*, 243, 4722 (1968).
684. Schwartz and Reed, *J. Biol. Chem.*, 243, 639 (1968).
685. Frisell and Mackenzie, *J. Biol. Chem.*, 237, 94 (1962).
686. Rene and Campbell, *J. Biol. Chem.*, 244, 1445 (1969).
687. Chu, *J. Biol. Chem.*, 243, 4342 (1968).
688. Borja, *Biochemistry*, 8, 71 (1969).
689. Huseby and Murray, *Biochem. Biophys. Res. Commun.*, 35, 169 (1969).
690. Marder, Shulman, and Carroll, *J. Biol. Chem.*, 244, 2111 (1969).
691. Iizuka and Yang, *Biochemistry*, 7, 2218 (1968).
692. Takasaki, Kosugi, and Kambayashi, *Agric. Biol. Chem.*, 33, 1527 (1969).
693. Rattazzi, *Biochim. Biophys. Acta*, 181, 1 (1969).
694. Bruni, Auricchio, and Covelli, *J. Biol. Chem.*, 244, 4735 (1969).
695. Legler, *Hoppe-Seylers Z. Physiol. Chem.*, 349, 1755 (1970).
696. Legler, *Hoppe-Seylers Z. Physiol. Chem.*, 348, 1359 (1967).
697. Herman-Boussier, Moretti, and Jayle, *Bull. Soc. Chim. Biol.*, 42, 837 (1960).
698. Guinand, Tonnelat, Boussier, and Jayle, *Bull. Soc. Chim. Biol.*, 38, 329 (1956).
699. Dobryszycka, Elwyn, and Kukral, *Biochim. Biophys. Acta*, 175, 220 (1969).
700. Okazaki, Wittenberg, Briehl, and Wittenberg, *Biochim. Biophys. Acta*, 140, 258 (1967).
701. Webster, *Biochim. Biophys. Acta*, 207, 371 (1970).
702. Bucci, Fronticelli, and Ragatz, *J. Biol. Chem.*, 243, 241 (1968).
703. Chiancone, Curell, Vecchini, Antonini, and Wyman, *J. Biol. Chem.*, 245, 4105 (1970).
704. Peacock, Pastewka, Reed, and Ness, *Biochemistry*, 9, 2275 (1970).
705. Atassi, Brown, and McEwan, *Immunochemistry*, 2, 379 (1965).
706. Benesch and Benesch, *J. Biol. Chem.*, 236, 405 (1961).
707. Bucci, Fronticelli, Bellelli, Antonini, Wyman, and Rossi-Fanelli, *Arch. Biochem. Biophys.*, 100, 364 (1963).
708. Li and Johnson, *Biochemistry*, 3, 2083 (1969).
709. Antonini, Bucci, Fronticelli, Chiancone, Wyman, and Rossi-Fanelli, *J. Mol. Biol.*, 17, 29 (1966).
710. Ueda, Shiga, and Tyuma, *Biochem. Biophys. Acta*, 207, 18 (1970).
711. Ozawa, *Biochemistry*, 9, 2158 (1970).
712. Harrison and Garratt, *Biochem. J.*, 113, 733 (1969).
713. Moller, Castleman, and Terhorst, *FEBS Lett.*, 8, 192 (1970).
714. Frank and Veros, *Biochem. Biophys. Res. Commun.*, 32, 155 (1968).
715. Chung and Franzen, *Biochemistry*, 8, 3175 (1969).
716. Illingworth and Tipton, *Biochem. J.*, 118, 253 (1970).
717. Malmstrom, Reinhammar, and Vanngard, *Biochim. Biophys. Acta*, 205, 48 (1970).
718. Tang, Coleman, and Myer, *J. Biol. Chem.*, 243, 4286 (1968).
719. Lin, *Biochemistry*, 9, 984 (1970).
720. Ebner, Denton, and Brodbeck, *Biochem. Biophys. Res. Comm.*, 22, 232 (1966).
721. Stevens, Brown, and Smith, *Arch. Biochem. Biophys.*, 136, 413 (1970).
722. Eriksson and Svensson, *Biochim. Biophys. Acta*, 198, 449 (1970).
723. Soda and Misone, *Biochemistry*, 7, 4110 (1968).
724. Sher and Mallette, *Arch. Biochem. Biophys.*, 53, 354 (1954).
725. Soda and Moriguchi, *Biochem. Biophys. Res. Comm.*, 34, 34 (1969).
726. Takeda, Yamamoto, Kojima, and Hayaishi, *J. Biol. Chem.*, 244, 2935 (1969).
727. Petek and Villarroya, *Bull. Soc. Chim. Biol.*, 50, 725 (1968).
728. Kajita, Kerwar, and Huennekens, *Arch. Biochem. Biophys.*, 130, 662 (1969).
729. Takesue and Omura, *J. Biochem.* (Tokyo), 67, 267 (1970).
730. Nishibayashi-Yamashita and Sato, *J. Biochem.* (Tokyo), 67, 199 (1970).
731. Karlsson, Eaker, and Porath, *Biochim. Biophys. Acta*, 127, 505 (1966).
732. Pho, Olomucki, Huc, and Thoai, *Biochim. Biophys. Acta*, 206, 46 (1970).
733. Webster, *Biochim. Biophys. Acta*, 207, 371 (1970).
734. Willumsen, *C. R. Trav. Lab. Carlsberg*, 37, 21 (1969).
735. Cheesman, *Proc. R. Soc. Lond. (Biol.)*, 149, 571 (1958).
736. Peraino, Bunville, and Tahmisian, *J. Biol. Chem.*, 244, 2241 (1969).
737. Willick, Schonbaum, and Kay, *Biochemistry*, 8, 3729 (1969).
738. Hardin Strickland, Kay, and Shannon, *J. Biol. Chem.*, 243, 3560 (1968).
739. Hamaguchi, Ikeda, Yoshida, and Morita, *J. Biochem.* (Tokyo), 66, 191 (1969).

740. Mazza, Charles, Bouchet, Ricard, and Raynaud, *Biochim. Biophys. Acta,* 167, 89 (1968).
741. Blasi, Fragomele, and Corelli, *J. Biol. Chem.,* 244, 4866 (1969).
742. Simpson, Vallee, and Taft, *Biochemistry,* 7, 4336 (1968).
743. Schlesinger, *J. Biol. Chem.,* 243, 3877 (1968).
744. Rothman and Byrne, *J. Mol. Biol.,* 6, 330 (1963).
745. Morton, *Biochem. J.,* 60, 573 (1955).
746. Alvarez and Lora-Tamayo, *Biochem. J.,* 69, 312 (1958).
747. Dennert and Hoglund, *Eur. J. Biochem.,* 12, 502 (1970).
748. Hunsley and Suelter, *J. Biol. Chem.,* 244, 4185 (1969).
749. Richardson, *Proc. Natl. Acad. Sci., U.S.A.,* 55, 1616 (1966).
750. Neuhoff, Schill, and Sternbach, *Hoppe-Seylers Z. Physiol. Chem.,* 350, 767 (1969).
751. Brown, Canellakis, Lundin, Reichard, and Thelander, *Eur. J. Biochem.,* 9, 561 (1969).
752. Brown, Eliasson, Reichard, and Thelander, *Eur. J. Biochem.,* 9, 512 (1969).
753. Lee, Patrick, and Barnes, *J. Biol. Chem.,* 245, 1357 (1970).
754. Matsubara and Nishimura, *J. Biochem.* (Tokyo), 45, 503 (1958).
755. Yamada, Kumagi, Nagate, and Yoshida, *Biochem. Biophys. Res. Commun.,* 39, 53 (1970).
756. Brewer and Edelhoch, *J. Biol. Chem.,* 245, 2402 (1970).
757. Salvatore, Vecchio, Salvatore, Cahnmann, and Robbins, *J. Biol. Chem.,* 240, 2935 (1965).
758. Han and Benson, *Biochem. Biophys. Res. Commun.,* 38, 378 (1970).
759. Mikola and Suolinna, *Eur. J. Biochem.,* 9, 555 (1969).
760. Poillon, Maeno, Koike, and Feigelson, *J. Biol. Chem.,* 244, 3447 (1968).
761. Henning, Helinski, Chao, and Yanofsky, *J. Biol. Chem.,* 237, 1523 (1962).
762. Hathaway and Crawford, *Biochemistry,* 9, 1801 (1970).
763. Wilson and Crawford, *J. Biol. Chem.,* 240, 4801 (1965).
764. Lemaire, van Rapenbusch, Gros, and Labouesse, *Eur. J. Biochem.,* 10, 334 (1969).
765. Paul and Stigbrand, *Biochim. Biophys. Acta,* 221, 255 (1970).
766. Haas, Lumfrom, and Goldblatt, *Arch. Biochem. Biophys.,* 44, 79 (1953).
767. Malhorta and Roni, *Indian J. Biochem.,* 6, 15 (1969).
768. Blakeley, Webb, and Zerner, *Biochemistry,* 8, 1984 (1969).
769. Bailey and Boulter, *Biochem. J.,* 112, 669 (1969).
770. George and Phillip, *J. Biol. Chem.,* 245, 529 (1970).
771. Kawano, Morimoto, and Uemura, *J. Biochem.* (Tokyo), 67, 333 (1970).
772. Scraba, Hostvedt, and Colter, *Can. J. Biochem.,* 47, 165 (1969).
773. Hull, Hills, and Markham, *Virology,* 37, 416 (1969).
774. Stevens and Lauffer, *Biochemistry,* 4, 31 (1965).
775. Reichman, *J. Biol. Chem.,* 235, 2959 (1966).
776. Crewther and Harrap, *J. Biol. Chem.,* 242, 4310 (1967).
777. Mega, Ikenaka, and Matsushima, *J. Biochem.* (Tokyo), 68, 109 (1970).
778. Villafranca and Mildvan, *J. Biol. Chem.,* 246, 772 (1971).
779. Kimura and Huang, *Arch. Biochem. Biophys.,* 137, 357 (1963).
780. Shimomura and Johnson, *Biochemistry,* 8, 3991 (1969).
781. Rosso, Takashima, and Adams, *Biochem. Biophys. Res. Commun.,* 34, 134 (1969).
782. Hamlin, *Exp. Gerontol.,* 4, 189 (1969).
783. Von Tigerstron and Razzell, *J. Biol. Chem.,* 243, 2691 (1968).
784. Von Tigerstron and Razzell, *J. Biol. Chem.,* 243, 6495 (1968).
785. Dimroth, Guchhait, Stoll, and Lane, *Proc. Natl. Acad. Sci., U.S.A.,* 67, 1353 (1970).
786. Mayer and Miller, *Anal. Biochem.,* 36, 91 (1970).
787. Anastasi, Erspamer, and Endean, *Arch. Biochem. Biophys.,* 125, 57 (1968).
788. Keutmann, Parsons, Potts, Jr., and Schlueter, *J. Biol. Chem.,* 245, 1491 (1970).
789. Marshall and Cohen, *J. Biol. Chem.,* 241, 4197 (1970).
790. Lacko and Neurath, *Biochemistry,* 9, 4680 (1970).
791. Visuri, Mikola, and Enari, *Eur. J. Biochem.,* 7, 193 (1969).
792. Boston and Prescott, *Arch. Biochem. Biophys.,* 128, 88 (1968).
793. Yue, Palmieri, Olson, and Kuby, *Biochemistry,* 6, 3204 (1967).
794. Gosselin-Rey and Gerday, *Biochim. Biophys. Acta,* 221, 241 (1970).
795. Yue, Jacobs, Okabe, Keutel, and Kuby, *Biochemistry,* 7, 4291 (1968).
796. Kumudavalli, Moreland, and Watts, *Biochem. J.,* 117, 513 (1970).
797. Hass and Hill, *J. Biol. Chem.,* 244, 6080 (1969).
798. Fraenkel-Conrat and Singer, *Arch. Biochem. Biophys.,* 60, 64 (1956).
799. Zimmerman and Coleman, *J. Biol. Chem.,* 246, 309 (1971).
800. Tai and Sih, *J. Biol. Chem.,* 245, 5062 (1970).
801. Shotton, *Methods Enzymol.,* 19, 113 (1970).

802.   Wang and Keen, *Arch. Biochem. Biophys.*, 141, 749 (1970).
803.   Seto, Sato, and Tamiya, *Biochim. Biophys. Acta*, 214, 483 (1970).
804.   Fanelli, Chiancone, Vecchini, and Antonini, *Arch. Biochem. Biophys.*, 141, 278, (1970).
805.   Hartz and Deutsch, *J. Biol. Chem.*, 244 4565 (1969).
806.   Svedberg and Sjogren, *J. Am. Chem. Soc.*, 52, 279 (1930).
807.   Lee, Travis, and Black, Jr., *Arch. Biochem. Biophys.*, 141, 676 (1970).
808.   Newman, Ihle, and Dure, III, *Biochem. Biophys. Res. Commun.*, 36, 947 (1969).
809.   Mayhew, Petering, Palmer, and Foust, *J. Biol. Chem.*, 244, 2830 (1969).
810.   Devanathan, Akagi, Hersh, and Himes, *J. Biol. Chem.*, 244, 2846 (1969).
811.   Lovenberg, Buchanan, and Rabinowitz, *J. Biol. Chem.*, 238, 3899 (1963).
812.   Bayer, Eckstein, Hagenmaier, Josef, Koch, Krauss, Roder, and Schretzmann, *Eur. J. Biochem.*, 8, 33 (1969).
813.   Hong and Rabinowitz, *J. Biol. Chem.*, 245, 4982 (1970).
814.   Moss, Petering, and Palmer, *J. Biol. Chem.*, 244, 2275 (1969).
815.   Petering, Fee, and Palmer, *J. Biol. Chem.*, 246, 643 (1971).
816.   Hormann and Gollwitzer, *Z. Physiol. Chem.*, 346, 21 (1966).
817.   Mihalyi, *Biochemistry*, 7, 208 (1968).
818.   Budzynski, *Biochim. Biophys. Acta*, 229, 663 (1971).
819.   Huseby, Mosesson, and Murray, *Physiol. Chem. Phys.*, 2, 374 (1970).
820.   Fuller and Doolittle, *Biochemistry*, 10, 1305 (1971).
821.   Holloway, Antonini, and Brunori, *FEBS Lett.*, 4, 299 (1969).
822.   Kramer and Whitaker, *J. Biol. Chem.*, 239, 2178 (1964).
823.   Brewer, Ljungdahl, Spencer, and Neece, *J. Biol. Chem.*, 245, 4798 (1970).
824.   Pontremoli, *Methods Enzymol.*, 9, 625 (1966).
825.   Mosesson and Umfleet, *J. Biol. Chem.*, 245, 5728 (1970).
826.   Danno, *Agric. Biol. Chem.*, 34, 1795 (1970).
827.   Strausbauch and Fischer, *Biochemistry*, 9, 226 (1970).
828.   Fahien and Cohen, *Methods Enzymol.*, 17A, 839 (1970).
829.   Dessen and Pantaloni, *Eur. J. Biochem.*, 8, 292 (1969).
830.   Winnacker and Barker, *Biochim. Biophys. Acta*, 212, 225 (1970).
831.   Barker, *Methods Enzymol.*, 13, 319 (1969).
832.   Rowe, Ronzio, Wellner, and Meister, *Methods Enzymol.*, 17A, 900 (1970).
833.   Woodin and Segel, *Biochim. Biophys. Acta*, 167, 64 (1968).
834.   Kohn, Warner, and Carroll, *J. Biol. Chem.*, 245, 3820 (1970).
835.   Edelhoch and Lippoldt, *J. Biol. Chem.*, 245, 4199 (1970).
836.   Gould and Scheinberg, *Arch. Biochem. Biophys.*, 137, 1 (1970).
837.   Sage and Connett, *J. Biol. Chem.*, 244, 4713 (1969).
838.   Ferrell and Kitto, *Biochemistry*, 9, 3053 (1970).
839.   Keresztes-Nagy, Ph.D. thesis, *Northwestern University, 1962; cited in* Subramanian, Holleman, and Klotz, *Biochemistry*, 7, 2859 (1968).
840.   Morimoto and Kegeles, *Arch. Biochem. Biophys.*, 142, 247 (1971).
841.   Gruber, in *Physiology and Biochemistry of Hemocyanins*, Ghiretti, Ed., Academic Press, New York, 1968, 49.
842.   Konings, Dijk, Wichertjes, Beuvery, and Gruber, *Biochim. Biophys. Acta*, 188, 43 (1969).
843.   Ellerton, Carpenter, and Van Holde, *Biochemistry*, 9, 2225 (1970).
844.   Joniau, Grossberg, and Pressman, *Immunochemistry*, 7, 755 (1970).
845.   Amkraut, personal communication.
846.   Shaklai and Daniel, *Biochemistry*, 9, 564 (1970).
847.   Maki, Yamamoto, Nozaki, and Hayaishi, *J. Biol. Chem.*, 244, 2942 (1969).
848.   Leslie and Cohen, *Biochem. J.*, 120, 787 (1970).
849.   Herskovits, *Arch. Biochem. Biophys.*, 130, 19 (1969).
850.   Meachum, Jr., Colvin, Jr., and Braymer, *Biochemistry*, 10, 326 (1971).
851.   Yokobayashi, Misaki, and Harada, *Biochim. Biophys. Acta*, 212, 458 (1970).
852.   McFadden, *Methods Enzymol.*, 13, 163 (1969).
853.   Parsons and Burns, *J. Biol. Chem.*, 244, 996 (1969).
854.   Yu, Harmon, Wachter, and Blank, *Arch. Biochem. Biophys.*, 135, 363 (1969).
855.   Alfsen, Baulieu, Claquin, and Falcoz-Kelly, *Proc. 2nd Int. Congr. Hormonal Steroids*, 1967, 508.
856.   Malkin and Malmstrom, *Adv. Enzymol.*, 33, 178 (1970).
857.   Nakamura and Ogura, *J. Biochem.* (Tokyo), 59, 449 (1966).
858.   Malkin, Malmstrom, and Vanngard, *Eur. J. Biochem.*, 10, 324 (1969).
859.   Malkin, Malmstrom, and Vanngard, *Eur. J. Biochem.*, 7, 253 (1969).
860.   Karkhanis and Cormier, *Biochemistry*, 10, 317 (1971).
861.   Haino, *Biochim. Biophys. Acta*, 229, 459 (1971).
862.   Chibata, Ishikawa, and Tosa, *Methods Enzymol.*, 19, 675 (1970).

863. Nagasawa, Sugihara, Han, and Suzuki, *J. Biochem.* (Tokyo), 67, 809 (1970).
864. Schonenberger, Schmidtberger, and Schultze, *Z. Naturforsch.,* 136, 761 (1958).
865. Jacquot-Armand and Guinand, *Biochim. Biophys. Acta,* 133, 289 (1967).
866. Berggard and Bearn, *J. Biol. Chem.,* 243, 4095 (1968).
867. Gregory and Harrison, *Biochem. Biophys. Res. Commun.,* 40, 995 (1970).
868. Covelli, Consiglio, and Varrone, *Biochim. Biophys. Acta,* 184, 678 (1969).
869. Gerding and Wolfe, *J. Biol. Chem.,* 244, 1164 (1969).
870. Kitto, *Methods Enzymol.,* 13, 106 (1969).
871. Mann and Vestling, *Biochemistry,* 8, 1105 (1969).
872. Spina, Jr., Bright, and Rosenbloom, *Biochemistry,* 9, 3794 (1970).
873. Ramachandran, *Biochem. Biophys. Res. Commun.,* 41, 353 (1970).
874. Lawrence, *Eur. J. Biochem.,* 15, 436 (1970).
875. Agner, *Acta Chem. Scand.,* 12, 89 (1958).
876. Schirmer, Schirmer, Schulz, and Thuma, *FEBS Lett.,* 10, 333 (1970).
877. Leuzinger, *Biochem. J.,* 123, 139 (1971).
878. Muruyama, *J. Biochem.* (Tokyo), 69, 369 (1971).
879. Minato, *J. Biochem.* (Tokyo), 64, 813 (1969).
880. Zielke and Suelter, *J. Biol. Chem.,* 246, 2179 (1971).
881. Zielke and Suelter, *Fed. Proc.,* 28, 2624 (1969).
882. Tweedie and Segel, *J. Biol. Chem.,* 246, 2438 (1971).
883. Lebeault, Zevaco, and Hermier, *Bull. Soc. Chim. Biol.,* 52, 1073 (1970).
884. Sterman and Foster, *J. Am. Chem. Soc.,* 78, 3656 (1956).
885. Frigerio and Hettinger, *Biochim. Biophys. Acta,* 59, 228 (1962).
886. Emery, *Biochemistry,* 8, 877 (1969).
887. Mayer and Miller, *Anal. Biochem.,* 36, 91 (1970).
888. King and Spencer, *J. Biol. Chem.,* 245, 6134 (1970).
889. Webster, *Biochim. Biophys. Acta,* 207, 371 (1970).
890. Pederson and Foster, *Biochemistry,* 8, 2357 (1969).
891. Noelken, *Biochemistry,* 9, 4117 (1970).
892. Noelken, *Biochemistry,* 9, 4122 (1970).
893. Van Kley and Stahmann, *J. Am. Chem. Soc.,* 81, 4374 (1959).
894. Janatova, Fuller, and Hunter, *J. Biol. Chem.,* 243, 3612 (1968).
895. Lerner and Barnum, *Arch. Biochem. Biophys.,* 10, 417 (1946).
896. Groulade, Chicault, and Waltzinger, *Bull. Soc. Chim. Biol.,* 49, 1609 (1967).
897. Warner and Schumaker, *Biochemistry,* 9, 451 (1970).
898. Petersen and Foster, *J. Biol. Chem.,* 240, 3861 (1965).
899. Laggner, Kratky, Palm, and Holasek, *FEBS Lett.,* 15, 220 (1971).
900. Joniau, Grossberg, and Pressman, *Immunochemistry,* 7, 755 (1970).
901. Polis, Shmukler, and Custer, *J. Biol. Chem.,* 187, 349 (1950).
902. Zigman and Lerman, *Biochim. Biophys. Acta,* 154, 423 (1968).
903. Lerman, *Can. J. Biochem.,* 47, 1115 (1969).
904. Bonnichsen, *Acta Chem. Scand.,* 4, 715 (1950).
905. Ehrenberg and Dalziel, *Acta Chem. Scand.,* 12, 465 (1958).
906. Green and McKay, *J. Biol. Chem.,* 244, 5034 (1969).
907. Cannon and McKay, *Biochem. Biophys. Res. Commun.,* 35, 403 (1969).
908. Drum, Li, and Vallee, *Biochemistry,* 8, 3783 (1969).
909. Drum, Harrison, Li, Bethune, and Vallee, *Proc. Natl. Acad. Sci., U.S.A.,* 57, 1434 (1967).
910. Drum and Vallee, *Biochem. Biophys. Res. Commun.,* 41, 33 (1970).
911. Von Wartburg, Bethune, and Vallee, *Biochemistry,* 3, 1775 (1964).
912. Negelein and Wulff, *Biochem. Z.,* 293, 351 (1937).
913. Koberstein, Weber, and Jaenicke, *Z. Naturforsch. B,* 23, 474 (1968).
914. Buhner and Sund, *Eur. J. Biochem.,* 11, 73 (1969).
915. Swaisgood and Pattee, *J. Food Sci.,* 33, 400 (1968).
916. Sofer and Ursprung, *J. Biol. Chem.,* 243, 3110 (1968).
917. Biszku, Boross, and Szabolcsi, *Acta Physiol. Acad. Sci. Hung.,* 25, 161 (1964).
918. Kawahara and Tanford, *Biochemistry,* 5, 1578 (1966).
919. Sine and Hass, *J. Biol. Chem.,* 244, 430 (1969).
920. Reisler and Eisenberg, *Biochemistry,* 8, 4572 (1969).
921. Hass, *Biochemistry,* 3, 535 (1964).
922. Gracy, Lacko, and Horecker, *J. Biol. Chem.,* 244, 3913 (1969).
923. Marquardt, *Can. J. Biochem.,* 47, 517 (1969).
924. Marquardt, *Can. J. Biochem,* 49, 647 (1971).

925. Marquardt, *Can. J. Biochem.,* 49, 658 (1971).
926. Suh, *Fed. Proc.,* 30, 1157 (1971).
927. Lai and Chen, *Arch. Biochem. Biophys.,* 144, 467 (1971).
928. Kobes, Simpson, Vallee, and Rutter, *Biochemistry,* 8, 585 (1969).
929. Harris, Kobes, Teller, and Rutter, *Biochemistry,* 8, 2442 (1969).
930. Rapoport, Davis, and Horecker, *Arch. Biochem. Biophys.,* 132, 286 (1969).
931. DeGraaf, Goedvolk-DeGroot, and Stouthamer, *Biochem. Biophys. Acta,* 221, 566 (1971).
932. Dimroth, Guchhait, and Lane, *Hoppe-Seylers Z. Physiol. Chem.,* 352, 351 (1971).
933. Doyle, Pittz, and Woodside, *Carbohydr. Res.,* 8, 89 (1968).
934. Agrawal and Goldstein, *Biochim. Biophys. Acta,* 133, 376 (1967).
935. Olson and Liener, *Biochemistry,* 6, 105 (1967).
936. Yariv, Kalb, and Levitzki, *Biochim. Biophys. Acta,* 165, 303 (1968).
937. Cheesman, Zagalsky, and Ceccaldi, *Proc. R. Soc. Lond. (Biol.),* 164, 130 (1966).
938. Cummings, *Biochem. Biophys. Res. Commun.,* 33, 165 (1968).
939. Hamlin, *Exp. Gerentol.,* 4, 189 (1969).
940. Cavallini, Scandurra, and Dupre, in *Biological and Chemical Aspects of Oxygenases,* Bloch and Hayaishi, Eds., Maruzen, Tokyo, 1966, 73.
941. Mochan, *Biochim. Biophys. Acta,* 216, 80 (1970).
942. Poulos and Price, *J. Biol. Chem.,* 246, 4041 (1971).
943. Jargenson, *J. Biol. Chem.,* 240, 1064 (1965).
944. Kaplan and Dugas, *Biochem. Biophys. Res. Commun.,* 34, 681 (1969).
945. Wasi and Hofmann, *Biochem. J.,* 106, 926 (1968).
946. Gertler and Trop, *Eur. J. Biochem.,* 19, 90 (1971).
947. Kimura and Kubata, *Bull. Jap. Soc. Sci. Fish.,* 34, 535 (1968).
948. Vulpis, Vulpis, and Santoro, *Ital. J. Biochem.,* 15, 189 (1966).
949. Ruth, Soja, and Wold, *Arch. Biochem. Biophys.,* 140, 1 (1970).
950. Shethna, Stombaugh, and Burris, *Biochem. Biophys. Res. Commun.,* 42, 1108 (1971).
951. Buchanan, Matsubara, and Evans, *Biochim. Biophys. Acta,* 189, 46 (1969).
952. Vetter and Knappe, *Hoppe-Seylers Z. Physiol. Chem.,* 352, 433 (1970).
953. Mortenson, *Biochim. Biophys. Acta,* 81, 71 (1964).
954. Sobel and Lovenberg, *Biochemistry,* 5, 6 (1966).
955. Hong and Rabinowitz, *J. Biol. Chem.,* 245, 4982 (1970).
956. Aggarwal, Rao, and Matsubara, *J. Biochem. (Tokyo),* 69, 601 (1971).
957. Buchanan and Rabinowitz, *J. Bacteriol.,* 88, 806 (1964).
958. Lode and Coon, *J. Biol. Chem.,* 246, 791 (1971).
959. Dubourdieu and Le Gall, *Biochem. Biophys. Res. Commun.,* 38, 965 (1970).
960. Vetter, Jr. and Knappe, *Hoppe-Seylers Z. Physiol. Chem.,* 352, 433 (1970).
961. Mayhew, *Biochim. Biophys. Acta,* 235, 276 (1971).
962. Mayhew, *Biochim. Biophys. Acta,* 235, 289 (1971).
963. Mayhew and Massey, *J. Biol. Chem.,* 244, 794 (1969).
964. Zak, Steczko, and Ostrowski, *Bull. Soc. Chim. Biol.,* 51, 1065 (1969).
965. Schade and Reinhart, *Biochem. J.,* 118, 181 (1970).
966. Hauge, *Arch. Biochem. Biophys.,* 94, 308 (1961).
967. Yoshimura and Isemura, *J. Biochem. (Tokyo),* 69, 839 (1971).
968. Olive and Levy, *J. Biol. Chem.,* 246, 2043 (1971).
969. Flohe, Eisele, and Wendel, *Hoppe-Seylers Z. Physiol. Chem.,* 352, 151 (1971).
970. Harrington and Karr, *J. Mol. Biol.,* 13, 885 (1965).
971. Cseke and Boross, *Acta Biochim. Biophys. Acad. Sci. Hung.,* 2, 39 (1967).
972. Parker and Allison, *J. Biol. Chem.,* 244, 180 (1969).
973. Koberstein, Weber, and Jaenicke, *Z. Naturforsch. (B),* 23, 474 (1968).
974. Murdock and Koeppe, *J. Biol. Chem.,* 239, 1983 (1964).
975. McMurray and Trentham, *Biochem. J.,* 115, 913 (1959).
976. Wassarman, Watson, and Major, *Biochim. Biophys. Acta,* 191, 1 (1969).
977. Davidson, Sajgo, Noller, and Harris, *Nature,* 216, 1181 (1962).
978. Allison, *Methods Enzymol.,* 9, 210 (1966).
979. Devijlder, Boers, and Slater, *Biochim. Biophys. Acta,* 191, 214 (1969).
980. Warburg and Christian, *Biochem. Z.,* 303, 40 (1939).
981. Jaenicke, in *Pyridine Nucleotide Dependent Dehydrogenases,* Sund, Ed., Springer-Verlag, Berlin, 1970, 70.
982. Durchschlag, Puchwein, Kratky, Schuster, and Kirschner, *Eur. J. Biochem.,* 19, 9 (1971).
983. Oguchi, *J. Biochem. (Tokyo),* 68, 427 (1970).
984. Heinz and Kulbe, *Hoppe-Seylers Z. Physiol. Chem.,* 351, 249 (1970).
985. Baliga, Bhatnagar, and Jagannathan, *Indian J. Biochem.,* 1, 86 (1964).

986. Thorner and Paulus, *J. Biol. Chem.,* 246, 3385 (1971).
987. Fondy, Ross, and Sollohub, *J. Biol. Chem.,* 244, 1631 (1969).
988. Beisenherz, Bucher, and Gorbade, *Methods Enzymol.,* 1, 397 (1955).
989. Ankel, Bucher, and Czok, *Biochem. Z.,* 332, 315 (1960).
990. Catsimpoolas, Berg, and Meyer, *Int. J. Protein Res.,* 3, 63 (1971).
991. Wolf and Briggs, *Arch. Biochem. Biophys.,* 63, 40 (1959).
992. Hrkal and Muller-Eberhard, *Biochemistry,* 10, 1746 (1971).
993. Dus, DeKlerk, Sletten, and Bartsch, *Biochem. Biophys. Acta,* 140, 291 (1967).
994. Buffoni and Blaschko, *Proc. R. Soc. Lond. (Biol.),* 161, 153 (1965).
995. Klee, *J. Biol. Chem.,* 245, 3143 (1970).
996. Rechler, *J. Biol. Chem.,* 244, 551 (1969).
997. Riley and Snell, *Biochemistry,* 9, 1485 (1970).
998. Rechler and Bruni, *J. Biol. Chem.,* 246, 1806 (1971).
999. Champagne, Pouyet, Ouellet, and Garel, *Bull. Soc. Chim. Biol.,* 52, 377 (1970).
1000. Oh, *J. Biol. Chem.,* 245, 6404 (1970).
1001. Matsuo and Greenberg, *J. Biol. Chem.,* 230, 545 (1958).
1002. Datta, *J. Biol. Chem.,* 245, 5779 (1970).
1003. Janin, van Rapenbusch, Truffa-Bachi, and Cohen, *Eur. J. Biochem.,* 8, 128 (1969).
1004. Rhodes, Dodgson, Olavesen, and Hogberg, *Biochem. J.,* 122, 575 (1971).
1005. Newman, Berenson, Mathews, Goldwasser, and Dorfman, *J. Biol. Chem.,* 217, 31 (1955).
1006. Haschke and Campbell, *J. Bacteriol.,* 105, 249 (1971).
1007. Nakos and Mortenson, *Biochemistry,* 10, 2442 (1971).
1008. Yano, Morimoto, Higashi, and Arima, in *Biological and Chemical Aspects of Oxygenases,* Bloch and Hayaishi, Eds., Maruzen, Tokyo, 1966, 331.
1009. Hesp, Calvin, and Hosokawa, *J. Biol. Chem.,* 244, 5644 (1968).
1010. Aschhoff and Pfeil, *Hoppe-Seylers Z. Physiol. Chem.,* 351, 818 (1970).
1011. Finlay and Adams, *J. Biol. Chem.,* 245, 5248 (1970).
1012. Martin, *Arch. Biochem. Biophys.,* 138, 239 (1970).
1013. Hamaguichi and Migita, *J. Biochem.* (Tokyo), 56, 512 (1964).
1014. Ruffilli and Givol, *Eur. J. Biochem.,* 2, 429 (1967).
1015. Heimburger, Heide, and Haupt, *Clin. Chim. Acta,* 10, 293 (1964).
1016. Tomasi and Bienenstock, *Adv. Immunol.,* 9, 1 (1968).
1017. Lerner and Barnum, *Arch. Biochem.,* 10, 417 (1946).
1018. Lewis, Bergsagel, Bruce-Robertson, Schachter, and Connell, *Blood,* 32, 189 (1968).
1019. Connell, Dorington, Lewis, and Parr, *Can. J. Biochem.,* 48, 784 (1970).
1020. Schultze and Heremans, *Molecular Biology of Human Protein,* Vol. I, Elsevier, Amsterdam, 1966, 234.
1021. Bennich and Johansson, in *Gamma Globulins, Nobel Symposium No. 3,* Killander, Ed., Interscience, New York, 1967, 200.
1022. Orlans, *Immunology,* 14, 61 (1968).
1023. Yamashita, Franek, Skvaril, and Simek, *Eur. J. Biochem.,* 6, 34 (1968).
1024. Clem, *J. Biol. Chem.,* 246, 9 (1971).
1025. Acton, Weinheimer, Dupree, Evans, and Bennett, *Biochemistry,* 10, 2028 (1971).
1026. Freedman, Grossberg, and Pressman, *Biochemistry,* 7, 1941 (1968).
1027. Porter, *Biochem. J.,* 66, 677 (1957).
1028. Knight and Dray, *Biochemistry,* 7, 3830 (1968).
1029. Van Dalen, Seijen, and Gruber, *Biochim. Biophys. Acta,* 147, 421 (1967).
1030. Day, Sturtevant, and Singer, *Ann. N.Y. Acad. Sci.,* 103, 611 (1963).
1031. Onoue, Yagi, Grossberg, and Pressman, *Immunochemistry,* 2, 401 (1965).
1032. Freedman, Grossberg, and Pressman, *Immunochemistry,* 5, 367 (1968).
1033. Givol and Hurwitz, *Biochem. J.,* 115, 371 (1969).
1034. Haimovich, Schechter, and Sela, *Eur. J. Biochem.,* 4, 537 (1969).
1035. Sokol, Hana, and Albrecht, *Folia Microbiol.* (Praha), 6, 145 (1961).
1036. Montgomery, Dorrington, and Rockey, *Biochemistry,* 8, 1247 (1969).
1037. Eisen, Simms, and Potter, *Biochemistry,* 7, 4126 (1968).
1038. Tur-Sinai, Birk, Gertler, and Rigbi, *Isr. J. Chem.,* 8, 176 (1970).
1039. Mikola and Suolinna, *Arch. Biochem. Biophys.,* 144, 566 (1971).
1040. Malmström, Agro, and Antonini, *Eur. J. Biochem.,* 9, 383 (1969).
1041. Phillips and Jenness, *Biochim. Biophys. Acta,* 229, 407 (1971).
1042. Schmidt and Ebner, *Biochim. Biophys. Acta,* 243, 273 (1971).
1043. Bell, Hopper, McKenzie, Murphy, and Shaw, *Biochim. Biophys. Acta,* 214, 437 (1970).
1044. Kuwabara and Lloyd, *Biochem. J.,* 124, 215 (1971).
1045. Móra and Elödi, *Eur. J. Biochem.,* 5, 574 (1968).

1046. Gutfreund, Cantwell, McMurray, Criddle, and Hathaway, *Biochem. J.,* 106, 683 (1968).
1047. Reeves and Fimognari, *Methods Enzymol.,* 9, 289 (1966).
1048. Jaenicke, in *Pyridine Dependent Dehydrogenases,* Sund, Ed., Springer-Verlag, New York, 1970, 70.
1049. Koberstein, Weber, and Jaenicke, *Z. Naturforsch. (B),* 23, 474 (1968).
1050. Foye and Solis, *J. Pharm. Sci.,* 58, 352 (1969).
1051. Pfleiderer and Jeckel, *Biochem. Z.,* 329, 370 (1957).
1052. Velick, *J. Biol. Chem.,* 233, 1455, (1958).
1053. Hakala, Glaid, and Schwert, *J. Biol. Chem.,* 221, 191 (1956).
1054. Neilands, *J. Biol. Chem.,* 199, 373 (1952).
1055. Nisselbaum and Bodansky, *J. Biol. Chem.,* 236, 323 (1961).
1056. Symons and Burgoyne, *Methods Enzymol.,* 9, 314 (1966).
1057. Kaloustian, Stolzenbach, Everse, and Kaplan, *J. Biol. Chem.,* 244, 2891 (1969).
1058. Vestling and Kunsch, *Arch. Biochem. Biophsy.,* 127, 568 (1968).
1059. Kubowitz and Ott, *Biochem. Z.,* 314, 94 (1943).
1060. Castellino, Fish, and Mann, *J. Biol. Chem.,* 245, 4269 (1970).
1061. Ghosh, Chaudhuri, Roy, Sinha, and Sen, *Arch. Biochem. Biophys.,* 144, 6 (1971).
1062. Joniau, Bloemmen, and Lontie, *Biochim. Biophys. Acta,* 214, 468 (1970).
1063. Groves, in *Milk Proteins, Chemistry and Molecular Biology,* McKenzie, Ed., Vol. 2, Academic Press, New York, 1971, 367.
1064. Hou and Perlman, *J. Biol. Chem.,* 245, 1289 (1970).
1065. Carlstrom, *Acta Chem. Scand.,* 23, 185 (1969).
1066. Theorell and Pedersen, in *The Svedberg,* Tiselius and Pedersen, Eds., Almqvist and Wiksells Boktryckeri, Uppsala and Stockholm, 1944, 523.
1067. Polis and Shmukler, *J. Biol. Chem.,* 201, 475 (1953).
1068. Appleby, *Biochim. Biophys. Acta,* 188, 222 (1969).
1069. Sjögren and Svedberg, *J. Am. Chem. Soc.,* 52, 3279 (1930).
1070. Hermier, Lebeault, and Zevaco, *Bull. Soc. Chim. Biol.,* 52, 1089 (1970).
1071. Verger, Sarda, and Desnuelle, *Biochim. Biophys. Acta,* 242, 580 (1971).
1072. Vandermeers and Christophe, *Biochim. Biophys. Acta,* 154, 110 (1968).
1073. Brown, Levy, and Fredrickson, *J. Biol. Chem.,* 245, 6588 (1970).
1074. Koga, Horwitz, and Scanu, *J. Lipid Res.,* 10, 577 (1969).
1075. Chlumecká, Tigerstrom, D'Obrenan, and Smith, *J. Biol. Chem.,* 245, 5481 (1969).
1076. Hayaishi, in *Oxidases and Related Systems,* King, Mason, and Morrison, Eds., John Wiley & Sons, New York, 1965, 286.
1077. Cassio and Waller, *Eur. J. Biochem.,* 20, 283 (1971).
1078. Dalton, Morris, Ward, and Mortensen, *Biochemistry,* 10, 2066 (1971).
1079. Keilin and Hartree, *Biochem. J.,* 61, 153 (1953).
1080. Harrison and Blout, *J. Biol. Chem.,* 240, 299 (1965).
1081. Brown, Martinez, Johnstone, and Olcott, *J. Biol. Chem.,* 237, 81 (1962).
1082. Rossi-Fanelli, Antonini, and Povoledo, in *Symposium on Protein Structure,* Neuberger, Ed., Methuen and Co., London, 1958, 144.
1083. Atassi and Saplin, *Biochem. J.,* 98, 82 (1966).
1084. Herskovits, *Arch. Biochem. Biophys.,* 130, 19 (1969).
1085. Crumpton and Wilkinson, *Biochem. J.,* 94, 545 (1965).
1086. Atassi, *Biochim. Biophys. Acta,* 221, 612 (1970).
1087. Cameron, Azzam, Kotite, and Awad, *J. Lab. Clin. Med.,* 65, 883 (1965).
1088. Crumpton and Polson, *J. Mol. Biol.,* 11, 722 (1965).
1089. Boegman and Crumpton, *Biochem. J.,* 120, 373 (1970).
1090. Breslow, *J. Biol. Chem.,* 239, 486 (1964).
1091. Hermans, Jr., *Biochemistry,* 1, 193 (1962).
1092. Edmundson and Hirs, *Nature,* 190, 663 (1961).
1093. Ray and Gurd, *J. Biol. Chem.,* 242, 2062 (1967).
1094. Terwilliger and Read, *Comp. Biochem. Physiol.,* 29, 551 (1969).
1095. Terwilliger and Read, *Comp. Biochem. Physiol.,* 31, 55 (1969).
1096. Kritcher, Thyrum, and Luchi, *Biochim. Biophys. Acta,* 221, 264 (1970).
1097. Gazith, Himmelfarb, and Harrington, *J. Biol. Chem.,* 245, 15 (1970).
1098. Gershman, Stracher, and Dreizen, *J. Biol. Chem.,* 244, 2726 (1969).
1099. Shimizu, Morita, and Yagi, *J. Biochem.* (Tokyo), 69, 447 (1971).
1100. Young, Blanchard, and Brown, *Proc. Natl. Acad. Sci. U.S.A.,* 61, 1087 (1968).
1101. Otaiza and Jaenicke, *Hoppe-Seylers Z. Physiol. Chem.,* 352, 385 (1971).
1102. Bocchini, *Eur. J. Biochem.,* 15, 127 (1970).
1103. Karlsson, Arnberg, and Eaker, *Eur. J. Biochem.,* 21, 1 (1971).

1104.   Miranda, Kupeyan, Rochat, Rochat, and Lissitzky, *Eur. J. Biochem.*, 17, 477 (1970).
1105.   Miranda, Kupeyan, Rochat, Rochat, and Lissitzky, *Eur. J. Biochem.*, 16, 514 (1970).
1106.   Holcenberg and Stadtman, *J. Biol. Chem.*, 244, 1194 (1969).
1107.   Forget, *Eur. J. Biochem.*, 18, 442 (1971).
1108.   Burns, Holsten, and Hardy, *Biochem. Biophys. Res. Commun.*, 39, 90 (1970).
1109.   Nossal and Hershfield, *J. Biol. Chem.*, 246, 541 (1971).
1110.   Omenn, Onjes, and Anfinsen, *Biochemistry*, 9, 304 (1970).
1111.   Albrecht, *Biochemistry*, 9, 2462 (1970).
1112.   Glazer and Smith, *J. Biol. Chem.*, 235, PC43 (1960).
1113.   Weintraub and Schlamowitz, *Comp. Biochem. Physiol.*, 38B, 513 (1971).
1114.   Weintraub and Schlamowitz, *Comp. Biochem. Physiol.*, 37, 49 (1970).
1115.   Tomimatsu, Clary, and Bartulovich, *Arch. Biochem. Biophys.*, 115, 536 (1966).
1116.   Liu, Means, and Feeny, *Biochim. Biophys. Acta*, 229, 176 (1971).
1117.   Donovan, Mapes, Davis, and Hamburg, *Biochemistry*, 8, 4190 (1969).
1118.   Donovan, Davis, and White, *Biochim. Biophys. Acta*, 207, 190 (1970).
1119.   Davis, Mapes, and Donovan, *Biochemistry*, 10, 39 (1971).
1120.   Bender and Brubacher, *J. Am. Chem. Soc.*, 86, 5333 (1964).
1121.   Hinkle and Kirsch, *Biochemistry*, 9, 4633 (1970).
1122.   Arnon and Shapira, *J. Biol. Chem.*, 244, 1033 (1969).
1123.   Olander, *Biochemistry*, 10, 601 (1971).
1124.   Woods, *Biochem. J.*, 113, 39 (1969).
1125.   Milstein, *Biochem. J.* 103, 634 (1967).
1126.   Citri and Pollock, *Adv. Enzymol.*, 28, 237 (1966).
1127.   Imsande, Gillin, Tanis, and Atherly, *J. Biol. Chem.*, 245, 2205 (1970).
1128.   Lindstrom, Boman, and Steele, *J. Biol. Chem.*, 101, 218 (1970).
1129.   Sodek and Hofmann, *Methods Enzymol.*, 19, 372 (1970).
1130.   Morita, Toshida, and Maeda, *Agric. Biol. Chem.*, 35, 1074 (1971).
1131.   Staprans and Watanabe, *J. Biol. Chem.*, 245, 5962 (1970).
1132.   Peterson, *J. Biol. Chem.*, 246, 34 (1971).
1133.   Peterson and Berggard, *J. Biol. Chem.*, 246, 25 (1971).
1134.   Kuehn and McFadden, *Biochemistry*, 8, 2394 (1968).
1135.   Nicholson, *Biochem. J.*, 123, 117 (1971).
1136.   Labow and Robinson, *J. Biol. Chem.*, 241, 1239 (1966).
1137.   Dowhan, Jr. and Snell, *J. Biol. Chem.* 245, 4618 (1970).
1138.   Matsubara, *Methods Enzymol.*, 19, 642 (1970).
1139.   Berglund and Sjöberg, *J. Biol. Chem.*, 245, 6030 (1970).
1140.   Fedberg and Datta, *Eur. J. Biochem.*, 21, 438 (1971).
1141.   Hug and Roth, *Biochemistry*, 10, 1397 (1971).
1142.   White and Barlow, *Methods Enzymol.*, 19, 665 (1970).
1143.   Gumpf and Hamilton, *Virology*, 35, 87 (1968).
1144.   Yamazaki and Kaesberg, *J. Mol. Biol.*, 6, 465 (1963).
1145.   Stubbs and Kaesberg, *J. Mol. Biol.*, 8, 314 (1964).
1146.   Bachrach and Van den Woude, *Virology*, 34, 282 (1968).
1147.   Rueckert, *Virology*, 26, 345 (1965).
1148.   Ghabrial, Shepherd, and Grogan, *Virology*, 33, 17 (1967).
1149.   Damirdagh and Shepherd, *Virology*, 40, 84 (1970).
1150.   Budzynski and Fraenkel-Conrat, *Biochemistry*, 9, 3300 (1970).
1151.   Tremaine and Stace-Smith, *Virology*, 35, 102 (1968).
1152.   Dorne, Jonard, Witz, and Hirth, *Virology*, 43, 279 (1971).
1153.   Miki and Knight, *Virology*, 31, 55 (1967).
1154.   Wallace, *Biochim. Biophys. Acta*, 215, 176 (1970).
1155.   Bray, Chisholm, Hart, Meriwether, and Watts, in *Flavins and Flavoproteins*, Slater, Ed., Elsevier, Amsterdam, 1966, 117.
1156.   West, Nagy, and Gergely, in *Symposium on Fibrous Proteins*, Crewther, Ed., Plenum Press, New York, 1968, 164.
1157.   Kimura and Ting, *Biochem. Biophys. Res. Commun.*, 45, 1227 (1971).
1158.   Fushimi, Hamison, and Ravin, *J. Biochem.* (Tokyo), 69, 1041 (1971).
1159.   Shinowara, in *Blood Platelets, Henry Ford International Symposium No. 10*, Johnson, Monto, Rebuck, and Horn, Jr., Eds., Little Brown, Boston, 1961, 347.
1160.   Reynolds and Johnson, *Biochemistry*, 10, 2821 (1971).
1161.   Wu, Cluskey, Krull, and Friedman, *Can. J. Biochem.*, 49, 1042 (1971).
1162.   Berrens and Bleumink, *Int. Arch. Allergy*, 28, 150 (1965).

1163.   Feldman and Weiner, *J. Biol. Chem.*, 247, 260 (1972).
1164.   Guha, Lai, and Horecker, *Arch. Biochem. Biophys.*, 147, 692 (1971).
1165.   Caban and Hass, *J. Biol. Chem.*, 246, 6807 (1971).
1166.   Penhoet, Kochman and Rutter, *Biochemistry*, 8, 4396 (1969).
1167.   Chiu and Feingold, *Biochemistry*, 8, 98 (1969).
1168.   Robertson, Hammerstedt, and Wood, *J. Biol. Chem.*, 246, 2075 (1971).
1169.   'S-Gravenmade, Drift, Van Der, and Vogels, *Biochim. Biophys. Acta*, 251, 393 (1971).
1170.   King and Norman, *Biochemistry*, 1, 709 (1962).
1171.   Underdown and Goodfriend, *Biochemistry*, 8, 980 (1969).
1172.   Mazelis and Crews, *Biochem. J.*, 108, 725 (1968).
1173.   Svedberg and Sjögren, *J. Am. Chem. Soc.*, 52, 279 (1930).
1174.   Eady and Large, *Biochem. J.*, 123, 757 (1971).
1175.   Henn and Ackers, *J. Biol. Chem.*, 244, 465 (1969).
1176.   Miyake and Yamano, *Biochim. Biophys. Acta*, 198, 438 (1970).
1177.   Soda and Osumi, *Biochem. Biophys. Res. Commun.*, 35, 363 (1969).
1178.   McKeehan and Hardesty, *J. Biol. Chem.*, 244, 4330 (1969).
1179.   Yamada, Adachi, and Ogata, in *Pyridoxyl Catalysis: Enzymes and Model Systems,* Snell, Braunstein, Severin, and Torchinsky, Eds., Interscience, New York, 1968, 347.
1180.   Wang, Achee, and Yasunobu, *Arch. Biochem. Biophys.*, 128, 106 (1968).
1181.   Prescott, Wilkes, Wagner, and Wilson, *J. Biol. Chem.*, 246, 1756 (1971).
1182.   Hanson, Hutter, Mansfeldt, Kretschmer, and Sohr, *Hoppe-Seylers Z. Physiol. Chem.*, 348, 680 (1967).
1183.   Wacker, Lehky, Fischer, and Stein, *Helv. Chim. Acta*, 54, 473 (1971).
1184.   Roncari and Zuber, *Int. J. Protein Res.*, 1, 45 (1969).
1185.   Pfleiderer and Femfert, *FEBS Lett.*, 4, 265 (1969).
1186.   Auricchio and Bruni, *Biochem. Z.*, 340, 321 (1964).
1187.   Yaron and Mlynar, *Biochem. Biophys. Res. Commun.*, 32, 658 (1968).
1188.   Grszkiewicz, *Acta Biochem. Pol.*, 9, 301 (1962).
1189.   Vandermeers and Christophe, *Biochim. Biophys. Acta*, 154, 110 (1968).
1190.   Menzi, Stein, and Fischer, *Helv. Chim. Acta*, 40, 534 (1957).
1191.   Yutani, Yutani, and Isemura, *J. Biochem.* (Tokyo), 66, 823 (1969).
1192.   Nishida, Fukumoto, and Yamamoto, *Agric. Biol. Chem.*, 31, 682 (1967).
1193.   Nishida, Ph. D. thesis; cited in Nishida, Fukumoto, and Yamamoto, *Agric. Biol. Chem.*, 31, 682 (1967).
1194.   Ogasahara, Imanishi, and Isemura, *J. Biochem.* (Tokyo), 67, 65 (1970).
1195.   Junge, Stein, Neurath, and Fischer, *J. Biol. Chem.*, 234, 556 (1959).
1196.   Sanders and Rutter, *Biochemistry*, 11, 130 (1972).
1197.   Krysteva and Erodi, *Acta Biochim. Biophys. Acad. Sci. Hung.*, 3, 275 (1968).
1198.   Yoshida, Hiroshi, and Ono, *J. Biochem.* (Tokyo), 65, 741 (1969).
1199.   Yutani, Yutani, and Isemura, *J. Biochem.* (Tokyo), 65, 201 (1969).
1200.   Shainkin and Berk, *Biochim. Biophys. Acta*, 221, 502 (1970).
1201.   Takeda and Hizukuri, *Biochim. Biophys. Acta*, 185, 469 (1969).
1202.   Englard, Sorof, and Singer, *J. Biol. Chem.*, 189, 217 (1951).
1203.   Nagano and Zalkin, *J. Biol. Chem.*, 245, 3097 (1970).
1204.   Ito, Cox, and Yanofsky, *J. Bacteriol.*, 97, 725 (1969).
1205.   Henderson and Zalkin, *J. Biol. Chem.*, 246, 6891 (1971).
1206.   Warner and Schumaker, *Biochemistry*, 9, 451 (1970).
1207.   Bezkorovainy, Springer, and Dese, *Biochemistry*, 10, 3761 (1971).
1208.   Kim, *Vox Sang*, 20, 461 (1971).
1209.   Bezkorovainy, Springer, and Hotti, *Biochim. Biophys. Acta*, 115, 501 (1966).
1210.   Preer, *J. Immunol.*, 33, 385 (1959).
1211.   Kaji and Tagawa, *Biochim. Biophys. Acta*, 207, 456 (1970).
1212.   Harell and Sokolovsky, *Eur. J. Biochem.*, 25, 102 (1972).
1213.   Sakai and Murachi, *Physiol. Chem. Phys.*, 1, 31 (1969).
1214.   Blethen, Boeker, and Snell, *J. Biol. Chem.*, 243, 1671 (1968).
1215.   Regnouf, Pradel, Kassab, and Thoai, *Biochim. Biophys. Acta*, 194, 540 (1969).
1216.   Oriol-Audit, Landon, Robin, and Thoai, *Biochim. Biophys. Acta*, 188, 132 (1969).
1217.   Kassab, Fattoum, and Pradel, *Eur. J. Biochem.*, 12, 264 (1970).
1218.   Kassab, Roustan, and Pradel, *Biochim. Biophys. Acta*, 167, 308 (1968).
1219.   Landon, Oriol, and Thoai, *Biochim. Biophys. Acta*, 214, 168 (1970).
1220.   Yorifuji, Ogata, and Soda, *J. Biol. Chem.*, 246, 5085 (1971).
1221.   Bray and Ratner, *Arch. Biochem. Biophys.*, 146, 531 (1971).
1222.   Nakano, Tsutsumi, and Danowski, *J. Biol. Chem.*, 245, 4443 (1974).
1223.   Nakamura, Makino, and Ogura, *J. Biochem.* (Tokyo), 64, 189 (1969).

1224. **Penton and Dawson,** in *Oxidases and Related Redox Systems,* King, Mason, and Morrison, Eds., John Wiley and Sons, New York, 1965, 221.
1225. **Tosa, Sano, Yamamoto, Nakamura, and Chibata,** *Biochemistry,* 11, 21 (1972).
1226. **Lu and Handschumacher,** *J. Biol. Chem.,* 247, 66 (1972).
1227. **Nishumara, Makino, Takenaka, and Inada,** *Biochim. Biophys. Acta,* 227, 171 (1971).
1228. **Shifrin and Grochowski,** *J. Biol. Chem.,* 247, 1048 (1972).
1229. **Kakimoto, Kato, Shibatani, Nishimura, and Chibata,** *J. Biol. Chem.,* 244, 353 (1969).
1230. **Banks, Doonan, Lawrence, and Vernon,** *Eur. J. Biochem.,* 5, 528 (1968).
1231. **Banks and Vernon,** *J. Chem. Soc.,* p. 1968 (1961).
1232. **Arrio-Dupont, Cournil, and Duie,** *FEBS Lett.,* 11, 144 (1970).
1233. **Martinez-Carrion, Kuczenski, Tiemeier, and Peterson,** *J. Biol. Chem.,* 245, 799 (1970).
1234. **Bergami, Marino, and Scardi,** *Biochem. J.,* 110, 471 (1968).
1235. **Bertlund and Kaplan,** *Biochemistry,* 9, 2653 (1970).
1236. **Magee and Phillips,** *Biochemistry,* 10, 3397 (1971).
1237. **Marino, Greco, Scardi, and Zito,** *Biochem. J.,* 99, 589 (1966).
1238. **Scardi,** in *Pyridoxal Catalysis: Enzymes and Model Systems,* Snell, Braunstein, Severin, and Torchinsky, Eds., Interscience, New York, 1968, 179.
1239. **Nelbach, Pigiet, Gerhart, and Schachman,** *Biochemistry,* 11, 315 (1972).
1240. **Meighen, Pigiet, and Schachman,** *Proc. Natl. Acad. Sci., U.S.A.,* 65, 234 (1970).
1241. **Benisek,** *J. Biol. Chem.,* 246, 3151 (1971).
1242. **Vanaman and Stark,** *J. Biol. Chem.,* 245, 3565 (1970).
1243. **Biswas, Gray, and Paulus,** *J. Biol. Chem.,* 245, 4900 (1970).
1244. **Lafuma, Gros, and Patte,** *Eur. J. Biochem.,* 15, 111 (1970).
1245. **Janin, van Rapenbusch, Truffa-Bachi, Cohen, and Gros,** *Eur. J. Biochem.,* 8, 128 (1969).
1246. **Truffa-Bachi, van Rapenbusch, Janin, Gros, and Cohen,** *Eur. J. Biochem.,* 7, 401 (1969).
1247. **Falcoz-Kelly, van Rapenbusch, and Cohen,** *Eur. J. Biochem.,* 8, 146 (1969).
1248. **Inoue, Suzuki, Fukunishi, Adachi, and Takeda,** *J. Biol. Chem.,* 60, 543 (1966).
1249. **Melamed and Green,** *Biochem. J.,* 89, 591 (1963).
1250. **Green,** *Biochem. J.,* 89, 599 (1963).
1251. **Scrutton,** *Biochemistry,* 10, 3897 (1971).
1252. **Brill, Bryce, and Maria,** *Biochim. Biophys. Acta,* 154, 342 (1968).
1253. **Haupt, Heimburger, Krantz, and Schwick,** *Eur. J. Biochem.,* 17, 254 (1970).
1254. **Vonemasu, Stroud, Niedermeier, and Butler,** *Biochem. Biophys. Res. Commun.,* 43, 1388 (1971).
1255. **Nilsson and Lindskog,** *Eur. J. Biochem.,* 2, 309 (1967).
1256. **Carter,** *Biochim. Biophys. Acta,* 235, 222 (1971).
1257. **Bernstein and Schraer,** *J. Biol. Chem.,* 247, 1306 (1972).
1258. **Bernstein and Schraer,** *Fed. Proc.,* Abstr. 1387, 30 (1291).
1259. **Maynard and Coleman,** *J. Biol. Chem.,* 246, 4455 (1971).
1260. **Carter and Parsons,** *Biochem. J.,* 120, 797 (1970).
1261. **Funakoshi and Deutsch,** *J. Biol. Chem.,* 245, 4913 (1970).
1262. **Edsall, Mehta, Myers, and Armstrong,** *Biochem. Z.,* 345, 9 (1966).
1263. **Ashworth, Spencer, and Brewer,** *Arch. Biochem. Biophys.,* 142, 122 (1971).
1264. **Tanis, Tashian, and Yu,** *J. Biol. Chem.,* 245, 6003 (1970).
1265. **Rossi, Chersi, and Cortivo,** in CO_2: *Chemical, Biochemical and Physiological Aspects,* Forster, Edsall, Otis, and Roughton, Eds., NASA SP-188, 1969, 131.
1266. **Runnegar, Scott, Webb, and Zerner,** *Biochemistry,* 8, 2013 (1969).
1267. **Runnegar, Webb, and Zerner,** *Biochemistry,* 8, 2018 (1969).
1268. **Franz and Krisch,** *Hoppe-Seylers Z. Physiol. Chem.,* 149, 575 (1968).
1269. **Chase and Tubbs,** *Biochem. J.,* 111, 225 (1969).
1270. **Zagalsky, Cheesman, and Ceccaldi,** *Comp. Biochem. Physiol.,* 22, 851 (1967).
1271. **Zagalsky, Ceccaldi, and Daumas,** *Comp. Biochem. Physiol.,* 34, 579 (1970).
1272. **Dumas and Garnier,** *J. Dairy Res.,* 37, 269 (1970).
1273. **Thompson and Pepper,** *J. Dairy Sci.,* 47, 633 (1964).
1274. **Herskovits,** *Arch. Biochem. Biophys.,* 130, 19 (1969).
1275. **Carrico and Deutsch,** *J. Biol. Chem.,* 244, 6087 (1969).
1276. **Porter and Folch,** *J. Neurochem.,* 1, 260 (1957).
1277. **Ashwell and Morell,** in *Red Cross Scientific Symposium on Glycoproteins of Blood Cells and Plasma,* Jamieson and Greenwalt, Eds., J. B. Lippincott, Philadelphia, 1971, 173.
1278. **Blumberg, Eisinger, Aisen, Morell, and Scheinberg,** *J. Biol. Chem.,* 238, 1675 (1963).
1279. **Ryden,** *Int. J. Protein Res.,* 3, 131 (1971).
1280. **Matsunaga and Nosoh,** *Biochim. Biophys. Acta,* 215, 280 (1970).
1281. **Osaki,** *J. Biochem.* (Tokyo), 48, 190 (1960).

1282. Milne and Matrone, *Biochim. Biophys. Acta,* 212, 43 (1970).
1283. Holtzman and Gaumnitz, *J. Biol. Chem.,* 245, 2350 (1970).
1284. Lospalluto and Finkelstein, *Biochim. Biophys. Acta,* 257, 158 (1972).
1285. Kover, Szaboic, and Csabal, *Arch. Biochem. Biophys.,* 106, 333 (1964).
1286. Schumberger, *Z. Naturforsch. (B),* 23, 1412 (1968).
1287. Bahl, *J. Biol. Chem.,* 244, 567 (1969).
1288. Koch, Shaw, and Gibson, *Biochim. Biophys. Acta,* 229, 805 (1971).
1289. Koch, Shaw, and Gibson, *Biochim. Biophys. Acta,* 229, 795 (1971).
1290. Koch, Shaw, and Gibson, *Biochim. Biophys. Acta,* 212, 375 (1970).
1291. Koch, Shaw, and Gibson, *Biochim. Biophys. Acta,* 212, 387 (1970).
1292. Nakagawa and Bender, *Biochemistry,* 9, 259 (1970).
1293. Singh, Thornton, and Westheimer, *J. Biol. Chem.,* 237, PC3006 (1962).
1294. Webster, *Biochim. Biophys. Acta,* 207, 371 (1970).
1295. Krausz and Becker, *J. Biol. Chem.,* 243, 4606 (1968).
1296. Oliver, Viswanatha, and Whish, *Biochem. Biophys. Res. Commun.,* 27, 107 (1967).
1297. Babul and Stellwagen, *Anal. Biochem.,* 28, 216 (1969).
1298. Rovery, *Methods Enzymol.,* 11, 231 (1967).
1299. Coan, Roberts, and Travis, *Biochemistry,* 10, 2711 (1971).
1300. Jackson and Brandts, *Biochemistry,* 9, 2294 (1970).
1301. Chervenka and Wilcox, *J. Biol. Chem.,* 222, 621 (1956).
1302. Glazer and Smith, *J. Biol. Chem.,* 235, PC43 (1960).
1303. Vandermeers and Christophe *Biochim. Biophys. Acta,* 188, 101 (1969).
1304. Brandts and Lumry, *J. Phys. Chem.,* 67, 1484 (1963).
1305. Nichol, *J. Biol. Chem.,* 243, 4065 (1968).
1306. Prahl and Neurath, *Biochemistry,* 5, 2131 (1966).
1307. Gratecos, Guy, Rovery, and Desnuelle, *Biochim. Biophys. Acta,* 175, 82 (1969).
1308. Singh, Brooks, and Srere, *J. Biol. Chem.,* 245, 4636 (1970).
1309. Berger, Kafatos, Felsted, and Law, *J. Biol. Chem.,* 246, 4131 (1971).
1310. Hruska and Law, *Methods Enzymol.,* 19, 221 (1970).
1311. Schwartz and Helinski, *J. Biol. Chem.,* 246, 6318 (1971).
1312. Grant and Alburn, *Arch. Biochem. Biophys.,* 82, 245 (1959).
1313. Rhoads and Udenfriend, *Arch. Biochem. Biophys.,* 139, 329 (1970).
1314. Shin and Mayer, *Biochemistry,* 7, 2991 (1968).
1315. Emery, *Biochemistry,* 8, 877 (1969).
1316. Ehrenpreis and Warner, *Arch. Biochem. Biophys.,* 61, 38 (1956).
1317. Grant-Greene and Friedberg, *Int. J. Protein Res.,* 2, 235 (1970).
1318. Lerman, *Can. J. Biochem.,* 47, 1115 (1969).
1319. Augusteyn and Spector, *Biochem. J.,* 124, 345 (1971).
1320. Gregory, Holdsworth, and Ottesen *C. R. Trav. Lab. Carlsberg Ser. Chim.,* 30, 147 (1957).
1321. Holdsworth, *Biochim. Biophys. Acta,* 51, 295 (1961).
1322. Rosenberry, Chang, and Chen, *J. Biol. Chem.,* 247, 1555 (1972).
1323. Verpoorte, *J. Biol. Chem.,* 247, 4787 (1972).
1324. West, Nagy, and Gergely, in *Symposium on Fibrous Proteins,* Crewther, Ed., Plenum Press, New York, 1968, 164.
1325. Schramm and Hochstein, *Biochemistry,* 11, 2777 (1972).
1326. Day, Franklin, Pettersson, and Philipson, *Eur. J. Biochem.,* 29, 537 (1972).
1327. Ronca-Testoni, Ranieri, Raggi, and Ronca, *Ital. J. Biochem.,* 19, 262 (1970).
1328. Suhara, Takemori, and Katagiri, *Biochim. Biophys. Acta,* 263, 272 (1972).
1329. Levine, Kaplan, and Greenaway, *Biochem. J.,* 129, 847 (1972).
1330. Claesson, *Ark. Kem.,* 10, 4 (1956).
1331. Wallenfels and Herrmann, *Methods Enzymol.,* 9, 608 (1966).
1332. Berrens, in *The Chemistry of Atopic Allergens,* Karger, Basel, 1971, 205.
1333. Marsuzawa and Segal, *J. Biol. Chem.,* 243, 5929 (1968).
1334. Visuri and Nummi, *Eur. J. Biochem.,* 28, 555 (1972).
1335. Wissler, *Eur. J. Immunol.,* 2, 73 (1972).
1336. Stellwagen, Rysavy, and Babul, *J. Biol. Chem.,* 247, 8074 (1972).
1337. Lux, John, and Brewer, *J. Biol. Chem.,* 247, 7510 (1972).
1338. Tombs, *Biochem. J.,* 96, 119 (1965).
1339. Grazi and Magri, *Biochem. J.,* 126, 667 (1972).
1340. Nakamura, Makino and Ogura, *J. Biochem.* (Tokyo), 64, 189 (1968).
1341. Cammack, Marlborough, and Miller, *Biochem. J.,* 126, 316 (1972).
1342. Laboureur, Langlois, Labrousse, Boudon, Emeraud, Samain, Ageron, and Dumesnil, *Biochimie,* 53, 1147 (1971).

1343.   Scandurra and Cannella, *Eur. J. Biochem.,* 27, 196 (1972).
1344.   D'Aniello and Rocca, *Comp. Biochem. Physiol.,* B41, 625 (1972).
1345.   Falcoz-Kelly, Janin, Saari, Veron, Truffa-Bachi, and Cohen, *Eur. J. Biochem.,* 28, 507 (1972).
1346.   Thuma, Schirmer, and Schirmer, *Biochim. Biophys. Acta,* 268, 81 (1972).
1347.   Tweedie and Segel, *Prep. Biochem.,* 1, 91 (1971).
1348.   Wolff and Siegel, *J. Biol. Chem.,* 247, 4180 (1972).
1349.   Virden, *Biochem. J.,* 127, 503 (1972).
1350.   Brundell, Falkbring, and Nyman, *Biochim. Biophys. Acta,* 284, 311 (1972).
1351.   Kang, Storm, and Carson, *Biochim. Biophys. Res. Commun.,* 49, 621 (1972).
1352.   Ihle and Dure, III, *J. Biol. Chem.,* 247, 5034 (1972).
1353.   Johansen, Livingston, and Vallee, *Biochemistry,* 11, 2584 (1972).
1354.   Zagalsky and Herring, *Comp. Biochem. Physiol.,* B41, 397 (1972).
1355.   Bonaventura, Schroeder, and Fang, *Arch. Biochem. Biophys.,* 150, 606 (1972).
1356.   Price, Sterling, Tarantola, Hartley, and Rechcigl, *J. Biol. Chem.,* 237, 3468 (1962).
1357.   Otto and Bhakdi, *Hoppe-Seyler's Z. Physiol. Chem.,* 350, 1577 (1969).
1358.   Skujins, Pukite, and McLaren, *Enzymologia,* 39, 353 (1970).
1359.   Timmis, *J. Bacteriol.,* 109, 12 (1972).
1360.   Maylie, Charles, Gache, and Desnuelle, *Biochim. Biophys. Acta,* 229, 286 (1971).
1361.   Barth, Bunnenberg, and Djerassi, *Anal. Biochem.,* 48, 471 (1972).
1362.   Waterson, Castellino, Hass, and Hill, *J. Biol. Chem.,* 247, 5266 (1972).
1363.   Wissler, *Eur. J. Immunol.,* 2, 84 (1972).
1364.   Goldberger, Smith, Tisdale, and Bomstein, *J. Biol. Chem.,* 236, 2788 (1961).
1365.   Lederer and Simon, *Eur. J. Biochem.,* 20, 469 (1971).
1366.   Groudinsky, *Eur. J. Biochem.,* 18, 480 (1971).
1367.   Mevel-Ninio, Pajot, and Labeyrie, *Biochimie,* 53, 35 (1971).
1368.   Monteilhet and Risler, *Eur. J. Biochem.,* 12, 165 (1970).
1369.   Strittmatter and Velick, *J. Biol. Chem.,* 221, 253 (1956).
1370.   Itagaki and Hager, *J. Biol. Chem.,* 241, 3687 (1966).
1371.   Iwasaki and Shidara, *Plant Cell Physiol.,* 10, 291 (1969).
1372.   Webster, *Biochim. Biophys. Acta,* 207, 371 (1970).
1373.   Flatmark and Sletten, *J. Biol. Chem.,* 243, 1623 (1968).
1374.   Mayer and Miller, *Anal. Biochem.,* 36, 91 (1970).
1375.   Herskovits, *Arch. Biochem. Biophys.,* 130, 19 (1969).
1376.   Herskovits, Jaillet, and Gadegbeku, *J. Biol. Chem.,* 245, 4544 (1970).
1377.   Bolard and Garnier, *Biochim. Biophys. Acta,* 263, 535 (1972).
1378.   Scholes, McLain, and Smith, *Biochemistry,* 10, 2072 (1971).
1379.   Schejter, Grosman, and Sokolovsky, *Isr. J. Chem.,* 10, 37 (1972).
1380.   Yu, Yu, and King, *J. Biol. Chem.,* 247, 1012 (1972).
1381.   Horio and Kamen, *Biochim. Biophys. Acta,* 48, 266 (1961).
1382.   Cusanovich, Tedro, and Kamen, *Arch. Biochem. Biophys.,* 141, 557 (1970).
1383.   Iwasaki and Matsubara, *J. Biochem.* (Tokyo), 69, 847 (1971).
1384.   Miki and Okunuki, *J. Biochem.* (Tokyo), 66, 831 (1969).
1385.   Clark-Walker and Lascelles, *Arch. Biochem. Biophys.,* 136, 153 (1970).
1386.   Yamanaka, Takenami, Akijama, and Okunuki, *J. Biochem.* (Tokyo), 70, 349 (1971).
1387.   Shioi, Takamiya, and Nishimura, *J. Biochem.* (Tokyo), 71, 285 (1972).
1388.   Yong and King, *J. Biol. Chem.,* 245, 1331 (1970).
1389.   Sugimura and Yakushiji, *J. Biochem.* (Tokyo), 63, 281 (1968).
1390.   Laycock and Craigie, *Can. J. Biochem.,* 49, 641 (1971).
1391.   Kusel, Suriano, and Weber, *Arch. Biochem. Biophys.,* 133, 293 (1969).
1392.   Stripp, Greene, and Gillette, *Pharmacology,* 6, 56 (1971).
1393.   Kuronen and Ellfolk, *Biochim. Biophys. Acta,* 275, 308 (1972).
1394.   Ellfolk and Soininen, *Acta Chem. Scand.,* 25, 1535 (1971).
1395.   Hiraoka, Fukumoto, and Tsuru, *J. Biochem.* (Tokyo), 71, 57 (1972).
1396.   D'Souza, Warwick, and Freisheim, *Biochemistry,* 11, 1528 (1972).
1397.   Gunderson, Dunlap, Harding, Freisheim, Otting, and Huennekens, *Biochemistry,* 11, 1018 (1972).
1398.   Greenfield, Williams, Poe, and Hoogsteen, *Biochemistry,* 11, 4706 (1972).
1399.   Erickson and Mathews, *Biochemistry,* 12, 372 (1973).
1400.   Butzow, *Biochim. Biophys. Acta,* 168, 490 (1968).
1401.   Setlow, Brutlag, and Kornberg, *J. Biol. Chem.,* 247, 224 (1972).
1402.   Ueda, Lode, and Coon, *J. Biol. Chem.,* 247, 2109 (1972).
1403.   Robinson and Maxwell, *J. Biol. Chem.,* 247, 7023 (1972).
1404.   Rexova-Benkova and Slezarik, *Collect. Czech. Chem. Commun.,* 35, 1255 (1970).

1405. Spring and Wold, *J. Biol. Chem.,* 246, 6797 (1971).
1406. Winstead, *Biochemistry,* 11, 1046 (1972).
1407. Malmstrom, *Arch. Biochem. Biophys. Suppl.,* 1, 247 (1962).
1408. Maroux, Baratti, and Desnuelle, *J. Biol. Chem.,* 246, 5031 (1971).
1409. Schantz, Roessler, Woodburn, Lyach, Jacoby, Silverman, Gorman, and Spero, *Biochemistry,* 11, 360 (1972).
1410. Borja, Fanning, Huang, and Bergdoll, *J. Biol. Chem.,* 247, 2456 (1972).
1411. Markland, and Damus, *J. Biol. Chem.,* 246, 6460 (1971).
1412. Swaney, and Klotz, *Arch. Biochem. Biophys.,* 147, 475 (1971).
1413. Weser, Bunnenberg, Cammack, Djerassi, Flohe, Thomas, and Voelter, *Biochim. Biophys. Acta,* 243, 203 (1971).
1414. Weser, and Hartmann, *Fed. Eur. Biochem. Soc. Lett.,* 17, 78 (1971).
1415. Weser, Barth, Djerassi, Hartmann, Krauss, Voelker, Voelter, and Voetsch, *Biochem. Biophys. Acta,* 278, 28 (1972).
1416. Espada, Langton, and Dorado, *Biochim. Biophys. Acta,* 285, 427 (1972).
1417. Omori-Satoh, Sadahiro, Ohsaka, and Murata, *Biochim. Biophys. Acta,* 285, 414 (1972).
1418. Aloof-Hirsch, DeVries, and Berger, *Biochim. Biophys. Acta,* 154, 53 (1968).
1419. Taylor, Mitchell, and Cohen, *J. Biol. Chem.,* 247, 5928 (1972).
1420. Fujikawa, Legaz, and Davie, *Biochemistry,* 11, 4882 (1972).
1421. Radcliffe, and Barton, *J. Biol. Chem.,* 247, 7735 (1972).
1422. Takagi, and Konishi, *Biochim. Biophys. Acta,* 271, 363 (1972).
1423. Monahan, Rivier, Vale, Guillemin, and Burgus, *Biochim. Biophys. Res. Commun.,* 47, 551 (1972).
1424. Dutler, Coon, Kull, Vogel, Waldvogel, and Prelog, *Eur. J. Biochem.,* 22, 203 (1971).
1425. Yoch, and Arnon, *J. Biol. Chem.,* 247, 4514 (1972).
1426. Shanmugan, Buchanan, and Arnon, *Biochim. Biophys. Acta,* 256, 477 (1972).
1427. Fee, and Palmer, *Biochim. Biophys. Acta,* 245, 175 (1971).
1428. Gersonde, Trittelvitz, Schlaak, and Stabel, *Eur. J. Biochem.,* 22, 57 (1971).
1429. Nakamura, and Kimura, *J. Biol. Chem.,* 245, 6235 (1971).
1430. Jackson, Munro, and Korner, *Biochim. Biophys. Acta,* 91, 666 (1964).
1431. Murray, Oikawa, and Kay, *Biochim. Biophys. Acta,* 175, 331 (1969).
1432. Graham, in *Glycoproteins,* Gottschalk, Ed., Elsevier, Amsterdam, 1966, 361.
1433. Blomback, *Ark, Kem.,* 12, 99 (1958).
1434. Pisano, Finlayson, Peyton, and Nagai, *Proc. Natl. Acad. Sci. U.S.A.,* 68, 770 (1971).
1435. Gollwitzer, Timpl, Becker, and Furthmayr, *Eur. J. Biochem.,* 28, 497 (1972).
1436. Gollwitzer, Karges, Hormann, and Kuhn, *Biochim. Biophys. Acta,* 207, 445 (1970).
1437. Kazal, Amsel, Miller, and Tocantins, *Proc. Soc. Exp. Biol. Med.,* 113, 989 (1963).
1438. Marker, Budzynski, and James, *J. Biol. Chem.,* 247, 4775 (1972).
1439. Bion, Marguérie, Hudry, and Chagniel, *C.R. Acad. Sci. (D)* (Paris), 273, 901 (1971).
1440. Whitaker, *Biochemistry,* 8, 1896 (1969).
1441. Cusanovich, and Edmondson, *Biochem. Biophys. Res. Commun.,* 45, 327 (1971).
1442. D'Anna, and Tollin, *Biochemistry,* 11, 1073 (1972).
1443. Zumft, and Spiller, *Biochem. Biophys. Res. Commun.,* 45, 112 (1971).
1444. Wickner, and Tabor, *J. Biol. Chem.,* 247, 1605 (1972).
1445. Curthoys, and Rabinowitz, *J. Biol. Chem.,* 246, 6942 (1971).
1446. Welch, Buttlaire, Hersh, and Himes, *Biochim. Biophys. Acta,* 236, 599 (1971).
1447. Sanchez, Gonzalez, and Pontis, *Biochim. Biophys. Acta,* 227, 67 (1971).
1448. Olson, and Marquardt, *Biochim. Biophys. Acta,* 268, 453 (1972).
1449. Mendicino, Kratowich, and Oliver, *J. Biol. Chem.,* 247, 6643 (1972).
1450. Traniello, Melloni, Pontremoli, Sia, and Horecker, *Arch. Biochem. Biophys.,* 149, 222 (1972).
1451. Fernando, Pontremoli, and Horecker, *Arch. Biochem. Biophys.,* 129, 370 (1969).
1452. Tashima, Tholey, Drummond, Bertrand, Rosenberg, and Horecker, *Arch. Biochem. Biophys.,* 149, 118 (1972).
1453. Frieden, Bock, and Alberty, *J. Am. Chem. Soc.,* 76, 2482 (1953).
1454. Loontiens, Wallenfels, and Weil, *Eur. J. Biochem.,* 14, 138 (1970).
1455. Klee, and Klee, *J. Biol. Chem.,* 247, 2336 (1972).
1456. Barth, Bunnenberg, and Djerassi, *Anal. Biochem.,* 48, 471 (1972).
1457. Tang, Wolf, Caputto, and Trucco, *J. Biol. Chem.,* 234, 1174 (1959).
1458. Platt, and Kasarda, *Biochim. Biophys. Acta,* 243, 407 (1971).
1459. Konieczny and Domanski, *Acta Biochim. Pol.,* 10, 325 (1963).
1460. Vodrazka, Hrkal, Kodicek, and Jandova, *Eur. J. Biochem.,* 31, 296 (1972).
1461. Amiconi, Antonini, Brunori, Formaneck, and Huber, *Eur. J. Biochem.,* 31, 52 (1972).
1462. Nashi, *Cancer Res.,* 30, 2507 (1970).
1463. Yip, Waks, and Beychok, *J. Biol. Chem.,* 247, 7237 (1972).
1464. Marshall and Pensky, *Arch. Biochem. Biophys.,* 146, 76 (1971).
1465. Kohsiyama, *Int. J. Pept. Protein Res.,* 4, 167 (1972).

1466. Swann, and Hammes, *Biochemistry,* 8, 1 (1969).
1467. Kay, and Marsh, *Biochim. Biophys. Acta,* 33, 251 (1959).
1468. D'Anna, Jr. and Tollin, *Biochemistry,* 11, 1073 (1972).
1469. Legler, von Radloff, and Kempfle, *Biochim. Biophys. Acta,* 257, 40 (1971).
1470. Legler, *Hoppe-Seyler's Z. Physiol. Chem.,* 348, 1359 (1967).
1471. Malcolm, *Hoppe-Seyler's Z. Physiol. Chem.,* 352, 883 (1971).
1472. Smith, Langdon, Piszkiewicz, Brattin, Langley, and Melamed, *Proc. Natl. Acad. Sci. U.S.A.,* 67, 724 (1970).
1473. Winnacker, and Barker, *Biochim. Biophys. Acta,* 212, 225 (1970).
1474. Egan, and Dalziel, *Biochim. Biophys. Acta,* 250, 47 (1971).
1475. Sund, and Akeson, *Biochem. Z.,* 340, 421 (1964).
1476. Miller, and Stadtman, *J. Biol. Chem.,* 247, 7407 (1972).
1477. Roberts, Holcenberg, and Dolowy., *J. Biol. Chem.,* 247, 84 (1972).
1478. Stahl, and Jaenicke, *Eur. J. Biochem.,* 29, 401 (1972).
1479. Cooper, and Meister, *Biochemistry,* 11, 661 (1972).
1480. Orlowski, and Meister, *J. Biol. Chem.,* 246, 7095 (1971).
1481. Ida, and Morita, *Agric. Biol. Chem.,* 35, 1542 (1971).
1482. Wu, Cluskey, Krull, and Friedman, *Can. J. Biochem.,* 49, 1042 (1971).
1483. Jaenicke, in *Pyridine Nucleotide-dependent Dehydrogenases,* Sund, Ed., Springer-Verlag, Berlin, 1970, 70.
1484. Malhotra, and Bernhard, *J. Biol. Chem.,* 243, 1243 (1968).
1485. Aune and Timasheff, *Biochemistry,* 9, 1481 (1970).
1486. D'Alessio and Josse, *J. Biol. Chem.,* 246, 4326 (1971).
1487. Ross, Curry, Schwartz, and Fondy, *Arch. Biochem. Biophys.,* 145, 591 (1971).
1488. White, III, and Kaplan, *J. Biol. Chem.,* 244, 6031 (1969).
1489. Barel, Turneer, and Dolmans, *Eur. J. Biochem.,* 30, 26 (1972).
4190. Frigerio and Harbury, *J. Biol. Chem.,* 231, 135 (1958).
1491. Tamm and Horsfall, *J. Exp. Med.,* 95, 71 (1952).
1492. Maxfield, *Arch. Biochem. Biophys.,* 85, 382 (1959).
1493. Fletcher, Neuberger, and Ratcliffe, *Biochem. J.,* 120, 417 (1970).
1494. Frot-Coutaz, Louisot, and Got, *Biochim. Biophys. Acta,* 264, 362 (1972).
1495. Nisselbaum and Bernfeld, *J. Am. Chem. Soc.,* 78, 687 (1965).
1496. Li and Li, *J. Biol. Chem., Soc.,* 245, 825 (1970).
1497. Patrito and Martin, *Hoppe-Seyler's Z. Physiol. Chem.,* 352, 89 (1971).
1498. Ryan and Westphal, *J. Biol. Chem.,* 247, 4050 (1972).
1499. Heimburger, Haupt, Kranz, and Baudner, *Hoppe-Seyler's Z. Physiol. Chem.,* 353, 1133 (1972).
1500. Iwasaki and Schmid, *J. Biol. Chem.,* 245, 1814 (1970).
1501. Haupt, Baudner, Kranz, and Heimburger, *Eur. J. Biochem.,* 23, 242 (1971).
1502. Boenisch and Alper, *Biochim. Biophys. Acta,* 221, 529 (1970).
1503. Labat, Ishiguro, Fujisaki, and Schmid, *J. Biol. Chem.,* 244, 4975 (1969).
1503a. Kredich, Keenan, and Foote, *J. Biol. Chem.,* 247, 7157 (1972).
1503b. Schwartz, Pizzo, Hill, and McKee, *J. Biol. Chem.,* 248, 1395 (1973).
1503c. Haupt, Heimburger, Kranz, and Boudner, *Hoppe-Seyler's Z. Physiol. Chem.,* 353, 1841 (1972).
1503d. Boenisch and Alper, *Biochim. Biophys. Acta,* 214, 135 (1970).
1503e. Marr, Neuberger, and Ratcliffe, *Biochem. J.,* 122, 623 (1971).
1503f. Rambhar and Ramachandran, *Indian J. Biochem. Biophys.,* 9, 21 (1972).
1503g. Waks, Kahn, and Beychok, *Biochem. Biophys. Res. Commun.,* 45, 1232 (1971).
1504. Huprikar and Sohonie, *Enzymologia,* 28, 333 (1965).
1505. Dahlgren, Porath, and Lindahl-Kiessling, *Arch. Biochem. Biophys.,* 37, 306 (1970).
1506. Howard and Sage, *Biochemistry,* 8, 2436 (1969).
1507. Howard, Sage, Stein, Yound, Leon, and Dyckes, *J. Biol. Chem.,* 246, 1590 (1971).
1508. Garbett, Darnall, Klotz, and Williams, *Arch. Biochem. Biophys.,* 135, 419 (1969).
1509. Makino, *J. Biochem.* (Tokyo), 70, 149 (1971).
1510. Giamberardino, *Arch. Biochem. Biophys.,* 118, 273 (1967).
1511. Bannister and Wood, *Comp. Biochem. Physiol.,* B40, 7 (1971).
1512. Nickerson and Van Holde, *Comp. Biochem. Physiol.,* B39, 855 (1971).
1513. Wittenberg, Briehl, and Wittenberg, *Biochem. J.,* 96, 363 (1965).
1514. Wittenberg, Wittenberg, and Noble, *J. Biol. Chem.,* 247, 4008 (1971).
1515. Wittenberg, Ozazaki, and Wittenberg, *Biochim. Biophys. Acta,* 111, 485 (1965).
1516. Figueiredo, Gomez, Heneine, Santos, and Hargreaves, *Comp. Biochem. Physiol.,* B44, 481 (1973).
1517. Seamonds, Forster, and George, *J. Biol. Chem.,* 246, 5391 (1971).
1518. Terwilliger and Read, *Comp. Biochem. Physiol.,* 36, 339 (1970).
1519. Yamaguchi, Kochiyama, Hashimoto, and Matsuura, *Bull. Jap. Soc. Sci. Fish.,* 28, 184 (1962).
1520. Yamaguchi, Kochiyama, Hashimoto, and Matsuura, *Bull. Jap. Soc. Sci. Fish.,* 29, 174 (1963).

1521. Buhler, *J. Biol. Chem.*, 238, 1665 (1963).
1522. Mohr, Scheler, Schumann, and Muller, *Eur. J. Biochem.*, 3, 158 (1967).
1523. Boyer, Hathaway, Pascasio, Bordley, Orton, and Naughton, *J. Biol. Chem.*, 242, 2211 (1967).
1524. Babul and Stellwagen, *Anal. Biochem.*, 28, 216 (1969).
1525. Inada, Kurozumi, and Shibata, *Arch. Biochem. Biophys.*, 93, 30 (1961).
1526. Herskovits, Gabegbeku, and Jzillet, *J. Biol. Chem.*, 245, 2588 (1970).
1527. Mayer and Miller, *Anal. Biochem.*, 36, 91 (1970).
1528. Javahezian and Beychok, *J. Mol. Biol.*, 37, 1 (1968).
1529. Herskovits, *Arch. Biochem. Biophys.*, 130, 19 (1969).
1530. Malchy and Dixon, *Can. J. Biochem.*, 48, 192 (1970).
1531. Sidwell, Munch, Guzman Barron, and Hogness, *J. Biol. Chem.*, 123, 335 (1938).
1532. Jones and Schroeder, *Biochemistry*, 2, 1357 (1963).
1533. Olson and Gibson, *J. Biol. Chem.*, 246, 5241 (1971).
1534. Allis and Steinhardt, *Biochemistry*, 9, 2286 (1970).
1535. Antonini, Brunori, Caputo, Chiancone, Rossi-Fanelli, and Wyman, *Biochim. Biophys. Acta*, 79, 284 (1964).
1536. Beaven, Hoch, and Holiday, *Biochem. J.*, 49, 374 (1951).
1537. Itano, Fogarty, Jr., and Alford, *Am. J. Clin. Pathol.*, 55, 135 (1971).
1538. Morningstar, Williams, and Suutarinen, *Am. J. Clin. Pathol.*, 46, 603 (1966).
1539. Zettner and Mensch, *Am. J. Clin. Pathol.*, 49, 196 (1968).
1540. Tentori, Vivaldi, and Salvati, *Clin. Chim. Acta*, 14, 276 (1966).
1541. Sugita and Yoneyama, *J. Biol. Chem.*, 246, 389 (1971).
1542. Bucci and Fronticelli, *J. Biol. Chem.*, 240, PC 551 (1965).
1543. DeBruin and Bucci, *J. Biol. Chem.*, 246, 5228 (1971).
1544. Bucci and Fronticelli, *Biochim. Biophys. Acta*, 243, 170 (1971).
1545. Waterman and Yonetani, *J. Biol. Chem.*, 245, 5842 (1970).
1546. Oshino, Asakura, Tamura, Oshino, and Chance, *Biochem. Biophys. Res. Commun.*, 46, 1055 (1972).
1547. Seery, Hathaway, and Eberhard, *Arch. Biochem. Biophys.*, 150, 269 (1972).
1548. Chou and Wilson, *Arch. Biochem. Biophys.*, 151, 48 (1972).
1549. Easterby and Rosemeyer, *Eur. J. Biochem.*, 28, 241 (1972).
1550. Klee, *J. Biol. Chem.*, 247, 1398 (1972).
1551. Wickett, Li, and Isenberg, *Biochemistry*, 11, 2952 (1972).
1552. Pieri and Kergueris, *C. R. Acad. Sci., Paris*, 2740, 2366 (1973).
1553. Mori and Hollands, *J. Biol. Chem.*, 246, 7223 (1971).
1554. Bewley and Li, *Arch. Biochem. Biophys.*, 144, 589 (1971).
1555. Donini, Puzzuoli, D'Alessio, and Donini, in *Pharmacology of Hormonal Polypeptides and Proteins*, Beck, Martini, and Paoletti, Eds., Plenum Press, New York, 1968, 229.
1556. Bewley and Li, *Biochemistry*, 11, 927 (1972).
1557. Bewley, Brovetto-Cruz, and Li, *Biochemistry*, 8, 4701 (1969).
1558. Shome and Friesen, *Endocrinology*, 89, 631 (1971).
1559. Bewley and Li, *Biochemistry*, 11, 884 (1972).
1560. Ma, Brovetto-Cruz, and Li, *Biochemistry*, 9, 2302 (1970).
1561. Bewley, Sairam, and Li, *Biochemistry*, 11, 932 (1971).
1562. Woodhead, O'Riordan, Keutmann, Stolz, Dawson, Niall, Robinson, and Potts, Jr., *Biochemistry*, 10, 2787 (1971).
1563. Bewley and Li, *Int. J. Protein Res.*, 1, 117 (1969).
1564. Jackson and Lovenberg, *J. Biol. Chem.*, 246, 4280 (1971).
1565. Gerstner and Pfeil, *Hoppe-Seyler's Z. Physiol. Chem.*, 353, 271 (1972).
1566. Corneil and Wofsy, *Immunochemistry*, 4, 183 (1967).
1567. Pollet, Rossi, and Edelhoch, *J. Biol. Chem.*, 247, 5921 (1972).
1568. Anders Karlsson, Peterson, and Berggard, *J. Biol. Chem.*, 247, 1065 (1972).
1569. Barth, Bunnenberg, and Djerassi, *Anal. Biochem.*, 48, 471 (1972).
1570. Evans, Herron, and Goldstein, *J. Immunol.*, 101, 915 (1968).
1571. Grey, Abel, and Zimmerman, *Ann. N.Y. Acad. Sci.*, 190, 37 (1972).
1572. Kochwa, Terry, Capra, and Yang, *Ann. N.Y. Acad. Sci.*, 190, 49 (1971).
1573. Kaygorodova and Kaversneva, *Mol. Biol. USSR*, 1, 224 (1967); cited in Egaroy, Chernyak, Dunaevsky, Gavrilova, and Moiseev, *Immunochemistry*, 8, 157 (1971).
1574. Stevenson and Dorrington, *Biochem. J.*, 118, 703 (1970).
1575. O'Daly and Cebra, in *Protides of the Biological Fluids*, Peeters, Ed., Pergamon Press, New York, 1969, 205.
1576. Levine and Levytska, *J. Immunol.*, 102, 647 (1969).
1577. Painter, Sage, and Tanford, *Biochemistry*, 11, 1327 (1972).
1578. Underdown, Simms, and Eisen, *Biochemistry*, 10, 4359 (1971).
1579. Weintraub and Schlamowitz, *Comp. Biochem. Physiol.*, B38, 513 (1971).

1580.   Helms and Allen, *Comp. Biochem. Physiol.*, B38, 439 (1971).
1581.   Reynolds and Johnson, *Biochemistry*, 10, 2821 (1971).
1582.   Butler, *Biochim Biophys. Acta.*, 251, 435 (1971).
1583.   Narayana, Shurpalekab, and Sundarvalli, *Indian J. Biochem.*, 7, 241 (1970).
1584.   Vidal and Stoppani, *Arch. Biochim. Biophys.*, 147, 66 (1971).
1585.   Kiyohara, Iwasaki, and Yoshikawa, *J. Biochem.* (Tokyo), 73, 89 (1972).
1586.   Iwasaki, Kiyohara, and Yoshikawa, *J. Biochem.* (Tokyo), 70, 817 (1971).
1587.   Chu and Chi, *Sci. Sin.*, 14, 1441 (1965).
1588.   Markussen, *Int. J. Protein Res.*, 3, 201 (1971).
1589.   Brandenberg, Gattner, and Wollmer, *Hoppe-Seyler's Z. Physiol. Chem.*, 353, 599 (1972).
1590.   Holdsworth, *Biochim. Biophys. Acta*, 51, 295 (1961).
1591.   Illingworth, *Biochem. J.*, 129, 1119 (1972).
1592.   Sanderman, Jr., and Strominger, *J. Biol. Chem.*, 247, 5123 (1972).
1593.   Kutzbach and Schmidt-Kastner, *Hoppe-Seyler's Z. Physiol. Chem.*, 353, 1099 (1972).
1594.   Fielder, Muller, and Werle, *Fed. Eur. Biochem. Soc. Lett.*, 22, 1 (1972).
1595.   Crewther, Fraser, Lennox, and Lindley, *Adv. Protein Chem.*, 20, 191 (1965).
1596.   Gillespie, *Comp. Biochem. Physiol.*, B41, 723 (1972).
1597.   Barel, Prieels, Maes, Looze, and Leonis, *Biochim. Biophys. Acta*, 257, 288 (1972).
1598.   Cowburn, Brew, and Gratzer, *Biochemistry*, 11, 1228 (1972).
1599.   Dagleish and Peacocke, *Biochem. J.*, 125, 155 (1971).
1600.   Apella and Markert, *Biochem. Biophys. Res. Commun.*, 6, 171 (1961).
1601.   Thuwissen, Masson, Osinski, and Heremans, *Eur. J. Biochem.*, 31, 239 (1972).
1602.   Masson, *La Lactoferrine*, Arsica, Brussels and Maloine, Paris, 1970.
1603.   Aisen and Leibman, *Biochim. Biophys. Acta*, 257, 314 (1972).
1604.   Berrens and Bleumink, *Int. Arch. Allergy*, 28, 150 (1965).
1605.   Trayer and Hill, *J. Biol. Chem.*, 246, 6666 (1971).
1606.   Peive, Atanasov, Zhiznevskaya, and Krasnobaeva, *Dokl. Akad. Nauk SSR Biochem. Sect. (Transl.)*, 202, 39 (1972).
1607.   Atanasov, Bulgarian Academy of Sciences, Bulgaria, submitted.
1608.   Soda, Misono, Mori, and Sakato, *Biochem. Biophys. Res. Commun.*, 44, 931 (1971).
1609.   Garner, Jr. and Smith, *J. Biol. Chem.*, 247, 561 (1972).
1610.   Edelstein, Lim, and Scanu, *J. Biol. Chem.*, 247, 5842 (1972).
1611.   Scanu, Lim, and Edelstein, *J. Biol. Chem.*, 247, 5850 (1972).
1612.   Smith, Dawson, and Tanford, *J. Biol. Chem.*, 247, 3376 (1972).
1613.   Dejmal and Brookes, *J. Biol. Chem.*, 247, 869 (1972).
1614.   Christopher, Pistorius, and Axelrod, *Biochim. Biophys. Acta*, 198, 12 (1970).
1615.   Bellisario, Spencer, and Cormier, *Biochemistry*, 11, 2256 (1972).
1616.   Webster, *Biochim. Biophys. Acta*, 207, 371 (1970).
1617.   Bradshaw and Deranleau, *Biochemistry*, 9, 3310 (1970).
1618.   Kravchenko and Lapuk, *Biokhimia*, 34, 832 (1969).
1619.   Davies, Neuberger, and Wilson, *Biochim. Biophys. Acta*, 178, 294 (1969).
1620.   Blake, Johnson, Mair, North, Phillips, and Sarma, *Proc R. Soc. Lond. (Biol.)*, 167, 378 (1967).
1621.   Donovan, Davis, and White, *Biochim. Biophys. Acta*, 270, 190 (1970).
1622.   Roxby and Tanford, *Biochemistry*, 10, 3348 (1971).
1623.   Ehrenpreis and Warner, *Arch Biochem. Biophys.*, 61, 38 (1956).
1624.   Lin and Koshland, *J. Biol. Chem.*, 244, 505 (1969).
1625.   Teichberg, Kay, and Sharon, *Eur. J. Biochem.*, 16, 55 (1970).
1626.   Hayashi, Imoto, Funatsu, and Funatsu, *J. Biochem.*, (Tokyo), 58, 227 (1965).
1627.   Franek and Pechan, *Scr. Fac. Sci. Nat. Ujep. Brunensis Chem.*, 1, 67 (1971).
1628.   Riblet and Herzenberg, *Science*, 168, 45 (1970).
1629.   Barel, Prieels, Maes, Looze, and Leonis, *Biochim. Biophys. Acta*, 257, 288 (1972).
1630.   Fawcett, Limbird, Oliver, and Borders, *Can. J. Biochem.*, 49, 816 (1971).
1631.   Latovitzki, Halper, and Beychok, *J. Biol. Chem.*, 246, 1457 (1971).
1632.   Parry, Jr., Chandan, and Shahani, *Arch. Biochem. Biophys.*, 130, 59 (1969).
1633.   Cowburn, Brew, and Gratzer, *Biochemistry*, 11, 1228 (1972).
1634.   Mitchell and Hash, *J. Biol. Chem.*, 244, 17 (1969).
1635.   Greene, Damian, and Hubbard, *Biochim. Biophys. Acta*, 236, 659 (1971).
1636.   Humphries, Rohrbach, and Harrison, *Biochem. Biophys. Res. Commun.*, 50, 493 (1973).
1637.   Allen, *Eur. J. Biochem.*, 35, 338 (1973).
1638.   Yamaguchi, Tokushige, and Katsuki, *J. Biochem.* (Tokyo), 73, 169 (1973).
1639.   Saita, Ikenaka, and Mutsushima, *J. Biochem.* (Tokyo), 70, 827 (1971).
1640.   Vachek and Wood, *Biochim. Biophys. Acta*, 258, 133 (1972).

1641. Heazlitt, Conway, and Montag, *Biochim. Biophys. Acta,* 317, 316 (1973).
1642. Lymn and Taylor, *Biochemistry,* 10, 4617 (1971).
1643. Weser, Donay, and Rupp, *FEBS Lett.,* 32, 171 (1973).
1644. Hirata, Nakazawa, Nozaki, and Hayaishi, *J. Biol. Chem.,* 246, 5882 (1971).
1645. Kuma and Inomata, *J. Biol. Chem.,* 247, 556 (1972).
1646. Oh and Conrad, *Arch. Biochem. Biophys.,* 146, 525 (1971).
1647. Morris, Martenson, Deibler, and Cagan, *J. Biol. Chem.,* 248, 534 (1973).
1648. Tamura, Asakura, and Yonetani, *Biochim. Biophys. Acta,* 295, 467 (1973).
1649. Tamura, Woodrow, and Yonetani, *Biochim. Biophys. Acta,* 317, 34 (1973).
1650. Goldbloom and Brown, *Arch. Biochem. Biophys.,* 147, 367 (1971).
1651. Deconinck, Peiffer, Schnek, and Leonis, *Biochimie,* 54, 969 (1972).
1652. Bolard and Garnier, *Biochim. Biophys. Acta,* 263, 535 (1972).
1653. Harrington and Himmelfarb, *Biochemistry,* 11, 2945 (1972).
1654. Kakol, *Biochem. J.,* 125, 261 (1972).
1655. Katoh, Kubo, and Takahashi, *J. Biochem.* (Tokyo), 74, 771 (1973).
1656. Bjorkman and Janson, *Biochim. Biophys. Acta,* 276, 508 (1972).
1657. Suhara, Ikeda, Takemori, and Katagiri, *FEBS Lett.,* 28, 45 (1972).
1658. Yuan, Barnett, and Anderson, *J. Biol. Chem.,* 247, 511 (1972).
1659. Siegel, Murphy, and Kamin, *J. Biol. Chem.,* 248, 251 (1973).
1660. Frazier, Hogue-Angeletti, Sherman, and Bradshaw, *Biochemistry,* 12, 328 (1973).
1661. Angeletti, *Biochim. Biophys. Acta,* 214, 478 (1970).
1662. Pearce, Banks, Banthorpe, Berry, Davies, and Vernon, *Eur. J. Biochem.,* 29, 417 (1972).
1663. Kopeyan, vanRietschoten, Martinez, Rochat, and Miranda, *Eur. J. Biochem.,* 35, 244 (1973).
1664. Karlsson, Eaker, and Ponterius, *Biochim. Biophys. Acta,* 257, 235 (1972).
1665. Karlsson, Eaker, and Ryden, *Toxicon,* 10, 405 (1972).
1666. Iwasaki and Matsubara, *J. Biochem.* (Tokyo), 71, 645 (1972).
1667. Prakash and Sadana, *Arch. Biochem. Biophys.,* 148, 614 (1972).
1668. Zumft, *Biochim. Biophys. Acta,* 276, 363 (1972).
1669. Burns and Hardy, *Methods Enzymol.,* 24B, 480 (1972).
1670. Eady, Smith, Cook, and Postgate, *Biochem. J.,* 128, 655 (1972).
1671. Luisi, Olomucki, Baici, and Karlovic, *Biochemistry,* 12, 4100 (1973).
1672. Marshall and Cohen, *J. Biol. Chem.,* 247, 1641 (1972).
1673. Marcais, Nicot, and Moretti, *Bull. Soc. Chim. Biol.,* 52, 741 (1970).
1674. Joniau, Bloemmen, and Lontie, *Biochim. Biophys. Acta,* 214, 468 (1970).
1675. Berrens and Bleumink, *Int. Arch. Allergy,* 28, 150 (1965).
1676. Holt and Creeth, *Biochem. J.,* 129, 665 (1972).
1677. Willumsen, *C.R. Trav. Lab. Carlsberg,* 36, 247 (1967).
1678. Ifft, *C.R. Trav. Lab. Carlsberg,* 38, 315 (1971).
1679. Babul and Stellwagen, *Anal. Biochem.,* 28, 216 (1969).
1680. Ketterber, *Biochem. J.,* 96, 372 (1965).
1681. Barth, Bunnenberg, and Djerassi, *Anal. Biochem.,* 48, 471 (1972).
1682. Davis, Zahnley, and Donovan, *Biochemistry,* 8, 2044 (1966).
1683. Kay, Strickland, and Billups, *J. Biol. Chem.,* 249, 797 (1974).
1684. Sjoberg, and Feeney, *Biochim. Biophys. Acta,* 168, 79 (1968).
1685. Warner and Weber, *J. Biol. Chem.,* 191, 173 (1951).
1686. Rhodes, Azari, and Feeney, *J. Biol. Chem.,* 230, 399 (1958).
1687. Azari, and Phillips, *Arch. Biochem. Biophys.,* 138, 32 (1970).
1688. Azari, and Baugh, *Arch. Biochem. Biophys.,* 118, 138 (1967).
1689. Weintraub, Vincent, Baulieu, and Alfsen, *FEBS Lett.,* 37, 82 (1973).
1690. Jolley, Evans, Makino, and Mason, *J. Biol. Chem.,* 249, 335 (1974).
1691. Darby, *J. Biol. Chem.,* 139, 721 (1941).
1692. Lauwers, in *West European Symp. on Clin. Chem. – Symp. on Enzymes in Clin. Chem.,* Ruyssen and Vandendriessche, Eds., Elsevier, New York, 1965, 19.
1693. Piront and Gerday, *Comp. Biochem. Physiol.,* 46B, 349 (1973).
1694. Pechere, Capony, and Ryden, *Eur. J. Biochem.,* 23, 421 (1971).
1695. Parello and Pechere, *Biochimie,* 53, 1079 (1971).
1696. Jones, and Hofmann, *Can. J. Biochem.,* 50, 1297 (1972).
1697. Glazer and Smith, *J. Biol. Chem.,* 235, PC43 (1960).
1698. Webster, *Biochim. Biophys. Acta,* 207, 371 (1970).
1699. Blumenfeld and Perlmann, *J. Gen. Physiol.,* 42, 563 (1959).
1700. Lang and Kassell, *Biochemistry,* 10, 2296 (1971).
1701. Mayer and Miller, *Anal. Biochem.,* 36, 91 (1970).

1702.   Bohak, *J. Biol. Chem.*, 244, 4638 (1969).
1703.   Meitner and Kassell, *Biochem. J.*, 121, 249 (1971).
1704.   Marcinszyn, and Kassell, *J. Biol. Chem.*, 246, 6560 (1971).
1705.   Kikuchi and Sakaguchi, *Agric. Biol. Chem.*, 37, 827 (1973).
1706.   Ternynck and Avrameas, *FEBS Lett.*, 23, 24 (1972).
1707.   Herskovits, *Arch. Biochem. Biophys.*, 130, 19 (1969).
1708.   Shih, Shannon, Kay, and Lew, *J. Biol. Chem.*, 246, 4546 (1971).
1709.   Roman and Dunford, *Biochemistry*, 11, 2076 (1972).
1710.   Paul and Stigbrand, *Acta Chem. Scand.*, 24, 3607 (1970).
1711.   Ohlsson and Paul, *Biochim. Biophys. Acta*, 315, 293 (1973).
1712.   Morita and Yoshida, *Agric. Biol. Chem.*, 34, 590 (1970).
1713.   Ex-Fekih and Kertesz, *Bull. Soc. Chim. Biol.*, 50, 547 (1968).
1714.   Stelmaszynska and Zgliczynski, *Eur. J. Biochem.*, 19, 56 (1971).
1715.   Neujahr and Gaal, *Eur. J. Biochem.*, 35, 386 (1973).
1716.   Havir and Hanson, *Biochemistry*, 12, 1583 (1973).
1717.   Fisher, Kirkwood, and Kaufman, *J. Biol. Chem.*, 247, 5161 (1972).
1718.   Huang, Max, and Kaufman, *J. Biol. Chem.*, 248, 4235 (1973).
1719.   Hulett-Cowling and Campbell, *Biochemistry*, 10, 1364 (1971).
1720.   Yoshizumi and Coleman, *Arch. Biochem. Biophys.*, 160, 255 (1974).
1721.   Halford, Benneti, Trentham, and Gutfreund, *Biochem. J.*, 114, 243 (1969).
1722.   Taylor, Lau, Applebury, and Coleman, *J. Biol. Chem.*, 248, 6216 (1973).
1723.   Glew and Heath, *J. Biol. Chem.*, 246, 1556 (1971).
1724.   Igarashi, Takahashi, and Tsuyama, *Biochim. Biophys. Acta*, 220, 85 (1970).
1725.   Igarashi and Hollander, *J. Biol. Chem.*, 243, 6084 (1968).
1726.   Uehara, Fujimoto, and Taniguchi, *J. Biochem.* (Tokyo), 70, 183 (1971).
1727.   Jacobs, Nyc, and Brown, *J. Biol. Chem.*, 246, 1419 (1971).
1728.   Shimada and Sugino, *Biochim. Biophys. Acta*, 185, 367 (1969).
1729.   Cannata, *J. Biol. Chem.*, 245, 792 (1970).
1730.   Smith, *J. Biol. Chem.*, 246, 4234 (1971).
1731.   Wohl and Markus, *J. Biol. Chem.*, 237, 5785 (1972).
1732.   Maeba and Sanwal, *J. Biol. Chem.*, 244, 2549 (1969).
1733.   Kopperschlager, Lorenz, Diezel, Marquardt, and Hofmann, *Acta Biol. Med. Ger.*, 29, 561 (1972).
1734.   Najjar, in *The Enzymes* Vol. 6, Boyer, Lardy, and Myrbach, Eds., Academic Press, New York, 1962, 161.
1735.   Filmer and Koshland, *Biochim. Biophys. Acta*, 77, 334 (1963).
1736.   Hirose, Sugimoto, and Chiba, *Biochim. Biophys. Acta*, 250, 514 (1971).
1737.   Silverberg and Dalziel, *Eur. J. Biochem.*, 38, 229 (1973).
1738.   Tsuboi, Fukunaga, and Chervenka, *J. Biol. Chem.*, 246, 7586 (1971).
1739.   Krietsch and Bucher, *Eur. J. Biochem.*, 17, 568 (1970).
1740.   Scopes, *Biochem. J.*, 113, 551 (1969).
1741.   Scopes, *Biochem. J.*, 122, 89 (1971).
1742.   Sasaki, Sugimoto, and Chiba, *Biochim. Biophys. Acta*, 227, 584 (1971).
1743.   James, Hurst, and Flynn, *Can. J. Biochem.*, 49, 1183 (1971).
1744.   Kawauchi, Iwanaga, Samejima, and Suzuki, *Biochim. Biophys. Acta*, 236, 142 (1971).
1745.   Janssen, deBruin, and Haas, *Eur. J. Biochem.*, 28, 156 (1972).
1746.   Boman, *Ark. Kem.* 12, 453 (1958).
1747.   Felenbok, *Eur. J. Biochem.*, 17, 165 (1970).
1748.   Rutner, *Biochemistry*, 9, 178 (1970).
1749.   Hartman, *J. Biol. Chem.*, 238, 3024 (1963).
1750.   Blasi, Aloj, and Goldberger, *Biochemistry*, 10, 1409 (1971).
1751.   Sevilla and Fischer, *Biochemistry*, 8, 2161 (1969).
1752.   Kamogawa and Fukui, *Biochim. Biophys. Acta*, 242, 55 (1971).
1753.   Cohen, Duewer, and Fischer, *Biochemistry*, 10, 2683 (1971).
1754.   Fosset, Muir, Nielsen, and Fischer, *Biochemistry*, 10, 4105 (1971).
1755.   Assaf and Yunis, *Biochemistry*, 12, 1423 (1973).
1756.   Assaf and Graves, *J. Biol. Chem.*, 244, 5544 (1969).
1757.   Kamogawa, Fukui, and Nikuni *J. Biochem.* (Tokyo), 63, 361 (1968).
1758.   Cohen, *Eur. J. Biochem.*, 34, 1 (1972).
1759.   Hayakawa, Perkins, Walsh, and Krebs, *Biochemistry*, 12, 567 (1973).
1760.   Mano and Yoshida, *J. Biochem.* (Tokyo), 66, 105 (1969).
1761.   Walinder, *Biochim. Biophys. Acta*, 258, 411 (1972).
1762.   Glazer, Fang, and Brown, *J. Biol. Chem.*, 248, 5679 (1973).
1763.   Maccoll, Habig, and Berns, *J. Biol. Chem.*, 248, 7080 (1973).

1764.　Glazer and Fang, *J. Biol. Chem.,* 248, 659 (1973).
1765.　Boucher, Crespi, and Katz, *Biochemistry,* 5, 3796 (1968).
1766.　Kao, Berns, and Town, *Biochem. J.,* 131, 39 (1973).
1767.　Anderson, Jenner, and Mumford, *Biochim. Biophys. Acta,* 221, 69 (1970).
1768.　Tobin and Briggs, *Photochem. Photobiol.,* 18, 487 (1973).
1769.　Bourrillon and Font, *Biochim. Biophys. Acta,* 154, 28 (1968).
1770.　Berrens and Maesen, *Clin. Exp. Immunol.,* 10, 383 (1972).
1771.　Kochiyama, Yamaguchi, Hashimoto, and Matsuura, *Bull. Jap. Soc. Sci. Fish.,* 32, 867 (1966).
1772.　Sjoholm, Wiman, and Wallen, *Eur. J. Biochem.,* 39, 471 (1973).
1773.　Wallen and Wiman, *Biochim. Biophys. Acta,* 221, 20 (1970).
1774.　Ramshaw, Brown, Scawen, and Boulter, *Biochim. Biophys. Acta,* 303, 269 (1973).
1775.　Katoh, Shiratori, and Takamiya, *J. Biochem.* (Tokyo), 51, 32 (1962).
1776.　Milne and Wells, *J. Biol. Chem.,* 245, 1566 (1970).
1777.　Klee, *J. Biol. Chem.,* 244, 2558 (1969).
1778.　Yurewicz, Ghalambor, Duckworth, and Heath, *J. Biol. Chem.,* 246, 5607 (1971).
1779.　Treffry and Ainsworth, *Biochem. J.,* 137, 319 (1974).
1780.　Cejka and Fleischmann, *Arch. Biochem. Biophys.,* 157, 168 (1973).
1781.　Heide, Haupt, and Schultze, *Nature,* 201, 1218 (1964).
1782.　Peterson, *J. Biol. Chem.,* 246, 34 (1971).
1783.　Schultze, Schonenberger, and Schwick, *Biochem. Z.,* 328, 267 (1956).
1784.　Raz and Goodman, *J. Biol. Chem.,* 244, 3230 (1969).
1785.　Seal and Doe, in *Proc. 2nd Int. Congr. Endocrinology,* Vol. 19, Sect. 3, Part 1, Excerpta Medica, Amsterdam, 1964, 229.
1786.　Schultze and Heremans, *Molecular Biology of Human Proteins,* Vol. 1, Elsevier, Amsterdam, 1966, 234.
1787.　Tritsch, *J. Med.* (Basel), 3, 129 (1972).
1788.　Oppenheimer, Surks, Smith, and Squef, *J. Biol. Chem.,* 240, 173 (1965).
1789.　Vahlquist and Peterson, *Biochemistry,* 11, 4526 (1972).
1790.　Jaarsveld, Branch, Robbins, Morgan, Kanda, and Canfield, *J. Biol. Chem.,* 248, 7898 (1973).
1791.　Stratil, *Animal Blood Groups Biochem. Genet.,* 3, 63 (1972).
1792.　Ashida, *Arch. Biochem. Biophys.,* 144, 749 (1973).
1793.　Van der Wel, *FEBS Lett.,* 21, 88 (1972).
1794.　Behnke and Vallee, *Fed. Proc.,* 31, 435 (1972).
1795.　Gates and Travis, *Biochemistry,* 12, 1867 (1973).
1796.　Lacko and Neurath, *Biochemistry,* 9, 4680 (1970).
1797.　Uren and Neurath, *Biochemistry,* 11, 4483 (1972).
1798.　Reeck and Neurath, *Biochemistry,* 11, 3947 (1972).
1799.　Gertler and Birk, *Eur. J. Biochem.,* 12, 170 (1970).
1800.　Uram and Lamy, *Biochim. Biophys. Acta,* 194, 102 (1969).
1801.　Lea, *Biochim. Biophys. Acta,* 317, 351 (1973).
1802.　Frank, Veros, and Pekar, *Biochemistry,* 11, 4926 (1972).
1803.　Markussen, *Int. J. Protein Res.,* 3, 201 (1971).
1804.　Dixon, Schmidt, and Pankov, *Arch. Biochem. Biophys.,* 141, 705 (1970).
1805.　Murao, Funakoshi, and Oda, *Agric. Biol. Chem.,* 36, 1327 (1972).
1806.　Tsuru, Hattori, Tsuji, and Fukumoto, *J. Biochem.* (Tokyo), 67, 15 (1970).
1807.　Sodek and Hofmann, *Methods Enzymol.,* 19, 372 (1970).
1808.　Oda, Kamada, and Murao, *Agric. Biol. Chem.,* 36, 1103 (1972).
1809.　McDowell, *Eur. J. Biochem.,* 14, 214 (1970).
1810.　Sekine, *Agric. Biol. Chem.,* 36, 198 (1972).
1811.　Yoshimoto, Fukumoto, and Tsuru, *Int. J. Protein Res.,* 3, 285 (1971).
1812.　Tsuru, Yoshimoto, Yoshida, Kira, and Fukumoto, *Int. J. Protein Res.,* 2, 75 (1970).
1813.　Shinano and Fukushima, *Agric. Biol. Chem.,* 33, 1236 (1969).
1814.　Levy, Fishman, and Schenkein, *Methods Enzymol.,* 19, 672 (1970).
1815.　Jurasek and Whitaker, *Can. J. Biochem.,* 43, 1955 (1965).
1816.　Kaplan and Whitaker, *Can. J. Biochem.,* 47, 305 (1969).
1817.　Paterson and Whitaker, *Can. J. Biochem.,* 47, 317 (1969).
1818.　Sauer and Senft, *Comp. Biochem. Physiol.,* 42B, 205 (1972).
1819.　Merkel and Sipos, *Arch. Biochem. Biophys.,* 145, 126 (1971).
1820.　Siegel, Brady, and Awad, *J. Biol. Chem.,* 247, 4155 (1972).
1821.　Liu, Neumann, Elliott, Moore, and Stein, *J. Biol. Chem.,* 238, 251 (1963).
1822.　Hata, Hayashi, and Doi, *Agric. Biol. Chem.,* 31, 357 (1967).
1823.　Aibard, Hayashi, and Hata, *Agric. Biol. Chem.,* 35, 658 (1971).
1824.　Hiramatsu and Ouchi, *J. Biochem.* (Tokyo), 71, 676 (1972).

1825.   Pangburn, Burstein, Morgan, Walsh, and Neurath, *Biochem. Biophys. Res. Commun.,* 54, 371 (1973).
1826.   Van Heyningen, *Eur. J. Biochem.,* 27, 436 (1972).
1827.   Nasuno and Ohara, *Agric. Biol. Chem.,* 36, 1791 (1972).
1828.   Turkova, Mikes, Gancev, and Boublik, *Biochim. Biophys. Acta,* 178, 100 (1969).
1829.   Gertler and Hayashi, *Biochim. Biophys. Acta,* 235, 378 (1971).
1830.   Mizusawa and Yoshida, *J. Biol. Chem.,* 247, 6978 (1972).
1831.   Lees, Leston, and Marfey, *J. Neurochem.,* 16, 1025 (1969).
1832.   Cox and Hanahan, *Biochim. Biophys. Acta,* 207, 49 (1970).
1833.   Fujisawa and Hayaishi, *J. Biol. Chem.,* 243, 2673 (1968).
1834.   Ono, Nozaki, and Hayaishi, *Biochim. Biophys. Acta,* 220, 224 (1970).
1835.   Berg and Prockop, *J. Biol. Chem.,* 248, 1175 (1973).
1836.   DeSa, *J. Biol. Chem.,* 247, 5527 (1972).
1837.   Yui, *J. Biochem.* (Tokyo), 69, 101 (1971).

PROTEIN pK VALUES

Lynne H. Botelho and Frank R. N. Gurd

The general techniques for determining individual pK values in proteins usually depend on NMR,[1,2] absorption,[3] or kinetic[4] procedures. Effects of neighboring groups may be evident in chemical shift influences[5-7] or in electrostatic influences on the hydrogen ion equilibria proper.[8]

1. Markley, Finkenstadt, Dugas, Leduc, and Drapeau, *Biochemistry,* 14, 998 (1975).
2. Markley, *Acc. Chem. Res.,* 8, 70 (1975).
3. Tanford, Hanenstein, and Rands, *J. Am. Chem. Soc.,* 77, 6409 (1955).
4. Garner, Bogardt, and Gurd, *J. Biol. Chem.,* in press.
5. Sachs, Schechter, and Cohen, *J. Biol. Chem.,* 246, 6576 (1971).
6. Shrager, Cohen, Heller, Sachs, and Schechter, *Biochemistry,* 11, 541 (1972).
7. Deslauriers, McGregor, Sarantakis, and Smith, *Biochemistry,* 13, 3443 (1974).
8. Roxby and Tanford, *Biochemistry,* 10, 3348 (1971).

Table 1A
SPECIFIC His pK ASSIGNMENTS
IN RIBONUCLEASE A

Protein	pK values	Reference
Bovine		
His 12	6.3	1
His 48	5.8	1
His 73	6.4	1
His 105	6.7	1
Rat		
His 12	6.6	1
His 48	6.2	1
His 73	7.6	1
His 105	6.3	1
His 119	6.1	1

Compiled by Lynne H. Botelho and Frank R. N. Gurd.

REFERENCES

1. Migchelsen and Beintema, *J. Mol. Biol.,* 79, 25 (1973).

Table 1B
NONSPECIFIC His pK VALUES
FOR RIBONUCLEASE A

Protein	pK values				Reference
Coypu	5.8	6.3	6.3	8.0	1
Chinchilla	4.9	6.0	6.1	7.2	1
Bovine	6.01	6.17	6.72	6.9	2

Compiled by Lynne H. Botelho and Frank R. N. Gurd.

REFERENCES

1. Migchelsen and Beintema, *J. Mol. Biol.,* 79, 25 (1973).
2. Markley, *Acc. Chem. Res.,* 8, 70 (1975).

Table 1C
pK VALUES FOR HISTIDINE RESIDUES
IN MYOGLOBIN

Species	pK observed		Reference
Sperm whale	5.37	5.33	1, 2
	5.53	5.39	
	6.34	6.21	
	6.44	6.31	
	6.65	6.55	
	6.83	6.72	
	8.05	7.97	
Horse	5.7	5.5	1, 2
	6.0	5.8	
	6.6	6.5	
	6.9	6.8	
	7.0	6.9	
	7.6	7.6	
California grey whale	5.7		2
	6.2		
	6.6		
	6.8		
	7.8		
Inia geoffrensis	5.53		2
	5.95		
	6.17		
	6.31		
	6.45		
	6.66		
	8.05		
Tursiops truncatus	5.50		2
	5.95		
	6.24		
	6.26		
	6.42		
	6.60		
	7.82		
Balaenoptera acutorostrata	5.46		2
	5.65		
	6.10		
	6.23		
	6.41		
	6.59		
	7.86		

Compiled by Lynne H. Botelho and Frank R. N. Gurd.

REFERENCES

1. Cohen, Hagenmaier, Pollard, and Schechter, *J. Mol. Biol.*, 71, 513 (1972).
2. Botelho, Hanania, and Gurd, unpublished observations.

Table 1D
HISTIDINE pK VALUES IN
HUMAN HEMOGLOBIN

Hemoglobin	pK values	Reference
Human	6.8	1, 2
	7.0	
	7.0	
	7.1	
	7.2	
	7.13	
	7.7	
	8.1	
	8.1	

Compiled by Lynne H. Botelho and Frank R. N. Gurd.

REFERENCES

1. Donovan, *Methods Enzymol.*, 27, 497 (1973).
2. Mandel, *Proc. Natl. Acad. Sci. U.S.A.*, 52, 736 (1964).

Table 1E
pK VALUE FOR HUMAN Hb His 146 β

Protein	His 146 β pK value	Reference
Human hemoglobin Deoxy, His 146 β	8.0	1
	8.1	2
	7.4	3
Human hemoglobin Carboxy, His 146 β	7.1	1
	6.8	2

Compiled by Lynne H. Botelho and Frank R. N. Gurd.

REFERENCES

1. Kilmartin, Breen, Roberts, and Ho, *Proc. Natl. Acad. Sci. U.S.A.*, 70, 1246 (1973).
2. Greenfield and Williams, *Biochim. Biophys. Acta*, 257, 187 (1972).
3. Huestis and Raftery, *Biochemistry*, 11, 1648 (1972).

Table 1F
SPECIFIC HISTIDINE pK VALUE FOR CYTOCHROME *c*

Species	pK values	Reference
Horse		
His 33	6.41	1
Yeast		
His 33	6.74	2
His 39	6.56	2

Compiled by Lynne H. Botelho and Frank R. N. Gurd.

REFERENCES

1. Cohen, Fisher, and Schechter, *J. Biol. Chem.*, 249, 1113 (1974).
2. Cohen and Hayes, *J. Biol. Chem.*, 249, 5472 (1974).

Table 1G
pK VALUES FOR HISTIDINE RESIDUES IN CARBONIC ANHYDRASE

Species	pK values			Reference
Human, B	5.91	5.88	6	1–3
	6.04	6.09	6.98	
	7.00	6.93	7.23	
	7.23	7.23	8.2	
		8.2	8.24	
Human, C	5.87	5.74		2, 3
	5.96	6.43		
	6.10	6.49		
	6.20	6.5		
	6.63	6.57		
	7.20	6.63		
	7.28	7.25		

Compiled by Lynne H. Botelho and Frank R. N. Gurd.

REFERENCES

1. King and Roberts, *Biochemistry,* 10, 558 (1971).
2. Cohen, Yem, Kandel, Gornall, Kandell, and Friedman, *Biochemistry,* 11, 327 (1972).
3. Pesando, *Biochemistry,* 14, 675 (1975).

Table 1H
HISTIDINE pK VALUES FROM NUCLEASE

Protein	pK values		Reference
Staphylococcus aureus			
Foggi	5.46	5.37	1, 2
	5.76	5.71	
	5.66, 5.74, 6.54[a]	5.74	
	6.57	6.50	
Staphylococcus aureus			
V8	5.55		1
	5.80, 6.10[a]		
	6.50		

Compiled by Lynne H. Botelho and Frank R. N. Gurd.

[a]pK values of one histidine existing in multiple conformational forms of the enzyme which slowly interconvert.

REFERENCES

1. Markley, *Acc. Chem. Res.,* 8, 70 (1975).
2. Cohen, Shrager, McNeel, and Schechter, *Nature,* 228, 642 (1970).

Table 1I
HISTIDINE pK VALUES IN VARIOUS PROTEINS

Protein	Number of His resolved	pK	Specific assignment	Reference
Adenylate kinase (pig)	2 of 2	<5.5		1
		6.3		
Chymotrypsin A$_\delta$ (cow)	1 of 2	7.2	His 57	2, 3
Chymotrypsinogen A (cow)	1 of 2	7.2	His 57	2, 3
Lysozyme				
chicken		5.8		4, 5
human		7.1		4, 6
Neurophysin II (cow)	1 of 1	6.87		7
Ovomucoid (chicken)	4 of 4	5.94		8
		6.71		
		6.75		
		8.07		
Protease				
α-Lyter (Myxobacter 495)	1 of 1	<4		9, 10
(*Staphylococcus aureus,* V8)	3 of 3	6.69		11
		6.85		
		7.19		
Ribonuclease T$_1$ (*Aspergillus oryzae*)	2 of 3	7.9		12
		8.0		
Serine esterase		6.5–7.5		13
Trypsin (pig, β form)	4 of 4	5.0		14
		6.54		
		6.66		
		7.20		
Trypsin inhibitor (soybean, Kunitz)	2 of 2	5.27		15
		7.00		

Compiled by Lynne H. Botelho and Frank R. N. Gurd.

REFERENCES

1. Cohn, Leigh, and Reed, *Cold Spring Harbor Symp. Quant. Biol.,* 36, 533 (1972).
2. Robillard and Shulman, *J. Mol. Biol.,* 71, 507 (1972).
3. Robillard and Shulman, *Ann. N.Y. Acad. Sci.,* 69, 599 (1972).
4. Meadows, Markley, Cohen, and Jardetsky, *Proc. Natl. Acad. Sci. U.S.A.,* 58, 1307 (1967).
5. Cohen, Hagenmaier, Pollard, and Schechter, *J. Mol. Biol.,* 71, 513 (1972).
6. Cohen, *Nature* (Lond.), 223, 43 (1969).
7. Cohen, Griffen, Camier, Caizergues, Fromageot, and Cohen, *FEBS Lett.,* 25, 282 (1972).
8. Markley, *Ann. N.Y. Acad. Sci.,* 222, 347 (1973).
9. Hunkapiller, Smallcombe, Whitaker, and Richards, *J. Biol. Chem.,* 248, 8306 (1973).
10. Hunkapiller, Smallcombe, Whitaker, and Richards, *Biochemistry,* 12, 4732 (1973).
11. Markley, Finkenstadt, Dugas, Leduc, and Drapeau, *Biochemistry,* 14, 998 (1975).
12. Riterjans and Pongs, *Eur. J. Biochem.,* 18, 313 (1971).
13. Polgar and Bender, *Proc. Natl. Acad. Sci. U.S.A.,* 69, 599 (1972).
14. Markley, *Acc. Chem. Res.,* 8, 70 (1975).
15. Markley, *Biochemistry,* 12, 2245 (1973).

Table 2
pK VALUES FOR α-AMINO GROUPS
IN PROTEINS

Protein	pK′ value	Reference
Human hemoglobin		
Carboxy		
α chain	6.72	1
	6.95	2
β chain	7.05	2
Cyano		
α chain	6.74	2
β chain	6.93	2
Deoxy		
α chain	7.79	2
β chain	6.84	2
Myoglobin		
Sperm whale	7.77	2
	7.96	3
California grey whale	7.74	3
Pilot whale	7.43	3
Dall porpoise	7.22	3
Harbor seal	7.66	3
Bovine pancreatic	8.14	4
Ribonuclease A		
Horse hemoglobin		
Oxy		
α chain	7.3	5
Deoxy		
α chain	7.7	5

Compiled by Lynne H. Botelho and Frank R. N. Gurd.

REFERENCES

1. **Hill and Davis,** *J. Biol. Chem.,* 242, 2005 (1967).
2. **Garner, Bogardt, Jr., and Gurd,** *J. Biol. Chem.,* in press.
3. **Garner, Garner, and Gurd,** *J. Biol. Chem.,* 248, 5451 (1973).
4. **Carty and Hirs,** *J. Biol. Chem.,* 243, 5254 (1968).
5. **Kilmartin and Rossi-Bernardi,** *Biochem. J.,* 124, 31 (1971).

Table 3
pK VALUES FOR ε-AMINO GROUPS IN PROTEINS

Protein	pK' value	Method	Reference
Bovine pancreatic			
Ribonuclease A			
Lys-41	9.11	Kinetic	1
Other Lys	10.1	Kinetic	1
All Lys	10.2	Titration	2
Sperm Whale			
Myoglobin			
All Lys	10.6	Titration	3
except one		(intrinsic pK)	
Hen egg white			
Lysozyme			
Lys 97	10.1	NMR	4
Lys 116	10.2		
Lys 13	10.3		
Lys 33	10.4		
Lys 1	10.6		
Lys 96	10.7		

Compiled by Lynne H. Botelho and Frank R. N. Gurd.

REFERENCES

1. Carty and Hirs, *J. Biol. Chem.*, 243, 5254 (1968).
2. Tanford and Hanenstein, *J. Am. Chem. Soc.*, 78, 5287 (1956).
3. Shire, Hanania, and Gurd, *Biochemistry*, 13, 2967 (1974).
4. Bradbury and Brown, *Eur. J. Biochem.*, 40, 565 (1973).

Table 4A
TYROSINE pK VALUES

Protein	No of groups	pK	Reference
Ribonuclease A	3 of 6	9.9	1
Insulin		9.7	2
Pepsin		9.5	3
Serum Albumin		10.35	3
Lysozyme		10.8	4
Trypsin inhibitor			
(BPTI)			
Bovine, tyrosines		10.6	5
		10.8	
		11.1	
		11.6	

Compiled by Lynne H. Botelho and Frank R. N. Gurd.

REFERENCES

1. Tanford, Hauenstein, and Rands, *J. Am. Chem. Soc.*, 77, 6409 (1955).
2. Tanford and Epstein, *J. Am. Chem. Soc.*, 76, 2163 (1954).
3. Tanford and Roberts, Jr., *J. Am. Chem. Soc.*, 74, 2509 (1952).
4. Fromageot and Schnek, *Biochem. Biophys. Acta*, 6, 113 (1950); Tanford and Wagner, *J. Am. Chem. Soc.*, 76, 2331 (1954).
5. Karplus, Snyder, and Sykes, *Biochemistry*, 12, 1323 (1973).

Table 4B
TYROSINE pK VALUES IN Hb

Species	No. of residues	pK	Specific residue pK	Reference
Horse	8 of 12	10.6		1
	4 of 12	>12		
Human A	8 of 12	10.6		1
	4 of 12	>12		
Human A carboxy	8 of 12	10.60	β145 10.6	2
			β130 10.6	
	4 of 12	>10.6	β35 >10.6	
Human a deoxy	6 of 12	10.77		2
Human F carboxy	6 of 10	10.45		2
Human F deoxy	4 of 10	10.65		2

Compiled by Lynne H. Botelho and Frank R. N. Gurd.

REFERENCES

1. **Hermans, Jr.,** *Biochemistry,* 1, 193 (1962).
2. **Nagel, Ranney, and Kucinskis,** *Biochemistry,* 5, 1934 (1966).

Table 4C
TYROSINE pK VALUES IN Mb

Species	pK	Reference
Sperm whale	10.3	1
	11.5	
	>12.8	
Horse	10.3	1
	11.5	
	>12.8	

Compiled by Lynne H. Botelho and Frank R. N. Gurd.

REFERENCE

1. **Hermans, Jr.,** *Biochemistry,* 1, 193 (1962).

Table 5
pK VALUES FOR HUMAN Hb Cys β 93 SH

Human hemoglobin	pK	Reference
Deoxy, cys β 93 SH	>11	1
	>10	2
	>9.5	3
Carboxy, cys β 93 SH	>11	1

Compiled by Lynne H. Botelho and Frank R. N. Gurd.

REFERENCES

1. Janssen, Willekens, De Bruin, and van Os, *Eur. J. Biochem.*, 45, 53 (1974).
2. Snow, *Biochem. J.*, 84, 360 (1962).
3. Guidotti, *J. Biol. Chem.*, 242, 3673 (1967).

Table 6
CARBOXYL SIDE CHAIN pK VALUES ESTIMATED IN LYSOZYMES

Residue	Range of pK values[1]
Gln 35	6–6.5
Asp 101	4.2–4.7
Asp 66	1.5–2
Asp 52	3–4.6

Compiled by Lynne H. Botelho and Frank R. N. Gurd.

REFERENCE

1. Imoto, Johnson, North, Phillips, and Rupley, in *The Enzymes,* Vol. VII, 3rd ed., Bayer, Ed., Academic Press, New York, 1972, 665.

GEL ELECTROPHORESIS OF PROTEINS

SDS ELECTROPHORESIS

PROTEINS STUDIED BY SDS-ELECTROPHORESIS

Protein	Mol. wt. of polypeptide chain	Ref.
Myosin*	220,000	9, 10, 11
β-Galactosidase*	130,000	12
Paramyosin	100,000	13
Phosphorylase a*	94,000	12, 14
Serum albumin	68,000	15
L-Amino acid oxidase	63,000	16
Catalase*	60,000	17, 18
Pyruvate kinase*	57,000	19
Glutamate dehydrogenase*	53,000	12, 20, 12
Leucine amino peptidase	53,000	12, 20, 21
γ-Globulin, H chain*	50,000	22
Fumarase*	49,000	23
Ovalbumin	43,000	24
Alcohol dehydrogenase (liver)*	41,000	24, 25[b]
Enolase*	41,000	24, 26
Aldolase*	40,000	24, 27, 28
Creatine kinase*	40,000	29
D-Amino acid oxidase*	37,000	30
Alcohol dehydrogenase (yeast)*	37,000	31[c]
Glyceraldehyde phosphate dehydrogenase*	36,000	32, 33
Tropomyosin*	36,000	13, 34
Lactate dehydrogenase*	36,000	24
Pepsin	35,000	35
Aspartate transcarbamylase, C chain*	34,000	2, 3
Carboxypeptidase A	34,600	36
Carbonic anhydrase	29,000	37
Subtilisin[d]	27,600	38
γ-Globulin, L chain*	23,500	22
Chymotrypsinogen	25,700	e
Trypsin	23,300	e
Papain (carboxymethyl)	23,000	e,f
β-Lactoglobulin*	18,400	15
Myoglobin	17,200	e
Aspartate transcarbamylase, R chain*	17,000	2, 3
Hemoglobin*	15,500	e
Qβ coat protein	15,000	40[g]
Lysozyme	14,300	e
R17 coat protein	13,750	e
Ribonuclease	13,700	e
Cytochrome c	11,700	e
Chymotrypsin*, 2 chains	11,000 and 13,000	e

Note: The table lists molecular weights of the polypeptide chains taken from the literature. Proteins which under native conditions exist as oligomers are indicated by an asterisk.

PROTEINS STUDIED BY SDS-ELECTROPHORESIS (continued)

a Weber, K., unpublished results.
b Jornvall, H. and Harris, J. I., FEBS, 1968, abst. 759.
c Butler, P. J. G. and Harris, J. I., FEBS, 1968, abst. 741.
d After performic acid oxidation.
e Calculated from the amino acid sequences given in Dayhoff and Eck.[39]
f Corrected according to the X-ray structure (J. Drenth, personal communication)
g Konigsberg, W., personal communication.

REFERENCES

1. Shapiro, A. L., Vinuela, E., and Maizel, J. V., Jr., *Biochem. Biophys. Res. Comm.*, 28, 815, 1967.
2. Weber, K., *Nature*, 218, 1116, 1968.
3. Wiley, D. C. and Lipscomb, W. N., *Nature*, 218, 1119, 1968.
4. Weber, K., *Biochemistry*, 6, 3144, 1967.
5. Hirs, C. H. W., *J. Biol. Chem.*, 219, 611, 1956.
6. Sela, M., White, F. H., and Anfinsen, C. B., *Biochim. Biophys. Acta.* 31, 1959.
7. Ornstein, L., *Ann. N.Y. Acad. Sci.*, 121, 404, 1964.
8. Davis. B. J., *Ann. N.Y. Acad. Sci.*, 121, 404, 1964.
9. Woods, E. F., Himmelfarb, S., and Harrington, W. F., *J. Biol. Chem.*, 238, 2374, 1964.
10. Gershman, L. C. and Dreizen, P., *Biophys. J.*, 9, A235, 1969.
11. Slayter, H. S. and Lowey, S., *Proc. Natl. Acad. Sci. U.S.A.*, 58, 1611, 1967.
12. Ullman, A., Goldberg, M. E., Perrin, D., and Monod, J., *Biochemistry*, 7, 261, 1968.
13. Olander, J., Emerson, M. F., and Holtzer, A., *J. Am. Chem. Soc.*, 89, 3058, 1967.
14. Seery, V. L., Fischer, E. H., and Teller, D. C., *Biochemistry*, 6, 3315, 1967.
15. Tanford, C., Kawahara, K., and Lapanje, S., *J. Am. Chem. Soc.*, 89, 729, 1967.
16. Singer, T. P. and Kearney, E. B., *Arch. Biochem.*, 29, 190, 1950.
17. Sund, H., Weber, K., and Molbert, E., *Eur. J. Biochem.*, 1, 40, 1967.
18. Schroeder, W. A., Shelton, J. R., Shelton, J. B., and Olson, B. M., *Biochim. Biophys. Acta*, 89, 47, 1964.
19. Steinmetz, M. A. and Deal, W. C., Jr., *Biochemistry*, 5, 1399, 1966.
20. Marler, E. and Tanford, C., *J. Biol. Chem.*, 239, 4217, 1964.
21. Eisenberg, H. and Tomkins, G. M., *J. Mol. Biol.*, 31, 37, 1968.
22. Rutishauser, U., Cunningham, B. A., Bennett, C., Konigsberg, W. H., and Edelman, G. M., *Proc. Natl. Acad. Sci. U.S.A.* 61, 1414, 1968.
23. Kanarek, L., Marler, E., Bradshaw, R. A., Fellows, R. E. and Hill, R. L., *J. Biol. Chem.*, 239, 4207, 1964.
24. Castellino, F. J. and Barker, R., *Biochemistry*, 7, 2207, 1968.
25. Branden, C.-I., *Arch. Biochem. Biophys.*, 112, 215, 1965.
26. Winstead, J. A. and Wold, F., *Biochemistry*, 5, 1578, 1966.
27. Kawahara, K. and Tanford, C., *Biochemistry*, 5, 1578, 1966.
28. Morse, D. E., Chan. W. and Horecker, B. L., *Proc. Natl. Acad. Sci. U.S.A.* 58, 628, 1967.
29. Dawson, D. M., Eppenberger, H. M., and Kaplan, O., *J. Biol. Chem.*, 242, 210, 1967.
30. Henn, S. W. and Ackers, G. K., *J. Biol. Chem.*, 244, 465, 1969.
31. Sund, H., *Biochemistry Z.*, 333, 205, 1060.
32. Harrington, W. F. and Karr, G. M., *J. Mol. Biol.*, 13, 885, 1965.
33. Davidson, B. E., Sajgo, M., Noller, H. F., and Harris, J. I., *Nature*, 216, 1181, 1967.
34. Woods, E. F., *Nature*, 207, 82, 1965.
35. Bovey, F. A. and Yanari, S. S., in *The Enzymes*, Vol. 4, Boyer, P. D., Lardy, H., and Myrback, K., Eds., Academic Press, New York, 1960, 63.
36. Bargetzi, J.-P., Sampath Kumar, K. S. V., Cox, D. J., Walsh, K. A., and Neurath, H., *Biochemistry*, 2, 1468, 1963.
37. Armstrong, J. M., Myers, D. V., Verpoorte, J. A., and Edsall, J. T., *J. Biol. Chem.*, 241, 5137, 1966.
38. Smith, E. L., Markland, F. S., Kasper, C. B., DeLange, R. J., Landon, M., and Evans, W. H., *J. Biol. Chem.*, 241, 5974, 1966.
39. Dayhoff, M. O. and Eck, R. V., *Atlas of Protein Sequence and Structure*, National Biomedical Research Foundation, Silver Spring, Md., 1967—1968.
40. Overby, L. R., Barlow, G. H., Dor, R. H., Jacob, M., and Spiegelman, S., *J. Bacteriol.* 92, 739, 1966.
41. Tanford, C., *Adv. Prot. Chem.*, 23, 121, 1968.
42. DeCrombrugghe, B., Pitt-Rivers, R., and Edelhoch, H., *J. Biol. Chem.*, 241, 2766, 1966.

PROTEINS STUDIED BY SDS-ELECTROPHORESIS (continued)

43. **Craven, G. R., Steers, E., Jr., and Anfinsen, C. B.,** *J. Biol. Chem.,* 240, 2468, 1965.
44. **Kirby Hade, E. P. and Tanford, C.,** *J. Am. Chem. Soc.,* 83, 5034, 1967.
45. **Reithel, F. J. and Sakura, J. D.** *J. Phys. Chem.,* 67, 2497, 1963.
46. **Drum, D. E., Harrison, J. H., Li, T.-K., Bethune, J. L. and Vallee, B. L.,** *Proc. Natl. Acad. Sci. U.S.A.,* 57, 1434, 1967.
47. **Schachman, H. K. and Edelstein, S. J.,** *Biochemistry,* 5, 2681, 1966.
48. **Fonda, M. L. and Anderson, B. M.,** *J. Biol. Chem.* 243, 5635, 1968; 244, 666, 1969.
49. **Nathans, N. D.,** *J. Mol Biol.,* 13, 521, 1965.

From Weber, K. and Osborn, M., *J. Biol. Chem.,* 244, 4406, 1969. With permission.

COMPARISON OF MOBILITIES OF PROTEINS OF DIFFERENT
MOLECULAR WEIGHTS

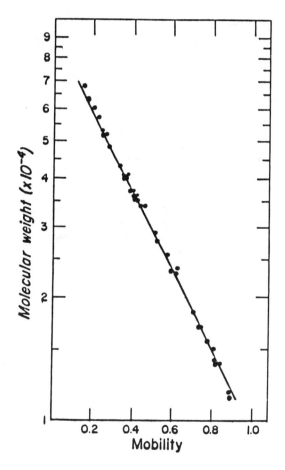

Comparison of the molecular weights of 37 different polypeptide chains in the molecular weight range from 11,000 to 70,000 with their electrophoretic mobilities on gels with the normal amount of cross-linker. The references to the molecular weights are given in the preceding table. (From Weber, K. and Osborn, M., *J. Biol. Chem.,* 244, 4406, 1969. With permission.)

A NONUREA ELECTROPHORETIC GEL SYSTEM FOR RESOLUTION OF POLYPEPTIDES OF M_r 2000 to M_r 200,000

Molecular Weights of Some Proteins or Peptides

Protein or peptide	Molecular weight
Myosin	200,000
Gel Code molecular weight markers	92,500
	81,000
	57,000
	35,000
	29,000
Myoglobin	17,000
Myoglobin I and II (dimer)	14,400
Myoglobin I	8,200
Myoglobin II	6,200
ACTH	4,500
ACTH peptide 1—24	2,900
Myoglobin III	2,500
ACTH peptide 1—17	2,100
Myoglobin fragment[a]	1,400

[a] This species has not been characterized, and although it is reproducible, the manufacturer suggests that it be ignored.

REFERENCES

1. **Swank, R. T. and Munkres, K. D.,** *Anal. Biochem,* 39, 462, 1971.
2. **Burr, F. A. and Burr, D.,** in *Methods in Enzymology,* Fleischer, S. and Fleischer, B., Eds., Vol. 96, Academic Press, New York, 1983, 239.
3. **Kyte, J. and Rodriguez, H.,** *Anal. Biochem,* 133, 515, 1983.
4. **Glyn, M. C. P., Bull, J., and Wright, J.,** *S. Afr. J. Sci.,* 78, 36, 1982.
5. **Anderson, B. L., Berry, R. W., and Telser, A.,** *Anal. Biochem.,* 132, 365, 1983.
6. **Hashimoto, F., Horigome, T., Kanbayashi, M., Yoshida, K., and Sugano, H.,** *Anal. Biochem.,* 129, 192, 1983.
7. **Anderson, N. L. and Anderson, N. G.,** *Anal. Biochem.,* 85, 341, 1978.
8. **Tollaksen, S. L., Anderson, N. L., and Anderson, N. G.,** in Operation of the ISO-DALT System, 7th ed., ANL-BIM-84-1, Argonne National Laboratory, Argonne, Ill., 1984.
9. **Anderson, N. G. and Anderson, N. L.,** *Anal. Biochem.,* 85, 331, 1978.
10. **Merril, C. R., Goldman, D., Sedman, S. A., and Ebert, M. H.,** *Science,* 211, 1437, 1981.
11. **Sammons, D. W., Adams. L. D., Vidmar, T. J., Hatfield, C. A., Jones, D. H., Chuba, P. J., and Crooks, S. W.,** in *Two-Dimensional Gel Electrophoresis of Proteins,* Celis, J. E. and Bravo, R., Eds., Academic Press, Orlando, Fla., 1984, 112.
12. **Johnson, G.,** *Biochem. Genet.,* 17, 499, 1979.
13. **Moore, D., Sowa, B. A., and Ippen-Ihler, K.,** *J. Bacteriol.,* 146, 251, 1981.
14. **Sowa, B. A., Moore, D., and Ippen-Ihler, K.,** *J. Bacteriol.,* 153, 962, 1983.
15. **Westermann, R.,** *Electrophoresis,* 6, 136, 1985.
16. **DeWald, D. B., Adams. L. D., and Pearson, J. D.,** *Proc. Symp. Am. Protein Chem., 1st.,* in press.
17. **Ochs, D. C., McConkey, E. H., and Sammons, D. W.,** (1981) *Electrophoresis,* 2, 304, 1981.
18. **Adams. L. D. and Sammons, D. W.,** in *Electrophoresis 81: Advanced Methods, Biochemical and Clinical Applications,* Allen, R. C. and Arnaud, P., Eds., De Gruyter, New York, 1981, 155.
19. **Guevara, J., Johnston, D. A., Ramagali, L. S., Martin, B. A., Capetillo, S., and Rodriguez, L. V.,** *Electrophoresis,* 3, 197, 1982.

From De Wald, D. B., Adams, L. D., and Pearson, J. D., *Anal. Biochem.,* 154, 502, 1986. With permission.

MOLECULAR WEIGHTS OF PHAGE PROTEINS DETERMINED BY COMPARING THEIR MOBILITY IN SDS GELS WITH THOSE OF MARKER PROTEINS WITH KNOWN MOLECULAR WEIGHTS

Gene product	Observed value in SDS gels	Published value
P18	69,000	50,000
P20	63,000	
P23	56,000	
P23	46,500	46,000
P24	45,000	
P24	43,500	
P22	31,000	
IP	23,500	
IP	21,000	
P19	18,000	

Note: The following marker proteins were used: serum albumin (68,000); γ-globulin, heavy chain (50,000); ovalbumin (48,000); γ-globulin, light chain (23,500); and TMV (17,000), to calibrate 10% acrylamide gels. In this improved gel system the distances of migration of the various marker proteins relative to the distance of migration of bromophenol blue are also a linear function of the logarithm of the molecular weight of the marker proteins, as has been described. Radioactively labeled phage proteins were mixed with unlabeled marker proteins before electrophoresis; the distance of migration of the phage proteins was determined from the autoradiogram and those of the marker proteins from the stained gel. It is assumed that the phage proteins also separate in SDS gels solely according to their molecular weight.

MOLECULAR WEIGHT ANALYSIS OF OLIGOPEPTIDES BY ELECTROPHORESIS IN POLYACRYLAMIDE GEL WITH SODIUM DODECYL SULFATE

MOLECULAR WEIGHTS OF POLYPEPTIDES EMPLOYED IN ELECTROPHORESIS

Protein	Molecular weight
1. Ovalbumin	46,000[a]
2. Carboxypeptidase A	34,500[b]
3. Myoglobin	17,200[c]
4. Myoglobin I + II	14,900[c]
5. Cytochrome *c*	12,300[c]
6. Myoglobin I	8,270[c]
7. Cytochrome *c* I	7,760[c]
8. Myoglobin II	6,420[c]
9. Bovine trypsin inhibitor	6,160[c]
10. Adrenocorticotropic hormone	4,550[c]
11. Insulin	5,700[c]
12. Insulin B	3,400[c]
13. Insulin A	2,300[c]
14. Glucagon	3,460[c]
15. Cytochrome *c* II	2,780[c]
16. Myoglobin III	2,550[c]
17. Cytochrome *c* III	1,810[c]
18. Bacitracin	1,400[d]
19. Polymyxin B	1,225[d]

Note: Myoglobin I, myoglobin II, cytochrome *c* I, etc. refer to the cyanogen bromide cleavage products of myoglobin and cytochrome *c* numbered in order of decreasing size. Myoglobin I + II is the peptide composed of fragments myoglobin I and myoglobin II apparently resulting from incomplete cyanogen bromide cleavage.

[a] Neurath, H., Ed., *The Proteins,* Vol. 1, Academic Press, New York, 1963.

[b] Bradshaw, R. A., Ericsson, L. H., Walsh, K. A., and Neurath, H., *Proc. Natl. Acad. Sci. U.S.A.,* 63, 1389, 1969.

[c] Dayhoff, M. O., Ed., *Atlas of Protein Sequence and Structure,* Vol. 4, National Biomedical Research Foundation, Silver Spring, Md., 1969.

[d] Bodansky, M. and Perlman, D. *Nature (London),* 204, 840, 1964.

From Swank, R. T. and Munkres, K. D., *Anal. Biochem.,* 39, 462, 1971. With permission.

USE OF DIMETHYL SUBERIMIDATE, A CROSS-LINKING REAGENT, IN STUDYING THE SUBUNIT STRUCTURE OF OLIGOMERIC PROTEINS

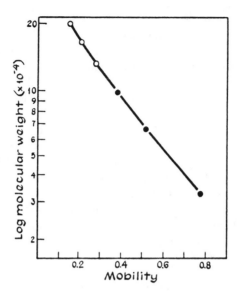

Semilog plot of molecular weight vs. migration relative to bromphenol blue for the covalently linked species produced by amidination of the catalytic subunit of aspartate transcarbamylase. (From Davies, G. E. and Stark, G. R., in *Proc. Natl. Acad. Sci. U.S.A.*, 66(3), 651, 1970. With permission.)

PRESTAINED ELECTROPHORESIS STANDARDS*

Polyacrylamide gel electrophoresis (PAGE) in the presence of sodium dodecylsulfate (SDS) has gained wide use in the separation of protein subunits and the determination of molecular weights. An estimate of the molecular weight of a given protein can be determined by comparing its electrophoretic mobility with that of known protein standards. An approximate linear relationship is obtained if the logarithms of the molecular weight of standard proteins are plotted against their respective electrophoretic mobility. The protein standards provided in the kits from Diversified Biotech have a molecular weight range common to most proteins and their subunits. Each kit contains a total volume of 500 μl. As little as 3 μl per application may be used to achieve optimal staining intensity. The concentration of individual components in the unstained kits have been determined to yield bands of approximately equal color intensity following staining with Brilliant Blue R (Coomassie Brilliant Blue). Prestained kits and individual prestained standards are also provided for monitoring protein migraiton in SDS-PAGE, as well as for use as internal reference markers in preparative electrophoresis. Prestained proteins can also be transferred to nitrocellulose, using Western and Southern blotting techniques, and provide convenient molecular weight markers without the necessity of additional protein stain.

LOW-RANGE KITS

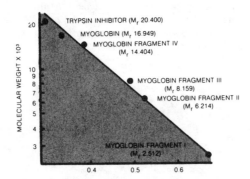

Figure 1. Relative mobilities determined with the gel system of Swank and Munkres.[3]

Kit contains trypsin inhibitor, myoglobin, and four peptide fragments prepared from myoglobin by cyanogen bromide cleavage. Relative mobilities determined with the gel system of Swank and Munkres were identical for the unstained and prestained components. Proteins are suspended in 500 μl of 50 mM Tris·Cl (pH 7.5) containing 1% SDS, 1% beta mercaptoethanol, and 10 mM EDTA. Use as little as 3 μl per application.

Myoglobin fragment I	2,512 daltons
Myoglobin fragment II	6,224 daltons
Myoglobin fragment III	8,159 daltons
Myoglobin fragment IV	14,400 daltons
Myoglobin	16,950 daltons
Tryspin inhibitor	20,400 daltons

* Courtesy of Diversified Biotech, 46 Marcellus Drive, Newton Centre, MA 02159.

MID-RANGE KITS

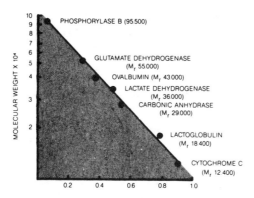

FIGURE 2. Relative mobility of prestained proteins using the gel system of Laemmli.[1]

Kit contains seven homogenous proteins at concentrations yielding equal staining intensity with Coomassie Blue or as prestained components. Relative mobilities as determined with the Laemmli gel system were identical for both the unstained and prestained kits. Proteins are suspended in 50 mM Tris·Cl (pH 7.5 with 1% SDS, 1% beta mercaptoethanol, and 10 mM DTA. Use as little as 3 μl per application.

Cytochrome c	12,400 daltons
Lactoglobulin	18,400 daltons
Carbonic anhydrase	29,000 daltons
Lactate dehydrogenase	36,000 daltons
Ovalbumin	43,000 daltons
Glutamate dehydrogenase	55,000 daltons
Phosphorylase B	95,500 daltons

Standards	Relative Mobility
Insulin	3,000 daltons
Cytochrome c	12,400 daltons
Myoglobin	16,950 daltons
Trypsin inhibitor	20,400 daltons
Carbonic anhydrase	29,000 daltons
Lactate dehjydrogenase	36,000 daltons
Alcohol dehydrogenase	39,000 daltons
Ovalbumin	43,000 daltons
Glutamate dehydrogenase	55,000 daltons
Bovine serum albumin	68,000 daltons
Phosphorylase b	95,500 daltons
B Galactosidase	116,000 daltons
Myosin	200,000 daltons

REFERENCES

1. **Laemmli, U.,** *Nature,* 227, 680, 1970.
2. **Weber, K. and Osborn, M. J.,** *Anal. Biochem,* 244, 4406, 1970.
3. **Swank, R. and Munkres, K.,** *Anal. Biochem.,* 39, 462, 1971.
4. **Davies, G. and Stark, G.,** *Proc. Natl. Acad. Sci. U.S.A.,* 66, 651, 1970.

POLYACRYLAMIDE GELS

RELATIONSHIP OF GEL CONCENTRATION AND MOBILITY FOR A NUMBER OF DIFFERENT PROTEINS

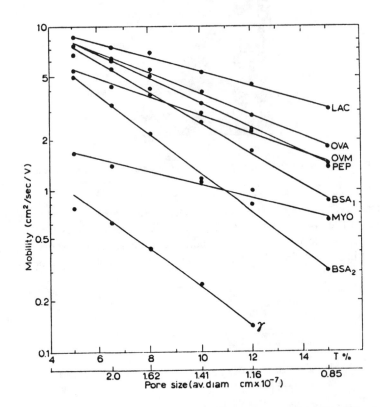

With increased total concentration of acrylamide plus bisacrylamide (T), the pore size decreases as shown above. For all the proteins examined, log mobility is found to be proportional to pore size. All mobilities have been measured in gels with 5% of bisacrylamide at pH 8.83. LAC: β-lactoglobulin; OVA: ovalbumin; OVM: ovomucoid; PEP: pepsin; MYO: myoglobin; γ: bovine γ-globulin; BSA₁, BSA₂: bovine serum albumin monomer and dimer.

From Morris, C. J. O. R., in Peeters, H., Ed., *Protides of the Biological Fluids,* Elsevier, Amsterdam, 1966, 543. Reprinted from Gordon, A. H., Ed., *Electrophoresis of Proteins in Polyacrylamide and Starch Gels* (in Work, T. S. and Work, E., General Eds., *Laboratory Techniques in Biochemistry and Molecular Biology*), North-Holland, New York, 1971, 37. With permission of Elsevier Science Publishers.

BUFFERS FOR ACRYLAMIDE GELS (SINGLE-GEL SYSTEMS)

In gel			In electrode vessels			Characteristics of system	Apparatus type[a]	Examples of materials investigated
Buffer ions	M	pH	Buffer ions	M	pH			
Veronal	0.09	8.6	Same as in gel			—	VP	Serum[1]
Tris	0.070	8.7				Albumin band too narrow for acurate quantitation	VP	Serum[1]
EDTA	0.007							
Borate[b]	0.010							
Tris	0.070	8.65				Mobilities are low in this buffer	VP	Serum[1]
EDTA	0.007							
Ca lactate	0.0016							
Tris	0.006	8.5				—	VP	
Veronal	0.022							
Na lactate	0.017							
Acetate	0.05	5.7				Rapid; cooled in petroleum ether	FP	Lactic dehydrogenase isoenzymes[3]
Tris	0.068	7.2	Borate[b]	0.05	8.6	12 M urea in gel	VP	Myosin subunits[4]
Citrate	0.025							
Phenol:acetic acid:water 2:1:1			Same as in gel			—	FP	Ribosomal proteins[5]
Tris	0.010	9.2	Tris	0.04		Sample unstabilized	D	Serum[6]
Glycine	0.005		Glycine	0.022	9.2			
Tris borate	0.1	8.7	Same as in gel			Preliminary soaking of gel possible	VP	Serum[7]
Potassium acetate	0.02	2.9	Valine (anode)	0.3	4.0 (acetic)	Sucrose used to stabilize sample	D	Histones[8]
			Glycine (cathode)	0.3				
Tris glycine	0.37	9.5	Tris glycine	0.37	9.5	Sucrose use to stabilize sample	D	Serum[12]
Veronal	0.05	8.4	—		—	7 M urea in gel	FP	Lactic dehydrogenase isoenzymes[3]
Tris	0.083	8.9	Same as in gel			Sample concentrates as it enters gel	D	α-Crystallin[14]
Tris	0.37	8.8	Borate[b]	0.05	9.2		D	Serum[15]

[a] D, Disc; FP, flat plate (in the horizontal; VP, vertical plate.
[b] Molarity based on weight of orthoboric acid (H_3BO_3) taken.

From *Electrophoresis of Proteins in Polyacrylamide and Starch Gels*, Gordon, A. H., Ed. (*Laboratory Techniques in Biochemistry and Molecular Biology*, Work, T. S. and Work, E., Gen. Eds.), North-Holland, New York, 1971, 40. With permission of Elsevier Science Publishers.

BUFFERS FOR ACRYLAMIDE GELS WITH MORE THAN ONE LAYER

Gel layer	Buffer solutions						Characteristics of system	Apparatus type[a]	Ref.
	In gel			In electrode vessels					
	Buffer ions	M	pH	Buffer ions	M	pH			
Sample (large pore)	Tris / Chloride	0.062 / 0.06	6.7	Tris / Glycine	0.005 / 0.038	8.3	The sample is set in the upper-most gel and concentrates in the spacer gel; the actual separation occurs in the bottom gel	D	9
Spacer (large pore)	Tris / Chloride	0.062 / 0.06	6.7						
Small-pore gel	Tris / Chloride	0.37 / 0.06	8.9						
Same gel layers as above and buffers				Tris / Glycine	0.05 / 0.38	8.3	Same as Ref. 9 except that improved resolution is obtained if carried out at 4—5°C	D	10
Spacer (large pore)	Tris / Chloride	0.075 / 0.075	6.7	Tris / Glycine	0.05 / 0.38	8.3	Same as Ref. 9 except that sample gel is omitted; the sample is stabilized with sucrose	D	2
Small-pore gel	Tris / Chloride	0.046 / 0.075	8.7						
Spacer (large pore)	Tris / H_3PO_4	0.049 / 0.026	6.7	Tris / Glycine	0.025 / 0.19	8.4	Same as Ref. 9 except that sample gel is omitted; the sample is stabilized with acrylamide monomer	VP	11
Small-pore gel	Tris / Chloride	0.37 / 0.06	8.9						
Sample (large pore)	Acetic acid / KOH	0.062 / 0.06	6.8	Acetic acid / β-Alanine	0.14 / 0.35	4.5	An acid gel suitable for separation of bases	D	13
Spacer (large pore)	Acetic acid / KOH	0.062 / 0.06	6.8						
Small-pore gel	Acetic acid / KOH	0.75 / 0.12	4.3						

[a] D: disc; FP: flat plate (in the horizontal); VP: vertical plate.

[b] Molarity based on weight of orthoboric acid (H_4BO_3) taken. For additional buffers see Starch Gels. All buffers recommended for starch gel can be used in acrylamide gels.

From *Electrophoresis of Proteins in Polyacrylamide and Starch Gels*, Gordon, A. H., Ed. (*Laboratory Techniques in Biochemistry and Molecular Biology*, Work, T. S. and Work, E., Gen. Eds.), North-Holland, New York, 1971. 41. With permission of Elsevier Science Publishers.

REFERENCES FOR ACRYLAMIDE GEL TABLES

1. **Ferris, T. G., Easterling, R. E., and Budd, R. E.,** *Anal. Biochem.,* 8, 477, 1964.
2. **Pun, J. Y. and Lombarozo, L.,** *Anal. Biochem.,* 9, 9, 1964.
3. **Jensen, K.,** *Scand. J. Clin. Lab. Invest.,* 17, 192, 1965.
4. **Small, P. A., Harrington, W. F., and Keilley, W. W.,** *Biochim. Biophys. Acta.* 49, 462, 1961.
5. **Work, T. S.,** *J. Mol. Biol.,* 10, 544, 1964.
6. **Matson, C. F.,** *Anal. Biochem.,* 13, 294, 1965.
7. **Lorber, A.,** *J. Lab. Clin. Med.,* 64, 133, 1964.
8. **Shepherd, G. R. and Gurley, L. R.,** *Anal. Biochem.,* 14, 356, 1966.
9. **Davies, B. J.,** *Ann. N.Y. Acad. Sci.,* 121, 404, 1964.
10. **Pastewka, J. V., Neass, A. T., and Peacock, A. C.,** *Clin. Chim. Acta,* 14, 219, 1966.
11. **Ritchie, R. F., Harter, J. G., and Bayles, T. B.,** *J. Lab. Clin. Med.,* 68, 842, 1966.
12. **Hjerten, S., Jerstedt, S., and Tiselius, A.,** *Anal. Biochem.,* 11, 219, 1965.
13. **Reisfeld, R. A., Lewis, U. J., and Williams, D. E.,** *Nature,* 195, 281, 1962.
14. **Bloemendal, H., Bout, W. S., Jongkind, J. F., and Wisse, J. H.,** *Exp. Eye Res.,* 1, 300, 1962.
15. **Gordon, A. H. and Louis, L. N.,** *Anal. Biochem.,* 21, 190, 1967.

DETECTION METHODS FOR ENZYMES AFTER SEPARATION IN STARCH OR ACRYLAMIDE GEL

Enzyme	Detection method used in		Procedures	Ref.
	Starch gel	Acrylamide gel[a]		
Red cell acid phosphatase	+		Phenolphthalein diphosphate followed by alkali	1,5
Phosphoglucomutase	+		Reduction of the tetrazolium salt of MTT	2
Placental alkaline phosphatase	+		β-Naphthyl phosphate and diazo salt of fast blue RR	3,4,5,11
β-Glucuronidase	+		8-*HO*-quinoline glucuronide + blue RR salt	5
Cholinesterase	+		6-Bromo-s-naphthylcarbonaphthoxycholine iodide + blue B salt	5
Glucose-6-phosphate dehydrogenase	+		Application of an agar overlay containing a tetrazolium salt	6
6-Phosphogluconate dehydrogenase	+		Application of an agar overlay containing a tetrazolium salt	7
Cytochrome oxidase	+		α-Naphthol + dimethylparaphenylenediamine	5
Esterases	+	+	α-Naphthyl butyrate + diazo salt of fast blue RR	5,11
Lactic and malic dehydrogenases	+	+	Reduction of a tetrazolium salt (nitro blue tetrazolium)	5,9
Acid phosphatase	+	+	Sodium α-naphthyl acid phosphate + diazo salt of 5-chloro-*o*-toluidine	11
Phosphorylase		+	Silver phosphate is reduced to metallic silver by UV light	10
Succinic dehydrogenase	+		Reduction of a tetrazolium salt (nitro blue tetrazolium)	9
β-*HO*-butyric dehydrogenase	+		Reduction of a tetrazolium salt (nitro blue tetrazolium)	9
Glutamic dehydrogenase	+		Reduction of a tetrazolium salt (nitro blue tetrazolium)	9
Catalase	+		Inhibition of starch iodide reaction	8
Ceruloplasmin	+		*O*-Dianisidine	8,12
Leucine aminopeptidase	+		Alanyl-β-naphthylamide + diazotized *O*-aminoazotoluene	9,13
Hemoglobin as peroxidase	+		Benzidine + H_2O_2	

[a] Detection methods used in starch gel may be expected to work equally well in acrylamide gel.

REFERENCES

1. **Hopkinson, D. A., Spencer, N., and Harris, H.,** *Nature,* 199, 969, 1963.
2. **Spencer, N., Hopkinson, D. A., and Harris, H.,** *Nature,* 204, 742, 1964.
3. **Boyer, S. H.,** *Science,* 134, 1002, 1961.
4. **Robson, E. B. and Harris, H.,** *Nature* 207, 1257, 1965.
5. **Lawrence, S. H., Melnick, P. J., and Weimer, H. E.,** *Proc. Soc. Exp. Biol. Med.,* 105, 572, 1960.
6. **Fildes, R. A. and Parr, C. W.,** *Biochem. J.,* 87, 45P, 1963.
7. **Fildes, R. A. and Parr, C. W.,** *Nature,* 200, 890, 1963.
8. **Baur, E. W.,** *J. Lab. Clin. Med.,* 61, 166, 1963.
9. **Goldberg, E.,** *Science* 139, 602, 1963.
10. **Frederick, J. F.,** *Phytochemistry,* 2, 413, 1963.
11. **Allen, J. M. and Hunter, R. L.,** *J. Histochem. Cytochem.,* 8, 50, 1960.
12. **Owen, J. A. and Smith, H.,** *Clin. Chim. Acta,* 6, 441, 1961.
13. **Dubbs, C. A., Vivonia, C., and Hilburn, J. M.,** *Nature,* 191, 1203, 1961.

From *Electrophoresis of Proteins in Polyacrylamide and Starch Gels,* Gordon, A. H., Ed. (*Laboratory Techniques in Biochemistry and Molecular Biology,* Work, T. S. and Work, E., Gen. Eds.). North-Holland, New York, 1971, 62. With permission of Elsevier Science Publishers.

STARCH GELS

Table 1
STARCH GELS: BUFFER SOLUTIONS

Buffer solutions						Materials investigated (examples only)	Results
In gel			**In electrode vessels**				
Buffer ions	*M*	*p*H	Buffer ions	*M*	*p*H		
Borate[a,b]	0.025	8.5	Borate	0.3	8.5	Serum[1]	
Tris	0.076					Serum,[2,3] diphtheria toxin,[2] horse radish peroxidase,[2] haptoglobins,[5] and tranferins[6]	Citrate borate boundary is seen as a brown band
Citrate[a]	0.005	8.6	Borate	0.3	8.5	Hemoglobins,[4] human and animal sera,[7] staphylococcal enterotoxin B[16]	The electrophoresis is over in less time compared to borate only
Tris	0.045		Tris	0.013			
Borate[a]	0.025	8.4	Borate	0.075	8.4	Hemoglobins[12-14]	Hb A and F are separated
EDTA	0.002		EDTA	0.006		Serum[17]	
Tris	0.1		Tris	0.01			
Malate[a]	0.01	7.4	Malate	0.1	7.4	Hemolysates invested for phosphoglucomutase[20]	
EDTA	0.001		EDTA	0.01			
MgCl₂	0.001		MgCl₂	0.01			
Glycine	0.05	8.9	Borate	0.3	8.9	Myeloma γ-globulin[8]	Five bands
Citrate[a]	0.0027		—	—			
Borate[a]	0.0076		Borate	0.38		Anterior pituitary hormones[15]	
Tris	0.0144	8.0	—	—			
Lithium	0.0020		Lithium	0.10			
Tris	0.03	8.4	Tris	0.05		Tissue dehydrogenases[21]	
HCl	—		HCl	—			
Acetate	0.1	5.0	—	—		Pepsin and gastricsin[23]	The enzymes were recovered from the gel

[a] Molarities based on weight of orthoboric (H_4BO_3) or other acid taken. When no cation is given, the acids have been adjusted to the pH shown by addition of NaOH.

[b] For optimum results, use borate concentration recommended for each individual batch of starch.

From Gordon, A. H., Ed., *Electrophoresis of Proteins in Polyacrylamide and Starch Gels* (in *Laboratory Techniques in Biochemistry and Molecular Biology,* Work, T. S. and Work, E., Gen. Eds.), North-Holland, New York, 1971, 100. With permission of Elsevier Science Publishers.

Table 2
STARCH GELS CONTAINING UREA BUFFER SOLUTIONS

Buffer solutions							Materials investigated (examples only)	Results
In gel				In electrode vessels				
Buffer ions	*M*	*p*H	Urea *M*	Buffer ions	*M*	*p*H		
Formate	0.05	3.0	8	Formate	0.2	2—3	γ-Myeloma, heavy chains,[10] antibodies[18]	Slightly sharper bands were obtained if alkylation was done as well as reduction
Aluminum lactate	0.05	3.1	3	Aluminum lactate			Gliadin[11]	
Glycine	0.035	7—8	8	Borate	0.3	8.2	γ-Globulin heavy and light chains[9]	Light chains well separated
Borate[a]	0.025	8.5	8	Borate	0.3	8.5	Haptoglobins[19]	Without prior reduction urea has no effect
Tris Citrate[a]	0.076 0.005	8.6	7	Borate	0.3	8.6	α- and β-casein[22,25]	Urea treatment of α-crystallin, without reduction, leads to formation of many bands
Acetate	0.0365	5.6	6	—	—	—	*E. coli* ribosomal proteins[24]	At least 14 bands

[a] Molarities based on weight of orthoboric (H_4BO_3) or other acid taken. When no cation is given, the acids have been adjusted to the pH shown by addition of NaOH.

[b] For optimum results, use borate concentration recommended for each individual batch of starch.

From Gordon, A. H., Ed., *Electrophoresis of Proteins in Polyacrylamide and Starch Gels* (in *Laboratory Techniques in Biochemistry and Molecular Biology*, Work, T. S. and Work, E., Gen. Eds.), North-Holland, New York, 1971. 101. with permission of Elsevier Science Publishers.

REFERENCES TO TABLES 1 AND 2

1. **Smithies, O.,** *Biophys. J.,* 61, 629, 1955.
2. **Poulik, M.D.,** *Nature,* 180, 1477, 1957.
3. **Poulik, M. D.,** *J. Immunol.,* 82, 502, 1959
4. **De Grouchy, J.,** *Rev. Fr. Etud. Clin. Biol.,* 3, 877, 1958.
5. **Giblett, E. R., Motulsky, A. G., and Frazer, G. R.,** *Am. J. Human Genet.,* 18, 553, 1966.
6. **Harris, H., Robson, E. B., and Siniscalco, M.,** *Nature,* 182, 452, 1958.
7. **Krotski, W. A., Benjamin, D. C., and Weimer, H. E.,** *Can. J. Biochem.,* 44, 545, 1966.
8. **Askonas, I.,** *Biochem. J.,* 79, 33, 1961.
9. **Cohen, S. and Porter, R. R.,** *Biochem. J.,* 90, 278, 1964.
10. **Edelman, G. M. and Poulik, M. D.,** *J. Exp. Med.,* 113, 861, 1961.
11. **Woychick, J. H., Boundy, J. A., and Dimler, R. J.,** *Arch. Biochem. Biophys.,* 94, 477, 1961.
12. **Winterhalter, K. H. and Huehns, E. R.,** *J. Biol. Chem.,* 239, 3699, 1964.
13. **Huehns, E. R., Dance, N., Beaven, G. H., Keil, J. V., Hecht, F., and Motulsky, A. G.,** *Nature,* 201, 1095, 1964.
14. **Chernoff, A. I. and Pettit, N. M.,** *Blood,* 25, 646, 1965.
15. **Ferguson, K. A. and Wallace, A. L. C.,** *Nature,* 190, 629, 1961.
16. **Baird-Parker, A. C. and Jospeh, R. L.,** *Nature,* 202, 570,1964
17. **Poulik, M. D.,** in *Methods of Biochemical Analysis,* Vol. 14, Glick, D., Ed., Interscience, 1966, 455.
18. **Edelman, G. M., Benacerraf, B., and Ovary, Z.,** *J. Exp. Med.,* 118, 229, 1963.
19. **Smithies, O. and Connell, G. E.,** in *Biochemistry of Human Genetics,* Churchill, London, 1959, 178.
20. **Spencer, N., Hopkinson, D. A., and Harris, H.,** *Nature,* 204, 742, 1964.
21. **Tsao, M. U.,** *Arch. Biochem. Biophys.,* 90, 234, 1960
22. **Wake, R. G. and Baldwin, R. L.,** *Biochim. Biophys. Acta,* 47, 227, 1961.
23. **Tang, J., Wolf, S., Caputto, R., and Trucco, R. E.,** *J. Biol. Chem.,* 234, 1174, 1959.
24. **Waller, J. P. and Harris, J. I.,** *Proc. Nat. Acad. Sci. U.S.A.* 47, 18, 1961.
25. **Bloemendal, H., Bout, W. S., Jonkind, J. F., and Wisse, J. H.,** *Nature,* 193, 437, 1962.

Section 2

Nucleosides, Nucleotides, and Nucleic Acids

ALKYL BASES, NUCLEOSIDES, AND NUCLEOTIDES

UV SPECTRAL CHARACTERISTICS AND ACIDIC DISSOCIATION CONSTANTS OF 280 ALKYL BASES, NUCLEOSIDES, AND NUCLEOTIDES

B. Singer

The λ_{max}(nm), in those cases where more than one value has been reported, are either the most frequent value or an average of several values, the range being $\pm$ 1 to 2 nm for the λ_{max}. Since the λ_{min} is more sensitive than the λ_{max} to impurities in the sample, the values of λ_{min} in the table are generally the lowest reported. Values in parentheses are shoulders or inflexions. The cationic and anionic forms are either so stated by the authors or are arbitrarily taken at pH 1 and pH 13. Individual values are given for pK_a except when there are more than two values. In that case, a range is given.

Complete spectra representing a range of derivatives are shown in the figures, and reference to these is made in the table with an asterisk and number preceding the name of the compound. All spectra were obtained in the author's laboratory from samples isolated from paper chromatograms. It is recognized that pH 1 or pH 13 is not ideal for obtaining the cationic or anionic forms when these pHs are close to a pK. Nevertheless, these conditions are useful for purposes of identification since the spectra are reproducible.

Additional data not quoted here are available in many of the references. These data include spectral characteristics in other solvents than H_2O and at other pH values, extinction coefficients, R_F values in various paper chromatographic systems, column chromatographic systems, methods of synthesis or preparation of alkyl derivatives, mass spectra, and NMR, optical rotatory dispersion and infrared spectra.

UV SPECTRAL CHARACTERISTICS AND ACIDIC DISSOCIATION CONSTANTS OF ALKYL BASES, NUCLEOSIDES, AND NUCLEOTIDES[a]

	Acidic		Basic			
	λ_{max}(nm)	λ_{min}(nm)	λ_{max}(nm)	λ_{min}(nm)	pK_a	References
[1] Adenine						
Monoalkylated						
*[2] 1-Methyl-	259	228	270	234	7.2	1−7
1-Ethyl-	260	233	271	242	6.9, 7.0	5, 8, 9
1-Isopropyl-	259		269			10
1-Benzyl-	260		271		7.0	11, 12
1-(2-Hydroxypropyl)-	259		271		7.2	13
1-(2-Hydroxyethylthioethyl)-	262		271		7.2	6
2-Methyl-	267	228	270 (280)	239	~5.1	3, 4
*[3] 3-Methyl-	274	235	273	244	6.1, 6.1	2, 5, 6, 13−15
3-Ethyl-	274	240	273	247	6.5	5, 8, 16
3-Isopropyl-	274		273			10
3-Benzyl-	275		272		5.1	17, 18
3-(2-Hydroxypropyl)-	274		273		6.0	13

[a]Much of the data and some of the spectra were published in a review by B. Singer in *Prog. Nucleic Acids Res. Mol. Biol.*, 15, 219−284, 330−332 (1975).

UV SPECTRAL CHARACTERISTICS AND ACIDIC DISSOCIATION CONSTANTS OF ALKYL BASES, NUCLEOSIDES, AND NUCLEOTIDES[a] (continued)

	Acidic		Basic			
	λ_{max}(nm)	λ_{min}(nm)	λ_{max}(nm)	λ_{min}(nm)	pK_a	References
*[1] **Adenine** (continued)						
3-(2-Diethylaminoethyl)-	275	236	274	245		19
*[4] N^6-Methyl-	267	231	273 (280)	238	4.2, 4.2	1–4, 7, 15, 20, 21
N^6-Ethyl-	268	231	274 (281)	241		8, 21
N^6-Butyl-	270		275			21
N^6-(2-Hydroxyethyl)-	272	233	273	236	3.7	72
N^6-(2-Diethylaminoethyl)-	275	233	274	239		19
*[5] 7-Methyl-	273	237	270 (280)	230	~3.5, 3.6	8, 15, 18, 23, 24
7-Ethyl-	272	239	270 (280)	234		8, 16
7-Isopropyl-	272		272			10
9-Methyl-	261	230	262	228	3.9	5, 13, 15
9-Ethyl-	258	230	262	228	4.1	5, 13, 25
9-Isopropyl-	260		262			10
9-(2-Hydroxypropyl)-	259		261			13
9-(2-Diethylaminoethyl)-	258	227	261	229		19
9-Benzyl-	259		261			18
Dialkylated						
1,N^6-Dimethyl-	261	230	273	245		1, 8, 20
1,7-Dimethyl-	270					26
1-Butyl, 7-methyl-	268					26
1,9-Dimethyl-	260	235	260 (265)	235	9.08	13, 20, 27
1,9-Di-(2-hydroxypropyl)-	260		260			13
1,9-Di-(2-diethylaminoethyl)-	257	233	261	232		19
1-Ethyl-9-methyl-	261		261		9.16	27
1-Propyl-9-methyl-	261		261		9.15	27
3,N^6-Dimethyl-	281		287			28
3,N^6-Di-(2-diethylaminoethyl)-	282	243	282	249		19
3,7-Dimethyl-	276	246	225, 280	221, 247	11	2, 13, 20
3,7-Dibenzyl-	278		281		9.6	18
3,7-Di-(2-hydroxypropyl)-	278		281			13
3,7-Di-(2-diethylaminoethyl)-	276	237	279	245		19
N^6,N^6-Dimethyl-	276	236	282	245		7, 21
N^6,N^6-Diethyl-	278		282			21, 29
N^6,7-Dimethyl-	279		275			24, 26
*[6] N^6-Methyl-7-ethyl-	277 (285)	241	276	244		8
N^6-Propyl-7-methyl-	281		277			24
N^6-Butyl-7-methyl-	279					26
N^6,9-Dimethyl-	265		268		4.02	27, 29
N^6,9-Di-(2-diethylaminoethyl)-	266	229	270	233		19
N^6-Ethyl-9-methyl-	265		268		4.08	27, 29
N^6-Propyl-9-methyl-	266		270		4.14	27
N^6-Butyl-9-ethyl-	266		269			27
Trialkylated						
1,N^6,N^6-Trimethyl-	221, 293	246	232, 301	262		30
3,N^6,N^6-Trimethyl-	290	243	293	250		30
N^6,N^6,7-Trimethyl-	233, 293	250	291	246		30
N^6,N^6,9-Trimethyl-	269	234	276	237		29, 30
N^6,N^6-Dimethyl-9-ethyl-	270		277			25
3,N^6,7-Tribenzyl-	289		Unstable		9.4	18

UV SPECTRAL CHARACTERISTICS AND ACIDIC DISSOCIATION CONSTANTS OF ALKYL BASES, NUCLEOSIDES, AND NUCLEOTIDES[a] (continued)

	Acidic		Basic			
	λ_{max}(nm)	λ_{min}(nm)	λ_{max}(nm)	λ_{min}(nm)	pK_a	References

[7]Guanine

	λ_{max}(nm)	λ_{min}(nm)	λ_{max}(nm)	λ_{min}(nm)	pK_a	References
Monoalkylated						
1-Methyl-	250 (270)	228	277 (262)	242	3.1	4, 7, 15, 23, 31–34
*[8] 1-Ethyl-	251 (274)	229	278 (260)	243		35, 36
1-Isopropyl-	253					10
1-(2-Diethylaminoethyl)-	255, 275		257, 268			19
N^2-Methyl-[b]	251, 279	228	245–255, 278	238	3.3	4, 7, 23, 33, 34, 37, 38
*[9] N^2-Ethyl-	253 (280)	229	245, 279	263		37, 39
N^2-Isopropyl-	252		277			10
3-Methyl-	263 (244)	227	273	246		15, 33, 40, 41
*[10] 3-Ethyl-	263 (244)	233	273	248		39
3-Isopropyl-	263		273			10
3-Benzyl-	263 (243)		274		4.00	42
O^6-Methyl-	286		246, 284			43, 44
*[11] O^6-Ethyl-	286	253	284 (246)	259		35, 44
O^6-Propyl-	286		246, 283			44
O^6-Isopropyl-	285 (230)		283 (245)			10, 45
O^6-Butyl-	286		246, 285			44
O^6-Isobutenyl-	286 (232)		283 (245)			45
7-Methyl-	249 (272)	226	281 (240)	255	3.5	4, 23, 31, 32, 46
*[12] 7-Ethyl-	249 (274)	233	280	258	3.7	5, 35
7-Isopropyl-	249 (274)		278 (240)			10, 45
7-Benzyl-	250		281		3.2	11
7-(2-Hydroxyethyl)-	250	229	281	261		47
7-(2-Hydroxypropyl)-	250, 272	229	280	257		13
7-(2-Diethylaminoethyl)-	253, 270		275 (250)			19
7-(β-Hydroxyethylthioethyl)-	250		281			48
8-Propyl-	249, 276		276			45
8-Isobutyl-	249, 278		276			45
8-(3-Methylbutyl)-	249, 277		275			45
9-Methyl	251, 276		268 (258)		2.9	33
*[13] 9-Ethyl-	252, 277	230	253, 268	238		35
9-Isopropyl-	253, 276		256, 258			10, 45
Dialkylated						
1,7-Dimethyl-	252 (272)	230	284 (251)	262		23, 33, 35
*[14] 1,7-Diethyl-	252 (275)	232	285 (250)	263		49
1,9-Dimethyl-	254 (277)	229				33
N^2,N^2-Dimethyl-[b]	255 (289)	229	277–283			4, 34, 37, 38
7,8-Dimethyl-	249, 277		280 (235)		4.4	50
7,9-Dimethyl	254 (278)	229	c			28, 33
7,9-Di-(2-diethylaminoethyl)-	257, 278		c			19

[b] N^2-Alkyl guanines and guanosines do not exhibit sharp maxima or minima, particularly in basic solution, as shown in the spectra published by Hall,[4] Smith and Dunn,[34] and Singer and Fraenkel-Conrat.[39] Therefore, some of the data are given as a range of values.

[c] All 7-alkyl purine nucleosides and nucleotides and 7,9-dialkyl purines are unstable in alkali and the imidazole ring opens at varying rates. For this reason spectral data obtained in alkaline solution do not represent the original compound and thus such data are omitted. The opening of the imidazole ring in alkali can be used as a means of identifying this class of alkyl compounds. Ring opening can lead to a number of derivatives.[39]

UV SPECTRAL CHARACTERISTICS AND ACIDIC DISSOCIATION CONSTANTS OF ALKYL BASES, NUCLEOSIDES, AND NUCLEOTIDES[a] (continued)

	Acidic		Basic			
	λ_{max}(nm)	λ_{min}(nm)	λ_{max}(nm)	λ_{min}(nm)	pK_a	References
[*7] Guanine (continued)						
7-Methyl-9-ethyl-	254, 281		c		7.3	14
8,9-Dimethyl-	252, 277 (289)		280 (252)		4.11	50
Trialkylated						
1,7,9-Trimethyl-	254, 280		c			33
[*15] Cytosine						
Monoalkylated						
1-Methyl-	213, 283	241	274	250	4.55–4.61	51–53
O^2-Methyl-	260	241	270	246	5.41	53
3-Methyl-	273	240	294	251	7.4, 7.49	4, 53, 54
[*16] 3-Ethyl-	275	241	296	257		55
N^4-Methyl-	278	240	286 (230)	256	4.55	4, 56, 57
N^4-Ethyl-	277	244	284	253	4.58	55, 57
5-Methyl-	211, 284	242	288	254	4.6	58
6-Methyl-					5.13	59
Dialkylated						
1,3-Dimethyl-	281	243	272	247	9.29–9.4	51, 53, 54
1,N^4-Dimethyl-	218, 285	244	274 (235)	250	4.38–4.47	51, 53, 56
N^4,N^4-Dimethyl-	283	242	290 (235)	259	4.15, 4.25	56, 57
1,5-Dimethyl-	291	244			4.76	60
Trialkylated						
1,3,N^4-Trimethyl-	212, 287	248	280	247	9.65	19, 51, 53
1,N^4,N^4-Trimethyl-	220, 288	248	283	242	4.2	56
5-Hydroxymethylcytosine						
Unmodified						
	279		284			7
3-Methyl-	278		296		7.1	14
[*17] Xanthine						
[*18] 1-Methyl-[d]	260–265 (235)	239	283 (245)	257	1.3	35, 61, 62
3-Methyl-[d]	266		275 (232)		0.8	61, 62
[*19] 7-Methyl-[d]	267	233	289 (237)	255	0.8	32, 35, 61
9-Methyl-[d]	260		245, 278		2.0	61, 62
1,3-Dimethyl-	266		275		0.7	61, 62
1,7-Dimethyl-	~260		233, 289		0.5	61, 62
1,9-Dimethyl-	262		248, 277		2.5	61, 62
3,7-Dimethyl-	265		234, 274		0.3	61, 62
3,9-Dimethyl-	265		270 (240)		1.0	61, 62
7,9-Dimethyl-	239, 262		c			63
8,9-Dimethyl-	238, 265		245, 278			63
1,3,7-Trimethyl-	266				0.5	61
1,3,9-Trimethyl-	266				0.6	61
1,7,9-Trimethyl-	232, 262		c			63

[d] Basic values are those of the dianion (pH 14).

UV SPECTRAL CHARACTERISTICS AND ACIDIC DISSOCIATION CONSTANTS OF ALKYL BASES, NUCLEOSIDES, AND NUCLEOTIDES[a] (continued)

	Acidic		Basic			
	λ_{max}(nm)	λ_{min}(nm)	λ_{max}(nm)	λ_{min}(nm)	pK_a	References
*[20] **Uracil**						
1-Methyl-	208, 268	241	265	242	~1.8	58, 64
O^2-Ethyl-	218, 260		221, 265			58
*[21] 3-Methyl-	258	230	218, 283	245		4, 58, 65
O^4-Methyl-	267		276			65a
O^4-Ethyl-	269		220, 278			58
1,3-Dimethyl-	266	234	266	234		58
1,6-Dimethyl-	208, 268	234	266	241		65
O^2,3-Dimethyl-	213, 269					65b
3,6-Dimethyl-	205, 259	231	281	245		65
5,6-Dimethyl-	267	236	275	245		65
1,5,6-Trimethyl-	206, 276	240	273	245		65
*[22] **Thymine**						
1-Methyl-			269	244		66
1-(2-Diethylaminoethyl)-			265	248		19
3-Methyl-	266	237	290	248		31, 35, 66
*[23] 3-Ethyl-	265	237	289	247		35
3-(2-Diethylaminoethyl)-			288	244		19
1,3-Di-(2-diethylaminoethyl)-			269	245		19
1,O^4-Di-(2-diethylaminoethyl)-			274	245		19
O^2,3-Dimethyl-	217, 272					65b
Hypoxanthine						
1-Methyl-	249		260			15, 66a
3-Methyl-	253		265		2.61	15, 66a, 67
3-Ethyl-	254		266			68
3-Benzyl-	254		264 (277)			69
3-(2-Diethylaminoethyl)-	260	234	262	243		19
O^6-(3-Methyl-2-butenyl)-	247		262			45
7-Methyl-	250		262		2.12	15
9-Methyl-	250		254			15
1,7-Dimethyl-	252	232	267	237		70
1,7-Dibenzyl-	255		256			69
1,9-Dibenzyl-	263		259			69
3,7-Dimethyl-			267			20
3,7-Dibenzyl-	256		267			69
N^6,7-Dimethyl-	256		258			24
7,9-Dimethyl-	251		c			28
*[24] **Adenosine**						
*[25] 1-Methyl-	257	231	258 (265)	233	8.8, 8.3	3, 4, 8, 16, 71
1-Methyldeoxy-	257	239	258	242		8, 72
*[26] 1-Ethyl-	259	235	261 (268, 300)	237		8
1-Ethyldeoxy-	259	231	260 (268)	236		8
1-Benzyldeoxy-	259		259			11
2-Methyl-	258	230	264	227		3, 4, 73
*[27] N^6-Methyl-	262	231	266	223	4.0	1–4, 8, 74, 74a
N^6-Methyldeoxy-	262	231	266	226		4, 8, 72, 74

UV SPECTRAL CHARACTERISTICS AND ACIDIC DISSOCIATION CONSTANTS OF ALKYL BASES, NUCLEOSIDES, AND NUCLEOTIDES[a] (continued)

	Acidic		Basic			
	λ_{max}(nm)	λ_{min}(nm)	λ_{max}(nm)	λ_{min}(nm)	pK_a	References
			[24] Adenosine (continued)			
N^6-Ethyl-	264	239	268	243		8
N^6-Ethyldeoxy-	263	237	268	241		8
N^6-Benzyl-	265	235	268	236		69, 75
N^6-Benzyldeoxy-	264		268			11
N^6-Butyl-	263		267			76
N^6-(2-Hydroxyethyl)-	263	233	267	232	3.1	22
*[28] 7-Ethyl-	268	239	c			8
*[29] 1,N^6-Dimethyl-	261	234	263 (300)	234		1, 8, 20
N^6,N^6-Dimethyl-	268	233	276	237	4.5	3, 4, 30
*[30] N^6,7-Dimethyl-	276	241	c			8
N^6-Methyl-7-ethyl-	276	242	c			8
			[31] Guanosine			
*[32] 1-Methyl-	258 (280)	230	255 (270)	228	2.6	4, 7, 20, 34
1-Methyldeoxy-	257 (278)	232	255 (270)	229		20, 31, 77
1-Ethyl-	261 (272)	232	258 (270)	239	2.8	49
1-Ethyldeoxy-	256		257			77
1-Butyldeoxy-	258 (282)		257 (280)			77
N^2-Methyl-[b]	251–258 (280–290)	222–234	248–258 (270–275)	227–238		4, 34, 38
O^6-Methyl-	284 (243)	259	243, 277	239, 261	2.4	49
O^6-Methyldeoxy-	284 (230)	252	243, 278	233, 261		31, 78
*[33] O^6-Ethyl-	244, 286	239, 260	247, 278	233, 261	2.5	49
O^6-Ethyldeoxy-	286	252	248, 280	261		77
O^6-Butyldeoxy-	246, 287	260	248, 280	261		77
7-Methyl-	257 (275)	230	c		6.7–7.3	4, 14, 46, 49, 74
7-Methyldeoxy-	256 (275)	229	c			31, 32, 74, 77
*[34] 7-Ethyl-	258 (277)	238	c		7.2, 7.4	14, 49
7-Isopropyl-	256 (275)		c			45
7-Benzyl-	258		c		7.2	11
7-Butyldeoxy-	257 (280)		c			77
8-Methyl-	260 (273)		256		3.01	50
1,7-Dimethyl-	260 (270)	236	c			20, 49
*[35] 1,7-Diethyl-	263 (270)	237	c			49
1-Methyl-7-ethyl-	259 (277)	233	c			49
1-Ethyl-7-methyl-	261 (275)	235	c			49
N^2,N^2-Dimethyl-	265 (290)	237	262 (283)	240		4, 7, 34, 38
N^2,O^6-Dimethyldeoxy-	288		249, 284			77
*[36] N^2,O^6-Diethyl-	246, 292	239, 267	252, 281	237, 268		35, 49
N^2,N^2,7-Trimethyl-	267, 300	239, 286	c			79
			[37] Cytidine			
O^2-Methyl-	233, 262	221, 243	Unstable		>8.6	35
*[37a] O^2-Ethyl-	233, 262	221, 243	Unstable		>8.6	35
3-Methyl-	278	243	225, 267	212, 244	8.3–8.9	4, 54–56, 71
*[38] 3-Ethyl-	280	247	267	248	8.4	55
3-Ethyldeoxy-	280	245	268	247	8.6	55
3-Benzyl-	281		266		7.7	11

UV SPECTRAL CHARACTERISTICS AND ACIDIC DISSOCIATION CONSTANTS OF ALKYL BASES, NUCLEOSIDES, AND NUCLEOTIDES[a] (continued)

	Acidic		Basic			
	λ_{max}(nm)	λ_{min}(nm)	λ_{max}(nm)	λ_{min}(nm)	pK$_a$	References
[*37] Cytidine (continued)						
3-(2-Diethylaminoethyl)deoxy-	284	243	271			19
N^4-Methyl-	217, 281	207, 243	237, 273	250	3.85, 3.92	56, 57
N^4-Methyldeoxy-	282	243	236, 270	229, 248	4.01	57
[*39] N^4-Ethyl-	281	244	272	253	4.2	55
N^4-Ethyldeoxy-	279	247	272	253	4.2	55
2'-O,N^4-Dimethyl-[e]	281	242	271	250		80
2',3',5'-Tri-O-methyl-N^4-methyl-[e]	281	242	271	247		81
5-Methyl-	288	245	278	255	4.28	60
5-Methyldeoxy-	287	246	278	255	4.40	60, 82
5-Ethyldeoxy-			278	255		83
6-Methyl	278	241	273	252	4.42	59, 84
6-Methyldeoxy-	278	241	273	252		84
3,N^4-Dimethyl-	286	249	277	249		55
[*40] 3,N^4-Diethyl-	287	252	277	253		55
3,N^4-Di-(2-diethylaminoethyl)-deoxy-	284	245				19
N^4,N^4-Dimethyl-	219, 285	245	279	238	3.7, 3.62	56, 57
N^4,N^4-Dimethyldeoxy-	287	245	278	238	3.79	57
[*41] N^4,N^4-Diethyl-	286	249	276	249		55
2'-O,N^4,N^4-Trimethyl-[e]	287	246	278	238		80
N^4,5-Dimethyl-			275 (234)	252		60
N^4,5-Dimethyldeoxy-	218, 287	246	275 (235)		4.04	60
[*42] 3,N^4,N^4-Triethyl-	287	252	289	253		55
[*43] Uridine						
O^2-Methyl-	229, 251	213, 238				35, 86, 87
O^2-Ethyl-	228, 253	213, 237				87
3-Methyl-	263	233	262	233		4, 14, 55, 85
[*44] 3-Ethyl-	262	235	264	237		35, 55
3-Benzyl-			264	235		83
3-(2-Hydroxyethyl)-	261	235	262			88
[*44a] O^4-Methyl-	271	235	274	239		80
6-Methyl-	261	230	264	242		65, 84
5,6-Dimethyl-	206, 269	236	270	248		65
[*45] Thymidine (deoxy-)						
3-Methyl-	266	238	267	239		31, 35
[*46] 3-Ethyl-	269	239	270	240		35
3-(2-Diethylaminoethyl)-			270	242		19
O^4-Methyl-(α)	279	245	279	243		89
O^4-Methyl-(β)	279	241	280	243		89
[*47] Inosine						
1-Methyl-	250	223	249			4
1-Methyldeoxy-	250		250 (265)			31
1-Benzyl-	251		249			69

[e]Alkylation of ribose does not cause any change in spectrum.

UV SPECTRAL CHARACTERISTICS AND ACIDIC DISSOCIATION CONSTANTS OF ALKYL BASES, NUCLEOSIDES, AND NUCLEOTIDES[a] (continued)

	Acidic		Basic			
	λ_{max}(nm)	λ_{min}(nm)	λ_{max}(nm)	λ_{min}(nm)	pK$_a$	References
[*47] Inosine (continued)						
1-(2-Hydroxyethyl)-	250	226	250	226		88
O^6-Methyl-	250		250			66a, 74a
[*48] 7-Methyl-	252	221	c		6.4	35, 70
7-Ethyl-	252		c			68
1,7-Dimethyl-	265		c			70
Xanthosine						
7-Methyl-	262	237	c			32, 35
Adenylic Acid or ADP[e,f]						
1-Methyl-	258	232	259 (268)	230		6, 35, 90
1-(2-Hydroxyethylthioethyl)-	261		261			6
2-Methyl-	259		263			73
N^6-Methyl-	261	231	265	229		6, 90
N^6-(2-Hydroxyethylthioethyl)-			268			6
N^6-(2-Hydroxyethyl)-	263	233	266	230		91
2,N^6-Dimethyl	263		269			92
Guanylic Acid or GDP[e,f]						
1-Methyl-	258 (280)	230	256 (273)	230	<3	7, 93
7-Methyl-	259, 279	230	c		6.9−7.2	46, 94
O^6-Methyl-	245, 288	262	249, 281	263		95
N^2-Methyl-	263	237	263	240		96
N^2,N^2-Dimethyl-	265	232	263	237	~3	7, 96
N^2,N^2,7-Trimethyl-	262 (290)	237	c		7.4	96
Cytidylic Acid or CDP[e,f]						
3-Methyl-	276	242	223, 266	243	9.0, 9.2	97−99
N^4-Methyl-	217, 281	242	271	249	4.25	98, 99
N^4,N^4-Dimethyl-	219, 287	245	225, 278	238	4.0	98, 99
5-Methyl-	284		279			100
5-Methyldeoxy-	287	243	277	254	4.5	101
Uridylic Acid or UDP[e,f]						
3-Methyl-	261	235	262	235		35, 91
3-Ethyl-	261		263			35
Thymidylic Acid (deoxy-)[f]						
[*49] 3-Methyl-	267	240	268	241		35
3-Ethyl-	267	241	268	242		35

Compiled by B. Singer.

[f]Alkylated nucleoside diphosphates have the same spectral characteristics as alkylated nucleotides and the data are not separated. Alkylation of the phosphate group does not cause any change in spectrum.

UV SPECTRAL CHARACTERISTICS AND ACIDIC DISSOCIATION CONSTANTS OF ALKYL BASES, NUCLEOSIDES, AND NUCLEOTIDES[a] (continued)

REFERENCES

1. Wacker and Ebert, *Z. Naturforsch.*, 14b, 709 (1959).
2. Brookes and Lawley, *J. Chem. Soc.*, (Lond.), p. 539 (1960).
3. Garrett and Mehta, *J. Am. Chem. Soc.*, 94, 8532 (1972).
4. Hall, *The Modified Nucleosides in Nucleic Acids*, Columbia University Press, New York, 1971.
5. Pal, *Biochemistry*, 1, 558 (1962).
6. Shooter, Edwards, and Lawley, *Biochem. J.*, 125, 829 (1971).
7. Venkstern and Baer, *Absorption Spectra of Minor Bases*, Plenum Press, New York, 1965.
8. Singer, Sun, and Fraenkel-Conrat, *Biochemistry*, 13, 1913 (1974).
9. Ludlum, *Biochim. Biophys. Acta*, 174, 773 (1969).
10. Lawley, Orr, and Jarman, *Biochem. J.*, 145, 73 (1975).
11. Brookes, Dipple, and Lawley, *J. Chem. Soc. C*, p. 2026 (1968).
12. Leonard and Fujii, *Proc. Natl. Acad. Sci. U.S.A.*, 51, 73 (1964).
13. Lawley and Jarman, *Biochem. J.*, 126, 893 (1972).
14. Lawley and Brookes, *Biochem. J.*, 89, 127 (1963).
15. Elion, *J. Org. Chem.*, 27, 2478 (1962).
16. Lawley and Brookes, *Biochem. J.*, 92, 19c (1964).
17. Montgomery and Thomas, *J. Am. Chem. Soc.*, 85, 2672 (1963).
18. Montgomery and Thomas, *J. Heterocycl. Chem.*, 1, 115 (1964).
19. Price, Gaucher, Koneru, Shibakawa, Sowa, and Yamaguchi, *Biochim. Biophys. Acta*, 166, 327 (1968).
20. Broom, Townsend, Jones, and Robins, *Biochemistry*, 3, 494 (1964).
21. Elion, Burgi, and Hitchings, *J. Am. Chem. Soc.*, 74, 411 (1952).
22. Windmueller and Kaplan, *J. Biol. Chem.*, 236, 2716 (1961).
23. Reiner and Zamenhof, *J. Biol. Chem.*, 228, 475 (1957).
24. Prasad and Robins, *J. Am. Chem. Soc.*, 79, 6401 (1957).
25. Montgomery and Temple, *J. Am. Chem. Soc.*, 79, 5238 (1957).
26. Taylor and Loeffler, *J. Am. Chem. Soc.*, 82, 3147 (1960).
27. Itaya, Tanaka, and Fujii, *Tetrahedron*, 28, 535 (1972).
28. Jones and Robins, *J. Am. Chem. Soc.*, 84, 1914 (1962).
29. Robins and Lin, *J. Am. Chem. Soc.*, 79, 490 (1957).
30. Townsend, Robins, Loeppky, and Leonard, *J. Chem. Soc.* (Lond.), p. 5320 (1964).
31. Friedman, Mahapatra, Dash, and Stevenson, *Biochim. Biophys. Acta*, 103, 286 (1965).
32. Haines, Reese, and Todd, *J. Chem. Soc.* (Lond.), p. 5281 (1962).
33. Shapiro, *Prog. Nucleic Acid Res. Mol. Biol.*, 8, 73 (1968).
34. Smith and Dunn, *Biochem. J.*, 72, 294 (1959).
35. Singer, unpublished.
36. Kriek and Emmelot, *Biochemistry*, 2, 733 (1963).
37. Elion, Lange, and Hitchings, *J. Am. Chem. Soc.*, 78, 217 (1956).
38. Gerster and Robins, *J. Am. Chem. Soc.*, 87, 3752 (1965).
39. Singer and Fraenkel-Conrat, *Biochemistry*, 14, 772 (1975).
40. Lawley, Orr, and Shah, *Chem. Biol. Interact.*, 4, 431 (1971/72).
41. Townsend and Robins, *J. Chem. Soc.* (Lond.), p. 3008 (1962).
42. Miyaki and Shimizu, *Chem. Pharm. Bull.* (Tokyo), 18, 1446 (1970).
43. Lawley and Thatcher, *Biochem. J.*, 116, 693 (1970).
44. Balsiger and Montgomery, *J. Am. Chem. Soc.*, 25, 1573 (1960).
45. Leonard and Frihart, *J. Am. Chem. Soc.*, 96, 5894 (1974).
46. Hendler, Fürer, and Srinivasan, *Biochemistry*, 9, 4141 (1970).
47. Brookes and Lawley, *J. Chem. Soc.* (Lond.), p. 3923 (1961).
48. Brookes and Lawley, *Biochem. J.*, 77, 478 (1960).
49. Singer, *Biochemistry*, 11, 3939 (1972).
50. Pfleiderer, Shanshal, and Eistetter, *Chem. Ber.*, 105, 1497 (1972).
51. Kenner, Reese, and Todd, *J. Chem. Soc.* (Lond.), p. 855 (1955).
52. Fox and Shugar, *Biochim. Biophys. Acta*, 9, 369 (1952).
53. Sukhorukov, Gukovskaya, Sukhoruchkina, and Lavrrenova, *Biophysica*, 17, 5 (1972).
54. Brookes and Lawley, *J. Chem. Soc.* (Lond.), p. 1348 (1962).
55. Sun and Singer, *Biochemistry*, 13, 1905 (1974).
56. Szer and Shugar, *Acta Biochim. Pol.*, 13, 177 (1966).
57. Wempen, Duschinsky, Kaplan, and Fox, *J. Am. Chem. Soc.*, 83, 4755 (1961).

UV SPECTRAL CHARACTERISTICS AND ACIDIC DISSOCIATION CONSTANTS OF ALKYL BASES, NUCLEOSIDES, AND NUCLEOTIDES[a] (continued)

58. Shugar and Fox, *Biochim. Biophys. Acta,* 9, 199 (1952).
59. Notari, Witiak, DeYoung, and Lin, *J. Med. Chem.,* 15, 1207 (1972).
60. Fox, Praag, Wempen, Doerr, Cheong, Knoll, Eidinoff, Bendich, and Brown, *J. Am. Chem. Soc.,* 81, 178 (1959).
61. Lichtenberg, Bergmann, and Neiman, *J. Chem. Soc. C,* p. 1676 (1971).
62. Pfleiderer and Nübel, *Justus Liebigs Ann. Chem.,* 647, 155 (1961).
63. Pfleiderer, *Justus Liebigs Ann. Chem.,* 647, 161 (1961).
64. Brown, Hoerger, and Mason, *J. Chem. Soc.,* (Lond.), p. 211 (1955).
65. Wittenburg, *Collect. Czech. Chem. Commun.,* 36, 246 (1971).
65a. Wong and Fuchs, *J. Org. Chem.,* 35, 3786 (1970).
65b. Wong and Fuchs, *J. Org. Chem.,* 36, 848 (1971).
66. Wierzchowski, Litonska, and Shugar, *J. Am. Chem. Soc.,* 87, 4621 (1965).
66a. Miles, *J. Org. Chem.,* 26, 4761 (1961).
67. Bergmann, Levin, Kalmus, and Kwietny-Govrin, *J. Am. Chem. Soc.,* 26, 1504 (1961).
68. Rajabalee and Hanessian, *Can. J. Chem.,* 49, 1981 (1971).
69. Montgomery and Thomas, *J. Am. Chem. Soc.,* 28, 2304 (1963).
70. Michelson and Pochon, *Biochim. Biophys. Acta,* 114, 469 (1966).
71. Haines, Reese, and Todd, *J. Chem. Soc.* (Lond.), p. 1406 (1964).
72. Coddington, *Biochim. Biophys. Acta,* 59, 472 (1962).
73. Saneyoshi, Ohashi, Harada, and Nishimura, *Biochim. Biophys. Acta,* 262, 1 (1972).
74. Jones and Robins, *J. Am. Chem. Soc.,* 85, 193 (1963).
74a. Johnson, Thomas, and Schaeffer, *J. Am. Chem. Soc.,* 80, 699 (1958).
75. Kissman and Weiss, *J. Am. Chem. Soc.,* 21, 1053 (1956).
76. Fleysher, *J. Med. Chem.,* 15, 187 (1972).
77. Farmer, Foster, Jarman, and Tisdale, *Biochem. J.,* 135, 203 (1973).
78. Loveless, *Nature,* 223, 206 (1969).
79. Saponara and Enger, *Nature,* 223, 1365 (1969).
80. Robins and Naik, *Biochemistry,* 10, 3591 (1971).
81. Kusmierek, Giziewica, and Shugar, *Biochemistry,* 12, 194 (1973).
82. Dekker and Elmore, *J. Chem. Soc.* (Lond.), p. 2864 (1951).
83. Imura, Tsuruo, and Ukita, *Chem. Pharm. Bull.,* 16, 1105 (1968).
84. Winkley and Robins, *J. Org. Chem.,* 33, 2822 (1968).
85. Miles, *Biochim. Biophys. Acta,* 22, 247 (1956).
86. Brown, Todd, and Varadarajan, *J. Chem. Soc.* (Lond.), p. 868 (1957).
87. Kimura, Fujisawa, Sawada, and Mitsunobu, *Chem. Lett.,* 691 (1974).
88. Holy, Bald, and Hong, *Collect. Czech. Chem. Commun.,* 36, 2658 (1971).
89. Lawley, Orr, Shah, Farmer, and Jarman, *Biochem. J.,* 135, 193 (1973).
90. Griffin and Reese, *Biochim. Biophys. Acta,* 68, 185 (1963).
91. Michelson and Grunberg-Manago, *Biochim. Biophys. Acta,* 91, 92 (1964).
92. Hattori, Ikehara, and Miles, *Biochim. Biophys. Acta,* 13, 2754 (1974).
93. Pochon and Michelson, *Biochim. Biophys. Acta,* 145, 321 (1967).
94. Lawley and Shah, *Biochem. J.,* 128, 117 (1972).
95. Gerchman, Dombrowski, and Ludlum, *Biochim. Biophys. Acta,* 272, 672 (1972).
96. Pochon and Michelson, *Biochim. Biophys. Acta,* 182, 17 (1969).
97. Ludlum and Wilhelm, *J. Biol. Chem.,* 243, 2750 (1968).
98. Brimacombe and Reese, *J. Chem. Soc. C,* p. 588 (1966).
99. Brimacombe, *Biochim. Biophys. Acta,* 142, 24 (1967).
100. Szer, *Biochem. Biophys. Res. Commun.,* 20, 182 (1965).
101. Cohn, *J. Am. Chem. Soc.,* 73, 1539 (1951).

ULTRAVIOLET ABSORBANCE OF OLIGONUCLEOTIDES CONTAINING 2'-O-METHYLPENTOSE[a] RESIDUES

Compound	Min	Max	Absorbance ratios[b]				
			240	250	270	280	290
			pH 12				
Gm-Cp	230	268	0.74	0.89	1.02	0.71	0.20
Gm-Up	226	255	0.63	0.97	0.81	0.52	0.19
Um-Ap	230	258	0.56	0.84	0.69	0.21	0.04
Um-Gp	233	260	0.65	0.87	0.87	0.48	0.09
Um-Up	241	260	0.79	0.86	0.79	0.29	0.04
Am-Am-Up	229	258	0.53	0.86	0.71	0.24	0.04
Am-Gm-Cp	228	261	0.59	0.86	0.83	0.46	0.12
Am-Um-Gp	230	259	0.55	0.85	0.79	0.36	0.14

[a]The pentose is presumed to be 2'-O-methylribose since the dinucleotides were obtained from yeast ribonucleic acid. Evidence for the chemical constitution of the modified pentose is given in Howlett et al.[1] and Trim and Parker.[2]

[b]Absorbance ratios were calculated from optical densities at 240, 250, 270, 280, and 290 nm relative to that at 260 nm.

Compiled by A. R. Trim.

Values for trinucleotides are from Trim and Parker.[3] Data on dinucleotides are reprinted by permission from Trim and Parker, *Biochem. J.*, Vol. 116, p. 589, copyright© 1970 by the Biochemical Society, London.

REFERENCES

1. **Howlett, Johnson, Trim, Eagles, and Self,** *Anal. Biochem.*, 39, 429 (1971).
2. **Trim and Parker,** *Anal. Biochem.*, 46, 482 (1972).
3. **Trim and Parker,** unpublished data.

THE OPTICAL ROTATORY DISPERSION PARAMETERS OF OLIGONUCLEOTIDES OF RNA

Myron M. Warshaw

Explanation of Table

The λ_{x1}, λ_{x2}, . . . , are the first and second crossover wavelengths, respectively. They are given in the order of decreasing λ_x. λ_1, $[\Phi]_1$, . . . λ_4, $[\Phi]_4$ are the wavelength and rotation, respectively, of the first extremum . . . out to the fourth extremum. The value of $[\Phi]$ is given as rotation per mole of residue measured in a 1 cm pathlength cell:

$$[\Phi] = \phi \; cm \times \frac{1}{M} \times 100$$

Where ϕ is the measured rotation in degrees in a 1 cm path and M is the concentration in moles of residue (base) per liter. Unless otherwise indicated, the Lorenz field correction, $3/(n^2 + 2)$, was not applied to the values of $[\Phi]$. The wavelength, λ, is in units of $m\mu$.

Unless otherwise indicated, all spectra were measured in solutions of dilute buffer at an ionic strength of 0.1 and at a temperature of 25°C.[1,2] Reference to each group of compounds appears in parentheses beside each category. Reference to a particular compound appears in parentheses next to the compound's name. The conditions are approximate, and the pH values are nominal. For exact values see references.

pH 1.0

Compound[a]	λ_1	$[\phi]_1$ $\times 10^{-4}$	λ_{x1}	λ_2	$[\phi]_2$ $\times 10^{-4}$	λ_{x2}	λ_3	$[\phi]_3$ $\times 10^{-4}$	λ_{x3}	λ_4	$[\phi]_4$ $\times 10^{-4}$	λ_{x4}
Mononucleotides[1,3]												
pA	267	−0.40	251	240	0.12	233	215	−0.52				
pC	299	0.74	282	257	−0.82		248	−0.77		277	−0.96	214
pG	265	0.24	252	218	−0.58	209						
pU	283	0.45	272	253	−1.04	218						
Dinucleoside Phosphates[1,3]												
ApA	276	−0.09		258	−0.51	246	236	0.13	230	216	−1.01	205
ApC	298	0.38	284	268	−0.79		238	−0.24		219	−0.60	
ApG	288	−0.23		274	−0.54	259	256	0.05	254	218	−0.89	210
ApU	284	0.25	276	260	−0.70	238	235	0.04	230	215	−0.57	208
CpC	300	0.83	286	268	−0.74		249	−0.65		230	−0.86	214
CpA	299	0.32	284	266	−0.67		242	−0.31		230	−0.54	212
CpG	300	0.24	288	277	−0.17		265	−0.10		266	−0.47	209
CpU	296	0.74	282	268	−1.12							
GpG[b]	270	0.48	261	251	−0.59		236	−0.18				
GpA	296	0.24	285	275	−0.41	265	260	0.19	256	236	−0.35	
GpC	305	0.61	294	285	−0.56	276	265	0.74	255	238	−0.58	210
GpU	300	0.50	289	277	−0.84	268	258	1.19	248	239	−0.53	
UpU	286	0.78	275	260	−1.41		235	−0.04		222	−0.20	213
UpA	284	0.25	275	261	−0.75		232	−0.08		217	−0.45	
UpC	297	0.66	282	264	−0.95		231	−0.35				
UpG	297	0.14	282	270	−0.18		260	−0.08		248	−0.38	209
Trinucleoside Diphosphates[2,4]												
ApApC	300	0.16	290	258	−0.60		242	−0.42				
ApApU				257	−0.49	237	232	0.02				
GpGpC	302	0.46	291	283	−0.32	277	265	0.62				

[a]The mononucleotides are the 5′ phosphoric esters. The nomenclature ApA, ApC, . . . represents adenylyl(3′-5′)adenosine, adenylyl(3′-5′)cytidine, . . . the unnatural 2′ → 5′ linkage is explicitly indicated in naming the compound.

[b]Calculated from trinucleoside diphosphates and the 5′ mononucleotides, see Reference 1.

pH 1.0 (continued)

Compound[a]	λ_1	$[\phi]_1$ $\times 10^{-4}$	λ_{x1}	λ_2	$[\phi]_2$ $\times 10^{-4}$	λ_{x2}	λ_3	$[\phi]_3$ $\times 10^{-4}$	λ_{x3}	λ_4	$[\phi]_4$ $\times 10^{-4}$	λ_{x4}
Trinucleoside Diphosphates[2,4] (continued)												
GpApU	287	0.19	279	251	−0.39	...	232	−0.14				
ApGpU	295	0.19	290	277	−0.46	268	259	0.58				
GpGpU	300	0.23	290	280	−0.32	271	263	0.55				
Trinucleotides[5]												
ApApGp	272	0.07	268	250	−0.29							
ApCpGp	286	0.16	271	265	−0.26							
CpApGp	286	0.30	276	270	−0.09	...						
ApUpGp	280	0.19	267	261	−0.20							
UpApGp	281	0.32	265	250	−0.18							
CpCpGp	297	0.76	283	275	−0.59	...						
UpUpGp	280	0.31	272	255	−0.78	227						

pH 7.0

Compound[a]	λ_1	$[\phi]_1$ $\times 10^{-4}$	λ_{x1}	λ_2	$[\phi]_2$ $\times 10^{-4}$	λ_{x2}	λ_3	$[\phi]_3$ $\times 10^{-4}$	λ_{x3}	λ_4	$[\phi]_4$ $\times 10^{-4}$	λ_{x4}
Mononucleotides[1,3]												
pA	274	−0.35	258	246	0.18	234	219	−0.20	—	—	—	—
pC	288	0.85	273	239	−1.45	214	—	—	—	—	—	—
pG	281	−0.25	—	268	−0.16	—	259	−0.22	246	233	0.18	224
pU	282	0.43	273	255	−1.20	220	210	1.03	202	—	—	—
Dinucleoside Phosphates[1,3]												
ApA	282	0.88	274	260	−2.94	250	240	1.43	222	215	−3.12	208
ApC	286	1.06	276	263	−2.06	238	227	0.90	218	—	—	—
ApG	291	0.24	286	270	−1.30	257	247	0.42	218	...	—	—
ApU	278	0.58	271	260	−1.26	244	230	0.39	218	212	−0.80	—
CpC	292	1.90	280	267	−1.92	—	225	−0.56	—	220	−0.76	214
CpA	288	0.80	277	262	−1.64	230	226	0.20	222	215	−0.85	—
CpG	293	0.65	284	266	−1.00	—	213	−0.50	210	—	—	—
CpU	288	1.53	277	260	−2.04	—	222	−0.32	211	...	—	—
GpG[b]	271	0.40	260	250	−0.74	235	230	0.17	226	—	—	—
GpA	276	0.50	267	255	−0.63	—	234	−0.03	...	—	—	—
GpC	293	0.65	272	257	−0.70	226	—	—	—	—	—	—
GpU	290	0.20	278	274	−0.07	269	258	0.12	231	241	−0.36	...
UpU	284	0.81	274	260	−1.53	234	228	0.12	222	218	−0.29	214
UpA	280	0.20	272	261	−0.75	—	—	—	—	—	—	—
UpC	288	1.03	276	258	−1.37	—	228	−0.43	—	220	−0.54	213
UpG	271	0.78	282	270	−0.70	—	252	−0.11	—	210	−0.64	...
2′→5′ ApC[6]	291	0.73	281	266	−1.90	253	236	0.94	220	—	—	—
2′→5′ GpA[7]	281	0.35	266	250	−0.61	—	225	−0.52	—	—	—	—
C-2′-O-Me-pC	292	2.25	280	268	−2.27	—	242	−1.58	214	—	—	—
Trinucleoside Diphosphates[2,4]												
ApApA	283	0.90	272	259	−3.08	247	240	1.08	—	—	—	—
ApApC	285	1.68	275	260	−4.43	241	225	0.50	—	—	—	—
ApApU	282	1.11	273	259	−2.92	247	237	0.80	—	—	—	—
GpGpC	287	0.46	265	248	−0.78	—	218	−0.10	—	—	—	—
GpApU	278	0.82	269	258	−1.17	240	230	0.31	—	—	—	—
ApGpU	295	0.18	283	268	−0.68	259	254	0.16	—	—	—	—
GpGpU	270	0.45	259	248	−0.56	232	228	0.09	—	—	—	—
Trinucleotides[5]												
ApApGp	284	0.55	274	260	−1.89	246	—	—	—	—	—	—
ApCpGp	286	0.68	274	260	−1.17	235	—	—	—	—	—	—
CpApGp	288	0.56	278	265	−1.20	225	—	—	—	—	—	—
ApUpGp	283	0.22	271	259	−0.32	244	—	—	—	—	—	—
UpApGp	282	0.72	270	267	−0.09	264	—	—	—	—	—	—
CpCpGp	289	1.21	277	260	−1.63	—	—	—	—	—	—	—
UpUpGp	283	0.32	274	257	−0.83	—	—	—	—	—	—	—

pH 11.5

Compound[a]	λ_1	$[\phi]_1$ $\times 10^{-4}$	λ_{x1}	λ_2	$[\phi]_2$ $\times 10^{-4}$	λ_{x2}	λ_3	$[\phi]_3$ $\times 10^{-4}$	λ_{x3}	λ_4	$[\phi]_4$ $\times 10^{-4}$	λ_{x4}
Mononucleotides[1],[3]												
pA	277	−0.34	258	248	0.15	235	218	−0.40				
pC	289	0.86	274	240	−1.54	215						
pG	284	−0.39	261	239	0.10	228	220	−0.18	214			
pU	278	0.65	268	249	−1.72	219						
Dinucleoside Phosphates[1],[3]												
ApA	282	0.87	275	260	−2.77	250	240	1.07	222	215	−2.75	
ApC	288	1.05	276	262	−1.98	238	228	0.94	216			
ApG	288	0.00		270	−0.93	250	239	0.53		230	0.32	220
ApU	280	0.48	271	258	−0.81	238	225	0.30	214			
CpC	292	2.00	280	268	−2.06		253	−1.52		245	−1.62	215
CpA	288	0.80	276	263	−1.57	228	226	0.16	222	214	−0.91	
CpG	290	0.53	282	267	−0.86	228	223	0.15	218			
CpU	287	1.26	275	252	−1.80	217						
GpG[b]	283	−0.18	270	260	0.15	228						
GpA	279	−0.40	254	236	0.49		230	0.41		225	0.49	220
GpC	290	0.38	282	254	−0.58	230						
GpU	282	0.24	269	250	−0.81							
UpU	281	0.77	268	250	−1.60	220						
UpA	278	0.26	270	246	−0.89		218	−0.26				
UpC	286	0.75	273	247	−1.56	214						
UpG	281	0.18	274	250	−0.80							
Trinucleoside Disphosphates[2],[3]												
ApApC	285	1.56	275	260	−4.47	226	225	0.04				
ApApU	282	0.82	273	259	−2.40	245	240	0.36				
GpGpC	290	0.19	282	255	−0.48		226	−0.02				
GpApU				255	−0.55	240	226	0.43				
ApGpU	285	0.14	281	265	−0.40	235	230	0.08				
GpGpU				250	−0.57		225	−0.11				
Trinucleotides[c] [5]												
ApApGp	282	0.41	273	258	−1.68	247						
ApCpGp	286	0.70	274	258	−1.03	232						
CpApGp	284	0.58	272	260	−1.10	228						
ApUpGp	278	0.16	267	247	−0.60	218						
UpApGp	284	0.24	263	244	−0.43	228						
CpCpGp	289	1.23	275	259	−1.23	222						
UpUpGp	275	0.47	266	245	−1.02	222						

[c]The pH was 11.0, unbuffered.[5]

Compiled by Myron M. Warshaw.

This table originally appeared in Sober, H., *Handbook of Biochemistry and Selected Data for Molecular Biology,* 2nd ed., Chemical Rubber Co., Cleveland, 1970.

REFERENCES

1. **Warshaw and Tinoco**, *J. Mol. Biol.,* 13, 54 (1965).
2. **Cantor and Tinoco**, *J. Mol. Biol.,* 13, 65 (1965).
3. **Warshaw**, Ph.D. Dissertation, University of California, Berkeley, 1966.
4. **Cantor**, Ph.D. Dissertation, University of California, Berkeley, 1966.
5. **Inoue, Aoyagi, and Nakanishi**, *J. Am. Chem. Soc.,* 89, 5701 (1967).
6. **Warshaw and Cantor**, *Biopolymers,* in press.
7. **Warshaw and Cantor**, unpublished results.

CIRCULAR DICHROISM PARAMETERS OF OLIGONUCLEOTIDES OF RNA

Myron M. Warshaw

Explanation of Table

The λ_1, $[\Theta]_1$, ... λ_4, $[\Theta]_4$ are the wavelength and circular dichroism, respectively, for each Cotton effect in the order of decreasing wavelength. The λ_{x1} ... λ_{x4} are the crossover points given in order of decreasing wavelength. The parameters are tabulated such that two consecutive Cotton effects have the same sign when there is no crossover between them. The value of $[\Theta]$ is given as ellipticity per mole of residue measured in a 1 cm pathlength cell:

$$[\Theta] = \theta/\text{cm} \times \frac{1}{M} \times 100$$

Where θ/cm is the measured ellipticity in degrees per 1 cm path and M is the concentration in moles of residue (base) per liter. The wavelength, λ, is given in units of $m\mu$.

Unless otherwise indicated, these results are those of Warshaw and Cantor.[1] The approximate conditions are dilute buffer at an ionic strength of 0.1 at a temperature of $26°$[1] unless otherwise stated.

pH 7

Compound[a]	λ_1	$[\Theta]_1$ $\times 10^{-4}$	λ_{x1}	λ_2	$[\Theta]_2$ $\times 10^{-4}$	λ_{x2}	λ_3	$[\Theta]_3$ $\times 10^{-4}$	λ_{x3}	λ_4	$[\Theta]_4$ $\times 10^{-4}$	λ_{x4}
					Mononucleosides[1]							
A	263	−0.38	238	227	0.11	215	—	—	—	—	—	—
U	267	0.92	252	238	−0.51	—	217	−0.56	205	—	—	—
C	272	1.04	240	218	−0.94	—	—	—	—	—	—	—
G	252	−0.29	233	215	0.81	208	—	—	—	—	—	—
					Arabinose Mononucleotides[3]							
araCMP-5′	272	2.47	235	217	−1.59	—	—	—	—	—	—	—
araCMP-3′	271	2.55	232	216	−1.39	—	—	—	—	—	—	—
araCMP-2′	271	2.78	233	217	−1.68	—	—	—	—	—	—	—
					Dinucleoside Phosphates							
ApA	271	2.03	261	252	−2.51	234	220	2.00	215	207	−3.31	—
2′→5′ApA	272	1.97	260	252	−1.65	238	218	3.45	213	207	−4.41	—
ApU	269	1.35	258	248	−0.68	231	218	0.82	—	—	—	—
ApC	275	2.35	260	238	−1.15	226	220	0.74	—	205	0.85	—
2′→5′ApC	279	1.68	266	255	−1.36	238	220	2.35	213	203	−1.50	—
ApG	280	0.75	271	260	−1.28	236	216	0.43	211	—	—	—
UpA	268	0.76	257	244	−0.42	—	—	—	—	—	—	—
UpU	271	1.61	256	242	−0.75	—	210	—	—	—	—	—
UpC	275	1.94	251	238	−0.40	—	214	−0.32	206	—	—	—
UpG	280	0.59	267	258	−0.30	—	235	−0.84	204	—	—	—
CpA	274	1.88	256	233	−1.09	—	207	−0.20	—	—	—	—
CpU	275	2.72	253	235	−0.90	—	210	−1.56	202	—	—	—
CpC	280	3.08	244	230	−0.74	—	215	−0.93	209	—	—	—
C-2′-O-Me-pC	280	3.34	243	230	−0.81	—	214	−0.99	209	—	—	—

[a]Nomenclature: ApA, araCparaC, etc. represent adenylyl(3′ → 5′)adenosine, 1-β-D-arabinofuranosylcytidylyl(3′ → 5′)-1-β-D-arabinofuranosylcytidine, respectively. AraCMP-5′ represents 1-β-D-arabinofuranosylcytidine-5′-monophosphate, etc. The unnatural 2′ → 5′ phosphodiester linkage is explicitly indicated.
[b]Measured in 0.01M Tris buffer, pH 7.4 – no added salt.

pH 7 (continued)

Compound[a]	λ_1	$[\phi]_1$ $\times 10^{-4}$	λ_{x1}	λ_2	$[\phi]_2$ $\times 10^{-4}$	λ_{x2}	λ_3	$[\phi]_3$ $\times 10^{-4}$	λ_{x3}	λ_4	$[\phi]_4$ $\times 10^{-4}$	λ_{x4}
CpG	282	1.28	258	229	−0.60	—	212	−0.95	—	—	—	—
GpA	285	−0.24	279	265	0.84	252	247	−0.16	237	219	0.45	208
2′→5′GpA	295	−0.04	289	263	0.61	—	215	0.69	210	—	—	—
GpU	280	0.29	—	251	0.46	235	231	−0.02	228	217	0.34	209
GpC	269	0.94	244	227	−0.60	—	212	−0.79	208	—	—	—
GpG	280	−0.13	272	259	0.79	248	240	−0.31	230	218	0.63	209
A-2′-*O*-Me-pA[b2]	270	1.15	261	252	−1.23	238	218	1.4	—	—	—	—
Arabinose Dinucleoside Phosphates[3]												
AraCparaC	274	2.28	234	218	−1.55	—	—	—	—	—	—	—
2′→5′ araCparaC	268	2.02	233	216	−1.14	—	—	—	—	—	—	—
Dinucleotides												
ApAp[b2]	271	1.00	261	252	−1.84	—	—	—	—	—	—	—
A-2′-*O*-Me-pAp[b2]	273	0.62	264	252	−1.08	239	216	1.5	—	—	—	—

[b]Measured in 0.01*M* Tris buffer, pH 7.4-no added salt.

This table originally appeared in Sober, H., *Handbook of Biochemistry and Selected Data for Molecular Biology,* 2nd ed., Chemical Rubber Co., Cleveland, 1970.

CIRCULAR DICHROISM PARAMETERS OF OLIGONUCLEOTIDES OF DNA

Myron M. Warshaw and A. F. Drake

Explanation of Table

The λ_1, $[\Theta]_1$, ... λ_4, $[\Theta]_4$ are the wavelength and circular dichroism, respectively, for each Cotton effect in the order of decreasing wavelength. The λ_{x1}, ... λ_{x4} are the crossover points given in order of decreasing wavelength. The parameters are tabulated such that two consecutive Cotton effects have the same sign when there is no crossover between them. The value of $[\Theta]$ is given as ellipticity per mole of residue measured in a 1 cm pathlength cell:

$$[\Theta] = \theta/\text{cm} \times \frac{1}{M} \times 100$$

Where θ/cm is the measured ellipticity in degrees per 1 cm path and M is the concentration in moles of residue (base) per liter. The wavelength, λ, is given in units of $m\mu$.

Unless otherwise indicated, these results are those of Cantor, Warshaw and Shapiro.[1] The approximate conditions are dilute buffer at an ionic strength of 0.1 at a temperature of $26°$[1] unless otherwise stated.

pH 7

Compound[a]	λ_1	$[\Theta]_1$ $\times 10^{-4}$	λ_{x1}	λ_2	$[\Theta]_2$ $\times 10^{-4}$	λ_{x2}	λ_3	$[\Theta]_3$ $\times 10^{-4}$	λ_{x3}	λ_4	$[\Theta]_4$ $\times 10^{-4}$	λ_{x4}
Deoxymononucleosides[1]
dA	264	−0.21	242	219	0.26	207	---	---	---		---	---
dU	268	0.75	251	239	−0.43	---	218	−0.37	207		---	---
dT	274	0.38	257	240	−0.38	---	217	−0.63	207		---	---
dC	272	0.90	236	216	−1.11	207	---	---	---		---	---
dG	250	−0.32	233	212	1.15	205	---	---	---		---	---
Deoxydinucleoside Phosphates[1]
d(ApT)	273	1.62	263	252	−1.43	234	219	1.18	212		---	---
d(ApG)	281	−0.64	---	272	−0.63	---	260	−0.56	238	214	1.71	---
d(TpT)	279	0.72	264	250	−0.62	---	215	−0.08	211		---	---
d(TpC)	279	0.85	255	239	−0.32	---	216	−0.54	206		---	---
d(CpA)	273	1.17	255	212	−1.47	---	---	---	---		---	---
d(CpT)	280	1.55	258	235	−0.36	---	213	−0.92	---		---	---
d(CpC)	277	1.42	239	215	−0.84	---	---	---	---		---	---
d(CpG)	278	0.69	263	247	−0.88	---	207	−2.09	---		---	---
d(GpA)	270	−0.41	261	248	0.88	---	215	0.76	210		---	---
d(GpT)	284	0.56	273	265	−0.35	254	246	0.20	---	210	0.20	---
d(GpC)	278	0.60	258	249	−0.25	---	228	−0.15	216		---	---
dAprA[b2]	267	1.65	259	250	−1.75	237	217	2.60	211		---	---

[a]d(ApG), d(ApAp) represents deoxyadenylyl(3′ → 5′)deoxyguanosine and deoxyadenylyl(3′ → 5′)deoxyadenylyl-3′-monophosphate, respectively. dAprA represents deoxyadenylyl(3′ → 5′)riboadenosine. d(pTpT) represents thymidylyl(3′ → 5′)-thymidylyl(3′ → 5′)cyclic monophosphate.

[b]Measured in 0.01M Tris buffer, pH 7.4 – no added salt. These are dimers of mixed sugars.

pH 7 (continued)

Compound[a]	λ_1	$[\Theta]_1$ $\times 10^{-4}$	λ_{x1}	λ_2	$[\Theta]_2$ $\times 10^{-4}$	λ_{x2}	λ_3	$[\Theta]_3$ $\times 10^{-4}$	λ_{x3}	λ_4	$[\Theta]_4$ $\times 10^{-4}$	λ_{x4}
Deoxydinucleotides[1]												
d(ApAp)	269	1.66	260	250	−2.26	237	218	3.90	213	—	—	—
d(pApA)[3]	272	1.23	261	252	−1.98	241	219	4.56	214	—	—	—
d(ApCp)	273	1.66	257	241	−1.25	224	218	0.71	213	—	—	—
d(TpAp)	270	0.85	262	250	−1.07	230	218	0.65	211	—	—	—
d(TpGp)	285	0.35	274	260	−0.73	—	213	−0.03	—	—	—	—
d(GpGp)	280	−0.46	265	255	0.38	244	234	−0.12	228	213	1.05	207
pdAprA[b2]	268	1.81	260	250	−2.14	—	—	—	—	—	—	—
dp(TpT)[4]	278	0.71	256	248	−0.64	—	216	−0.37	208	—	—	—
d(pTpT)	280	−0.80	270	259	0.90	240	235	−0.06	—	215	−0.22	209
Deoxytrinucleoside diphosphates[1]												
d(ApCpC)	273	0.96	255	239	−0.64	224	216	0.34	—	—	—	—
d(ApTpT)	273	1.27	263	250	−1.01	232	220	0.54	213	—	—	—
d(GpTpT)	280	0.69	266	255	−0.24	230	225	0.09	—	—	—	—
Deoxytrinucleotides[1]												
d(pTpTpTp)	276	0.96	263	248	−0.70	—	218	−0.10	—	—	—	—
d(pTpTpT)	277	0.99	263	249	−0.77	—	214	−0.21	209	—	—	—
d(pTpTpT)[4]	291	0.03	—	261	0.52	—	223	0.96	216	207	−0.92	198
Deoxypentanucleotides[5]												
d(pG)$_s$ [c]	277	−0.93	266	255	+1.45	240	233	−0.33	255	210	+0.93	206
Polyhomodioxynucleotides[6]												

d[C]$_n$
d[A]$_n$
d[T]$_n$

[c] Conditions: 0.02M Na phosphate buffer, pH 7.0, 20°C.

Compiled by Myron M. Warshaw and A. F. Drake.

FUNCTIONAL CHARACTERIZATION OF NUCLEOPHOSPHODIESTERASES*†

David Kowalski and Michael Laskowski, Sr.

This compilation was influenced by previously published patterns[1-3] as well as the tables from the earlier editions of this Handbook. The major classification into deoxyribonucleases, ribonucleases, and nucleases nonspecific to sugar is retained because common names emphasize this division. Since they require a different tabulation, restriction nucleases are listed separately and precede DNases. Also, RNases H from various species are listed together (rather than by their source) under RNases because of their obvious functional similarity. Otherwise, each of the three groups is subdivided into (A) 5'-monoester formers and (B) 3'-monoester formers. In the case of ribonucleases, the formation of the cyclic intermediate is mentioned, and when possible, relative rates of the formation and hydrolysis of the cyclic intermediate are compared. RNases for which hydrolysis of the cyclic derivative is either extremely slow, or possibly nonexistent, are listed in a third subdivision, (C) 2',3'-cyclic diester formers. The terms exo- and endonuclease are still used even though the sharpness of this delineation is decreased by the discovery of enzymes with intermediate properties. A further complication is caused by the two different definitions of exonuclease. Originally, exonucleases were defined as enzymes that consecutively liberate mononucleotides from the same terminus.[10] Recently, the definition has been expanded to include all enzymes that require a free terminus (or termini) as determined by the inability to open closed-circular polynucleotide substrates.[242,416] In addition, the expanded definition is not restricted to mononucleotide formation, and oligonucleotide products of varying chain lengths are included. In describing the mode of attack of a specific enzyme (Column 2 of the table), the term *exo* is used in its original meaning,[10] while for nucleases called exonucleases by the expanded definition, the terminus required (5',3', or both) is listed. The term *processive* is used as originally defined by Singer[124] to mean that once the chain has been attacked, it is degraded to the final products prior to the attack on the next chain. The terms α, ω, and 5',3' termini are used interchangeably. The term *nick* refers to a single phosphodiester interruption in one of the two strands of a duplex DNA molecule.[415]

*During the preparation of this review, the authors had grants from the Atomic Energy Commission, the National Science Foundation, the National Institutes of Health, and the American Cancer Society.
†Acknowledgments: We are indebted to Drs. P. H. Johnson, I. R. Lehman, C. C. Richardson, M. F. Singer, R. L. Sinsheimer, and C. M. Wilson for their criticism and suggestions.

Table 1
DEOXYRIBONUCLEASES*

Enzyme	Mode of attack	Linkages attacked and products formed	Additional characteristics
		A. 5'-Monoester formers	
DNase I of bovine pancreas Crystalline.[4,20] Preparation contains 4 isoenzymes, A to D,[131,230-237] which are glycoproteins. Primary structure.[234,235] For a review see Reference 229.	Endo[4]	Preference for dsDNA.[5,13] In the early stages, dsDNA is nicked[31] preferentially at dA and dT;[227] at pseudo-equilibrium, the preferentially hydrolyzed bond is dR↓pdY and the products have an average size of a tetranucleotide.[4-8,10] With d(pApT)$_n$, only even numbered products d(pTpA)$_n$ are formed.[130] Bivalent cation affects the preferentially hydrolyzed linkage.[9,228,240,247] After exhaustive digestion, products are d(pN), 5%; d(pNpN), 60%; d(pNpNpN), 25%.[11] The linkage N$^\alpha$↓ pN$^\beta$ . . . N$^\omega$ is resistant and no nucleoside is formed, but the linkage N$^\alpha$ . . . N$^\psi$↓pN$^\omega$p is susceptible and pN$^\omega$p is produced.[12]	Crystalline DNase[4] is a mixture of two thirds of four isoenzymes, (of which DNase A predominates) and about one third of chymotrypsin B + chymotrypsinogen B.[14,131] A specific protein inhibitor of DNase is widely distributed in tissues of vertebrates.[17,20] It was crystallized and shown to form a 1:1 complex.[18,19] The inhibitor is actin.[413] Kinetics of DNase acting on native DNA shows autoretardation.[2,229] Hydroxy biphenyls inhibit.[239] The only known exception among 5'-monoester formers because NO$_2$PhpdTp↓PhNO$_2$ is hydrolyzed as shown.[238]

*Abbreviations: AdoMet, S-adenosylmethionine; ss, single-stranded; ds, double-stranded; MW, molecular weight; R, purine nucleoside; and Y, pyrimidine nucleoside.

Table 1 (continued)
DEOXYRIBONUCLEASES

A. 5'-Monoester Formers (continued)

Enzyme	Mode of attack	Linkages attacked and products formed	Additional characteristics
DNase III (mammalian)[146-149,186] The predominant DNA exonuclease in many mammalian tissues; localized in nuclei.[186]	Exo	Denatured DNA is hydrolyzed four times faster than native. Attacks from the 3'-terminus liberating predominantly 5'-mononucleotides.	DNA terminated in 3'-P is digested at half the rate of DNA terminated in 3'-OH. MW ~52,000. Requires Mg^{2+} or Mn^{2+}. Optimal pH 8.5.[186]
DNase IV (mammalian)[150,273,274]	Requires a free 5'-terminus, partly processive[274]	Specific for native DNA. Attacks from the 5'-terminus liberating 5'-mono- and oligonucleotides. Removes oligonucleotides containing pyrimidine dimers. Synthetic ds-polydeoxyribonucleotides of simple repeating base sequence are degraded more rapidly than naturally occurring DNAs.[273]	Attacks both 5'-P and 5'-OH termini. Can act at nicks.[274] Requires Mg^{2+} or Mn^{2+}. MW ~42,000. Activity is very similar to the 5'→3' exonuclease activity of *E. coli* DNA polymerase I; however DNase IV has no polymerase activity.[273]
Mammaliam endonuclease (thymus)[275,276]	Endo	Specific for dsDNA containing apurinic sites and analogous lesions involving loss of a base. Produces single-strand breaks on the 3'-side of apurinic residues.	Does not attack native DNA, alkali-denatured DNA, or ssDNA containing apurinic sites MW ~32,000. Strongly stimulated by Mg^{2+} and Mn^{2+}. Optimal pH 8.5.
Octopus DNase[5]	Endo	Preference for N-pC. Products formed are di- and trinucleotides. Dinucleotides are resistant.	

Table 1 (continued)
DEOXYRIBONUCLEASES

A. 5'-Monoester Formers (continued)

Enzyme	Mode of attack	Linkages attacked and products formed	Additional characteristics
Streptococcal DNase (Streptodornase)[91-96] At least 4 isoenzymes, of which B and D also hydrolyze RNA.[241]	Endo	Preference for N↓pG. Products are predominantly tri- to hexanucleotides. Dinucleotides are resistant. Isoenzyme B shows a preference for r(U↓pN).	Attacks native DNA much faster than denatured. Requires Mg^{2+}. Optimal pH ~7.[95,96]
Escherichia coli Endonuclease I[3,118-121,414]	Endo	Attacks ssDNA and dsDNA. Average chain length of products is 7. No preference for bases.[118-120]	Hydrolyzes native DNA seven times faster than denatured.[118] Produces ds-breaks in native DNA.[121] Poly(A) · poly(U) and tRNA are strong competitive inhibitors.[3,119]
Endonuclease II[138,414,417,424]	Endo	Specific for apurinic sites in dsDNA.[424] Apparent hydrolysis of alkylated DNA[417] may be caused by depurination.[276]	Appears to have no effect on native DNA, alkylated DNA, or ssDNA[424]
Exonuclease I[3,106-109,244]	Exo	Attack starts from the 3'-hydroxyl terminus.[108] Terminal dinucleotide $pN^{\alpha}pN^{\beta}$ is resistant.[108] Produces 5'-mononucleotides.	Denatured DNA is hydrolyzed 40,000 times faster than native.[107] $d(A-T)_n$, however, is a good substrate. Degrades glucosylated DNA.

Table 1 (continued)
DEOXYRIBONUCLEASES

Enzyme	Mode of attack	Linkages attacked and products formed	Additional characteristics
		A. 5'-Monoester Formers (continued)	
3'→5' Exonuclease activity of DNA polymerase I[3,110, 111,245,246,414,416]	Exo	Specific for ssDNA. Absolute requirement for 3'-hydroxyl terminus from which the attack begins. Specifically removes unpaired or mismatched nucleotides from the template-primer.[246] Produces 5'-mononucleotides.	Oligonucleotides bearing 3'-phosphoryl groups are resistant.[111] After tryptic digestion of polymerase I, this activity remains in the large fragment together with the polymerase activity.
5'→3' Exonuclease activity of DNA polymerase I[190,191, 245,246,249,250,414,416]	Requires a free 5'-terminus	Cleaves the 5'-base-paired end of ds DNA Nonspecific for terminus: 5'-hydroxyl and mono-, di-, or triphosphates are hydrolyzed. Can excise oligonucleotides up to 10 residues long from the 5'-end. Concurrent polymerization increases the rate of hydrolysis tenfold. Produces 5'-mono- and oligonucleotides. Mononucleotides predominate in the absence of polymerization.[416]	Can remove the RNA priming fragment of enzymatically synthesized DNA.[416] ssDNA and small oligonucleotides are resistant. The activity is associated with the small tryptic fragment of DNA polymerase I. The hydrolytic activity of DNA polymerase, comprised of this activity and the 3'→5' exonuclease, has been called exonuclease II (see Reviews 414 416). The 5'→3' activity[256] was also called exonuclease VI.

Table 1 (continued)
DEOXYRIBONUCLEASES

A. 5'-Monoester Formers (continued)

Enzyme	Mode of attack	Linkages attacked and products formed	Additional characteristics
Exonuclease III or DNA phosphatase[113,114,414]	Exo	Specific for native DNA[113,114] Attack starts from 3'-hydroxyl terminus.[114] Hydrolysis ceases once the bihelical structure is lost as a result of sustained exonucleolytic attack. If the chain is terminated in 3'-phosphoryl, P_i is liberated. Produces 5'-mononucleotides and resistant ss-fragments.	Nicked bihelical DNA is attacked at those ss-breaks as well as at the 3'-terminus.[415] ssDNA and small oligonucleotides are resistant.
Exonuclease IV Two enzymes, IVA and IVB, are separable on DEAE-cellulose.[115-117,151,414,418]	Exo	Strong preference for oligonucleotide chains shorter than 100. Attack starts from the 3'-hydroxyl terminus but is not processive. Produces 5'-mononucleotides exclusively.	Polynucleotides bearing 3'-phosphate are resistant. $N^\alpha pN^\beta$ are 9 fold more resistant than $pN^\alpha pN^\beta$. Requires Mg^{2+}. Optimal pH 8–9.
Exonuclease V, or rec BC DNase[251-255,421,422]	Requires a free 3'- or 5'-terminus, processive; endo on circular ssDNA	Attacks linear ds- or ssDNA from either terminus. Requires ATP for activity on ds- and single-stranded linear DNAs, but not on circular ssDNA. 2–40 ATP hydrolyzed per one internucleotide bond. Initially produces ss-fragments several hundred nucleotides in length, which are ultimately hydrolyzed to oligo- and mononucleotides.[421]	No activity against intact or nicked circular dsDNA. MW ~270,000 (two nonidentical subunits).[422] Duplex circles with gaps as small as 5 residues are degraded.[255] Optimal pH 9.2, requires Mg^{2+}.

Table 1 (continued)
DEOXYRIBONUCLEASES

A. 5'-Monoester Formers (continued)

Enzyme	Mode of attack	Linkages attacked and products formed	Additional characteristics
Exonuclease VII[257,258]	Requires a free 3'- or 5'-terminus, processive	Specific for ssDNA or dsDNA with ss-termini. Attacks processively from either 5'- or 3'-terminus yielding large fragments, which are later hydrolyzed to oligonucleotides. Circular ss-molecules are not attacked.	Excises UV induced dimers after pretreatment with *M. luteus* nuclease. Stimulated by Mg^{2+}, but almost fully active in its absence. Optimal pH 7.9, MW = 88,000.
Exonuclease VIII[259]	Unknown	Attacks native DNA	Does not require ATP; is inactive on T_4 DNA.
λ-Induced exonuclease crystalline enzyme; homogeneous by several criteria[179,414]	Exo	Attacks from the 5'-terminus.[180] Shows a strong preference for 5'-P over 5'-OH termini.[180] Native T_7 DNA is attacked 350-fold faster than denatured. Produces 5'-mononucleotides.	Does not act at nicks. MW ~24,000. β protein forms a 1:1 complex with the exonuclease but has no demonstrable influence on its activity in vitro.[420]
T_2- and T_4-Induced exonuclease[115-117,151,414,418]	Exo	Catalytically very similar to exonuclease IV	
T_4-Induced endonuclease II[260]	Endo	Attacks native λ DNA by predominantly making single-strand breaks. At the 5'-terminus, C and G are most frequent. ds-fragments produced from λ DNA are about 1,000 nucleotides long.[260]	Tenfold greater activity on native DNA than on denatured. T_4 DNA (glucosylated or nonglucosylated) is resistant.[260]

Table 1 (continued)
DEOXYRIBONUCLEASES

A. 5'-Monoester Formers (continued)

Enzyme	Mode of attack	Linkages attacked and products formed	Additional characteristics
T_4-Induced endonuclease III[145,263]	Endo	Preference for denatured DNA. Produces oligonucleotides. 5'-terminal nucleotides are not specific.[263]	Does attack denatured T_4 DNA. No acid soluble material produced from poly(dC).[263]
T_4-Induced endonuclease IV[262,263]	Endo	Products average 50 nucleotides long and contain exclusively 5'-pC; cleaves fd and denatured λ DNA but has no effect on denatured T_4 DNA (glucosylated or nonglucosylated).[262]	About 200-fold greater activity against ssDNA than against dsDNA.[263]
T_4-Induced endonuclease V[264]	Endo	Does not act on either native or denatured DNA; induces single-strand breaks in UV-irradiated DNA. Cleaves on the 5'-side of pyrimidine dimers. The cleavage with respect to P is unknown.	Shows optimal activity at pH 7. No requirement for bivalent cation.
T_5-Induced exonuclease[242,243]	Requires a free 5'-terminus	Attacks ss- or dsDNA. Hydrolysis starts from the 5'-terminus and produces 5'-phosphonucleotides of an average length of 3 residues.[242]	Does not attack circular DNA. Attacks UV-irradiated DNA. Can initiate hydrolysis at ss-breaks.[242]

Table 1 (continued)
DEOXYRIBONUCLEASES

A. 5'-Monoester Formers (continued)

Enzyme	Mode of attack	Linkages attacked and products formed	Additional characteristics
T$_7$-Induced endonuclease I[265-267,419]	Endo	Hydrolyzes ssDNA 150-fold faster than dsDNA; acid soluble material is produced only from ss-substrates. 5'-end groups are predominantly pyrimidine residues.[267,419]	Extensive hydrolysis of denatured T$_7$, native T$_7$, and native *E. coli*. DNA yields products of average single-strand MW of 6×10^3, 1×10^5, and 6×10^3, respectively.[267] Inhibited by high concentration of RNA. Requires Mg^{2+}. MW ~18,000. Optimal pH 8–9.
T$_7$-Induced endonuclease II[268]	Endo	Exhaustive digestion of native T$_7$ DNA produces 13 single-strand cleavages. Has no activity toward ssDNA.	5'-termini produced contain all four bases. Requires Mg^{2+} and sulfhydryl reagent. Optimal pH 6.5.
T$_7$-Induced exonuclease[269,270]	Requires a free 5'-terminus	Hydrolyzes duplex DNA producing 50% acid soluble material, composed almost entirely of 5'-mononucleotides. The attack begins at 5'-terminus. The rate of hydrolysis is not affected by the presence of 5'-P. From a 5'-hydroxyl terminus, a dinucleoside monophosphate is cleaved.	Marked preference for dsDNA. Attacks at nicks as well as external termini. Absolute requirement for bivalent cations and sulfhydryl reagents. K$^+$ stimulates.

Table 1 (continued)
DEOXYRIBONUCLEASES

A. 5'-Monoester Formers (continued)

Enzyme	Mode of attack	Linkages attacked and products formed	Additional characteristics
Micrococcus luteus (*M. lysodeikticus*)			
3'→5' Exonuclease activity of DNA polymerase[423,437−439]	Exo	Specific for ssDNA. Produces 5'-P-mono-nucleotides.[440]	Similar to activity in *E. coli* DNA polymerase I.
5'→3' Exonuclease activity of DNA polymerase[423,437−439]	Requires a free 5'-terminus	Specific for dsDNA. Produces mono-, di-, and trinucleotides.[440] Stimulated by NTP without concurrent polymerization.[441] Excises pyrimidine dimers after treatment of damaged DNA with a *M. luteus* UV endonuclease.[423]	MW ~105,000.[423] Optimal pH 7.5—9.0.[440] Similar to activity in *E. coli* DNA polymerase I.
UV-endonuclease[423,443] An *M. luteus* UV-endonuclease of different specificity is listed under 3'-monoester formers.	Endo	Specific for UV-irradiated native DNA.[443] Makes single-strand breaks on the 5'-side of pyrimidine dimers forming 3'-OH and 5'-P termini.[423]	Inactive on unirradiated native or denatured DNA and irradiated denatured DNA. MW ~15,000.[443]
Exonuclease[306]	Requires a free 3'- or 5'-terminus	Specific for denatured DNA. Attacks from the 3'- or 5'-terminus producing 5'-mono-nucleotides. UV-irradiated and unirradiated denatured DNA are equally susceptible to attack. Pyrimidine dimers are excised as part of an oligonucleotide fragment.	Can remove pyrimidine dimers from irradiated duplex DNA incised with the 3'-monoester-forming *M. luteus* endonuclease.[304,305]

Table 1 (continued)
DEOXYRIBONUCLEASES

Enzyme	Mode of attack	Linkages attacked and products formed	Additional characteristics
A. 5'-Monoester Formers (continued)			
ATP-dependent DNase[280,281]	Endo	Native DNA is degraded 40-fold faster than denatured. Predominant products are oligonucleotides of an average length of 5.5 residues at the limit of digestion.	Absolute requirement for Mg^{2+} and ATP, which is hydrolyzed to ADP + P_i (three molecules of ATP for each internucleotide bond). Optimal pH 9.4.
Hemophilus influenzae			
Exonuclease[271]	Exo	Specific for dsDNA. Attack starts from the 3'-terminus. Produces 5'-mononucleotides and resistant ss-fragments.	Optimal pH 8.3. Requires Mg^{2+} or Mn^{2+}. Resembles exonuclease III of *E. coli*.
ATP-dependent DNase[444-446]	Requires a free 3'- and 5'-terminus, processive; endo on circular ssDNA	dsDNA is degraded ten times faster than ssDNA. Attacks linear duplex DNA to give products that are initially several hundred bases in length but that are further degraded to oligonucleotides of an average size of 6 residues.	Possesses DNA-dependent ATPase activity. Requires ATP and Mg^{2+}. Does not attack circular dsDNA or duplex circles containing nicks or gaps. MW ~270,000. Optimal pH 7.5–9.0.
Aspergillus DNase K_2[139-144]	Endo	Preference for $dR\downarrow pdY$ linkage and denatured DNA.	Fragments rich in dT are produced.[144]
Bacillus subtilus			
ATP-dependent DNase[447]	Unknown	Attacks duplex DNA and denatured DNA producing oligonucleotides.	Possesses DNA-dependent ATPase activity. Requires ATP and Mg^{2+}.

Table 1 (continued)
DEOXYRIBONUCLEASES

A. 5'-Monoester Formers (continued)

Enzyme	Mode of attack	Linkages attacked and products formed	Additional characteristics
Phage SP3-induced DNase[152,272]	Sequential cleavage of dinucleotides	Specific for denatured DNA. Attacks from the 5'-terminus[272] by cleaving 5'-phospho-dinucleotides. Among the products, some trinucleotides are present.	Requires Mg^{2+}, optimal pH 8–9.
Vaccinia virus DNase (V_1)[277-279]	Mixed, endo, and exo	Highly specific for ssDNA. Produces 5'-P mono- and oligonucleotides.	The enzyme is associated with the viral core. Properties resemble that of nuclease S_1 (*Asperillus*) except that V_1 is DNA specific. Optimal pH 4.4. Two subunits, each of MW ~50,000.
Yeast DNase[282]	Endo	750-fold preference for ssDNA over native DNA. Predominant products are oligonucleotides of an average length of 4 residues.	Requires Me^{2+}. Carries some RNase activity, presumably as a contaminant.
Chlorella DNase[283]	Probably endo	Strong preference for denatured DNA. Acid-soluble products varying from mono- to trinucleotides (not characterized).	The enzyme appears in synchronized cultures at the time of DNA synthesis. After this stage a nondialyzable, heat-stable inhibitor appears. Optimal pH 8.5.

Table 1 (continued)
DEOXYRIBONUCLEASES

B. 3'-Monoester Formers

Enzyme	Mode of attack	Linkages attacked and products formed	Additional characteristics
DNase II spleen and thymus;[21-24,135-137,284-293] Probably a similar activity in brain[296] and several other tissues.[425] For reviews see References 136 and 425.	Endo	Attacks native DNA faster than denatured. Cuts two strands simultaneously.[31] Pronounced autoretardation.[2,26] In early stages of reaction, preference for dGp↓dC;[28] at pseudo-equilibrium, dYp↓dR.[17] Bivalent anions at 10^{-5} M accelerate, at 10^{-2} M inhibit. Anion activation increases the number of single-strand breaks.[295]	Localized in the lysosomes. Glycoprotein of MW ~38,000. Naturally occurring inhibitor.[187] Trinucleotides bearing 5'-monophosphate are resistant.[29,30] dCpdT appears to be relatively resistant.[12,28,29] The frequency of a nucleoside at α and ω terminus is proportional to the C + G content of the bacterial DNA used as substrate. The enzyme can be purified from exonuclease and nonspecific phosphodiesterase.[290-293] Optimal pH 4.5–5.5.
Human acid DNases of gastric mucosa and cervix[297]	Endo	Very similar enzymes from both sources. Resembles the spleen emzyme; 10–20-fold preference for dsDNA. Average chain length of products is 10 bases.	MW ~38,000. Optimal pH 5.2–5.5.
Crab testes DNase[60,298]	Endo	Attacks native DNA by making predominantly ds-breaks.[298] Preference for dNp↓dT; dNp↓dC rather resistant.[60] Produces di- and trinucleotides.	$S_{20}w = 5.7$.[298] Requires bivalent cations, optimal pH 7.5.

Table 1 (continued)
DEOXYRIBONUCLEASES

B. 3'-Monoester Formers (continued)

Enzyme	Mode of attack	Linkages attacked and products formed	Additional characteristics
Snail DNase[299,300]	Endo	Hydrolyzes DNA and d(A-T)n but not (dG)n · (dC)n. Predominantly 4–10 mers formed. At the 3'-terminus 78% dA, 1% dC; at 5'-terminus 45% dG.	Plotting frequency of a nucleoside at either terminus versus dG + dC content of substrate DNA gives a straight line.
Aspergillus DNase K$_1$[139-144]	Mixed, endo, and exo	Preference for dRp↓dR and denatured DNA. Predominant product dNp.	
Salmon testes DNase[301-303]	Endo	Products formed are oligonucleotides. Gp predominates at the 3'-terminus.	dsDNA is a better substrate than ssDNA.
M. luteus UV-endonuclease[304,305]	Endo	Introduces nicks into a strand containing UV damage. Cleaves on 3'-side of dipyrimidine.	

Compiled by David Kowalski and Michael Laskowski, Sr.

Table 2
RIBONUCLEASES*

A. 5'-Monoester Formers

Enzyme	Mode of attack	Linkages attacked and products formed	Additional characteristics
RNase H (hybrid), cellular: Hydrolyzes the RNA strand of an RNA · DNA hybrid (the five entries below)			
Calf thymus[308-310]	Endo[309,310]	No significant base specificity. Products are 5'-P-mono- and oligonucleotides (2–9 bases).[310]	Requires divalent cation.[309,310] Native and denatured DNAs and ssRNAs are not hydrolyzed.[310] MW ~64,000,[309] ~74,000.[310]
E. coli[311-314]	Endo[311-313]	Products are 5'-P-oligonucleotides (2–6 bases).[312,313]	Requires divalent cation. Degrades circular (rA)$_n$ in the presence of (dT)$_n$.[312,313] Cannot cleave the phosphodiester bond linking RNA to DNA.[313] MW ~40,000.[311]
Chick embryo and human KB cells[315,316]	Endo[315]	Products are 5'-P-mono- and oligonucleotides (mono << oligo).[315]	Requires divalent cation. Degrades Col E$_1$ DNA, a chloramphenicol induced, closed-circular, RNA · DNA hybrid. MW = 70,000–90,000.[315]

*Abbreviations: AdoMet, S-adenosylmethionine; ss, single-stranded; ds, double-stranded; MW, molecular weight; R, purine nucleoside; and Y, pyrimidine nucleoside.

Table 2 (continued)
RIBONUCLEASES

A. 5'-Monoester Formers (continued)

Enzyme	Mode of attack	Linkages attacked and products formed	Additional characteristics
Rat liver[317]	Endo[317]	With hybrid homopolymer substrates, hydrolyzes only $(rR)_n$. Products are 5'-P-oligonucleotides of < 15 bases (no mononucleotides).[317]	Requires divalent cation. Native and denatured DNAs and natural and synthetic polyribonucleotides are not hydrolyzed. MW ~125,000. Subunits 85,000 and 43,000.[317]
Ustilago maydis[318]	Endo[318]	Products are 5'-P-oligonucleotides.[318]	Requires divalent cation. $(rU)_n \cdot (dA)_n$ is a poor substrate. Polyribonucleotides are not degraded except for $(rU)_n$ and $(rI)_n$.[318]
RNase H (hybrid), viral: Hydrolyzes the RNA strand of an RNA · DNA hybrid (the two entries below)			
Avian myeloblastosis virus[312,319-323]	Requires a free 3'- or 5'-terminus[312,322]	Attacks from both the 3'- and 5'-termini in a processive fashion.[312,322] Products are 5'-P-oligonucleotides of 2–8 bases.[312] Does not attack $(rU)_n \cdot (dA)_n$.[312]	Requires divalent cation. Present in the viral core.[319] Copurifies with the RNA dependent DNA polymerase.[312,319,320,323] Differential inactivation of the two activities is possible.[312,319] Does not hydrolyze closed-circular hybrids.[312,322] Two subunits: α, MW = 65,000 and β, MW = 105,000.[323]

Table 2 (continued)
RIBONUCLEASES

A. 5'-Monoester Formers (continued)

Enzyme	Mode of attack	Linkages attacked and products formed	Additional characteristics
Avian myeloblastosis virus[312,319–323] (continued)			RNase H and DNA polymerase activities reside on a single polypeptide subunit, α. α is a random exonuclease whereas the α-β complex is a processive exonuclease.[322]
Rous associated virus[315]	Requires a free terminus	Products are 5'-P-mono- and oligonucleotides (mono << oligo).[315]	Requires divalent cation. Does not hydrolyze closed-circular RNA · DNA hybrids.[315] Copurifies with RNA dependent DNA polymerase.
Pig liver nuclei RNase[45]	Endo	Hydrolyzes rRNA and all ribohomopolymers except $(G)_n$. Products are oligonucleotides of 2—6 bases.	Requires divalent cation. Hydrolysis of $(I)_n$ requires additional Mg^{2+}
Nuclear RNase (from mouse liver, kidney, embryo, mammary tumor, and Ehrlich ascites tumor)[161,162]	Exo, processive	Degrades from the 3'-OH end producing 5'-mononucleotides.	Requires divalent cation. Strongly inhibited by terminal 3'-P or cyclic P.[162]
Leukemic cell RNase[48]	Exo	Products are 5'-mononucleotides.	Does not attack NO_2PhpdT.
Escherichia coli			
RNase P[324]	Endo	Cleaves only one bond of the 129 nucleotide tRNA^{Tyr} precursor removing all of the extra nucleotides at the 5'-terminus as a 41 nucleotide fragment.	Requires mono- and divalent cation for optimal activity.

Table 2 (continued)
RIBONUCLEASES

A. 5'-Monoester Formers (continued)

Enzyme	Mode of attack	Linkages attacked and products formed	Additional characteristics
RNase II[124,163]	Exo, processive	Attacks from the 3'-OH end producing 5'-mononucleotides. Oligonucleotides of less than 8 bases are fairly resistant to hydrolysis.	Requires mono- and divalent cation. Specific for ss-polyribonucleotides.
RNase III[159,325]	Endo	Specific for dsRNA producing 5'-P-oligonucleotides of about 14 bases.[325,326]	Does not degrade DNA · RNA hybrids.[325] Requires mono- and divalent cation.
RNase V[327] The existence of this activity is questionable and may be due to residual RNase II associated with ribosomes.[426,427]	Exo	Attacks polyribonucleotides $(U)_n$, $(A)_n$, T_4 phage and *E. coli* mRNAs from the 5'-terminus producing 5'-mononucleotides.	Does not degrade rRNA. Activity requires G and T_4 factors, tRNA, K^+, Mg^{2+}, GTP, and reducing compound.
Anacystis nidulans			
RNase α[328]	Endo	Specific for $O^{2'}$-methylated RNA Does not attack homoribopolymers.	
RNase β[328]	Endo	Hydrolyzes yeast RNA; slow degrades $(U)_n$.	
Bacillus AU$_2$ RNase[329]	Exo	Degrades RNA and synthetic polyribonucleotides from the 3'-terminus first to 5'-mononucleotides, then to N + p due to associated 5'-nucleotidase activity.	Does not hydrolyze N > p or $(A)_nA$ > p. Ribonuclease and 5'-nucleotidase activities are inseparable by gel electrophoresis or by density gradient centrifugation. Both activities require divalent cation.

Table 2 (continued)
RIBONUCLEASES

Enzyme	Mode of attack	Linkages attacked and products formed	Additional characteristics
A. 5'-Monoester Formers (continued)			
Lactobacillus plantarum RNase II[164]	Exo, processive	Hydrolyzes ss-polyribonucleotides to 5'-mononucleotides.	Requires mono- and divalent cation Similar to *E. coli* RNase II.
Physarum polycephalum RNase PP4[330]	Endo	Hydrolyzes RNA to 5'-mono- and oligonucleotides. Preference for R↓pN.	
B. 3'-Monoester formers			
Bovine Pancreatic RNase (RNase A),[32-34] primary structure,[331] three dimensional structure[154,332,333] synthesis[155,156] RNase B is a glycoprotein of identical primary structure.[334] RNase S is RNase A after subtilisin cleavage between residues 20 and 21.[335] For reviews see References 336 and 337.	Endo[32,33]	Specific for Yp↓N,[35,36] Cp cleaved faster than Up.[35-37] Products are 2',3'-cyclic, then 3'-monophosphates.	MW = 13,680. Polyadenylate[38] and polyribophosphate[39] are slowly hydrolyzed. RNase[40] and RNase S-protein[41] are capable of transesterification, resulting in oligonucleotide synthesis.
Venom RNase[58]	Endo	Preference for Yp↓R. Products are 3'-P-oligonucleotides (12 bases and higher).	Similar to, but not identical with, pancreatic RNase in specificity.
Plant RNases I[62-66,338,340,368,369,428,432] For a review see Reference 339.	Endo	Produces 2',3'-cyclic phosphates. R > p are slowly hydrolyzed to Rp.	Y> p not hydrolyzed. MW = 20,000-25,000. Optimal pH near 5.[339]
Plant RNases II[62,66-68,368,429-431] For review see Reference 339.	Endo	Produces 2',3'-cyclic phosphates. N > p slowly hydrolyzed to Np.	MW = 17,000-20,000. Optimal pH near 6.[339]

Table 2 (continued)
RIBONUCLEASES

B. 3'-Monoester Formers (continued)

Enzyme	Mode of attack	Linkages attacked and products formed	Additional characteristics
Escherichia coli			
RNase I[123]	Endo	Preference for Ap↓N and Up↓N in the early stages. Produces 2',3'-cyclic and then 3'-monophosphates.	A > p and C > p are hydrolyzed five times faster than G > p and U > p.
RNase IV[341]		Nature of the bonds cleaved is not known. Specifically cleaves R17 RNA into two large fragments: 15S and 21S.	
Aspergillus oryzae			
RNase T$_1$[73,74] primary structure[157,158,342]	Endo[74]	Specific for Gp↓N. Produces cyclic, then 3'-monophosphates. I > p and X > p are hydrolyzed. A > p, C > p, and U > p are resistant.[74]	MW = 11,085.[342]
RNase T$_2$[75,76,158,342]	Endo[75]	Preference for Ap↓N in the early stages. Produces cyclic, then 3'-monophosphates.[75,76]	Glycoprotein of MW ~36,000. Also cleaves ψp-N and hUp-N.[158]
Neurospora crassa			
RNase N$_1$[158,342]	Endo	Specific for Gp↓N. Produces cyclic, then 3'-monophosphates (slowly).	MW = 11,000. Extracellular. Slow hydrolysis of G > p makes N$_1$ useful for synthetic reactions.[342]
RNase N$_2$[158,342]	Endo	No absolute base specificity. Produces cyclic, then 3'-monophosphates.	MW ~36,000.[342]

Table 2 (continued)
RIBONUCLEASES

Enzyme	Mode of attack	Linkages attacked and products formed	Additional characteristics
		B. 3'-Monoester Formers (continued)	
RNase N$_3$[158,342]	Endo	Specific for Gp↓N. Produces cyclic, then 3'-monophosphates.	MW of N$_3$ > MW of N$_1$. Intracellular.[342]
Ustilago sphaerogena			
RNase U$_1$[158,181–184, 342,343]	Endo	Specific for Gp↓N. Produces cyclic, then 3'-monophosphates.	MW = 11,000.[342,343] Extracellular.[158]
RNase U$_2$[158,183,184, 342,344]	Endo	Specific for Rp↓N, however, not absolutely.[342,344] Exhaustive digestion yields mono-, di-, and trinucleotides terminated in all four nucleosides.[344] 2', 3'-cyclic phosphate intermediates are hydrolyzed slowly.[184]	MW = 10,000. Extracellular.[342]
RNase U$_3$[158,183,184]	Endo	An enzyme similar to the above.	
RNase U$_4$[183,184,345]	Exo	No base specificity. Attacks from the 5'-terminus producing 2', 3'-cyclic mononucleotides, which are degraded to 3'-mononucleotides.[345]	Little or no activity towards denatured DNA and bis(p-nitrophenyl) phosphate.[345]
Chalaropsis sp. RNase[346,347]	Endo	Specific for Gp↓N. Produces cyclic, then 3'-monophosphates.	MW = 11,842. Primary sequence of first 18 residues closely resembles RNase T$_1$.[347]
Microbial RNases I (for reviews see References 158 and 342.)	Endo	Specific from Gp↓N. Produces cyclic, then 3'-monophosphates.	

Table 2 (continued)
RIBONUCLEASES

B. 3'-Monoester Formers (continued)

Enzyme	Mode of attack	Linkages attacked and products formed	Additional characteristics
Acrocylindrium sp.[342] *Actinomyces areoventicillatus*[342] *Aspergillus fumigatus*[348] *Aspergillus* sp.[128] *Bacillus pumilus*[128] *Mucor genevensis*[128] *Penicillium brevicompactum*[370] *Physarum polycephalum* (PP₁)[349] *Streptomyces albogriseolus*[350] *Streptomyces erythreus*[99] *Ustilago zeae*[342]	Endo	No specificity toward bases. Produces cyclic, then 3'-mono-phosphates.	
Microbial RNases II (for reviews, see References (158 and 342.) *Aspergillus saitoi*[342] *Bacillus subtilis* Marburg[351] *Penicillium janthinellum*[352] *Physarum polycephalum* (PP₂, PP₃)[353] *Salmonella typhimurium*[354] *Tricoderma koningi* (I, II)[342]			
Enterobacter sp. RNase[355]	Endo	Strong preference for Cp↓N. Products from yeast RNA are C > p, Cp, and 3'-P-oligonucleo-tides terminated in C.	$(A)_n$, $(U)_n$, and $(G)_n$ are not degraded. Cleaves 5S RNA of *B. stearothermophilus* mainly at Cp↓A.[356]
Thiobacillus thioparus RNase[357-359] several similar enzymes	Endo	Specific for Yp↓N.	

Table 2 (continued)
RIBONUCLEASES

Enzyme	Mode of attack	Linkages attacked and products formed	Additional characteristics
		B. 3'-Monoester Formers (continued)	
Xenopus laevis RNase[360]	Endo	Preferential for Yp↓N. $(U)_n$ is hydrolyzed 30 times faster than $(C)_n$. Does not hydrolyze U > p.	Requires divalent cation. The 3'-ends formed from 4S RNA from sea urchin embryo cytoplasm are primarily U. $(A)_n$ and $(G)_n$ are not degraded.
Bacillus subtilis RNase (*B. amyloliquefaciens*)[100]	Endo	Preference for Gp↓R. Produces cyclic, then 3'-monophosphates.[100-102]	Crystalline enzyme is free of DNase, phosphodiesterases and phosphomonoesterases. MW ~10,700. Extracellular.[100]
Pleospora RNase[342]	Endo	Specific for Rp↓N. Produces cyclic, then 3'-monophosphates.	
Rhizopus oligosparus RNases[361]	Endo	Preference for Yp↓N. Produces cyclic, then 3'-monophosphates.	Two RNases with very similar properties. MW = 30,000.
Tricoderma koningi RNase III[342]	Endo	Preference for Rp↓N. Produces cyclic, then 3'-monophosphates.	MW = 10,000
Yeast RNase[362]	Exo	No specificity towards bases. Produces 3'-mononucleotides.	Does not hydrolyze N > p.

Table 2 (continued)
RIBONUCLEASES

Enzyme	Mode of attack	Linkages attacked and products formed	Additional characteristics
		C. 2′, 3′-Cyclic Diester Formers**	
Azotobacter agilis RNase[363]	Endo	Digests RNA completely to the four 2′, 3′-cyclic mono-nucleotides.	
Bacillus subtilis RNase[364-366]	Endo	Digests RNA completely to the four 2′, 3′-cyclic mono-nucleotides.	Strongly inhibited by ATP and dATP;[365,366] intracellular.
Proteus mirabilis RNase II[367]	Endo	Produces 2′, 3′-cyclic nucleotides. Cp-N bonds are fairly resistant.	

**See the introductory remarks.

Compiled by David Kowalski and Michael Laskowski, Sr.

Table 3
NUCLEASES NONSPECIFIC FOR THE SUGAR MOIETY*

Enzyme	Mode of attack	Linkages attacked and products formed	Additional characteristics
A. 5'-Monoester Formers			
Sheep kidney nuclease[165]	Endo	Specific for single stranded. Produces 5'-monophosphates.	Native DNA and $(A)_n$ $(U)_n$ are resistant.
Lamb brain nuclease[47]	Endo	Specific for single stranded. Produces 5'-P-hexanucleotides and higher.	
Chicken pancreas nuclease[44]	Endo	Produces 5'-P-di to tetra-nucleotides.	
Neurospora crassa mitochondrial nuclease[126]	Endo	Produces 5'-oligo- and mono-nucleotides. Of the mono-nucleotides, pG and pC pre-dominate.	Requires divalent cation. Optimal pH 6.5–7.5.
N. crassa nuclease[127,371-376]	Endo	Produces 5'-oligo- and mononucleotides. Of the mononucleotides, pG predom-inates. Attacks superhelical DNA to produce nicked circles.[373]	Cleaves λ DNA in the (A + T)-rich middle.[376] Optimal pH 7.5–8.5.
Aspergillus nuclease S$_1$[377,378,453]	Mixed, endo, and exo	Produces predominantly 5'-P-mononucleotides.[377] Attacks superhelical DNA to produce linear molecules.[379]	Cleaves nicked duplex DNA opposite nicks.[307] Inflicts a limited number of cleavages in native viral DNA.[380] Optimal pH 4.5.

The next four enzymes have many properties in common, particularly a very high specificity toward single-stranded structures, mismatched regions, and supercoiled DNA. They cleave native viral DNA into a limited number of fragments and in this respect resemble restriction enzymes.

*Abbreviations: AdoMet, S-adenosylmethionine; ss, single-stranded; ds, double-stranded; MW, molecular weight; R, purine nucleoside; and Y, pyrimidine nucleoside.

Table 3 (continued)
NUCLEASES NONSPECIFIC FOR THE SUGAR MOIETY

Enzyme	Mode of attack	Linkages attacked and products formed	Additional characteristics
A. 5'-Monoester Formers (continued)			
Mung bean nuclease[70-72,166,351] (Purified to apparent homogeneity, inseparable from ω-monophosphatase activity.)	Endonucleolytically attacks RNA and DNA. Also has ω-exonucleolytic activity on deoxy-, ribo- and arabinose oligonucleotides.	30,000:1 preference for T_4 ssDNA over T_4 dsDNA. Produces 5'-P-mono- and oligonucleotides. From various 3'-phosphoesters, liberates nucleoside and phosphoester $(N{\downarrow}pPhNO_2)$.[383]	Inflicts a limited number of cleavages in native viral DNA. Splits λ DNA into two halves.[72] Excises heat denatured regions from various DNAs.[382] Cleaves off the sticky ends of λ DNA.[384] Optimal pH 4.8–5.5.
Wheat seedling nuclease, inseparable from ω-monophosphatase[385-387]	Endonucleolytically attacks DNA. $(rA)_n$ is attacked predominantly exonucleolytically.	Highly preferential for ssDNA. Produces 5'-mono- and oligonucleotides.	Cleaves native phage DNA into fragments of MW approximately 2×10^6.[387] Probably very similar to mung bean nuclease. Optimal pH 4.8–5.5.
Silk worm nuclease[61,388]	Endo	$A{\downarrow}pN$ preferential; $G{\downarrow}pN$ discriminated against. Produces 5'-oligo- and a small amount of mononucleotides.	Requires Mg^{2+}. Optimal pH ~10. A similar enzyme exists in larval and pupal stages.[388]
Potato nuclease I[69,389]	Endo	5'-Terminus 75% G; 3'-terminus 55% A, 40% T. Produces 5'-P-oligonucleotides.	Requires Mg^{2+}. Optimal pH 6–7. Inseparable from the 3'-nucleotidase activity.[390]
Azotobacter nuclease[97,98]	Endo	Preferentially attacked linkage appears to be $A{\downarrow}pA$; linkage discriminated against is $G{\downarrow}pG$. Produces 5'-oligonucleotides of predominantly 2–4 bases.	Autoretardation

Table 3 (continued)
NUCLEASES NONSPECIFIC FOR THE SUGAR MOIETY

Enzyme	Mode of attack	Linkages attacked and products formed	Additional characteristics
A. 5'-Monoester Formers (continued)			
Hog kidney phosphodiesterase[46]	Exo	Attacks from the 3'-terminus producing 5'-mononucleotides. ω-Monophosphatase activity is present.	Inactive on high-MW RNA, but active on native DNA
Lactobacillus exonuclease[391]	Exo	Produces 5'-mononucleotides.	
Yeast nuclease[435,436]	Endo	Hydrolyzes yeast ribosomal RNA and denatured DNA. Products are di- and trinucleotides with 5'-P.	Purified 30-fold. Optimal pH 7.6, Mg^{2+} is required for optimal activity.
Carrot exonuclease[172,189]	Exo	Attacks from the 3'-terminus and produces 5'-mononucleotides.	dsDNA is resistant. Requires divalent cation.
Phoma cucurbitacearum nuclease[433]	Endo	Products terminated in 5'-P	Carries 3'-nucleotidase activity. Inhibited by EDTA.
Venom exonuclease (syn. venom phosphodiesterase).[52] Purified to apparent homogeneity, inseparable from ATPase;[392] immobilized preparation available.[393] Removal of 5'-nucleotidase from commercial preparations was described.[396] For reviews see References 2, 52, and 448.	Attacks ribo-deoxyribo- and arabinose derivatives[52-55] exonucleolytically from the 3'-terminus[56] regardless of the location of monophosphate. Attacks poly(adenosine diphosphate ribose) endonucleolytically.[397]	Chains terminated in 5'-phosphate are best substrates. Removal of phosphate decreases susceptibility tenfold; introduction of 3'-phosphate decreases another 50-fold. Enzyme is useful for quantitative determination of contaminating RNA in preparations of DNA.[394]	Requires Mg^{2+} or Ca^{2+} for activity. Optimal pH ~9, except pH 6 for chains bearing 3'-phosphates.[56] With chains terminated in 3'-phosphate, both termini can be quantitatively determined as $N\alpha$ and $pN\omega p$.[394,395]

Table 3 (continued)
NUCLEASES NONSPECIFIC FOR THE SUGAR MOIETY

Enzyme	Mode of attack	Linkages attacked and products formed	Additional characteristics
A. 5'-Monoester Formers (continued)			
Avena leaf phosphodiesterase[398]	Exo	Attacks from the 3'-terminus. producing 5'-mononucleotides. Resembles venom exonuclease.	dsDNA is resistant.
Avena leaf nuclease[434]	Endo	Preference for ssDNA. Produces 5'-P-mono- and oligonucleotides.	MW approximately 33,000. Zn^{2+} selectively inhibits RNase activity. EDTA inhibits DNase and RNase activities.
Tobacco culture nuclease (extracellular)[399]	Endo	Produces 5'-mono- and oligonucleotides (predominantly of 5 residues or less).	Highly preferential for ssDNA.
Penicillium citrium P_1 nuclease[400-403]	Mixed, endo, and exo	Initially produces oligonucleotides which are further degraded to 5'-mononucleotides. Hydrolyzes $dTpPhNO_2$ but not NO_2PhpdT.	Requires Zn^{2+}. Optimal temp. 70°. Optimal pH 4.5–6.0. The same enzyme hydrolyzes 2'-P-mononucleotides 3,000 times slower than 3'-mononucleotides to nucleosides and P_i. Ribo 3'-mononucleotides are hydrolyzed 30 times faster than deoxyribo.
Serratia marcescens nuclease (extracellular)[404,405]	Endo	No preference for bases or for native vs. denatured DNA. Produces mainly di- to tetranucleotides.	DNA and RNA are equally susceptible.

Table 3 (continued)
NUCLEASES NONSPECIFIC FOR THE SUGAR MOIETY

Enzyme	Mode of attack	Linkages attacked and products formed	Additional characteristics
		B. 3'-Monoester formers	
Micrococcal (staphylococcal) nuclease; homogeneous, crystalline,[77-80,89] amino-acid sequence,[168-170] 3-dimensional structure.[185,449] For a review see Reference 450.	Original attack endonucleolytic; soon thereafter, ω-exonucleolytic action starts.[83,406]	Attacks denatured DNA faster than native.[90,171] Shows autoacceleration.[171,407] Proximal 5'-monophosphate inhibits while proximal 3'-monophosphate enhances.[89]	Absolute requirement for Ca^{2+};[409] as fragments become shorter, higher Ca^{2+} concentration allows regulation of the size of products.[408] Strongly inhibited by pAp and pTp,[409] which limits usefulness of this enzyme for identification of $pN^{\alpha}p$ and N^{ω} in 5'-oligonucleotides to tetranucleotides and shorter fragments. Optimal pH 9.
Spleen exonuclease[49-51,410] For a review see Reference 451.	Exo	Preference for ss-substrates. Attacks from the 5'-hydroxyl terminus producing 3'-mononucleotides; chains bearing 5'-phosphate are resistant at pH 4.8,[50] but slowly hydrolyzed at pH 6.2.[452]	Glucosylated oligonucleotides (T_4 DNA) and 2'-O-acetylated tRNA are resistant. Widely used for the identification of the 3'-hydroxyl terminus after treatment with monophosphatase.
L. acidophilus phosphodiesterase[105]	Exo	Produces 3'-monophosphates from the 5'-hydroxyl terminus.	

Table 3 (continued)
NUCLEASES NONSPECIFIC FOR THE SUGAR MOIETY

Enzyme	Mode of attack	Linkages attacked and products formed	Additional characteristics
B. 3'-Monoester Formers (continued)			
B. subtilis exonuclease[103],[104]	Exo	Attacks ssDNA from the 5'-terminus, whereas dsDNA is initially attacked from 3'-terminus.[104] Produces 3'-mononucleotides.	Hydrolyzes denatured DNA faster than native
Salmon testes exonuclease[411]	Exo	Attacks from the 5'-terminus, producing 3'-mononucleotides.	
Chlamydomonas endonuclease[412]	Endo	Denatured DNA is hydrolyzed 200–300-fold faster than native.	Optimum pH ~8.8; activated by Ca^{2+}, inhibited by NaCl but not by KCl

Compiled by David Kowalski and Michael Laskowski, Sr.

REFERENCES

1. Privat de Garilhe, *Les Nucleases,* Herman, Paris, France, 1964.
2. Laskowski, Sr., *Adv. Enzymol.,* 29, 165 (1967).
3. Lehman, *Prog. Nucleic Acid Res.,* 2, 83 (1963).
4. Kunitz, *J. Gen. Physiol.,* 33, 349 (1950).
5. Laskowski, Sr., in *The Enzymes,* 2nd ed., Boyer, Lardy, and Myrback, Eds., Academic Press, New York, 1961, 123.
6. Potter, Brown, and Laskowski, Sr., *Biochim. Biophys. Acta,* 9, 150 (1952).
7. Sinsheimer and Koerner, *Science,* 114, 42 (1951).
8. Sinsheimer and Koerner, *J. Biol. Chem.,* 198, 293 (1952).
9. Bollum, *J. Biol. Chem.,* 240, 2599 (1965).
10. Laskowski, *Ann. N.Y. Acad. Sci.,* 81, 776 (1959).
11. Vanecko and Laskowski, Sr., *J. Biol. Chem.,* 236, 3312 (1961).
12. Vanecko and Laskowski, Sr., *J. Biol. Chem.,* 236, 1135 (1961).
13. Kurnick, *J. Am. Chem. Soc.,* 76, 417, 4040 (1954).
14. Potter, personal communication.
15. Palatnick and Bachrach, *Anal. Biochem.,* 2, 168 (1961).
16. Sandeen and Zimmerman, *Anal. Biochem.,* 14, 269 (1966).
17. Dabrowska, Cooper, and Laskowski, Sr., *J. Biol. Chem.,* 177, 991 (1949).
18. Lindberg, *J. Biol. Chem.,* 241, 1246 (1966).
19. Lindberg, *Biochemistry,* 6, 323, 343 (1967).
20. Laskowski, Sr., in *Procedures in Nucleic Acid Research,* Cantoni and Davies, Eds., Harper and Row, New York, 1966, 85.
21. Maver and Greco, *J. Biol. Chem.,* 181, 853, 861 (1949).
22. Bernardi, in *Procedures in Nucleic Acid Research,* Cantoni and Davies, Eds., Harper and Row, New York, 1966, 102.
23. Bernardi, Bernardi, and Chersi, *Biochim. Biophys. Acta,* 129, 1 (1966).
24. Slor and Hodes, personal communication.
25. Privat de Garilhe, Cunningham, Laurila, and Laskowski, Sr., *J. Biol. Chem.,* 224, 751 (1957).
26. Koerner and Sinsheimer, *J. Biol. Chem.,* 228, 1039, 1049 (1957).
27. Laurila and Laskowski, Sr., *J. Biol. Chem.,* 228, 49 (1957).
28. Doskočil and Šorm, *Biochim. Biophys. Acta,* 48, 211 (1961).
29. Vanecko and Laskowski, Sr., *Biochim. Biophys. Acta,* 61, 547 (1962).
30. Privat de Garilhe and Laskowski, Sr., *Biochim. Biophys. Acta,* 14, 147 (1954).
31. Young and Sinsheimer, *J. Biol. Chem.,* 240, 1274 (1965).
32. Kunitz, *J. Gen. Physiol.,* 24, 15 (1940).
33. Hirs, Moore, and Stein, *J. Biol. Chem.,* 200 493 (1953).
34. Klee, in *Procedures in Nucleic Acid Research,* Cantoni and Davies, Eds., Harper and Row, New York, 1966, 20.
35. Schmidt, Cubiles, Zöllner, Hecht, Strickler, Saraidarian, and Thannhauser, *J. Biol. Chem.,* 192, 715 (1951).
36. Markham and Smith, *Biochem. J.,* 52, 552 (1952).
37. Rushizky, Knight, and Sober, *J. Biol. Chem.,* 236, 2732 (1961).
38. Beers, Jr., *J. Biol. Chem.,* 235, 2393 (1960).
39. Rosenberg and Zamenhof, *J. Biol. Chem.,* 236, 2845 (1961).
40. Heppel and Whitfeld, *Biochem. J.,* 60, 1 (1955).
41. Bernfield, *J. Biol. Chem.,* 240, 4753 (1965).
42. Maver and Greco, *J. Biol. Chem.,* 211, 907 (1962).
43. Bernardi, in *Procedures in Nucleic Acid Research,* Cantoni and Davies, Eds., Harper and Row, New York, 1966, 37.
44. Eley and Roth, *J. Biol. Chem.,* 241, 3063, 3070 (1966).
45. Heppel, in *Procedures in Nucleic Acid Research,* Cantoni and Davies, Eds., Harper and Row, New York, 1966, 31.
46. Razzell, *J. Biol. Chem.,* 236, 3031 (1961).
47. Healy, Stollar, and Levine, in *Procedures in Nucleic Acid Research,* Cantoni and Davies, Eds., Harper and Row, New York, 1966, 188.
48. Anderson and Heppel, *Biochim. Biophys. Acta,* 43, 79 (1960).
49. Hilmoe, *J. Biol. Chem.,* 235, 2117 (1960).
50. Razzell and Khorana, *J. Biol. Chem.,* 236, 1144 (1961).
51. Bernardi and Bernardi, in *Procedures in Nucleic Acid Research,* Cantoni and Davies, Eds., Harper and Row, New York, 1966, 144.
52. Laskowski, Sr., in *Procedures in Nucleic Acid Research,* Cantoni and Davies Eds., Harper and Row, New York 1966, 154.
53. Wechter, *Abstr. Am. Chem. Soc. 152 Meet.,* New York, Sept. 1966, p. P17.

54. Wechter, *J. Med. Chem.,* 10, 762 (1967).
55. Richards, Tutas, Wechter, and Laskowski, Sr., *Biochemistry,* 6, 2908 (1967).
56. Richards and Laskowski, Sr., *Biochemistry,* 8, 1786 (1969).
57. Georgatsos and Laskowski, Sr., *Biochemistry,* 1, 288 (1962).
58. McLennan and Lane, *Can. J. Biochem.,* 46, 93 (1968).
59. Georgatsos and Antonoglou, *J. Biol. Chem.,* 241, 2151 (1966).
60. Georgatsos, *Biochim. Biophys. Acta,* 95, 544 (1965).
61. Mukai, *Biochem. Biophys. Res. Commun.,* 21, 562 (1965).
62. Reddi, in *Procedures in Nucleic Acid Research,* Cantoni and Davies, Eds., Harper and Row, New York, 1966, 71.
63. Markham and Strominger, *Biochem. J.,* 64, 46P (1956).
64. Reddi, *Biochim. Biophys. Acta,* 28, 386 (1958).
65. Reddi, *Biochim. Biophys. Acta,* 30, 638 (1958).
66. Shuster, Khorana, and Heppel, *Biochim. Biophys. Acta,* 33, 452 (1959).
67. Tuve and Anfinsen, *J. Biol. Chem.,* 235, 3437 (1960).
68. Reddi and Mauser, *Proc. Natl. Acad. Sci. U.S.A.,* 53, 607 (1965).
69. Bjork, *Biochim. Biophys. Acta,* 95, 652 (1965).
70. Masui, Hara, and Hiromatsu, *Biochim. Biophys. Acta,* 30, 215 (1958).
71. Sung and Laskowski, Sr., *J. Biol. Chem.,* 237, 506 (1962).
72. Johnson and Laskowski, Sr., *J. Biol. Chem.,* 243, 3421 (1968).
73. Sato and Egami, *J. Biochem.* (Tokyo), 44, 753 (1957).
74. Uchida and Egami, in *Procedures in Nucleic Acid Research,* Cantoni and Davies, Eds., Harper and Row, New York, 1966, 3.
75. Uchida and Egami, in *Procedures in Nucleic Acid Research,* Cantoni and Davies, Eds., Harper and Row, New York, 1966, 46.
76. Rushizky and Sober, *J. Biol. Chem.,* 238, 371 (1963).
77. Cunningham, Catlin, and Privat de Garilhe, *J. Am. Chem. Soc.,* 78, 4642 (1956).
78. Taniuchi and Anfinsen, *J. Biol. Chem.,* 241, 4366 (1966).
79. Cotton, Hazen, and Richardson, *J. Biol. Chem.,* 241, 4389 (1966).
80. Sulkowski and Laskowski, Sr., *J. Biol. Chem.,* 241, 4386 (1966).
81. Smith quoted by Cunningham, *Ann. N.Y. Acad. Sci.,* 81, 788 (1959).
82. Alexander, Heppel, and Hurwitz, *J. Biol. Chem.,* 236, 3014 (1961).
83. Sulkowski and Laskowski, Sr., *J. Biol. Chem.,* 237, 2620 (1962).
84. de Meuron-Landolt and Privat de Garilhe, *Biochim. Biophys. Acta,* 91, 433 (1964).
85. Rushizky, Knight, Roberts, and Dekker, *Biochem. Biophys. Res. Commun.,* 2, 153 (1960).
86. Reddi, *Biochim. Biophys. Acta,* 47, 47 (1961).
87. Dekker, *Annu. Rev. Biochem.,* 29, 453 (1960).
88. Pochon and Privat de Garilhe, *Bull. Soc. Chim. Biol.,* 42, 795 (1960).
89. Mikulski, Sulkowski, Stasiuk, and Laskowski, Sr., *J. Biol. Chem.,* 244, 6559 (1969).
90. von Hippel and Felsenfeld, *Biochemistry,* 3, 27 (1964).
91. Winter and Bernheimer, *J. Biol. Chem.,* 239, 215 (1964).
92. Yasmineh, Gray, and Wannamaker, *Fed. Proc.,* 22, 348 (1963).
93. Yasmineh, Gray, and Wannamaker, *Fed. Proc.,* 24, 227 (1965).
94. Potter and Laskowski, Sr., *J. Biol. Chem.,* 234, 1263 (1959).
95. Stone and Burton, *Biochem. J.,* 83, 492 (1962).
96. Georgatsos, Unterholzner, and Laskowski, Sr., *J. Biol. Chem.,* 237, 2626 (1962).
97. Stevens and Hilmoe, *J. Biol. Chem.,* 235, 3016 (1960).
98. Stevens and Hilmoe, *J. Biol. Chem.,* 235, 3023 (1960).
99. Tanaka, in *Procedures in Nucleic Acid Research,* Cantoni and Davies, Eds., Harper and Row, New York, 1966, 14.
100. Nishimura, in *Procedures in Nucleic Acid Research,* Cantoni and Davies, Eds., Harper and Row, New York, 1966, 56.
101. Rushizky, Greco, Hartley, Jr., and Sober, *Biochemistry,* 2, 787 (1962).
102. Whitfeld and Witzel, *Biochim. Biophys. Acta,* 72, 362 (1963).
103. Kerr, Pratt, and Lehman, *Biochem. Biophys. Res. Commun.,* 20, 154 (1965).
104. Okazaki, Okazaki, and Sakabe, *Biochem. Biophys. Res. Commun.,* 22, 611 (1966).
105. Fiers and Khorana, *J. Biol. Chem.,* 238, 2780, 2789 (1963).
106. Lehman, *J. Biol. Chem.,* 235, 1479 (1960).
107. Lehman and Nussbaum, *J. Biol. Chem.,* 239, 2628 (1964).
108. Adler, Lehman, Bessman, Simms, and Kornberg, *Proc. Natl. Acad. Sci. U.S.A.,* 44, 641 (1958).
109. Schachman, Adler, Radding, Lehman, and Kornberg, *J. Biol. Chem.,* 235, 3242 (1960).
110. Lehman, *Fed. Proc.,* 21, 378 (1962).
111. Lehman and Richardson, *J. Biol. Chem.,* 239, 233 (1964).

112. Richardson, Sample, Shildkraut, Lehman, and Kornberg, *Fed. Proc.*, 22, 349 (1963).
113. Richardson, Lehman, and Kornberg, *J. Biol. Chem.*, 239, 251 (1964).
114. Richardson and Kornberg, *J. Biol. Chem.*, 239, 242 (1964).
115. Jorgensen and Koerner, *J. Biol. Chem.*, 241, 3090 (1966).
116. Oleson and Koerner, *J. Biol. Chem.*, 239, 2935 (1964).
117. Short, Jr., and Koerner, *Proc. Natl. Acad. Sci. U.S.A.*, 54, 595 (1965).
118. Lehman, Roussos, and Pratt, *J. Biol. Chem.*, 237, 819 (1962).
119. Lehman, Roussos, and Pratt, *J. Biol. Chem.*, 237, 829 (1962).
120. De Waard and Lehman, in *Procedures in Nucleic Acid Research*, Cantoni and Davies, Eds., Harper and Row, 1966, 122.
121. Studier, *J. Mol. Biol.*, 11, 373 (1965).
122. Paul and Lehman, *J. Biol. Chem.*, 241, 3441 (1966).
123. Spahr, in *Procedures in Nucleic Acid Research*, Cantoni and Davies, Eds., Harper and Row, New York, 1966, 64.
124. Singer, in *Procedures in Nucleic Acid Research*, Cantoni and Davies, Eds., Harper and Row, New York, 1966, 192.
125. Anraku, in *Procedures in Nucleic Acid Research*, Cantoni and Davies, Eds., Harper and Row, New York, 1966, 130.
126. Linn and Lehman, *J. Biol. Chem.*, 241, 2694 (1966).
127. Linn and Lehman, *J. Biol. Chem.*, 240, 1287 (1965).
128. Rushizky, Greco, Hartley, and Sober, *J. Biol. Chem.*, 239, 2165 (1964).
129. Wechter, Mikulski, and Laskowski, Sr., *Biochem. Biophys. Res. Commun.*, 30, 318 (1968).
130. Scheffler, Elson, and Baldwin, *J. Mol. Biol.*, 36, 291 (1968).
131. Price, Liu, Stein, and Moore, *J. Biol. Chem.*, 244, 917 (1969).
132. Salnikow, Stein, and Moore, *Fed. Proc.*, 28, 344 (1969).
133. Melgar and Goldthwait, *J. Biol. Chem.*, 243, 4401 (1968).
134. Melgar and Goldthwait, *J. Biol. Chem.*, 243, 4409 (1968).
135. Carrara and Bernardi, *Biochemistry*, 7, 1121 (1968).
136. Bernardi, *Adv. Enzymol.*, 31, 1 (1968).
137. Swenson and Hodes, *J. Biol. Chem.*, 244, 1803 (1969).
138. Frieberg and Goldthwait, *Proc. Natl. Acad. Sci. U.S.A.*, 62, 934 (1969).
139. Kato and Ikeda, *J. Biochem.* (Tokyo), 64, 321 (1968).
140. Kato, Ando, and Ikeda, *J. Biochem.* (Tokyo), 64, 329 (1968).
141. Kato and Ikeda, *J. Biochem.* (Tokyo), 65, 43 (1968).
142. Shishido, Kato, and Ikeda, *J. Biochem.* (Tokyo), 65, 49 (1968).
143. Kato, Ikawa, Ikeda, and Fuke, *J. Biochem.* (Tokyo), 65, 185 (1968).
144. Shishido, Kato, and Ikeda, *J. Biochem.* (Tokyo), 65, 479 (1968).
145. Meselson and Yuan, *Nature*, 217, 1110 (1968).
146. Georgatsos, *Biochim. Biophys. Acta*, 129, 204 (1966).
147. Ip and Sung, *Can J. Biochem.*, 46, 1121 (1968).
148. Eron and McAuslan, *Biochem. Biophys. Res. Commun.*, 22, 518 (1966).
149. Morrison and Keir, *Biochem. J.*, 98, 37c (1966).
150. Lindahl, Gally, and Edelman, *Proc. Natl. Acad. Sci. U.S.A.*, 62, 597 (1969).
151. Short and Koerner, *J. Biol. Chem.*, 244, 1487 (1969).
152. Trilling and Aposhian, *Proc. Natl. Acad. Sci. U.S.A.*, 60, 214 (1968).
153. Lacks and Greenberg, *J. Biol. Chem.*, 242, 3108 (1967).
154. Kartha, Bello, and Harker, *Nature*, 213, 862 (1967).
155. Gutte and Merrifield, *J. Am. Chem. Soc.*, 91, 501 (1969).
156. Denkewalter, Veber, Holly, and Hirschmann, *J. Am. Chem. Soc.*, 91, 502 (1969).
157. Takahashi, *J. Biol. Chem.*, 240, 4117 (1965).
158. Egami and Nakamura, *Microbial Ribonucleases*, Springer-Verlag, New York, 1969.
159. Robertson, Webster, and Zinder, *J. Biol. Chem.*, 243, 82 (1968).
160. Dvorak and Heppel, *J. Biol. Chem.*, 243, 2647 (1968).
161. Lazarus and Sporn, *Proc. Natl. Acad. Sci. U.S.A.*, 57, 1387 (1967).
162. Sporn, Lazarus, Smith, and Henderson, *Biochemistry*, 8, 1698 (1969).
163. Nossal and Singer, *J. Biol. Chem.*, 243, 913 (1968).
164. Logan and Singer, *J. Biol. Chem.*, 243, 6161 (1968).
165. Kasai and Grunberg-Manago, *Eur. J. Biochem.*, 1, 152 (1967).
166. Mikulski and Laskowski, Sr., *J. Biol. Chem.*, 245, 5026 (1970).
167. Johnson and Laskowski, Sr., *J. Biol. Chem.*, 245, 891 (1970).
168. Taniuchi, Anfinsen, and Sodja, *J. Biol. Chem.*, 242, 4736 1967.
169. Taniuchi, Cusumano, Anfinsen, and Cone, *J. Biol. Chem.*, 243, 4475 (1968).
170. Cuatrecasas, Fuchs, and Anfinsen, *J. Biol. Chem.*, 244, 406 (1969).
171. Bernardi and Bernardi, *Biochim. Biophys. Acta*, 155, 360 (1968).
172. Harvey, Malsman, and Nussbaum, *Biochemistry*, 6, 3689 (1967).

173. Jovin, Englund, and Bertsch, *J. Biol. Chem.*, 244, 2996 (1969).
174. Jovin, Englund, and Kornberg, *J. Biol. Chem.*, 244, 3009 (1969).
175. Deutscher and Kornberg, *J. Biol. Chem.*, 244, 3019 (1969).
176. Deutscher and Kornberg, *J. Biol. Chem.*, 244, 3029 (1969).
177. Englund, Huberman, Jovin, and Kornberg, *J. Biol. Chem.*, 244, 3038 (1969).
178. Englund, Kelly, and Kornberg, *J. Biol. Chem.*, 244, 3045 (1969).
179. Little, Lehman, and Kaiser, *J. Biol. Chem.*, 242, 672 (1967).
180. Little, *J. Biol. Chem.*, 242, 679 (1967).
181. Glitz and Decker, *Biochemistry*, 3, 1391 (1964).
182. Glitz and Decker, *Biochemistry*, 3, 1399 (1964).
183. Arima, Uchida, and Egami, *Biochem. J.*, 106, 601 (1968).
184. Arima, Uchida, and Egami, *Biochem. J.*, 106, 609 (1968).
185. Cotton, Hazen, Richardson, Richardson, Arnone, and Bier, *Acta Crystallogr. Suppl.*, A25, S188 (1969).
186. Lindahl, Gally, and Edelman, *J. Biol. Chem.*, 244, 5014 (1969).
187. Lesca and Paoletti, *Proc. Natl. Acad. Sci. U.S.A.*, 64, 913 (1969).
188. Friedberg, Hadi, and Goldthwait, *J. Biol. Chem.*, 244, 5879 (1969).
189. Harvey, Olson, and Wright, *Biochemistry*, 9, 921 (1970).
190. Brutlag, Atkinson, Setlow, and Kornberg, *Biochem. Biophys. Res. Commun.*, 37, 982 (1969).
191. Klenow and Henningsen, *Proc. Natl. Acad. Sci. U.S.A.*, 65, 168 (1970).
192. Arber and Linn, *Annu. Rev. Biochem.*, 38, 467 (1969).
193. Boyer, *Annu. Rev. Microbiol.*, 25, 153 (1971).
194. Meselson, Yuan, and Heywood, *Annu. Rev. Biochem.*, 41, 447 (1972).
195. Arber, *Prog. Nucleic Acid Res.*, 14, 1 (1974).
196. Smith and Nathans, *J. Mol. Biol.*, 81, 419 (1973).
197. Linn and Arber, *Proc. Natl. Acad. Sci. U.S.A.*, 59, 1300 (1968).
198. Roulland-Dussoix and Boyer, *Biochim. Biophys. Acta*, 195, 219 (1969).
199. Eskin and Linn, *J. Biol. Chem.*, 247, 6183 (1972).
200. Linn, Lautenberger, Eskin, and Lackey, *Fed. Proc.*, 33, 1128 (1974).
201. Van Ormondt, *FEBS Lett.*, 33, 177 (1973).
202. Boyer, Scibienski, Slocum, and Roulland-Dussoix, *Virology*, 46, 703 (1971).
203. Horiuchi and Zinder, *Proc. Natl. Acad. Sci. U.S.A.*, 69, 3220 (1972).
204. Adler and Nathans, *Biochim. Biophys. Acta*, 299, 177 (1973).
205. Meselson and Yuan, *Nature*, 217, 1110 (1968).
206. Murray, Batten, and Murray, *J. Mol. Biol.*, 81, 395 (1973).
207. Brockes, Brown, and Murray, *J. Mol. Biol.*, 88, 437 (1974).
208. Smith and Wilcox, *J. Mol. Biol.*, 51, 379 (1970).
209. Kelly and Smith, *J. Mol. Biol.*, 51, 393 (1970).
210. Roy and Smith, *J. Mol. Biol.*, 81, 445 (1973).
211. Roy and Smith, *J. Mol. Biol.*, 81, 427 (1973).
212. Landy, Ruedisueli, Robinson, Foeller, and Ross, *Biochemistry*, 13, 2134 (1974).
213. Murray and Old, *Prog. Nucleic Acid Res.*, 14, 117, (1974).
214. Yoshimori, Ph.D. thesis, University of California, San Francisco, 1971.
215. Mulder and Delius, *Proc. Natl. Acad. Sci. U.S.A.*, 69, 3215 (1972).
216. Hedgepeth, Goodman, and Boyer, *Proc. Natl. Acad. Sci. U.S.A.*, 69, 3448 (1972).
217. Dugaiczyk, Hedgepeth, Boyer, and Goodman, *Biochemistry*, 13, 503 (1974).
218. Bigger, Murray, and Murray, *Nat. New Biol.*, 244, 7 (1973).
219. Boyer, Chow, Dugaiczyk, Hedgepeth, and Goodman, *Nat. New Biol.*, 244, 40 (1973).
220. Middleton, Edgell, and Hutchison, *J. Virol.*, 10, 42 (1972).
221. Boyer, *Fed. Proc.*, 33, 1125 (1974).
222. Gromkova and Goodgal, *J. Bacteriol.*, 109, 987 (1972).
223. Sharp, Snyder, and Sambrook, *Biochemistry*, 12, 3055 (1973).
224. Sack and Nathans, *Virology*, 51, 517 (1973).
225. Johnson, Lee, and Sinsheimer, *J. Virol.*, 11, 596 (1973).
226. Sugisaki and Takanami, *Nat. New Biol.*, 246, 138 (1973).
227. Weiss, Jaquemin-Sablon, Live, Fareed, and Richardson, *J. Biol. Chem.*, 243, 4543 (1968).
228. Junowicz and Spencer, *Biochim. Biophys. Acta*, 312, 85 (1973).
229. Laskowski, Sr., in *The Enzymes*, Vol. 4, 3rd ed., 1971, 289.
230. Price, Moore, and Stein, *J. Biol. Chem.*, 244, 924 (1969).
231. Price, Stein, and Moore, *J. Biol. Chem.*, 244, 929 (1969).
232. Catley, Moore, and Stein, *J. Biol. Chem.*, 244, 933 (1969).
233. Salnikow, Moore, and Stein, *J. Biol. Chem.*, 245, 5685 (1970).
234. Salnikow, Liao, Moore, and Stein, *J. Biol. Chem.*, 248, 1480 (1973).

235. Liao, Salnikow, Moore, and Stein, *J. Biol. Chem.*, 248, 1489 (1973).
236. Salnikow and Murphy, *J. Biol. Chem.*, 248, 1499 (1973).
237. Liao, *J. Biol. Chem.*, 249, 2354 (1974).
238. Liao, *J. Biol. Chem.*, 250, 3721 (1975).
239. Gottesfeld, Adams, El-Bardy, Moses, and Calvin, *Biochim. Biophys. Acta*, 228, 365 (1971).
240. Clark and Eichhorn, *Biochemistry*, 13, 5098 (1974).
241. Tavernier and Gray, *Biochemistry*, 9, 2846 (1970).
242. Frenkel and Richardson, *J. Biol. Chem.*, 246, 4839 (1971).
243. Frenkel and Richardson, *J. Biol. Chem.*, 246, 4848 (1971).
244. Ray, Reuben, and Molineux, *J. Biol. Chem.*, 249, 5379 (1974).
245. Setlow, Brutlag, and Kornberg, *J. Biol. Chem.*, 247, 224 (1972).
246. Brutlag and Kornberg, *J. Biol. Chem.*, 247, 241 (1972).
247. Eichhorn, Clark, and Tarien, *J. Biol. Chem.*, 244, 937 (1969).
248. Wovcha and Warner, *J. Biol. Chem.*, 248, 1746 (1973).
249. Setlow and Kornberg, *J. Biol. Chem.*, 247, 232 (1972).
250. Friedberg and Lehman, *Biochem. Biophys. Res. Commun.*, 58, 132 (1974).
251. Oishi, *Proc. Natl. Acad. Sci. U.S.A.*, 64, 1292 (1969).
252. Goldmark and Linn, *Proc. Natl. Acad. Sci. U.S.A.*, 67, 434 (1970).
253. Wright, Buttin, and Hurwitz, *J. Biol. Chem.*, 246, 6543 (1971).
254. Goldmark and Linn, *J. Biol. Chem.*, 247, 1849 (1972).
255. Karu, McKay, Goldmark, and Linn, *J. Biol. Chem.*, 248, 4874 (1973).
256. Klett, Ceraini, and Reich, *Proc. Natl. Acad. Sci. U.S.A.*, 60, 943 (1968).
257. Chase and Richardson, *J. Biol. Chem.*, 249, 4545 (1974).
258. Chase and Richardson, *J. Biol. Chem.*, 249, 4553 (1974).
259. Barbour, Nagaishi, Temphlin, and Clark, *Proc. Natl. Acad. Sci. U.S.A.*, 67, 128 (1970).
260. Sadowski and Hurwitz, *J. Biol. Chem.*, 244, 6182 (1969).
261. Altman and Meselson, *Proc. Natl. Acad. Sci. U.S.A.*, 65, 716 (1970).
262. Sadowski and Hurwitz, *J. Biol. Chem.*, 244, 6192 (1969).
263. Sadowski and Bakyta, *J. Biol. Chem.*, 247, 405 (1972).
264. Yasuda and Sekiguchi, *Proc. Natl. Acad. Sci. U.S.A.*, 67, 1839 (1970).
265. Center, Studier, and Richardson, *Proc. Natl. Acad. Sci. U.S.A.*, 65, 242 (1970).
266. Center and Richardson, *J. Biol. Chem.*, 245, 6285 (1970).
267. Center and Richardson, *J. Biol. Chem.*, 245, 6292 (1970).
268. Center, *J. Biol. Chem.*, 247, 146 (1972).
269. Kerr and Sadowski, *J. Biol. Chem.*, 247, 305 (1972).
270. Kerr and Sadowski, *J. Biol. Chem.*, 247, 311 (1972).
271. Gunther and Goodgal, *J. Biol. Chem.*, 245, 5341 (1970).
272. Aposhian, Friedman, Nishihara, Heimer, and Nussbaum, *J. Mol. Biol.*, 49, 367 (1970).
273. Lindahl, *Eur. J. Biochem.*, 18, 407 (1971).
274. Lindahl, *Eur. J. Biochem.*, 18, 415 (1971).
275. Ljungquist and Lindahl, *J. Biol. Chem.*, 249, 1530 (1974).
276. Ljungquist, Anderson, and Lindahl, *J. Biol. Chem.*, 249, 1536 (1974).
277. Pogo and Dales, *Proc. Natl. Acad. Sci. U.S.A.*, 63, 820 (1969).
278. Rosemond-Hornbeak and Moss, *J. Biol. Chem.*, 249, 3292 (1974).
279. Rosemond-Hornbeak, Paoletti, and Moss, *J. Biol. Chem.*, 249, 3287 (1974).
280. Anai, Hirashi, and Takagi, *J. Biol. Chem.*, 245, 767 (1970).
281. Anai, Hirashi, and Takagi, *J. Biol. Chem.*, 245, 775 (1970).
282. Pinon, *Biochemistry*, 9, 2839 (1970).
283. Schonherr, Wanka, and Kuyper, *Biochim. Biophys. Acta*, 224, 74 (1970).
284. Devillers-Thiery, Ehrlich, and Bernardi, *Eur. J. Biochem.*, 38, 416 (1973).
285. Soave, Thiery, Ehrlich, and Bernardi, *Eur. J. Biochem.*, 38, 423 (1973).
286. Thiery, Ehrlich, Devillers-Thiery, and Bernardi, *Eur. J. Biochem.*, 38, 434 (1973).
287. Ehrlich, Bertaazzoni, and Bernardi, *Eur. J. Biochem.*, 40, 143 (1973).
288. Ehrlich, Bertaazzoni, and Bernardi, *Eur. J. Biochem.*, 40, 149 (1973).
289. Bernardi, Ehrlich, and Thiery, *Nat. New Biol.*, 246, 36 (1973).
290. Slor, *Biochem. Biophys. Res. Commun.*, 38, 1084 (1970).
291. Slor and Hodes, *Arch. Biochem. Biophys.*, 139, 72 (1970).
292. Ehrlich, Devillers-Thiery, and Bernardi, *Eur. J. Biochem.*, 40, 139 (1973).
293. Ehrlich, Torti, and Bernardi, *Biochemistry*, 10, 2000 (1971).
294. Oshima and Price, *J. Biol. Chem.*, 248, 7522 (1973).
295. Oshima and Price, *J. Biol. Chem.*, 249, 4435 (1974).
296. Rosenbluth and Sung, *Can. J. Biochem.*, 47, 1081 (1969).

297. Yamanaka, Tsubota, Anai, Ishimatsu, Okomura, Katsuki, and Takagi, *J. Biol. Chem.*, 249, 3884 (1974).
298. Sabeur, Sicard, and Aubel-Sadron, *Biochemistry*, 13, 3203 (1974).
299. Laval and Paoletti, *Biochemistry*, 11, 3604 (1972).
300. Laval, Thiery, Ehrlich, Paoletti, and Bernardi, *Eur. J. Biochem.*, 40, 133 (1973).
301. Yamamoto, *Biochim. Biophys. Acta*, 228, 95 (1971).
302. Yamamoto and Bicknell, *Arch. Biochem. Biophys.*, 151, 261 (1972).
303. Sieliwanowicz, Yamamoto, Stasiuk, and Laskowski, Sr., *Biochemistry*, 14, 39 (1975).
304. Kaplan, Kushner, and Grossman, *Proc. Natl. Acad. Sci. U.S.A.*, 63, 144 (1969).
305. Kushner and Grossman, *Methods Enzymol.*, 21, 244 (1971).
306. Kaplan and Grossman, *Methods Enzymol.*, 21, 249 (1971).
307. Germond, Vogt, and Hirt, *Eur. J. Biochem.*, 43, 591 (1974).
308. Stein and Hausen, *Science*, 166, 393 (1969).
309. Stavrianopoulos and Chargaff, *Proc. Natl. Acad. Sci. U.S.A.*, 70, 1959 (1973).
310. Haberkern and Cantoni, *Biochemistry*, 12, 2389 (1973).
311. Miller, Riggs, and Gill, *J. Biol. Chem.*, 248, 2621 (1973).
312. Leis, Berkower, and Hurwitz, *Proc. Natl. Acad. Sci. U.S.A.*, 70, 466 (1973).
313. Berkower, Leis, and Hurwitz, *J. Biol. Chem.*, 248, 5914 (1973).
314. Henry, Ferdinand, and Knippers, *Biochem. Biophys. Res. Commun.*, 50, 603 (1973).
315. Keller and Crouch, *Proc. Natl. Acad. Sci. U.S.A.*, 69, 3360 (1972).
316. Keller, *Proc. Natl. Acad. Sci. U.S.A.*, 69, 1560 (1972).
317. Roewekamp and Sekeris, *Eur. J. Biochem.*, 43, 405 (1974).
318. Banks, *Eur. J. Biochem.*, 47, 499 (1974).
319. Mölling, Bolognesi, Bauer, Büsen, Plassman, and Hausen, *Nat. New Biol.*, 234, 240 (1971).
320. Watson, Mölling, and Bauer, *Biochem. Biophys. Res. Commun.*, 51, 232 (1973).
321. Baltimore and Smoler, *J. Biol. Chem.*, 247, 7282 (1972).
322. Grandgenett and Green, *J. Biol. Chem.*, 249, 5148 (1974).
323. Grandgenett, Gerard, and Green, *Proc. Natl. Acad. Sci. U.S.A.*, 70, 230 (1973).
324. Robertson, Altman, and Smith, *J. Biol. Chem.*, 247, 5243 (1972).
325. Crouch, *J. Biol. Chem.*, 249, 1314 (1974).
326. Schweitz and Ebel, *Biochimie*, 53, 585 (1971).
327. Kuwano, Kwan, Apirion, and Schlessinger, *Proc. Natl. Acad. Sci. U.S.A.*, 64, 693 (1969).
328. Norton and Roth, *J. Biol. Chem.*, 242, 2029 (1967).
329. Jacobson and Rodwell, *J. Biol. Chem.*, 247, 5811 (1972).
330. Hiramaru, Uchida, and Egami, *J. Biochem.* (Tokyo), 65, 701 (1969).
331. Smyth, Stein, and Moore, *J. Biol. Chem.*, 238, 227 (1963).
332. Wyckoff, Hardman, Allewell, Inagami, Johnson, and Richards, *J. Biol. Chem.*, 242, 3984 (1967).
333. Wyckoff, Tsernoglou, Hanson, Knox, Lee, and Richards, *J. Biol. Chem.*, 245, 305 (1970).
334. Plummer and Hirs, *J. Biol. Chem.*, 239, 2530 (1964).
335. Richards and Vithayathil, *J. Biol. Chem.*, 234, 1459 (1959).
336. Barnard, *Annu. Rev. Biochem.*, 38, 677 (1969).
337. Richards and Wyckoff, in *The Enzymes*, Vol. 4, 3rd ed., Boyer, Ed., Academic Press, New York, 1971, 647.
338. Wilson, *J. Biol. Chem.*, 242, 2260 (1967).
339. Wilson, *Annu. Rev. Plant Physiol.*, 26, 187 (1975).
340. Wyen, Udvardy, Solymosy, Marre, and Farkas, *Biochim. Biophys. Acta*, 191, 588 (1969).
341. Spahr and Gesteland, *Proc. Natl. Acad. Sci. U.S.A.*, 59, 876 (1968).
342. Uchida and Egami, in *The Enzymes*, Vol. 4, 3rd ed., 1971, 205.
343. Kenney and Dekker, *Biochemistry*, 10, 4963 (1971).
344. Rushizky, Mozejko, Rogerson, and Sober, *Biochemistry*, 9, 4966 (1970).
345. Blank and Dekker, *Biochemistry*, 11, 3956 (1972).
346. Fetcher and Hash, *Biochemistry*, 11, 4274 (1972).
347. Fletcher and Hash, *Biochemistry*, 11, 4281 (1972).
348. Glitz, Angel, and Eichler, *Biochemistry*, 11, 1746 (1972).
349. Hiramaru, Uchida, and Egami, *J. Biochem.* (Tokyo), 65, 701 (1969).
350. Yoneda, *J. Biochem.* (Tokyo), 55, 469 (1964).
351. Nikai, Minami, Yamasaki, and Tsugita, *J. Biol. Chem.*, 57, 96 (1965).
352. Ghosh and Thangamani, *Biochim. Biophys. Acta*, 361, 321 (1974).
353. Hiramaru, Uchida, and Egami, *J. Biochem.* (Tokyo), 65, 697 (1969).
354. Chakraburtty and Burma, *J. Biol. Chem.*, 243, 1133 (1968).
355. Levy and Goldman, *J. Biol. Chem.*, 245, 3257 (1970).
356. Marotta, Levy, Weissman, and Varricchio, *Biochemistry*, 12, 2901 (1973).
357. Ostrowski and Walczak, *Acta Biochim. Pol.*, 8, 345 (1961).
358. Walczak and Ostrowski, *Acta Biochim. Pol.*, 11, 241 (1964).

359. Ostrowski, Walczak, Stasiuk, and Laskowski, *Acta Biochim. Pol.,* 17, 73 (1970).
360. Berridge, Loewensteiner, and Aronson, *Biochemistry,* 13, 2520 (1974).
361. Woodroof and Glitz, *Biochemistry,* 10, 1532 (1971).
362. Ohtaka, Uchida, and Sakai, *J. Biochem.* (Tokyo), 54, 322 (1963).
363. Shiio, Ishii, and Shimizu, *J. Biochem.* (Tokyo), 59, 363 (1966).
364. Nishimura and Maruo, *Biochim. Biophys. Acta,* 40, 355 (1960).
365. Yamasaki and Arima, *Biochim. Biophys. Acta,* 139, 202 (1967).
366. Yamasaki and Arima, *Biochem. Biophys. Res. Commun.,* 37, 430 (1969).
367. Center and Behal, *Biochim. Biophys. Acta,* 151, 698 (1968).
368. Jervis, *Phytochemistry,* 13, 709 (1974).
369. Torti, Mapelli, and Soave, *Biochim. Biophys. Acta,* 324, 254 (1973).
370. Bezbarodova, Sukhodolskaya, and Gulyaeva, *Prikl. Biokhim. Mikrobial.,* 10, 432 (1974).
371. Linn, *Methods Enzymol.,* 12, 247 (1967).
372. Rabin and Fraser, *Can. J. Biochem.,* 48, 389 (1970).
373. Kato, Bartok, Fraser, and Denhardt, *Biochim. Biophys. Acta,* 308, 68 (1973).
374. Bartok, Garon, Berry, Fraser, and Rose, *J. Mol. Biol.,* 87, 437 (1974).
375. Bartok, Fraser, and Fareed, *Biochem. Biophys. Res. Commun.,* 60, 507 (1974).
376. Roizes, *Nucleic Acid Res.,* 1, 443 (1974).
377. Ando, *Biochim. Biophys. Acta,* 114, 158 (1966).
378. Sutton, *Biochim. Biophys. Acta,* 240, 522 (1971).
379. Beard, Marrow, and Berg, *J. Virol.,* 12, 1303 (1973).
380. Godson, *Biochim. Biophys. Acta,* 308, 59 (1973).
381. Ardelt and Laskowski, Sr., *Biochem. Biophys. Res. Commun.,* 44, 1205 (1971).
382. Kedzierski, Laskowski, Sr., and Mandel, *J. Biol. Chem.,* 248, 1227 (1973).
383. Kole, Sierakowska, Szemplinska, and Shugar, *Nucleic Acid Res.,* 1, 699, (1974).
384. Ghangas and Wu, *J. Biol. Chem.,* 249, 7550 (1974).
385. Hanson and Fairley, *J. Biol. Chem.,* 244, 2440 (1969).
386. Kroeker, Hanson, and Fairley, *J. Biol. Chem.,* 250, 3767 (1975).
387. Kroeker and Fairley, *J. Biol. Chem.,* 250, 3773 (1975).
388. Funaguma and Mukai, *Comp. Biochem. Physiol.,* 44B, 633 (1973).
389. Björk, *Ark. Kemi,* 27, 539 (1967).
390. Nomura, Suno, and Mizuno, *J. Biochem.* (Tokyo), 70, 993 (1971).
391. Sabatini and Hotchkiss, *Biochemistry,* 8, 4831 (1969).
392. Dolapchiev, Sulkowski, and Laskowski, Sr., *Biochem. Biophys. Res. Commun.,* 61, 273 (1974).
393. Sulkowski and Laskowski, Sr., *Biochem. Biophys. Res. Commun.,* 57, 463 (1974).
394. Duch and Laskowski, Sr., *Anal. Biochem.,* 44, 42 (1971).
395. Duch, Borkowska, Stasiuk, and Laskowski, Sr., *Anal. Biochem.,* 53, 459 (1973).
396. Sulkowski and Laskowski, Sr., *Biochim. Biophys. Acta,* 240, 443 (1971).
397. Matsubara, Haregawa, Fujimura, *J. Biol. Chem.,* 245, 3606 (1970).
398. Udvardy, Marre, and Farkas, *Biochim. Biophys. Acta,* 206, 392 (1970).
399. Oleson, Janski, and Clark, *Biochim. Biophys. Acta,* 366, 89 (1974).
400. Fujimoto, Kuninaka, and Yoshino, *Agric. Biol. Chem.,* 38, 777 (1974).
401. Fujimoto, Kuninaka, and Yoshino, *Agric. Biol. Chem.,* 38, 785 (1974).
402. Fujimoto, Kuninaka, and Yoshino, *Agric. Biol. Chem.,* 38, 1555 (1974).
403. Fujimoto, Fujiyama, Kuninaka, and Yoshino, *Agric. Biol. Chem.,* 38, 2141 (1974).
404. Nestle and Roberts, *J. Biol. Chem.,* 244, 5213 (1969).
405. Nestle and Roberts, *J. Biol. Chem.,* 244, 5219 (1969).
406. Sulkowski and Laskowski, Sr., *J. Biol. Chem.,* 244, 3818 (1969).
407. Sulkowski and Laskowski, Sr., *J. Biol. Chem.,* 243, 4917 (1968).
408. Sulkowski, Odlyzko, and Laskowski, Sr., *Anal. Biochem.,* 38, 393 (1970).
409. Sulkowski and Laskowski, Sr., *J. Biol. Chem.,* 243, 651 (1968).
410. Bernardi and Bernardi, *Biochim. Biophys. Acta,* 155, 360 (1968).
411. Menon and Smith, *Biochemistry,* 9, 1584 (1970).
412. Small and Sparks, *Arch. Biochem. Biophys.,* 153, 171 (1972).
413. Lazarides and Lindberg, *Proc. Natl. Acad. Sci. U.S.A.,* 71, 4742 (1974).
414. Lehman, in *The Enzymes,* Vol. 4, 3rd ed., Boyer, Ed., Academic Press, New York, 1971, 251.
415. Masamune, Fleischman, and Richardson, *J. Biol. Chem.,* 246, 2680 (1971).
416. Kornberg, in *DNA Synthesis,* W. H. Freeman & Co., San Francisco, 1974, pp. 1–399.
417. Hadi, Kirtikar, and Goldthwait, *Biochemistry,* 12, 2747 (1973).
418. Koerner, *Annu. Rev. Biochem.,* 39, 291 (1970).
419. Sadowski, *J. Biol. Chem.,* 246, 209 (1971).
420. Carter and Radding, *J. Biol. Chem.,* 246, 2502 (1971).

421. Mackay and Linn, *J. Biol. Chem.,* 249, 4286 (1974).
422. Lieberman and Oishi, *Nat. New Biol.,* 243, 75 (1973).
423. Hamilton, Mahler, and Grossman, *Biochemistry,* 13, 1886 (1974).
424. Verly, Paquette, and Thibodeau, *Nat. New Biol.,* 244, 68 (1973).
425. Bernardi, in *The Enzymes,* Vol. 4, 3rd ed., Boyer, Ed., Academic Press, New York, 1971, 271.
426. Holmes and Singer, *Biochem. Biophys. Res. Commun.,* 44, 837 (1971).
427. Bothwell and Apirion, *Biochem. Biophys. Res. Commun.,* 44, 844 (1971).
428. Lantero and Klosterman, *Phytochemistry,* 12, 775 (1973).
429. Wilson, *Plant Physiol.,* 43, 1332 (1968).
430. Wilson, *Plant Physiol.,* 43, 1339 (1968).
431. Wilson, *Plant Physiol.,* 48, 64 (1971).
432. Tang and Maretzki, *Biochim. Biophys. Acta,* 212, 300 (1970).
433. Tone and Okazaki, *Enzymology,* 34, 101 (1968).
434. Wyen, Erdei, and Farras, *Biochim. Biophys. Acta,* 232, 472 (1971).
435. Nakao, Lee, Halverson, and Bock, *Biochim. Biophys. Acta,* 151, 114 (1968).
436. Lee, Nakao, and Bock, *Biochim. Biophys. Acta,* 151, 126 (1968).
437. Zimmerman, *J. Biol. Chem.,* 241, 2035 (1966).
438. Litman, *J. Biol. Chem.,* 243, 6222 (1968).
439. Harwood, Schendel, and Wells, *J. Biol. Chem.,* 245, 5614 (1970).
440. Miller and Wells, *J. Biol. Chem.,* 247, 2667 (1972).
441. Miller and Wells, *J. Biol. Chem.,* 247, 2675 (1972).
442. Litman, *Biochem. Biophys. Res. Commun.,* 41, 91 (1970).
443. Carrier and Setlow, *J. Bacteriol.,* 102, 181 (1970).
444. Friedman and Smith, *J. Biol. Chem.,* 247, 2846 (1972).
445. Smith and Friedman, *J. Biol. Chem.,* 247, 2854 (1972).
446. Friedman and Smith, *J. Biol. Chem.,* 247, 2859 (1972).
447. Ohi and Sueoka, *J. Biol. Chem.,* 248, 7336 (1973).
448. Laskowski, Sr., in *The Enzymes,* Vol. 4, 3rd ed., 1971, 313.
449. Cotton and Hazen, in *The Enzymes,* Vol. 4, 3rd ed., 1971, 153.
450. Anfinsen, Cuatrecasas, and Taniuchi, in *The Enzymes,* Vol. 4, 3rd ed., 1971, 177.
451. Bernardi and Bernardi, in *The Enzymes,* Vol. 4, 3rd ed., 1971, 329.
452. Bernardi and Cantoni, *J. Biol. Chem.,* 244, 1468 (1969).
453. Vogt, *Eur. J. Biochem.,* 33, 192 (1973).

DEOXYNUCLEOTIDE POLYMERIZING ENZYMES

Lucy M. S. Chang

Three kinds of deoxynucleotide polymerizing enzymes are known to exist; they include terminal deoxynucleotidyl transferase, DNA polymerase, and reverse transcriptase. All three enzymes are found in mammalian and avian systems while only DNA polymerase is present in procaryotes. Terminal transferase is found only in thymus and some leukemic cells and is present in trace amounts in bone marrow cells. DNA polymerase is present in all mammalian and avian cells. The subclasses of DNA polymerase include the two major enzymes (the high molecular weight and the low molecular weight DNA polymerases), R-DNA polymerase, and mitochondrial DNA polymerase. The two major enzymes (constituting over 90% of total DNA polymerase activity in the cell) have been identified in all plants and in avian and mammalian systems. R-DNA polymerase and mitochondrial DNA polymerase have been identified in relatively few sources. The reverse transcriptase is found in RNA tumor viruses and in cancerous cells with possible viral etiology. The data compiled in this table do not include all information in the literature concerning these enzymes, but only provide information concerning the presence of these enzymes in various cells and some biochemical properties distinguishing these enzymes.

The reaction catalyzed by terminal transferase can be written as follows:

$$d(pY)_m + n(dXTP) \xrightarrow{M^{++}} d(pY)_{m+n} + nPP_i.$$

where $m \geq 3$. The reaction catalyzed by DNA polymerase can be written as:

$$\text{initiated single-stranded DNA} + n \begin{bmatrix} dATP \\ dCTP \\ dGTP \\ dTTP \end{bmatrix} \xrightleftharpoons{M^{++}} \text{double-stranded DNA} + nPP_i$$

The reaction catalyzed by mammalian low molecular weight DNA polymerase is irreversible. Two reactions are catalyzed by the reverse transcriptase:

and

$$\text{initiated single-stranded RNA} + n \begin{bmatrix} dATP \\ dCTP \\ dGTP \\ dTTP \end{bmatrix} \xrightleftharpoons{M^{++}} \text{DNA-RNA hybrid} + nPP_i$$

$$\text{initiated single-stranded DNA} + n \begin{bmatrix} dATP \\ dCTP \\ dGTP \\ dTTP \end{bmatrix} \xrightleftharpoons{M^{++}} \text{double-stranded DNA} + nPP_i$$

Due to the complex structures of natural DNA and RNA templates, model homopolymer templates are often used to study DNA polymerases and reverse transcriptases. A general structure of the template system for DNA polymerase can be written as, $\alpha(pX)_n \cdot d(pY)_m$ representing 1:1 complex of $\alpha(pX)_n$ and $\alpha(pY)_m$. In this complex, X represents the template nucleotide, Y represents the initiator nucleotide, $n \geq 100$ represents the chain length of the template, and $7 \leq m \leq 20$ represents the chain length of initiator. The base sequence of template chain $d(pX)_n$ directs the polymerization of nucleotide and the 3'-hydroxyl of the initiator chain $d(pY)_m$ provides the growing point for polymerization. A general structure of model template system for reverse transcriptase can be written as $r(pX)_n \cdot d(pY)_m$. Since some deoxynucleotide polymerizing enzymes use ribose initiator, such model systems can be written as $d(pX)_n \cdot r(pY)_m$. Although the model template systems are easily available and simple to use, nondiscriminating uses of these templates often lead to misleading results since some DNA polymerases can use some $r(pX)_n \cdot d(pY)_m$ templates under certain reaction conditions. The criterion used to distinguish a reverse transcriptase from DNA polymerase is the ability of replication of an initiated natural RNA template.

In addition to the polymerization reaction, some deoxynucleotide polymerizing enzymes catalyze the reverse reaction. The reverse reaction of polymerization is known as pyrophosphorolysis. A combination of polymerization and pyrophosphorolysis reactions results in pyrophosphate exchange reactions. In addition to polymerization and pyrophosphorolysis functions, procaryotic enzymes have associated nuclease activities. Thus far, all DNA polymerases isolated from procaryotic sources were found to have an associated 3'-exonuclease activity that degrades DNA from 3'-end and produces mononucleotides. A combination of polymerase and 3'-exonuclease activities results in a net degradation of deoxynucleoside triphosphate. Some procaryotic enzymes also have an associated 5'-exonuclease activity that degrades DNA from 5'-end and produces mononucleotides and oligonucleotides. The number and the kinds of enzymatic functions associated with purified DNA polymerases perhaps reflect the in vivo functions of these enzymes and differences in detailed mechanism of actions of these enzymes.

RIBONUCLEOTIDE POLYMERIZING ENZYMES

Michael I. Goldberg and William J. Rutter

The following table contains the common or literature name for each polymerizing enzyme, its molecular weight, number of subunits, and the molecular weight of each subunit. Molecular weights have either been taken directly from the text of the referenced article or calculated from an "S" value (assuming a globular molecule) or the molecular weights of the individual subunits. In some cases, there may be a discrepancy between the molecular weight of the oligomer and the sum of the subunit molecular weights.

The stoichiometry of individual subunits is based on the authors' best estimates. Usually the native enzymes have been sedimented in glycerol gradients or chromatographed and then sodium dodecyl sulfate electropherograms run of individual fractions. The subunits associated with the native enzyme have been found to maintain their stoichiometry in most of these experiments. However, there are some cases where the stoichiometry has been extrapolated from the molecular weight of the native enzyme and the number of subunits. With a few enzymes, only the molecular weight of the oligomers is known.

The enzymes described are considered to be representative of their class. No attempt has been made to list all the ribonucleotide polymerizing enzymes whose subunit structure is known. Other ribonucleotide polymerizing enzymes such as the postulated reticulocyte RNA-dependent RNA polymerase,[36] whose molecular weights have not been determined, are also not included. All the enzymes are proteins having subunits held together by noncovalent bonds.

RIBONUCLEOTIDE POLYMERIZING ENZYMES

Protein	Mol wt	Subunits		Reference
		Number	Mol wt	

Procaryotic Ribonucleotide Polymerizing Enzymes

Protein	Mol wt	Number	Mol wt	Reference
E. coli DNA-dependent RNA polymerase[a]	490,000	1	165,000	1, 2
		1	155,000	
		1	85,000	
		2	39,000	
E. coli DNA-dependent RNA polymerase	456,000	1	165,000	3
		1	155,000	
		1	56,000	
		2	39,000	
B. subtilis DNA-dependent RNA polymerase[a]	470,000	1	160,000	4, 5
		1	150,000	
		1	56,000	
		2	41,000	
		1	11,000	
		1	9,500	
H. cutirubrum DNA-dependent RNA polymerase	36,000	1	18,000	6
		1	18,000	
T7 DNA-dependent RNA polymerase	105,000	1	105,000	7
T3 DNA-dependent RNA polymerase	105,000	1	105,000	8
P4 polydC·polydG-dependent polyriboguanylate polymerase	~65,000	–	–	9
Qβ RNA-dependent RNA polymerase	215,000	1	70,000	10
		1	65,000	
		1	45,000	
		1	35,000	
f2 RNA-dependent RNA polymerase	213,000	1	70,000	11
		1	63,000	
		1	45,000	
		1	35,000	
H. cutirubrum RNA-dependent RNA polymerase	17,400	–	–	12

[a]The five proteins tabulated above have been shown by reconstitution experiments to be necessary for RNA polymerase activity.[37,38] Several other proteins are also known to associate with polymerase following either T4 infection of *E. coli* or SP infection of *B. subtilis*. These proteins appear subsequent to bacteriophage mediated alterations in the host polymerase subunits.[39-41]

Phage-induced subunits

Protein	Number	Mol wt	Reference
T4-infected *E. coli* DNA-dependent RNA polymerase	1	22,000	42–44
	1	15,000	
	1	12,000	
	1	12,000	
SPO1-infected *B. subtilis* DNA-dependent RNA polymerase	1	28,000	45
	1	13,000	
SP82 infected *B. subtilis* DNA-dependent RNA polymerase		16,000	46

RIBONUCLEOTIDE POLYMERIZING ENZYMES (continued)

Protein	Mol wt	Subunits		Reference
		Number	Mol wt	

Procaryotic Ribonucleotide Polymerizing Enzymes (continued)

Protein	Mol wt	Number	Mol wt	Reference
E. coli polynucleotide polymerase	190,000	6	32,000	13
M. luteus polynucleotide polymerase	260,000	4	67,000	14
H. cutirubrum polynucleotide polymerase	11,000–12,000	–	–	12
E. coli polyriboadenylate polymerase (α subunit of DNA-dependent RNA polymerase)	39,000	–	–	15
E. coli polyriboadenylate polymerase (also DNA-dependent RNA polymerase)	456,000	1	165,000	3
		1	155,000	
		1	56,000	
		2	39,000	
E. coli DNAG gene product: DNA-dependent RNA polymerase	64,000	1	64,000	54
E. coli DNA-dependent RNA polymerase III	~550,000	1	165,000	55
		1	155,000	
		1	90,000	
		1	<64,000	
		2	39,000	

Eucaryotic Ribonucleotide Polymerizing Enzymes

Protein	Mol wt	Number	Mol wt	Reference
Calf thymus DNA-dependent RNA polymerase I[b]	550,000	1	197,000	16
		1	126,000	
		1	51,000	
		1	44,000	
		2	25,000	
		2	16,000	
Mouse myeloma DNA-dependent RNA polymerase I	500,000	1	195,000	17
		1	117,000	
		1	60,000	
		1	50,000	
		1	27,000	
		1–2	16,500	
S. cerevisiae DNA-dependent RNA polymerase I		1	185,000	18, 53
		1	137,000	
		2	48,000	
		1	44,000	
		1	41,000	
		1–2	36,000	
		2	28,000	
		1	24,000	
		1	20,000	
		1	14,500	
		1	12,000	
Calf thymus DNA-dependent RNA polymerase IIA[c]	600,000	1	214,000	16
		1	140,000	
		1–2	34,000	
		2	25,000	
		3–4	16,500	

[b]The subunit structure of polymerase I has also been determined in *Physarum polycephalum*,[47] Krebs II ascites cells,[48] *Mucor rouxii*,[49] and Novikoff ascites cells.[50]
[c]The subunit structure of polymerase II has also been determined in rat liver,[51] KB cells,[52] and *Mucor rouxii*.[49]

RIBONUCLEOTIDE POLYMERIZING ENZYMES (continued)

Protein	Mol wt	Subunits		Reference
		Number	Mol wt	
Eucaryotic Ribonucleotide Polymerizing Enzymes (continued)				
Calf thymus DNA-dependent	570,000	1	180,000	16
RNA polymerase IIB		1	140,000	
		1–2	34,000	
		2	25,000	
		3–4	16,500	
Mouse myeloma DNA-dependent	500,000	1	205,000	19,20
RNA polymerase IIA		1	140,000	
		1	41,000	
		1	29,000	
		1	27,000	
		1	22,000	
		2	19,000	
		1	16,000	
Mouse myeloma DNA-dependent	500,000	1	170,000	19, 20
RNA polymerase IIB		1	140,000	
		1	41,000	
		1	29,000	
		1	27,000	
		1	22,000	
		2	19,000	
		1	16,000	
S. cerevisiae DNA-dependent	~550,000	1	173,000	21
RNA polymerase II		1	145,000	
		0.5	41,000	
		1	33,500	
		2	28,000	
		1–2	24,000	
		1	18,000	
		1	14,500	
		1	12,500	
Maize nuclear DNA-dependent	500,000	1	180,000	22
RNA polymerase IIA		1	160,000	
		1	35,000	
		1	25,000	
		1	20,000	
		1	17,000	
Mouse myeloma DNA-dependent	620,000	1	155,000	19
RNA polymerase IIA + IIB		1	138,000	
		1	89,000	
		1	70,000	
		1	52,000	
		1	43,000	
		1	34,000	
		1	29,000	
		1	19,000	
Wheat germ DNA-dependent		1	220,000	23
RNA polymerase II		1	140,000	
		1	40,000	
		1	25,000	
		1	20,000	
		2	17,000	
		2	13,000	

RIBONUCLEOTIDE POLYMERIZING ENZYMES (continued)

Protein	Mol wt	Subunits		Reference
		Number	Mol wt	
Eucaryotic Ribonucleotide Polymerizing Enzymes (continued)				
Xenopus laevis DNA-dependent RNA polymerase III	620,000	1	155,000	19
		1	137,000	
		1	92,000	
		1	68,000	
		1	52,000	
		1	42,000	
		1	33,000	
		1	29,000	
		1	19,000	
Mouse myeloma (MOPC$_{315}$) DNA-dependent RNA polymerase III	650,000	1	155,000	24
		1	138,000	
		1	89,000	
		1	70,000	
		1	53,000	
		1	49,000	
		1	41,000	
		1	32,000	
		1	29,000	
		1–3	19,000	
S. cerevisiae DNA-dependent RNA polymerase III	~650,000	1	160,000	25
		1	128,000	
		1	82,000	
		1	41,000	
		1	40,000	
		1	37,000	
		1	34,000	
		2	28,000	
		1	24,000	
		1	20,000	
		1	14,500	
		1	11,000	
Xenopus laevis mitochondrial DNA-dependent RNA polymerase	46,000	1	46,000	26
Neurospora crassa mitochondrial DNA-dependent RNA polymerase	64,000	1	64,000	27
Wheat leaf soluble DNA-dependent RNA polymerase	64,000	1	64,000	28
Maize chloroplast DNA-dependent RNA polymerase	500,000	1	180,000	29
		1	140,000	
		1	40,000	
Encephalomyocarditis virus RNA-dependent RNA polymerase	249,000	1	72,000	30
		1	65,000	
		1	57,000	
		1	45,000	
		1	35,000	
Vaccinia polyriboadenylate polymerase	80,000	1	51,000	31
		1	35,000	
Yeast polyriboadenylate polymerase I (RNA primer dependent)	100,000	1	–	32
Yeast polyriboadenylate polymerase II (polyA primer dependent)	100,000	–	–	32
Calf thymus polyriboadenylate polymerase (Mg^{++} dependent)	140,000–160,000	–	–	33
Calf thymus polyriboadenylate polymerase (Mn^{++} dependent)	62,000	1	62,000	34, 35

REFERENCES

1.   Burgess, *J. Biol. Chem.*, 244, 6168 (1969).
2.   Burgess, Travers, Dunn, and Bautz, *Nature*, 221, 43 (1969).
3.   Iwakura, Fukuda, and Ishihama, *J. Mol. Biol.*, 83, 369 (1974).
4.   Avila, Hermoso, Vinuela, and Salas, *Eur. J. Biochem.*, 21, 526 (1971).
5.   Duffy and Geiduschek, *J. Biol. Chem.*, 250, 4530 (1975).
6.   Louis and Fitt, *Biochem. J.*, 127, 69 (1972).
7.   Chamberlin, McGrath, and Waskell, *Nat. New Biol.*, 228, 227 (1970).
8.   Chakraborty, Sarkar, Huang, and Maitra, *J. Biol. Chem.*, 248, 6637 (1973).
9.   Barrett, Gibbs, and Calendar, *Proc. Natl. Acad. Sci. U.S.A.*, 69, 2986 (1972).
10.   Kondo, Gallerani, and Weissman, *Nature*, 228, 525 (1970).
11.   Federoff and Zinder, *Proc. Natl. Acad. Sci. U.S.A.*, 68, 1838 (1971).
12.   Louis, Peterkin, and Fitt, *Biochem. J.*, 121, 635 (1971).
13.   Ullman, Goldberg, Perrin, and Monod, *Biochemistry*, 7, 261 (1968).
14.   Klee, *J. Biol. Chem.*, 244, 2558 (1969).
15.   Ohasa and Tsugita, *Nat. New Biol.*, 240, 35 (1972).
16.   Kedinger, Gissinger, and Chambon, *Eur. J. Biochem.*, 44, 421 (1974).
17.   Schwartz and Roeder, *J. Biol. Chem.*, 249, 5898 (1974).
18.   Valenzuela, Weinberg, Bell, and Rutter, *J. Biol. Chem.*, 251, 1464 (1976).
19.   Sklar, Schwartz, and Roeder, *Proc. Natl. Acad. Sci. U.S.A.*, 72, 348 (1975)
20.   Schwartz and Roeder, *J. Biol. Chem.*, 250, 3221 (1975)
21.   Hager, Holland, and Rutter, in preparation.
22.   Mullinix, Strain, and Bogorad, *Proc. Natl. Acad. Sci. U.S.A.*, 70, 2386 (1973).
23.   Jendrisak and Burgess, personal communication.
24.   Sklar and Roeder, *J. Biol. Chem.*, in press, 1976.
25.   Valenzuela, Hager, Weinberg, and Rutter, *Proc. Natl. Acad. Sci. U.S.A.*, 73, 1024 (1976).
26.   Wu and Dawid, *Biochemistry*, 11, 3589 (1972).
27.   Kuntzel and Schafer, *Nat. New Biol.*, 231, 265 (1971).
28.   Polya, *Arch. Biochem. Biophys.*, 155, 125 (1973).
29.   Smith and Bogorad, *Proc. Natl. Acad. Sci. U.S.A.*, 71, 4839 (1974).
30.   Rosenberg, Diskin, Dron, and Traub, *Proc. Natl. Acad. Sci. U.S.A.*, 69, 3815 (1972).
31.   Moss, Rosenblum, and Gershowitz, *J. Biol. Chem.*, 250, 4722 (1975).
32.   Haff and Keller, *Biochem. Biophys. Res. Commun.*, 51, 704 (1973).
33.   Winters and Edmonds, *J. Biol. Chem.*, 248, 4756 (1973).
34.   Tsiapalis, Dorson, and Bollum, *J. Biol. Chem.*, 250, 4486 (1975).
35.   Tsiapalis, Dorson, DeSante, and Bollum, *Biochem. Biophys. Res. Commun.*, 50, 737 (1973).
36.   Downey, Byrnes, Jurmark, and So, *Proc. Natl. Acad. Sci. U.S.A.*, 70, 3400 (1973).
37.   Heil and Zillig, *FEBS Lett.*, 11, 165 (1972).
38.   Ishihama and Ito, *J. Mol. Biol.*, 72, 111 (1972).
39.   Goff and Weber, *Cold Spring Harbor Symp. Quant. Biol.*, 35, 101 (1972).
40.   Schachner and Zillig, *Eur. J. Biochem.*, 22, 531 (1971).
41.   Goff, *J. Biol. Chem.*, 249, 6181 (1974).
42.   Stevens, *Proc. Natl. Acad. Sci. U.S.A.*, 69, 603.
43.   Stevens, *Biochemistry*, 13, 493 (1974).
44.   Horvitz, *Nat. New Biol.*, 244, 137 (1973).
45.   Duffy and Geiduschek, *J. Biol. Chem.*, 250, 4530 (1975).
46.   Spiegelman and Whiteley, *J. Biol. Chem.*, 249, 1476 (1974).
47.   Gornicki, Vuturo, West, and Weaver, *J. Biol. Chem.*, 249, 1792 (1974).
48.   Goldberg, Perriard, and Rutter, in preparation.
49.   Young and Whiteley, *J. Biol. Chem.*, 250, 479 (1975).
50.   Froehner and Bonner, *Biochemistry*, 12, 3064 (1973).
51.   Weaver, Blatti, and Rutter, *Proc. Natl. Acad. Sci. U.S.A.*, 68, 2994 (1971).
52.   Sugden and Keller, *J. Biol. Chem.*, 248, 3777 (1973).
53.   Buhler, Sentenac, and Fromageot, *J. Biol. Chem.*, 249, 5963 (1974).
54.   Bouché, Zechel, and Kornberg, *J. Biol. Chem.*, 250, 5995 (1975).
55.   Wickner and Kornberg, *Proc. Natl. Acad. Sci. U.S.A.*, 71, 4425 (1974).

GEL ELECTROPHORESIS OF NUCLEIC ACIDS

Table 1
TWO-DIMENSIONAL GEL SYSTEMS FOR RNA SEPARATION

Main difference first/second dimension	First dimension			Second dimension			RNA mixture separated	Ref.
	Gel conc (%)	pH range	Urea conc (M)	Gel conc (%)	pH range	Urea conc (M)		
Urea shift	12.5	Neutral	0	12.5	Neutral	8	5S RNA fragments	1
	15—16	Neutral	6—7	16	Neutral	0	tRNAs	2.3
	6	Neutral	0	6	Neutral	5	mRNAs	4
Concentration shift	10	Neutral	0	20	Neutral	0	Small RNAs	5
	10.4	Neutral	4	20.8	Neutral	4	tRNAs and precursors	6
pH shift Urea shift Concentration shift	10.3	Acid	6	20.6	Neutral	0	Viral RNA fragments	8
	10.3	Acid	6	20.6	Neutral	0	Viral RNA oligonucleotides	9

Note: Gel concentration is expressed as % weight/volume including the crosslinker. The "neutral pH range" is between pH 4.5 and pH 8.5; acid range is below pH 4.5. Most neutral gels are run at pH 8.0 or 8.3, and the acid ones at pH 3.3. The term "RNA fragment" is used for products of partial RNase hydrolysis. Most oligonucleotide mixtures are products of complete T_1RNase hydrolysis. The references are limited to those introducing a new procedure or applying it to a new type of RNA mixture.

From De Wachter, R. and Fiers, W., in *Gel Electrophoresis of Nucleic Acids. A Practical Approach,* Rickwood, D. and Hames, B.D., Eds., IRL Press, Arlington, Va., 1982, 80. With permission.

Table 2
CONDITIONS FOR THE ELECTROPHORESIS OF OLIGONUCLEOTIDES: COMPARISON OF THE ORIGINAL PROCEDURE AND SOME MODIFICATIONS

Procedure: authors and reference	Dimension	Gel concentration (% w/v)[a]		Buffer	Gel size — height: width: thickness (mm)	Electrophoresis conditions		
		Acrylamide	Bisacrylamide			Voltage (V)	Time (h)	Marker distance[b] (cm)
1. De Watcher and Fiers (8)	First	10.0	0.325	0.025 M citric acid, 6 M urea	400:200:2	900	6	20
	Second	20.0	0.65	0.04 M Tris-citrate (pH 8.0)	300:250:2	350	15	13.5
2. Colfin and Billeter (17)	First	10.0	0.33	0.025 M citric acid, 6 M urea	360:170:2	500	20	21
	Second	21.8	0.65	0.04 M Tris-citrate (pH 8.0)	250:215:2	350	16	15
3. Frisby et al. (15)	First	10.0	0.35	0.025 M citric acid, 6 M urea	400:200:?	400		19
	Second	20.0	0.66	0.1 M Tris-borate, 0.0025 M EDTA (pH 8.3)	400:400:?	400		19
4. Kennedy (19)	First	10.0	0.325	0.025 M citric acid, 6 M urea	380:170:3	500		21
	Second	20.0	0.65	0.2 M Tris-borate, 0.005 M EDTA (pH 8.3)	380:380:3	400		18.5
5. Pedersen and Haseltine (14)	First	10.0	0.3	0.025 M citric acid, 6 M urea[c]	400:200:0.75	1700	2	18
	Second	22.8	0.8	0.05 M Tris-borate (pH 8.3)	400:330:0.38	750	9	25

[a] Amounts of catalyst used are nearly the same in all procedures.

[b] Distance moved by the bromophenol blue marker when the electrophoresis is stopped.

[c] In this procedure, only the gel contains urea. The buffer compartments contain 0.025 M citric acid.

From De Wachter, R. and Fiers, W., in *Gel Electrophoresis of Nucleic Acids. A Practical Approach*, Rickwood, D. and Hames, B. D., Eds., IRL Press, Arlington, Va., 1982, 94. With permission.

Table 3
CONDITIONS FOR THE ELECTROPHORESIS OF RNA FRAGMENTS

Procedure: authors; reference: fragment length	Dimension	Gel concentration (% w/v)		Buffer	Gel size — height: width: thickness (mm)	Electrophoresis conditions		
		Acrylamide	Bisacrylamide			Voltage (V)	Time (h)	Marker distance[a] (cm)
1. Vigne and Jordan; (1) 5S RNA fragments 13≤N≤79	First	12.1	0.4	0.04 M Tris-acetate (pH 8.4)	180:130:3	300	3—4	
	Second	12.1	0.4	0.04 M Tris-acetate, 8 M urea (pH 8.4)	400:200:4	400	14—18	
2. De Wachter and Fiers; (8) phage RNA fragments 10≤N≤80	First	10.0	0.325	0.025 M citric acid, 6 M urea	400:200:2	900	6	20
	Second	20.0	0.65	0.04 M Tris-acetate (pH 8.0)	300:250:2	350	15	13.5
3. De Wachter and Fiers; (8) phage RNA fragments $N>80$	First	8.0	0.26	0.025 M citric acid, 6 M urea	400:200:2	900	4	19
	Second	16.0	0.52	0.04 M Tris-citrate (pH 8.0)	300:250:2	350	15	20

Note: In the first procedure the gel-buffer mixture for both dimensions is polymerized by addition of 0.1 ml TEMED and 40 mg $(NH_4)_2S_2O_8$/100 ml of solution. The second and third procedures are described in the original text. N designates the chain length of separated RNA fragments.

[a] Distance moved by the bromophenol blue when electrophoresis is stopped.

From De Wachter, R. and Fiers, W., in *Gel Electrophoresis of Nucleic Acids. A Practical Approach*, Rickwood, D. and Hames, B. D., Eds., IRL Press, Arlington, Va., 1982, 99. With permission.

Table 4
CONDITIONS FOR THE ELECTROPHORESIS OF SMALL RNA MOLECULES

Procedure: authors (reference)	Dimension	Gel concentration (% w/v)		Buffer	Gel size — height: width: thickness (mm)	Electrophoresis conditions	
		Acrylamide	Bisacrylamide			Voltage (V)	Time (h)
1. Ikemura and Dahlberg (5)[a]	First	9.5	0.5	0.089 *M* Tris base 0.0028 *M* EDTA 0.089 *M* H$_3$BO$_3$ (pH 8.3)	170:130:4	340	3
	Second	19.0	1.0	0.089 *M* Tris base 0.0028 *M* EDTA 0.089 *M* H$_3$BO$_3$ (pH 8.3)	170:130:4	340	17
2. Varricchio and Ernst (3)[b]	First	15.2	0.8	0.089 *M* Tris base 0.0028 *M* EDTA 0.089 *M* H$_3$BO$_3$ 7 *M* urea	150:φ = 6[6]	100	16
	Second	15.2	0.8	0.089 *M* Tris base 0.0028 *M* EDTA 0.089 *M* H$_3$BO$_3$ (pH 8.3)	180:150:1.6	225	23

[a] Both the first- and second-dimensional gel are polymerized by addition of 0.4 ml diaminopropionitrile (DMAPN) and 40 mg (NH$_4$)$_2$S$_2$O$_8$/100 ml of acrylamide-buffer solution. Both runs are carried out using a commercial apparatus from E.C. Corporation.

[b] The first dimension is run in tubes, the gel cylinders being 15 cm long and 6 mm in diameter. The second-dimensional slab gel is run in a "Buchler" vertical starch gel electrophoresis apparatus. The catalyst for polymerization consists of 0.42 ml DMAPN and 56 mg (NH$_4$)$_2$S$_2$O$_8$/100 ml of gel. In both dimensions, a layer of approximately 1.5 cm of "stacking gel" is cast on top of the resolving gel. It contains 4.75 g acrylamide and 0.25 g bisacrylamide/100 ml, and 0.05 *M* tris-HCl (pH 3.6). Immediately before use, 0.25 ml of 0.02% riboflavin is added to 5 ml of this solution, and the gels are polymerized by exposure to a fluorescent light source for 2 h. After the first dimension, the gel cylinder is sliced lengthwise, and a 1.5 mm-thick slice is pushed between the glass plates containing the prepared second-dimensional slab so that it lies horizontally on top of the stacking gel layer.

[c] Gel cyclinder of 15 cm length and 0.6 cm diameter.

From De Wachter, R. and Fiers, W., in *Gel Electrophoresis of Nucleic Acids. A Practical Approach,* Rickwood, D. and Hames, B. D., Eds., IRL Press, Arlington, Va., 1982, 99. With permission.

Table 5
CONDITIONS FOR THE ELECTROPHORESIS OF mRNA[4]

| Dimension | Gel concentration (% w/v) | | Catalyst (amount/100 ml) | Buffer | Gel size — height: width: thickness (mm) | Electrophoresis conditions | |
	Acrylamide	Bisacrylamide				Voltage (V)	Time (h)
First	6.0	0.16	$(NH_4)_2S_2O_8$, 200 mg; TEMED, 200 μl	0.04 M Tris base 0.02 M sodium acetate 0.001 M EDTA Acetic acid to pH 7.2	90:φ = 2.2[a]	60	4.5—9
Second	6.0	0.16	$(NH_4)_2S_2O_8$, 200 mg: TEMED, 200 μl	0.04 M Tris base 0.02 M sodium acetate 0.001 M EDTA Acetic acid to pH 7.2 5 M urea	90:170:2	60	4.5—9

Note: The first dimension is run in a cylindrical gel using the apparatus described by Davis.[11] The 9 cm long resolving gel is surmounted by a 1 cm long 3% "stacking gel" (half the amount of catalyst is used in casting the latter gel). After electrophoresis, the gel cylinder is immersed for 1 h in a solution containing half the buffer concentration used for electrophoresis, plus 5 M urea. The cylinder is then aligned horizontally near the top side between the glass plates assembled for casting the second-dimensional gel slab. The gel solution is poured between the plates to a height of 5 mm below the cylinder and overlayered with water during polymerization to obtain a flat surface. The space remaining between the slab and the cylinder is filled with 1% agarose solution containing half the buffer concentration used for electrophoresis, plus 0.2% SDS and 5 M urea. This solution is allowed to set at 4°C for 1 h. Temperature control (in the range 21°C to 35°C) is achieved by carrying out the electrophoresis in an incubator after 2 h of equilibration.

[a] Cylinder height and diameter.

From De Wachter, R. and Fiers, W., in *Gel Electrophoresis of Nucleic Acids. A Practical Approach*, Rickwood, D. and Hames, B. D., Eds., IRL Press, Arlington, Va., 1982, 104. With permission.

Table 6
CONDITIONS FOR THE ELECTROPHORESIS OF DNA RESTRICTION FRAGMENTS ON CONVENTIONAL GELS[27]

Dimension	Gel concentration[a,b] (% w/v)			Catalyst (amount/100 ml)	Buffer	Gel size — height: width: thickness (mm)	Electrophoresis conditions		
	Acrylamide	Bisacrylamide	Agarose				Voltage (V)	Time (h)	Temp (°C)
First	8.0	—	1	$(NH_4)_2S_2O_8$, 20 mg; TEMED, 40 µl	0.036 M Tris base 0.03 M NaH$_2$PO$_4$ 0.001 M EDTA NaOH to pH 7.7	400:200:2	370	16—20	4[c]
	12.0	—	1						
	15.0	—	1						
Second	4.0	0.2	—	$(NH_4)_2S_2O_8$, 50 mg; TEMED, 100 µl	0.089 M Tris base 0.089 M H$_3$BO$_3$ 0.0025 M EDTA (pH 8.3)	300:250:2	100	17—22	50
	5.0	0.25	—						
	6.0	0.3	—						

Note: The gels can be made up from the following stock solutions of gelling agent, catalysts, and buffer: acrylamide, 400 g/l; acrylamide, 400 g/l; bisacrylamide, 2 g/l; $(NH_4)_2S_2O_8$, 5 mg/ml; buffer for first dimension: 10 times the final concentration; buffer for second dimension: twice the final concentration.

[a] The concentration stages indicated for both dimensions are only examples; the concentration can vary from 8 to 2% in the first dimension and from 4 to 10% in the second dimension, according to the molecular weight of the fragments to be separated.

[b] Only the composition of the resolving gels is given. Consult the text for the composition of the "bottom gel" (first dimension) and "stacking gel" (second dimension).

[c] The electrophoresis is run in the coldroom at this temperature. The temperature in the gel rises due to electrical heating and may actually be closer to 20°C since there is no vigorous cooling by circulating liquid, as is the case for the second dimension.

From De Wachter, R. and Fiers, W., in *Gel Electrophoresis of Nucleic Acids. A Practical Approach*, Rickwood, D. and Hames, B. D., Eds., IRL Press, Arlington, Va., 1982, 107. With permission.

Table 7

CONDITIONS FOR THE ELECTROPHORESIS OF DNA RESTRICTION FRAGMENTS ON DENATURING GRADIENT GELS[28]

| Dimension | Gel concentration (% w/v) | | Agarose | Catalyst (amount/ml) | Buffer | Gel size — height: width: thickness (nn) | Electrophoresis conditions | | |
	Acrylamide	Bisacrylamide					Voltage (V)	Time (h)	Temp (°C)
First	—	—	1	—	0.04 M Tris base 0.02 M sodium acetate 0.001 M EDTA Acetic acid to pH 8.0	177:177:2.5	34	20	20
Second	4.0	0.1067	—	$(NH_4)_2S_2O_8$, 100 mg; TEMED, 10µL	0.04 M Tris base 0.02 M sodium acetate 0.001 M EDTA Acetic acid to pH 8.0 Plus gradient: 0—7 M urea 0—40% (v/v) Formamide	177:177:2.5	150	20—30	60

Note: The gels can be made up from the following stock solutions of gelling agent, catalyst, and buffer: acrylamide, 300 g/l, bisacrylamide, 8 g/l; $(NH_4)_2S_2O_8$, 200 mg/ml; buffer at 20 times the final concentration. The gel plug, L, cast in the trough, B, contains 9% acrylamide, 0.24% bisacrylamide, and the same buffer as the first-dimensional gel.

From De Wachter, R. and Fiers, W. in *Gel Electrophoresis of Nucleic Acids. A Practical Approach*, Rickwood, D. and Hames, B. D., Eds., IRL Press, Arlington, Va., 1982, 113. With permission.

REFERENCES TO TABLES

1. **Vigne, R. and Jordan, B. R.,** *Biochemie,* 53, 981, 1971.
2. **Stein, M. and Varricchio, F.,** *Anal. Biochem.,*61, 112, 1974.
3. **Varricchio, F. and Ernst, H. J.,** *Anal. Biochem.,* 68, 485, 1975.
4. **Burckhardt, J. and Birnstiel, M. L.,** *J. Mol. Biol.,* 118, 61, 1978.
5. **Ikemura, T. and Dahlberg, J. E.,** *J. Biol. Chem.,* 248, 5024, 1973.
6. **Fradin, A., Gruhl, H., and Feldman, H.,***FEBS Lett.,* 50, 185, 1975.
7. **De Wachter, R., Merregaert, J., Vandenberghe, A., Contreras, R., and Fiers, W.,** *Eur. J. Biochem.,* 22, 400, 1971.
8. **De Wachter, R. and Fiers, W.,** *Anal. Biochem.,* 49, 184, 1972.
9. **Billeter, M. A., Parsons, J. T., and Coffin, J. M.,** *Proc. Natl. Acad. Sci. U.S.A.,* 71, 3560, 1974.
10. **Studier, W.,** *J. Mol. Biol.,* 79, 237, 1973.
11. **Davis, B. J.,** *Ann. N.Y. Acad. Sci.,* 121, 404, 1964.
12. **Laskey, R. A. and Mills, A. D.,** *Eur. J. Biochem.,* 56, 335, 1975.
13. **Hassur, R. A. and Whitlock, H. W.,** *Anal. Biochem.,* 59, 162, 1974.
14. **Pederson, F. S. and Haseltine, W. A.,** in *Methods in Enzymology,* Vol. 65, Grossman, L. and Moldave, K., Eds.; Academic Press, Inc., New York, 1980, 680.
15. **Sanger, F., Brownlee, G. G., and Barrell, B. G.,** *J. Mol. Biol.,* 13, 373, 1965.
16. **Brownlee, G. G. and Sanger, F.,** *Eur. J. Biochem.,* 11, 395, 1969.
17. **Coffin, J. M. and Billeter, M. A.,** *J. Mol. Biol.,* 100, 293, 1976.
18. **Frisby, D. P., Newton, C., Carey, N. H., Fellner, P., Newman, J. F. E., Harris, T. J. R., and Brown, F.,** *Virology,* 71, 379, 1976.
19. **Kennedy, S. I. T.,** *J. Mol. Biol.,* 108, 491, 1976.
20. **Frisby, D. P.,** *Nucleic Acids Res.,* 4, 2975, 1977.
21. **Clewly, J., Gentsch, J., and Bishop, D. H. L.,** *J. Virology,* 22, 459, 1977.
22. **Rommelaere, J., Faller, D. V., and Hopkins, N.,** *Proc. Natl. Acad. Sci. U.S.A.* 75, 495, 1978.
23. **Young, J. F., Desselberger, U., and Palese, P.,** *Cell,* 18, 73, 1979.
24. **Reddy, R., Sitz, T. O., Ro-Choi, T. S., and Busch, H.,** *Biochem. Bophys. Res. Comm.,* 56, 1017, 1974.
25. **Ikemura, T. and Ozeki, H.,** *Eur. J. Biochem.,* 51, 117, 1975.
26. **Peters, G., Harada, F., Dahlberg, J. E., Panet, A., Haseltine, W. A., and Baltimore, D.,** *J. Virol.,* 21, 1031, 1977.
27. **Derynck, R. and Fiers, W.,** *J. Mol. Biol.,* 110, 387, 1977.
28. **Fischer, S. G. and Lerman, L. S.,** *Cell,* 16, 191, 1979.
29. **Fischer, S. G. and Lerman, L. S.,** in *Methods in Enzymology,* Vol. 68, Wu, R., Ed., Academic Press, New York, 1980, 183.
30. **Chamberlain, J. P.,** *Anal. Biochem.,* 98, 132, 1979.

ELECTROPHORESIS OF NUCLEOPROTEINS

RECIPES OF GEL MIXTURES USED FOR SEPARATING POLYSOMES AND RIBOSOMES

Gel (%)	Buffer	20% Acrylamide-bis-acrylamide soln	Agarose (g)	Water (ml)	6.4% DMAPN[b] (ml)	Buffer stocks				1.6% Ammonium persulfate (ml)	Time of electrophoresis	Voltage (V)	Temp. (°C)
						10 × Tris-borate-EDTA (ml)	1 M Tris-HCl (ml)	3 M KCl (ml)	1 M MgCl$_2$ (ml)				
2.25/0.5	25 mM Tris-HCl 60 mM KCl 10 mM MgCl$_2$ (pH 7.6)	18	0.8	118	10	—	4	3.2	1.6	5	6 h (buffer change every 2 h)	120	4
2.5/0.5	25 mM Tris-acetate 60 mM potassium acetate 10 mM magnesium acetate (pH 7.6)	20	0.8	116	10	—	4[a]	3.2[a]	1.6[a]	5	6 h (buffer change every 2 h)	120	10
2.25/0.5	25 mM Tris-HCl 6 mM KCl 2 mM MgCl$_2$ (pH 7.6)	18	0.8	122	10	—	4	0.32	0.32	5	4 h (buffer change every 2 h)	200	4
2.75/0.5	25 mM Tris-HCl 0.2 mM MgCl$_2$ (pH 7.6)	22	0.8	119	10	—	4	—	0.032	5	6 h (buffer change every 2 h)	200	4
3/0.5	Tris-borate-EDTA (pH 8.3)	24	0.8	105	10	16	—	—	—	5	6 h (no buffer change)	200	4

Note: First dimension as in original text. Second dimension: 2h, 200V, 3°C in 1 × Tris-borate-EDTA buffer.

[a] In this case replace chloride with acetate.
[b] 3-Dimethylaminopropionitrile.

REFERENCES

1. **Kornberg, R. D.,** *Annu. Rev. Biochem.,* 46, 931, 1977.
2. **Varshavsky, A. J., Bakayeveva, J. G., and Georgiev, G. P.,** *Nucleic Acids Res.,* 3, 477, 1976.
3. **Mathew, C. G. P., Goodwin, G. H., and Johns, E. W.,** *Nucleic Acids Res.,* 6, 167, 1979.
4. **Goodwin, G. H., Woodhead, L., and Johns, E. W.,** *FEBS Lett.,* 73, 85, 1977.
5. **Johns, E. W.,** *J. Chromatog.,* 42, 152, 1969.
6. **Panyim, S. and Chalkley, R.,** *Arch. Biochem. Biophys.,* 130, 337, 1969.
7. **Weintraub, H., Palter, K., and Van Lente, F.,** *Cell,* 6, 85, 1975.
8. **Maniatis, T., Jeffrey, A., and van de Sande, H.,** *Biochemistry,* 14, 3787, 1975.
9. **Bakayev, V. V., Bakayeva, T. G., and Varshavsky, A. J.,** *Cell,* 11, 619, 1977.
10. **Bakayev, V. V., Bakayeva, T. G., Sehmatchenko, V. V., and Georgiev, G. P.,** *Eur. J. Biochem.,* 91, 291, 1978.
11. **Todd, R. D. and Garrard, W. T.,** *J. Biol. Chem.,* 254, 3074, 1979.
12. **Studier, F. W.,** *J. Mol. Biol.,* 79, 237, 1973.
13. **O'Farrell, P. and O'Farrell, P.,** in *Methods in Cell Biology,* Vol. 16, Stein, J. and Stein, G. S., Eds., Academic Press, New York, 1977, 407.
14. **Goodwin, G. H., Wright, C. A., and Johns, E. W.,** *Nucleic Acids Res.,* 9, 2761, 1977.
15. **Shaw, B. R. and Richards, R. G.,** in *Chromatin Structure and Function, Part A,* Nicolini, C. A., Ed., Plenum, New York, 125, 1979.
16. **Spiker, S.,** *Ann. Biochem.,* 108, 263, 1980.
17. **Todd, R. D. and Garrad, R. D.,** *J. Biol. Chem.,* 252, 4729, 1977.
18. **Weisbrod, S. and Weintraub, H.,** *Proc. Natl. Acad. Sci. U.S.A.,* 76, 630, 1979.
19. **Bloom, K. S. and Anmderson, J. H.,** *Cell,* 15, 141, 1978.
20. **Peacock, A. C. and Dingman, C. W.,** *Biochemistry,* 7, 668, 1968.
21. **Lindahl, L. and Forchhammer, J.,** *J. Mol. Biol.,* 43, 593, 1969.
22. **Hobden, A. N. and Cundliffe, E.,** *Biophys. J.,* 190, 765, 1980
23. **Dahlberg, A. E., Dingman, C. W., and Peacock, A. C.,** *J. Mol. Biol.,* 41, 139, 1969.
24. **Talens, A., Van Diggelen, O. P., Brongers, M., Popa, L. M., and Bosch, L.,** *Eur. J. Biochem.,* 37, 121, 1973.
25. **Dahlberg, A. E., Lund, E., and Kjeldgaard, N. O.,** *J. Mol. Biol.,* 78, 627, 1973.
26. **Helser, L., Baan, R. A., and Dahlberg, A. E.,** *Mol. Cell. Biol.* 1, 51, 1981.
27. **Dahlberg, A. E.,** *J. Biol. Chem.,* 249, 7673, 1974.
28. **Szer, W. and Leffler, S.,** *Proc. Natl. Acad. Sci. U.S.A.* 71, 3611, 1974.
29. **Tokimatsu, H., Strycharz, W., and Dahlberg, A. E.,** *J. Mol. Biol.,* 152, 397, 1981.
30. **Bowman, C. M., Dahlberg, J. E., Ikemura, T. Konisky, J., and Nomura, M.,** *Proc. Natl. Acad. Sci. U.S.A.,* 68, 964, 1971.
31. **Senior, B. W. and Holland, I. B.,** *Proc. Natl. Acad. Sci. U.S.A.* 68, 959, 1971.
32. **Dahlberg, A. E., Dahlberg, J. E., Lund, E., Tokimatsu, H., Rabson, A. B., Calvert, P. C., Reynolds, F., and Zahalak, M.,** *Proc. Natl. Acad. Sci. U.S.A.* 75, 3598, 1978.
33. **Dahlberg, A. E. and Peacock, A. C.,** *J. Mol. Biol.,* 55, 61, 1971.
34. **Dahlberg, A. E. and Peacock, A. C.,** *J. Mol. Biol.,* 60, 409, 1971.

From Goodwin, G. H. and Dahlberg, A. F., in *Gel Electrophoresis of Nucleic Acids, A Practical Approach,* Rickwood, D. and Hames, B. D., Eds., IRL Press, Arlington, Va., 1982, 199. With permission.

NUCLEIC ACID MOLECULAR WEIGHT MARKERS

DNA SIZE MARKERS

pBR322

The plasmid pBR322 is an extremely convenient plasmid for use as a molecular weight marker. By using a relatively small range of restriction endonucleases, a wide range of molecular weight markers can be generated. The linear 4362 bp molecule, useful as a marker for agarose gels, can be obtained using *Bam* HI, *Eco* RI, *Pst* I, *Ava* I, *Hind* III, and *Pvu* II, each of which cleaves at a single site in the plasmid. Other enzymes which have multiple restriction sites, together with the sizes of the fragments obtained, are listed in Table I.

Table 1
SIZES OF THE RESTRICTION FRAGMENTS OF pBR322[a,b]

Acc I	Rsa I	Hae III	Hpa II	Alu I	Hinf I	Taq I	Tha I (Fnu DII)	Hha I	Hae II	Mbo I	Mnl I
2767	2117	587	622	910	1631	1444	581	393	1876	1374	611
1595	1565	540	527	659	517	1307	493	347	622	665	400
	680	504	404	655	506	475	452	337	439	358	314
		458	309	521	396	368	372	332	430	341	267
		434	242	403	344	315	355	270	370	317	262
		267	238	281	298	312	341	259	227	272	218
		234	217	257	221	141	332	206	181	258	206
		213	201	226	220		330	190	83	207	204
		192	190	136	154		145	174	60	105	199
		184	180	100	75		129	153	53	91	186
		124	160	63			129	152	21	78	186
		123	160	57			122	151		75	136
		104	147	49			115	141		46	130
		89	147	19			104	132		36	116
		80	122	15			97	131		31	111
		64	110	11			68	109		27	81
		57	90				66	104		18	68
		51	76				61	100		17	61
		21	67				27	93		15	60
		18	34				26	83		12	58
		11	34				10	75		11	57
		7	26				5	67		8	56
			26				2	62			38
			15					60			30
			9					53			27
			9					40			
								36			
								33			
								30			
								28			
								21			

[a] Data from Reference 1.
[b] These sizes (in base pairs) do not include any single-stranded extensions which may be left by the restriction nuclease.

From Minter, S. and Sealey, P., in *Gel Electrophoresis of Nucleic Acids. A Practical Approach,* Rickwood, D. and Hames, B. B., Eds., IRL Press, Arlington, Va., 1982, 227. With permission.

SV40

The SV40 genome is a double-stranded circular DNA 5241 bp long. It can be linearized using *Msp* I (*Hpa* II), *Taq* I, *Eco* RI, *Bam* HI, *Kpn* I, *Hae* II, or *Bcl* I.

Table 2
SIZES OF RESTRICTION FRAGMENTS
FROM SV40[a,b]

Hind III	*Hinf* I	*Mbo* II	*Hae* III	*Alu* I	
1768	1845	1350	1661	775	53
1169	1085	756	765	483	50
1116	766	687	540	329	49
526	543	645	373	286	46
447	525	409	329	275	41
215	237	395	323	253	38
	109	381	300	253	30
	83	375	299	243	29
	24	69	227	224	28
	24	65	179	223	27
		31	49	177	12
		30	45	157	10
		14	41	154	8
		13	33	153	7
		11	30	146	
		10	29	144	
			14	123	
			9	75	
			6	54	

[a] There are no cleavage sites for *Fnu* DII, *Pst* I, or *Ava* I in SV40. *Mnl* I gives 51 fragments.
[b] From sequence data reported in References 2 and 3.

From Minter, S. and Sealey, P., in *Gel Electrophoresis of Nucleic Acids. A Practical Approach,* Rickwood, D. and Hames, B. B., Eds., IRL Press, Arlington, Va., 1982, 228. With permission.

Bacteriophage Lambda (λ)

Bacteriophage λ DNA is double-stranded; strain C 72 is 49,502 bp long, whereas strain cI *ts* 857 is 48,540 bp long. It is perhaps the most extensively studied bacteriophage of *E. coli.*

Table 3 lists restriction endonuclease fragments of λ cI *ts* 857, whereas Table 4 lists similar data for λ C 72 DNA.

Table 3
SIZES OF RESTRICTION FRAGMENTS OF PHAGE λ cI *ts* 857[a]

Eco RI	Hind III	Sal I	Bam HI	Kpn I	Hpa I	Pvu I	Pvu II	Xho I	Sma I	Ava I	Avr II	Bgl II	Xba I	Sst I	Sst II
21240	23150	32760	16840	29960	8680	14310	21110	33520	19420	14670	24340	22010	24530	24790	20340
7420	9420	15270	7230	17070	6910	12730	4775	15020	12220	8630	24120	13290	24010	22550	18780
5810	6560	500	6785	1500	5410	11960	4420		8630	6890	70	9700		1190	8130
5650	4380		6530		4540	9540	4270		8270	4740		2390			1080
4880	2320		5620		4490		3980			4720		650			210
3540	2020		5530		4350		3910			3730		435			
	560				3400		2300			1880		60			
	125				3380		1710			1670					
					3040		640			1600					
					2240		530								
					750		470								
					440		230								
					410		140								
					250		60								
					230										

[a]   Fragment sizes calculated by R. Hayward from sequence data provided by F. Sanger.

From Minter, S. and Sealey, P., in *Gel Electrophoresis of Nucleic Acids. A Practical Approach*, Rickwood, D. and Hames, B. B., Eds., IRL Press, Arlington, Va., 1982, 228. With permission.

Table 4
SIZES OF RESTRICTION FRAGMENTS OF PHAGE
λC 72[a]

Hind III	Bam HI	Bgl I	Eco RI	Ava I	Kpn I	Pvu I
28329	17177	22308	25538	14898	47981	25312
9593	12782	13340	7536	13964	1521	14554
6764	7346	10707	5902	6909		9636
2286	6556	2441	5681	4781		
1945	5641	651	4845	3712		
581		60		1930		
				1671		
				1637		

[a] Data from Reference 4.

From Minter, S. and Sealey, P., in *Gel Electrophoresis of Nucleic Acids. A Practical Approach,* Rickwood, D. and Hames, B. B., Eds., IRL Press, Arlington, Va., 1982, 228. With permission.

φX174

This is a bacteriophage of *E. coli* with a single-stranded DNA molecule. The RF form is double-stranded and 5386 bp long. It can be linearized using *Ava* II, *Sac* II, *Pst* I, or *Mst* II. The restriction fragment sizes are given in Table 5.

Table 5
SIZES OF RESTRICTION FRAGMENTS OF φX174[a,b]

Eco RII	Acc I	Hpa II	Hae II	Taq I	Hae III	Rsa I	Mbo II	Fnu DII	Mnl I
2767	3034	2748	2314	2914	1353	1560	1130	1553	1050
2619	2352	1697	1565	1175	1078	964	1064	640	870
		374	786	404	878	645	857	614	718
		348	269	327	605	525	812	532	695
		219	185	231	310	472	396	305	530
			125	141	281	392	394	300	496
			93	87	271	247	324	269	259
			54	54	232	197	224	201	176
				33	194	157	118	192	156
				20	118	138	89	145	127
					72	89	3	143	114
								123	103
								101	79
								93	19
								84	
								54	
								35	
								2	

[a] *Hinf* I gives 21 fragments, *Alu* I gives 24 fragments. There are no cleavage sites for *Bam* HI, *Eco* RI, *Ava* I, *Pvu* II, or *Hind* III.

[a] Data from Reference 5.

From Minter, S. and Sealey, P., in *Gel Electrophoresis of Nucleic Acids. A Practical Approach,* Rickwood, D. and Hames, B. B., Eds., IRL Press, Arlington, Va., 1982, 228. With permission.

fd DNA

This is a male-specific coliphage closely related to the phage M13. The virion is a single-stranded circular DNA molecule 6408 nucleotides long. The fd DNA can be linearized using either *Acc* I or *Hinc* II. The restriction fragment sizes are given in Table 6.

Table 6
SIZES OF RESTRICTION FRAGMENTS FOR fd DNA[a,b]

Bam HI	*Hae* II	*Mbo* II	*Hae* III	*Taq* I	*Msp* I (*Hpa* II)	*Alu* I
3425	3550	4349	2528	2019	1596	1446
2983	2033	666	1633	850	829	1330
	817	384	849	703	819	705
	8	332	352	652	652	554
		318	311	579	648	484
		196	309	441	501	366
		163	154	381	454	314
			106	357	381	257
			103	287	156	220
			69	139	129	204
					123	201
					60	166
					42	111
					12	29
					6	27
						24

[a] Data from reference 6.

[a] *Hinf* I gives 24 fragments and *Mnl* I gives 51 fragments. There are no cut sites for *Hind* III, *Bcl* I, *Eco* RI or *Kpn* I.

From Minter, S. and Sealey, P., in *Gel Electrophoresis of Nucleic Acids. A Practical Approach,* Rickwood, D. and Hames, B. B., Eds., IRL Press, Arlington, Va., 1982, 228. With permission.

RNA Size Markers

RNA can be electrophoresed on nondenaturing agarose or polyacrylamide gels, or on denaturing gels, either agarose containing methyl mercuric hydroxide or polyacrylamide gels containing 98% formamide or 8 M urea. Hence, the apparent molecular weight will depend upon the conditions of electrophoresis. In addition to the RNA molecular weight markers listed in Table 7, restriction endonuclease fragments of DNA (see Tables 1 to 6) may also be useful as size markers in certain situations.

Table 7
MOLECULAR WEIGHT MARKERS FOR GEL
ELECTROPHORESIS OF RNA

RNA species	Molecular weight[a]	Number of nucleotides	Ref.
Fibroin mRNA (silkworm)	57×10^6	19000	10[b]
Myosin heavy chain mRNA (chicken)	2.02×10^6	6500	11[c]
28S rRNA (Hela)[c]	1.9×10^6	6333	7
25S rRNA (*Aspergillus*)[c]	1.24×10^6	4000	18
23S rRNA (*E. coli*)[c]	1.07×10^6	3566	8
18S rRNA (Hela)[c]	0.71×10^6	2366	7
17S RNA (*Aspergillus*)[c]	0.62×10^6	2000	18
16S rRNA (*E. coli*)[c]	0.53×10^6	1776	9
A2 crystallin (calf lens)	0.45×10^6	1460	13[c]
Immunoglobulin light chain (mouse)	0.39×10^6	1250	14[c]
β Globin mRNA (mouse)	0.24×10^6	783	15
β Globin mRNA (rabbit)	0.22×10^6	710	16[c]
α Globin mRNA (mouse)	0.22×10^6	696	15
α Globin mRNA (rabbit)	0.20×10^6	630	16[c]
Histone H4 (sea urchin)	0.13×10^6	410	17[b]
5.8S RNA (*Aspergillus*)	4.89×10^4	158	18
5S (*E. coli*)	3.72×10^4	120	19
4S *Aspergillus*	2.63×10^4	85	18

[a] Molecular weights are approximate only and are based upon an average "molecular weight" of 310 for each nucleotide.
[c] Nondenaturing gel system.
[c] Denaturing formamide gel system.
[d] A more extensive list of sizes of rRNAs is given in Reference 20.

From Minter, S. and Sealey, P., in *Gel Electrophoresis of Nucleic Acids. A Practical Approach,* Rickwood, D. and Hames, B. B., Eds., IRL Press, Arlington, Va., 1982, 228. With permission.

REFERENCES TO TABLES

1. Sutcliffe, G., *Cold Spring Harbor Symp. Quant. Biol.*, 43, 77, 1979.
2. Fiers, W., Contreras, R., Haegeman, G., Rogiers, R., Van De Voorde, A., Van Heuverswyn, H., Van Herreweghe, J., Volckaert, G., and Ysebaert, M., *Nature*, 273, 113, 1978.
3. Van Henversuryn, H. and Fiers, W., *Eur. J. Biochem.*, 100, 50, 1979.
4. Schroeder, J. L. and Blattner, F. R., *Gene* 4, 167, 1978.
5. Sanger, F., Coulson, A. R., Friedmann, T., Air, G. M., Barrell, B. G., Brown, N. L., Fiddes, J. C., Hutchison, C. A., Slocombe, P. M., and Smith, M., *J. Mol. Biol.*, 125, 225, 1978.
6. Beck, F., Sommer, R., Auerswald, E. A., Kurz, Ch., Zink, B., Osterburg, G., and Schaller, H., *Nucleic Acids Res.*, 5, 4495, 1978.
7. McConkey, E. and Hopkins, J., *J. Mol. Biol.*, 29, 545, 1969.
8. Stanley, W. M. and Bock, R. M., *Biochemistry*, 4, 1302, 1965.
9. Pearce, T. C., Rowe, A. J., and Turnock, G., *J. Mol. Biol.*, 97, 193, 1975.
10. Lizardi, P. M., Williamson, R., and Brown, D. D., *Cell*, 4, 199, 1975.
11. Mondal, H., Sutton, A., Chen, C. J., and Sarker, S., *Biochem. Biophys. Res. Comm.*, 56, 988, 1974.
12. Zelenka, P. and Platigorsky, J., *Proc. Natl. Acad. Sci. U.S.A.* 71, 1896, 1974.
13. Berns, A., Jansson, P., and Bloemendal, H., *Biochem. Biophys. Res. Comm.*, 59, 1157, 1974.
14. Stravnezer, J., Huang, R. C. C., Stravnezer, E., and Bishop, J. M., *J. Mol. Biol.*, 88, 43, 1974.
15. Morrison, M. R. and Lingrel, J. B., *Biochim. Biophys. Acta*. 447, 104, 1976.
16. Hamlyn, P. H. and Gould, H. J., *J. Mol. Biol.*, 94, 101, 1973.
17. Grunstein, M. and Schedl, P., *J. Mol. Biol.*, 104, 323, 1976.
18. Scazzochio, C., personal communication.
19. Brownlee, G., Sanger, F., and Barrel, B. G., *J. Mol. Biol.*, 34, 379, 1968.
20. Loening, U. E., *J. Mol. Biol.*, 38, 355, 1968.

NUCLEASES

ENDONUCLEASES SPECIFIC FOR SINGLE-STRANDED POLYNUCLEOTIDES

PROPERTIES OF SINGLE-STRANDED POLYNUCLEOTIDE-SPECIFIC ENDONUCLEASES

Enzyme source	Degree of purity (%)	Molecular weight	Ratio DNase/RNase	Final products of hydrolysis	pH optimum	Divalent cations	Inhibitors	Comments
N. crassa (1,2)[a]	Unknown	55,000	1	>90% Nucleoside 5'-phosphates	7.0—9.0	Zn^{2+} Co^{2+} Mg^{2+}	2-Mercaptoethanol, potassium phosphate, ATP	Preferential attack at G and dG residues
U. maydis (7,8)	Unknown	42,000	3	>90% Nucleoside 5'-phosphates	8.0	Mg^{2+} Ca^{2+} Co^{2+} Zn^{2+}	2-Mercaptoethanol, potassium phosphate, ATP	Preferential attack at G and dG residues
A. oryzae (3,4)	>90	32,000	5	>90% Nucleoside 5'-phosphates	4.0—4.3	Zn^{2+} Co^{2+}	—	—
P. citrinum (5,6)	>90	Unknown	0.7	>90% Nucleoside 5'-phosphates	5.0	Zn^{2+}	—	Possesses 3'-nucleotidase activity; cold labile
Mung bean (9,10)	>90	39,000	1.2	>90% Nucleoside 5'-phosphates	5.0	Zn^{2+}	—	Possesses 3'-nucleotidase activity
Wheat seedlings (11)	>90	43,000	2	>90% Nucleoside 5'-phosphates	4.8—5.5	Zn^{2+} required for stabilization	Potassium phosphate, NaF	Possesses 3'-nucleotidase activity; preferential attack at A and dA residues

[a] Numbers in parentheses are reference numbers.

REFERENCES

1. **Linn, S. and Lehman, I. R.**, *J. Biol. Chem.*, 240, 1287, 1965.
2. **Linn, S. and Lehman, I. R.**, *J. Biol. Chem.*, 240, 1294, 1965.
3. **Ando, T.**, *Biochim. Biophys. Acta*, 114, 158, 1966.
4. **Vogt, V. M.**, *Eur. J. Biochem.*, 33, 192, 1973.
5. **Fujimoto, M., Kuninaka, A., and Yoshino, H.**, *Agric. Biol. Chem.* 38, 777, 1974.
6. **Fujimoto, M., Kuninaka, A., and Yoshino, H.**, *Agric. Biol. Chem.* 38, 785, 1974.
7. **Holloman, W. K. and Holliday, R.**, *J. Biol. Chem.*, 248, 8107, 1973.
8. **Holloman, R.**, *J. Biol. Chem.*, 248, 8114, 1973.
9. **Kowalski, D., Kroeker, W. D., and Laskowski, M., Sr.**, *Biochemistry*, 15, 4457, 1976.
10. **Kroeker, W. D., Kowalski, D., and Laskowski, M., Sr.**, *Biochemistry*, 15, 4463, 1976.
11. **Kroeker, W. D. and Fairley, J. L.**, *J. Biol. Chem.*, 250, 3773, 1975.

EXODEOXYRIBONUCLEASES OF *ESCHERICHIA COLI*

EXONUCLEASES OF *E. COLI*

Enzyme	Gene	Characteristics[a]
Exonuclease I	*sbcB*	Single-strand specific, processive; leaves 5'-terminal dinucleotide intact
(Exonuclease II)		Former name for $3' \rightarrow 5'$ exonuclease activity of DNA polymerase 1 (see below)
Exonuclease III	*xthA*	Double-strand-specific; associated activities: DNA-3'-phosphatase, AP endonuclease, and RNase H
Exonucleases IVA and IVB		Single-strand-specific; degrades oligonucleotides completely; not well characterized[b]
Exonuclease V (*recBC* DNase)	*recBC*	ATP-dependent; both $3' \rightarrow 5'$ and $5' \rightarrow 3'$; double-strand specific, processive; inactive at a nick; releases large oligonucleotides initially; also a single-strand specific endonuclease; a recombination enzyme
(Exonuclease VI)		Former name for $5' \rightarrow 3'$ exonuclease activity of DNA polymerase I (see below)
Exonuclease VII	*xse*	Single-strand specific, processive; EDTA- resistant; both $3' \rightarrow 5'$ and $5' \rightarrow 3'$; oligonucleotide products
Exonuclease VIII	*recE*	Double-strand specific; not well characterized[b]; product of the cryptic Rac prophage of *E. coli* K12; detected only in *xbcA* mutants
DNA polymerase I	*polA*	$3' \rightarrow 5'$ activity that is single-strand specific; $5' \rightarrow 3'$ activity: double-strand specific; mono- and oligonucleotide products; RNase H activity; repair functions
DNA polymerase II	*polB*	$3' \rightarrow 5'$ only; single-strand specific
DNA polymerase III	*polC*	$3' \rightarrow 5'$ activity: single-strand specific; does not attack dinucleotides; $5' \rightarrow 3'$ activity: single- strand specific but can attack a duplex after initiating hydrolysis at a single-stranded end

[a] Unless otherwise noted, directionality is $3' \rightarrow 5'$ and products are 5'-mononucleotides.

[b] Directionality is unknown.

From Weiss, B., in *The Enzymes*, 3rd ed., Boyer, P., Ed., Vol. XIVA, Academic Press, New York, 1981, 231. With permission.

recBC-LIKE ENZYMES

EXONUCLEASE V: DEOXYNUCLEASES

PURIFICATION AND PROPERTIES OF ExoV ENZYMES

Source of enzyme	Purification (-fold)	Specific activity[a] (U/mg)	$s_{20,w}$ (S)	Molecular weight	Subunit molecular weight
E. coli	17,000	57,000	12	270,000	140,000
					128,000
E. coli	3,500	21,600	12.4	350,000	170,000
					60,000
H. influenzae	2,000	28,000	12	290,000	115,000
					107,000
					68,000
B. subtilis	350	237	—	270,000	81,000
					70,000
					62,000
					52,500
					42,500
B. subtilis	5,000	4,000	13—14	300,000	155,000
					140,000
M. luteus	2,300	67,000	—	—	—
B. laterosporus	150	54,500	—	—	—
M. smegmatis	100	8,700	—	—	—
D. pneumoniae	743	28,000	—	—	—
B. cereus	4,200	2,450	—	—	—
P. aeruginosa	97,000	91,000	13.6	300,000	—
A. faecalis	820	1,900	—	—	—
S. intermedius	200	4,700	15	450,000	—

All values have been adjusted so that 1 unit is the amount of enzyme needed to render acid-soluble 1 nmol of duplex DNA-nucleotides in 30 min.

From Telander Muskavitch, K. M. and Linn, S., in *The Enzymes,* Vol. XIVA, Boyer, P., Ed., Academic Press, Orlando, 233, 1981. With permission.

RESTRICTION AND MODIFICATION ENZYMES AND THEIR RECOGNITION SEQUENCES

Richard J. Roberts*

Microorganism	Source	Enzyme[a]	Sequence[b]	Number of cleavage sites[c]					Ref.
				λ	Ad2	SV40	ΦX	pBR	
Acetobacter aceti	IFO 3281	AatI (StuI)	AGG↑CCT	6	11	7	1	0	289
		AatII	GACGT↑C	10	3	0	1	1	289
Acetobacter aceti sub. liquefaciens	IFO 12388	AacI (BamHI)	GGATCC	5	3	1	0	1	262
Acetobacter aceti sub. liquefaciens	M. Van Montagu	AaeI (BamHI)	GGATCC	5	3	1	0	1	262
Acetobacter aceti sub. orleanensis	NCIB 8622	AorI (EcoRII)	CC↑(A/T)GG	71	136	17	2	6	262
Acetobacter liquefaciens	IFO 12257	Ali12257I (BamHI)	GGATCC	5	3	1	0	1	445
Acetobacter liquefaciens	IFO 12258	Ali12258I (BamHI)	GGATCC	5	3	1	0	1	445
Acetobacter liquefaciens	K. Komagata	Ali2882I (PstI)	CTGCAG	28	30	2	1	1	445
Acetobacter liquefaciens	IFO 12388	AliI (BamHI)	G↑GATCC	5	3	1	0	1	335
Acetobacter liquefaciens AJ 2881	K. Komagata	AliAJI (PstI)	CTGCA↑G	28	30	2	1	1	246
Acetobacter pasteurianus	IFO 13753	ApaLI	G↑TGCAC	4	7	0	1	3	346
Acetobacter pasteurianus sub. pasteurianus	NCIB 7215	ApaI	GGG*CC↑C	1	12	1	0	0	261;306
Acetobacter xylinus	IFO 3288	AxyI (SauI)	CC↑TNAGG	2	7	0	0	0	339
Achromobacter eurydice	ATCC 39312	AeuI (EcoRII)	CC↑(A/T)GG	71	136	17	2	6	482
Achromobacter immobilis	ATCC 15934	AimI	?	?	?	?	?	?	66
Achromobacter pestifer	ATCC 15445	ApeI (MluI)	ACGCGT	7	5	0	2	0	475
Achromobacter species	U. Mayr	AspHI (HgiAI)	G(A/T)GC(A/T)↑C	28	38	0	3	8	519
Achromobacter species 697	C. Kessler	Asp697I (AvaII)	GG(A/T)CC	35	73	6	1	8	143
Achromobacter species 700	C. Kessler	Asp700I (XmnI)	GAANN↑NNTTC	24	5	0	3	2	143
Achromobacter species 703	C. Kessler	Asp703I (XhoI)	CTCGAG	1	6	0	1	0	143
Achromobacter species 707	C. Kessler	Asp707I (ClaI)	ATCGAT	15	2	0	0	1	143
Achromobacter species 708	C. Kessler	Asp708I (PstI)	CTGCAG	28	30	2	1	1	143
Achromobacter species 718	C. Kessler	Asp718I (KpnI)	G↑GTACC	2	8	1	0	0	19
Achromobacter species 742	C. Kessler	Asp742I (HaeIII)	GGCC	149	216	18	11	22	381
Achromobacter species 748	C. Kessler	Asp748I (HpaII)	CCGG	328	171	1	5	26	381
Achromobacter species 763	C. Kessler	Asp763I (ScaI)	AGTACT	5	5	0	0	1	381
Acinetobacter calcoaceticus	R.J. Roberts	AccI	GT↑(A/C)(G/T)AC	9	17	1	2	2	342
		AccII (FnuDII)	CG↑CG	157	303	0	14	23	342;150
		AccIII (BspMII)	T↑CCGGA	24	8	0	0	1	211;357;357
Acinetobacter calcoaceticus EBF 65/65	A. Vivian	AccEBI (BamHI)	G↑GATCC	5	3	1	0	1	367

*Cold Spring Harbor Laboratory, Cold Spring Harbor, N. Y.

RESTRICTION AND MODIFICATION ENZYMES AND THEIR RECOGNITION SEQUENCES (continued)

Microorganism	Source	Enzyme[a]	Sequence[b]	Number of cleavage sites[c]					Ref.
				λ	Ad2	SV40	ΦX	pBR	
Acinetobacter lwoffi	NEB 402	*Alw*I (*Bin*I)	GGATC (4/5)	58	35	6	0	12	202
Acinetobacter lwoffi	NEB 419	*Alw*NI	CAGNNN↑CTG	41	25	2	0	1	457
Acinetobacter lwoffi X	NEB 413	*Alw*XI (*Bbv*I)	GCAGC (8/12)	199	179	22	14	21	389
Actinomadura madurae	ATCC 15904	*Ama*I (*Nru*I)	TCGCGA	5	5	0	2	1	116
Actinosynnema pretiosum	ATCC 31281	*Apr*I (*Nae*I)	GCCGGC	1	13	1	0	4	361
Agmenellum quadruplicatum PR-6	ATCC 27264	*Aqu*I (*Ava*I)	*C↑PyCGPuG	8	40	?	?	1	163;398
Agrobacterium tumefaciens	ATCC 15955	*Atu*AI	?	>30	>30	?	?	?	260
Agrobacterium tumefaciens	IAM B-26-1	*Atu*AMI	?	?	?	?	?	?	290
Agrobacterium tumefaciens B6806	E. Nester	*Atu*BI (*Eco*RII)	CC(A/T)GG	71	136	17	2	6	241
Agrobacterium tumefaciens C58	E. Nester	*Atu*CI (*Bcl*I)	TGATCA	8	5	1	0	0	260
Agrobacterium tumefaciens ID 135	C. Kado	*Atu*II (*Eco*RII)	CC(A/T)GG	71	136	17	2	6	170
Agrobacterium tumefaciens IIBV7	G. Roizes	*Atu*BVI	?	>14	?	1	2	?	240
Agrobacterium tumefaciens RFL1	A.A. Janulaitis	*Atu*II (*Eco*RII)	CC(A/T)GG	71	136	17	2	6	420
Alcaligenes species	N. Brown	*Asp*AI (*Bst*EII)	G↑GTNACC	13	10	0	0	0	27
Alcaligenes species 47	C. Kessler	*Asp*47I (*Xho*I)	CTCGAG	1	6	0	1	0	381
Alcaligenes species 52	C. Kessler	*Asp*52I (*Hind*III)	AAGCTT	6	12	6	0	0	381
Alcaligenes species 78	C. Kessler	*Asp*78I (*Stu*I)	AGGCCT	6	11	7	1	0	381
Alcaligenes species RFL36	A.A. Janulaitis	*Asp*36I (*Pst*I)	CTGCAG	28	30	2	1	1	122
Alteromonas putrefaciens	M. Sargent	*Apu*I (*Asu*I)	GGNCC	74	164	11	2	15	368
Anabaena catenula	CCAP 1403/1	*Aca*I (*Asu*II)	TTCGAA	7	1	1	0	0	114,514
		*Aca*II (*Bam*HI)	GGATCC	5	3	1	0	1	514
		*Aca*III (*Mst*I)	TGCGCA	15	17	0	1	4	514
		*Aca*IV (*Hae*III)	GGCC	149	216	18	11	22	514
Anabaena cylindrica	CCAP 1403/2a	*Acy*I	GPu↑CGPyC	40	44	0	7	6	55
		*Acy*II	?	?	?	?	?	?	55
Anabaena flos-aquae	CCAP 1403/13f	*Afl*I (*Ava*II)	G↑G(A/T)CC	35	73	6	1	8	328
		*Afl*II	C↑TTAAG	3	4	1	2	0	328
Anabaena inqualis	CCAP 1446/1a	*Afl*III	A↑CPuPyGT	20	25	0	2	1	328
		*Ain*I (*Pst*I)	CTGCAG	28	30	2	1	1	514
		*Ain*II (*Bam*HI)	GGATCC	5	3	1	0	1	514
Anabaena oscillarioides	CCAP 1403/11	*Aos*I (*Mst*I)	TGC↑GCA	15	17	0	1	4	56
		*Aos*II (*Acy*I)	GPu↑CGPyC	40	44	0	7	6	56
		*Aos*III (*Sac*II)	CCGCGG	4	33	0	1	0	53
Anabaena species	CCAP 1403/9	*Aoc*I (*Sau*I)	CC↑TNAGG	2	7	0	0	0	53
		*Aoc*II (*Sdu*I)	G(A/G/T)GC(A/C/T)↑C	38	105	4	3	10	53
Anabaena species	ATCC 29151	*Asp*BI (*Ava*I)	CPyCGPuG	8	40	0	1	1	63
		*Asp*BII (*Ava*I)	CPyCGPuG	8	40	0	1	1	63

RESTRICTION AND MODIFICATION ENZYMES AND THEIR RECOGNITION SEQUENCES (continued)

Microorganism	Source	Enzyme[a]	Sequence[b]	λ	Ad2	SV40	ΦX	pBR	Ref.
Anabaena species	ATCC 27893	AspCI (*Ava*I)	CPyCGPuG	8	40	0	1	1	63
		AspCII (*Ava*II)	GG(A/T)CC	35	73	6	1	8	63
Anabaena species	ATCC 27898	AspDI (*Ava*I)	CPyCGPuG	8	40	0	1	1	63
		AspDII (*Ava*II)	GG(A/T)CC	35	73	6	1	8	63
Anabaena species	ATCC 29208	AsrWI (*Acy*I)	GPu↑CGPyC	40	44	0	7	6	54
Anabaena species TA 1	P. Wolk	AspTI (*Pst*I)	CTGCAG	28	30	2	1	1	514
		AspTII (*Bam*HI)	GGATCC	5	3	1	0	1	514
		AspTIII (*Hae*III)	GGCC	149	216	18	11	22	514
Anabaena subcylindrica	CCAP 1403/4b	AsuI	G↑GNCC	74	164	11	2	15	113
		AsuII	TT↑CGAA	7	1	0	0	0	211,54;54
		AsuIII (*Acy*I)	GPu↑CGPyC	40	44	0	7	6	54
Anabaena variabilis	ATCC 27892	AvaI	C↑PyCGPuG	8	40	0	1	1	208;115
		AvaII	G↑G(A/T)CC	35	73	6	1	8	208;296,115,71
		AvaIII	ATGCAT	14	9	3	0	0	239;239,268
Anabaena variabilis (halle)	CCAP 1403/12	AviI (*Asu*II)	TT↑CGAA	7	1	0	0	0	53
		AviII (*Mst*I)	TGC↑GCA	15	17	0	1	4	53
Anabaena variabilis uw	E.C. Rosenvold	AvrI (*Ava*I)	CPyCGPuG	8	40	0	1	1	243
		AvrII	C↑CTAGG	2	2	2	0	0	243;242
Anabaenopsis circularis	ATCC 27895	AcrI (*Ava*I)	CPyCGPuG	8	40	0	1	1	514
		AcrII (*Bst*EII)	G↑GTNACC	13	10	0	0	0	514
Anacystis nidulans R2	L. Sherman	AniI	?	?	?	?	?	0	464
Aphanothece halophytica	ATCC 29534	AhaI (*Cau*II)	CC↑(C/G)GG	114	97	0	1	10	404
		AhaII (*Acy*I)	GPu↑CGPyC	40	44	0	7	6	404
		AhaIII	TTT↑AAA	13	12	12	2	3	326
Aquaspirillum itersonii	ATCC 12639	AitAI (*Xho*II)	PuGATCPy	21	22	3	0	8	477
Aquaspirillum itersonii	NEB 443	AitI (*Eco*47III)	AGC↑GCT	2	13	1	0	4	478
		AitII (*Xho*II)	PuGATCPy	21	22	3	0	8	478
Aquaspirillum metamorphum	ATCC 15280	AmeI (*Apa*LI)	GTGCAC	4	7	0	1	3	477
		AmeII (*Nae*I)	GCCGGC	1	13	1	0	4	477
Aquaspirillum peregrinum	ATCC 15387	ApeAI (*Nae*I)	GCCGGC	1	13	1	0	4	477
Aquaspirillum serpens	NEB 448	AseI (*Vsp*I)	AT↑TAAT	17	3	3	2	1	482
		AseII (*Cau*II)	CC↑(C/G)GG	114	97	0	1	10	482
Arthrobacter luteus	ATCC 21606	AluI	AG↑*CT	143	158	34	24	16	236;236;336,376,376;337,376
Arthrobacter nicotianae	ATCC 15236	AniMI (*Nae*I)	GCCGGC	1	13	1	0	4	202
Arthrobacter pyridinolis	R. DiLauro	ApyI (*Eco*RII)	CC↑(A/T)GG	71	136	17	2	6	58
Arthrobacter species RFL1	A.A. Janulaitis	Asp11 (*Cau*II)	CC(C/G)GG	114	97	0	1	10	127

RESTRICTION AND MODIFICATION ENZYMES AND THEIR RECOGNITION SEQUENCES (continued)

Microorganism	Source	Enzyme[a]	Sequence[b]	λ	Ad2	SV40	ΦX	pBR	Ref.
Azospirillum amazonense	G. Schwabe	AamI	?	?	?	?	?	?	259
Azospirillum brasilense	ATCC 29711	AbrI (*XhoI*)	C↑TCGAG	1	6	0	1	0	259
Bacillus acidocaldarius	ATCC 27009	BacI (*SacII*)	CCGCGG	4	33	0	1	0	190,211
Bacillus alcalophilus 36	M.V. Jones	Bac36I (*AsuI*)	G↑GNCC	74	164	11	2	15	368
Bacillus alvei	ATCC 6348	BavI (*PvuII*)	CAG↑CTG	15	24	3	0	1	192
Bacillus amyloliquefaciens F	ATCC 23350	BamFI (*BamHI*)	GGATCC	5	3	1	0	1	267
Bacillus amyloliquefaciens H	F.E. Young	BamHI	G↑GAT*CC	5	3	1	0	1	329;238;103;103
Bacillus amyloliquefaciens K	T. Kaneko	BamKI (*BamHI*)	GGATCC	5	3	1	0	1	267
Bacillus amyloliquefaciens N	T. Ando	BamNI (*BamHI*)	GGATCC	5	3	1	0	1	266
		BamNxI (*AvaII*)	G↑G(A/T)CC	35	73	6	1	8	265,266;119
Bacillus aneurinolyticus	IAM 1077	BanI (*HgiCI*)	G↑GPyPuCC	25	57	1	3	9	289;289,463
		BanII (*HgiJII*)	GPuGCPy↑C	7	57	2	0	2	289
		BanIII (*ClaI*)	ATCGAT	15	2	0	0	1	289
Bacillus brevis	ATCC 9999	BbvI	GCAGC (8/12)	199	179	22	14	21	85;83,250;103;103
Bacillus brevis 80	V.M. Kramarov	BbvII	GAAGAC (2/6)	24	27	3	3	3	187
Bacillus brevis S	A.P. Zarubina	BbvSI	*GC(A/T)GC	Specific methylase					314
Bacillus caldolyticus	A. Atkinson	BclI	T↑GATCA	8	5	1	0	0	15
Bacillus centrosporus RFL1	A.A. Janulaitis	BcnI (*CauII*)	CC↑(C/G)GG	114	97	0	1	10	129,130,546;124,430,489
Bacillus cereus	IAM 1229	Bce1229I	?	>10	?	?	?	?	267
Bacillus cereus	ATCC 14579	Bce14579I	?	>10	?	?	?	?	267
Bacillus cereus	T. Ando	Bce170I (*PstI*)	CTGCAG	28	30	2	1	1	267
Bacillus cereus	IOC 243	Bce243I (*MboI*)	↑GATC#	116	87	8	0	22	52
Bacillus cereus	ATCC 31293	BceFI (*FnuDII*)	CGCG	157	303	0	14	23	219
Bacillus cereus	NEB 442	BcrI (*NlaIV*)	GGNNCC	82	178	16	6	24	478
Bacillus cereus 71	S.F. Ye	Bce711 (*HaeIII*)	GGCC	149	216	18	11	22	439
Bacillus cereus Rf sm st	T. Ando	BceRI (*FnuDII*)	CGCG	157	303	0	14	23	267
Bacillus cereus sub. *fluorescens*	P. Venetianer	BcefI	ACGGC (12/13)	115	80	11	11	3	526
Bacillus coagulans RFL33	A.A. Janulaitis	Bco33I (*HaeIII*)	GGCC	149	216	18	11	22	423
Bacillus coagulans RFL35	A.A. Janulaitis	Bco35I (*GsuI*)	CTGGAG	25	32	6	3	4	538
Bacillus globigii	G.A. Wilson	BglI	GCCNNNN↑NGGC	29	20	1	0	3	60,330;11,310
		BglII	A↑GATCT	6	11	0	0	3	60,330;224
Bacillus licheniformis	P.L. Manachini	BliI (*HaeIII*)	GGCC	149	216	18	11	22	462
Bacillus macerans	ATCC 8244	BmaAI (*PvuI*)	CGATCG	3	7	0	0	1	477
Bacillus macerans	ATCC 8510	BmaBI (*PvuI*)	CGATCG	3	7	0	0	1	477
Bacillus macerans	ATCC 7069	BmaCI (*PvuI*)	CGATCG	3	7	0	0	1	477
Bacillus macerans	ATCC 8509	BmaDI (*PvuI*)	CGATCG	3	7	0	0	1	477

RESTRICTION AND MODIFICATION ENZYMES AND THEIR RECOGNITION SEQUENCES (continued)

Microorganism	Source	Enzyme[a]	Sequence[b]	λ	Ad2	SV40	ΦX	pBR	Ref.
				Number of cleavage sites[c]					
Bacillus macerans	ATCC 8513	*BmaI (PvuI)*	CGATCG	3	7	0	0	1	479
Bacillus megaterium	J. Upcroft	*BmeI*	?	>10	>20	4	?	?	78
Bacillus megaterium 216	V.M. Kramarov	*Bme216I (AvaII)*	G↑G(A/T)CC	35	73	6	1	8	155;465
Bacillus megaterium 899	B899	*Bme899I*	?	>5	?	?	?	?	267
Bacillus megaterium B205-3	T. Kaneko	*Bme205I*	?	>10	?	?	?	?	267
Bacillus pumilus AHU1387A	T. Ando	*BpuI*	?	6	>30	2	?	?	118
Bacillus species	N.V. Tsvetkova	*BcmI (ClaI)*	AT↑CGAT	15	2	0	0	1	509
Bacillus species	P. Eastlake	*BscI (ClaI)*	AT↑CGAT	15	2	0	0	1	380;27
Bacillus species H	NEB 394	*BspHI*	T↑CATGA	8	3	2	3	4	388
Bacillus species M	NEB 356	*BspMI*	ACCTGC (4/8)	41	39	0	3	1	202;203
		BspMII	T↑CCGGA	24	8	0	0	1	202;203
Bacillus species RFL12	A.A. Janulaitis	*Bsp12I (SacII)*	CCGCGG	4	33	0	1	0	423
		Bsp12II	?	?	?	?	?	?	423
Bacillus species RFL16	A.A. Janulaitis	*Bsp16I (EcoRV)*	GATATC	21	9	1	0	1	423
Bacillus species RFL17	A.A. Janulaitis	*Bsp17I (PstI)*	CTGCAG	28	30	2	1	1	423
Bacillus species RFL18	A.A. Janulaitis	*Bsp18I (MboI)*	GATC	116	87	8	0	22	423
Bacillus species RFL2	A.A. Janulaitis	*Bsp2I (ClaI)*	ATCGAT	15	2	0	0	1	423
Bacillus species RFL21	A.A. Janulaitis	*Bsp21I (Cfr10I)*	PuCCGGPy	61	40	1	0	7	420
Bacillus species RFL22	A.A. Janulaitis	*Bsp22I (GsuI)*	CTGGAG	25	32	6	3	4	420
Bacillus species RFL28	A.A. Janulaitis	*Bsp28I (GsuI)*	CTGGAG	25	32	6	3	4	423
Bacillus species RFL29	A.A. Janulaitis	*Bsp29I (NlaIV)*	GGNNCC	82	178	16	6	24	423
Bacillus species RFL30	A.A. Janulaitis	*Bsp30I (BamHI)*	GGATCC	5	3	1	0	1	423
Bacillus species RFL4	A.A. Janulaitis	*Bsp4I (ClaI)*	ATCGAT	15	2	0	0	1	423
Bacillus species RFL5	A.A. Janulaitis	*Bsp5I (HpaII)*	CCGG	328	171	1	5	26	423
Bacillus species RFL6	A.A. Janulaitis	*Bsp6I (Fnu4HI)*	GC↑NGC	379	411	24	31	42	547
		Bsp6II (Eco57I)	CTGAAG	40	23	3	0	2	420
Bacillus species RFL7	A.A. Janulaitis	*Bsp7I (CauII)*	CC(C/G)GG	114	97	0	1	10	423
Bacillus species RFL8	A.A. Janulaitis	*Bsp8I (CauII)*	CC(C/G)GG	114	97	0	1	10	423
Bacillus species RFL9	A.A. Janulaitis	*Bsp9I (MboI)*	GATC	116	87	8	0	22	423
Bacillus sphaericus	WHO/CCBC 1691	*BshAI (HaeIII)*	GGCC	149	216	18	11	22	495
Bacillus sphaericus	WHO/CCBC 1881	*BshBI (HaeIII)*	GGCC	149	216	18	11	22	495
Bacillus sphaericus	WHO/CCBC 2013-6	*BshCI (HaeIII)*	GGCC	149	216	18	11	22	495
Bacillus sphaericus	WHO/CCBC 2117-2	*BshDI (HaeIII)*	GGCC	149	216	18	11	22	495
Bacillus sphaericus	WHO/CCBC 2362	*BshEI (HaeIII)*	GGCC	149	216	18	11	22	495
Bacillus sphaericus	WHO/CCBC 2500	*BshFI (HaeIII)*	GGCC	149	216	18	11	22	495
Bacillus sphaericus	WHO/CCBC 1593	*BshI (HaeIII)*	GGCC	149	216	18	11	22	495
Bacillus sphaericus	IAM 1286	*Bsp1286I (SduI)*	G(A/G/T)GC(A/C/T)↑C	38	105	4	3	10	267;211,252
Bacillus sphaericus 105	Q-L. Li	*Bsp105I (MboI)*	↑GATC	116	87	8	0	22	439

RESTRICTION AND MODIFICATION ENZYMES AND THEIR RECOGNITION SEQUENCES (continued)

Microorganism	Source	Enzyme[a]	Sequence[b]	Number of cleavage sites[c]					Ref.
				λ	Ad2	SV40	ΦX	pBR	
Bacillus sphaericus 106	B. Zhou	*Bsp*106I (*Cla*I)	AT↑CGAT	15	2	0	0	1	440
Bacillus sphaericus 211	P.F. Yan	*Bsp*211I (*Hae*III)	GG↑CC	149	216	18	11	22	439
Bacillus sphaericus 226	P.F. Yan	*Bsp*226I (*Hae*III)	GGCC	149	216	18	11	22	439
Bacillus sphaericus 63	S.Y. Ye	*Bsp*63I (*Pst*I)	CTGCA↑G	28	30	2	1	1	439
Bacillus sphaericus 64	S.Y. Ye	*Bsp*64I (*Mbo*I)	GATC	116	87	8	0	22	439
Bacillus sphaericus 67	S.Y. Ye	*Bsp*67I (*Mbo*I)	↑GATC#	116	87	8	0	22	439
Bacillus sphaericus 71	P.F. Yan	*Bsp*71I (*Hae*III)	GGCC	149	216	18	11	22	439
Bacillus sphaericus 74	S.Y. Ye	*Bsp*74I (*Mbo*I)	GATC	116	87	8	0	22	439
Bacillus sphaericus 76	S.Y. Ye	*Bsp*76I (*Mbo*I)	GATC	116	87	8	0	22	439
Bacillus sphaericus 78	S.Y. Ye	*Bsp*78I (*Pst*I)	CTGCAG	28	30	2	1	1	439
Bacillus sphaericus JL14	R. Mullings	*Bsp*BI (*Pst*I)	CTGCA↑G	28	30	2	1	1	365
		*Bsp*BII (*Asu*I)	G↑GNCC	74	164	11	2	15	365
Bacillus sphaericus JL4B	R. Mullings	*Bsp*AI (*Mbo*I)	↑GATC#	116	87	8	0	22	365
Bacillus sphaericus R	P. Venetianer	*Bsp*RI (*Hae*III)	GG↑CC	149	216	18	11	22	149;316;153
Bacillus sphaericus X	D. Dupret	*Bsp*XI (*Cla*I)	AT↑CGAT	15	2	0	0	1	460
		*Bsp*XII (*Bcl*I)	T↑GATCA	8	5	1	0	0	460
Bacillus stearothermophilus	NEB 447	*Bsr*I	ACTGG (1/-1)	110	86	11	9	19	531
Bacillus stearothermophilus	D. Comb	*Bsr*NI (*Eco*RII)	CC↑(A/T)GG	71	136	17	2	6	253
Bacillus stearothermophilus	ATCC 12980	*Bst*PI (*Bst*EII)	G↑GTNACC	13	10	0	0	0	226
Bacillus stearothermophilus 1503-4R	N. Welker	*Bst*I (*Bam*HI)	G↑GATCC	5	3	1	0	1	38;41
Bacillus stearothermophilus 240	A. Atkinson	*Bst*AI	?	?	?	?	?	?	17
Bacillus stearothermophilus A664	Z. Chen	*Bsm*AI	GTCTC	37	60	2	4	3	490
Bacillus stearothermophilus B225	Z. Chen	*Bst*BI (*Asu*II)	TTCGAA	7	1	0	0	0	451
Bacillus stearothermophilus C1	N. Welker	*Bst*CI (*Hae*III)	GGCC	149	216	18	11	22	162
Bacillus stearothermophilus C11	N. Welker	*Bss*CI (*Hae*III)	GGCC	149	216	18	11	22	451
Bacillus stearothermophilus D428	Z. Chen	*Bst*DI (*Bst*EII)	GGTNACC	13	10	0	0	?	194
Bacillus stearothermophilus ET	N. Welker	*Bst*EI	?	?	?	?	?	?	194
		*Bst*EII	G↑GTNACC	13	10	0	0	0	194;165
Bacillus stearothermophilus		*Bst*EIII (*Mbo*I)	GATC	116	87	8	0	22	194;88;211
Bacillus stearothermophilus FH58	Z. Chen	*Bst*FI (*Hind*III)	A↑AGCTT	6	12	6	0	1	473
Bacillus stearothermophilus G3	N. Welker	*Bst*GI (*Bcl*I)	TGATCA	8	5	1	0	0	162
		*Bst*GII (*Eco*RII)	CC(A/T)GG	71	136	17	2	6	162
Bacillus stearothermophilus G6	N. Welker	*Bss*GI (*Bst*XI)	CCANNNNN↑NTGG	13	10	1	3	0	162
		*Bss*GII (*Mbo*I)	GATC	116	87	8	0	22	162
Bacillus stearothermophilus H1	N. Welker	*Bst*HI (*Xho*I)	CTCGAG	1	6	0	1	0	162

RESTRICTION AND MODIFICATION ENZYMES AND THEIR RECOGNITION SEQUENCES (continued)

Microorganism	Source	Enzyme[a]	Sequence[b]	Number of cleavage sites[c] λ	Ad2	SV40	ΦX	pBR	Ref.
Bacillus stearothermophilus H3	N. Welker	BssHI (XhoI)	CTCGAG	1	6	0	1	0	162
		BssHII (BsePI)	G↑CGCGC	6	52	0	1	0	162;255
Bacillus stearothermophilus H4	N. Welker	BsrHI (BsePI)	GCGCGC	6	52	0	1	0	162
Bacillus stearothermophilus K1460	Z. Chen	BstJI (HaeIII)	GGCC	149	216	18	11	22	451
Bacillus stearothermophilus K554	Z. Chen	BstK1 (BclI)	TGATCA	8	5	1	0	0	451
Bacillus stearothermophilus L95	Z. Chen	BstLI (XhoI)	CTCGAG	1	6	0	1	1	451
Bacillus stearothermophilus M571	Z. Chen	BstMI (ScaI)	AGTACT	5	5	0	0	1	451
		BsmI	GAATGC (1/-1)	46	10	4	3	1	211
Bacillus stearothermophilus NUB 36	N. Welker	Bst31I (BstEII)	GGTNACC	13	10	0	0	0	105
Bacillus stearothermophilus NUB31	Z. Chen	BsrOI (EcoRII)	CC(A/T)GG	71	136	17	2	6	451
Bacillus stearothermophilus O22	N. Welker	BssPI	?	>30	?	?	?	?	162
Bacillus stearothermophilus P1	N. Welker	BsrPI	?	11	>20	?	?	0	162
Bacillus stearothermophilus P5	N. Welker	BsrPII (MboI)	GATC#	116	87	8	0	22	162
Bacillus stearothermophilus P6	N. Welker	BsePI	GCGCGC	6	52	0	1	0	162
Bacillus stearothermophilus P8	N. Welker	BsaPI (MboI)	GATC	116	87	8	0	22	162
Bacillus stearothermophilus P9	N. Welker	BsoPI (BsePI)	GCGCGC	6	52	0	1	0	162
Bacillus stearothermophilus Q407	Z. Chen	BstQI (BamHI)	GGATCC	5	3	1	0	1	451
Bacillus stearothermophilus R463	Z. Chen	BsrRI (EcoRV)	GATATC	21	9	1	0	1	451
Bacillus stearothermophilus S183	Z. Chen	BstSI (AvaI)	C↑PyCGPuG	8	40	0	1	1	451
Bacillus stearothermophilus T12	N. Welker	BstTI (BstXI)	CCANNNNN↓NTGG	13	10	1	3	0	162
Bacillus stearothermophilus U458	Z. Chen	BstUI (FnuDII)	CG↑CG	157	303	0	14	23	451
Bacillus stearothermophilus V	C. Vasquez	BstVI (XhoI)	CTCGAG	1	6	1	1	0	382
Bacillus stearothermophilus W574	Z. Chen	BstWI (EcoNI)	CCTNNNNNAGG	9	10	2	0	1	490
Bacillus stearothermophilus X1	N. Welker	BstXI	CCANNNNN↑NTGG	13	10	1	3	0	162;256
		BstXII (MboI)	GATC	116	87	8	0	22	162
Bacillus stearothermophilus Y406	Z. Chen	BstYI (XhoII)	Pu↑GATCPy	21	22	3	0	8	451
Bacillus stearothermophilus Z130	Z. Chen	BstZI (XmaIII)	CGGCCG	2	19	0	0	1	451
Bacillus stearothermophilus strain 822	T. Oshima	BseI (HaeIII)	GGCC	149	216	18	11	22	271
Bacillus stearothermophilus strain 881	Z. Chen	BseII (HpaI)	GTTAAC	14	6	4	3	0	271
		BssI (NlaIV)	GGNNCC	82	178	16	6	24	490
Bacillus subtilis	IAM 1076	Bsu1076I (HaeIII)	GGCC	149	216	18	11	22	267
Bacillus subtilis	IAM 1114	Bsu1114I (HaeIII)	GGCC	149	216	18	11	22	267
Bacillus subtilis	ATCC 14593	Bsu1145I	?	>20	?	?	?	?	267
Bacillus subtilis	IAM 1192	Bsu1192I (HpaII)	CCGG	328	171	1	5	26	267;228
		Bsu1192II (FnuDII)	CGCG	157	303	0	14	23	228
Bacillus subtilis	IAM 1193	Bsu1193I (FnuDII)	CGCG	157	303	0	14	23	267;228
Bacillus subtilis	IAM 1259	Bsu1259I	?	>8	?	?	?	?	267

RESTRICTION AND MODIFICATION ENZYMES AND THEIR RECOGNITION SEQUENCES (continued)

Microorganism	Source	Enzyme[a]	Sequence[b]	Number of cleavage sites[c]					Ref.
				λ	Ad2	SV40	ΦX	pBR	
Bacillus subtilis	ATCC 6633	Bsu6633I (*FnuDII*)	CGCG	157	303	0	14	23	267;120
Bacillus subtilis	IAM 1247	BsuBI (*PstI*)	CTGCAG	28	30	2	1	1	267;110
Bacillus subtilis	IAM 1231	BsuEII (*FnuDII*)	*CGCG	157	303	0	14	23	228;228;527;136,527
Bacillus subtilis 1532	A.A. Prozorov	BsuFI (*HpaII*)	*CCGG	328	171	1	5	26	267;228,136
Bacillus subtilis 1532	A.A. Prozorov	Bsu15321 (*FnuDII*)	CG↑CG	157	303	0	14	23	529
Bacillus subtilis 1854	A.A. Prozorov	Bsi18541 (*HgiJII*)	GPuGCPy↑C	7	57	2	0	2	529
Bacillus subtilis 36	B. Zhou	Bsu36I (*SauI*)	CC↑TNAGG	2	7	0	0	0	441
Bacillus subtilis Marburg 168	T. Ando	BsuMI (*XhoI*)	*CTCGAG	1	6	0	1	0	267;136;136
Bacillus subtilis strain R	T. Trautner	BsuRI (*HaeIII*)	GG*CC	149	216	18	11	22	26;25;95
Bacillus thuringiensis	R.R. Azizbekyan	BtiI (*AvaII*)	GG(A/T)CC	35	73	6	1	8	4
Bacillus vulgatis	OSB816	BvuI (*HgiJII*)	GPuGCPy↑C	7	57	2	0	2	8
Bifidobacterium bifidum YIT4007	T. Khosaka	BbiI (*PstI*)	CTGCAG	28	30	2	1	1	147
		BbiII (*AcyI*)	GPu↑CGPyC	40	44	0	7	6	147
		BbiIII (*XhoI*)	CTCGAG	1	6	0	1	0	147
		BbiIV	?	?	?	?	?	?	147
Bifidobacterium breve S1	ATCC 15700	BbeSI	?	?	?	?	?	?	145
Bifidobacterium breve S50	ATCC 15698	BbeAI (*NarI*)	GGCGCC	1	20	2	2	4	145
		BbeAII	?	?	?	?	?	?	145
Bifidobacterium breve YIT4006	H. Takahashi	BbeI (*NarI*)	GGCGC↑C	1	20	2	2	4	148
		BbeII	?	?	?	?	?	?	145
Bifidobacterium infantis 659	ATCC 25962	BinI	GGATC (4/5)	58	35	6	0	12	146
Bifidobacterium infantis S76e	ATCC 15702	BinSI (*EcoRII*)	CC(A/T)GG	71	136	17	2	6	145
		BinSII (*NarI*)	GGCGCC	1	20	0	2	4	145
Bifidobacterium longum E194b	ATCC 15707	BloI	?	?	?	?	?	?	145
Bifidobacterium thermophilum RU326	ATCC 25866	BthI (*XhoI*)	CTCGAG	1	6	0	1	0	145
		BthII (*BinI*)	GGATC	58	35	6	0	12	145
Bordetella bronchiseptica	ATCC 19395	BbrI (*HindIII*)	AAGCTT	6	12	6	0	1	211
Bordetella pertussis	P. Novotny	BpeI (*HindIII*)	AAGCTT	6	12	6	0	1	90,141
Branhamella catarrhalis	ATCC 25240	BcaI (*HhaI*)	GCGC	215	375	2	18	31	479
Brevibacterium albidum	ATCC 15831	BalI	TGG*CCA	18	17	0	0	1	79;79;306
Brevibacterium epidermidis	P. Venetianer	BepI (*FnuDII*)	CG↑CG	157	303	0	14	23	506
Brevibacterium luteum	ATCC 15830	BluI (*XhoI*)	C↑TCGAG	1	6	0	1	0	84
		BluII (*HaeIII*)	GGCC	149	216	18	11	22	311
Brevibacterium protophormiae	IFO 12128	BprI	?	?	?	?	?	?	290
Calothrix scopulorum	CCAP 1410/5	CscI (*SacII*)	CCGC↑GG	4	33	0	1	0	62

RESTRICTION AND MODIFICATION ENZYMES AND THEIR RECOGNITION SEQUENCES (continued)

Microorganism	Source	Enzyme[a]	Sequence[b]	Number of cleavage sites[c]					Ref.
				λ	Ad2	SV40	ΦX	pBR	
Caryophanon latum	ATCC 15219	*ClmI (HaeIII)*	GGCC	149	216	18	11	22	279
		ClmII (AvaII)	GG(A/T)CC	35	73	6	1	8	279
Caryophanon latum	DSM 484	*CltI (HaeIII)*	GG↑CC	149	216	18	11	22	190
Caryophanon latum H7	W.C. Trentini	*CalI*	?	14	?	?	?	?	190
Caryophanon latum L	H. Mayer	*ClaI*	AT↑CGAT	15	2	0	0	1	188
Caryophanon latum RII	H. Mayer	*ClaI*	?	>20	?	?	?	?	190
Caseobacter polymorphus	ATCC 33010	*CpoI (RsrII)*	CGG(A/T)CCG	5	2	0	0	0	477
Caulobacter crescentus CB-13	J. Poindexter	*CcrI (XhoI)*	C↑TCGAG	1	6	0	1	0	297
Caulobacter fusiformis	A.A. Janulaitis	*CfuI (DpnI)*	GA*↑TC	Only cleaves methylated DNA					125;425
Caulobacter subvibroides	ATCC 15264	*CsuI*	?	?	?	?	?	?	480
Cellulomonas flavigena	IFO 3753	*CflI (PstI)*	CTGCA↑G	28	30	2	1	1	356
Chlorella strain NC64A (A1-2C)	J.L. Van Etten	*CviDI (HinfI)*	GANTC	148	72	10	21	10	401
Chlorella strain NC64A (CA-1A)	J.L. Van Etten	*CviKI (CviJI)*	PuGCPy	692	680	87	69	73	511
Chlorella strain NC64A (CA-1D)	J.L. Van Etten	*CviGI (HinfI)*	GANTC	148	72	10	21	10	401
Chlorella strain NC64A (CA-2A)	J.L. Van Etten	*CviLI (CviJI)*	PuGCPy	692	680	87	69	73	511
Chlorella strain NC64A (IL-2A)	J.L. Van Etten	*CviMI (CviJI)*	PuGCPy	692	680	87	69	73	511
Chlorella strain NC64A (IL-2B)	J.L. Van Etten	*CviNI (CviJI)*	PuGCPy	692	680	87	69	73	511
Chlorella strain NC64A (IL-3A)	J.L. Van Etten	*CviJI*	PuG↑CPy	692	680	87	69	73	467
Chlorella strain NC64A (IL-3D)	J.L. Van Etten	*CviOI (CviJI)*	PuGCPy	692	680	87	69	73	511
Chlorella strain NC64A (MA-1E)	J.L. Van Etten	*CviEI (HinfI)*	GANTC	148	72	10	21	10	401
Chlorella strain NC64A (NC-1A)	J.L. Van Etten	*CviBI (HinfI)*	G↑ANTC	148	72	10	21	10	401
		CviBII (MboI)	GATC	Specific methylase					499
		CviBIII (TaqI)	TCGA	Specific methylase					499
Chlorella strain NC64A (NE-8A)	J.L. Van Etten	*CviHI (MboI)*	GATC	116	87	8	0	22	511
Chlorella strain NC64A (NE-8A)	J.L. Van Etten	*CviCI (HinfI)*	GANTC	148	72	10	21	10	401
Chlorella strain NC64A (NY-2A)	J.L. Van Etten	*CviQI (RsaI)*	GT↑TAC	113	83	12	11	3	496
		CviQII	?	?	?	?	?	?	510
Chlorella strain NC64A (NY-2F)	J.L. Van Etten	*CviFI (HinfI)*	GANTC	148	72	10	21	10	401
Chlorella strain NC64A (NYs-1)	J.L. Van Etten	*CviPI*	?	?	?	?	?	?	510
Chlorella strain NC64A (PBCV-1)	J.L. Van Etten	*CviAI (MboI)*	↑GA*TC	116	87	8	0	22	399;400
Chloroflexus aurantiacus	A. Bingham	*CauI (AvaII)*	G↑TG(A/T)CC	35	73	6	1	8	16;201
		CauII	CC↑(C/G)GG	114	97	0	1	10	16;173,201
		CauIII (PstI)	CTGCAG	28	30	2	1	1	10
Chloroflexus aurantiacus B3	V.M. Kramarov	*CauB3I (BspMII)*	T↑CCGGA	24	8	0	0	1	524
Chromatium vinosum	G.C. Grosveld	*CvnI (SauI)*	CC↑TNAGG	2	7	0	0	0	94,435
Chromobacterium violaceum	ATCC 12472	*CviI*	?	?	?	?	?	?	66

RESTRICTION AND MODIFICATION ENZYMES AND THEIR RECOGNITION SEQUENCES (continued)

Microorganism	Source	Enzyme[a]	Sequence[b]	Number of cleavage sites[c]					Ref.
				λ	Ad2	SV40	ΦX	pBR	
Citrobacter diversus RFL27	A.A. Janulaitis	Cdi27I (*EcoRII*)	CC(A/T)GG	71	136	17	2	6	122
Citrobacter freundii	ATCC 11102	CfrNI (*AsuI*)	GGNCC	74	164	11	2	15	481
Citrobacter freundii A4	CAMB 2537	Cfr A4I (*PstI*)	CTGCA↑G	28	30	2	1	1	484
Citrobacter freundii RFL10	A.A. Janulaitis	Cfr10I	Pu↑CCGGPy	61	40	1	0	7	132,134,134;131;489,134
Citrobacter freundii RFL11	A.A. Janulaitis	Cfr11I (*EcoRII*)	CC(A/T)GG	71	136	17	2	6	132,134
Citrobacter freundii RFL13	A.A. Janulaitis	Cfr13I (*AsuI*)	G↑TGNCC	74	164	11	2	15	132,132;18
Citrobacter freundii RFL14	A.A. Janulaitis	Cfr14I (*CfrI*)	PyGGCCPu	39	70	0	0	6	132
Citrobacter freundii RFL19	A.A. Janulaitis	Cfr19I (*BstEII*)	GGTNACC	13	10	0	0	0	428
Citrobacter freundii RFL2	A.A. Janulaitis	CfrI	Py↑GG*CCPu	39	70	0	2	6	133,412;134,489,489
Citrobacter freundii RFL20	A.A. Janulaitis	Cfr20I (*EcoRII*)	CC(A/T)GG	71	136	17	2	6	428
Citrobacter freundii RFL22	A.A. Janulaitis	Cfr22I (*EcoRII*)	CC(A/T)GG	71	136	17	2	6	428
Citrobacter freundii RFL23	A.A. Janulaitis	Cfr23I (*AsuI*)	GGNCC	74	164	11	2	15	429
Citrobacter freundii RFL24	A.A. Janulaitis	Cfr24I (*EcoRII*)	CC(A/T)GG	71	136	17	2	6	428
Citrobacter freundii RFL25	A.A. Janulaitis	Cfr25I (*EcoRII*)	CC(A/T)GG	71	136	17	2	6	428
Citrobacter freundii RFL27	A.A. Janulaitis	Cfr27I (*EcoRII*)	CC(A/T)GG	71	136	17	2	6	428
Citrobacter freundii RFL28	A.A. Janulaitis	Cfr28I (*EcoRII*)	CC(A/T)GG	71	136	17	2	6	428
Citrobacter freundii RFL29	A.A. Janulaitis	Cfr29I (*EcoRII*)	CC(A/T)GG	71	136	17	2	6	428
Citrobacter freundii RFL30	A.A. Janulaitis	Cfr30I (*EcoRII*)	CC(A/T)GG	71	136	17	2	6	428
Citrobacter freundii RFL31	A.A. Janulaitis	Cfr31I (*EcoRII*)	CC(A/T)GG	71	136	17	2	6	428
Citrobacter freundii RFL32	A.A. Janulaitis	Cfr32I (*HindIII*)	AAGCTT	6	12	6	0	1	428
Citrobacter freundii RFL33	A.A. Janulaitis	Cfr33I (*AsuI*)	GGNCC	74	164	11	2	15	428
Citrobacter freundii RFL35	A.A. Janulaitis	Cfr35I (*EcoRII*)	CC(A/T)GG	71	136	17	2	6	428
Citrobacter freundii RFL37	A.A. Janulaitis	Cfr37I (*SacII*)	CCGCGG	4	33	0	1	0	429
Citrobacter freundii RFL38	A.A. Janulaitis	Cfr38I (*CfrI*)	PyGGCCPu	39	70	0	2	6	429
Citrobacter freundii RFL39	A.A. Janulaitis	Cfr39I (*CfrI*)	PyGGCCPu	39	70	0	2	6	428
Citrobacter freundii RFL4	A.A. Janulaitis	Cfr4I (*AsuI*)	GGNCC	74	164	11	2	15	132,134
Citrobacter freundii RFL40	A.A. Janulaitis	Cfr40I (*CfrI*)	PyGGCCPu	39	70	0	2	6	428
Citrobacter freundii RFL41	A.A. Janulaitis	Cfr41I (*SacII*)	CCGCGG	4	33	0	1	0	429
Citrobacter freundii RFL42	A.A. Janulaitis	Cfr42I (*SacII*)	CCGC↑GG	4	33	0	1	0	537
Citrobacter freundii RFL43	A.A. Janulaitis	Cfr43I (*SacII*)	CCGCGG	4	33	0	1	0	428
Citrobacter freundii RFL45	A.A. Janulaitis	Cfr45I (*AsuI*)	GGNCC	74	164	11	2	15	428
Citrobacter freundii RFL45	A.A. Janulaitis	Cfr45II (*SacII*)	CCGCGG	4	33	0	1	0	428
Citrobacter freundii RFL46	A.A. Janulaitis	Cfr46I (*AsuI*)	GGNCC	74	164	11	2	15	428
Citrobacter freundii RFL47	A.A. Janulaitis	Cfr47I (*AsuI*)	GGNCC	74	164	11	2	15	428
Citrobacter freundii RFL48	A.A. Janulaitis	Cfr48I (*HgiJII*)	GPuGCPyC	7	57	2	0	2	428
Citrobacter freundii RFL5	A.A. Janulaitis	Cfr5I (*EcoRII*)	CC(A/T)GG	71	136	17	2	6	132,134
Citrobacter freundii RFL51	A.A. Janulaitis	Cfr51I (*PvuI*)	CGATCG	3	7	0	0	1	423
Citrobacter freundii RFL52	A.A. Janulaitis	Cfr52I (*AsuI*)	GGNCC	74	164	11	2	15	538
Citrobacter freundii RFL54	A.A. Janulaitis	Cfr54I (*AsuI*)	GGNCC	74	164	11	2	15	539

RESTRICTION AND MODIFICATION ENZYMES AND THEIR RECOGNITION SEQUENCES (continued)

Microorganism	Source	Enzyme[a]	Sequence[b]	λ	Ad2	SV40	ΦX	pBR	Ref.
Citrobacter freundii RFL6	A.A. Janulaitis	Cfr6I (PvuII)	CAG↑CTG	15	24	3	0	1	132,134;488;488,134
Citrobacter freundii RFL7	A.A. Janulaitis	Cfr7I (BstEII)	GGTNACC	13	10	0	0	0	132,134
Citrobacter freundii RFL8	A.A. Janulaitis	Cfr8I (AsuI)	GGNCC	74	164	11	2	15	132,134
Citrobacter freundii RFL9	A.A. Janulaitis	Cfr9I (SmaI)	C↑CCGGG	3	12	0	0	0	132,134;472;472,134
Citrobacter freundii S39	W. Piepersberg	CfrS37I (EcoRII)	CC(A/T)GG	71	136	17	2	6	381
Clostridium formicoaceticum	ATCC 23439	CfoI (HhaI)	GCGC	215	375	2	18	31	181
Clostridium histolyticum	R. Hansen	ChI	?	?	?	?	?	?	100
Clostridium pasteurianum	NRRC 33011	CpaI (MboI)	GATC	116	87	8	0	22	316
Clostridium perfringens	R. Hansen	CpfI (MboI)	↑GATC#	116	87	8	0	22	100
Clostridium thermocellum	ATCC 27405	CthI (BclI)	TGATCA	8	5	1	0	0	446
		ChII (EcoRII)	CC↑(A/T)GG	71	136	17	2	6	446
Coccochloris elabens 17a	ATCC 27265	CelI (BamHI)	GGATCC	5	3	1	0	1	514
		CelII (EspI)	GCTNAGC	7	8	1	0	0	514
Corynebacterium equii	E.G. Duda	CeqI (EcoRV)	GAT↑ATC	21	9	1	0	1	454
Corynebacterium humiferum	ATCC 21108	ChuI (HindIII)	AAGCTT	6	12	6	1	1	66
		ChuII (HindII)	GTPyPuAC	35	25	7	13	2	66
Corynebacterium hydrocarboclastum	ATCC 21628	ChyI (StuI)	AGGCCT	6	11	7	1	0	479
Corynebacterium petrophilum	ATCC 19080	CpeI (BclI)	TGATCA	8	5	1	0	0	68
Corynebacterium species RFL2	A.A. Janulaitis	Csp2I (HaeIII)	GGCC	149	216	18	11	22	539
Cylindrospermum lichenforme	ATCC 29412	ClcI (PstI)	CTGCAG	28	30	2	1	1	514
		ClcII (MstI)	TGCGCA	15	17	0	1	4	514
Cylindrospermum lichenforme	UTEX 2014	CliI (AvaII)	GG(A/T)CC	35	73	6	1	8	535
		CliII (MstI)	TGCGCA	15	17	0	1	4	535
		CliIII	?	?	?	?	?	?	535
Cystobacter velatus Plv9	H. Reichenbach	CveI	?	?	?	?	?	?	190
Dactylococcopsis salina	A.E. Walsby	DsaI	C↑CPuPyGG	46	82	3	3	2	364
		DsaII (HaeIII)	GG↑CC	149	216	18	11	22	364
Deinococcus radiophilus	ATCC 27603	DraI (AhaIII)	TTT↑AAA	13	12	12	2	3	227
		DraII	PuG↑GNCCPy	3	44	3	0	4	100,93;57,93
		DraIII	CACNNN↑GTG	10	10	0	1	0	100,93;57,93
Deinococcus species RFL1	A.A. Janulaitis	Dspl1 (SacII)	CCGCGG	4	33	0	1	0	538
Desulfococcus mobilus	W. Zillig	DmoI	?	?	?	?	?	?	193
Desulfovibrio desulfuricans	ATCC 27774	DdsI (BamHI)	GGATCC	5	3	1	0	1	181
Desulfovibrio desulfuricans Norway	H. Peck strain	DdeI	*C↑TNAG	104	97	20	14	8	182;80;392
		DdeII (XhoI)	CTCGAG	1	6	0	1	0	211

RESTRICTION AND MODIFICATION ENZYMES AND THEIR RECOGNITION SEQUENCES (continued)

Microorganism	Source	Enzyme[a]	Sequence[b]	Number of cleavage sites[c]					Ref.
				λ	Ad2	SV40	ΦX	pBR	
Diplococcus pneumoniae	S. Lacks	DpnI	GA↑T*C	Only cleaves methylated DNA					159;76,160
		DpnII (MboI)	GATC	116	87	8	0	22	159;160
Enterobacter aerogenes	P.R. Whitehead	EaeI (CfrI)	Py↑GGCCPu	39	70	0	2	6	327;449
Enterobacter aerogenes	ATCC 15038	EaePI (PstI)	CTGCAG	28	30	2	1	1	219
Enterobacter aerogenes	NEB 450	EarI (Ksp632I)	CTCTTC	34	29	1	2	2	477
Enterobacter agglomerans	NEB 368	EagI (XmaIII)	C↑GGCCG	2	19	0	0	1	355
Enterobacter agglomerans	L. Kauc	EagKI (EcoRII)	CC(A/T)GG	71	136	17	2	6	474
Enterobacter agglomerans M3	CAMB 2541	EagMI (AvaII)	G↑G(A/T)CC	35	73	6	1	8	485
Enterobacter cloacae	DSM 30056	EcaI (BstEII)	G↑GTNACC	13	10	0	0	0	109
		EcaII (EcoRII)	CC(A/T)GG	71	136	17	2	6	211
Enterobacter cloacae	DSM 30060	EccI (SacII)	CCGCGG	4	33	0	1	0	189;211
Enterobacter cloacae	H. Hartmann	EclI	?	14	?	?	?	?	101
		EclII (EcoRII)	CC(A/T)GG	71	136	17	2	6	101
Enterobacter cloacae	CAMB 2542	EclJI (PvuI)	CGAT↑CG	3	7	0	0	1	486
Enterobacter cloacae	W. Piepersberg	EclS39I (EcoRII)	CC(A/T)GG	71	136	17	2	6	381
Enterobacter cloacae 593	C. Kessler	Ecl593I (PstI)	CTGCAG	28	30	2	1	1	381
Enterobacter cloacae RFL133	A.A. Janulaitis	Ecl133I (PstI)	CTGCAG	28	30	2	1	1	422
Enterobacter cloacae RFL136	A.A. Janulaitis	Ecl136I (EcoRII)	CC(A/T)GG	71	136	17	2	6	422
		Ecl136II (SacI)	GAG↑CTC	2	16	0	0	0	421
Enterobacter cloacae RFL137	A.A. Janulaitis	Ecl137I (SacI)	GAGCTC	2	16	0	0	0	408
		Eco137II (EcoRII)	CC(A/T)GG	71	136	17	2	6	408
Enterobacter cloacae RFL28	A.A. Janulaitis	Ecl28I (SacII)	CCGCGG	4	33	0	1	0	122
Enterobacter cloacae RFL37	A.A. Janulaitis	Ecl37I (SacII)	CCGCGG	4	33	0	1	0	121
Enterobacter cloacae RFL66	A.A. Janulaitis	Ecl66I (EcoRII)	CC(A/T)GG	71	136	17	2	6	414
Enterobacter cloacae RFL77	A.A. Janulaitis	Ecl77I (PstI)	CTGCAG	28	30	2	1	1	414
Enterobacter species RFL141	A.A. Janulaitis	Esp141I (PstI)	CTGCAG	28	30	2	1	1	408
Erwinia herbicola 9/5	A.A. Janulaitis	EheI (NarI)	GGC↑GCC	1	20	0	2	4	553,554
Erwinia rhaponici	D. Jones	ErpI (AvaII)	G↑G(A/T)CC	35	73	6	1	8	368
Escherichia coli	ATCC 26	EcoHI (CfrI)	PyGGCCPu	39	70	0	2	6	479
Escherichia coli	NEB 441	EcoNI	CCTNN↑NNNAGG	9	10	2	0	1	475
Escherichia coli	S. Glover	EcoR124/3I	GAANNNNNNNPuT*G[j]	Type I enzyme					387
		EcoR124I	GAANNNNNNNPuTCG	Type I enzyme					387
Escherichia coli (PI)	K. Murray	EcoPI	AG*ACC[e]	Type III enzyme					96;5;23,24;5,102
Escherichia coli 15T-	T. Bickle	EcoAI	G*AGNNNNNNNGTCA[i]	Type I enzyme					293,156
Escherichia coli 2bT	ICR 0020	EcoICRI (SacI)	GAGCTC	2	16	0	0	0	290

RESTRICTION AND MODIFICATION ENZYMES AND THEIR RECOGNITION SEQUENCES (continued)

Microorganism	Source	Enzyme[a]	Sequence[b]	λ	Ad2	SV40	ΦX	pBR	Ref.
Escherichia coli B	W. Arber	*Eco*BI	TG*ANNNNNNNTGCT	Type I enzyme					67;166,231;167;313
Escherichia coli CK	S.S. Debov	*Eco*CKI	?	4	?	?	?	?	308
Escherichia coli E1585-68	M. Yoshikawa	*Eco*VIII (*Hind*III)	A↑AGCTT	6	12	6	0	1	198
Escherichia coli E166	N.E. Murray	*Eco*DI	TTANNNNNNNGTCPy	Type I enzyme					212
Escherichia coli H304	K. Mise	*Eco*O34I	?	?	?	?	?	?	507
Escherichia coli H709c	I. Orskov	*Eco*O109I (*Dra*II)	PuG↑GNCCPy	3	44	3	0	4	200
Escherichia coli J62 pLG74	L.I. Glatman	*Eco*RV	GAT↑ATC	21	9	1	0	1	144;144,251
Escherichia coli K	M. Meselson	*Eco*KI	AACNNNNNNNGTGC[d]	Type I enzyme					195;12,139;97;139
Escherichia coli K11a	K. Mise	*Eco*O65I (*Bst*EII)	G↑GTNACC	13	10	0	0	0	507,508
Escherichia coli P15	W. Arber	*Eco*P15I	CAGCAG	Type III[e] enzyme					233;98
Escherichia coli R245	R.N. Yoshimori	*Eco*RII	↑CC(A/T)GG	71	136	17	2	6	340;13,22;340;22
Escherichia coli RFL100	A.A. Janulaitis	*Eco*100I (*Sac*II)	CCGCGG	4	33	0	1	0	408
Escherichia coli RFL101	A.A. Janulaitis	*Eco*101I (*Eco*31I)	GGTCTC	2	18	0	0	1	408
Escherichia coli RFL104	A.A. Janulaitis	*Eco*104I (*Sac*II)	CCGCGG	4	33	0	1	0	415
Escherichia coli RFL105	A.A. Janulaitis	*Eco*105I (*Sna*BI)	TAC↑GTA	1	0	0	0	0	421
Escherichia coli RFL113	A.A. Janulaitis	*Eco*113I (*Hgi*JII)	GPuGCPyC	7	57	2	0	2	408
Escherichia coli RFL115	A.A. Janulaitis	*Eco*115I (*Sau*I)	CCTNAGG	2	7	0	0	0	408
Escherichia coli RFL118	A.A. Janulaitis	*Eco*118I (*Sau*I)	CCTNAGG	2	7	0	0	0	408
Escherichia coli RFL120	A.A. Janulaitis	*Eco*120I (*Eco*31I)	GGTCTC	2	18	0	0	1	408
Escherichia coli RFL121	A.A. Janulaitis	*Eco*121I (*Cau*II)	CC(C/G)GG	114	97	0	1	10	408
Escherichia coli RFL125	A.A. Janulaitis	*Eco*125I (*Eco*57I)	CTGAAG	40	23	3	0	2	422
Escherichia coli RFL127	A.A. Janulaitis	*Eco*127I (*Eco*31I)	GGTCTC	2	18	0	0	1	422
Escherichia coli RFL128	A.A. Janulaitis	*Eco*128I (*Eco*RII)	CC(A/T)GG	71	136	17	2	6	422
Escherichia coli RFL129	A.A. Janulaitis	*Eco*129I (*Eco*31I)	GGTCTC	2	18	0	0	1	422
Escherichia coli RFL130	A.A. Janulaitis	*Eco*130I (*Sty*I)	C↑C(A/T)(A/T)GG	10	44	8	0	1	423,544
Escherichia coli RFL133	A.A. Janulaitis	*Eco*133I (*Pst*I)	CTGCAG	28	30	2	1	1	422
Escherichia coli RFL134	A.A. Janulaitis	*Eco*134I (*Sac*II)	CCGCGG	4	33	0	1	0	422
Escherichia coli RFL135	A.A. Janulaitis	*Eco*135I (*Sac*II)	CCGCGG	4	33	0	1	0	422
Escherichia coli RFL136	A.A. Janulaitis	*Eco*136I (*Eco*RII)	CC(A/T)GG	71	136	17	2	6	422
	A.A. Janulaitis	*Eco*136II (*Sac*I)	GAGCTC	2	16	0	0	1	422
Escherichia coli RFL141	A.A. Janulaitis	*Eco*141I (*Pst*I)	CTGCAG	28	30	2	1	1	408
Escherichia coli RFL143	A.A. Janulaitis	*Eco*143I (*Bss*HII)	GCGCGC	6	52	0	0	0	127
Escherichia coli RFL147	A.A. Janulaitis	*Eco*147I (*Stu*I)	AGG↑CCT	6	11	7	1	0	423,545
Escherichia coli RFL149	A.A. Janulaitis	*Eco*149I (*Kpn*I)	GGTACC	2	8	1	0	0	422
Escherichia coli RFL153	A.A. Janulaitis	*Eco*153I (*Scr*FI)	CCNGG	185	233	17	3	16	408
Escherichia coli RFL155	A.A. Janulaitis	*Eco*155I (*Eco*31I)	GGTCTC	2	18	0	0	1	408
Escherichia coli RFL156	A.A. Janulaitis	*Eco*156I (*Eco*31I)	GGTCTC	2	18	0	0	1	408

RESTRICTION AND MODIFICATION ENZYMES AND THEIR RECOGNITION SEQUENCES (continued)

Microorganism	Source	Enzyme[a]	Sequence[b]	Number of cleavage sites[c]					Ref.
				λ	Ad2	SV40	ΦX	pBR	
Escherichia coli RFL157	A.A. Janulaitis	Eco157I (Eco31I)	GGTCTC	2	18	0	0	1	408
Escherichia coli RFL158	A.A. Janulaitis	Eco158I (SacII)	CCGCGG	4	33	0	1	0	408
	A.A. Janulaitis	Eco158II (SnaBI)	TACGTA	1	0	0	0	0	408
Escherichia coli RFL159	A.A. Janulaitis	Eco159I (EcoRI)	GAATTC	5	5	1	0	1	422
Escherichia coli RFL161	A.A. Janulaitis	Eco161I (PstI)	CTGCAG	28	30	2	1	1	422
Escherichia coli RFL162	A.A. Janulaitis	Eco162I (Eco31I)	GGTCTC	2	18	0	0	1	422
Escherichia coli RFL164	A.A. Janulaitis	Eco164I (CfrI)	PyGGCCPu	39	70	0	2	6	127
Escherichia coli RFL165	A.A. Janulaitis	Eco165I (EcoRII)	CC(A/T)GG	71	136	17	2	6	127
Escherichia coli RFL167	A.A. Janulaitis	Eco167I (PstI)	CTGCAG	28	30	2	1	1	422
Escherichia coli RFL168	A.A. Janulaitis	Eco168I (HgiCI)	GGPyPuCC	25	57	1	3	9	408
Escherichia coli RFL169	A.A. Janulaitis	Eco169I (HgiCI)	GGPyPuCC	25	57	1	3	9	127
Escherichia coli RFL170	A.A. Janulaitis	Eco170I (EcoRII)	CC(A/T)GG	71	136	17	2	6	127
Escherichia coli RFL171	A.A. Janulaitis	Eco171I (HgiCI)	GGPyPuCC	25	57	1	3	9	422
Escherichia coli RFL173	A.A. Janulaitis	Eco173I (HgiCI)	GGPyPuCC	25	57	1	3	9	422
Escherichia coli RFL178	A.A. Janulaitis	Eco178I (EcoRV)	GATATC	21	9	1	0	1	422
Escherichia coli RFL179	A.A. Janulaitis	Eco179I (CauII)	CC(C/G)GG	114	97	0	1	10	422
Escherichia coli RFL180	A.A. Janulaitis	Eco180I (HgiJII)	GPuGCPyC	7	57	2	0	2	422
Escherichia coli RFL182	A.A. Janulaitis	Eco182I (SacII)	CCGCGG	4	33	0	1	0	422
Escherichia coli RFL185	A.A. Janulaitis	Eco185I (Eco31I)	GGTCTC	2	18	0	0	1	408
Escherichia coli RFL188	A.A. Janulaitis	Eco188I (HindIII)	AAGCTT	6	12	6	0	1	408
Escherichia coli RFL190	A.A. Janulaitis	Eco190I (CauII)	CC(C/G)GG	114	97	0	1	10	408
Escherichia coli RFL191	A.A. Janulaitis	Eco191I (Eco31I)	GGTCTC	2	18	0	0	1	408
Escherichia coli RFL193	A.A. Janulaitis	Eco193I (EcoRII)	CC(A/T)GG	71	136	17	2	6	408
Escherichia coli RFL195	A.A. Janulaitis	Eco195I (HgiCI)	GGPyPuCC	25	57	1	3	9	408
Escherichia coli RFL196	A.A. Janulaitis	Eco196I (SacII)	CCGCGG	4	33	0	1	0	422
	A.A. Janulaitis	Eco196II (AsuI)	GGNCC	74	164	11	2	15	422
Escherichia coli RFL200	A.A. Janulaitis	Eco200I (ScrFI)	CCNGG	185	233	17	3	16	422
Escherichia coli RFL201	A.A. Janulaitis	Eco201I (AsuI)	GGNCC	74	164	11	2	15	422
Escherichia coli RFL203	A.A. Janulaitis	Eco203I (Eco31I)	GGTCTC	2	18	0	0	1	408
Escherichia coli RFL204	A.A. Janulaitis	Eco204I (Eco31I)	GGTCTC	2	18	0	0	1	408
Escherichia coli RFL205	A.A. Janulaitis	Eco205I (Eco31I)	GGTCTC	2	18	0	0	1	408
Escherichia coli RFL24	A.A. Janulaitis	Eco24I (HgiJII)	GPuGCPy↑C	7	57	2	0	2	122, 122;540
Escherichia coli RFL25	A.A. Janulaitis	Eco25I (HgiJII)	GPuGCPyC	7	57	2	0	2	122
Escherichia coli RFL26	A.A. Janulaitis	Eco26I (HgiJII)	GPuGCPyC	7	57	2	0	2	134
Escherichia coli RFL31	A.A. Janulaitis	Eco31I	GGTCTC (1/5)	2	18	0	0	1	384
Escherichia coli RFL32	A.A. Janulaitis	Eco32I (EcoRV)	GAT↑ATC	21	9	1	0	1	134;407
Escherichia coli RFL35	A.A. Janulaitis	Eco35I (HgiJII)	GPuGCPyC	7	57	2	0	2	122
Escherichia coli RFL38	A.A. Janulaitis	Eco38I (EcoRII)	CC(A/T)GG	71	136	17	2	6	121
Escherichia coli RFL39	A.A. Janulaitis	Eco39I (AsuI)	GGNCC	74	164	11	2	15	121
Escherichia coli RFL40	A.A. Janulaitis	Eco40I (EcoRII)	CC(A/T)GG	71	136	17	2	6	121
Escherichia coli RFL41	A.A. Janulaitis	Eco41I (EcoRII)	CC(A/T)GG	71	136	17	2	6	121

RESTRICTION AND MODIFICATION ENZYMES AND THEIR RECOGNITION SEQUENCES (continued)

Microorganism	Source	Enzyme[a]	Sequence[b]	λ	Ad2	SV40	ΦX	pBR	Ref.
Escherichia coli RFL42	A.A. Janulaitis	*Eco*42I (*Eco*31I)	GGTCTC	2	18	0	0	1	408
Escherichia coli RFL43	A.A. Janulaitis	*Eco*43I (*Scr*FI)	CCNGG	185	233	17	3	16	409
Escherichia coli RFL47	A.A. Janulaitis	*Eco*47I (*Ava*II)	G↑G(A/T)CC	35	73	6	1	8	128;410
		*Eco*47II (*Asu*I)	GGNCC	74	164	11	2	15	128
		*Eco*47III	AGC↑GCT	2	13	1	0	4	128
Escherichia coli RFL48	A.A. Janulaitis	*Eco*48I (*Pst*I)	CTGCAG	28	30	2	1	1	127
Escherichia coli RFL49	A.A. Janulaitis	*Eco*49I (*Pst*I)	CTGCAG	28	30	2	1	1	127
Escherichia coli RFL50	A.A. Janulaitis	*Eco*50I (*Hgi*CI)	GGPyPuCC	25	57	1	3	9	127
Escherichia coli RFL51	A.A. Janulaitis	*Eco*51I (*Eco*31I)	GGTCTC	2	18	0	0	1	127
		*Eco*51II (*Scr*FI)	CCNGG	185	233	17	3	16	127
Escherichia coli RFL52	A.A. Janulaitis	*Eco*52I (*Xma*III)	C↑GGCCG	2	19	0	0	1	134;410
Escherichia coli RFL55	A.A. Janulaitis	*Eco*55I (*Sac*II)	CCGCGG	4	33	0	1	0	127
Escherichia coli RFL56	A.A. Janulaitis	*Eco*56I (*Nae*I)	G↑CCGGC	1	13	0	1	4	134,541
Escherichia coli RFL57	A.A. Janulaitis	*Eco*57I	CTGAAG (16/14)	40	23	3	0	2	411
Escherichia coli RFL60	A.A. Janulaitis	*Eco*60I (*Eco*RII)	CC(A/T)GG	71	136	17	2	6	134
Escherichia coli RFL61	A.A. Janulaitis	*Eco*61I (*Eco*RII)	CC(A/T)GG	71	136	17	2	6	134
Escherichia coli RFL64	A.A. Janulaitis	*Eco*64I (*Hgi*CI)	G↑GPyPuCC	25	57	1	3	9	413,542
Escherichia coli RFL65	A.A. Janulaitis	*Eco*65I (*Hind*III)	AAGCTT	6	12	6	0	1	414
Escherichia coli RFL67	A.A. Janulaitis	*Eco*67I (*Eco*RII)	CC(A/T)GG	71	136	17	2	6	413
Escherichia coli RFL68	A.A. Janulaitis	*Eco*68I (*Hgi*JII)	GPuGCPyC	7	57	2	2	2	414
Escherichia coli RFL70	A.A. Janulaitis	*Eco*70I (*Eco*RII)	CC(A/T)GG	71	136	17	2	6	413
Escherichia coli RFL71	A.A. Janulaitis	*Eco*71I (*Eco*RII)	CC(A/T)GG	71	136	17	2	6	415
Escherichia coli RFL72	A.A. Janulaitis	*Eco*72I (*Pma*CI)	CAC↑GTG	3	10	0	0	0	416
Escherichia coli RFL76	A.A. Janulaitis	*Eco*76I (*Sau*I)	CCTNAGG	2	7	0	0	0	414
Escherichia coli RFL78	A.A. Janulaitis	*Eco*78I (*Nar*I)	GGC↑GCC	1	20	0	2	4	417
Escherichia coli RFL80	A.A. Janulaitis	*Eco*80I (*Scr*FI)	CCNGG	185	233	17	3	16	414
Escherichia coli RFL81	A.A. Janulaitis	*Eco*81I (*Sau*I)	CC↑TNAGG	2	7	0	0	0	418
Escherichia coli RFL82	A.A. Janulaitis	*Eco*82I (*Eco*RI)	GAATTC	5	5	1	0	1	413
Escherichia coli RFL83	A.A. Janulaitis	*Eco*83I (*Pst*I)	CTGCAG	28	30	2	1	1	414
Escherichia coli RFL85	A.A. Janulaitis	*Eco*85I (*Scr*FI)	CCNGG	185	233	17	3	16	409
Escherichia coli RFL88	A.A. Janulaitis	*Eco*88I (*Ava*I)	CPyCGPuG	8	40	0	1	1	419
Escherichia coli RFL90	A.A. Janulaitis	*Eco*90I (*Cfr*I)	PyGGCCPu	39	70	0	2	6	420
Escherichia coli RFL91	A.A. Janulaitis	*Eco*91I (*Bst*EII)	G↑GTNACC	13	10	0	0	0	420,543
Escherichia coli RFL92	A.A. Janulaitis	*Eco*92I (*Sac*II)	CCGCGG	4	33	0	1	0	414
Escherichia coli RFL93	A.A. Janulaitis	*Eco*93I (*Scr*FI)	CCNGG	185	233	17	3	16	414
Escherichia coli RFL95	A.A. Janulaitis	*Eco*95I (*Eco*31I)	GGTCTC	2	18	0	0	1	413
Escherichia coli RFL96	A.A. Janulaitis	*Eco*96I (*Sac*II)	CCGCGG	4	33	0	1	0	414
Escherichia coli RFL97	A.A. Janulaitis	*Eco*97I (*Eco*31I)	GGTCTC	2	18	0	0	1	414
Escherichia coli RFL98	A.A. Janulaitis	*Eco*98I (*Hind*III)	AAGCTT	6	12	6	0	1	414

RESTRICTION AND MODIFICATION ENZYMES AND THEIR RECOGNITION SEQUENCES (continued)

Microorganism	Source	Enzyme[a]	Sequence[b]	λ	Ad2	SV40	ΦX	pBR	Ref.
						Number of cleavage sites[c]			
Escherichia coli RFL99	A.A. Janulaitis	*Eco*99I (*Sac*II)	CCGCGG	4	33	0	1	0	414
Escherichia coli RY13	R.N. Yoshimori	*Eco*RI	G↑A*ATTC	5	5	1	0	1	91;104;91;59
Escherichia coli TB104	N. Terakado	*Eco*T104I (*Sty*I)	CC(A/T)(A/T)GG	10	44	8	0	1	338
Escherichia coli TB14	N. Terakado	*Eco*T14I (*Sty*I)	C↑C(A/T)(A/T)GG	10	44	8	0	1	338
Escherichia coli TB22	N. Terakado	*Eco*T22I (*Ava*III)	ATGCA↑T	14	9	3	0	0	395
Escherichia coli TH38	K. Mise	*Eco*T38I (*Hgi*JII)	GPuGCPyC	7	57	2	0	2	395
Escherichia coli pDXX1	A. Piekarowicz	*Eco*DXXI	TCANNNNNNNATTC	Type I enzyme					385;391
Eucapsis species	PCC 6906	*Esp*I	GC↑TNAGC	7	8	1	0	0	34
		*Esp*II	?	?	?	?	?	?	514
Fischerella species	ATCC 29114	*Fsp*I (*Mst*I)	TGC↑GCA	15	17	0	1	4	298;406
		*Fsp*II (*Asu*II)	TT↑CGAA	7	1	0	0	0	298
Fischerella species	M. Sargent	*Fsp*MSI (*Ava*II)	G↑G(A/T)CC	35	73	6	1	8	368
Flavobacterium balustinum	ATCC 33487	*Fba*I (*Bcl*I)	TGATCA	8	5	1	0	0	359
Flavobacterium breve	NEB 379	*Fbr*I (*Fnu*4HI)	GC↑NGC	379	411	24	31	42	359
Flavobacterium indologenes	NEB 382	*Fin*I	GTCCC	38	59	8	2	4	202
		*Fin*II (*Hpa*II)	CCGG	328	171	1	5	26	202
Flavobacterium indoltheticum	ATCC 27950	*Fin*SI (*Hae*III)	GGCC	149	216	18	11	22	359
Flavobacterium okeanokoites	IFO 12536	*Fok*I	GGATG (9/13)	150	78	11	8	12	288
Flavobacterium species	NEB 380	*Fsp*MI (*Fnu*DII)	CGCG	157	303	0	14	23	202
Flavobacterium suaveolens	NEB 400	*Fsa*I	?	?	?	?	?	?	390
Flavobacterium suaveolens	ATCC 13718	*Fsu*I (*Tth*111I)	GACNNNGTC	2	12	0	0	1	390
Flavobacterium sulfureum	NEB 386	*Fsf*I (*Eco*57I)	CTGAAG	40	23	3	0	2	455
Fremyella diplosiphon	PCC 7601	*Fdi*I (*Ava*II)	G↑G(A/T)CC	35	73	6	1	8	309,286
		*Fdi*II (*Mst*I)	TGC↑GCA	15	17	0	1	4	309,286
Fusobacterium nucleatum 48	M. Smith	*Fnu*48I	?	>50	?	?	>10	?	176
Fusobacterium nucleatum 4H	M. Smith	*Fnu*4HI	GC↑NGC	379	411	24	31	42	172
Fusobacterium nucleatum A	M. Smith	*Fnu*AI (*Hinf*I)	G↑ANTC	148	72	10	21	10	177
		*Fnu*AII (*Mbo*I)	GATC	116	87	8	0	22	177;211
Fusobacterium nucleatum C	M. Smith	*Fnu*CI (*Mbo*I)	↑GATC	116	87	8	0	22	177
Fusobacterium nucleatum D	M. Smith	*Fnu*DI (*Hae*III)	GG↑CC	149	216	18	11	22	177
		*Fnu*DII	CG↑CG	157	303	0	14	23	177
		*Fnu*DIII (*Hha*I)	GCG↑C	215	375	2	18	31	177
Fusobacterium nucleatum E	M. Smith	*Fnu*EI (*Mbo*I)	↑GATC#	116	87	8	0	22	177
Gloeocapsa species	ATCC 29159	*Gse*I (*Asu*I)	GGNCC	74	164	11	2	15	533
		*Gse*II (*Pst*I)	CTGCAG	28	30	2	1	1	533
		*Gse*III (*Bam*HI)	GGATCC	5	3	1	0	1	533

RESTRICTION AND MODIFICATION ENZYMES AND THEIR RECOGNITION SEQUENCES (continued)

Microorganism	Source	Enzyme[a]	Sequence[b]	Number of cleavage sites[c]					Ref.
				λ	Ad2	SV40	ΦX	pBR	
Gloeothece species	ATCC 27152	GspI (PvuII)	CAGCTG	15	24	3	0	1	514
Gloeotricia species	D. Clark	GspAI (AvaII)	GG(A/T)CC	35	73	6	1	8	495
		GspAII (MstI)	TGCGCA	15	17	0	1	4	495
		GspAIII	?	1	4	?	?	?	495
Gluconobacter albidus	IFO 3251	GalI (SacII)	CCGC↑GG	4	33	0	1	0	356
Gluconobacter cerinus	IFO 3285	GceGLI (SacII)	CCGC↑GG	4	33	0	1	0	348
Gluconobacter cerinus	IFO 3262	GceI (SacII)	CCGC↑GG	4	33	0	1	0	356
Gluconobacter dioxyacetonicus	IAM 1814	GdiI (StuI)	AGG↑CCT	6	11	7	1	0	311
		GdiII	PyGGCCG (-5/-1)	21	53	0	2	5	311
Gluconobacter dioxyacetonicus	IAM 1840	GdoI (BamHI)	GGATCC	5	3	1	0	1	262
Gluconobacter gluconicus	IFO 3285	GgI	?	?	?	?	?	?	290
Gluconobacter industricus	IFO 3260	GinI (BamHI)	GGATCC	5	3	1	0	1	290
Gluconobacter oxydans sub. melonogenes	IAM 1836	GoxI (BamHI)	GGATCC	5	3	1	0	1	262
Gluconobacter suboxydans H-15T	M.S. Loytsianskaya	GsuI	CTGGAG (16/14)	25	32	6	3	4	123,411;411
	ATCC 11116	HaeI	(A/T)GG↑CC(A/T)	64	56	11	6	7	213
Haemophilus aegyptius		HaeII	PuGCGC↑Py	48	76	1	8	11	235;307
		HaeIII	GG↑*CC	149	216	18	11	22	196;25;184;184
Haemophilus aphrophilus	ATCC 19415	HapI	?	>30	?	?	?	?	211
		HapII (HpaII)	C↑CGG	328	171	1	5	26	301;291
Haemophilus gallinarum	ATCC 14385	HgaI	GACGC (5/10)	102	87	0	14	11	301;30,287
Haemophilus haemoglobinophilus	ATCC 19416	HhgI (HaeIII)	GGCC	149	216	18	11	22	211
Haemophilus haemolyticus	ATCC 10014	HhaI	*GCG↑C	215	375	2	18	31	237;237;403
		HhaII (HinfI)	G↑ANTC	148	72	10	21	10	183;183,402;276
	J. Stuy	Hin1056I (FnuDII)	CGCG	157	303	0	14	23	217
		Hin1056II	?	>30	>30	?	5	?	217
Haemophilus influenzae 1056									
Haemophilus influenzae 173	J. Chirikjian	Hin173I (HindIII)	AAGCTT	6	12	6	0	1	280
Haemophilus influenzae GU	J. Chirikjian	HinGUI (HhaI)	GCGC	215	375	2	18	31	280,40
		HinGUII (FokI)	GGATG	150	78	11	8	12	280;213,303
Haemophilus influenzae H-1	M. Takanami	HinHI (HaeII)	PuGCGCPy	48	76	1	8	11	301
Haemophilus influenzae JC9	A. Piekarowicz	HinJCI (HindII)	GTPy↑PuAC	35	25	7	13	2	223
		HinJCII (HindIII)	AAGCTT	6	12	6	0	1	223
Haemophilus influenzae P1	S. Shen	HinP1I (HhaI)	G↑CGC	215	375	2	18	31	264
Haemophilus influenzae RFL1	A.A. Janulaitis	HinII (AcyI)	GPu↑CGPyC	40	44	0	7	6	537
		HinIII (NlaIII)	CATG↑	181	183	17	22	26	548
Haemophilus influenzae RFL2	A.A. Janulaitis	Hin2I (HpaII)	CCGG	328	171	1	5	26	549
Haemophilus influenzae RFL3	A.A. Janulaitis	Hin3I (CauII)	CC(C/G)GG	114	97	0	1	10	549

RESTRICTION AND MODIFICATION ENZYMES AND THEIR RECOGNITION SEQUENCES (continued)

Microorganism	Source	Enzyme[a]	Sequence[b]	Number of cleavage sites[c]					Ref.
				λ	Ad2	SV40	ΦX	pBR	
Haemophilus influenzae RFL5	A.A. Janulaitis	*Hin*5I (*Hpa*II)	CCGG	328	171	1	5	26	550
		*Hin*5II (*Asu*I)	GGNCC	74	164	11	2	15	550
		*Hin*5III (*Hind*III)	AAGCTT	6	12	6	0	1	550
Haemophilus influenzae RFL6	A.A. Janulaitis	*Hin*6I (*Hha*I)	G↑CGC	215	375	2	18	31	551
Haemophilus influenzae RFL7	A.A. Janulaitis	*Hin*7I (*Hha*I)	GCGC	215	375	2	18	31	550
Haemophilus influenzae RFL8	A.A. Janulaitis	*Hin*8I (*Acy*I)	GPuCGPyC	40	44	0	7	6	550
		*Hin*8II (*Nla*III)	CATG	181	183	17	22	26	550
Haemophilus influenzae Rb	C.A. Hutchison	*Hinb*III (*Hind*III)	AAGCTT	6	12	6	0	1	197,211
Haemophilus influenzae Rc	A. Landy, G. Leidy	*Hinc*II (*Hind*II)	GTPy↑PuAC	35	25	7	13	2	161
Haemophilus influenzae Rd (exo-mutant)	S. H. Goodgal	*Hind*I	C$\overset{*}{A}$C	Specific methylase					244;245
		*Hind*II	GTPy↑Pu$\overset{*}{A}$C	35	25	7	13	2	278;142;244;245
		*Hind*III	$\overset{*}{A}$↑AGCTT	6	12	6	0	1	216;216;244;245
		*Hind*IV	G$\overset{*}{A}$T	Specific methylase					244;245
Haemophilus influenzae Rf	C.A. Hutchison	*Hinf*I	G↑ANTC	148	72	10	21	10	197;117,209
		*Hinf*II (*Hind*III)	AAGCTT	6	12	6	0	1	185
		*Hinf*III	CGAAT[e,f]	Type III enzyme					140;222
Haemophilus influenzae S1	S. Shen	*Hin*S11 (*Hha*I)	GCGC	215	375	2	18	31	264
Haemophilus influenzae S2	S. Shen	*Hin*S21 (*Hha*I)	GCGC	215	375	2	18	31	264
Haemophilus influenzae serotype b, 1076	J. Stuy	*Hin*1076II (*Hind*III)	AAGCTT	6	12	6	0	1	217
Haemophilus influenzae serotype c, 1160	J. Stuy	*Hin*1160II (*Hind*II)	GTPyPuAC	35	25	7	13	2	217
Haemophilus influenzae serotype c, 1161	J. Stuy	*Hin*1161II (*Hind*II)	GTPyPuAC	35	25	7	13	2	217
Haemophilus influenzae serotype e	A. Piekarowicz	*Hine*I (*Hind*III)	CGAAT[e,f]	Type III enzyme					220
Haemophilus parahaemolyticus	C.A. Hutchison	*Hph*I	GGTGA (8/7)	168	99	4	9	12	197;151
Haemophilus parainfluenzae	J. Setlow	*Hpa*I	GTT↑AAC	14	6	4	3	0	263;74;1
		*Hpa*II	C↑$\overset{*}{C}$GG	328	171	1	5	26	263;74,184;184
Haemophilus suis	ATCC 19417	*Hsu*I (*Hind*III)	A↑AGCTT	6	12	6	0	1	211
Hafnia species RFL2	A.A. Janulaitis	*Hsp*2I (*Ava*II)	GG(A/T)CC	35	73	6	1	8	538
Halobacterium cutirubrum	D. Oesterhelt	*Hcu*I	?	?	?	?	?	?	434
Halobacterium halobium NRC817	D. Oesterhelt	*Hhl*I	?	?	?	?	?	?	434
Halobacterium palinarium	D. Oesterhelt	*Hsa*I	?	?	?	?	?	?	434
Halococcus acetoinfaciens	IAM 12094	*Hac*I (*Mbo*I)	↑GATC	116	87	8	0	22	356
Halococcus agglomeratus	ATCC 25862	*Hag*I	?	?	?	?	?	?	228

RESTRICTION AND MODIFICATION ENZYMES AND THEIR RECOGNITION SEQUENCES (continued)

Microorganism	Source	Enzyme[a]	Sequence[b]	Number of cleavage sites[c]					Ref.
				λ	Ad2	SV40	ΦX	pBR	
Herpetosiphon giganteus HFS101	H. Foster	*HgiJI (AvaII)*	G↑G(A/T)CC	35	73	6	1	8	405
		HgiJII	GPuGCPy↑C	7	57	2	0	2	405
Herpetosiphon giganteus HP1023	J.H. Parish	*HgiAI*	G(A/T)GC(A/T)↑C	28	38	0	3	8	29
Herpetosiphon giganteus HP1049	J.H. Parish	*HgiHI (HgiCI)*	G↑GPyPuCC	25	57	1	3	9	405
		HgiHII (AcyI)	GPu↑CGPyC	40	44	0	7	6	405
		HgiHIII (AvaII)	G↑G(A/T)CC	35	73	6	1	8	405
Herpetosiphon giganteus Hpa1	H. Reichenbach	*HgiGI (AcyI)*	GPu↑CGPyC	40	44	0	7	6	157
Herpetosiphon giganteus Hpa2	H. Reichenbach	*HgiDI (AcyI)*	GPu↑CGPyC	40	44	0	7	6	157
Herpetosiphon giganteus Hpg14	H. Reichenbach	*HgiDII (SalI)*	G↑TCGAC	2	3	0	0	1	157
		HgiFI	?	?	15	?	?	?	190
Herpetosiphon giganteus Hpg24	H. Reichenbach	*HgiEI (AvaII)*	G↑G(A/T)CC	35	73	6	1	8	157
		HgiEII	ACCNNNNNNGGT	14	10	1	1	2	157
Herpetosiphon giganteus Hpg32	H. Reichenbach	*HgiKI*	?	>18	>20	?	?	?	190
Herpetosiphon giganteus Hpg5	H. Reichenbach	*HgiBI (AvaII)*	G↑G(A/T)CC	35	73	6	1	8	157
Herpetosiphon giganteus Hpg9	H. Reichenbach	*HgiCI*	G↑GPyPuCC	25	57	1	3	9	157;463
		HgiCII (AvaII)	G↑G(A/T)CC	35	73	6	1	8	157
		HgiCIII (SalI)	G↑TCGAC	2	3	0	0	1	157
Herpetosiphon giganteus S21	H. Foster	*HgiS21I (CauII)*	CC(C/G)GG	114	97	0	1	10	381
Hyphomonas jannaschiana	R. Weiner	*HjaI (EcoRV)*	GATATC	21	9	1	0	1	517
Klebsiella oxytoca	CAMB 2553	*KoxI (BstEII)*	G↑GTNACC	13	10	0	0	0	487
		KoxII (HgiJII)	GPuGCPy↑C	7	57	2	2	2	487
Klebsiella oxytoca RFL165	A.A. Janulaitis	*Kox165I (EcoRII)*	CC(A/T)GG	71	136	17	2	6	127
Klebsiella pneumoniae OK8	J. Davies	*KpnI*	GGTAC↑C	2	8	1	0	0	275;304
Klebsiella pneumoniae RFL10	A.A. Janulaitis	*Kpn10I (EcoRII)*	CC(A/T)GG	71	136	17	2	6	422
Klebsiella pneumoniae RFL12	A.A. Janulaitis	*Kpn12I (PstI)*	CTGCAG	28	30	2	1	1	422
Klebsiella pneumoniae RFL13	A.A. Janulaitis	*Kpn13I (EcoRII)*	CC(A/T)GG	71	136	17	2	6	422
Klebsiella pneumoniae RFL14	A.A. Janulaitis	*Kpn14I (EcoRII)*	CC(A/T)GG	71	136	17	2	6	408
Klebsiella pneumoniae RFL16	A.A. Janulaitis	*Kpn16I (EcoRII)*	CC(A/T)GG	71	136	17	2	6	408
Klebsiella pneumoniae RFL2	A.A. Janulaitis	*Kpn21I (BspMII)*	T↑CCGGA	24	8	0	0	1	423,552
Klebsiella pneumoniae RFL30	A.A. Janulaitis	*Kpn30I (BssHII)*	GCGCGC	6	52	0	1	0	122
Klebsiella pneumoniae mmK14	W. Piepersberg	*KpnK14I (KpnI)*	GGTACC	2	8	1	0	0	381
Kluyvera species 632	DSM 4196	*Ksp6321*	CTCTTC (1/4)	34	29	1	2	2	518
Lactobacillus species	P. Eastlake	*LspI (AsuII)*	TT↑CGAA	7	1	0	0	0	366;27
Legionella pneumophila	A. Brown	*LpnI*	?	>12	?	?	?	?	452
Legionella pneumophila Philadelphia 1	A. Brown	*LpnII*	?	>5	?	?	?	?	452

RESTRICTION AND MODIFICATION ENZYMES AND THEIR RECOGNITION SEQUENCES (continued)

Microorganism	Source	Enzyme[a]	Sequence[b]	Number of cleavage sites[c]					Ref.
				λ	Ad2	SV40	ΦX	pBR	
Mastigocladus laminosus	CCAP 1447/1	MlaI (AsuII)	TT↑CGAA	7	1	0	0	0	61
Methanococcus aeolicus PL-15/H	K.O. Stetter	MaeI	C↑TAG	13	54	12	3	5	258
		MaeII	A↑CGT	143	83	0	19	10	258
		MaeIII	↑GTNAC	156	118	14	17	17	258
Methanococcus jannashii	H. Escalante	MjaI (MaeI)	CTAG	13	54	12	3	5	344
		MjaII (AsuI)	GGNCC	74	164	11	2	15	344
Methanococcus vannielii	M. Thomm	MvnI (FnuDII)	CG↑CG	157	303	0	14	23	513
Methylophilus methylotrophus	W.J. Brammer	MmeI	TCCPuAC (20/18)	18	25	2	5	4	397
		MmeII (MboI)	GATC	116	87	8	0	22	397
Microbacterium flavum	IAM 1642	MflI (XhoII)	Pu↑GATCPy	21	22	3	0	8	108
Microbacterium thermosphactum	ATCC 11509	MthI (MboI)	GATC	116	87	8	0	22	162
Micrococcus aurantiacus	IFO 12422	MauI (PstI)	CTGCAG	28	30	2	1	1	290
Micrococcus euryhalis	ATCC 14389	MeuI (MboI)	GATC	116	87	8	0	22	361
Micrococcus kristinae	ATCC 27571	MkrI (PstI)	CTGCAG	28	30	2	1	1	361
Micrococcus luteus	ATCC 540	MleI (BamHI)	GGATCC	5	3	1	0	1	361
Micrococcus luteus	ATCC 400	MltI (AluI)	AG↑CT	143	158	34	24	16	361;362
Micrococcus luteus	IFO 12992	MluI	A↑CGCGT	7	5	0	2	0	288
Micrococcus radiodurans	ATCC 13939	MraI (SacII)	CCGCGG	4	33	0	1	0	322
Micrococcus roseus	F. Kato	MroI (BspMII)	T↑CCGGA	24	8	0	0	1	512
Micrococcus species	R. Meagher	MisI (NaeI)	GCCGGC	1	13	1	0	4	194
Micrococcus species	NEB 446	MseI	T↑TAA	195	115	47	35	15	523
Micrococcus varians RFL19	A.A. Janulaitis	MvaI (EcoRII)	CC↑(A/T)GG *	71	136	17	2	6	33,444;33
Microcoleus species	D. Comb	MstI	TGC↑GCA	15	17	0	1	4	49;83
		MstII (SauI)	CC↑TNAGG	2	7	0	0	0	250
Micromonospora carbonacea	C. Kessler	McaI (XhoI)	CTCGAG	1	6	0	1	0	143
Micromonospora echinospora subs. echinospora	ATCC 15837	MecI (XhoI)	CTCGAG	1	6	0	1	0	448
Micromonospora purpurea	ATCC 15835	MpuI (XhoI)	CTCGAG	1	6	0	1	0	448
Micromonospora zionensis	H. Lechevalier	MziI (PvuII)	CAGCTG	15	24	3	0	1	448
Moraxella bovis	ATCC 10900	MboI	↑GATC	116	87	8	0	22	77
Moraxella bovis	ATCC 17947	MboII	GAAGA (8/7) *	130	113	16	11	11	77;28,64;350;350
Moraxella glueidi LG1	J. Davies	MbvI	?	?	?	?	?	?	138
Moraxella glueidi LG1	J. Davies	MglI	?	?	?	?	?	?	275
Moraxella glueidi LG2	J. Davies	MglII	?	?	?	?	?	?	275
Moraxella kingae	ATCC 23331	MkiI (HindIII)	AAGCTT	6	12	6	0	1	138
Moraxella nonliquefaciens	ATCC 19966	MniI (HaeIII)	GGCC	149	216	18	11	22	138
		MniII (HpaII)	CCGG	328	171	1	5	26	138

RESTRICTION AND MODIFICATION ENZYMES AND THEIR RECOGNITION SEQUENCES (continued)

Microorganism	Source	Enzyme[a]	Sequence[b]	Number of cleavage sites[c]					Ref.
				λ	Ad2	SV40	ΦX	pBR	
Moraxella nonliquefaciens	ATCC 17953	MnlI	CCTC (7/7)	262	397	51	34	26	341;250
Moraxella nonliquefaciens	ATCC 17954	MnnI (HindII)	GTPyPuAC	35	25	7	13	2	99
		MnnII (HaeIII)	GGCC	149	216	18	11	22	99
		MnnIII	?	>10	>6	3	?	?	99
		MnnIV (HhaI)	GCGC	215	375	2	18	31	99
Moraxella nonliquefaciens	ATCC 19975	MnoI (HpaII)	C↑CGG	328	171	1	5	26	211;7
		MnoII (MnnIII)	?	>10	>6	3	?	?	211
		MnoIII (MboI)	GATC	116	87	8	0	22	211
Moraxella osloensis	ATCC 19976	MosI (MboI)	GATC	116	87	8	0	22	77
Moraxella phenylpyruvica	ATCC 19955	MphI (EcoRII)	CC(A/T)GG	71	136	17	2	6	138
Moraxella species	R.J. Roberts	MspI (HpaII)	*C↑CGG	328	171	1	5	26	312;312,250;137
Moraxella species MS67	M. Sargent	Msp67I (ScrFI)	CC↑NGG	185	233	17	3	16	368
		Msp67II (MboI)	GATC	116	87	8	0	22	368
Mycoplasma fermentans	N.F. Halden	MfeI	CAATTG	8	4	4	1	0	558
Myxococcus stipitatus Mxs2	H. Reichenbach	MsiI (XhoI)	CTCGAG	1	6	0	1	0	197,211
		MsiII	?	?	?	?	?	?	190
Myxococcus virescens V-2	H. Reichenbach	MviI	?	1	?	?	?	?	204
		MviII	?	?	?	?	?	?	204
Neisseria animalis	ATCC 19573	NanI (EcoRV)	GATATC	21	9	1	0	1	360
		NanII (DpnI)	GATC	Only cleaves methylated DNA					360
Neisseria caviae	NRCC 31003	NcaI (HinfI)	GANTC	148	72	10	21	10	318
Neisseria cinerea	NRCC 31006	NciI (CauII)	CC↑(C/G)GG[g]	114	97	0	1	10	323;112
Neisseria cuniculi	ATCC 14688	NcuI (MboII)	GAAGA	130	113	16	11	11	31
Neisseria denitrificans	NRCC 31009	NdeI	CA↑TATG	7	2	2	0	1	324
		NdeII (MboI)	↑GATC	116	87	8	0	22	318
Neisseria flavescens	ATCC 13120	NflAI (EcoRV)	GATATC	21	9	1	0	1	186
		NflAII (MboI)	GATC	116	87	8	0	22	186
Neisseria flavescens	ATCC 13115	NflBI (MboI)	GATC	116	87	8	0	22	480
Neisseria flavescens	NRCC 31011	NflI (MboI)	GATC	116	87	8	0	22	318
		NflII	?	?	?	?	?	?	318
		NflIII	?	?	?	?	?	?	318
Neisseria gonorrhoea	G. Wilson	NgoI (HaeII)	PuGCGCPy	48	76	1	8	11	331
Neisseria gonorrhoea	CDC 66	NgoII (HaeIII)	GGCC	149	216	18	11	22	42
Neisseria gonorrhoea KH 7764-45	L. Mayer	NgoIII (SacII)	CCGCGG	4	33	0	1	0	214
Neisseria gonorrhoea MS11	M. So	NgoMI (NaeI)	GCCGGC	1	13	1	0	4	491
Neisseria gonorrhoea PGH3-2	D.C. Stein	NgoSI (HaeIII)	GGCC	149	216	18	11	22	528
Neisseria gonorrhoeae	C. Korch	NgoVIII (HphI)	GGTGA[i]*	Specific methylase					393

RESTRICTION AND MODIFICATION ENZYMES AND THEIR RECOGNITION SEQUENCES (continued)

Microorganism	Source	Enzyme[a]	Sequence[b]	Number of cleavage sites[c]					Ref.
				λ	Ad2	SV40	ΦX	pBR	
Neisseria gonorrhoeae JKD211	J.K. Davies	NgoDI (*SacII*)	CCGCGG	4	33	0	1	0	492
		NgoDII	?	?	?	?	?	?	492
		NgoDIII (*DpnI*)	GATC	Only cleaves methylated DNA					492
Neisseria gonorrhoeae P9-2	J.R. Saunders	NgoPII (*HaeIII*)	GG↑CC	149	216	18	11	22	493
		NgoPIII (*SacII*)	CCGC↑GG	4	33	0	1	0	493
Neisseria gonorrhoeae WR302	D.C. Stein	NgoBI (*HphI*)	GGTGA i *	168	99	4	9	12	471;520
Neisseria lactamica	NRCC 2118	NlaI (*HaeIII*)	GGCC	149	216	18	11	22	371
		NlaII (*MboI*)	↑GATC	116	87	8	0	22	371
		NlaIII	CATG↑	181	183	17	22	26	371
		NlaIV	GGN↑NCC	82	178	16	6	24	371
Neisseria lactamica	NRCC 31016	NlaSI (*SacII*)	CCGCGG	4	33	0	1	0	36
		NlaSII (*AcyI*)	GPuCGPyC	40	44	0	7	6	36
Neisseria lactamica 5841	D.C. Stein	NlaDI (*MboI*)	GATC	116	87	8	0	22	471
		NlaDII (*AsuI*)	GGNCC	74	164	11	2	15	471
		NlaDIII (*SacII*)	CCGCGG	4	33	0	1	0	471
Neisseria meningitidis C114	C.A. Hart	NmeCI (*MboI*)	↑GATC	116	87	8	0	22	493
Neisseria meningitidis DRES-30	R. Sparling	NmeIV	?	?	?	?	?	?	282
Neisseria meningitidis DRES-W34	R. Sparling	NmeI	?	18	?	?	?	?	282
		NmeII	?	28	?	0	0	?	282
	R. Sparling	NmeIII	?	?	?	?	?	?	282
Neisseria meningitidis M1011	ATCC 25999	NheI	G↑CTAGC	1	4	0	0	1	48
Neisseria mucosa	ATCC 25997	NmuDI (*DpnI*)	GATC	Only cleaves methylated DNA					35
Neisseria mucosa	ATCC 25996	NmuEI (*DpnI*)	GATC	Only cleaves methylated DNA					30
Neisseria mucosa		NmuEII (*AsuI*)	GGNCC	74	164	11	2	15	35
Neisseria mucosa	ATCC 19697	NmuFI (*NaeI*)	GCCGGC	1	13	1	0	4	35
Neisseria mucosa	NRCC 31013	NmuI (*NaeI*)	GCCGGC	1	13	1	0	4	318
Neisseria mucosa	ATCC 19693	NmuSI (*AsuI*)	GGNCC	74	164	11	2	15	47
Neisseria ovis	NRCC 31020	NovI	?	?	?	?	?	?	318
		NovII (*HinfI*)	GANTC	148	72	10	21	10	318
Neisseria pharyngis C245	C.A. Hart	NphI (*MboI*)	↑GATC	116	87	8	0	22	493
Neisseria sicca	ATCC 9913	NsiAI (*MboI*)	GATC	116	87	8	0	22	44
Neisseria sicca	NRCC 31004	NsiHI (*HinfI*)	GANTC	148	72	10	21	10	317
Neisseria sicca	ATCC 29256	NsiI (*AvaIII*)	ATGCA↑T	14	9	3	0	0	46
Neisseria sicca C351	C.A. Hart	NsiCI (*EcoRV*)	GAT↑ATC	21	9	1	0	1	493
Neisseria subflava	ATCC 19243	NsuDI (*DpnI*)	GATC	Only cleaves methylated DNA					35
Neisseria subflava	ATCC 14221	NsuI (*MboI*)	GATC	116	87	8	0	22	35

RESTRICTION AND MODIFICATION ENZYMES AND THEIR RECOGNITION SEQUENCES (continued)

Microorganism	Source	Enzyme[a]	Sequence[b]	Number of cleavage sites[c]					Ref.
				λ	Ad2	SV40	ΦX	pBR	
Nocardia aerocolonigenes	ATCC 23870	*Nae*I	GCC↑GGC	1	13	1	0	4	51
Nocardia amarae	ATCC 27809	*Nam*I (*Nar*I)	GGCGCC	1	20	0	2	4	175
Nocardia argentinensis	ATCC 31306	*Nar*I	GG↑CGCC	1	20	0	2	4	50
Nocardia asteroides	ATCC 9970	*Nas*BI (*Bam*HI)	GGATCC	5	3	1	0	1	361
Nocardia asteroides	ATCC 7372	*Nas*I (*Pst*I)	CTGCAG	28	30	2	1	0	359
Nocardia asteroides	ATCC 9969	*Nas*SI (*Sac*I)	GAGCTC	2	16	0	0	0	359
Nocardia asteroides	ATCC 14759	*Nas*WI (*Nae*I)	GCCGGC	1	13	1	0	4	361
Nocardia blackwellii	ATCC 6846	*Nbl*I (*Pvu*I)	CGAT↑CG	3	7	0	0	1	250
Nocardia brasiliensis	ATCC 19296	*Nba*I (*Nae*I)	GCCGGC	1	13	1	0	4	228
Nocardia brasiliensis	ATCC 27936	*Nbr*I (*Nae*I)	GCCGGC	1	13	1	0	4	228
Nocardia corallina	ATCC 19070	*Nco*I	C↑CATGG	4	20	3	0	0	162
Nocardia dassonvillei	ATCC 21944	*Nda*I (*Nar*I)	GG↑CGCC	1	20	0	2	4	45
Nocardia globerula	ATCC 21292	*Ngb*I (*Pst*I)	CTGCAG	28	30	2	1	1	361
Nocardia minima	ATCC 19150	*Nmi*I (*Kpn*I)	GGTACC	2	8	1	0	0	48
Nocardia opaca	ATCC 21507	*Nop*I (*Sal*I)	G↑TCGAC	2	3	0	0	1	250
		*Nop*II	?	?	?	?	?	?	162
Nocardia otitidis-caviarum	ATCC 14629	*Noc*I (*Pst*I)	CTGCAG	28	30	2	1	1	48
Nocardia otitidis-caviarum	ATCC 14630	*Not*I	GC↑GGCCGC	0	7	0	0	0	21;257
Nocardia rubra	ATCC 15906	*Nrn*D	TCG↑CGA	5	5	0	2	1	48
Nocardia species	ATCC 19170	*Nsp*AI (*Mbo*I)	GATC	116	87	8	0	22	361
Nocardia species	ATCC 29100	*Nsp*WI (*Nae*I)	GCCGGC	1	13	1	0	4	361
Nocardia tartaricans	ATCC 31191	*Nta*I (*Tth*111I)	GACNNNGTC	2	12	0	0	0	359
Nocardia tartaricans	ATCC 31190	*Nta*SI (*Stu*I)	AGGCCT	6	11	7	1	0	359
		*Nta*SII (*Nae*I)	GCCGGC	1	13	1	0	4	359
Nocardia uniformis	ATCC 21806	*Nun*I	?	?	?	?	?	?	162
		*Nun*II (*Nar*I)	GG↑CGCC	1	20	0	2	4	162
Nostoc linckia	A. de Waard	*Nli*I (*Ava*I)	CPyCGPuG	8	40	0	1	1	63
		*Nli*II (*Ava*II)	GG(A/T)CC	35	73	6	1	8	63
Nostoc muscorum M-131-G	A. de Waard	*Nmu*AI (*Ava*I)	CPyCGPuG	8	40	0	1	8	63
		*Nmu*AII (*Ava*II)	GG(A/T)CC	35	73	6	1	8	63
Nostoc species	ATCC 29131	*Nsp*BI (*Asu*II)	TTCGAA	7	1	0	0	0	63
Nostoc species		*Nsp*BII	C(A/C)G↑C(G/T)G	75	95	4	5	6	63
Nostoc species	ATCC 29105	*Nsp*DI (*Ava*I)	CPyCGPuG	8	40	0	1	1	63
		*Nsp*DII (*Ava*II)	GG(A/T)CC	35	73	6	1	8	63
Nostoc species	ATCC 27896	*Nsp*EI (*Ava*I)	CPyCGPuG	8	40	0	1	1	534
		*Nsp*EII	?	?	?	?	?	?	534
Nostoc species	ATCC 29150	*Nsp*FI (*Asu*II)	TTCGAA	7	1	0	0	0	533

RESTRICTION AND MODIFICATION ENZYMES AND THEIR RECOGNITION SEQUENCES (continued)

Microorganism	Source	Enzyme[a]	Sequence[b]	Number of cleavage sites[c]					Ref.
				λ	Ad2	SV40	ΦX	pBR	
Nostoc species	ATCC 29106	*Nsp*HI (*Nsp*I)	PuCATG↓Py	32	41	2	0	4	63
		*Nsp*HII (*Ava*II)	GG(A/T)CC	35	73	6	1	8	63
		*Nsp*HIII (*Mst*I)	TGCGCA	15	17	0	1	4	53
Nostoc species	PCC 8009	*Nsp*MACI (*Bgl*II)	A↓GATCT	6	11	0	0	0	164
Nostoc species	ATCC 27897	*Nsp*MI (*Mst*I)	TGCGCA	15	17	0	1	4	514
Nostoc species 19-6C-C	C.P. Wolk	*Nsp*KI (*Ava*II)	GG(A/T)CC	35	73	6	1	8	535
Nostoc species 23-9B	C.P. Wolk	*Nsp*GI (*Ava*II)	GG(A/T)CC	35	73	6	0	8	514
Nostoc species 78-12B	C.P. Wolk	*Nsp*II (*Asu*II)	TTCGAA	7	1	0	0	0	514
Nostoc species C	ATCC 29411	*Nsp*I	PuCATG↓Py	32	41	2	0	4	232
		*Nsp*II (*Sdu*I)	G(A/G/T)GC(A/C/T)↓C	38	105	4	3	10	232
		*Nsp*III (*Ava*I)	C↓PyCGPuG	8	40	0	1	1	232
		*Nsp*IV (*Asu*I)	G↓GNCC	74	164	11	2	15	232
		*Nsp*V (*Asu*II)	TTCGAA	7	1	0	0	0	232
Nostoc species SA	D. Jones	*Nsp*SAI (*Ava*I)	C↓PyCGPuG	8	40	0	1	1	369
		*Nsp*SAII (*Bst*EII)	G↓GTNACC	13	10	0	0	0	369
		*Nsp*SAIII (*Nco*I)	CCATGG	4	20	3	0	0	369
		*Nsp*SAIV (*Bam*HI)	G↓GATCC	5	3	1	0	1	369
Nostoc species UM-3	C.P. Wolk	*Nsp*LI (*Mst*I)	TGCGCA	15	17	0	1	4	535
		*Nsp*LII (*Asu*I)	GGNCC	74	164	11	2	15	535
		*Nsp*LIII	?	?	?	?	?	?	535
		*Nsp*LIV	?	?	?	?	?	?	535
Oerskovia turbata	ATCC 27403	*Ota*I (*Alu*I)	AGCT	143	158	34	24	16	477
Oerskovia turbata	ATCC 25835	*Otu*NI (*Alu*I)	AGCT	143	158	34	24	16	479
Oerskovia xanthineolytica	R. Shekman	*Oxa*I (*Alu*I)	AGCT	143	158	34	24	16	285
		*Oxa*II	?	?	?	?	?	?	285
Oerskovia xanthineolytica N	NEB 401	*Oxa*NI (*Sau*I)	CC↓TNAGG	2	7	0	0	0	389
Proteus myxofaciens	ATCC 19692	*Pmy*I (*Pst*I)	CTGCAG	28	30	2	1	1	375
Proteus vulgaris	ATCC 13315	*Pvu*I	CGAT↓CG	3	7	0	0	1	82
		*Pvu*II	CAG↓CTG	15	24	3	0	1	82;488
Providencia alcalifaciens	ATCC 9886	*Pal*I (*Hae*III)	GG↓CC	149	216	18	11	22	78;436
Providencia stuartii 164	J. Davies	*Pst*I	CTGCA↓G	28	30	2	1	1	275;31;383
Pseudoanabaena species	ATCC 29541	*Pse*I (*Asu*I)	GGNCC	74	164	11	2	15	53
Pseudoanabaena species	ATCC 27263	*Psp*I (*Asu*I)	GGNCC	74	164	11	2	15	206
Pseudomonas aeruginosa	CAMB 2549	*Pae*AI (*Sac*II)	CCGC↓GG	4	33	0	1	0	483
Pseudomonas aeruginosa	N.N. Sokolov	*Pae*I (*Sph*I)	GCATG↓C	6	8	2	0	1	281

RESTRICTION AND MODIFICATION ENZYMES AND THEIR RECOGNITION SEQUENCES (continued)

Microorganism	Source	Enzyme[a]	Sequence[b]	Number of cleavage sites[c]					Ref.
				λ	Ad2	SV40	ΦX	pBR	
Pseudomonas aeruginosa	G.A. Jacoby	*PaeR7I* (*XhoI*)	C↓TCGAG	1	5[h]	0	0	0	107;81;81
Pseudomonas aeruginosa RFL177	A.A. Janulaitis	*Pae177I* (*BamHI*)	GGATCC	5	3	1	0	1	422
Pseudomonas aeruginosa RFL181	A.A. Janulaitis	*Pae181I* (*CauII*)	CC(C/G)GG	114	97	0	1	10	422
Pseudomonas alkaligenes	ATCC 12815	*PaiI* (*HaeIII*)	GGCC	149	216	18	11	22	290
Pseudomonas alkanolytica	IFO 12319	*PanI* (*XhoI*)	C↓TCGAG	1	6	0	1	0	290
Pseudomonas facilis	M. VanMontagu	*PfaI* (*MboI*)	GATC	116	87	8	0	22	312
Pseudomonas fluorescens	IFO 3507	*PflI*	?	?	?	?	?	?	290
Pseudomonas fluorescens	NEB 375	*PflMI*	CCANNNN↓NTGG	14	18	2	2	2	202
Pseudomonas fluorescens	ATCC 33512	*PflNI* (*XhoI*)	CTCGAG	1	6	0	1	0	479
Pseudomonas fluorescens	T.S. Wang	*PflWI* (*XhoI*)	CTCGAG	1	6	0	1	0	320
Pseudomonas fluorescens	NEB 431	*PfiaI* (*SplI*)	CGTACG	1	4	0	2	0	477
Pseudomonas fluorescens type A	ATCC 17582	*PflAI* (*FnuDII*)	CGCG	157	303	0	14	23	477
Pseudomonas glycinae	J.V. Leary	*PglI* (*NaeI*)	GCCGGC	0	13	0	0	4	169
		PglII	?	0	>25	?	?	1	169
Pseudomonas lemoignei	NEB 418	*PleI*	GAGTC (4/5)	61	40	5	10	4	458
Pseudomonas maltophila	D. Comb	*PmaI* (*PstI*)	CTGCAG	28	30	2	1	1	250
Pseudomonas maltophila CB50P	C.A. Hart	*PmaCI*	CAC↓GTG	3	10	0	0	0	363
Pseudomonas maltophilia	CAMB 2550	*Pma44I* (*PstI*)	CTGCA↓G	28	30	2	1	1	484
Pseudomonas mirabilis	N.L. Bakh	*PmiI*	?	9	?	?	?	1	372
Pseudomonas ovalis	S. Riazuddin	*PovI* (*BclI*)	T↓GATCA	8	5	1	0	0	378,525
Pseudomonas paucimobilis	NEB 376	*PpaI* (*Eco31I*)	GGTCTC	2	18	0	0	1	202
Pseudomonas putida C-83	Toyoboseki Co.	*PpuI* (*HaeIII*)	GGCC	149	216	18	11	22	290
Pseudomonas putida M	NEB 372	*PpuMI*	PuG↓G(A/T)CCPy	3	23	1	0	2	351;202
Pseudomonas species	R.A. Makula	*PssI* (*DraII*)	PuGGNC↓CPy	3	44	3	0	4	181;9
		PssII	?	?	?	?	?	?	181
Pseudomonas species MS61	M. Sargent	*Psp6II* (*NaeI*)	GCCGGC	1	13	1	0	4	368
Rhizobium leguminosarum 300	J. Beringer	*RleI*	?	6	>10	?	?	?	333
Rhizobium lupini 1	W. Heumann	*RluI* (*NaeI*)	GCCGGC	1	13	1	0	4	332,106;274
Rhizobium meliloti	J.L. Denarie	*RmeI*	?	8	>10	?	?	?	106
Rhodococcus luteus RFL1	A.A. Janulaitis	*Rlu1I* (*MboI*)	GATC	116	87	8	0	22	420
Rhodococcus luteus RFL3	A.A. Janulaitis	*Rlu3I* (*NlaIV*)	GGNNCC	82	178	16	6	24	538
Rhodococcus luteus RFL4	A.A. Janulaitis	*Rlu4I* (*BamHI*)	GGATCC	5	3	1	0	1	127
Rhodococcus rhodochrous	ATCC 14349	*RrhI* (*SalI*)	GTCGAC	2	3	0	0	1	228
		RrhII	?	?	?	?	?	?	228
Rhodococcus rhodochrous	ATCC 4276	*RroI* (*SalI*)	GTCGAC	2	3	0	0	1	228
Rhodococcus species	ATCC 21664	*RheI* (*SalI*)	GTCGAC	2	3	0	0	1	116
Rhodococcus species	ATCC 19148	*RhpI* (*SalI*)	GTCGAC	2	3	0	0	1	116
		RhpII	?	?	?	?	?	?	116
Rhodococcus species	ATCC 13259	*RhsI* (*BamHI*)	GGATCC	5	3	1	0	1	116

RESTRICTION AND MODIFICATION ENZYMES AND THEIR RECOGNITION SEQUENCES (continued)

Microorganism	Source	Enzyme[a]	Sequence[b]	λ	Ad2	SV40	ΦX	pBR	Ref.
Rhodococcus species	W. Tsui	RspXI (*Bsp*HI)	T↑CATGA	8	3	2	3	4	494
Rhodopseudomonas sphaeroides	S. Kaplan	*Rsa*I	GT↑AC	113	83	12	11	3	180
Rhodopseudomonas sphaeroides	V.M. Kramarov	*Rsh*I (*Pvu*I)	CGAT↑CG	3	7	0	0	1	179
Rhodopseudomonas sphaeroides	R. Lascelles	*Rsh*II (*Cau*II)	CC(C/G)GG	114	97	0	1	10	155
Rhodopseudomonas sphaeroides	S. Kaplan	*Rsp*I (*Pvu*I)	CGATCG	3	7	0	0	1	14
Rhodopseudomonas sphaeroides	S. Kaplan	*Rsr*I (*Eco*RI)	G↑AATTC	5	5	1	0	1	73;394
Rhodospirillum rubrum	J. Chirikjian	*Rsr*II	CG↑G(A/T)CCG	5	2	0	0	0	215
Salmonella anatum	Y. Yoshida	*Rrb*I	?	?	4	5	1	?	171
Salmonella bareilly	Y. Yoshida	*San*I	?	?	?	?	?	?	470
Salmonella blockley YY156	Y. Yoshida	*Sba*I	?	?	?	?	?	?	470
Salmonella blockley YY156	Y. Yoshida	*Sbl*AI (*Sty*I)	CC(A/T)(A/T)GG	10	44	8	0	1	469
Salmonella blockley YY176	Y. Yoshida	*Sbl*BI (*Sty*I)	CC(A/T)(A/T)GG	10	44	8	0	1	469
Salmonella blockley YY242	Y. Yoshida	*Sbl*CI (*Sty*I)	CC(A/T)(A/T)GG	10	44	8	0	1	469
Salmonella bredeney	Y. Yoshida	*Sbr*I	?	?	?	?	?	?	470
Salmonella hybrid	N.E. Murray	*Sty*SJI	GAGNNNNNNGTPuC	Type I enzyme					522
Salmonella hybrid	L.R. Bullas	*Sty*SQI	AACNNNNNNPuTAPyG	Type I enzyme					353
Salmonella infantis	A. deWaard	*Sin*I (*Ava*II)	G↑G(A/T)CC*	35	73	6	1	8	178;532;53
Salmonella infantis 85005	S. Horiuchi	*Sin*CI (*Ava*II)	GG(A/T)CC	35	73	6	1	8	468
Salmonella infantis 85020	S. Horiuchi	*Sin*DI (*Ava*II)	GG(A/T)CC	35	73	6	1	8	468
Salmonella infantis 85064	S. Horiuchi	*Sin*EI (*Ava*II)	GG(A/T)CC	35	73	6	1	8	468
Salmonella infantis 85084	S. Horiuchi	*Sin*FI (*Ava*II)	GG(A/T)CC	35	73	6	1	8	468
Salmonella infantis 85144	S. Horiuchi	*Sin*GI (*Ava*II)	GG(A/T)CC	35	73	6	1	8	468
Salmonella infantis 85166	S. Horiuchi	*Sin*HI (*Ava*II)	GG(A/T)CC	35	73	6	1	8	468
Salmonella infantis 85325	S. Horiuchi	*Sin*JI (*Ava*II)	GG(A/T)CC	35	73	6	1	8	468
Salmonella infantis YY163	Y. Yoshida	*Sin*AI (*Ava*II)	GG(A/T)CC	35	73	6	1	8	468
Salmonella infantis YY190	Y. Yoshida	*Sin*BI (*Ava*II)	GG(A/T)CC	35	73	6	1	8	468
Salmonella isangi	Y. Yoshida	*Sis*I	?	?	?	?	?	?	470
Salmonella potsdam	L.R. Bullas	*Sty*SPI	*AACNNNNNNG*TPuC[i]	Type I enzyme					352
Salmonella schwarzengrund	Y. Yoshida	*Ssc*I	?	?	?	?	?	?	470
Salmonella thompson YY106	Y. Yoshida	*Sth*BI (*Kpn*I)	GGTACC	2	8	1	0	0	468
Salmonella thompson YY148	Y. Yoshida	*Sth*NI (*Kpn*I)	GGTACC	2	8	1	0	0	470,469
Salmonella thompson YY150	Y. Yoshida	*Sth*CI (*Kpn*I)	GGTACC	2	8	1	0	0	468
Salmonella thompson YY17	Y. Yoshida	*Sth*AI (*Kpn*I)	GGTACC	2	8	1	0	0	468
Salmonella thompson YY197	Y. Yoshida	*Sth*DI (*Kpn*I)	GGTACC	2	8	1	0	0	468
Salmonella thompson YY200	Y. Yoshida	*Sth*EI (*Kpn*I)	GGTACC	2	8	1	0	0	468
Salmonella thompson YY209	Y. Yoshida	*Sth*FI (*Kpn*I)	GGTACC	2	8	1	0	0	468
Salmonella thompson YY217	Y. Yoshida	*Sth*GI (*Kpn*I)	GGTACC	2	8	1	0	0	468
Salmonella thompson YY224	Y. Yoshida	*Sth*HI (*Kpn*I)	GGTACC	2	8	1	0	0	468

RESTRICTION AND MODIFICATION ENZYMES AND THEIR RECOGNITION SEQUENCES (continued)

Microorganism	Source	Enzyme[a]	Sequence[b]	Number of cleavage sites[c]					Ref.
				λ	Ad2	SV40	ΦX	pBR	
Salmonella thompson YY225	Y. Yoshida	*Sth*JI (*Kpn*I)	GGTACC	2	8	1	0	0	468
Salmonella thompson YY228	Y. Yoshida	*Sth*KI (*Kpn*I)	GGTACC	2	8	1	0	0	468
Salmonella thompson YY247	Y. Yoshida	*Sth*LI (*Kpn*I)	GGTACC	2	8	1	0	0	468
Salmonella thompson YY347	Y. Yoshida	*Sth*MI (*Kpn*I)	GGTACC	2	8	1	0	0	468
Salmonella thompson YY356	Y. Yoshida	*Sth*I (*Kpn*I)	G↑GTACC	2	8	1	0	0	468,470
Salmonella typhi 27	E.S. Anderson	*Sty*I	C↑C(A/T)(A/T)GG	10	44	8	0	1	199
Salmonella typhimurium	Y. Yoshida	*Stm*I	?	?	?	?	?	?	470
Salmonella typhimurium	L.R. Bullas	*Sty*SBI	GA GNNNNNNPu TAPyG[i]	Type I enzyme					352
Serratia fonticola	ATCC 29938	*Sfn*I (*Ava*II)	GG(A/T)TCC	35	73	6	1	8	359
Serratia fonticola	NEB 369	*Sfo*I (*Nar*I)	GGCGCC	1	20	0	2	4	202
Serratia marcescens Sb	C. Mulder	*Sma*I	CC↑*GGG	3	12	0	0	0	92;65;472
Serratia species SAI	B. Torheim	*Ssp*XI	?	?	?	?	?	?	305
Shigella boydii 13	NCTC 9361	*Sbo*13I (*Nru*I)	TCG↑TCGA	5	5	0	2	1	395
Shigella sonnei 47	T.M. Uporova	*Sso*I (*Eco*RI)	G↑AATTC	5	5	1	0	1	374;516;516
		*Sso*II (*Scr*FI)	↑*CC NGG	185	233	17	3	16	374;396
Sphaerotilus natans	ATCC 15291	*Sna*BI	TAC↑GTA	1	0	0	0	0	20
Sphaerotilus natans	ATCC 13923	*Spe*I	A↑CTAGT	0	3	0	0	0	48
Sphaerotilus natans	ATCC 13925	*Ssp*I	AAT↑ATT	20	5	1	6	1	89
Sphaerotilus natans C	A. Pope	*Sna*I	GTATAC	3	3	0	0	1	225
Spiroplasma citri ASP2	M.A. Stephens	*Sci*NI (*Hha*I)	G↑CGC	215	375	2	18	31	283
Spirulina platensis	M. Kawamura	*Spl*I	C↑GTACG	1	4	0	2	0	370
		*Spl*II (*Tth*111I)	GACNNNGTC	2	12	0	0	1	370
		*Spl*III (*Hae*III)	GGCC	149	216	18	11	22	370
Staphylococcus aureus 3A	E.E. Stobberingh	*Sau*3AI (*Mbo*I)	↑GATC*#	116	87	8	0	22	294;432
Staphylococcus aureus 6782	E.E. Arutyunyan	*Sau*6782I (*Mbo*I)	GATC	116	87	8	0	22	437
Staphylococcus aureus PS96	E.E. Stobberingh	*Sau*96I (*Asu*I)	G↑GNCC	74	164	11	2	15	295
Staphylococcus intermedius	ATCC 29663	*Sin*MI (*Mbo*I)	GATC	116	87	8	0	22	35
		*Sin*MII	?	?	?	?	?	?	35
Staphylococcus saprophyticus	ATCC 13518	*Ssa*I	?	>10	?	?	?	?	162
Streptococcus cremoris F	C. Daly	*Scr*FI	CC↑NGG	185	233	17	3	16	69
Streptococcus durans RFL3	A.A. Janulaitis	*Sdu*I	G(A/G/T)GC(A/C/T)↑C	38	105	4	3	10	126,424
Streptococcus dysgalactiae	ATCC 9926	*Sdy*I (*Asu*I)	GGNCC	74	164	11	2	15	228
Streptococcus faecalis GU	J. Chirikjian	*Sfa*GUI (*Hpa*II)	CCGG	328	171	1	5	26	43

RESTRICTION AND MODIFICATION ENZYMES AND THEIR RECOGNITION SEQUENCES (continued)

Microorganism	Source	Enzyme[a]	Sequence[b]	Number of cleavage sites[c]					Ref.
				λ	Ad2	SV40	ΦX	pBR	
Streptococcus faecalis ND547	D. Clewell	*Sfa*NI	GCATC (5/9)	169	84	6	12	22	260
Streptococcus faecalis var. zymogenes	R. Wu	*Sfa*I (*Hae*III)	GG↑CC	149	216	18	11	22	334
Streptomyces achromogenes	ATCC 21353	*Sac*AI (*Nae*I)	GCCGGC	1	13	1	0	4	481
Streptomyces achromogenes	ATCC 12767	*Sac*I	GAGCT↑C	2	16	0	0	0	2
		*Sac*II	CCGC↑GG	4	33	0	1	0	2
		*Sac*III	?	>100	>100	?	?	?	2
Streptomyces akiyosinicus	ATCC 13480	*Sak*I (*Sac*II)	CCGCGG	4	33	0	1	0	481
Streptomyces alanosinicus	ATCC 15710	*Saa*I (*Sac*II)	CCGCGG	4	33	0	1	0	202
Streptomyces albofaciens	ATCC 25184	*Sao*I (*Nae*I)	GCCGGC	1	13	1	0	4	390
Streptomyces albohelvatus	ATCC 19820	*Sab*I (*Sac*II)	CCGCGG	4	33	0	1	0	390
Streptomyces albulus	ATCC 12757	*Sal*HI (*Mbo*I)	GATC	116	87	8	0	22	475
Streptomyces albus	ATCC 21725	*Sal*AI (*Mbo*I)	GATC	116	87	8	0	22	475
Streptomyces albus	ATCC 21290	*Sal*CI (*Nae*I)	GCCGGC	1	13	0	0	4	477
Streptomyces albus	ATCC 21132	*Sal*DI (*Nru*I)	TCGCGA	5	5	2	2	1	477
Streptomyces albus	CMI 52766	*Sal*PI (*Pst*I)	CTGCA↑G	28	30	2	1	1	39;37
Streptomyces albus G	J.M. Ghuysen	*Sal*I	G↑TCGAC	2	3	2	0	1	3
		*Sal*II	?	>20	?	?	?	?	3
Streptomyces albus subspecies pathocidicus	KCC S-0166	*Spa*I (*Xho*I)	CTCGAG	1	6	0	1	0	270
Streptomyces aureofaciens	CCM 3239	*Sau*3239I (*Xho*I)	C↑TCGAG	1	6	0	1	0	75,447
Streptomyces aureofaciens	ATCC 15852	*Sau*AI (*Nae*I)	GCCGGC	1	13	1	0	4	480
Streptomyces aureofaciens IKA 18/4	J. Timko	*Sau*I	CC↑TNAGG	2	7	0	0	0	302
Streptomyces aureofaciens strain BM-K	J. Timko	*Sau*BMKI (*Nae*I)	GCC↑GGC	1	13	1	0	4	515
Streptomyces bobili	ATCC 3310	*Sbo*I (*Sac*II)	CCGCGG	4	33	0	1	0	270;299
Streptomyces caespitosus	H. Takahashi	*Sca*I	AGT↑ACT	5	5	0	0	1	152;89
Streptomyces coelicolor	ATCC 10147	*Sco*I (*Sac*I)	GAGCTC	2	16	0	0	0	390
Streptomyces cupidosporus	KCC S0316	*Scu*I (*Xho*I)	CTCGAG	1	6	0	1	0	270
Streptomyces exfoliatus	KCC S0030	*Sex*I (*Xho*I)	CTCGAG	1	6	0	1	0	270
		*Sex*II	?	2	?	?	?	?	270
Streptomyces fimbriatus	ATCC 15051	*Sfi*I	GGCCNNNN↑NGGCC	0	3	1	0	0	229
Streptomyces fradiae	ATCC 3355	*Sfr*I (*Sac*II)	CCGCGG	4	33	0	1	1	270;299
Streptomyces ganmycicus	KCC S0759	*Sga*I (*Xho*I)	CTCGAG	1	6	0	1	0	270
Streptomyces goshikiensis	KCC S0294	*Sgo*I (*Xho*I)	CTCGAG	1	6	0	1	0	270
Streptomyces griseus	ATCC 23345	*Sgr*I	?	0	7	0	?	?	2
Streptomyces griseus Kr. 20	A.V. Orekhov	*Sgr*II (*Eco*RII)	CC(A/T)GG	71	136	17	2	6	218
Streptomyces hygroscopicus	F. Walter	*Shy*I (*Sac*II)	CCGCGG	4	33	0	1	0	319

RESTRICTION AND MODIFICATION ENZYMES AND THEIR RECOGNITION SEQUENCES (continued)

Microorganism	Source	Enzyme[a]	Sequence[b]	Number of cleavage sites[c]					Ref.
				λ	Ad2	SV40	ΦX	pBR	
Streptomyces hygroscopicus	T. Yamaguchi	*ShyTI*	?	2	?	?	?	?	270
Streptomyces karnatakensis	ATCC 25463	*SkaI* (*NaeI*)	GCCGGC	1	13	1	0	4	35
		SkaII (*PstI*)	CTGCAG	28	30	2	1	1	35
Streptomyces lavendulae	ATCC 8664	*SlaI* (*XhoI*)	C↑TCGAG	1	6	0	1	0	300
Streptomyces luteoreticuli	KCC S0788	*SluI* (*XhoI*)	CTCGAG	1	6	0	1	0	299
Streptomyces novocastria	P. Eastlake	*SnoI* (*ApaLI*)	G↑TGCAC	4	7	0	1	3	366;27
Streptomyces oderifer	ATCC 6246	*SodI*	?	?	?	?	?	?	162
		SodII	?	?	?	?	?	?	162
Streptomyces phaeochromogenes	IFO 3108	*SpaXI* (*SphI*)	GCATGC	6	8	2	0	1	290
Streptomyces phaeochromogenes	F. Bolivar	*SphI*	GCATG↑C	6	8	2	0	1	72
Streptomyces species RFL1	A.A. Janulaitis	*Ssp11* (*AsuII*)	TT↑CGAA	7	1	0	0	0	556
Streptomyces species RFL2	A.A. Janulaitis	*Ssp21* (*CauII*)	CC(C/G)GG	114	97	0	1	10	127
Streptomyces species RFL4	A.A. Janulaitis	*Ssp4I* (*XhoI*)	CTCGAG	1	6	0	1	0	538
Streptomyces stanford	S. Goff	*SstI* (*SacI*)	GAGCT↑C	2	16	0	0	0	86;205
		SstII (*SacII*)	CCGC↑GG	4	33	0	1	0	86
		SstIII (*SacIII*)	?	>100	>100	?	?	?	86
		SstIV (*BclI*)	T↑GATCA	8	5	1	0	0	111
Streptomyces tubercidicus	H. Takahashi	*StuI*	AGG↑CCT	6	11	7	1	0	269
Streptoverticillium cinnamonium	S.T. Williams	*SciI* (*XhoI*)	CTC↑GAG	1	6	0	1	0	368
Streptoverticillium flavopersicum	Upjohn UC 5066	*SflI* (*PstI*)	CTGCA↑G	28	30	2	1	1	138
Sulfolobus acidocaldarius	DSM 639	*SuaI* (*HaeIII*)	GG↑CC	149	216	18	11	22	377
Sulfolobus acidocaldarius	W. Zillig	*SuII* (*HaeIII*)	GGCC	149	216	18	11	22	193
Synechococcus cedrorum	ATCC 29140	*SceI* (*FnuDII*)	CGCG	157	303	0	14	23	514
Synechococcus species	ATCC 29140	*SseI* (*BclI*)	TGATCA	8	5	1	0	0	533
		SseII (*SacII*)	CCGCGG	4	33	0	1	0	533
Synechocystis species	ATCC 27175	*SciAI* (*BstEII*)	GGTNACC	13	10	0	0	0	514
		SciAII (*PvuII*)	CAGCTG	15	24	3	0	1	514
Synechocystis species 6701	ATCC 27170	*SecI*	C↑CNNGG	105	234	16	6	8	345
		SecII (*HpaII*)	CCGG	0	0	0	0	0	345
		SecIII (*SauI*)	CCTNAGG	2	6	0	0	0	345
Tatlockia micdadei	A. Brown	*TmiI*	?	>10	?	?	?	?	452
Thermococcus celer	W. Zillig	*TceI* (*MboII*)	GAAGA	130	113	16	11	11	193
Thermoplasma acidophilum	D. Searcy	*ThaI* (*FnuDII*)	CG↑CG	157	303	0	14	23	191
Thermopolyspora glauca	ATCC 15345	*TglI* (*SacII*)	CCGCGG	4	33	0	1	0	85
Thermus aquaticus	S.A. Grachev	*TaqXI* (*EcoRII*)	CC↑(A/T)GG	71	136	17	2	6	87

RESTRICTION AND MODIFICATION ENZYMES AND THEIR RECOGNITION SEQUENCES (continued)

Microorganism	Source	Enzyme[a]	Sequence[b]	Number of cleavage sites[c]					Ref.
				λ	Ad2	SV40	ΦX	pBR	
Thermus aquaticus YT1	J.I. Harris	*TaqI*	T↑CGA*	121	50	1	10	7	247;247;248
		TaqII	GACCGA (11/9) CACCCA (11/9)	28	36	1	2	6	211,6
Thermus flavus AT62	T. Oshima	*TfII* (*TaqI*)	TCGA	121	50	1	10	7	248
Thermus ruber strain 21	BKM B-1258	*TruI* (*AvaII*)	GG(A/T)CC	35	73	6	1	8	501
		TruII (*MboI*)	GATC	116	87	8	0	22	501
		TruIII	?	>10	?	?	?	?	501
Thermus species	ATCC 31674	*TspI* (*Tth*111I)	GACNNNGTC	2	12	0	0	1	490
Thermus species strain 1E	R.A.D. Williams	*TspEI*	AATT	189	87	39	25	9	503
Thermus species strain 2AZN	N.D.H. Raven	*TspZNI* (*HaeIII*)	GGCC	149	216	18	11	22	504
Thermus species strain YS45	R.J. Sharp	*Tsp45I*	GT(C/G)AC	81	73	6	8	9	505
Thermus thermophilus HB8	T. Oshima	*TthHB8I* (*TaqI*)	T↑CGA*	121	50	1	10	7	249;315;249,450;248;248
Thermus thermophilus strain 110	T. Oshima	*TteI* (*Tth*111I)	GACNNNGTC	2	12	0	0	1	273
Thermus thermophilus strain 111	T. Oshima	*Tth*111I	GACN↑NNGTC	2	12	0	0	1	273
		*Tth*111II	CAAPuCA (11/9)	49	53	11	11	5	272
Thermus thermophilus strain 23	T. Oshima	*TtrI* (*Tth*111I)	GACNNNGTC	2	12	0	0	1	273
Tolypothrix tenuis	ATCC 27914	*TteAI* (*HaeIII*)	GGCC	149	216	18	11	22	536
Tolypothrix tenuis	W. Siegelman	*TtnI* (*HaeIII*)	GGCC	149	216	18	11	22	286
Tuberoidobacter mutans RFL1	A.A. Janulaitis	*TmuII* (*CauII*)	CC(C/G)GG	114	97	0	1	10	420
Unidentified bacterium RFL21	A.A. Janulaitis	*Uba211I* (*HgiAI*)	G(A/T)GC(A/T)↑C	28	38	0	3	8	557
Unidentified bacterium RFL26	A.A. Janulaitis	*Uba26I* (*BsmAI*)	GTCTC (1/5)	37	60	2	4	3	555
Unidentified bacterium RFL44	A.A. Janulaitis	*Uba44I* (*ApaLI*)	G↑TGCAC	4	7	0	1	3	557
Ureaplasma urealyticum 960	ATCC 27618	*Uur960I* (*Fnu4HI*)	GC↑NGC	379	411	24	31	42	521
Vibrio anguillarum	P. Wood	*VanI* (*BglI*)	GCCNNNNN↑NGGC	29	20	1	0	3	456
Vibrio harveyi	ATCC 14126	*VhaI* (*HaeIII*)	GGCC	149	216	18	11	22	116
Vibrio nereis	P. Wood	*VneAI* (*DraII*)	PuG↑GNCCPy	3	44	3	0	4	456
Vibrio nereis	S.K. Degtyarev	*VneI* (*ApaLI*)	G↑TGCAC	4	7	0	1	3	498
Vibrio nigripulchritudo	P. Wood	*VniI* (*HaeIII*)	GGCC	149	216	18	11	22	456
Vibrio species	S.K. Degtyarev	*VspI*	AT↑TAAT	17	3	3	2	1	497
Xanthomonas amaranthicola	ATCC 11645	*XamI* (*SalI*)	G↑TCGAC	2	3	0	0	1	3
Xanthomonas badrii	ATCC 11672	*XbaI*	T↑CTAGA	1	5	0	0	0	343
Xanthomonas campestris	NEB 420	*XcaI* (*SmaI*)	GTA↑TAC	3	3	0	0	1	459
Xanthomonas citrii	IFO 3835	*XciI* (*SalI*)	G↑TCGAC	2	3	0	0	1	325
Xanthomonas cyanopsidis 13D5	C.I. Kado	*XcyI* (*SmaI*)	C↑CCGGG	3	12	0	0	0	70
Xanthomonas holcicola	ATCC 13461	*XhoI*	C↑TCGAG	1	6	0	1	0	84
		XhoII	Pu↑GATCPy	21	22	3	0	8	217;85,154
Xanthomonas malvacearum	ATCC 9924	*XmaI* (*SmaI*)	C↑CCGGG	3	12	0	0	0	65

RESTRICTION AND MODIFICATION ENZYMES AND THEIR RECOGNITION SEQUENCES (continued)

Microorganism	Source	Enzyme[a]	Sequence[b]	Number of cleavage sites[c]					Ref.
				λ	Ad2	SV40	ΦX	pBR	
Xanthomonas manihotis 7AS1	B-C. Lin	XmaII (PstI)	CTGCAG	28	30	2	1	1	65
		XmaIII	C↑GGCC*G	2	19	0	0	1	158;306
Xanthomonas nigromaculans	ATCC 23390	XmnI	GAANN↑NNTTC	24	5	0	3	2	174,228
		XniI (PvuI)	CGATCG	3	7	0	0	1	99
Xanthomonas oryzae	M. Ehrlich	XorI (PstI)	CTGCAG	28	30	2	1	1	321
		XorII (PvuI)	CGAT↑CG	3	7	0	0	1	321;82
Xanthomonas papavericola	ATCC 14180	XpaI (XhoI)	C↑TCGAG	1	6	0	1	0	84
Xanthomonas phaseoli	Z.F. Bunina	XphI (PstI)	CTGCAG	28	30	2	1	1	32
Yersinia enterocolitica 08 85-775	T. Maruyama	YenEI (PstI)	CTGCAG	28	30	2	1	1	500
Yersinia enterocolitica 08 A2635	H. Watanabe	YenI (PstI)	CTGCA↑G	28	30	2	1	1	500;500
Yersinia enterocolitica 08 Bi1212	T. Maruyama	YenBI (PstI)	CTGCAG	28	30	2	1	1	500
Yersinia enterocolitica 08 Bi3995	T. Maruyama	YenCI (PstI)	CTGCAG	28	30	2	1	1	500
Yersinia enterocolitica 08 Bi9534	T. Maruyama	YenDI (PstI)	CTGCAG	28	30	2	1	1	500
Yersinia enterocolitica 08 WA	T. Maruyama	YenAI (PstI)	CTGCAG	28	30	2	1	1	500
Zymomonas anaerobia	O.J. Yoo	ZanI (EcoRII)	CC↑(A/T)GG	71	136	17	2	6	292

a When two enzymes recognize the same sequence (i.e., are isoschizomers) the prototype (i.e., the first example isolated) is indicated in parentheses. Note that *Mbo*I is sensitive to *dam* methylation, whereas its isoschizomer *Sau*3AI is not, and *Eco*RII is sensitive to *dcm* methylation, whereas its isoschizomer *Bst*NI is not. For other isoschizomers of these two enzymes, those that are known to be similar to *Sau*3AI are indicated by #; those known to be similar to *Bst*NI are indicated by $.

b Recognition sequences are written from $5' \rightarrow 3'$, only one strand being given, and the point of cleavage is indicated by an arrow (↑). When no arrow appears, the precise cleavage site has not been determined. For example, G↑GATCC is an abbreviation for

$5'$ G↑G A T C C $3'$
$3'$ C C T A G↑G $5'$

For enzymes such as *Hga*I, *Mbo*II, etc., which cleave away from their recognition sequence, the sites of cleavage are indicated in parentheses. For example, *Hga*I GACGC(5/10) indicates cleavage as shown below

$5'$ GACGCNNNNN↑ $3'$
$3'$ CTGCGNNNNNNNNNN↑ $5'$

In all cases, the recognition sequences are oriented so that the cleavage sites lie on their $3'$ side. Bases appearing in parentheses signify that either base may occupy that position in the recognition sequence. Thus, *Acc*I cleaves the sequences GTAGAC, GTATAC, GTCGAC, and GTCTAC. Where known, the base modified by the corresponding specific methylase is indicated by an asterisk.
Å is N6-methyladenosine. Č is 5-methylcytosine, except in the case of *Bcn*I, *Cfr*6I, *Cfr*9I, *Mva*I, and *Pvu*II, where the modified base is N4-methylcytosine.

c These columns indicate the frequency of cleavage by the various specific endonucleases on bacteriophage lambda DNA (λ), adenovirus-2 DNA (Ad2), simian virus 40 DNA (SV40), φX174 Rf DNA, and pBR322 DNA (pBR). In all cases the sites were derived by computer search of the complete sequences obtained from GENBANK.

d Where more than one reference appears, the second concerns the recognition sequence for the restriction enzyme, the third describes the purification procedure for the methylase, and the fourth describes the recognition sequence of the methylase. In some cases, several references appear in one of these categories when independent groups have reached similar conclusions.

e *Eco*PI, *Eco*P15, *Hine*I and *Hin*fIII have characteristics intermediate between those of the Type I and Type II restriction endonucleases. They are designated Type III in accordance with the suggestion of Kauc and Piekarowicz.[140]

f Both *Hin*fIII and *Hine*I cleave about 25 bases $3'$ of the recognition sequence.

g *Nci*I leaves termini carrying a $3'$-phosphate group.[112]

h *Pae*RI fails to cleave the *Xho* site at 26.5% on the Ad2 genome.[81]

i The * placed under the T (or G) residue signifies that the A (or C) residue, on the complementary strand at that position, is modified by the cognate methylases.

REFERENCES

1. Agarwal, K. unpublished observations.
2. Arrand, J.R., Myers, P.A. and Roberts, R.J. unpublished observations.
3. Arrand, J.R., Myers, P.A. and Roberts, R.J. (1978) *J. Mol. Biol.* 118: 127-135.
4. Azizbekyan, R.R., Rebentish, B.A., Stepanova, T.V., Netyksa, E.M. and Buchkova, M.A. (1984) *Dokl. Akad. Nauk. SSSR* 274: 742-744.
5. Bachi, B., Reiser, J. and Pirrotta, V. (1979) *J. Mol. Biol.* 128: 143-163.
6. Barker, D., Hoff, M., Oliphant, A. and White, R. (1984) *Nucl. Acids Res.* 12: 5567-5581.
7. Baumstark, B.R., Roberts, R.J. and RajBhandary, U.L. (1979) *J. Biol. Chem.* 254: 8943-8950.
8. Beaty, J.S., McLean-Bowen, C.A. and Brown, L.R. (1982) *Gene* 18: 61-67.
9. Belle Isle, H. unpublished observations.
10. Bennett, S.P. and Halford, S.E. unpublished observations.
11. Bickle, T.A. and Ineichen, K. (1980) *Gene* 9: 205-212.
12. Bickle, T., Yuan, R., Pirrotta, V. and Ineichen, K. unpublished observations.
13. Bigger, C.H., Murray, K. and Murray, N.E. (1973) *Nature New Biology* 244: 7-10.
14. Bingham, A.H.A., Atkinson, A. and Darbyshire, J. unpublished observations.
15. Bingham, A.H.A., Atkinson, T., Sciaky, D. and Roberts, R.J. (1978) *Nucl. Acids Res.* 5: 3457-3467.
16. Bingham, A.H.A. and Darbyshire, J. (1982) *Gene* 18: 87-91.
17. Bingham, A.H.A., Sharp, R.J. and Atkinson, T. unpublished observations.
18. Bitinaite, J.B., Klimasauskas, S.J., Butkus, V.V. and Janulaitis, A.A. (1985) *FEBS Letters* 182: 509-513.
19. Bolton, B., Nesch, G., Comer, M., Wolf, W. and Kessler, C. (1985) *FEBS Letters* 182: 130-134.
20. Borsetti, R., Grandoni, R. and Schildkraut, I. unpublished observations.
21. Borsetti, R., Wise, D. and Schildkraut, I. unpublished observations.
22. Boyer, H.W., Chow, L.T., Dugaiczyk, A., Hedgpeth, J. and Goodman, H.M. (1973) *Nature New Biology* 244: 40-43.
23. Brockes, J.P. (1973) *Biochem. J.* 133: 629-633.
24. Brockes, J.P., Brown, P.R. and Murray, K. (1972) *Biochem. J.* 127: 1-10.
25. Bron, S. and Murray, K. (1975) *Mol. Gen. Genet.* 143: 25-33.
26. Bron, S., Murray, K. and Trautner, T.A. (1975) *Mol. Gen. Genet.* 143: 13-23.
27. Brown, N.L. unpublished observations.
28. Brown, N.L., Hutchison, C.A. III and Smith, M. (1980) *J. Mol. Biol.* 140: 143-148.
29. Brown, N.L., McClelland, M. and Whitehead, P.R. (1980) *Gene* 9: 49-68.
30. Brown, N.L. and Smith, M. (1977) *Proc. Natl. Acad. Sci. USA* 74: 3213-3216.
31. Brown, N.L. and Smith, M. (1976) *FEBS Letters* 65: 284-287.
32. Bunina, Z.F., Kramarov, V.M., Smolyaninov, V.V. and Tolstova, L.A. (1984) *Bioorg. Khim.* 10: 1333-1335.
33. Butkus, V., Klimasauskas, S., Kersulyte, D., Vaitkevicius, D., Lebionka, A. and Janulaitis, A. (1985) *Nucl. Acids Res.* 13: 5727-5746.
34. Calleja, F., Dekker, B.M.M., Coursin, T. and deWaard, A. (1984) *FEBS Letters* 178: 69-72.
35. Camp, R. and Schildkraut, I. unpublished observations.
36. Camp, R. and Visentin, L.P. unpublished observations.
37. Carter, J.A., Chater, K.F., Bruton, C.J. and Brown, N.L. (1980) *Nucl. Acids Res.* 8: 4943-4954.
38. Catterall, J.F. and Welker, N.E. (1977) *J. Bacteriol.* 129: 1110-1120.
39. Chater, K.F. (1977) *Nucl. Acids Res.* 4: 1989-1998.
40. Chirikjian, J.G., George, A. and Smith, L.A. (1978) *Fed. Proc.* 37: 1415.
41. Clarke, C.M. and Hartley, B.S. (1979) *Biochem. J.* 177: 49-62.
42. Clanton, D.J., Woodward, J.M. and Miller, R.V. (1978) *J. Bacteriol.* 135: 270-273.
43. Coll, E. and Chirikjian, J. unpublished observations.
44. Comb, D.G. unpublished observations.
45. Comb, D.G., Hess, E.J. and Wilson, G. unpublished observations.
46. Comb, D.G., Parker, P., Grandoni, R. and Schildkraut, I. unpublished observations.
47. Comb, D.G., Parker, P. and Schildkraut, I. unpublished observations.
48. Comb, D.G. and Schildkraut, I. unpublished observations.
49. Comb, D.G., Schildkraut, I. and Roberts, R.J. unpublished observations.
50. Comb, D.G., Schildkraut, I., Wilson, G. and Greenough, L. unpublished observations.
51. Comb, D.G. and Wilson, G. unpublished observations.
52. Cruz, A.K., Kidane, G., Pires, M.Q., Rabinovitch, L., Guaycurus, T.V. and Morel, C.M. (1984) *FEBS Letters* 173: 99-102.
53. de Waard, A. unpublished observations.
54. deWaard, A. and Duyvesteyn, M. (1980) *Arch. Microbiol.* 128: 242-247.
55. de Waard, A., Korsuize, J., van Beveren, C.P. and Maat, J. (1978) *FEBS Letters* 96: 106-110.
56. de Waard, A., van Beveren, C.P., Duyvesteyn, M. and van Ormondt, H. (1979) *FEBS Letters* 101: 71-76.

57. **de Wit, C.M., Dekker, B.M.M., Neele, A.C. and de Waard, A.** (1985) *FEBS Letters* 180: 219-223.
58. **DiLauro, R.** unpublished observations.
59. **Dugaiczyk, A., Hedgpeth, J., Boyer, H.W. and Goodman, H.M.** (1974) *Biochemistry* 13: 503-512.
60. **Duncan, C.H., Wilson, G.A. and Young, F.E.** (1978) *J. Bacteriol.* 134: 338-344.
61. **Duyvesteyn, M.G.C. and deWaard, A.** (1980) *FEBS Letters* 111: 423-426.
62. **Duyvesteyn, M.G.C., Korsuize, J. and deWaard, A.** (1981) *Plant Mol. Biol.* 1: 75-79.
63. **Duyvesteyn, M.G.C., Korsuize, J., deWaard, A., Vonshak, A. and Wolk, C.P.** (1983) *Arch. Microbiol.* 134: 276-281.
64. **Endow, S.A.** (1977) *J. Mol. Biol.* 114: 441-449.
65. **Endow, S.A. and Roberts, R.J.** (1977) *J. Mol. Biol.* 112: 521-529.
66. **Endow, S.A. and Roberts, R.J.** unpublished observations.
67. **Eskin, B. and Linn, S.** (1972) *J. Biol. Chem.* 247: 6183-6191.
68. **Fisherman, J., Gingeras, T.R. and Roberts, R.J.** unpublished observations.
69. **Fitzgerald, G.F., Daly, C., Brown, L.R. and Gingeras, T.R.** (1982) *Nucl. Acids Res.* 10: 8171-8179.
70. **Froman, B.E., Tait, R.C., Kado, C.I. and Rodriguez, R.L.** (1984) *Gene* 28: 331-335.
71. **Fuchs, C., Rosenvold, E.C., Honigman, A. and Szybalski, W.** (1978) *Gene* 4: 1-23.
72. **Fuchs, L.Y., Covarrubias, L., Escalante, L., Sanchez, S. and Bolivar, F.** (1980) *Gene* 10: 39-46.
73. **Gardner, J.F., Cohen, L.K., Lynn, S.P. and Kaplan, S.** unpublished observations.
74. **Garfin, D.E. and Goodman, H.M.** (1974) *Biochem. Biophys. Res. Comm.* 59: 108-116.
75. **Gasperik, J., Godany, A., Hostinova, E. and Zelinka, J.** (1983) *Biologia (Bratislava)* 38: 315-319.
76. **Geier, G.E. and Modrich, P.** (1979) *J. Biol. Chem.* 254: 1408-1413.
77. **Gelinas, R.E., Myers, P.A. and Roberts, R.J.** (1977) *J. Mol. Biol.* 114: 169-179.
78. **Gelinas, R.E., Myers, P.A. and Roberts, R.J.** unpublished observations.
79. **Gelinas, R.E., Myers, P.A., Weiss, G.A., Murray, K. and Roberts, R.J.** (1977) *J. Mol. Biol.* 114: 433-440.
80. **Gelinas, R.E. and Roberts, R.J.** unpublished observations.
81. **Gingeras, T.R. and Brooks, J.E.** (1983) *Proc. Natl. Acad. Sci. USA* 80: 402-406.
82. **Gingeras, T.R., Greenough, L., Schildkraut, I. and Roberts, R.J.** (1981) *Nucl. Acids Res.* 9: 4525-4536.
83. **Gingeras, T.R., Milazzo, J.P. and Roberts, R.J.** (1979) *Nucl. Acids Res.* 5: 4105-4127.
84. **Gingeras, T.R., Myers, P.A., Olson, J.A., Hanberg, F.A. and Roberts, R.J.** (1978) *J. Mol. Biol.* 118: 113-122.
85. **Gingeras, T.R. and Roberts, R.J.** unpublished observations.
86. **Goff, S.P. and Rambach, A.** (1978) *Gene* 3: 347-352.
87. **Grachev, S.A., Mamaev, S.V., Gurevich, A.I., Igoshin, A.V., Kolosov, M.N. and Slyusarenko, A.G.** (1981) *Bioorg. Khim.* 7: 628-630.
88. **Grandoni, R.P. and Comb, D.** unpublished observations.
89. **Grandoni, R.P. and Schildkraut, I.** unpublished observations.
90. **Greenaway, P.J.** (1980) *Biochem. Biophys. Res. Comm.* 95: 1282-1287.
91. **Greene, P.J., Betlach, M.C., Boyer, H.W. and Goodman, H.M.** (1974) *Methods Mol. Biol.* 7: 87-111.
92. **Greene, R. and Mulder, C.** unpublished observations.
93. **Grosskopf, R., Wolf, W. and Kessler, C.** (1985) *Nucl. Acids Res.* 13: 1517-1528.
94. **Grosveld, G.C.** unpublished observations.
95. **Gunthert, U., Storm, K. and Bald, R.** (1978) *Eur. J. Biochem.* 90: 581-583.
96. **Haberman, A.** (1974) *J. Mol. Biol.* 89: 545-563.
97. **Haberman, A., Heywood, J. and Meselson, M.** (1972) *Proc. Natl. Acad. Sci. USA* 69: 3138-3141.
98. **Hadi, S.M., Bachi, B., Shepherd, J.C.W., Yuan, R., Ineichen, K. and Bickle, T.A.** (1979) *J. Mol. Biol.* 134: 655-666.
99. **Hanberg, F., Myers, P.A. and Roberts, R.J.** unpublished observations.
100. **Hansen, R.** unpublished observations.
101. **Hartmann, H. and Goebel, W.** (1977) *FEBS Letters* 80: 285-287.
102. **Hattman, S., Brooks, J.E. and Masurekar, M.** (1978) *J. Mol. Biol.* 126: 367-380.
103. **Hattman, S., Keisler, T. and Gottehrer, A.** (1978) *J. Mol. Biol.* 124: 701-711.
104. **Hedgpeth, J., Goodman, H.M. and Boyer, H.W.** (1972) *Proc. Natl. Acad. Sci. USA* 69: 3448-3452.
105. **Hendrix, J.D. and Welker, N.E.** (1985) *J. Bacteriol.* 162: 682-692.
106. **Heumann, W.** (1979) *Curr. Top. Microbiol. Immunol.* 88: 1-24.
107. **Hinkle, N.F. and Miller, R.V.** (1979) *Plasmid* 2: 387-393.
108. **Hiraoka, N., Kita, K., Nakajima, H. and Obayashi, A.** (1984) *J. Ferment. Technol.* 62: 583-588.
109. **Hobom, G., Schwarz, E., Melzer, M. and Mayer, H.** (1981) *Nucl. Acids Res.* 9: 4823-4832.
110. **Hoshino, T., Uozumi, T., Horinouchi, S., Ozaki, A., Beppu, T. and Arima, K.** (1977) *Biochim. Biophys. Acta* 479: 367-369.
111. **Hu, A.W., Kuebbing, D. and Blakesley, R.J.** (1978) *Fed. Proc.* 38: 780.
112. **Hu, A.W. and Marschel, A.H.** (1982) *Fed. Proc.* 41: 119.
113. **Hughes, S.G., Bruce, T. and Murray, K.** (1980) *Biochem. J.* 185: 59-63.
114. **Hughes, S.G., Bruce, T. and Murray, K.** unpublished observations.
115. **Hughes, S.G. and Murray, K.** (1980) *Biochem. J.* 185: 65-75.
116. **Hurlin, P. and Schildkraut, I.** unpublished observations.

117. **Hutchison, C.A. and Barrell, B.G.** unpublished observations.
118. **Ikawa, S., Shibata, T. and Ando, T.** (1976) *J. Biochem. (Tokyo)* 80: 1457-1460.
119. **Ikawa, S., Shibata, T. and Ando, T.** (1979) *Agric. Biol. Chem.* 43: 873-875.
120. **Ikawa, S., Shibata, T., Ando, T. and Saito, H.** (1980) *Molec. Gen. Genet.* 177: 359-368.
121. **Janulaitis, A. and Adomaviciute, L.** unpublished observations.
122. **Janulaitis, A. and Bitinaite, J.** unpublished observations.
123. **Janulaitis, A., Bitinaite, J. and Jaskeleviciene, B.** (1983) *FEBS Letters* 151: 243-247.
124. **Janulaitis, A., Klimasauskas, S., Petrusyte, M. and Butkus, V.** (1983) *FEBS Letters* 161: 131-134.
125. **Janulaitis, A.A., Marcinkeviciene, L.Y. and Petrusyte, M.P.** (1982) *Dokl. Akad. Nauk. SSSR* 262: 241-244.
126. **Janulaitis, A., Marcinkeviciene, L., Petrusyte, M. and Mironov, A.** (1981) *FEBS Letters* 134: 172-174.
127. **Janulaitis, A.A. and Petrusyte, M.** unpublished observations.
128. **Janulaitis, A.A., Petrusyte, M. and Butkus, V.V.** (1983) *FEBS Letters* 161: 213-216.
129. **Janulaitis, A.A., Petrusite, M.A., Jaskelaviciene, B.P., Krayev, A.S., Skryabin, K.G. and Bayev, A.A.** (1981) *Dokl. Akad. Nauk. SSSR* 257: 749-750.
130. **Janulaitis, A.A., Petrusite, M.A., Jaskelavicene, B.P., Krayev, A.S., Skryabin, K.G. and Bayev, A.A.** (1982) *FEBS Letters* 137: 178-180.
131. **Janulaitis, A.A., Stakenas, P.S. and Berlin, Y.** (1983) *FEBS Letters* 161: 210-212.
132. **Janulaitis, A.A., Stakenas, P.S., Bitinaite, J.B. and Jaskeleviciene, B.P.** (1983) *Dokl. Akad. Nauk. SSSR* 271: 483-485.
133. **Janulaitis, A.A., Stakenas, P.S., Jaskeleviciene, B.P., Lebedenko, E.N. and Berlin, Y.A.** (1980) *Bioorg. Khim.* 6: 1746-1748.
134. **Janulaitis, A.A., Stakenas, P.S., Petrusyte, M.P., Bitinaite, J.B., Klimasauskas, S.J. and Butkus, V.V.** (1983) *Molekulyarnaya Biologiya* 18: 115-129.
135. **Janulaitis, A.A., Vaitkevicius, D., Puntezis, S. and Jaskeleviciene, B.** unpublished observations.
136. **Jentsch, S.** (1983) *J. Bacteriol.* 156: 800-808.
137. **Jentsch, S., Gunthert, U. and Trautner, T.A.** (1981) *Nucl. Acids Res.* 12: 2753-2759.
138. **Jiang, B.D. and Myers, P.** unpublished observations.
139. **Kan, N.C., Lautenberger, J.A., Edgell, M.H. and Hutchison, C.A. III.** (1979) *J. Mol. Biol.* 130: 191-209.
140. **Kauc, L. and Piekarowicz, A.** (1978) *Eur. J. Biochem.* 92: 417-426.
141. **Kazennova, E.V. and Tarasov, A.P., Mileikovskaya, M.M., Semina, I.E. and Tsvetkova, N.V.** (1982) *Zh. Mikrobiol. Epidemiol. Immunobiol.* 0: 56-57.
142. **Kelly, T.J., Jr. and Smith, H.O.** (1970) *J. Mol. Biol.* 51: 393-409.
143. **Kessler, C. Neumaier, P.S. and Wolf, W.** (1985) *Gene* 33: 1-102.
144. **Kholmina, G.V., Rebentish, B.A., Skoblov, Y.S., Mironov, A.A., Yankovsky, N.K., Kozlov, Y.I., Glatman, L.I., Moroz, A.F. and Debabov, V.G.** (1980) *Dokl. Akad. Nauk. SSSR* 253: 495-497.
145. **Khosaka, T.** unpublished observations.
146. **Khosaka, T. and Kiwaki, M.** (1984) *Gene* 31: 251-255.
147. **Khosaka, T. and Kiwaki, M.** (1984) *FEBS Letters* 177: 57-60.
148. **Khosaka, T., Sakurai, T., Takahashi, H. and Saito, H.** (1982) *Gene* 17: 117-122.
149. **Kiss, A., Sain, B., Csordas-Toth, E. and Venetianer, P.** (1977) *Gene* 1: 323-329.
150. **Kita, K., Hiraoka, N., Kimizuka, F. and Obayashi, A.** (1984) *Agric. Biol. Chem.* 48: 531-532.
151. **Kleid, D., Humayun, Z., Jeffrey, A. and Ptashne, M.** (1976) *Proc. Natl. Acad. Sci. USA* 73: 293-297.
152. **Takahashi, H., Kojima, H. and Saito, H.** (1985) *Biochem. J.* 231: 229-232.
153. **Koncz, C., Kiss, A. and Venetianer, P.** (1978) *Eur. J. Biochem.* 89: 523-529.
154. **Kramarov, V.M., Mazanov, A.L. and Smolyaninov, V.V.** (1982) *Bioorg. Khim.* 8: 220-223.
155. **Kramarov, V.M., Pachkunov, D.M. and Matvienko, N.I.,** in Gaziev, A.I.(Ed.), (1983) *Nek. Aspekty Fiziol. Mikroorg., Akad. Nauk SSSR,* Nauchn. Tsentr Biol. Issled., Puschino, USSR, 22-26.
156. **Kroger, M. and Hobom, G.** (1984) *Nucl. Acids Res.* 12: 887-899.
157. **Kroger, M., Hobom, G., Schutte, H. and Mayer, H.** (1984) *Nucl. Acids Res.* 12: 3127-3141.
158. **Kunkel, L.M., Silberklang, M. and McCarthy, B.J.** (1979) *J. Mol. Biol.* 132: 133-139.
159. **Lacks, S. and Greenberg, B.** (1975) *J. Biol. Chem.* 250: 4060-4066.
160. **Lacks, S. and Greenberg, B.** (1977) *J. Mol. Biol.* 114: 153-168.
161. **Landy, A., Ruedisueli, E., Robinson, L., Foeller, C. and Ross, W.** (1974) *Biochemistry* 13: 2134-2142.
162. **Langdale, J.A., Myers, P.A. and Roberts, R.J.** unpublished observations.
163. **Lau, R.H. and Doolittle, W.F.** (1980) *FEBS Letters* 121: 200-202.
164. **Lau, R.H., Visentin, L.P., Martin, S.M., Hofman, J.D. and Doolittle, W.F.** (1985) *FEBS Letters* 179: 129-132.
165. **Lautenberger, J.A., Edgell, M.H. and Hutchison, C.A. III.** (1980) *Gene* 12: 171-174.
166. **Lautenberger, J.A., Kan, N.C., Lackey, D., Linn, S., Edgell, M.H. and Hutchison, C.A. III.** (1978) *Proc. Natl. Acad. Sci. USA* 75: 2271-2275.
167. **Lautenberger, J.A. and Linn, S.** (1972) *J. Biol. Chem.* 247: 6176-6182.
168. **Lautenberger, J.A., White, C.T., Haigwood, N.L., Edgell, M.H. and Hutchinson, C.A. III.** (1980) *Gene* 9: 213-231.
169. **Leary, J.V.** unpublished observations.
170. **LeBon, J.M., Kado, C., Rosenthal, L.J. and Chirikjian, J.** (1978) *Proc. Natl. Acad. Sci. USA* 75: 4097-4101.

171. **LeBon, J.M., LeBon, T., Blakesley, R. and Chirikjian, J.** unpublished observations.
172. **Leung, D.W., Lui, A.C.P., Merilees, H., McBride, B.C. and Smith, M.** (1979) *Nucl. Acids Res.* 6: 17-25.
173. **Levi, C. and Bickle, T.** unpublished observations.
174. **Lin, B.-C., Chien, M.-C. and Lou, S.-Y.** (1980) *Nucl. Acids Res.* 8: 6189-6198.
175. **Lin, P-M. and Roberts, R.J.** unpublished observations.
176. **Lui, A.C.P., McBride, B.C. and Smith, M.** unpublished observations.
177. **Lui, A.C.P., McBride, B.C., Vovis, G.F. and Smith, M.** (1979) *Nucl. Acids Res.* 6: 1-15.
178. **Lupker, H.S.C. and Dekker, B.M.M.** (1981) *Biochim. Biophys. Acta* 654: 297-299.
179. **Lynn, S.P., Cohen, L.K., Gardner, J.F. and Kaplan, S.** (1979) *J. Bacteriol.* 138: 505-509.
180. **Lynn, S.P., Cohen, L.K., Kaplan, S. and Gardner, J.F.** (1980) *J. Bacteriol.* 142: 380-383.
181. **Makula, R.A.** unpublished observations.
182. **Makula, R.A. and Meagher, R.B.** (1980) *Nucl. Acids Res.* 8: 3125-3131.
183. **Mann, M.B., Rao, R.N. and Smith, H.O.** (1978) *Gene* 3: 97-112.
184. **Mann, M.B. and Smith, H.O.** (1977) *Nucl. Acids Res.* 4: 4211-4221.
185. **Mann, M.B. and Smith, H.O.** unpublished observations.
186. **Maratea, E. and Camp, R.R.** unpublished observations.
187. **Matvienko, N.I., Pachkunov, D.M. and Kramarov, V.M.** (1984) *FEBS Letters* 177: 23-26.
188. **Mayer, H., Grosschedl, R., Schutte, H. and Hobom,G.** (1981) *Nucl. Acids Res.* 9: 4833-4845.
189. **Mayer, H. and Klaar, J.** unpublished observations.
190. **Mayer, H. and Schutte, H.** unpublished observations.
191. **McConnell, D.J., Searcy, D.G. and Sutcliffe, J.G.** (1978) *Nucl. Acids Res.* 5: 1729-1739.
192. **McEvoy, S. and Roberts, R.J.** unpublished observations.
193. **McWilliam, P.** quoted in reference 143
194. **Meagher, R.B.** unpublished observations.
195. **Meselson, M. and Yuan, R.** (1968) *Nature* 217: 1110-1114.
196. **Middleton, J.H., Edgell, M.H. and Hutchison, C.A. III.** (1972) *J. Virol.* 10: 42-50.
197. **Middleton, J.H., Stankus, P.V., Edgell, M.H. and Hutchison, C.A. III** unpublished observations.
198. **Mise, K. and Nakajima, K.** (1984) *Gene* 30: 79-85.
199. **Mise, K. and Nakajima, K.** (1985) *Gene* 33: 357-361.
200. **Mise, K. and Nakajima, K.** (1985) *Gene* 36: 363-367.
201. **Molemans, F., van Emmelo, J. and Fiers, W.** (1982) *Gene* 18: 93-96.
202. **Morgan, R.** unpublished observations.
203. **Morgan, R. and Hoffman, L.** unpublished observations.
204. **Morris, D.W. and Parish, J.H.** (1976) *Arch. Microbiol.* 108: 227-230.
205. **Muller, F., Stoffel, S. and Clarkson, S.G.** unpublished observations.
206. **Mulligan, B.J. and Szekeres, M.** unpublished observations.
208. **Murray, K., Hughes, S.G., Brown, J.S. and Bruce, S.A.** (1976) *Biochem. J.* 159: 317-322.
209. **Murray, K. and Morrison, A.** unpublished observations.
210. **Murray, K., Morrison, A., Cooke, H.W. and Roberts, R.J.** unpublished observations.
211. **Myers, P.A. and Roberts, R.J.** unpublished observations.
212. **Nagaraja, V., Stieger, M., Nager, C., Hadi, S.M. and Bickle, T.** (1985) *Nucl. Acids Res.* 13: 389-399.
213. **Nardone, G. and Blakesley, R.** (1981) *Fed. Proc.* 40: 1848.
214. **Norlander, L., Davies, J.K., Hagblom, P., and Normark, S.** (1981) *J. Bacteriol.* 145: 788-795.
215. **O'Connor, C.D., Metcalf, E., Wrighton, C.J., Harris, T.J.R. and Saunders, J.R.** (1984) *Nucl. Acids Res.* 12: 6701-6708.
216. **Old, R., Murray, K. and Roizes, G.** (1975) *J. Mol. Biol.* 92: 331-339.
217. **Olson, J.A., Myers, P.A. and Roberts, R.J.** unpublished observations.
218. **Orekhov, A.V., Rebentish, B.A. and Debabov, V.G.** (1982) *Dokl. Akad. Nauk. SSSR* 263: 217-220.
219. **Parker, P. and Schildkraut, I.** unpublished observations.
220. **Piekarowicz, A.** (1982) *J. Mol. Biol.* 157: 373-381.
221. **Piekarowicz, A.** unpublished observations.
222. **Piekarowicz, A., Bickle, T.A., Shepherd, J.C.W. and Ineichen, K.** (1981) *J. Mol. Biol.* 146: 167-172.
223. **Piekarowicz, A., Stasiak, A. and Stanczak, J.** (1980) *Acta Microbiol. Pol.* 29: 151-156.
224. **Pirrotta, V.** (1976) *Nucl. Acids Res.* 3: 1747-1760.
225. **Pope, A., Lynn, S.P. and Gardner, J.F.** unpublished observations.
226. **Pugatsch, T. and Weber, H.** (1979) *Nucl. Acids Res.* 7: 1429-1444.
227. **Purvis, I.J. and Moseley, B.E.B.** (1983) *Nucl. Acids Res.* 11: 5467-5474.
228. **Qiang, B.-Q. and Schildkraut, I.** unpublished observations.
229. **Qiang, B.-Q. and Schildkraut, I.** (1984) *Nucl. Acids Res.* 12: 4507-4515.
230. **Qiang, B.-Q., Schildkraut, I. and Visentin, L.** unpublished observations.
231. **Ravetch, J.V., Horiuchi, K. and Zinder, N.D.** (1978) *Proc. Natl. Acad. Sci. USA* 75: 2266-2270.
232. **Reaston, J., Duyvesteyn, M.G.C. and deWaard, A.** (1982) *Gene* 20: 103-110.

233. **Reiser, J. and Yuan, R.** (1977) *J. Biol. Chem.* 252: 451-456.
234. **Roberts, R.J.** (1985) *Nucl. Acids Res.* 13: r165-r200.
235. **Roberts, R.J., Breitmeyer, J.B., Tabachnik, N.F. and Myers, P.A.** (1975) *J. Mol. Biol.* 91: 121-123.
236. **Roberts, R.J., Myers, P.A., Morrison, A., and Murray, K.** (1976) *J. Mol. Biol.* 102: 157-165.
237. **Roberts, R.J., Myers, P.A., Morrison, A. and Murray, K.** (1976) *J. Mol. Biol.* 103: 199-208.
238. **Roberts, R.J., Wilson, G.A. and Young, F.E.** (1977) *Nature* 265: 82-84.
239. **Roizes, G., Nardeux, P-C. and Monier, R.** (1979) *FEBS Letters* 104: 39-44.
240. **Roizes, G., Pages, M., Lecou, C., Patillon, M. and Kovoor, A.** (1979) *Gene* 6: 43-50.
241. **Roizes, G., Patillon, M. and Kovoor, A.** (1977) *FEBS Letters* 82: 69-70.
242. **Rosenvold, E.C.** unpublished observations.
243. **Rosenvold, E.C. and Szybalski, W.** unpublished observations, cited in *Gene* 7, 217-270 (1979).
244. **Roy, P.H. and Smith, H.O.** (1973) *J. Mol. Biol.* 81: 427-444.
245. **Roy, P.H. and Smith, H.O.** (1973) *J. Mol. Biol.* 81: 445-459.
246. **Sasaki, J. and Yamada, Y.** (1984) *Agric. Biol. Chem.* 48: 3027-3034.
247. **Sato, S., Hutchison, C.A. and Harris, J.I.** (1977) *Proc. Natl. Acad. Sci. USA* 74: 542-546.
248. **Sato, S., Nakazawa, K. and Shinomiya, T.** (1980) *J. Biochem.* 88: 737-747.
249. **Sato, S. and Shinomiya, T.** (1978) *J. Biochem.* 84: 1319-1321.
250. **Schildkraut, I.** unpublished observations.
251. **Schildkraut, I., Banner, D.B., Rhodes, C.S. and Parekh, S.** (1984) *Gene* 27: 327-329.
252. **Schildkraut, I. and Christ, C.** unpublished observations.
253. **Schildkraut, I. and Comb, D.** unpublished observations.
254. **Schildkraut, I., Grandoni, C. and Comb, D.** unpublished observations.
255. **Schildkraut, I. and Greenough, L.** unpublished observations.
256. **Schildkraut, I. and Wise, R.** unpublished observations.
257. **Schildkraut, I. Wise, R., Borsetti, R. and Qiang, B.-Q.** unpublished observations.
258. **Schmid, K., Thomm, M., Laminet, A., Laue, F.G., Kessler, C., Stetter, K. and Schmitt, R.** (1984) *Nucl. Acids Res.* 12: 2619-2628.
259. **Schwabe, G., Posseckert, G. and Klingmuller, W.** (1985) *Gene* 39: 113-116.
260. **Sciaky, D. and Roberts, R.J.** unpublished observations.
261. **Seurinck, J., van de Voorde, A. and van Montagu, M.** (1983) *Nucl. Acids Res.* 11: 4409-4415.
262. **Seurinck, J. and van Montagu, M.** unpublished observations.
263. **Sharp, P.A., Sugden, B. and Sambrook, J.** (1973) *Biochemistry* 12: 3055-3063.
264. **Shen, S., Li, Q., Yan, P., Zhou, B., Ye, S., Lu, Y. and Wang, D.** (1980) *Sci. Sin.* 23: 1435-1442.
265. **Shibata, T. and Ando, T.** (1975) *Mol. Gen. Genetics* 138: 269-379.
266. **Shibata, T. and Ando, T.** (1976) *Biochim. Biophys. Acta* 442: 184-196.
267. **Shibata, T., Ikawa, S., Kim, C. and Ando, T.** (1976) *J. Bacteriol.* 128: 473-476.
268. **Shimatake, H. and Rosenberg, M.** unpublished observations.
269. **Shimotsu, H., Takahashi, H. and Saito, H.** (1980) *Gene* 11: 219-225.
270. **Shimotsu, H., Takahashi, H. and Saito, H.** (1980) *Agric. Biol. Chem.* 44: 1665-1666.
271. **Shinomiya, T.** unpublished observations.
272. **Shinomiya, T., Kobayashi, M. and Sato, S.** (1980) *Nucl. Acids Res.* 8: 3275-3285.
273. **Shinomiya, T. and Sato, S.** (1980) *Nucl. Acids Res.* 8: 43-56.
274. **Sievert, U. and Rosch, A.** unpublished observations.
275. **Smith, D.I., Blattner, F.R. and Davies, J.** (1976) *Nucl. Acids Res.* 3: 343-353.
276. **Smith, H.O.** unpublished observations.
277. **Smith, H.O. and Nathans, D.** (1973) *J. Mol. Biol.* 81: 419-423.
278. **Smith, H.O. and Wilcox, K.W.** (1970) *J. Mol. Biol.* 51: 379-391.
279. **Smith, J. and Comb, D.** unpublished observations.
280. **Smith, L., Blakesley, R. and Chirikjian, J.** unpublished observations.
281. **Sokolov, N.N., Fitsner, A.B., Anikeitcheva, N.V., Choroshoutina, Yu.B., Samko, O.T., Kolosha, V.O., Fodor, I. and Votrin, I.I.** (1985) *Molec. Biol. Rep.* 10: 159-161.
282. **Sparling, R. and Bhatti, A.R.** (1984) *Microbios* 41: 73-79.
283. **Stephens, M.A.** (1982) *J. Bacteriol.* 149: 508-514.
284. **Stobberingh, E.E., Schiphof, R. and Sussenbach, J.S.** (1977) *J. Bacteriol.* 131: 645-649.
285. **Stotz, A. and Philippson, P.** unpublished observations.
286. **Streips, U. and Golemboski, B.** unpublished observations.
287. **Sugisaki, H.** (1978) *Gene* 3: 17-28.
288. **Sugisaki, H. and Kanazawa, S.** (1981) *Gene* 16: 73-78.
289. **Sugisaki, H., Maekawa, Y., Kanazawa, S. and Takanami, M.** (1982) *Nucl. Acids Res.* 10: 5747-5752.
290. **Sugisaki, H., Maekawa, Y., Kanazawa, S. and Takanami, M.** (1982) *Bull. Inst. Chem. Res. Kyoto Univ.* 60: 328-335.
291. **Sugisaki, H. and Takanami, M.** (1973) *Nature New Biology* 246: 138-140.

292. **Sun, D.K. and Yoo, O.J.** unpublished observations.
293. **Suri, B., Shepherd, J.C.W. and Bickle, T.A.** (1984) *EMBO J.* 3: 575-579.
294. **Sussenbach, J.S., Monfoort, C.H., Schiphof, R. and Stobberingh, E.E.** (1976) *Nucl. Acids Res.* 3: 3193-3202.
295. **Sussenbach, J.S., Steenbergh, P.H., Rost, J.A., van Leeuwen, W.J. and van Embden, J.D.A.** (1978) *Nucl. Acids Res.* 5: 1153-1163.
296. **Sutcliffe, J.G. and Church, G.M.** (1978) *Nucl. Acids Res.* 5: 2313-2319.
297. **Syddall, R. and Stachow, C.** (1985) *Biochim. Biophys. Acta* 825: 236-243.
298. **Szekeres, M.** unpublished observations.
299. **Takahashi, H.** unpublished observations.
300. **Takahashi, H., Shimizu, M., Saito, H., Ikeda, Y. and Sugisaki, H.** (1979) *Gene* 5: 9-18.
301. **Takanami, M.** (1974) *Methods in Mol. Biol.* 7: 113-133.
302. **Timko, J., Horwitz, A.H., Zelinka, J. and Wilcox,G.** (1981) *J. Bacteriol.* 145: 873-877.
303. **Tolstoshev, C.M. and Blakesley, R.W.** (1982) *Nucl. Acids Res.* 10: 1-17.
304. **Tomassini, J., Roychoudhury, R., Wu, R. and Roberts, R.J.** (1978) *Nucl. Acids Res.* 5: 4055-4064.
305. **Torheim, B.** unpublished observations.
306. **Trautner, T.A.,** unpublished observations. quoted in **Gunthert, U. and Trautner, T.A.,** in Trautner, T.A. (Ed.), *DNA Methyltransferases of Bacillus subtilis and Its Bacteriophages,* (1984) Springer-Verlag, (Curr. Top. Microbiol. Immunol.) 108: 11-22.
307. **Tu, C-P.D., Roychoudhury, R. and Wu,R.** (1976) *Biochem. Biophys. Res. Comm.* 72: 355-362.
308. **Uporova, T.M., Nikolskaya, I.I., Rubtsova, E.N. and Debov, S.S.** (1981) *Vestnik Akad. Medicin. Nauk SSSR* 2: 21-26.
309. **van den Hondel, C.A.M.J.J., van Leen, R.W., van Arkel, G.A., Duyvesteyn, M. and deWaard, A.** (1983) *FEMS Microbiology Letters* 16: 7-12.
310. **van Heuverswyn, H. and Fiers, W.** (1980) *Gene* 9: 195-203.
311. **van Montagu, M.** unpublished observations.
312. **van Montagu, M., Sciaky, D., Myers, P.A. and Roberts, R.J.** unpublished observations.
313. **van Ormondt, H., Lautenberger, J.A., Linn, S. and deWaard, A.** (1973) *FEBS Letters* 33: 177-180.
314. **Vanyushin, B.F. and Dobritsa, A.P.** (1975) *Biochim. Biophys. Acta* 407: 61-72.
315. **Venegas, A., Vicuna, R., Alonso, A., Valdes, F. and Yudelevich, A.** (1980) *FEBS Letters* 109: 156-158.
316. **Venetianer, P.** unpublished observations.
317. **Visentin, L.P.** unpublished observations.
318. **Visentin, L.P., Watson, R.J., Martin, S. and Zuker, M.** unpublished observations.
319. **Walter, F., Hartmann, M. and Roth, M.** (1978) *Abstracts of 12th FEBS Symposium, Dresden*
320. **Wang, T.-S.** (1981) *Ko Hsueh Tung Pao* 26: 815-817.
321. **Wang, R.Y.-H., Shedlarski, J.G., Farber, M.B., Kuebbing, D. and Ehrlich, M.** (1980) *Biochim. Biophys. Acta* 606: 371-385.
322. **Wani, A.A., Stephens, R.E., D'Ambrosio, S.M. and Hart, R.W.** (1982) *Biochim. Biophys. Acta* 697: 178-184.
323. **Watson, R., Zuker, M., Martin, S.M. and Visentin, L.P.** (1980) *FEBS Letters* 118: 47-50.
324. **Watson, R.J., Schildkraut, I., Qiang, B.-Q., Martin, S.M. and Visentin, L.P.** (1982) *FEBS Letters* 150: 114-116.
325. **Whang, Y. and Yoo, O.J.** unpublished observations.
326. **Whitehead, P.R. and Brown, N.L.** (1982) *FEBS Letters* 143: 296-300.
327. **Whitehead, P.R. and Brown, N.L.** (1983) *FEBS Letters* 155: 97-102.
328. **Whitehead, P.R. and Brown, N.L.** (1985) *J. Gen. Microbiol.* 131: 951-958.
329. **Wilson, G.A. and Young, F.E.** (1975) *J. Mol. Biol.* 97: 123-125.
330. **Wilson, G.A. and Young, F.E.,** in Schlessinger, D., (Ed.), *Restriction and modification in the Bacillus subtilis genospecies* Microbiology 1976. (1976) American Society for Microbiology, Washington, 350-357.
331. **Wilson, G.A. and Young, F.E.** unpublished observations.
332. **Winkler, K.** Diploma Dissertation (1979).
333. **Winkler, K. and Rosch, A.** unpublished observations.
334. **Wu, R., King, C.T. and Jay, E.** (1978) *Gene* 4: 329-336.
335. **Yamada, Y., Yoshioka, H., Sasaki, J. and Tahara, Y.** (1983) *J. Gen. Appl. Microbiol.* 29: 157-166.
336. **Yoon, H., Suh, H., Han, M.H. and Yoo, O.J.** (1985) *Korean Biochem. J.* 18: 82-87.
337. **Yoon, H., Suh, H., Kim, K., Han, M.H. and Yoo, O.J.** (1985) *Korean Biochem. J.* 18: 88-93.
338. **Yoshida, Y. and Mise, K.** unpublished observations.
339. **Yoshioka, H., Nakamura, H., Sasaki, J., Tahara, Y. and Yamada, Y.** (1983) *Agric. Biol. Chem.* 47: 2871-2879.
340. **Yoshimori, R.N.** PhD Thesis 1971
341. **Zabeau, M., Greene, R., Myers, P.A. and Roberts, R.J.** unpublished observations.
342. **Zabeau, M. and Roberts, R.J.** unpublished observations.
343. **Zain, B.S. and Roberts, R.J.** (1977) *J. Mol. Biol.* 115: 249-255.
344. **Zerler, B., Myers, P.A., Escalante, H. and Roberts, R.J.** unpublished observations.
345. **Calleja, F., Tandeau de Marsac, N., Coursin, T., van Ormondt, H. and de Waard, A.** (1985) *Nucl. Acids Res.* 13: 6745-6750.
346. **Yamada, Y. and Murakami, M.** (1985) *Agric. Biol. Chem.* 49: 3627-3629.
348. **Sasaki, J., Murakami, M. and Yamada, Y.** (1985) *Agric. Biol. Chem.* 49: 3107-3122.

350. McClelland, M., Nelson, M. and Cantor, C.R. (1985) *Nucl. Acids Res.* 13: 7171-7182.
351. Morgan, R. and Hempstead, S.K. unpublished observations.
352. Nagaraja, V., Shepherd, J.C.W., Pripfl, T. and Bickle, T.A. (1985) *J. Mol. Biol.* 182: 579-587.
353. Nagaraja, V., Shepherd, J.C.W. and Bickle, T.A. (1985) *Nature* 316: 371-372.
355. Morgan, R., Camp, R. and Soltis, A. unpublished observations.
356. Hiraoka, N., Kita, K., Nakajima, F., Kimizuka, F. and Obayashi, A. (1985) *J. Ferment. Technol.* 63: 151-157.
357. Kita, K., Hiraoka, N., Oshima, A., Kadonishi, S. and Obayashi, A. (1985) *Nucl. Acids Res.* 13: 8685-8694.
359. Stote, R. and Schildkraut, I. unpublishd observations.
360. Dingman, C. and Schildkraut, I. unpublished observations.
361. Wickberg, L. and Schildkraut, I. unpublished observations.
362. Christ, C. and Wickberg, L. unpublished observations.
363. Walker, J.N.B., Dean, P.D.G and Saunders, J.R. (1986) *Nucl. Acids Res.* 14: 1293-1301.
364. Evans, L.R. and Brown, N.L. unpublished observations.
365. Mullings, R., Evans, L.R. and Brown, N.L. (1986) *FEMS Microbiol. Letts.* 37: 237-240.
366. Eastlake, P. unpublished observations.
367. Walker, J.M., Vivian, A. and Saunders, J.R. unpublished observations.
368. Walker, J.M., Dean, P.G. and Saunders, J.R. unpublished observations.
369. Dean, P.D.C. and Walker, J.N.B. (1985) *Biochem. Soc. Transactions* 13: 1055-1058.
370. Kawamura, M., Sakakibara, M., Watanabe, T., Kita, K., Hiraoka, N., Obayashi, A., Takagi, M. and Yano, K. (1986) *Nucl. Acids Res.* 14: 1985-1989.
371. Qiang, B.-Q. and Schildkraut, I. (1986) *Nucl. Acids Res.* 14: 1991-1999.
372. Bakh, N.L., Tsvetkova, N.V., Semina, I.E., Tarasov, A.P., Mileikovskaya, M.M., Gruber, I.M., Polyachenko, V.M. and Romanenko, E.E. (1985) *Antibiot. Med. Biotek.* 30: 342-344.
374. Uporova, T.M., Kartasheva, I.M., Skripkin, E.A., Lopareva, E.N., Nikol'skaya, I.I. and Debov, S.S. (1985) *Vopr. Med. Khim.* 31: 131-136.
375. Weule, K. and Roberts, R.J. unpublished observations.
376. Kramarov, V.M. and Smolyaninov, V.V. (1981) *Biokhimiya* 46: 1526-1529.
377. Prangishvili, D.A., Vashakidze, R.P., Chelidze, M.G. and Gabriadze, I.Yu. (1985) *FEBS Letters* 192: 57-60.
378. Sohail, A., Khan, E., Riazuddin, S. and Roberts, R.J. unpublished observations.
380. Gordon, R. unpublished observations.
381. Bolton, B.J., Comer, M.J. and Kessler, C. unpublished observations.
382. Vasquez, C. (1985) *Biochem. Int.* 10: 655-662.
383. Walder, R.Y., Walder, J.A. and Donelson, J.E. (1984) *J. Biol. Chem.* 259: 8015-8026.
384. Butkus, V., Bitinaite, J., Kersulyte, D. and Janulaitis, A. (1985) *Biochim. Biophys Acta* 826: 208-212.
385. Piekarowicz, A., Goguen, J.D. and Skrzypek, E. (1985) *Eur. J. Biochem.* 152: 387-393.
387. Price, C., Shepherd, J.C.W. and Bickle, T.A. (1987) *EMBO J.* 6: 1493-1497.
388. Hall, D., Camp, R., Morgan, R. and Hoffman, L. unpublished observations.
389. Morgan, R. and Ingalls, D. unpublished observations.
390. Stote, R. and Morgan, R. unpublished observations.
391. Piekarowicz, A. and Goguen, J.D. (1986) *Eur. J. Biochem.* 154: 295-298.
392. Howard, K.A., Card, C., Benner, J.S., Callahan, H.L., Maunus, R., Silber, K., Wilson, G. and Brooks, J.E. (1986) *Nucl. Acids Res.* 14:7939-7951.
393. Korch, C., Hagblom, P. and Normark, S. (1985) *J. Bacteriol.* 161: 1236-1237.
394. Aiken, C., Milarski-Brown, K. and Gumport, R.I. (1986) *Fed. Proc.* 45: 1914.
395. Mise, K., Nakajima, K., Terakado, N. and Ishidate, M. (1986) *Gene* 44: 165-169.
396. Nikolskaya, I.I., Karpetz, L.Z., Kartashova, I.M., Lopatina, N.G., Skripkin, E.A., Suchkov, S.V., Uporova, T.M., Gruber, I.M. and Debov, S.S. (1983) *Molek. Genet. Mikrobiol. Virusol.* 12: 5-10.
397. Boyd, A.C., Charles, I.G., Keyte, J.W. and Brammer, W.J. (1986) *Nucl. Acids Res.* 14: 5255-5274.
398. Karreman, C., Tandeau de Marsac, N. and deWaard, A. (1986) *Nucl. Acids Res.* 14: 5199-5205.
399. Xia, Y., Burbank, D.E., Uher, L., Rabussay, D. and Van Etten, J.L. (1986) *Mol. Cell. Biol.* 6: 1430-1439.
400. Xia, Y. and Van Etten, J.L. (1986) *Mol. Cell. Biol.* 6: 1440-1445.
401. Xia, Y., Burbank, D.E. and Van Etten, J.L. (1986) *Nucl. Acids Res.* 14: 6017-6030.
402. Kelly, S., Kaddurah-Daouk, R. and Smith, H.O. (1985) *J. Biol. Chem.* 260: 15339-15344.
403. Mann, M.B. and Smith, H.O., Specificity of DNA methylases from *Haemophilus* sp., in *Proc. Conf. Methylation*, Usdin, E., Borchardt, R.T. and Greveling, C.R., Eds., (1979) Elsevier/North Holland, New York., 483-492.
404. Whitehead, P.R. and Brown, N.L. (1985) *Arch. Microbiol.* 141: 70-74.
405. Whitehead, P.R., Jacobs, D. and Brown, N.L. (1986) *Nucl. Acids Res.* 14: 7031-7045.
406. Christ, C. and Ingalls, D. unpublished observations.
407. Maneliene, Z., Bitinaite, J., Butkus, V. and Janulaitis, A. unpublished observations.
408. Janulaitis, A. and Gilvonauskaite, R. unpublished observations.
409. Janulaitis, A., Bagdonaviciute, V. and Petrusyte, M. unpublished observations.

410. **Butkus, V.V., Petrusyte, M.P. and Janulaitis, A.A.** (1985) *Bioorg. Khim.* 11: 987-988.
411. **Petrusyte, M., Bitinaite, J., Kersulyte, D., Menkevicius, S., Butkus, V. and Janulaitis, A.** (1987) *Dokl. Akad. Nauk SSSR* 295: 1250-1253.
412. **Janulaitis, A.A., Stakenas, P.S., Lebedenko, E.N. and Berlin, Yu.A.** (1982) *Nucl. Acids Res.* 10: 6521-6530.
413. **Janulaitis, A., Kazlauskiene, R. and Gilvonauskaite, R.** unpublished observations.
414. **Janulaitis, A. and Kazlauskiene, R.** unpublished observations.
415. **Janulaitis, A., Kazlauskiene, R. and Steponaviciene, D.** unpublished observations.
416. **Kazlauskiene, R., Maneliene, Z., Butkus, V., Petrusyte, M. and Janulaitis, A.** (1986) *Bioorg. Khim.* 12: 836-838.
417. **Butkus, V., Kazlauskiene, R., Gilvonauskaite, R., Petrusyte, M. and Janulaitis, A.** (1985) *Bioorg. Khim.* 11: 1572-1573.
418. **Janulaitis, A., Kazlauskiene, R., Butkus, V., Petrauskiene, L. and Petrusyte, M.** unpublished observations.
419. **Janulaitis, A., Kazlauskiene, R. and Bagdonaviciute, V.** unpublished observations.
420. **Janulaitis, A., Kazlauskiene, R. and Petrusyte, M.** unpublished observations.
421. **Janulaitis, A., Steponaviciene, D., Butkus, V., Maneliene, Z. and Petrusyte, M.** unpublished observations.
422. **Janulaitis, A. and Steponaviciene, D.** unpublished observations.
423. **Janulaitis, A., Gilvonauskaite, R. and Petrusyte, M.** unpublished observations.
424. **Krayev, A.S., Zimin, A.A., Mironova, M.V., Janulaitis, A.A., Tanyashin, V.I., Skryabin, K.G. and Bayev, A.A.** (1981) *Dokl. Akad. Nauk SSSR* 270: 1495-1500.
425. **Butkus, V., Padegimiene, A., Laucys, V. and Janulaitis, A.** unpublished observations.
426. **Bitinaite, J., Kersulyte, D., Butkus, V. and Janulaitis, A.** unpublished observations.
427. **Butkus, V.V., Stakenas, P.S. and Janulaitis A.A.** unpublished observations.
428. **Janulaitis, A. and Lazareviciute, L.** unpublished observations.
429. **Janulaitis, A., Lazareviciute, L. and Lebionka, A.** unpublished observations.
430. **Petrusyte, M.P. and Janulaitis, A.** (1981) *Bioorg. Khim.* 7: 1885-1887.
431. **Klimasauskas, S., Butkus, V. and Janulaitis, A.** unpublished observations.
432. **Klimasauskas, S., Lebionka, A., Butkus, V. and Janulaitis, A.** unpublished observations.
433. **Petrauskiene, L., Klimasauskas, S., Butkus, V. and Janulaitis, A.** unpublished observations.
434. **Schinzel, R.** unpublished observations.
435. **Goossens, M., Dumez, Y., Kaplan, L., Lupker, M., Chabret, C., Henrion, R. and Rosa, J.** (1983) *New England J. Med.* 309: 831-833.
436. **Rushizky, G.W.,** Purification of the sequence specific endonuclease PalI in Chirikjian, J.G., Ed., *Gene Amplification and Analysis.* (1981) Elsevier/North Holland, 1: 239-242.
437. **Arutyunyan, E.E., Gruber, I.M., Polyachenko, V.M., Kvachadze, L.J., Andriashvili, I.A., Chanishvili, T.G. and Nikol'skaya, I.I.** (1985) *Vopr. Med. Khim.* 31: 127-132.
438. **Calleja, F., van Ormondt, H. and de Waard, A.** unpublished observations.
439. **Yan, P-F., Ye, S-Y., Wang, P-Z., Li, Q-L, Lu, Y-Y and Zhou, B.** (1982) *Acta Biochim. Biophys. Sinica* 14: 151-158.
440. **Yan, P.F. and Zhou, B.** unpublished observations.
441. **Zhou, B. and Li, Q.** unpublished observations.
444. **Janulaitis, A.A. and Vaitkevicius, D.P.** (1985) *Biotechnologiya* 1: 39-51.
445. **Yamada,Y. and Sasaki, J.** (1984) *J. Gen. Appl. Microbiol.* 30: 309-312.
446. **Yoo, O.J. and Choi, K.D.** unpublished observations.
447. **Simbochova, G., Timko, J., Zelinkova, E. and Zelinka, J.** (1986) *Biologia (Bratislava)* 41: 357-365.
448. **Meyertons, J.L., Tilley, B.C., Lechevalier, M.P. and Lechevalier, H.A.** (1987) *J. Ind. Microbiol.* 2: 293-303.
449. **Jacobs, D. and Brown, N.L.** (1986) *Biochem. J.* 238: 613-616.
450. **Uchida, Y.** unpublished observations.
451. **Chen, Z. and Kong, H.** unpublished observations.
452. **Chen, G.C.C., Brown, A. and Lema, M.W.** (1986) *Can. J. Microbiol.* 32: 591-593.
454. **Duda, E.G., Izsvak, Z. and Orosz, A.** (1987) *Nucl. Acids Res.* 15: 1334.
455. **Morgan, R., Stote, R. and Schildkraut, I.** unpublished observations.
456. **Wood, P.** unpublished observations.
457. **Morgan, R.D., Dalton, M. and Stote, R.** (1987) *Nucl. Acids Res.* 15: 7201.
458. **Morgan, R., Stote, R. and Soltis, A.** unpublished observations.
459. **Morgan, R. and Ellard, J.** unpublished observations.
460. **Zieger, M., Patillon, M., Roizes, G., Lerouge, T., Dupret, D. and Jeltsch, J.M.** (1987) *Nucl. Acids Res.* 15: 3919.
462. **Manachini, P.L., Parini, C., Fortina, M.G. and Benazzi, L.** (1987) *FEBS Letts.* 214: 305-307.
463. **Schildkraut, I., Lynch, J. and Morgan, R.** (1987) *Nucl. Acids Res.* 15: 5492.
464. **Gallagher, M.L. and Burke, W.F.** (1985) *FEMS Microbiol. Letts.* 26: 317-321.
465. **Matvienko, N.I., Kramarov, V.M. and Pachkunov, D.M.** (1987) *Eur. J. Biochem.* 165: 565-570.
467. **Xia, Y., Burbank, D.E., Uher, L., Rabussay, D. and Van Etten, J.L.** (1987) *Nucl. Acids Res.* 15: 6075-6090.
468. **Miyahara, M. and Mise, K.** unpublished observations.
469. **Mise, K.** unpublished observations.
470. **Matsui, M., Mise, K., Yoshida, Y. and Ishidate, M.** (1986) *Bull. Natl. Inst. Hyg. Sci. (Tokyo)* 104: 92-96.

471. **Stein, D.C.** unpublished observations.
472. **Butkus, V., Petrauskiene, L., Maneliene, Z., Klimasauskas, S., Laucys, V. and Janulaitis, A.** (1987) *Nucl. Acids Res.* 15: 7091-7102.
473. **Kong, H. and Chen, Z.** (1987) *Nucl. Acids Res.* 15: 7205.
474. **Kauc, L. and Leszczynska, K.** (1986) *Acta Microbiol. Pol.* 35: 317-320.
475. **Hall, D. and Morgan, R.** unpublished observations.
476. **Morgan, R., Mattila, P., Pitkanen, K. and Robinson, D.** unpublished observations.
477. **Polisson, C.** unpublished observations.
478. **Polisson, C. and McMahon, M.** unpublished observations.
479. **Hall, D.** unpublished observations.
480. **Grandoni, R.** unpublished observations.
481. **Schneider, A.** unpublished observations.
482. **Polisson, C. and Morgan, R.** unpublished observations.
483. **Sohail, A., Khan, E. and Riazuddin, S.** unpublished observations.
484. **Sohail, A., Khan, E., Maqbool, T. and Riazuddin, S.** unpublished observations.
485. **Sohail, A., Mushtaq, R., Khan, E. and Riazuddin, S.** unpublished observations.
486. **Maqbool, T., Sohail, A., Chudary, S. and Riazuddin, S.** unpublished observations.
487. **Sohail, A., Khan, E., Mushtaq, R. and Riazuddin, S.** unpublished observations.
488. **Butkus, V., Klimasauskas, S., Petrauskiene, L., Maneliene, Z., Lebionka, A. and Janulaitis, A.** (1987) *Biochim. Biophys. Acta* 909: 201-207.
489. **Klimasauskas, S., Butkus, V. and Janulaitis, A.** (1987) *Molekul. Biolog.* 21: 87-92.
490. **Chen, Z.** unpublished observations.
491. **Chien, R.H., Stein, D.C., Seifert, H.S., Floyd, K. and So, M.** unpublished observations.
492. **Duff, M.K. and Davies, J.K.** unpublished observations.
493. **Sullivan, K.M., Macdonald, H.J. and Saunders, J.R.** (1987) *FEMS Microbiology Lett.* 44: 389-393.
494. **Tsui, W.-C., Elgar, G., Merrill, C. and Maunders, M.** (1988) *Nucl. Acids Res.* 16: 4178.
495. **Clark, D.** unpublished observations.
496. **Xia, Y., Narva,K.E. and Van Etten, J.L.** (1987) *Nucl. Acids Res.* 15: 10063.
497. **Degtyarev, S.K., Repin, V.E., Rechkunova, N.I., Tchigikov, V.E., Malygin, E.G., Mikhajlov, V.V. and Rasskazov, V.A.** (1987) *Bioorg. Khim.* 13: 420-421.
498. **Degtyarev, S.K., Rechkunova, N.I., Netesova, N.A., Tchigikov, V.E., Malygin, E.G., Kochkin, A.V., Mikhajlov, V.V. and Rasskazov, V.A.** (1987) *Bioorg. Khim.* 13: 422-423.
499. **Narva, K.E., Wendell. D.L., Skrdla, M.P. and Van Etten, J.L.** (1987) *Nucl. Acids Res.* 15: 9807-9823.
500. **Miyahara, M., Maruyama, T., Wake, A. and Mise, K.** *Appl. Environ. Microbiol.* in press.
501. **Bernal, W.M., Raven, N.D.H and Williams, R.A.D.** (1986) *Proc. 14th Int. Congress Microbiology.* 204.
503. **Raven, N.D.H., Ghufoor, K. and Williams, R.A.D.** unpublished observations.
504. **Ghofoor, K., Raven, N.D.H. and Williams, R.A.D.** unpublished observations.
505. **Raven, N.D.H., Sharp, R.J. and Williams, R.A.D.** unpublished observations.
506. **Venetianer, P. and Orosz, A.** (1988) *Nucl. Acids Res.* 16: 350.
507. **Yoshida, Y. and Mise, K.** (1986) *J. Bacteriol.* 165: 357-362.
508. **Tsui, W.-C.** unpublished observations.
509. **Tsvetkova, N.V., Mileikovskaya, M.M., Gruber, I.M., Polyachenko, V.M., Butkus, V.V., Janulaitis, A.A., Sudzhyuvene, O.F. and Tarasov, A.P.** (1987) *Mol. Genet. Mikrobiol. Virusol.* 4: 19-22.
510. **Xia, Y. and van Etten, J.L.** unpublished observations.
511. **Xia, Y., Burbank, D.E. and van Etten, J.L.** unpublished observations.
512. **Kato, F., Suetake, T., Murata, A., Mukai, T. and Maekawa, N.** unpublished observations.
513. **Thomm, M., Frey, G., Bolton, B.G., Lhue, F., Kessler, C. and Stetter, K.O.** *FEMS Microbiol. Letts.* (1988) 52:229-234.
514. **Calleja, F. and De Waard, A.** unpublished observations.
515. **Timko, J., Turna, J. and Zelinka, J.,** in *Metabolism and Enzymology of Nucleic Acids* Zelinka, J. and Balan, J., Eds., (1987) Publishing House of the Slovak Akademy of Sciences, 6: 107-118.
516. **Lopatina, N.G., Kirnos, M.D., Suchkov, S.V., Vanyushin, B.F., Nikol'skaya, I.I. and Debov, S.S.** (1985) *Biokhimiya* 50: 495-502.
517. **Danaher, R. and Stein, D.C.** unpublished observations
518. **Bolton, B.J., Schmitz, G.S., Jarsch, M., Comer, M.J. and Kessler, C.** (1988) *Gene* in press.
519. **Bolton, B.J., Holtke, H.J., Glados, S., Jarsch, M., Schmidt, G.G. and Kessler, C.** unpublished observations.
520. **Piekarowicz, A., Yuan, R. and Stein, D.C.**
521. **Cocks, B.G. and Finch, L.R.** (1987) *Int. J. Sys. Bacteriol.* 37: 451-453.
522. **Gann, A.A.F., Campbell, A.J.B., Collins, J.F., Coulson, A.F.W. and Murray, N.E.** (1987) *Mol. Microbiol.* 1: 13-22.
523. **Morgan, R.D.** (1988) *Nucl. Acids Res.* 16: 3104.
524. **Kramarov, V.M., Fomenkov, A.I., Matvienko, N.I., Ubieta, R.H., Smolianov, V.V. and Gorlenko, V.M.** (1987) *Bioorg. Khim.* 13: 773-776.

525. Faruqi, A.F. and Ahmed, N. (1987) *Pakistan. J. Sci. Ind. Res.* 30: 390-392.

526. Venetianer, P. and Orosz, A. (1988) *Nucl. Acids Res.* 16: 3053-3060.

527. Gaido, M.L., Prostkos, C.R. and Strobl, J.S. (1988) *J. Biol. Chem.* 263: 4832-4836.

528. Stein, D.C., Gregoire, S. and Piekarowicz, A. (1988) *Infect. Immun.* 56: 112-116.

529. Kholmina, G.V., Rebentish, B.A., Kozlovskii, Y.E., Sorokin, A.V., Gol'denberg, D.S. and Prozorov, A.A. (1986) *Molek. Genet. Mikrobiol. Virusol.* 11: 23-25.

530. Nwoso, V.U., Connolly, B.A., Halford, S.E. and Garnett, J. (1988) *Nucl. Acids Res.* 16: 3705-3720.

531. Polisson, C. and Morgan, R.D. (1988) *Nucl. Acids Res.* 16: 5205.

532. Karreman, C. and de Waard, A. (1988) *J. Bacteriol.* 170: 2533-2536.

533. Brand, P. and de Waard, A. unpublished observations.

534. Flores, E. and Wolk, P. unpublished observations.

535. Karreman, C. and de Waard, A. unpublished observations.

536. Siegelman, B. unpublished observations.

537. Lazareviciute, L., Laucys, V., Maneliene, Z., Poliachenko, V., Bitinaite, J., Butkus, V. and Janulaitis, A. unpublished observations.

538. Steponaviciene, D., Petrusyte, M. and Janulaitis, A. unpublished observations.

539. Jagelavicius, M., Bitinaite, J. and Janulaitis, A. unpublished observations.

540. Naureckiene, S., Bitinaite, J., Vaitkevicius, D., Menkevicius, S., Butkus, V. and Janulaitis, A. unpublished observations.

541. Petrusyte, M., Padegimiene, A., Butkus, V. and Janulaitis, A. unpublished observations.

542. Naureckiene, S., Kazlauskiene, R., Maneliene, Z., Vaitkevicius, D., Butkus, V. and Janulaitis, A. unpublished observations.

543. Naureckiene, S., Kazlauskiene, R., Vaitkevicius, D., Butkus, V., Kiuduliene, L. and Janulaitis, A. unpublished observations.

544. Gilvonauskaite, R., Padegimiene, A., Petrusyte, M., Butkus, V. and Janulaitis, A. unpublished observations.

545. Gilvonauskiete, R., Maneliene, Z., Padegimiene, A., Petrusyte, M. and Janulaitis, A. unpublished observations.

546. Petrusyte, M. and Janulaitis, A. (1982) *Eur. J. Biochem.* 121: 377-381.

547. Gilvonauskaite, R., Kazlauskiene, R., Padegimiene, A., Petrusyte, M., Butkus, V. and Janulaitis, A. unpublished observations.

548. Lazareviciute, L., Laucys, V., Maneliene, Z., Poliachenko, V., Butkus, V. and Janulaitis, A. unpublished observations.

549. Lazareviciute, L., Laucys, V., Poliachenko, V. and Janulaitis, A. unpublished observations.

550. Lazareviciute, L., Bitinaite, J., Laucys, V., Poliachenko, V. and Janulaitis, A. unpublished observations.

551. Lazareviciute, L., Bitinaite, J., Laucys, V., Poliachenko, V. Padegimiene, A, Butkus, V. and Janulaitis, A. unpublished observations.

552. Gilvonauskaite, R., Petrusyte, M., Butkus, V., Kersulyte, D. and Janulaitis, A. unpublished observations.

553. Vaitkevicius, D., Kulba, A., Chernov, S., Butkus, V., Fomichiov, J. and Janulaitis, A. unpublished observations.

554. Kulba, A.M., Abdel-Sabur, M.S., Butkus, V.V., Janulaitis, A.A. and Fomichiov, J. (1987) *Molek. Biol.* 21: 250-254.

555. Gilvonauskaite, R., Bitinaite, J., Petrusyte, M., Butkus, V., Maneliene, Z. and Janulaitis, A. unpublished observations.

556. Petrusyte, M., Maneliene, Z., Butkus, V. and Janulaitis, A. unpublished observations.

557. Jagelavicius, M., Bitinaite, J., Maneliene, Z., Butkus, V. and Janulaitis, A. unpublished observations.

558. Halden, N.F., Wolf, J.B., Cross, S.L. and Leonard, W.J. (1988) *Clin. Res.* 36: 404A.

Section 3

Lipids

FATTY ACIDS: PHYSICAL AND CHEMICAL CHARACTERISTICS

Acid		Chemical formula	Molec-ular weight	Melting point °C	Boiling point °C/mm[a]
No	Systematic name	Common name				

No	Systematic name	Common name	Chemical formula	Molecular weight	Melting point °C	Boiling point °C/mm[a]
		SATURATED FATTY ACIDS				
1	**Methanoic**	Formic	HCOOH	46.0	8.4	100.5
2	**Ethanoic**	Acetic	CH_3COOH	60.1	16.7	118.2
3	**Propanoic**	Propionic	C_2H_5COOH	74.1	−22.0	141.1
4	**Butanoic**	Butyric	C_3H_7COOH	88.1	−7.9	163.5
5	**Pentanoic**	Valeric	C_4H_9COOH	102.1	−34.5	187
6	**Hexanoic**	Caproic	$C_5H_{11}COOH$	116.2	−3.4	205.8
7	**Heptanoic**	Heptylic[g]	$C_6H_{13}COOH$	130.2	−10.5	223.0
8	**Octanoic**	Caprylic	$C_7H_{15}COOH$	144.2	16.7	239.7
9	**Nonanoic**	Pelargonic	$C_8H_{17}COOH$	158.2	12.5	255.6
10	**Decanoic**	Capric	$C_9H_{19}COOH$	172.3	31.6	270⁻
11	**Undecanoic**[h]	Undecylic	$C_{10}H_{21}COOH$	186.3	29.3	284
12	**Dodecanoic**	Lauric	$C_{11}H_{23}COOH$	200.3	44.2	225/100
13	**Tridecanoic**	Tridecylic	$C_{12}H_{25}COOH$	214.3	41.5	236/100
14	**Tetradecanoic**	Myristic	$C_{13}H_{27}COOH$	228.4	53.9	250/100
15	**Pentadecanoic**	Pentadecylic	$C_{14}H_{29}COOH$	242.2	52.3	202.5/10
16	**Hexadecanoic**	Palmitic	$C_{15}H_{31}COOH$	256.4	63.1	268/100
17	**Heptadecanoic**	Margaric	$C_{16}H_{33}COOH$	270.4	61.3	220/10
18	**Octadecanoic**	Stearic	$C_{17}H_{35}COOH$	284.5	69.6	213/5
19	**Nonadecanoic**	Nonadecyclic	$C_{18}H_{37}COOH$	298.5	68.6	299/10
20	**Eicosanoic**	Arachidic	$C_{19}H_{39}COOH$	312.5	76.5	204/1
21	**Docosanoic**	Behenic	$C_{21}H_{43}COOH$	340.6	81.5	306/60
22	**Tetracosanoic**	Lignoceric	$C_{23}H_{47}COOH$	368.6	86.0	272/10
23	**Hexacosanoic**	Cerotic	$C_{25}H_{51}COOH$	396.7	88.5	—
24	**Octacosanoic**	Montanic	$C_{27}H_{55}COOH$	424.7	90.9	—
25	**Triacontanoic**	Melissic	$C_{29}H_{59}COOH$	452.8	93.6	—
26	**Dotriacontanoic**	Lacceroic	$C_{31}H_{63}COOH$	480.0	96.2	—
27	**Tetratriacontanoic**	Gheddic	$C_{33}H_{67}COOH$	508.9	98.4	—
28	**Pentatriacontanoic**	Ceroplastic	$C_{34}H_{69}COOH$	522.9	98.4	—
		UNSATURATED FATTY ACIDS (MONOETHENOIC)				
29	*trans*-**2-Butenoic**	Crotonic	$C_4H_6O_2$	86.1	72	189.0
30	*cis*-**2-Butenoic**	Isocrotonic	$C_4H_6O_2$	86.1	15.5	169.3
31	**2-Hexenoic**	Isohydrosorbic	$C_6H_{10}O_2$	114.1	32	217
32	**4-Decenoic**	Obtusilic	$C_{10}H_{18}O_2$	170.2	—	149/13
33	**9-Decenoic**	Caproleic	$C_{10}H_{18}O_2$	170.2	—	142/4
34	**4-Dodecenoic**	Linderic	$C_{12}H_{22}O_2$	198.3	1.0–1.3	171/13
35	**5-Dodecenoic**	Denticetic	$C_{12}H_{22}O_2$	198.3	—	—
36	**9-Dodecenoic**	Lauroleic	$C_{12}H_{22}O_2$	198.3	—	142/4
37	**4-Tetradecenoic**	Tsuzuic	$C_{14}H_{26}O_2$	226.4	18.0–18.5	185–188/13
38	**5-Tetradecenoic**	Physeteric	$C_{14}H_{26}O_2$	226.4	—	190–195/15
39	**9-Tetradecenoic**	Myristoleic	$C_{14}H_{26}O_2$	226.4	−4	—
40	**9-Hexadecenoic**	Palmitoleic	$C_{16}H_{30}O_2$	254.4	−0.5 to +0.5	131/0.06
41	**6-Octadecenoic**	Petroselinic	$C_{18}H_{34}O_2$	282.5	32–33	237.5/18
42	*cis*-**9-Octadecenoic**	Oleic	$C_{18}H_{34}O_2$	282.5	13.4(α), 16.3(β)	234/15
43	*trans*-**9-Octadecenoic**	Elaidic	$C_{18}H_{34}O_2$	282.5	44.5	288/100
44	*trans*-**11-Octadecenoic**	Vaccenic	$C_{18}H_{34}O_2$	282.5	44	—
45	**9-Eicosenoic**	Gadoleic	$C_{20}H_{38}O_2$	310.5	24–24.5	220/6
46	**11-Eicosenoic**	Gondoic	$C_{20}H_{38}O_2$	310.5	23.5–24	267/15
47	**11-Docosenoic**	Cetoleic	$C_{22}H_{42}O_2$	338.6	32.5–33	—
48	**13-Docosenoic**	Erucic	$C_{22}H_{42}O_2$	338.6	34.7	242/5
49	**15-Tetracosenoic**	Nervonic[i]	$C_{24}H_{46}O_2$	366.6	42.5–43.0	—
50	**17-Hexacosenoic**	Ximenic	$C_{26}H_{50}O_2$	394.7	45–45.5	—
51	**21-Triacontenoic**	Lumequeic	$C_{30}H_{58}O_2$	450.8	—	—
		UNSATURATED FATTY ACIDS (DIENOIC)				
52	**2,4-Pentadienoic**	β-Vinylacrylic	$C_5H_6O_2$	98.1	80	110 d.
53	**2,4-Hexadienoic**	Sorbic	$C_6H_8O_2$	112.1	134.5	228 d.
54	**2,4-Decadienoic**	Stillingic	$C_{10}H_{16}O_2$	168.2	—	—
55	**2,4-Dodecadienoic**	—	$C_{12}H_{20}O_2$	196.3	—	—
56	**9,12-Hexadecadienoic**	—	$C_{16}H_{28}O_2$	252.4	—	—
57	*cis*-**9,**cis**-12-Octadecadienoic**	α-Linoleic	$C_{18}H_{32}O_2$	280.5	−5.2 to −5.0	202/1.4

FATTY ACIDS: PHYSICAL AND CHEMICAL CHARACTERISTICS (continued)

No	Specific gravity[b]	Refractive index[c] n_D^C	Neutral- ization value[d]	Iodine value (calcu- lated)[e]	Solubility[f]	Reference	No
			SATURATED FATTY ACIDS				
1	$1.220^{20°}$	$1.3714^{20°}$	1.219	----	s.w.	13, 28, 29	1
2	$1.049^{20°}$	$1.3718^{20°}$	934.2	—	s.w.	15, 28, 29	2
3	$0.992^{20°}$	$1.3874^{20°}$	757.3		s.al., chl., eth., w.	28, 29	3
4	$0.9587^{20°}$	$1.33906^{20°}$	636.8		s.al., eth., w.	13, 29, 36	4
5	$0.942^{20°}$	$1.4086^{20°}$	549.3		s.al., eth.; sl.s.w.	29, 36	5
6	$0.929^{20°}$	$1.41635^{20°}$	483.0		s.al., eth.; sl.s.w.	29, 36	6
7	$0.92215^{20°}$	$1.4230^{20°}$	431.0		s.al., eth.; v.sl.s.w.	29, 36	7
8	$0.910^{20°}$	$1.4285^{20°}$	389.1		s.al., bz., eth.; v.sl.s.w.	13, 29, 36	8
9	$0.907^{20°}$	$1.4322^{20°}$	354.6		s.al., chl., eth.; v.sl.s.w.	29, 36	9
10	$0.8858^{40°}$	$1.42855^{40°}$	325.7		s.al., eth., pet.eth.; v.sl.s.w.	13, 29, 36	10
11	$0.9905^{25°}$	$1.4202^{70°}$	301.2		s.al., chl., eth., pet.eth.	29, 36	11
12	$0.8690^{50°}$	$1.4261^{60°}$	280.1		s.acet., al., eth., pet.eth.	13, 29, 36	12
13	$0.8458^{80°}$	$1.4286^{60°}$	261.8		s.acet., al., eth., pet.eth.	14, 29, 31, 36	13
14	$0.8622^{54°}$	$1.4273^{70°}$	245.7		s.acet., al., eth., pet.eth.	13, 29, 36	14
15	$0.8423^{80°}$	$1.4292^{70°}$	231.5		s.acet., al., eth., pet.eth.	29, 36	15
16	$0.8487^{70°}$	$1.4309^{70°}$	218.8		s.acet., h.al., eth., pet.eth.	29, 36	16
17	$0.853^{60°}$	$1.4324^{70°}$	207.5		s.acet., h.al., eth., pet.eth.	29, 36	17
18	0.8390^{80}	1.4337^{70}	197.2		s.acet., h.al., eth., pet.eth.	13, 29, 36	18
19	$0.8771^{24°}$	$1.4512^{25°}$	188.0		s.acet., h.al., eth., pet.eth.	29	19
20	$0.8240^{100°}$	$1.4250^{100°}$	179.5		s.bz., chl., eth., pet.eth.	13, 29, 36	20
21	$0.8221^{100°}$	$1.4270^{100°}$	164.7		sl.s.al., eth.	13, 29, 36	21
22	$0.8207^{100°}$	$1.4287^{100°}$	152.2		s.ac.a., bz., CS_2, eth.	28, 29, 36, 39	22
23	$0.8198^{100°}$	$1.4301^{100°}$	141.4		s.h.acet., h.chl., h.me.al.	13, 29, 36	23
24	$0.8191^{100°}$	$1.4313^{100°}$	132.1		s.h.ac.a., h.bz., h.me.al.	13, 29, 36	24
25		$1.4323^{100°}$	123.9		s.chl., CS_2, h.me.al.	13, 29, 36	25
26			116.7		s.h.acet., h.bz., chl.	13, 17, 29, 36	26
27			110.2		s.h.acet., h.bz., chl.	13, 29, 32, 36	27
28			107.3		s.h.acet., h.bz., chl.	13, 29, 32, 36	28
			UNSATURATED FATTY ACIDS (MONOETHENOIC)				
29	$0.964^{80°}$	$1.4228^{80°}$	651.7	294.9	s.acet., al., tol., w.	29	29
30	$1.0312^{15°}$	$1.4457^{20°}$	651.7	294.9	s.al., pet.eth., w.	29	30
31	$0.965^{20°}$	$1.4460^{40°}$	491.5	222.5	s.CS_2, eth.	29, 36	31
32	$0.9197^{20°}$	$1.4497^{20°}$	329.6	149.1	s.bz., eth.	2, 13, 29, 36	32
33	$0.9238^{15°}$	$1.4507^{15°}$	329.6	149.1	s.al., eth.	13, 29, 36, 37	33
34	$0.9081^{20°}$	$1.4529^{20°}$	282.9	128.0	s.bz., chl., eth.	13, 29, 36	34
35	$0.9130^{15°}$	$1.4535^{15°}$	282.9	128.0	s.bz., chl., eth.	13, 29, 36, 39	35
36			282.9	128.0	s.bz., chl., eth.	13, 29, 36	36
37	$0.9024^{20°}$	$1.4557^{20°}$	247.9	112.2	s.bz., pet.eth.	2, 13, 29, 36	37
38	$0.9046^{20°}$	$1.4552^{20°}$	247.9	112.2	s.bz., eth., pet.eth.	13, 29, 36, 39	38
39	$0.9018^{20°}$	$1.4519^{20°}$	247.9	112.2	s.bz., eth., pet.eth.	2, 13, 29, 36	39
40			220.5	99.8	s.bz., eth., pet.eth.	2, 13, 18, 29	40
41	$0.8824^{35°}$	$1.4533^{40°}$	198.6	89.9	s.al., eth., pet.eth.	13, 21, 22, 29, 36	41
42	$0.8905^{20°}$	$1.45823^{20°}$	198.6	89.9	s.acet., eth., me.al.	13, 29, 36	42
43	$0.851^{79°}$	$1.4468^{50°}$	198.6	89.9	s.al., chl., eth., pet.eth.	29	43
44	$0.8563^{70°}$	$1.4406^{70°}$	198.6	89.9	s.acet., me.al.	13, 29, 36	44
45	0.8882^{25}	$1.4597^{25°}$	180.7	81.8	s.acet., me.al., pet.eth.	13, 16, 21, 29	45
46			180.7	81.8	s.al., me.al.	13, 16, 29	46
47			165.7	75.0	s.al.	13, 21, 29, 39	47
48	0.85321^{70}	$1.4444^{70°}$	165.7	75.0	v.s.eth., me.al.	13, 29, 30, 36	48
49			153.0	69.2	s.acet., al., eth.	13, 21, 29, 36	49
50			142.2	64.3	s.bz., chl., eth., pet.eth.	13, 29	50
51			124.5	56.3	s.bz., chl., eth., pet.eth.	13, 29	51
			UNSATURATED FATTY ACIDS (DIENOIC)				
52			572.0	517.5	v.s.al., eth.; s.h.w.	29, 36	52
53			500.4	452.7	s.al., eth.; sl.s.w.	29, 36	53
54			333.5	301.7	s.acet., eth., hex.	29	54
55			285.8	258.6	s.acet., eth., pet.eth.	29	55
56			222.3	201.1	s.acet., eth., pet.eth.	7	56
57	$0.9038^{18°}$	$1.4699^{20°}$	200.1	181.0	s.acet., al., eth., pet.eth.	13, 29, 36	57

FATTY ACIDS: PHYSICAL AND CHEMICAL CHARACTERISTICS (continued)

No	Acid — Systematic name	Acid — Common name	Chemical formula	Molecular weight	Melting point °C	Boiling point °C/mm[a]
		UNSATURATED FATTY ACIDS (DIENOIC)				
58	*trans*-9,*trans*-12-Octadeca-dienoic	Linolelaidic	$C_{18}H_{32}O_2$	280.5	28–29	—
59	*trans*-10,*trans*-12-Octadeca-dienoic	—	$C_{18}H_{32}O_2$	280.5	55.5–56	—
60	11,14-Eicosadienoic	—	$C_{20}H_{36}O_2$	308.4	—	—
61	13,16-Docosadienoic	—	$C_{22}H_{40}O_2$	336.6	—	—
62	17,20-Hexacosadienoic	—	$C_{26}H_{48}O_2$	392.7	61	—
		UNSATURATED FATTY ACIDS (TRIENOIC)				
63	6,10,14-Hexadecatrienoic	Hiragonic	$C_{16}H_{26}O_2$	250.4	—	180–190/15
64	7,10-13-Hexadecatrienoic	—	$C_{16}H_{26}O_2$	250.4	—	—
65	*cis*-6,*cis*-9,*cis*-12-Octadeca-trienoic	γ-Linolenic	$C_{18}H_{30}O_2$	278.4	—	—
66	*trans*-8,*trans*-10,*cis*-12-Octa-decatrienoic	α-Calendic	$C_{18}H_{30}O_2$	278.4	40–40.5	—
67	*trans*-8,*trans*-10,*trans*-12-Octa-decatrienoic	β-Calendic	$C_{18}H_{30}O_2$	278.4	77–78	—
68	*cis*-8,*trans*-10,*cis*-12-Octa-decatrienoic	—	$C_{18}H_{30}O_2$	278.4	—	—
69	*cis*-9,*cis*-12,*cis*-15-Octadeca-trienoic	α-Linolenic	$C_{18}H_{30}O_2$	278.4	−10 to −11.3	157/0.001
70	*trans*-9,*trans*-12,*trans*-15-Octa-decatrienoic	Linolenelaidic	$C_{18}H_{30}O_2$	278.4	29–30	—
71	*cis*-9,*trans*-11,*trans*-13-Octa-decatrienoic	α-Eleostearic	$C_{18}H_{30}O_2$	278.4	48–49	235/15
72	*trans*-9,*trans*-11,*trans*-13-Octadecatrienoic	β-Eleostearic	$C_{18}H_{30}O_2$	278.4	71.5	—
73	*cis*-9,*trans*-11,*cis*-13-Octadeca-trienoic	Punicic	$C_{18}H_{30}O_2$	278.4	43.5–44	—
74	*trans*-9,*trans*-11,*trans*-13-Octa-decatrienoic	—	$C_{18}H_{30}O_2$	278.4	—	—
75	5,8,11-Eicosatrienoic	—	$C_{20}H_{34}O_2$	306.5	—	—
76	8,11,14-Eicosatrienoic	—	$C_{20}H_{34}O_2$	306.5	—	—
		UNSATURATED FATTY ACIDS (TETRAENOIC)				
77	4,8,11,14-Hexadecatetraenoic	—	$C_{16}H_{24}O_2$	248.4	—	—
78	6,9,12,15-Hexadecatetraenoic	—	$C_{16}H_{24}O_2$	248.4	—	—
79	4,8,12,15-Octadecatetraenoic	Moroctic	$C_{18}H_{28}O_2$	276.4	—	208–213/15
80	6,9,12,15-Octadecatetraenoic	—	$C_{18}H_{28}O_2$	276.4	−57.4 to −56.6	—
81	9,11,13,15-Octadecatetraenoic	α-Parinaric	$C_{18}H_{28}O_2$	276.4	85–86	—
82	9,11,13,15-Octadecatetraenoic	β-Parinaric	$C_{18}H_{28}O_2$	276.4	95–96	—
83	9,12,15,18-Octadecatetraenoic	—	$C_{18}H_{28}O_2$	276.4	—	—
84	4,8,12,16-Eicosatetraenoic	—	$C_{20}H_{32}O_2$	304.5	—	217–220/10
85	5,8,11,14-Eicosatetraenoic	Arachidonic	$C_{20}H_{32}O_2$	304.5	−49.5	163/1
86	6,10,14,18-Eicosatetraenoic?	—	$C_{20}H_{32}O_2$	304.5	—	—
87	4,7,10,13-Docosatetraenoic	—	$C_{22}H_{36}O_2$	332.5	—	—
88	7,10,13,16-Docosatetraenoic	—	$C_{22}H_{36}O_2$	332.5	—	—
89	8,12,16,19-Docosatetraenoic	—	$C_{22}H_{36}O_2$	332.5	—	—
		UNSATURATED FATTY ACIDS (PENTA- AND HEXA-ENOIC)				
90	4,8,12,15,18-Eicosapentaenoic	Timnodonic?	$C_{20}H_{30}O_2$	302.5	—	—
91	5,8,11,14,17-Eicosapentaenoic	—	$C_{20}H_{30}O_2$	302.5	−54.4 to −53.8	—
92	4,7,10,13,16-Docosapentaenoic	—	$C_{22}H_{34}O_2$	330.5	—	—
93	4,8,12,15,19-Docosapentaenoic	Clupanodonic	$C_{22}H_{34}O_2$	330.5	—	207–212/2
94	7,10,13,16,19-Docosapentaenoic	—	$C_{22}H_{34}O_2$	330.5	—	—
95	4,7,10,13,16,19-Docosahexaenoic	—	$C_{22}H_{32}O_2$	328.5	−44.5 to −44.1	—
96	4,8,12,15,18,21-Tetracosahexa-enoic	Nisinic	$C_{24}H_{36}O_2$	356.6	—	—

FATTY ACIDS: PHYSICAL AND CHEMICAL CHARACTERISTICS (continued)

No	Specific gravity[b]	Refractive index[c] n_D^c	Neutral- ization value[d]	Iodine value (calcu- lated)[e]	Solubility[f]	Reference	No
					UNSATURATED FATTY ACIDS (DIENOIC)		
58	---	---	200.1	181.0	s.al., eth., me.al., pet.eth.	29, 36	58
59	---	---	200.1	181.0	s.acet., cyc., eth.	23	59
60	---	---	181.9	164.5	s.acet., eth., pet.eth.	29	60
61	---	---	166.7	150.8	s.acet., eth.	29	61
62	---	---	142.9	129.3	s.eth., pet.eth.	29	62
					UNSATURATED FATTY ACIDS (TRIENOIC)		
63	0.9296[20*]	1.4850[50*]	224.1	304.1	s.al., eth.	13, 29, 36, 38	63
64	---	---	224.1	304.1	s.al., eth.	29	64
65	---	---	201.5	273.5	s.acet., eth., me.al.	29, 36	65
66	---	---	201.5	273.5	s.acet., pent.	8, 29, 36	66
67	---	---	201.5	273.5	s.me.al., pet.eth.	8	67
68	---	---	201.5	273.5	v.s.acet., al., pent., pet.eth.	10, 29	68
69	0.914[20*]	1.4678[50*]	201.5	273.5	s.acet., al., eth., pet.eth.	13, 29, 36	69
70	---	---	201.5	273.5	s.me.al., pet.eth.	29, 36	70
71	---	1.5112[50*]	201.5	273.5	s.al., cyc., eth., pet.eth.	13, 28, 29, 36	71
72	---	1.5002[75*]	201.5	273.5	s.al., eth., me.al., pet.eth.	13, 29, 36	72
73	0.9027[50*]	1.5114[50*]	201.5	273.5	s.al., pent., pet.eth.	13, 29, 36	73
74	---	---	201.5	273.5	s.acet., al., CS_2, pent.	9, 29	74
75	---	---	183.1	248.3	s.CS_2, hept., me.al.	29	75
76	---	---	183.1	248.3	s.CS_2, hept., me.al.	29	76
					UNSATURATED FATTY ACIDS (TETRAENOIC)		
77	---	---	225.9	408.8	s.acet., al., eth., pet.eth.	29	77
78	---	1.4870[29*]	225.9	408.8	s.acet., al., CS_2, eth., pent.	29	78
79	0.9297[20*]	1.4911[20*]	203.0	367.3	s.acet., al., eth., pet.eth.	13, 29, 36, 38	79
80	---	1.4888[16*]	203.0	367.3	s.CS_2, me.al.	29	80
81	---	---	203.0	367.3	s.acet., al., eth., pet.eth.	13, 29, 36	81
82	---	---	203.0	367.3	s.eth., pet.eth.	13, 29, 36	82
83	---	---	203.0	367.3	s.CS_2, me.al.	29	83
84	0.9263[20*]	1.4915[20*]	184.3	333.4	s.acet., eth.	13, 29, 36	84
85	0.9082[20*]	1.4824[20]	184.3	333.4	s.acet., eth., me.al., pet.eth.	13, 29, 36	85
86	0.9263[20*]	1.4935[20*]	184.3	333.4	s.acet., me.al., pet.eth.	29	86
87	---	---	168.7	305.4	s.acet., me.al., pet.eth.	24, 29	87
88	---	---	168.7	305.4	s.CS_2, hept., me.al.	1, 29	88
89	---	---	168.7	305.4	s.acet., me.al., pet.eth.		89
					UNSATURATED FATTY ACIDS (PENTA- AND HEXA-ENOIC)		
90	0.9399[15*]	1.5109[15*]	185.5	419.6	s.bz., chl., eth., pet.eth.	13, 29, 36	90
91	---	1.4977[23*]	185.5	419.6	s.hept., me.al.	25	91
92	---	---	169.8	384.0	s.chl., hept., me.al.	24, 29	92
93	0.9356[20*]	1.5014[20]	169.8	384.0	s.acet., eth., pet.eth.	13, 29, 36	93
94	---	---	169.8	384.0	s.bz., chl., me.al., pet.eth.	26, 29	94
95	---	1.5017[26*]	170.8	463.6	s.bz., chl., me.al., pet.eth.	**19, 24**, 29	95
96	0.9452[20]	1.5122[20*]	157.4	427.1	s.bz., chl., eth., pet.eth.	13, 29, 36	96

FATTY ACIDS: PHYSICAL AND CHEMICAL CHARACTERISTICS (continued)

No	Acid — Systematic name	Common name	Chemical formula	Molec-ular weight	Melting point °C	Boiling point °C/mm[a]

HYDROXYALKANOIC ACIDS

No	Systematic name	Common name	Chemical formula	Molecular weight	Melting point °C	Boiling point °C/mm[a]
97	2-Hydroxydodecanoic	2-Hydroxylauric	$C_{12}H_{24}O_3$	216.3	73–74	—
98	12-Hydroxydodecanoic	Sabinic	$C_{12}H_{24}O_3$	216.3	84	—
99	2-Hydroxytetradecanoic	2-Hydroxymyristic	$C_{14}H_{28}O_3$	244.4	81.5–82	—
100	11-Hydroxypentadecanoic	Convolvulinolic	$C_{15}H_{30}O_3$	258.4	63.5–64	—
101	2-Hydroxyhexadecanoic	2-Hydroxypalmitic	$C_{16}H_{32}O_3$	272.4	86–87	—
102	11-Hydroxyhexadecanoic	Jalapinolic	$C_{16}H_{32}O_3$	272.4	68–69	—
103	16-Hydroxyhexadecanoic	Juniperic	$C_{16}H_{32}O_3$	272.4	95	—
104	2-Hydroxyoctadecanoic	2-Hydroxystearic	$C_{18}H_{36}O_3$	300.5	91	—
105	23-Hydroxydocosanoic	Phellonic	$C_{22}H_{44}O_3$	356.6	95–96	—
106	2-Hydroxytetracosanoic	Cerebronic	$C_{24}H_{48}O_3$	384.6	99.5–100.5	—
107	3,11-Dihydroxytetradecanoic	Ipurolic	$C_{14}H_{28}O_4$	260.4	100–101	—
108	2,15-Dihydroxypentadecanoic	Dihydroxypentade-cyclic	$C_{15}H_{30}O_4$	274.4	102–103	—
109	15,16-Dihydroxyhexadecanoic	Ustilic A	$C_{16}H_{32}O_4$	288.4	112–113	—
110	9,10-Dihydroxyoctadecanoic	9,10-Dihydroxy-stearic	$C_{18}H_{36}O_4$	316.5	141[9]	—
111	9,10-Dihydroxyoctadecanoic	9,10-Dihydroxy-stearic	$C_{18}H_{36}O_4$	316.5	90[10]	—
112	11,12-Dihydroxyeicosanoic	11,12-Dihydroxy-arachidic	$C_{20}H_{40}O_4$	344.5	130[9]	—
113	2,15,16-Trihydroxyhexadecanoic	Ustilic	$C_{16}H_{32}O_5$	304.4	140	—
114	9,10,16-Trihydroxyhexadecanoic	Aleuritic	$C_{16}H_{32}O_5$	304.4	100	—

KETO, EPOXY, AND CYCLO FATTY ACIDS

No	Systematic name	Common name	Chemical formula	Molecular weight	Melting point °C	Boiling point °C/mm[a]
115	4-Ketopentanoic	Levulinic	$C_5H_8O_3$	116.1	37.2	154/15
116	6-Ketooctadecanoic	Lactarinic	$C_{18}H_{34}O_3$	298.5	87	—
117	4-Keto-9,11,13,octadecatrienoic	α-Licanic	$C_{18}H_{28}O_3$	292.4	74–75	—
118	4-Keto-*trans*-9,-*trans*-11,-*trans*-13-octadecatrienoic	β-Licanic	$C_{18}H_{28}O_3$	292.4	99.5	—
119	*cis*-12,13-Epoxy-*cis*-9-octa-decenoic	Vernolic	$C_{18}H_{32}O_3$	296.5	31–32	—
120	*cis*-9,10-Epoxyoctadecanoic	Epoxystearic	$C_{18}H_{34}O_3$	298.5	57.5–58	—
121	ω-(2-*n*-Octylcycloprop-1-enyl)-octanoic	Sterculic	$C_{19}H_{34}O_2$	294.5	18	—
122	ω-(2-*n*-Octylcyclopropyl)-octanoic	Lactobacillic	$C_{19}H_{36}O_2$	296.5	28–29	—
123	13-(2-Cyclopentenyl)-tridecanoic	Chaulmoogric	$C_{18}H_{32}O_2$	280.2	68.5	247.5/20
124	11-(2-Cyclopentenyl)-hendec-anoic	Hydnocarpic	$C_{16}H_{28}O_2$	252.2	60.5	—
125	9-(2-Cyclopentenyl)-nonanoic	Alepric	$C_{14}H_{24}O_2$	224.2	48.0	—
126	7-(2-Cyclopentenyl)-heptanoic	Aleprylic	$C_{12}H_{20}O_2$	196.2	32.0	—
127	5-(2-Cyclopentenyl)-pentanoic	Aleprestic	$C_{10}H_{16}O_2$	168.1	Liquid	—
128	2-Cyclopentenyl-1-oic	Aleprolic	$C_6H_8O_2$	112.1	Liquid	—
129	13-(2-Cyclopentenyl)-6-tri-decenoic	Gorlic	$C_{18}H_{30}O_2$	278.2	6.0	232.5

HYDROXY UNSATURATED ACIDS

No	Systematic name	Common name	Chemical formula	Molecular weight	Melting point °C	Boiling point °C/mm[a]
130	16-Hydroxy-7-hexadecenoic	Ambrettolic	$C_{16}H_{30}O_3$	270.5	25	—
131	9-Hydroxy-12-octadecenoic	—	$C_{18}H_{34}O_3$	298.5	—	—
132	*d*-12-Hydroxy-*cis*-9-octadecenoic	Ricinoleic	$C_{18}H_{34}O_3$	298.5	5, 7.7, & 16	225/10
133	*d*-12-Hydroxy-*trans*-9-octa-decenoic	Ricinelaidic	$C_{18}H_{34}O_3$	298.5	52–53	—
134	2-Hydroxy-15-tetracosenoic	Hydroxynervonic	$C_{24}H_{46}O_3$	382.6	65	—
135	9-Hydroxy-10,12-octadeca-dienoic	—	$C_{18}H_{32}O_3$	296.5	—	—
136	13-Hydroxy-9,11-octadeca-dienoic	—	$C_{18}H_{32}O_3$	296.5	—	—
137	18-Hydroxy-*cis*-9,*trans*-11,-*trans*-13-octadecatrienoic	α-Kamlolenic	$C_{18}H_{30}O_3$	294.4	77–78	—
138	18-Hydroxy-*trans*-9,*trans*-11,-*trans*-13-octadecatrienoic	β-Kamlolenic	$C_{18}H_{30}O_3$	294.4	88–89	—

FATTY ACIDS: PHYSICAL AND CHEMICAL CHARACTERISTICS (continued)

No	Specific gravity[b]	Refractive index[c] $n_D^{°C}$	Neutral-ization value[d]	Iodine value (calcu-lated)[e]	Solubility[f]	Reference	No
					HYDROXYALKANOIC ACIDS		
97	—	—	259.4	—	s.al., me.al.	29	97
98	—	—	259.4	—	s.al., h.bz.	3, 13, 29	98
99	—	—	229.1	—	s.al., chl., eth.	29	99
100	—	—	217.1	—	s.al., chl., eth.	13, 29, 34	100
101	—	—	206.0	—	s.al., me.al.	29, 40	101
102	—	—	206.0	—	s.al., eth.	13, 29, 33	102
103	—	—	206.0	—	s.al., bz., eth.	13, 21, 29, 36	103
104	—	—	186.7	—	s.al., me.al.	29, 40	104
105	—	—	157.3	—	s.acet., chl., eth., glac.ac.a, pyr.	5, 13, 29, 36	105
106	—	—	145.9	—	s.acet., h.al., eth., pyr.	13, 29, 36	106
107	—	—	215.5	—	s.chl., eth.	13, 29, 33	107
108	—	—	204.5	—	s.me.al.	29	108
109	—	—	194.5	—	s.me.al.	29	109
110	—	—	177.3	—	s.h.al.; sl.s.eth.	13, 29	110
111	—	—	177.3	—	s.al., eth., h.w.	29	111
112	—	—	162.9	—	s.acet., eth.	13, 29	112
113	—	—	184.3	—	s.me.al.	29	113
114	—	—	184.3	—	s.me.al.	29	114
					KETO, EPOXY, AND CYCLO FATTY ACIDS		
115	1.1395[20°]	1.442[15.8°]	483.2	—	v.s.al., eth., w.	29	115
116	—	—	188.0	—	s.h.al., chl., eth.	13, 21, 29, 36	116
117	—	—	191.9	260.4	s.h.pet.eth.	4, 15, 29, 36	117
118	—	—	191.9	260.4	s.h.pet.eth.	4, 15, 29, 36	118
119	—	—	189.3	85.6	s.acet., al., hex.	27, 29	119
120	—	—	188.0	—	s.acet., al., hex.	6	120
121	—	—	190.5	86.2	s.eth.	29	121
122	—	—	189.2	—	s.acet., eth., pet.eth.	29	122
123	—	—	200.1	90.5	s.acet., chl., eth.	12, 13, 21, 29, 36	123
124	—	—	222.3	100.6	s.al., chl., pet.eth.	21, 29, 36	124
125	—	—	250.1	113.1	s.al., eth., pet.eth.	29	125
126	—	—	285.8	129.3	s.acet., eth., pet.eth.	29	126
127	—	—	333.5	150.8	s.acet., eth., pet.eth.	29	127
128	—	—	500.4	226.4	s.acet., eth., pet.eth.	29	128
129	0.9436[25°]	1.4782[25°]	201.5	182.5	s.h.al.	11, 13, 29, 36	129
					HYDROXY UNSATURATED ACIDS		
130	—	—	207.5	93.9	s.al., eth.	29, 36	130
131	—	—	188.0	85.0	s.acet., al., eth.	29	131
132	0.940[27.4°]	1.4716[20°]	188.0	85.0	s.acet., al., eth.	13, 29, 36	132
133	—	—	188.0	85.0	s.acet., al., eth.	13, 29, 36	133
134	—	—	146.6	66.3	s.acet., al., chl., eth., pyr.; sl.s.pet.eth.	13, 29, 36	134
135	—	—	189.2	171.2	s.acet., al., pent.	8	135
136	—	—	189.2	171.2	s.acet., al., pent.	8	136
137	—	—	190.5	258.6	—	29	137
138	—	—	190.5	258.6	—	29	138

FATTY ACIDS: PHYSICAL AND CHEMICAL CHARACTERISTICS (continued)

	Acid		Chemical formula	Molecular weight	Melting point °C	Boiling point °C/mm[a]
No	Systematic name	Common name				
	BRANCHED-CHAIN FATTY ACIDS					
139	**3-Methylbutanoic**	Isovaleric	$C_5H_{10}O_2$	102.1	−37.6	176.7
140	**d-6-Methyloctanoic**	—	$C_9H_{18}O_2$	158.2	—	—
141	**8-Methyldecanoic**	—	$C_{11}H_{22}O_2$	186.3	−18.5	—
142	**10-Methylhendecanoic**	Isolauric	$C_{12}H_{24}O_2$	200.3	41.2	—
143	**d-10-Methyldodecanoic**	—	$C_{13}H_{26}O_2$	214.3	6.2–6.5	—
144	**11-Methyldodecanoic**	Isoundecylic	$C_{13}H_{26}O_2$	214.3	39.4–40	—
145	**12-Methyltridecanoic**	Isomyristic	$C_{14}H_{28}O_2$	228.4	53.6	—
146	**d-12-Methyltetradecanoic**	—	$C_{15}H_{30}O_2$	242.4	25.8	—
147	**13-Methyltetradecanoic**	Isopentadecylic	$C_{15}H_{30}O_2$	242.4	52.2	—
148	**14-Methylpentadecanoic**	Isopalmitic	$C_{16}H_{32}O_2$	256.4	62.4	—
149	**d-14-Methylhexadecanoic**	—	$C_{17}H_{34}O_2$	270.4	38.0	—
150	**15-Methylhexadecanoic**	—	$C_{17}H_{34}O_2$	270.4	60.5	—
151	**10-Methylheptadecanoic**	—	$C_{18}H_{36}O_2$	284.5	33.5	—
152	**16-Methylheptadecanoic**	Isostearic	$C_{18}H_{36}O_2$	284.5	69.5	—
153	**l-D-10-Methyloctadecanoic**	Tuberculostearic	$C_{19}H_{38}O_2$	298.5	13.2	175–178/0.7
154	**d-16-Methyloctadecanoic**	—	$C_{19}H_{38}O_2$	298.5	49.9–50.7	—
155	**18-Methylnonadecanoic**	Isoarachidic	$C_{20}H_{40}O_2$	312.5	75.3	—
156	**d-18-Methyleicosanoic**	—	$C_{21}H_{42}O_2$	326.6	55.6	—
157	**20-Methylheneicosanoic**	Isobehenic	$C_{22}H_{44}O_2$	340.6	79.5	—
158	**d-20-Methyldocosanoic**	—	$C_{23}H_{46}O_2$	354.6	62.1	—
159	**22-Methyltricosanoic**	Isolignoceric	$C_{24}H_{48}O_2$	368.6	83.1	—
160	**d-22-Methyltetracosanoic**	—	$C_{25}H_{50}O_2$	382.7	67.8	—
161	**24-Methylpentacosanoic**	Isocerotic	$C_{26}H_{52}O_2$	396.7	86.9	—
162	**d-24-Methylhexacosanoic**	—	$C_{27}H_{54}O_2$	410.7	72.9	—
163	**26-Methylheptacosanoic**	Isomontanic	$C_{28}H_{56}O_2$	424.7	89.3	—
164	**d-28-Methyltriacontanoic**	—	$C_{31}H_{62}O_2$	466.8	80.7	—
165	**2,4,6-(D)-Trimethyloctacosanoic**	Mycoceranic(myco-cerosic)	$C_{31}H_{62}O_2$	466.8	27–28	—
166	**2-Methyl-*cis*-2-butenoic**	Angelic	$C_5H_8O_2$	100.1	45	185
167	**2-Methyl-*trans*-2-butenoic**	Tiglic	$C_5H_8O_2$	100.1	65.5	198.5
168	**4-Methyl-3-pentenoic**	Pyroterebic	$C_6H_{10}O_2$	114.1	—	207
169	**d-2,4(L),6(L)-Trimethyl-*trans*-2-tetracosenoic**	C₂₋-Phthienoic (mycolipenic)	$C_2\text{-}H_{52}O_2$	408.7	39.5–41	—

These data were compiled originally for the *Biology Data Book* by Klare S. Markley (1964) pp. 370–80. Data are reproduced here in modified form by permission of the copyright owners of the above publication, the Federation of American Societies for Experimental Biology, Washington, D.C.

[a]Boiling Point: d. = decomposes; 760 mm of mercury (atmospheric pressure), unless otherwise specified.

[b]At temperature indicated in superscript, referred to water at 4°C.

[c]Refractive index (*n*) is given for the sodium D-line at temperature shown in superscript.

[d]Milligrams KOH required to neutralize one gram of acid.

[e]Grams of iodine absorbed by 100 grams of acid.

[f]Solubility: a. = acid; acet. = acetone; ac. = acetic; al. = alcohol; bz. = benzene; chl. = chloroform; cyc. = cyclohexane; eth. = ether; glac. = glacial; hept. = heptane; hex. = hexane; h. = hot; me. = methyl; pent. = pentane; pet. = petroleum; pyr. = pyridine; s. = soluble; sl. = slightly; tol. = toluene; v. = very; w. = water.

[g]Also called enanthic acid.

[h]Also called hendecanoic acid.

[i]Also called selacholeic acid.

FATTY ACIDS: PHYSICAL AND CHEMICAL CHARACTERISTICS (continued)

No	Specific gravity[b]	Refractive index[c] n_D^C	Neutralization value[d]	Iodine value (calculated)[e]	Solubility[f]	Reference	No
			BRANCHED-CHAIN FATTY ACIDS				
139	0.937^{15°	$1.40178^{22.4^\circ}$	549.3		s.al., chl., eth.; sl.s.w.	13, 28, 29, 36	139
140			354.6		s.acet., eth., me.al., pet.eth.	13, 29, 36, 40	140
141			301.2		s.acet., eth., me.al., pet.eth.	13, 29, 36, 40	141
142			280.1		s.acet., eth., me.al., pet.eth.	13, 29, 36, 40	142
143		1.4424^{25°	261.8		s.bz., chl., me.al., pet.eth.	13, 29, 36, 40	143
144		1.4293^{60°	261.8		s.acet., al., me.al., pet.eth.	29	144
145			245.7		s.acet., me.al., pet.eth.	13, 29, 36, 40	145
146		1.4327^{59°	231.5		s.chl., eth., me.al., pet.eth.	13, 29, 36, 40	146
147		1.4312^{59°	231.5		s.me.al., pet.eth.	29	147
148		1.4293^{70°	218.8		s.acet., eth., me.al., pet.eth.	13, 29, 36, 40	148
149			207.5		s.acet., eth., me.al., pet.eth.	13, 29, 36, 40	149
150		1.4315^{70°	207.5		s.acet., eth., pet.eth.	13, 29, 36, 40	150
151			197.2		s.acet., glac.ac.a.	20	151
152			197.2		s.acet., eth., pet.eth.	13, 29, 36, 40	152
153	0.887^{25°	1.4512^{25°	188.0		s.acet., al., me.al., pent.	13, 29, 36	153
154			188.0		s.acet., me.al., pet.eth.	13, 29, 35, 36, 40	154
155			179.5		s.al., eth., pet.eth.	13, 29, 36, 40	155
156			171.8		s.acet., chl., pet.eth.	13, 29, 36, 40	156
157			164.7		s.chl., eth., me.al., pet.eth.	13, 29, 36, 40	157
158			158.2		s.acet., chl., eth., pet.eth.	13, 29, 36, 40	158
159			152.2		s.acet., chl., pet.eth.	13, 29, 36, 40	159
160			146.6		s.al., bz., chl., pet.eth.	13, 29, 36, 40	160
161			141.4		s.acet., chl., glac.ac.a	13, 29, 36, 40	161
162			136.6		s.bz., chl., glac.ac.a., pet.eth.	13, 29, 36, 40	162
163			132.1		s.bz., chl., glac.ac.a., pet.eth.	13, 29, 36, 40	163
164			120.2		s.bz., chl., glac.ac.a., pet.eth.	13, 29, 36, 40	164
165			120.2		s.ch., pet.eth.	13, 29	165
166	0.983^{47°	1.4434^{47°	560.4	253.6	v.s.eth.; s.al.; sl.s.w.	29	166
167		1.4342^{81°	560.4	253.6	v.s.h.w.; s.al., eth.	29	167
168			491.6	222.4	s.al., chl., eth.	29	168
169		1.4598^{25°	137.3	62.1	s.acet., me.al., pet.eth.	29	169

REFERENCES

1. Baudert, *Bull. Soc. Chim. Fr. Ser. 5*, 9, 922 (1942).
2. Bosworth and Brown, *J. Biol. Chem.*, 103, 115 (1933).
3. Bougault and Bourdier, *J. Pharm. Chim. Ser. 6*, 30, 10 (1909).
4. Brown and Farmer, *Biochem. J.*, 29, 631 (1935).
5. Chibnall, Piper, and Williams, *Biochem. J.*, 30, 100 (1936).
6. Chisholm and Hopkins, *Chem. Ind.* (London), p. 1154 (1959).
7. Chisholm and Hopkins, *Can. J. Chem.*, 38, 805 (1960).
8. Chisholm and Hopkins, *Can. J. Chem.*, 38, 2500 (1960).
9. Chisholm and Hopkins, *J. Chem. Soc.*, p. 573 (1962).
10. Chisholm and Hopkins, *J. Org. Chem.*, 27, 3137 (1962).
11. Cole and Cardoss, *J. Am. Chem. Soc.*, 60, 612 (1938).
12. Cole and Cardoss, *J. Am. Chem. Soc.*, 61, 2349 (1939).
13. Deuel, *The Lipids; Their Chemistry and Biochemistry,* Interscience, New York, 1951–57.
14. Dorinson, McCorkle, and Ralston, *J. Am. Chem. Soc.*, 64, 2739 (1942).
15. Dyson, *A Manual of Organic Chemistry,* Vol. I, Longmans, Green, London, 1950.
16. Foreman and Brown, *Oil Soap* (Chicago), 21, 183 (1944).
17. Francis and Piper, *J. Am. Chem. Soc.*, 61, 577 (1939).
18. Gupta, Grollman, and Niyogy, *Proc. Nat. Inst. Sci.* (India), 19, 519 (1953).
19. Hammond and Lundberg, *J. Am. Oil Chem. Soc.*, 30, 438 (1953).
20. Hansen, Shorland, and Cooke, *Chem. Ind.* (London), p. 839 (1951).
21. Heilbron, *Dictionary of Organic Compounds.* Eyre and Spottiswoods, Eds., London, 1934.

FATTY ACIDS: PHYSICAL AND CHEMICAL CHARACTERISTICS (continued)

22. Hilditch and Jones, *Biochem. J.,* 22, 326 (1928).
23. Hopkins and Chisholm, *Chem. Ind.* (London), p. 2064 (1962).
24. Klenk and Bongard, *Z. Phys. Chem.,* 291, 104 (1952).
25. Klenk and Montag, *Ann. Chem.,* 604, 4 (1957).
26. Klenk and Tomuschat, *Z. Phys. Chem.,* 308, 165 (1957).
27. Krewson, Ard, and Riemenschneider, *J. Am. Oil Chem. Soc.,* 39, 334 (1962).
28. Lange, *Handbook of Chemistry,* 6th ed., Handbook Publications, Sandusky, Ohio, 1946.
29. Markley, *Fatty Acids,* 2nd ed., Interscience, New York, 1960–61.
30. Noller and Talbot, *Organic Synthesis Collection,* Vol. 12, John Wiley & Sons, New York, 1943.
31. Nunn, *J. Chem. Soc.,* p. 313 (1952).
32. Piper et al., *Biochem. J.,* 28, 2175 (1934).
33. Power and Rogerson, *J. Am. Chem. Soc.,* 32, 106 (1910).
34. Power and Rogerson, *J. Chem. Soc.,* 101(T), 1 (1912).
35. Prout, Cason, and Ingersoll, *J. Am. Chem. Soc.,* 69, 1233 (1947).
36. Ralston, *Fatty Acids and Their Derivations,* John Wiley & Sons, New York, 1948.
37. Smedley, *Biochem. J.,* 6, 451 (1912).
38. Teresi, J. D., Unpublished. U.S. Naval Radiological Defense Laboratory, San Francisco, Calif.
39. Warth, *The Chemistry and Technology of Waxes,* 2nd ed., Reinhold, New York, 1956.
40. Weitkamp, *J. Am. Chem. Soc.,* 67, 447 (1945).

DIELECTRIC CONSTANTS OF SOME FATS, FATTY ACIDS, AND ESTERS

Compound	Dielectric constant, ε	Temperature °C.	Compound	Dielectric constant, ε	Temperature °C.
Linolenic acid	2.55	− 10	**Ethyl palmitate**	3.07	30
	2.76	20		2.88	69
	2.97	60		2.71	104
	3.01	100		2.57	144
				2.46	182
Tripalmitin	2.272	− 45			
	2.354	− 30	**Ethyl stearate**	2.92	48
	2.402	− 6		2.69	100
	2.444	5		2.56	138
	2.544	20		2.48	167
	2.901	55			
	2.954	80	**Butyl stearate**	3.30	25
	2.924	120	**Ethyl oleate**	3.17	28
Methyl acetate	6.7	25		3.00	60
Ethyl acetate	6.15	20		2.87	89
Bornyl acetate	4.6	21		2.72	122
				2.63	150
Ethyl laurate	3.44	20			
	3.16	60	**Butyl oleate**	4.0	25
	2.91	101			
	2.73	143	**Glyceryl triacetate**	9.4	24

SOLUBILITIES OF FATTY ACIDS IN WATER

Acid	Grams acid per 100 g water				
	0°C	20°C	30°C	45°C	60°C
Caproic	0.864	0.968	1.019	1.095	1.171
Heptanoic	0.190	0.244	0.271	0.311	0.353
Caprylic	0.044	0.068	0.079	0.095	0.113
Nonanoic	0.014	0.026	0.032	0.041	0.051
Capric	0.0095	0.015	0.018	0.023	0.027
Hendecanoic	0.0063	0.0093	0.011	0.013	0.015
Lauric	0.0037	0.0055	0.0063	0.0075	0.0087
Tridecanoic	0.0021	0.0033	0.0038	0.0044	0.0054
Myristic	0.0013	0.0020	0.0024	0.0029	0.0034
Pentadecanoic	0.00076	0.0012	0.0014	0.0017	0.0020
Palmitic	0.00046	0.00072	0.00083	0.0010	0.0012
Heptadecanoic	0.00028	0.00042	0.00055	0.00069	0.00081
Stearic	0.00018	0.00029	0.00034	0.00042	0.00050

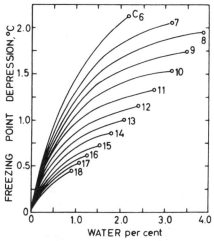

Effect of addition of water on the freezing point of the fatty
acids from C_6 to C_{18}

From Markley, Klare S., *Fatty Acids*, 2nd ed., Part 1, Interscience Publishers, Inc., New York, 1960, 616. Copyright
© 1960. Reprinted by permission of John Wiley & Sons, Inc.

APPROXIMATE SOLUBILITIES OF WATER IN SATURATED FATTY ACIDS AT VARIOUS TEMPERATURES

Acid	Temperature, °C.	Water, %	Acid	Temperature, °C.	Water, %
Caproic	− 5.4	2.21	**Lauric**	42.7	2.35
	12.3	4.73		75.0	2.70
	31.7	7.57		90.5	2.85
	46.3	9.70	**Tridecanoic**	40.8	2.00
Heptanoic	− 8.3	2.98	**Myristic**	53.2	1.70
	42.5	9.98	**Pentadecanoic**	51.8	1.46
Caprylic	14.4	3.88	**Palmitic**	61.8	1.25
Nonanoic	10.5	3.45	**Heptadecanoic**	60.4	1.06
Capric	29.4	3.12	**Stearic**	68.7	0.92
Hendecanoic	26.8	2.72		92.4	1.02
	57.5	4.21			

From Markley, Klare S., *Fatty Acids*, 2nd ed., Part 1, Interscience Publishers, Inc., New York, 1960, 617. Copyright © 1960.
Reprinted by permission of John Wiley & Sons, Inc.

SOLUBILITY OF SIMPLE SATURATED TRIGLYCERIDES[a]

Glyceride	Benzene		Diethyl ether		Chloroform		Ethanol	
	°C	Sol.	°C	Sol.	°C	Sol.	°C	Sol.
Tristearin	14.5	0.45	25.5	0.51	−3.4	0.46	59.8	0.074
	24.0	2.64	27.4	0.98	6.0	1.94	60.8	0.15
	26.7	5.07	31.2	2.11	11.9	4.52	62.4	0.22
	33.6	10.57	34.0	4.80	20.3	10.85	66.8	0.87
	38.1	18.91	40.3	14.25	25.8	15.71	66.9	1.55
	39.6	22.23	44.6	26.20	33.8	25.11		
	41.8	29.10	48.9	39.00	40.4	34.66		
	47.8	44.58	65.1	83.54	49.7	49.31		
	60.4	75.72			61.0	74.15		
Tripalmitin	13.7	7.74	19.0	0.75	2.5	3.08		
	18.3	4.61	24.8	2.12	8.8	7.38		
	22.8	9.69	28.0	4.49	13.5	12.13		
	26.4	16.43	31.0	8.10	18.6	18.40		
	29.5	25.30	33.8	14.06	23.3	24.24		
	34.3	37.11	37.2	21.86	28.3	31.60		
	38.4	48.07	40.3	32.62	32.3	37.67		
	48.0	56.43	43.2	42.07	41.9	52.38		
			47.5	58.12	52.4	71.08		
Trimyristin	11.3	7.56	6.7	0.57	−8.0	3.98		
	14.1	12.03	13.3	1.88	−4.7	6.12		
	15.5	14.50	18.0	4.76	0.9	10.79		
	19.1	21.94	22.2	9.54	5.5	15.04		
	21.2	27.02	25.3	15.12	10.6	21.46		
	24.9	35.85	28.0	24.46	14.8	26.36		
	30.9	49.76	31.0	34.73	19.2	32.25		
	37.2	64.69	35.7	50.53	22.6	38.04		
			40.3	65.16	24.3	40.29		
					38.1	62.90		
Trilaurin	9.0	23.50	−7.2	0.57	−12.5	13.09		
	11.6	28.25	0.8	2.25	−6.9	18.17		
	12.0	32.19	5.1	4.44	0.0	25.29		
	14.3	37.21	9.0	8.36	5.1	30.90		
	16.0	41.29	12.9	15.49	11.0	38.00		
	21.0	52.44	16.0	24.24	19.3	49.41		
	29.5	69.71	19.9	34.42	31.4	69.65		
			24.4	49.78				
			28.0	60.28				
Tricaprin	2.2	45.29	−14.2	3.46			9.5	0.44
	7.2	56.00	−9.5	6.67			13.8	0.99
	9.4	60.39	−3.7	14.64			17.6	2.01
	11.6	64.58	−0.3	21.57			23.4	5.81
			1.2	25.36			25.1	9.87
			3.0	30.40			25.2	14.36
			4.6	36.03			58.5	29.32
			8.5	48.43				
			15.9	69.20				

[a]All solubilities expressed in grams of triglyceride per 100 g solution.

From Markley, Klare S., *Fatty Acids*, 2nd ed., Part 1, Interscience Publishers, Inc., New York, 1960, 644. Copyright © 1960. Reprinted by permission of John Wiley & Sons, Inc.

SOLUBILITIES OF MIXED TRIACID TRIGLYCERIDES AT 25°C[a]

Material	2-Acyl equals	M.p., °C	Diethyl ether	Petroleum ether	Acetone	Ethanol
1-Stearyl-2-acyl-3-palmitin	Myristyl	59.5	10.97	7.59	0.18	0.03
	Lauryl	57.5	16.49	9.49	0.31	0.03
	Capryl	55.0	22.87	10.60	0.59	0.03
1-Stearyl-2-acyl-3-myristin	Palmityl	58.5	11.03	5.46	0.18	0.03
	Lauryl	55.0	30.68	16.26	0.68	0.04
	Capryl	52.5	53.75	37.03	1.96	0.08
1-Stearyl-2-acyl-3-Laurin	Palmityl	52.0	72.63	58.88	1.47	0.06
	Myristyl	49.5	112.59	81.44	2.53	0.07
	Capryl	41.8	192.13	179.56	13.49	0.39
1-Stearyl-2-acyl-3-caprin	Palmityl	50.0	---	89.97	2.38	0.09
	Myristyl	45.0	----	116.35	9.03	0.31
	Lauryl	44.0	----	148.42	26.04	0.36

[a]All solubilities expressed in grams of triglyceride per 100 g solution.

From Markley, Klare S., *Fatty Acids*, 2nd ed., Part 1, Interscience Publishers, Inc., New York, 1960, 664. Copyright © 1960. Reprinted by permission of John Wiley & Sons, Inc.

PROPERTIES AND FATTY ACID COMPOSITION OF FATS AND OILS

These data were compiled originally for Harwood, H. J. and Geyer, R. P., *Biology Data Book*, the Federation of American Societies for Experimental Biology, Washington, D.C., 1964, 380. By permission of the copyright owners.

Values are typical rather than average and frequently were derived from specific analyses for particular samples (especially the constituent fatty acids). Extreme variations may occur, depending on a number of variables such as source, treatment, and age of a fat or oil. Specific Gravity (column D) was calculated at the specified temperature (degrees centigrade) and referred to water at the same temperature, unless otherwise specified. Density, shown in parentheses (column D), was measured at the specified temperature (degrees centigrade). Refractive Index (column E) was measured at 50°C, unless otherwise specified.

			Constants				
	Fat or Oil	Source	Melting (or solidification) point, °C	Specific gravity (or density)	Refractive index $n_D^{40°C}$	Iodine value	Saponification value
No	(A)	(B)	(C)	(D)	(E)	(F)	(G)
	Land Animals						
1	Butterfat	*Bos taurus*	32.2	$0.911^{40/15°}$	1.4548	36.1	227
2	Depot fat	*Homo sapiens*	(15)	$0.918^{15°}$	1.4602	67.6	196.2
3	Lard oil	*Sus scrofa*	(30.5)	$0.919^{15°}$	1.4615	58.6	194.6
4	Neat's-foot oil	*B. taurus*	—	$0.910^{25°}$	$1.464^{25°}$	69–76	190–199
5	Tallow, beef	*B. taurus*	—	—	—	49.5	197
6	Tallow, mutton	*Ovis aries*	(42.0)	$0.945^{15°}$	1.4565	40	194
	Marine Animals						
7	Cod-liver oil	*Gadus morhua*	—	$0.925^{25°}$	$1.481^{25°}$	165	186
8	Herring oil	*Clupea harengus*	—	$0.900^{60°}$	$1.4610^{60°}$	140	192
9	Menhaden oil	*Brevoortia tyrannus*	—	$0.903^{60°}$	$1.4645^{60°}$	170	191
10	Sardine oil	*Sardinops caerulea*	—	$0.905^{60°}$	$1.4660^{60°}$	185	191
11	Sperm oil, body	*Physeter macrocephalus*		—	—	76–88	122–130
12	Sperm oil, head	*P. macrocephalus*	—	—	—	70	140–144
13	Whale oil	*Balaena mysticetus*	—	$0.892^{60°}$	$1.460^{60°}$	120	195
	Plants						
14	Babassu oil	*Attalea funifera*	22–26	$(0.893^{60°})$	$1.443^{60°}$	15.5	247
15	Castor oil	*Ricinus communis*	(−18.0)	$0.961^{15°}$	1.4770	85.5	180.3
16	Cocoa butter	*Theobroma cacao*	34.1	$0.964^{15°}$	1.4568	36.5	193.8
17	Coconut oil	*Cocos nucifera*	25.1	$0.924^{15°}$	1.4493	10.4	268
18	Corn oil	*Zea mays*	(−20.0)	$0.922^{15°}$	1.4734	122.6	192.0
19	Cotton seed oil	*Gossypium hirsutum*	(−1.0)	$0.917^{25°}$	1.4735	105.7	194.3
20	Linseed oil	*Linum usitatissimum*	(−24.0)	$0.938^{15°}$	$1.4782^{25°}$	178.7	190.3
21	Mustard oil	*Brassica hirta*	—	$0.9145^{15°}$	1.475	102	174
22	Neem oil	*Melia azadirachta*	−3	$0.917^{15°}$	1.4615	71	194.5
23	Niger-seed oil	*Guizotia abyssinica*	—	$0.925^{15°}$	1.471	128.5	190
24	Oiticica oil	*Licania rigida*	—	$0.974^{25°}$	—	140–180	—
25	Olive oil	*Olea europaea sativa*	(−6.0)	$0.918^{15°}$	1.4679	81.1	189.7
26	Palm oil	*Elaeis guineensis*	35.0	$0.915^{15°}$	1.4578	54.2	199.1
27	Palm-kernel oil	*E. guineensis*	24.1	$0.923^{15°}$	1.4569	37.0	219.9
28	Peanut oil	*Arachis hypogaea*	(3.0)	$0.914^{15°}$	1.4691	93.4	192.1
29	Perilla oil	*Perilla frutescens*	—	$(0.935^{15°})$	$1.481^{25°}$	195	192
30	Poppy-seed oil	*Papaver somniferum*	(−15)	$0.925^{15°}$	1.4685	135	194
31	Rapeseed oil	*Brassica campestris*	(−10)	$0.915^{15°}$	1.4706	98.6	174.7
32	Safflower oil	*Carthamus tinctorius*	—	$(0.900^{60°})$	$1.462^{60°}$	145	192
33	Sesame oil	*Sesamum indicum*	(−6.0)	$0.919^{25°}$	1.4646	106.6	187.9
34	Soybean oil	*Glycine soja*	(−16.0)	$0.927^{15°}$	1.4729	130.0	190.6
35	Sunflower-seed oil	*Helianthus annuus*	(−17.0)	$0.923^{15°}$	1.4694	125.5	188.7
36	Tung oil	*Aleurites fordi*	(−2.5)	$0.934^{15°}$	$1.5174^{25°}$	168.2	193.1
37	Wheat-germ oil	*Triticum aestivum*	—	—	—	125	—

PROPERTIES AND FATTY ACID COMPOSITION OF FATS AND OILS (continued)

Constituent fatty acids, g/100 g total fatty acids

	Saturated						Unsaturated					
	Lauric	Myris-tic	Palmi-tic	Stearic	Ara-chidic	Other	Palmit-oleic	Oleic	Lino-leic	Lino-lenic	Other	
No	(H)	(I)	(J)	(K)	(L)	(M)	(N)	(O)	(P)	(Q)	(R)	No
1	2.5	11.1	29.0	9.2	2.4	2.0[a]; 0.5[b]; 2.3[c]	4.6	26.7	3.6	—	3.6[d]; 0.1[e]; 0.1[f]; 0.9[g]; 1.4[h]; 1.0[i]; 1.0[j]; 0.4[k]	1
2	—	2.7	24.0	8.4	—	—	5	46.9	10.2	—	2.5	2
3	—	1.3	28.3	11.9	—	—	2.7	47.5	6	—	0.2[g]; 2.1[h]	3
4	—	—	17–18	2–3	—	—	—	74–76	—	—	—	4
5	—	6.3	27.4	14.1	—	—	—	49.6	2.5	—	—	5
6	—	4.6	24.6	30.5	—	—	—	36.0	4.3	—	—	6
7	—	5.8	8.4	0.6	—	—	20.0	←—29.1—→		—	25.4[l]; 9.6[m]	7
8	—	7.3	13.0	Trace	—	—	4.9	—	—	<1%	30.1[l]; 23.2[m]	8
9	—	5.9	16.3	0.6	0.6	—	15.5	—	—	<1%	19.0[l]; 11.7[m]; 0.8[n]	9
10	—	5.1	14.6	3.2	—	—	11.8	←—17.8—→		—	18.1[l]; 14.0[m]; trace[g]; 15.4	10
11	1	5	6.5	—	—	—	26.5	37	19	—	1[m]; 4[g]; 19[p]	11
12	16	14	8	2	—	3.5[c]	15	17	6.5	—	4[f]; 14[g]; 6.5[p]	12
13	0.2	9.3	15.6	2.8	—	—	14.4	35.2	—	—	13.6[l]; 5.9[m]; 2.5[g]; 0.2[q]	13
14	44.1	15.4	8.5	2.7	0.2	0.2[a]; 4.8[b]; 6.6[c]	—	16.1	1.4	—	—	14
15	←————	2.4		————→	—	—	—	7.4	3.1	—	87[r]	15
16	—	—	24.4	35.4	—	—	—	38.1	2.1	—	—	16
17	45.4	18.0	10.5	2.3	0.4[s]	0.8[a]; 5.4[b]; 8.4[c]	0.4	7.5	Trace	—	—	17
18	—	1.4	10.2	3.0	—	—	1.5	49.6	34.3	—	—	18
19	—	1.4	23.4	1.1	1.3	—	2.0	22.9	47.8	—	—	19
20	—	—	6.3	2.5	0.5	—	—	19.0	24.1	47.4	0.2[n]	20
21	—	1.3[t]	—	—	—	—	—	27.2[t]	16.6[t]	1.8[t]	1.1[n]; 1.0[u]; 51.0[v]	21
22	—	2.6[t]	14.1[t]	24.0[t]	0.8[t]	—	—	58.5[t]	—	—	—	22
23	—	3.3[t]	8.2[t]	4.8[t]	0.5[t]	—	—	30.3[t]	57.3[t]	—	—	23
24	←————	11.3[w]		————→	—	—	—	6.2	—	—	82.5[x]	24
25	—	Trace	6.9	2.3	0.1	—	—	84.4	4.6	—	—	25
26	—	1.4	40.1	5.5	—	—	—	42.7	10.3	—	—	26
27	46.9	14.1	8.8	1.3	—	2.7[b]; 7.0[c]	—	18.5	0.7	—	—	27
28	—	—	8.3	3.1	2.4	—	—	56.0	26.0	—	3.1[n]; 1.1[u]	28
29	←————	9.6[w]		————→	—	—	—	17.8	—	17.5	—	29
30	—	—	4.8[t]	2.9[t]	—	—	—	30.1[t]	62.2[t]	—	—	30
31	—	—	1	—	—	—	—	32	15	1	50[v]	31
32	←————	6.8[w]		————→	—	—	—	18.6	70.1	3.4	—	32
33	—	—	9.1	4.3	0.8	—	—	45.4	40.4	—	—	33
34	0.2	0.1	9.8	2.4	0.9	—	0.4	28.9	50.7	6.5	0.1[g]	34
35	—	—	5.6	2.2	0.9	—	—	25.1	66.2	—	—	35
36	←————	4.6[w]		————→	—	—	—	4.1	0.6	—	90.7[y]	36
37	←————	16.0[w]		————→	—	—	—	28.1	52.3	3.6	—	37

[a]Caproic.
[b]Caprylo.
[c]Capric.
[d]Butyric.
[e]Decenoic.
[f]C_{12} monoethenoic.
[g]C_{14} monoethenoic.
[h]Gadoleic plus erucic.
[i]C_{12} n-pentadecanoic.
[j]C_{17} margaric.

[k]12-Methyl tetradecanoic.
[l]C_{20} polyethenoic.
[m]C_{22} polyethenoic.
[n]Behenic.
[o]C_{14} polyethenoic.
[p]Gadoleic.
[q]C_{24} polyethenoic.
[r]Ricinoleic.
[s]Includes behenic and lignoceric.
[t]Percent by weight.

[u]Lignoceric.
[v]Erucic.
[w]Includes behenic.
[x]Licanic.
[y]Eleostearic.

Section 4

Physical-Chemical Data

THERMODYNAMIC TERMS RECOMMENDED

Table 1a
THERMODYNAMIC TERMS USED IN THESE RECOMMENDATIONS

$[A]$	Concentration of A
y_B	Activity coefficient of substance B (concentration basis)
a_B	Relative activity of substance B (*Note:* $a_B = y_B \cdot [B]$)
I	Ionic strength ($I_c = 1/2 \sum_1^S c_i \cdot z_i^2$; $I_m = 1/2 \sum_1^S m_i \cdot z_i^2$)
	where c_i is the concentration of the ith ion, m_i is the molality of the ith ion, and z_i is the charge number of the ith ion. S is the number of ion types present
K_c	pH-independent proper product of equilibrium concentrations (concentration equilibrium constant)
K	Thermodynamic equilibrium constant
K'(pH = x; etc.)	pH-dependent apparent proper product of summed equilibrium concentrations, constrained with respect to concentrations of stated species (apparent equilibrium constant)
K_{exp}	See K'
K_{app}	See K'
K_{obsd}	See K'
ΔG_c°	Standard Gibbs energy change corresponding to the pH-independent product of equilibrium concentrations ($\Delta G_c^\circ = -RT \ln K_c$)
$\Delta G_c^{\circ \prime}$ (pH = x)	Apparent standard Gibbs energy change corresponding to the apparent product of equilibrium concentrations at fixed pH in a buffered solution ($\Delta G_c^{\circ \prime} = -RT \ln K_c'$)
ΔG_{exp}°	See $\Delta G^{\circ \prime}$
ΔG_{app}°	See $\Delta G^{\circ \prime}$
ΔF	Formerly common symbol for free energy change; now more properly called Gibbs energy change and given the symbol ΔG
ΔF°	See ΔF, ΔG°, $\Delta G^{\circ \prime}$
ΔS	Entropy change
ΔH	Enthalpy change
C_p	Heat capacity at constant pressure

The units of molar enthalpy change and molar Gibbs energy change are generally expressed in joules per mole ($J \cdot mol^{-1}$) or kilojoules per mole ($kJ \cdot mol^{-1}$) of a reaction shown, but the units should be stated in every case. For particular purposes specific enthalpy change or specific Gibbs energy change may be given per unit mass of a given reactant or product (joules per kilogram) ($J \cdot kg^{-1}$).

The units of entropy change are generally expressed in joules per mole kelvin ($J \cdot mol^{-1} \cdot K^{-1}$) for a particular process, but the units should be stated in every case. For particular purposes specific entropy change may be given per unit mass of a given reactant or product, joule per kilogram kelvin, ($J \cdot kg^{-1} \cdot K^{-1}$) or the total entropy change for a stated process in which case the units are joules per kelvin ($J \cdot K^{-1}$).

Table 1a (continued)
THERMODYNAMIC TERMS USED IN THESE RECOMMENDATIONS

The units of absolute entropy and heat capacity are the same as those for entropy change ($J \cdot mol^{-1} \cdot K^{-1}$, $J \cdot kg^{-1} \cdot K^{-1}$ or $J \cdot K^{-1}$) but refer to a particular substance or aggregate of substances rather than to a process.

Some thermodynamic relations: The units for these quantities are joules, J, for the extensive quantity total energy, joules per mole ($J \cdot mol^{-1}$) for molar energy, and joule per kilogram ($J \cdot kg^{-1}$) for specific energy.

$\Delta U = Q + W$	First law of thermodynamics – the increase of internal energy of a system is the sum of heat supplied to the system and work done on the system.
$\Delta H = \Delta U + p \Delta V$	For a constant pressure system an increase in enthalpy of the system (ΔH) results in an increase in internal energy (ΔU) and work done by the system ($p \Delta V$).
$\Delta H = H(T_2) - H(T_1) = \int C_p dT$	For a system at constant pressure to which energy is supplied the increase of enthalpy is the integral of C_p over the range of the temperature change it causes.
$\Delta G = \Delta H - T \Delta S$	The increase in Gibbs energy of a system (ΔG) at constant temperature is the increase in enthalpy (ΔH) minus the increase in $T \Delta S$.

Table 1b
SUMMARY OF THERMODYNAMIC SYMBOLS AND OTHER RECOMMENDATIONS

Quantity	Apparent value at pH = constant (x)	pH-independent value
Equilibrium constant*	K'	K
Standard molar Gibbs energy change**	$\Delta G^{\circ\,\prime}$ (pH = x)	ΔG°
Concentration of reactant A	$[A]_{total}$	$[A]$

Recommended Conventions Concerning Factors in Expressions for Equilibrium Constants

In cases where water occurs as a reactant or product, state whether the factor for water is taken as unity or as 55.5 or other number.

State whether 10^{-pH} or some other measure is used for the hydrogen ion factor. The pH is not uncontested as an accurate measure of concentration or activity of the hydrogen ion.

Recommended Measurement Conditions

	Primary conditions	Secondary conditions
Temperature $t/^{\circ}C$ (or T/K)	25°C (298.15 K) (also, vary t)	37°C (310.15 K) (also, vary t)
Ionic strength, $I/mol \cdot dm^{-3}$	0.1 (made up with KCl)	0.1 (made up with KCl)
Hydrogen ion concentration	pH = 7	
Buffer concentration	Lowest effective	

*Approximately constant proper quotient of equilibrium concentrations.
**Formally calculated from the equilibrium constant.

Table 2
PHYSICAL QUANTITIES, SI UNITS, AND THEIR SYMBOLS

Name of unit   **Symbol for unit**

(a) Base Units

Length	metre	m
Mass	kilogram	kg
Time	second	s
Electric current	ampere	A
Temperature	kelvin	K
Amount of substance	mole	mol
Luminous intensity	candela	cd

(b) Derived Units (Examples)

Force	newton	N	$(kg \cdot m \cdot s^{-2})$
Pressure	pascal	Pa	$(N \cdot m^{-2})$
Energy	joule	J	$(kg \cdot m^2 \cdot s^{-2})$
Power	watt	W	$(J \cdot s^{-1})$
Electric charge	coulomb	C	$(A \cdot s)$
Electric potential difference	volt	V	$(J \cdot A^{-1} \cdot s^{-1})$
Electric resistance	ohm	Ω	$(V \cdot A^{-1})$
Frequency	hertz	Hz	(s^{-1})
Area	square metre	m^2	
Volume	cubic metre	m^3	
Density	kilogram per cubic metre	$kg \cdot m^{-3}$	

(c) SI Prefixes

Fraction	Prefix	Symbol	Multiple	Prefix	Symbol
10^{-1}	deci	d	10	deca	da
10^{-2}	centi	c	10^2	hecto	h
10^{-3}	milli	m	10^3	kilo	k
10^{-6}	micro	μ	10^6	mega	M
10^{-9}	nano	n	10^9	giga	G
10^{-12}	pico	p	10^{12}	tera	T
10^{-15}	femto	f	10^{15}	peta	P
10^{-18}	atto	a	10^{18}	exa	E

REFERENCES

1. Guide for the presentation in the primary literature of numerical data derived from experiments, *CODATA Bull.*, No. 9, December 1973.
2. A guide to procedures for the publication of thermodynamic data, *J. Chem. Thermodynamics*, 4, 511 (1972); *Pure Appl. Chem.*, 29, 395 (1972).
3. **Alberty,** *J. Biol. Chem.*, 244, 3290 (1969).
4. **McGlashan,** *Manual of Symbols and Terminology for Physiochemical Quantities and Units*, Butterworth and Co., London, 1970; also in *Pure Appl. Chem.*, 21, No. 1 (1970).
5. International Organization for Standardization (ISO), SI units and recommendations for the use of their multiples and of certain other units, International Standard ISO-1000, 1st ed., 1973-02-01, American National Standards Institute, New York. See also, for more detail: International Standard ISO-31/0-1974 and ISO-31/I-XII, 1965–1975 which deal with quantities, units, symbols, conversion factors, and conversion tables for various branches of science and technology. Copies are obtained through the ISO-member national standards organizations of various countries.
6. Bureau International des Poids et Mesures (BIPM), *Le Systéme International d'Unités*, OFFILIB, Paris, France, 1970. Authorized English translations are available: *The International System of Units (SI)*, National Bureau of Standards Publication 330, U.S. Gov. Print. Off. Washington, D.C., 1971, or *The International System of Units (SI)*, Her Majesty's Stationery Office, London.

DECI-NORMAL SOLUTIONS OF OXIDATION AND REDUCTION REAGENTS

Atomic and molecular weights in the following table are based upon the 1965 atomic weight scale and the isotope C-12. The weight in grams of the compound in 1 cc of the following deci-normal solutions is found by dividing the H equivalent in the last column by 1,000.

Name	Formula	Atomic or molecular weight	Hydrogen equivalent	0.1 Hydrogen equivalent in g
Antimony	Sb	121.75	$\frac{1}{3}$Sb	6.0875
Arsenic	As	74.9216	$\frac{1}{2}$As	3.7461
Arsenic trisulfide	As_2S_3	246.0352	$\frac{1}{4}As_2S_3$	6.1509
Arsenous oxide	As_2O_3	197.8414	$\frac{1}{4}As_2O_3$	4.9460
Barium peroxide	BaO_2	169.3388	$\frac{1}{2}BaO_2$	8.4669
Barium peroxide hydrate	$BaO_2 \cdot 8H_2O$	313.4615	$\frac{1}{2}BaO_2 \cdot 8H_2O$	15.6730
Calcium	Ca	40.08	$\frac{1}{2}$Ca	2.004
Calcium carbonate	$CaCO_3$	100.0894	$\frac{1}{2}CaCO_3$	5.0045
Calcium hypochlorite	$Ca(OCl)_2$	142.9848	$\frac{1}{4}Ca(OCl)_2$	3.5746
Calcium oxide	CaO	56.0794	$\frac{1}{2}$CaO	2.8040
Chlorine	Cl	35.453	Cl	3.5453
Chromium trioxide	CrO_3	99.9942	$\frac{1}{3}CrO_3$	3.3331
Ferrous ammonium sulfate	$FeSO_4(NH_4)SO_4 \cdot 6H_2O$	392.0764	$FeSO_4(NH_4)_2SO_4 \cdot 6H_2O$	39.2076
Hydroferrocyanic acid	$H_4Fe(CN)_6$	215.9860	$H_4Fe(CN)_6$	21.5986
Hydrogen peroxide	H_2O_2	34.0147	$\frac{1}{2}H_2O_2$	1.7007
Hydrogen sulfide	H_2S	34.0799	$\frac{1}{2}H_2S$	1.7040
Iodine	I	126.9044	I	12.6904
Iron	Fe	55.847	Fe	5.5847
Iron oxide (ferrous)	FeO	71.8464	FeO	7.1846
Iron oxide (ferric)	Fe_2O_3	159.6922	$\frac{1}{2}Fe_2O_3$	7.9846
Lead peroxide	PbO_2	239.1888	$\frac{1}{2}PbO_2$	11.9594
Manganese dioxide	MnO_2	86.9368	$\frac{1}{2}MnO_2$	4.3468
Nitric acid	HNO_3	63.0129	$\frac{1}{3}HNO_3$	2.1004
Nitrogen trioxide	N_2O_3	76.0116	$\frac{1}{4}N_2O_3$	1.9002
Nitrogen pentoxide	N_2O_5	108.0104	$\frac{1}{5}N_2O_5$	1.8001
Oxalic acid	$C_2H_2O_4$	90.0358	$\frac{1}{2}C_2H_2O_4$	4.5018
Oxalic acid hydrate	$C_2H_2O_4 \cdot 2H_2O$	126.0665	$\frac{1}{2}C_2H_2O_4 \cdot 2H_2O$	6.3033
Oxygen	O	15.9994	$\frac{1}{2}$O	0.8000
Potassium dichromate	$K_2Cr_2O_7$	294.1918	$\frac{1}{6}K_2Cr_2O_7$	4.9032
Potassium chlorate	$KClO_3$	122.5532	$\frac{1}{6}KClO_3$	2.0425
Potassium chromate	K_2CrO_4	194.1076	$\frac{1}{3}K_2CrO_4$	6.4733
Potassium ferrocyanide	$K_4Fe(CN)_6$	368.3621	$K_4Fe(CN)_6$	36.8362
Potassium ferrocyanide	$K_4Fe(CN)_6 \cdot 3H_2O$	422.4081	$K_4Fe(CN)_6 \cdot 3H_2O$	42.2408
Potassium iodide	KI	166.0064	KI	16.6006
Potassium nitrate	KNO_3	101.1069	$\frac{1}{3}KNO_3$	3.3702
Potassium perchlorate	$KClO_4$	138.5526	$\frac{1}{8}KClO_4$	1.7319
Potassium permanganate	$KMnO_4$	158.0376	$\frac{1}{5}KMnO_4$	3.1608
Sodium chlorate	$NaClO_3$	106.4410	$\frac{1}{6}NaClO_3$	1.7740
Sodium nitrate	$NaNO_3$	84.9947	$\frac{1}{3}NaNO_3$	2.8332
Sodium thiosulfate	$Na_2S_2O_3 \cdot 5H_2O$	248.1825	$Na_2S_2O_3 \cdot 5H_2O$	24.8183
Stannous chloride	$SnCl_2$	189.5960	$\frac{1}{2}SnCl_2$	9.4798
Stannous oxide	SnO	134.6894	$\frac{1}{2}$SnO	6.7345
Sulfur dioxide	SO_2	64.0628	$\frac{1}{2}SO_2$	3.2031
Tin	Sn	118.69	$\frac{1}{2}$Sn	5.935

This table originally appeared in Sober, Ed., *Handbook of Biochemistry and Selected Data for Molecular Biology*, 2nd ed., Chemical Rubber Co., Cleveland, 1970.

MEASUREMENT OF pH

Roger G. Bates and Maya Paabo

I. Definition of pH

The following definition of pH has received the endorsement of the International Union of Pure and Applied Chemistry.

1. *Operational definition.* In all existing national standards the definition of pH is an operational one. The electromotive force E_x of the cell:

Pt, H_2 |solution X|concentrated KCl solution| reference electrode

is measured and likewise the electromotive force E_s of the cell:

Pt, H_2 |solution S|concentrated KCl solution| reference electrode

both cells being at the same temperature throughout and the reference electrodes and bridge solutions being identical in the two cells. The pH of the solution X, denoted by pH(X), is then related to the pH of the solution S, denoted by pH(S), by the definition:

$$pH(X) = pH(S) + \frac{E_x - E_s}{(RT \ln 10)/F}$$

where R denotes the gas constant, T the thermodynamic temperature, and F the faraday constant. Thus defined the quantity pH is dimensionless.

To a good approximation, the hydrogen electrodes in both cells may be replaced by other hydrogen ionresponsive electrodes, e.g., glass or quinhydrone. The two bridge solutions may be of any molality not less than 3.5 mol kg^{-1}, provided they are the same (see *Pure Appl. Chem.*, 1, 163 1960).

2. *Standards.* The difference between the pH of two solutions having been defined as above, the definition of pH can be completed by assigning a value of pH at each temperature to one or more chosen solutions designated as standards. A series of pH(S) values for seven suitable standard reference solutions is given in Table 1. The constants for calculating pH(S) values over the temperature range for 0 to 95°C are given in Table 2.

If the definition of pH given above is adhered to strictly, then the pH of a solution may be slightly dependent on which standard solution is used. These unavoidable deviations are caused not only by imperfections in the response of the hydrogen ion electrodes but also by variations in the liquid-junction potentials resulting from the different ionic compositions and mobilities of the several standards and from differences in the structure of the liquid-liquid boundary. In fact such variations in measured pH are usually too small to be of practical significance. Moreover, the acceptance of several standards allows the use of the following alternative definition of pH.

Table 1
VALUES OF pH(S) FOR SEVEN PRIMARY STANDARD SOLUTIONS

$t/°C$	A	B	C	D	E	F	G	$t/°C$	A	B	C	D	E	F	G
0	—	3.863	4.003	6.984	7.534	9.464	10.317	40	3.547	3.753	4.035	6.838	7.380	9.068	9.889
5	—	3.840	3.999	6.951	7.500	9.395	10.245	45	3.547	3.750	4.047	6.834	7.373	9.038	9.856
10	—	3.820	3.998	6.923	7.472	9.332	10.179	50	3.549	3.749	4.060	6.833	7.367	9.011	9.828
15	—	3.802	3.999	6.900	7.448	9.276	10.118	55	3.554	—	4.075	6.834	—	8.985	—
20	—	3.788	4.002	6.881	7.429	9.225	10.062	60	3.560	—	4.091	6.836	—	8.962	—
25	3.557	3.776	4.008	6.865	7.413	9.180	10.012	70	3.580	—	4.126	6.845	—	8.921	—
30	3.552	3.766	4.015	6.853	7.400	9.139	9.966	80	3.609	—	4.164	6.859	—	8.885	—
35	3.549	3.759	4.024	6.844	7.389	9.102	9.925	90	3.650	—	4.205	6.877	—	8.850	—
38	3.548	3.755	4.030	6.840	7.384	9.081	9.903	95	3.674	—	4.227	6.886	—	8.833	—

The compositions of the standard solutions are:
A: KH tartrate (saturated at 25°C)
B: KH_2 citrate, $m = 0.05$ mol kg^{-1}
C: KH phthalate, $m = 0.05$ mol kg^{-1}
D: $KH_2 PO_4$, $m = 0.025$ mol kg^{-1}; $Na_2 HPO_4$, $m = 0.025$ mol kg^{-1}
E: $KH_2 PO_4$, $m = 0.008695$ mol kg^{-1}; $Na_2 HPO_4$, $m = 0.03043$ mol kg^{-1}
F: $Na_2 B_4 O_7$, $m = 0.01$ mol kg^{-1}
G: $NaHCO_3$, $m = 0.025$ mol kg^{-1}; $Na_2 CO_3$, $m = 0.025$ mol kg^{-1} where m denotes molality.

Table 2
VALUES OF THE CONSTANTS OF THE EQUATION: $pH(S) = \frac{A}{T} + B + CT + DT^2$, FOR SEVEN PRIMARY STANDARD BUFFER SOLUTIONS FROM 0 TO 95°C

Solution	Temperature range °C	A	B	C	$10^5 D$	Standard deviation of the fitted curves
A. Tartrate	25 to 95	−1727.96	23.7406	−0.075947	9.2873	0.0016
B. Citrate	0 to 50	1280.4	−4.1650	0.012230	0	0.0010
C. Phthalate	0 to 95	1678.30	−9.8357	0.034946	−2.4804	0.0027
D. Phosphate	0 to 95	3459.39	−21.0574	0.073301	−6.2266	0.0017
E. Phosphate	0 to 50	5706.61	−43.9428	0.154785	−15.6745	0.0011
F. Borax	0 to 95	5259.02	−33.1064	0.114826	−10.7860	0.0025
G. Carbonate	0 to 50	2557.1	−4.2846	0.019185	0	0.0026

The electromotive force E_x is measured, and likewise the electromotive forces E_1 and E_2, of two similar cells with the solution X replaced by the standard solutions S_1 and S_2 such that E_1 and E_2 values are on either side of, and as near as possible to, E_x. The pH of solution X is then obtained by assuming linearity between pH and E, that is to say

$$\frac{pH(X) - pH(S_1)}{pH(S_2) - pH(S_1)} = \frac{E_x - E_1}{E_2 - E_1}$$

This procedure is especially recommended when the hydrogen-ion-responsive electrode is a glass electrode.

II. Standard Solutions

The pH meter or other electrometric pH assembly does not, strictly speaking, measure the pH but rather indicates a difference between the pH of an unknown solution (X) and a standard solution (S), both of which are at the same temperature. The pH meter should always be standardized routinely with two reference solutions of assigned pH, chosen if possible to bracket the pH of the test solution. These standards are prepared as indicated in Table 3. For convenience, air weights of the buffer salts are given. A good grade of distilled or de-ionized water should be used; for the four solutions of highest pH, the water should be freed of dissolved carbon dioxide by boiling or purging. For a detailed discussion of the properties of the primary standard buffer solutions, the reader is referred to chapter 4 of R. G. Bates, *Determination of pH,* 2nd ed., (John Wiley and Sons, Inc., New York, 1973).

Highly pure buffer materials should be used. These materials are obtainable commercially; they are also distributed as certified standard reference materials by the National Bureau of Standards. It should be noted that individual lots show slight variations; hence, the values certified for a particular lot may differ slightly from those given in Table 1.

Table 3
PREPARATION OF PRIMARY STANDARD BUFFER SOLUTIONS

Standard solution	NBS SRM No.[a]	Buffer substance	Weight in air[b] (g)	Standard solution	NBS SRM No.[a]	Buffer substance	Weight in air[b] (g)
A. Tartrate	188	$KHC_4H_4O_6$	(Satd. at 25°C)	E. Phosphate	186Ic	KH_2PO_4	1.179
					186IIb	Na_2HPO_4	4.302
B. Citrate	190	$KH_2C_6H_5O_7$	11.41	F. Borax	187a	$Na_2B_4O_7 \cdot$	3.80
C. Phthalate	185d	$KHC_8H_4O_4$	10.12			$10\ H_2O$	
D. Phosphate	186Ic	KH_2PO_4	3.388	G. Carbonate	191	$NaHCO_3$	2.092
	186IIb	Na_2HPO_4	3.533		192	Na_2CO_3	2.640

[a]These materials may be ordered from the Office of Standard Reference Materials, National Bureau of Standards, Washington, D.C. 20234.
[b]This weight of salt to be dissolved in water and diluted to 1 liter at 25°C to provide concentrations indicated in Table 1.

The use of two or more standard reference solutions may disclose small inconsistencies in the standardization of the pH meter, depending on which standards are chosen. When this is the case, the best results are often obtained by assuming linearity between E and pH between the two calibrating points bracketing the pH of the unknown.

III. Electrodes

Although the hydrogen electrode is the ultimate standard on which the pH scale is based, in practice the convenient and versatile glass electrode is favored for the vast majority of pH measurements. New glass electrodes, or those that have been allowed to dry out, should be conditioned by soaking in water for several hours before use and after exposure to nonaqueous or dehydrating media. Some glass electrodes are designed especially for use at high temperatures, while others are best suited to low-temperature use. Special "high pH" electrodes are also available. For optimum results, careful attention should be paid to selection of the proper electrode for the problem at hand.

Glass electrodes of small dimensions are of great utility when sample volumes are limited. The pH-sensitive glasses are, however, moderately soluble, and small amounts of alkali are dissolved from the glass surface by the solutions in which the electrode is immersed. For this reason, the most accurate results are obtained when the ratio of the electrode area to sample volume is small.

The concentrated solution of potassium chloride that joins the reference electrode with the unknown or standard solution is reasonably effective in reducing the liquid-junction potential to small, fairly constant, values. It is important to assure that the flow of bridge solution into the test solution is neither excessive nor completely interrupted by crystallization of salt in the aperture where liquid-liquid contact is established.

Temperature gradients within the pH cell are a common source of difficulty, marked by variability and inaccuracy in the reading. Both of the electrodes, and the standard and test solutions as well, should be within a few degrees Celsius of the same temperature. For results of the highest reliability, temperature control should be provided. It is the function of the temperature compensator of the pH meter to adjust the pH-e.m.f. slope in such a manner that a difference of e.m.f. (in volts) is correctly converted to a difference of pH. This adjustment cannot compensate for inequalities of temperature through the cell or for differences between the temperature of the standard and test solutions.

IV. Techniques

Electrodes and sample cups should be washed carefully with distilled or de-ionized water and gently dried with clean absorbent tissue. The electrodes are immersed in the first standard solution and the temperature compensator of the measuring instrument is set at the temperature of the solutions whose pH is to be measured. The standardization control of the instrument is adjusted until the meter is balanced at the known pH of the standard, as given in Table 1. This procedure is repeated with successive portions of the same standard until replacement causes no change in the position of balance. The electrodes are then washed once more and dried.

A second standard solution is selected and the measurement repeated without altering the position of the standardization control. The pH reading of this second solution is noted and the sample replaced with a second portion of the same solution. This replacement is continued until successive readings agree within 0.02 pH unit, when the electrodes and meter may be judged to be functioning properly. It is advisable to make a final check with one of the buffers at the conclusion of a series of measurements.

After the instrument is properly standardized, a portion of the test solution is placed in the sample cup and the pH reading noted. Successive portions are again used until two measurements agree within the limits imposed by the reproducibility of the measuring instrument and the temperature control. With the best meters, measurements on buffered solutions should be reproducible to 0.01 unit or even better. With water or poorly buffered solutions, values agreeing to 0.1 unit may have to be accepted. Some improvement will result if poorly buffered solutions are protected from carbon dioxide of the atmosphere during the period of the measurements.

V. Interpretation of pH Numbers

The standard values of pH given in the table of an earlier section are based on hydrogen electrode potentials as measured in cells without a liquid junction. The uncertainty of the standard values is estimated at 0.005 unit. The accuracy of the results furnished by a given pH assembly adjusted with these primary standards is, however, further limited by inconsistencies which have their origin in defects of the glass electrode response and variations in the liquid-junction potential. For these reasons, the accuracy of experimental pH numbers can be considered to be better than 0.01 unit only under unusually favorable conditions.

The operational definition of pH fulfills adequately the need for an experimental scale capable of furnishing reproducible pH numbers. The interpretation of these numbers may be of secondary importance and should only be attempted when the standard and unknown solutions are matched so closely in composition that there is good reason to believe that the liquid-junction potential remains fairly constant when the standard is replaced by the unknown. In general, this will be the case when the unknowns are aqueous solutions of simple salts of total concentration not in excess of 0.2 M with pH values between 2.5 and 11.5.

When these "ideal" conditions prevail, the experimental pH can be considered to approach $-\log a_H$, where a_H is the conventional hydrogen ion activity defined in a manner consistent with the convention on which the standard values of pH(S) were based Bates, R. G., *J. Res. Natl. Bur. Standards,* 66A, 179 (1962)). All quantitative applications of pH measurements, when justifiable, should therefore be based on the approximation $pH(X) \approx -\log a_H = -\log m_H \gamma_H$, where m is molality and γ is the activity coefficient.

VI. Indicator Methods

Acid-base indicators have the property of altering the color of a solution in the region 1 to 2 pH units as the pH changes. They are therefore useful for pH measurements, although in general the accuracy is inferior to that obtainable by electrometric procedures. A list of suitable indicators, their pH ranges and color changes, is given in Table 4.

Table 4
ACID-BASE INDICATORS

Indicator	pH Range	Color change	Indicator	pH Range	Color change
Acid cresol red	0.2–1.8	Red–yellow	Metacresol purple	7.6–9.2	Yellow–purple
Acid metacresol purple	1.2–2.8	Red–yellow	Thymol blue	8.0–9.6	Yellow–blue
Acid thymol blue	1.2–2.8	Red–yellow	Phthalein red	8.6–10.2	Yellow–red
Bromophenol blue	3.0–4.6	Yellow–blue	Tolyl red	10.0–11.6	Red–yellow
Bromocresol green	3.8–5.4	Yellow–blue	Acyl red	10.0–11.6	Red–yellow
Methyl red	4.4–6.0	Red–yellow	Parazo orange	11.0–12.6	Yellow–orange
Chlorophenol red	5.2–6.8	Yellow–red	Acyl blue	12.0–13.6	Red–blue
Bromocresol purple	5.2–6.8	Yellow–purple	Benzo yellow	2.4–4.0	Red–yellow
Bromothymol blue	6.0–7.6	Yellow–blue	Benzo red	4.4–7.6	Red–blue
Phenol red	6.8–8.4	Yellow–red	Thymol red	8.0–11.2	Yellow–red
Cresol red	7.2–8.8	Yellow–red			

Courtesy of W. A. Taylor and Co.

Equal concentrations of the same indicator are added to the test solution and to each of a series of buffer solutions of known pH selected to bracket the pH of the test solution. Color comparisons are made with a colorimeter or spectrophotometer, and solutions of equal color are assumed to have the same pH. The pH of a series of suitable reference solutions can be determined in advance by electrometric methods. Alternatively, tables of pH as a function of composition can be utilized. The compositions and pH values of a set of useful solutions covering the range pH 1 to 13 are summarized in Table 5.

Table 5
BUFFER SOLUTIONS FOR INDICATOR MEASUREMENTS AND pH CONTROL

25 ml 0.2 M KCl, x ml 0.2 M HCl, diluted to 100 ml

pH	x	pH	x
1.00	67.0	1.50	20.7
1.10	52.8	1.60	16.2
1.20	42.5	1.70	13.0
1.30	33.6	1.80	10.2
1.40	26.6	1.90	8.1
—	—	2.00	6.5
—	—	2.10	5.1
—	—	2.20	3.9

50 ml 0.1 M KH Phthalate, x ml 0.1 M HCl, diluted to 100 ml

pH	x	pH	x
2.20	49.5	3.20	15.7
2.30	45.8	3.30	12.9
2.40	42.2	3.40	10.4
2.50	38.8	3.50	8.2
2.60	35.4	3.60	6.3
2.70	32.1	3.70	4.5
2.80	28.9	3.80	2.9
2.90	25.7	3.90	1.4
3.00	22.3	4.00	0.1
3.10	18.8	—	—

50 ml 0.1 M KH Phthalate, x ml 0.1 M NaOH, diluted to 100 ml

pH	x	pH	x
4.10	1.3	5.10	25.5
4.20	3.0	5.20	28.8
4.30	4.7	5.30	31.6
4.40	6.6	5.40	34.1
4.50	8.7	5.50	36.6
4.60	11.1	5.60	38.8
4.70	13.6	5.70	40.6
4.80	16.5	5.80	42.3
4.90	19.4	5.90	43.7
5.00	22.6	—	—

50 ml 0.1 M KH_2PO_4, x ml 0.1 M NaOH, diluted to 100 ml

pH	x	pH	x
5.80	3.6	6.80	22.4
5.90	4.6	6.90	25.9
6.00	5.6	7.00	29.1
6.10	6.8	7.10	32.1
6.20	8.1	7.20	34.7
6.30	9.7	7.30	37.0
6.40	11.6	7.40	39.1
6.50	13.9	7.50	41.1
6.60	16.4	7.60	42.8
6.70	19.3	7.70	44.2
—	—	7.80	45.3
—	—	7.90	46.1
—	—	8.00	46.7

50 ml of a mixture 0.1 M with respect to both KCl and H_3BO_3, x ml 0.1 M NaOH, diluted to 100 ml

pH	x	pH	x
8.00	3.9	9.00	20.8
8.10	4.9	9.10	23.6
8.20	6.0	9.20	26.4
8.30	7.2	9.30	29.3
8.40	8.6	9.40	32.1
8.50	10.1	9.50	34.6
8.60	11.8	9.60	36.9
8.70	13.7	9.70	38.9
8.80	15.8	9.80	40.6
8.90	18.1	9.90	42.2
—	—	10.00	43.7
—	—	10.10	45.0
—	—	10.20	46.2

50 ml 0.1 M Tris(hydroxmethyl)-aminomethane, x ml 0.1 M HCl, diluted to 100 ml

pH	x	pH	x
7.00	46.6	8.00	29.2
7.10	45.7	8.10	26.2
7.20	44.7	8.20	22.9
7.30	43.4	8.30	19.9
7.40	42.0	8.40	17.2
7.50	40.3	8.50	14.7
7.60	38.5	8.60	12.4
7.70	36.6	8.70	10.3
7.80	34.5	8.80	8.5
7.90	32.0	8.90	7.0
—	—	9.00	5.7

Table 5 (continued)
BUFFER SOLUTIONS FOR INDICATOR MEASUREMENTS AND pH CONTROL

50 ml 0.025 M Borax, x ml 0.1 M HCl, diluted to 100 ml

pH	x	pH	x
8.00	20.5	8.50	15.2
8.10	19.7	8.60	13.5
8.20	18.8	8.70	11.6
8.30	17.7	8.80	9.4
8.40	16.6	8.90	7.1
—	—	9.00	4.6
—	—	9.10	2.0

50 ml 0.025 M Borax, x ml 0.1 M NaOH, diluted to 100 ml

pH	x	pH	x
9.20	0.9	10.20	20.5
9.30	3.6	10.30	21.3
9.40	6.2	10.40	22.1
9.50	8.8	10.50	22.7
9.60	11.1	10.60	23.3
9.70	13.1	10.70	23.80
9.80	15.0	10.80	24.25
9.90	16.7	—	—
10.00	18.3	—	—
10.10	19.5	—	—

50 ml 0.05 M NaHCO$_3$, x ml 0.1 M NaOH, diluted to 100 ml

pH	x	pH	x
9.60	5.0	10.60	19.1
9.70	6.2	10.70	20.2
9.80	7.6	10.80	21.2
9.90	9.1	10.90	22.0
10.00	10.7	11.00	22.7
10.10	12.2	—	—
10.20	13.8	—	—
10.30	15.2	—	—
10.40	16.5	—	—
10.50	17.8	—	—

50 ml 0.05 M Na$_2$HPO$_4$, x ml 0.1 M NaOH, diluted to 100 ml

pH	x	pH	x
10.90	3.3	11.40	9.1
11.00	4.1	11.50	11.1
11.10	5.1	11.60	13.5
11.20	6.3	11.70	16.2
11.30	7.6	11.80	19.4
—	—	11.90	23.0
—	—	12.00	26.9

25 ml 0.2 M KCl, x ml 0.2 M NaOH, diluted to 100 ml

pH	x	pH	x
12.00	6.0	12.50	20.4
12.10	8.0	12.60	25.6
12.20	10.2	12.70	32.2
12.30	12.8	12.80	41.2
12.40	16.2	12.90	53.0
—	—	13.00	66.0

Source: Bower and Bates, *J. Res. Natl. Bur. Standards,* 55, 197 (1955); Bates and Bower, *Anal. Chem.,* 28, 1322 (1956).

Table 6

pH VALUES FOR MISCELLANEOUS BUFFER SOLUTIONS OVER A RANGE OF TEMPERATURE

Composition of the buffer solution	m^a	$t/°C$										
		0	5	10	15	20	25	30	35	40	45	50
Potassium dihydrogen phosphate (m)	0.005	—	—	—	—	—	6.251	—	—	—	—	—
Sodium succinate (m) (1)	0.015	—	—	—	—	—	6.162	—	—	—	—	—
	0.025	—	—	—	—	—	6.109	—	—	—	—	—
Piperazine phosphate (m) (2)	0.02	6.580	6.515	6.453	6.394	6.338	6.284	6.234	6.185	6.140	6.097	6.058
	0.05	6.589	6.525	6.463	6.404	6.348	6.294	6.243	6.195	6.149	6.106	6.066
2,2-Bis(hydroxymethyl)-2,2''-nitrilotriethanol (2m) Hydrochloric acid (m) (3)	0.02	7.000	6.905	6.812	6.722	6.635	6.551	6.469	6.390	6.312	6.237	6.165
	0.04	7.029	6.932	6.839	6.748	6.662	6.577	6.495	6.415	6.336	6.262	6.190
	0.06	7.050	6.953	6.859	6.767	6.681	6.595	6.513	6.434	6.353	6.280	6.208
	0.08	7.067	6.969	6.876	6.783	6.696	6.610	6.528	6.448	6.367	6.294	6.222
	0.10	7.082	6.983	6.889	6.796	6.710	6.623	6.540	6.460	6.378	6.306	6.235
Morpholine (1.5m) Hydrochloric acid (m) (4)	0.10	8.963	8.828	8.702	8.579	8.458	8.343	8.231	8.120	8.013	7.908	7.806
Tris(hydroxymethyl)aminomethane ("Tris") (m), Tris.HCl (m) (5)	0.05	8.946	8.774	8.614	8.461	8.313	8.173	8.036	7.904[b]	7.777	7.654	7.537
Tris (m), Tris.HCl (3m) (6)	0.01667	8.471	8.303	8.142	7.988	7.840	7.698	7.563	7.433	7.307	7.186	7.070

[a] mol kg^{-1}
[b] 7.851 at 37°C

Table 6 (continued)
pH VALUES FOR MISCELLANEOUS BUFFER SOLUTIONS OVER A RANGE OF TEMPERATURE

Composition of the buffer solution	m^a	$t/°C$											
		0	5	10	15	20	25	30	35	40	45	50
N-Tris(hydroxymethyl)methyl-glycine ("Tricine") (m), Na Tricinate (m) (7)	0.05	—	8.485	8.375	8.271	8.175	8.079	7.988	7.902	7.817	7.740	7.663
Tricine ($3m$), Na Tricinate (m) (7)	0.02	—	8.023	7.916	7.813	7.713	7.621	7.527	7.437[c]	7.355	7.275	7.197

[c]7.407 at 37°C

Compiled by Roger G. Bates and Maya Paabo.

Contribution from the National Bureau of Standards, not subject to copyright.

REFERENCES

1. **Paabo, Bates, and Robinson,** *J. Res. Natl. Bur. Standards,* 67A, 573 (1963).
2. **Hetzer, Robinson, and Bates,** *Anal. Chem.,* 40, 634 (1968).
3. **Paabo and Bates,** *J. Phys. Chem.,* 74, 702 (1970).
4. **Hetzer, Bates, and Robinson,** *J. Phys. Chem.,* 70, 2869 (1966).
5. **Bates and Robinson,** *Anal. Chem.,* 45, 420 (1973).
6. **Durst and Staples,** *Clin. Chem.,* 18, 206 (1972).
7. **Bates, Roy and Robinson,** *Anal. Chem.,* 45, 1663 (1973).

BUFFER SOLUTIONS[a]

Buffer solutions (or buffers) are solutions whose pH value is to a large degree insensitive to the addition of other substances. It is important to realize, however, that the pH value of a buffer solution not only changes when acids or bases are added or on dilution, but also when the temperature changes or neutral salts are added. In accurate work, therefore, it is important to check the pH value electrometrically after all the ingredients have been added. The extent to which the pH values of buffer solutions vary when acids or bases are added or the temperature changes is shown in the following tables. In general, dilution to half the concentration changes the pH value by only some hundredths of a unit (buffer No. 1 in the table is an exception in that the change amounts to a pH value of ca. 0.15); addition of neutral salt 0.1 mol l^{-1} may change the pH value by ca. 0.1.

In the table below, the solutions are classified into general buffers (mostly in use for the last 75 years), universal buffers with a low buffering capacity but a wide pH range, and buffers for biological media with a moderate pH range but containing stable ingredients (phosphate and borate, for example, often undergo secondary reactions with biological media). An important property is often the transparency to ultraviolet light. Occasionally it is desirable to have a volatile buffer which can be readily removed[1] (examples are buffers Nos. 20 and 21), but the use of very volatile systems makes a close control of the pH essential. Most of the older pH data found in the literature relate to the Sørensen scale, and it should be noted that the values given in the following table of buffers are on the conventional pH scale.

Both stock and buffer solutions should be made up with distilled water free of CO_2. Only standard reagents should be used. If there is any doubt as to the purity or water content of solutions, their amount of substance concentration must be checked by titration. The volumes x (in ml) of stock solutions required to make up a buffer solution of the desired pH value are given in the last table.

REFERENCES

1. For a list of volatile buffers, see Michl, H., in Heftmann, E., Ed., *Chromatography,* 3rd. ed., van Nostrand Reinhold, New York, 1975, page 288; and Perrin and Dempsey, *Buffers for pH and Metal Ion Control,* Chapman & Hall, London, 1974.
2. **Clark and Lubs,** *J. Bacteriol.,* 2, 1, 1917.
3. **Sørensen, S. P.,** *Biochem. Z.,* 21, 131, 1909, and 22, 352, 1909; *Ergebn. Physiol.,* 12, 393, 1912; Walbum, L. E., *Biochem. Z.,* 107, 219, 1920.
4. **Michaelis, L.,** *J. Biol. Chem.,* 87, 33, 1930.
5. **McIlvaine, T. C.,** *J. Biol. Chjem.,* 49, 183, 1921.
6. **Teorell and Stenhagen,** *Biochem. Z.,* 299, 416, 1938.
7. **Britton and Welford,** *J. Chem. Soc.,* 1, 1848, 1937.
8. **Walpole, G. S.,** *J. Chem. Soc.,* 105, 2501, 1914.
9. **Gomori, G.,** in *Methods in Enzymology,* Vol. 1, Colowick and Kaplan, Eds., Academic Press, New York, 1955, 138.
10. **Green, A. A.,** *J. Am. Chem. Soc.,* 55, 2331, 1933.
11. **Stafford et al.,** *Biochim. Biophys. Acta,* 18, 318, 1955; Krebs, H. A., unpublished, 1957.
12. **Smith and Smith,** *Biol. Bull.,* 96, 233, 1949.
13. **Semenza et al.,** *Helv. Chim. Acta,* 45, 2306, 1962.
14. **Gomori, G.,** *Proc. Soc. Exp. Biol. (N.Y.),* 68, 354, 1948.
15. **Mertz and Owen,** *Proc. Soc. Exp. Biol. (N.Y.),* 43, 204, 1940.
16. **Beisenherz et al.,** *Z. Naturforsch.,* 8b, 555, 1953.
17. **Leonis, J. C. R.,** *Lab Carlsberg. Ser. Chim.,* 26, 357, 1948.
18. **Delory and King,** *Biochem. J.,* 39, 245, 1945.

[a] This chapter on "Buffer Solutions" has been compiled by F. Kohler, Institut für Thermo- und Fluiddynamik, Ruhr-Universität, Bochum, FRG.

BUFFER SOLUTIONS

No.	Name	Range of pH value	Temperature (°C)	ΔpH/K
General buffers				
1	KCl/HCl (Clark and Lubs)[2]	1.0—2.2	Room	0
2	Glycine/HCl (Sørensen)[3]	1.2—3.4	Room	0
3	Na citrate/HCl (Sørensen)[3]	1.2—5.0	Room	0
4	K biphthalate/HCl (Clark and Lubs)[2]	2.4—4.0	20	+0.001
5	K biphthalate/NaOH (Clark and Lubs)[2]	4.2—6.2	20	
6	Na citrate/NaOH (Sørensen)[3]	5.2—6.6	20	+0.004
7	Phosphate (Sørensen).[3]	5.0—8.0	20	−0.003
8	Barbital-Na/HCl (Michaelis)[4]	7.0—9.0	18	
9	Na borate/HCl (Sørensen)[3]	7.8—9.2	20	−0.005
10	Glycine/NaOH (Sørensen)[3]	8.6—12.8	20	−0.025
11	Na borate/NaOH (Sørensen)[3]	9.4—10.6	20	−0.01
Universal buffers				
12	Citric acid/phosphate (McIlvaine)[5]	2.2—7.8	21	
13	Citrate-phosphate-borate/HCl (Teorell and Stenhagen)[6]	2.0—12.0	20	
14	Britton-Robinson[7]	2.6—11.8	25	At low pH:0 At high pH: −0.02
Buffers for biological media				
15	Acetate (Walpole)[8-10]	3.8—5.6	25	
16	Dimethylglutaric acid/NaOH[11]	3.2—7.6	21	
17	Piperazine/HCl[12,13]	4.6—6.4	20	
		8.8—10.6		
18	Tetraethylethylenediamine[a,13]	5.0—6.8	20	
		8.2—10.0		
19	Tris maleate[9,14]	5.2—8.6	23	
20	Dimethylaminoethylamine[a,13]	5.6—7.4	20	
		8.6—10.4		
21	Imidazole/HCl[15]	6.2—7.8	25	
22	Triethanolamine/HCl[16]	7.9—8.8	25	
23	N-Dimethylaminoleucylglycine/NaOH[17]	7.0—8.8	23	−0.015
24	Tris/HCl[9]	7.2—9.0	23	−0.02
25	2-Amino-2-methylpropane-1,3-diol/HCl[9,14]	7.8—10.0	23	
26	Carbonate (Delory and King)[9,18]	9.2—10.8	20	

[a] Can be combined with tris buffer to give a cationic universal buffer (see Semenza et al.[13]).

BUFFER SOLUTIONS (continued)

pH	1	2	3	4	5	6	7	8	9	10	11	12	13	14	15	16a	16b	17	18	19	20	21	22	23	24	25	26	pH
1.0	54.2																											1.0
1.2	36.0	11.1	9.0																									1.2
1.4	23.2	26.4	17.9																									1.4
1.6	14.7	36.2	23.6																									1.6
1.8	9.3	43.9	27.6																									1.8
2.0	5.9	50.7	30.2																									2.0
2.2	3.8	56.5	32.2									98.8	74.4															2.2
2.4		62.3	34.1	41.0								94.5	68.8															2.4
2.6		68.4	36.0	34.3								90.0	64.6															2.6
2.8		74.7	37.9	27.8								85.1	63.3	1.6														2.8
3.0		81.0	39.9	21.6								80.3	58.9	3.6														3.0
3.2		86.2	42.1	15.9								76.0	56.9	5.7														3.2
3.4		90.3	44.8	10.9								72.0	55.2	7.8														3.4
3.6			47.8	6.7								68.4	53.9	9.9		7.0	14.4											3.6
3.8			51.2	3.3								65.1	52.9	11.7		13.3	20.9											3.8
4.0			55.1	0.0								62.0	51.8	13.5	10.9	20.7	26.8											4.0
4.2			60.0		3.0							59.1	50.7	15.3	16.6	26.3	32.4											4.2
4.4			66.4		6.7							56.4	49.7	17.5	23.9	32.4	36.6											4.4
4.6			74.9		11.1							53.7	48.6	19.7	33.5	36.2	40.3											4.6
4.8			85.6		16.5		99.2					51.2	47.5	21.9	44.9	39.3	43.1											4.8
5.0			100.0		22.6	87.1	98.4					49.0	46.4	24.1	56.6	41.3	45.7											5.0
5.2					28.8	78.0	97.3					46.9	45.4	26.3	67.8	43.5	48.3											5.2
5.4					34.4	70.3	95.5					44.7	44.3	28.6	76.8	45.7	51.5											5.4
5.6					39.1	64.5	92.8					42.4	43.2	31.0	84.0	48.4	53.6											5.6
5.8					42.4	60.3	88.9					40.0	42.0	33.4	89.3	51.3	58.2											5.8
6.0					45.0	57.2	83.0					37.4	40.8	35.8		55.0	63.6											6.0
6.2					46.7	54.8	75.4					34.5	39.7	38.3		58.8	68.7											6.2
6.4						53.2	65.3					31.4	38.4	40.8		63.9	73.6											6.4
6.6							53.4	53.3				27.9	37.0	43.3		69.5	78.5	94.3	94.3									6.6
6.8							41.3	55.0				23.5	35.6	45.8		74.1	83.3	91.5	91.5									6.8
7.0							29.6	57.6				19.0	34.2	48.3		83.5	87.4	87.8	87.6									7.0
7.2							19.7	60.8				13.8	32.9	50.9		87.4	91.0	83.6	83.1									7.2
7.4							12.8	65.2				9.8	31.7	53.4		90.0	93.2	77.6	77.6	3.2	94.3							7.4
7.6							7.4	70.6				6.8	30.6	55.8		91.8	94.9	71.8	71.7	5.0	91.7							7.6
7.8							3.7	75.9	53.0			4.6	29.6	58.2		93.0	95.8	66.5	66.4	7.3	88.0							7.8
8.0								81.2	55.4				28.8	60.5		93.8	96.8	61.8	61.7	9.7	83.3							8.0
8.2								86.2	58.0	94.7			28.1	62.8				58.2	58.0	12.4	77.9							8.2
8.4								90.1	62.1	92.0			27.6	65.0				55.5	55.3	15.2	72.0							8.4
8.6								93.2	66.9	88.4			27.0	67.2						17.9	66.6		86.2					8.6
8.8									73.6	84.0	87.0		26.3	69.3						20.8	61.9	43.4	80.6	86.4				8.8
9.0									83.5	78.9	75.5		25.2	71.3				45.5	46.4	22.2	58.1	40.4	73.1	80.6	44.7	43.9		9.0
9.2									95.6	73.2	65.1		24.0	73.2				43.2	43.9	23.7	55.3	36.5	62.0	72.8	42.0	41.6	10.0	9.2
9.4										67.2	59.6		22.6	75.1				40.0	40.9	25.2	45.4	31.4	52.0	63.2	39.3	38.4	18.4	9.4
9.6										62.5	56.4		21.4	77.0				35.8	36.8	26.7	42.8	25.4	41.1	52.1	33.7	34.8	29.3	9.6
9.8										58.8	54.1		20.2	78.8				30.8	31.8	28.6	39.2	19.6	31.9	41.1	27.9	30.7	42.0	9.8
10.0										55.7	52.3		19.0	80.4				25.0	26.2	31.2	34.7	14.6	22.5	31.4	22.9	17.7	53.4	10.0
10.2										53.6			18.1	81.8				19.4	20.4	33.9	29.3	10.2	16.0	23.0	17.3	13.3	63.7	10.2
10.4													17.1	83.1				14.3	15.2	36.9	23.6	6.6	11.7	15.9	13.0	9.2	73.1	10.4
10.6													16.0	85.4				10.0	10.8	39.9	19.0			10.3	8.8	5.2	81.2	10.6

BUFFER SOLUTIONS (continued)

pH	1	2	3	4	5	6	7	8	9	10	11	12	13	14	15	16a	16b	17	18	19	20	21	22	23	24	25	26	pH
10.8	—	—	—	—	—	—	—	—	—	52.2	—	—	15.5	86.5	—	—	—	—	—	—	—	—	—	—	—	—	87.9	10.8
11.0	—	—	—	—	—	—	—	—	—	51.2	—	—	14.7	87.8	—	—	—	—	—	—	—	—	—	—	—	—	—	11.0
11.2	—	—	—	—	—	—	—	—	—	50.4	—	—	13.5	89.3	—	—	—	—	—	—	—	—	—	—	—	—	—	11.2
11.4	—	—	—	—	—	—	—	—	—	49.5	—	—	11.7	91.3	—	—	—	—	—	—	—	—	—	—	—	—	—	11.4
11.6	—	—	—	—	—	—	—	—	—	48.7	—	—	9.1	94.5	—	—	—	—	—	—	—	—	—	—	—	—	—	11.6
11.8	—	—	—	—	—	—	—	—	—	47.6	—	—	5.5	99.0	—	—	—	—	—	—	—	—	—	—	—	—	—	11.8
12.0	—	—	—	—	—	—	—	—	—	46.0	—	—	1.3	—	—	—	—	—	—	—	—	—	—	—	—	—	—	12.0
12.2	—	—	—	—	—	—	—	—	—	43.2	—	—	—	—	—	—	—	—	—	—	—	—	—	—	—	—	—	12.2
12.4	—	—	—	—	—	—	—	—	—	39.1	—	—	—	—	—	—	—	—	—	—	—	—	—	—	—	—	—	12.4
12.6	—	—	—	—	—	—	—	—	—	31.8	—	—	—	—	—	—	—	—	—	—	—	—	—	—	—	—	—	12.6
12.8	—	—	—	—	—	—	—	—	—	21.4	—	—	—	—	—	—	—	—	—	—	—	—	—	—	—	—	—	12.8

Note: The table gives the volumes *x* (in ml) of the stock solutions listed that are required to make up a buffer solution of the desired pH value.

BUFFER SOLUTIONS (continued)

Stock solutions and their amount of substance concentrations or mass and/or volume contents of the solutes

No.	A	B	Composition of the buffer
1	KCl 0.2 mol l^{-1} (14.91 g l^{-1})	HCl 0.2 mol l^{-1}	25 ml A + x ml B made up to 100 ml
2	Glycine 0.1 mol l^{-1} + NaCl 0.1 mol l^{-1} (1 l solution contains 7.507 g glycine + 5.844 g NaCl)	HCl 0.1 mol l^{-1}	x ml A + $(100 - x)$ ml B
3	Disodium citrate 0.1 mol l^{-1} (1 l solution contains 21.01 g citric acid monohydrate + 200 ml NaOH 1 mol l^{-1})	HCl 0.1 mol l^{-1}	x ml A + $(100 - x)$ ml B
4	Potassium biphthalate 0.1 mol l^{-1} (20.42 g l^{-1})	HCl 0.1 mol l^{-1}	50 ml A + x ml B made up to 100 ml
5	As No. 4	NaOH 0.1 mol l^{-1}	50 ml A + x ml B made up to 100 ml
6	As No. 3	NaOH 0.1 mol l^{-1}	x ml A + $(100 - x)$ ml B
7	Potassium dihydrogen phosphate $^1/_{15}$ mol l^{-1} (9.073 g l^{-1})	Disodium phosphate $^1/_{15}$ mol l^{-1} (Na$_2$HPO$_4$·2 H$_2$O, 11.87 g l^{-1})	x ml A + $(100 - x)$ ml B
8	Barbital-Na 0.1 mol l^{-1} (20.62 g l^{-1})	HCl 0.1 l^{-1}	x ml A + $(100 - x)$ ml B
9	Boric acid, half-neutralized, 0.2 mol l^{-1} (corresponds to 0.05 mol l^{-1} borax solution; 1 l solution contains 12.37 g boric acid + 100 ml NaOH 1 mol l^{-1})	HCl 0.1 mol l^{-1}	x ml A + $(100 - x)$ ml B
10	As No. 2	NaOH 0.1 mol l^{-1}	x ml A + $(100 - x)$ ml B
11	As No. 9	NaOH 0.1 mol l^{-1}	x ml A + $(100 - x)$ ml B
12	Citric acid 0.1 mol l^{-1} (citric acid monohydrate 21.01 g l^{-1})	Disodium phosphate 0.2 mol l^{-1} (Na$_2$HPO$_4$·2 H$_2$O, 35.60 g l^{-1})	x ml A + $(100 - x)$ ml B
13	To 100 ml citric acid and 100 ml phosphoric acid solution, each equivalent to 100 ml NaOH 1 mol l^{-1}, add 3.54 g boric acid and 343 ml NaOH 1 mol l^{-1} and make up to 1 l of solution	HCl 0.1 l^{-1}	20 ml A + x ml B made up to 100 l
14	Citric acid, potassium hydrogen phosphate, barbital, and boric acid, all 0.02857 mol l^{-1} (1 l solution contains 6.004 g citric acid monohydrate, 3.888 g potassium hydrogen phosphate, 5.263 g barbital, 1.767 g boric acid)	NaOH 0.2 mol l^{-1}	100 ml A + x ml B
15	Sodium acetate 0.1 mol l^{-1} (1 l solution contains 8.204 g C$_2$H$_3$O$_2$Na or 13.61 g C$_2$H$_3$O$_2$Na·3 H$_2$O)	Acetic acid 0.1 mol l^{-1} (6.005 g l^{-1})	x ml A + $(100 - x)$ ml B
16a	Dimethylglutaric acid 0.1 mol l^{-1} (16.02 g l^{-1})	NaOH 0.2 mol l^{-1}	(a) 100 ml A + x ml B made up to 1000 ml
16b	Dimethylglutaric acid 0.1 mol l^{-1} (16.02 g l^{-1})	NaOH 0.2 mol l^{-1}	(b) 100 ml A + x ml B + 5.844 g NaCl made up to 1000 ml NaCl - 0.1 mol l^{-1}

BUFFER SOLUTIONS (continued)

No.	A	B	Composition of the buffer
17	Piperazine 1 mol l^{-1} (86.14 g l^{-1})	HCl 0.1 mol l^{-1}	5 ml A + x ml B made up to 100 ml
18	Tetraethylethylenediamine 1 mol l^{-1} (172.32 g l^{-1})	HCl 0.1 mol l^{-1}	5 ml A + x ml B made up to 100 ml
19	Tris acid maleate 0.2 mol l^{-1} [1 l solution contains 24.23 g tris(hydroxymethyl)aminomethane + 23.21 g maleic acid or 19.61 g maleic anhydride]	NaOH 0.2 mol l^{-1}	25 ml A + x ml B made up to 100 ml
20	Dimethylaminoethylamine 1 mol l^{-1} (88 g l^{-1})	HCl 0.1 mol l^{-1}	5 ml A + x ml B made up to 100 ml
21	Imidazole 0.2 mol l^{-1} (13.62 g l^{-1})	HCl 0.1 mol l^{-1}	25 ml A + x ml B made up to 100 ml
22	Triethanolamine 0.5 mol l^{-1} + ethylenediamine-tetraacetic acid disodium salt (1 l solution contains 74.60 g $C_6H_{15}O_3N$ + 20 g $C_{10}H_{14}O_8N_2 \cdot 2\ H_2O$)	HCl 0.05 mol l^{-1}	10 ml A + x ml B made up to 100 ml
23	N-Dimethylaminoleucylglycine 0.1 mol l^{-1} + NaCl 0.2 mol l^{-1} (1 l solution contains 24.33 g $C_{10}H_{20}O_3N_2 \cdot \frac{1}{2}\ H_2O$ + 11.69 g NaCl)	NaOH 1 mol l^{-1} 100 ml made up to 1 l with solution A	x ml A + $(100 - x)$ ml B
24	Tris 0.2 mol l^{-1} [tris(hydroxymethyl)aminomethane 24.23 g l^{-1}]	HCl 0.1 mol l^{-1}	25 ml A + x ml B made up to 100 ml
25	2-Amino-2-methylpropane-1,3-diol 0.1 mol l^{-1} (10.51 g l^{-1})	HCl 0.1 mol l^{-1}	50 ml A + x ml B made up to 100 ml
26	Sodium carbonate anhydrous 0.1 mol l^{-1} (10.60g l^{-1})	Sodium bicarbonate 0.1 mol l^{-1} (8.401 g l^{-1})	x ml A + $(100 - x)$ ml B

Note: When not otherwise specified, both stock and buffer solutions should be made up with distilled water free of CO_2. Only standard reagents should be used. If there is any doubt as to the purity or water content of solutions, their amount of substance concentration must be checked by titration. The volumes x (in ml) of stock solutions required to make up a buffer solution of the desired pH value are given in the table on the next page.

From Lenter, C., Ed., *Geigy Scientific Tables*, 8th ed., volume 3, Ciba-Geigy, Basel, 1984, pages 58—60. With permission.

AMINE BUFFERS USEFUL FOR BIOLOGICAL RESEARCH

Norman Good

All of these amines are highly polar, water-soluble substances. Their advantages and disadvantages must be determined empirically for each biological reaction system. For best buffering performance they should be used at pH's close to the pKa, preferably within ±0.5 pH units of the pKa and never more than ±1.0 unit from the pKa. Note that the pKa's, and therefore the pH's of buffered solutions, change with temperature in the manner indicated.

AMINE BUFFERS USEFUL FOR BIOLOGICAL RESEARCH

Chemical name	Trivial name or acronym	Structure	pKa at 20°C	ΔpKa/°C
2-(N-Morpholino)ethanesulfonic acid	MES		6.15	−0.011
Bis(2-Hydroxyethyl)imino-tris-(hydroxymethyl)methane	Bistris	$(HOCH_2CH_2)_2=N-C\equiv(CH_2OH)_3$	6.5	—
N-(2-Acetamido)iminodiacetic acid	ADA[a]		6.6	−0.011
Piperazine-N,N'-bis(2-ethanesulfonic acid)	PIPES		6.8	−0.0085
1,3-Bis[tris(hydroxymethyl)-methylamino]propane	Bistrispropane	$(HOCH_2)_3=C-NH(CH_2)_3NH-C\equiv(CH_2OH)_3$	6.8 (9.0)	—
N-(Acetamido)-2-aminoethanesulfonic acid	ACES	$H_2NCOCH_2\ N^+H_2CH_2CH_2SO_3^-$	6.9	−0.020
3-(N-Morpholino)propanesulfonic acid	MOPS		7.15	−0.013
N,N'-Bis(2-Hydroxyethyl)-2-amino-ethanesulfonic acid	BES	$(HOCH_2CH_2)_2=N^+HCH_2CH_2SO_3^-$	7.15	−0.016
N-Tris(hydroxymethyl)methyl-2-amino-ethanesulfonic acid	TES	$(HOCH_2)_3=C-N^+H_2CH_2CH_2SO_3^-$	7.5	−0.020
N-2-Hydroxyethylpiperazine-N'-ethanesulfonic acid	HEPES[b]		7.55	−0.014
N-2-Hydroxyethylpiperazine-N'-propanesulfonic acid	HEPPS[b]		8.1	−0.015

[a]These substances may bind certain di- and polyvalent cations and therefore they may sometimes be useful for providing constant, low level concentrations of free heavy metal ions (heavy metal buffering).
[b]These substances interfere with and preclude the Folin protein assay.

AMINE BUFFERS USEFUL FOR BIOLOGICAL RESEARCH (continued)

Chemical name	Trivial name or acronym	Structure	pKa at 20°C	ΔpKa/°C
N-Tris(hydroxymethyl)methylglycine	Tricine[a]	$(HOCH_2)_3 \equiv C - N^+H_2CH_2COO^-$	8.15	-0.021
Tris(hydroxymethyl)aminomethane	Tris	$(HOCH_2)_3 \equiv CNH_2$	8.3	-0.031
N,N-Bis(2-hydroxyethyl)glycine	Bicine[a]	$(HOCH_2CH_2)_2 = N^+HCH_2COO^-$	8.35	-0.018
Glycylglycine	Glycylglycine[a]	$H_3N^+CH_2CONHCH_2COO^-$	8.4	-0.028
N-Tris(hydroxymethyl)methyl-3-aminopropanesulfonic acid	TAPS	$(HOCH_2)_3 \equiv C - N^+H_2(CH_2)_3SO_3^-$	8.55	-0.027
1,3-Bis[tris(hydroxymethyl)-methylamino]propane	Bistrispropane	$(HOCH_2)_3 \equiv C - NH(CH_2)_3NH - C \equiv (CH_2OH)_3$	9.0 (6.8)	—
Glycine	Glycine[a]	$H_3N^+CH_2COO^-$	9.9	—

Compiled by Norman Good.

For further information on these and other buffers, see Good and Izawa, in *Methods in Enzymology*, *Part B*, Vol. 24, Pietro, Ed., Academic Press, New York, 1972, 53.

PREPARATION OF BUFFERS FOR USE IN ENZYME STUDIES*

G. Gomori

The buffers described in this section are suitable for use either in enzymatic or histochemical studies. The accuracy of the tables is within ±0.05 pH at 23°. In most cases the pH values will not be off by more than ±0.12 pH even at 37° and at molarities slightly different from those given (usually 0.05 M).

The methods of preparation described are not necessarily identical with those of the original authors. The titration curves of the majority of the buffers recommended have been redetermined by the writer. The buffers are arranged in the order of ascending pH range. For more complete data on phosphate and acetate buffers over a wide range of concentrations, see Vol. I [10].*

*From Gomori, in *Methods in Enzymology,* Vol. 1, Colowick and Kaplan, Eds., Academic Press, New York, 1955, 138. With permission.

Table 1
HYDROCHLORIC ACID-POTASSIUM CHLORIDE BUFFER*

x	pH
97.0	1.0
78.0	1.1
64.5	1.2
51.0	1.3
41.5	1.4
33.3	1.5
26.3	1.6
20.6	1.7
16.6	1.8
13.2	1.9
10.6	2.0
8.4	2.1
6.7	2.2

*Stock solutions

A:0.2 M solution of KCl (14.91 g in 1,000 ml)
B:0.2 M HCl 50 ml of A + x ml of B, diluted to a total of 200 ml

REFERENCE

1. **Clark and Lubs,** *J. Bacteriol.,* 2, 1 (1917).

Table 2
GLYCINE-HCl BUFFER*

x	pH	x	pH
5.0	3.6	16.8	2.8
6.4	3.4	24.2	2.6
8.2	3.2	32.4	2.4
11.4	3.0	44.0	2.2

*Stock solutions

A:0.2 M solution of glycine (15.01 g in 1,000 ml)
B:0.2 M HCl 50 ml of A + x ml of B, diluted to a total of 200 ml

REFERENCE

1. **Sørensen,** *Biochem. Z.,* 21, 131 (1909); 22, 352 (1909).

Table 3
PHTHALATE-HYDROCHLORIC ACID BUFFER*

x	pH	x	pH
46.7	2.2	14.7	3.2
39.6	2.4	9.9	3.4
33.0	2.6	6.0	3.6
26.4	2.8	2.63	3.8
20.3	3.0		

*Stock solutions

A:0.2 *M* solution of potassium acid phthalate (40.84 g in 1,000 ml)
B:0.2 *M* HCl 50 ml of A + *x* ml of B, diluted to a total of 200 ml

REFERENCE

1. **Clark** and **Lubs,** *J. Bacteriol.*, 2, 1 (1917).

Table 4
ACONITATE BUFFER*

x	pH	x	pH
15.0	2.5	83.0	4.3
21.0	2.7	90.0	4.5
28.0	2.9	97.0	4.7
36.0	3.1	103.0	4.9
44.0	3.3	108.0	5.1
52.0	3.5	113.0	5.3
60.0	3.7	119.0	5.5
68.0	3.9	126.0	5.7
76.0	4.1		

*Stock solutions

A:0.5 *M* solution of aconitic acid (87.05 g in 1,000 ml)
B:0.2 *M* NaOH 20 ml of A + *x* ml of B, diluted to a total of 200 ml

REFERENCE

1. **Gomori,** unpublished data.

Table 5
CITRATE BUFFER*

x	y	pH
46.5	3.5	3.0
43.7	6.3	3.2
40.0	10.0	3.4
37.0	13.0	3.6
35.0	15.0	3.8
33.0	17.0	4.0
31.5	18.5	4.2
28.0	22.0	4.4
25.5	24.5	4.6
23.0	27.0	4.8
20.5	29.5	5.0
18.0	32.0	5.2
16.0	34.0	5.4
13.7	36.3	5.6
11.8	38.2	5.8
9.5	41.5	6.0
7.2	42.8	6.2

*Stock solutions

A:0.1 *M* solution of citric acid (21.01 g in 1,000 ml)
B:0.1 *M* solution of sodium citrate (29.41 g $C_6H_5O_7Na_3 \cdot 2H_2O$ in 1,000 ml; the use of the salt with 5½ H_2O is not recommended). *x* ml of A + *y* ml of B, diluted to a total of 100 ml

REFERENCE

1. **Lillie,** *Histopathologic Technique,* Blakiston, Philadelphia and Toronto, 1948.

Table 6
ACETATE BUFFER*

x	y	pH
46.3	3.7	3.6
44.0	6.0	3.8
41.0	9.0	4.0
36.8	13.2	4.2
30.5	19.5	4.4
25.5	24.5	4.6
20.0	30.0	4.8
14.8	35.2	5.0
10.5	39.5	5.2
8.8	41.2	5.4
4.8	45.2	5.6

*Stock solutions

A:0.2 M solution of acetic acid (11.55 ml in 1,000 ml)

B:0.2 M solution of sodium acetate (16.4 g of $C_2H_3O_2Na$ or 27.2 g of $C_2H_3O_2Na \cdot 3H_2O$ in 1,000 ml) x ml of A + y ml of B, diluted to a total of 100 ml

REFERENCE

1. **Walpole,** *J. Chem. Soc.,* 105, 2501 (1914).

Table 7
CITRATE-PHOSPHATE BUFFER*

x	y	pH
44.6	5.4	2.6
42.2	7.8	2.8
39.8	10.2	3.0
37.7	12.3	3.2
35.9	14.1	3.4
33.9	16.1	3.6
32.3	17.7	3.8
30.7	19.3	4.0
29.4	20.6	4.2
27.8	22.2	4.4
26.7	23.3	4.6
25.2	24.8	4.8
24.3	25.7	5.0
23.3	26.7	5.2
22.2	27.8	5.4
21.0	29.0	5.6
19.7	30.3	5.8
17.9	32.1	6.0
16.9	33.1	6.2
15.4	34.6	6.4
13.6	36.4	6.6
9.1	40.9	6.8
6.5	43.6	7.0

*Stock solutions

A:0.1 M solution of citric acid (19.21 g in 1,000 ml)

B:0.2 M solution of dibasic sodium phosphate (53.65 g of $Na_2\text{-}HPO_4 \cdot 7H_2O$ or 71.7 g of $Na_2HPO_4 \cdot 12H_2O$ in 1,000 ml) x ml of A + y ml of B, diluted to a total of 100 ml

REFERENCE

1. **McIlvaine,** *J. Biol. Chem.,* 49, 183 (1921).

Table 8
SUCCINATE BUFFER*

x	pH	x	pH
7.5	3.8	26.7	5.0
10.0	4.0	30.3	5.2
13.3	4.2	34.2	5.4
16.7	4.4	37.5	5.6
20.0	4.6	40.7	5.8
23.5	4.8	43.5	6.0

*Stock solutions

A:0.2 *M* solution of succinic acid (23.6 g in 1,000 ml)
B:0.2 *M* NaOH 25 ml of A + *x* ml of B, diluted to a total of 100 ml

REFERENCE

1. **Gomori,** unpublished data.

Table 10
MALEATE BUFFER*

x	pH	x	pH
7.2	5.2	33.0	6.2
10.5	5.4	38.0	6.4
15.3	5.6	41.6	6.6
20.8	5.8	44.4	6.8
26.9	6.0		

*Stock solutions

A:0.2 *M* solution of acid sodium maleate (8 g of NaOH + 23.2 g of maleic acid or 19.6 g of maleic anhydride in 1,000 ml)
B:0.2 *M* NaOH 50 ml of A + *x* ml of B, diluted to a total of 200 ml

REFERENCE

1. **Temple,** *J. Am. Chem. Soc.,* 51, 1754 (1929).

Table 9
PHTHALATE-SODIUM HYDROXIDE BUFFER*

x	pH	x	pH
3.7	4.2	30.0	5.2
7.5	4.4	35.5	5.4
12.2	4.6	39.8	5.6
17.7	4.8	43.0	5.8
23.9	5.0	45.5	6.0

*Stock solutions

A:0.2 *M* solution of potassium acid phthalate (40.84 g in 100 ml)
B:0.2 *M* NaOH 50 ml of A + *x* ml of B, diluted to a total of 200 ml

REFERENCE

1. **Clark and Lubs,** *J. Bacteriol.,* 2, 1 (1917).

Table 11
CACODYLATE BUFFER*

x	pH	x	pH
2.7	7.4	29.6	6.0
4.2	7.2	34.8	5.8
6.3	7.0	39.2	5.6
9.3	6.8	43.0	5.4
13.3	6.6	45.0	5.2
18.3	6.4	47.0	5.0
23.8	6.2		

*Stock solutions

A:0.2 *M* solution of sodium cacodylate (42.8 g of $Na(CH_3)_2 AsO_2 \cdot 3H_2O$ in 1,000 ml)
B:0.2 *M* HCl 50 ml of A + *x* ml of B, diluted to a total of 200 ml

REFERENCE

1. **Plumel,** *Bull. Soc. Chim. Biol.,* 30, 129 (1949).

Table 12
PHOSPHATE BUFFER*

x	y	pH	x	y	pH
93.5	6.5	5.7	45.0	55.0	6.9
92.0	8.0	5.8	39.0	61.0	7.0
90.0	10.0	5.9	33.0	67.0	7.1
87.7	12.3	6.0	28.0	72.0	7.2
85.0	15.0	6.1	23.0	77.0	7.3
81.5	18.5	6.2	19.0	81.0	7.4
77.5	22.5	6.3	16.0	84.0	7.5
73.5	26.5	6.4	13.0	87.0	7.6
68.5	31.5	6.5	10.5	90.5	7.7
62.5	37.5	6.6	8.5	91.5	7.8
56.5	43.5	6.7	7.0	93.0	7.9
51.0	49.0	6.8	5.3	94.7	8.0

*Stock solutions

A:0.2 *M* solution of monobasic sodium phosphate (27.8 g in 1,000 ml)

B:0.2 *M* solution of dibasic sodium phosphate (53.65 g of $Na_2HPO_4 \cdot 7H_2O$ or 71.7 g of $Na_2HPO_4 \cdot 12H_2O$ in 1,000 ml) x ml of A + y ml of B, diluted to a total of 200 ml

REFERENCE

1. Sφrensen, *Biochem. Z.*, 21, 131 (1909); 22, 352 (1909).

Table 13
TRIS(HYDROXYMETHYL)-
AMINO METHANE-MALEATE
(TRIS-MALEATE) BUFFER*[†]

x	pH	x	pH
7.0	5.2	48.0	7.0
10.8	5.4	51.0	7.2
15.5	5.6	54.0	7.4
20.5	5.8	58.0	7.6
26.0	6.0	63.5	7.8
31.5	6.2	69.0	8.0
37.0	6.4	75.0	8.2
42.5	6.6	81.0	8.4
45.0	6.8	86.5	8.6

*Stock solutions

A:0.2 *M* solution of Tris acid maleate (24.2 g. of tris(hydroxy-methyl)aminomethane + 23.2 g of maleic acid or 19.6 g of maleic anhydride in 1,000 ml)

B:0.2 *M* NaOH 50 ml of A + x ml of B, diluted to a total of 200 ml

[†] A buffer-grade Tris can be obtained from the Sigma Chemical Co., St. Louis, MO., or From Matheson Coleman & Bell, East Rutherford, NJ.

REFERENCE

1. **Gomori**, *Proc. Soc. Exp. Biol. Med.*, 68, 354 (1948).

Table 14
BARBITOL BUFFER*†

x	pH
1.5	9.2
2.5	9.0
4.0	8.8
6.0	8.6
9.0	8.4
12.7	8.2
17.5	8.0
22.5	7.8
27.5	7.6
32.5	7.4
39.0	7.2
43.0	7.0
45.0	6.8

*Stock solutions

A:0.2 *M* solution of sodium barbital (veronal) (41.2 g in 1,000 ml)
B:0.2 *M* HCl 50 ml of A + *x* ml of B, diluted to a total of 200 ml

†Solutions more concentrated than 0.05 *M* may crystallize on standing, especially in the cold.

REFERENCE

1. **Michaelis**, *J. Biol. Chem.*, 87, 33 (1930).

Table 15
TRIS(HYDROXYMETHYL)-AMINOMETHANE (TRIS) BUFFER*†

x	pH
5.0	9.0
8.1	8.8
12.2	8.6
16.5	8.4
21.9	8.2
26.8	8.0
32.5	7.8
38.4	7.6
41.4	7.4
44.2	7.2

*Stock solutions

A:0.2 *M* solution of tris(hydroxymethyl)aminomethane (24.2 g in 1,000 ml)
B:0.2 *M* HCl 50 ml of A + *x* ml of B, diluted to a total of 200 ml

†A buffer-grade Tris can be obtained from the Sigma Chemical Co., St. Louis, MO., or from Matheson Coleman & Bell, East Rutherford, NJ.

Table 16
BORIC ACID-BORAX BUFFER*

x	pH	x	pH
2.0	7.6	22.5	8.7
3.1	7.8	30.0	8.8
4.9	8.0	42.5	8.9
7.3	8.2	59.0	9.0
11.5	8.4	83.0	9.1
17.5	8.6	115.0	9.2

*Stock solutions

A:0.2 *M* solution of boric acid (12.4 g in 1,000 ml)
B:0.05 *M* solution of borax (19.05 g in 1,000 ml; 0.2 *M* in terms of sodium borate) 50 ml of A + *x* ml of B, diluted to a total of 200 ml

REFERENCE

1. **Holmes,** *Anat. Rec.,* 86, 163 (1943).

Table 17
2-AMINO-2-METHYL-1,3-PROPANEDIOL (AMMEDIOL) BUFFER*

x	pH	x	pH
2.0	10.0	22.0	8.8
3.7	9.8	29.5	8.6
5.7	9.6	34.0	8.4
8.5	9.4	37.7	8.2
12.5	9.2	41.0	8.0
16.7	9.0	43.5	7.8

*Stock solutions

A:0.2 *M* solution of 2-amino-2-methyl-1,3-propanediol (21.03 g in 1,000 ml)
B:0.2 *M* HCl 50 ml of A + *x* ml of B, diluted to a total of 200 ml

REFERENCE

1. **Gomori,** *Proc. Soc. Exp. Biol. Med.,* 62, 33 (1946).

Table 18
GLYCINE-NaOH BUFFER*

x	pH	x	pH
4.0	8.6	22.4	9.6
6.0	8.8	27.2	9.8
8.8	9.0	32.0	10.0
12.0	9.2	38.6	10.4
16.8	9.4	45.5	10.6

*Stock solutions

A:0.2 *M* solution of glycine (15.01 g in 1,000 ml)
B:0.2 *M* NaOH 50 ml of A + *x* ml of B, diluted to a total of 200 ml.

REFERENCE

1. **Sφrensen,** *Biochem. Z.,* 21, 131 (1909); 22, 352 (1909).

Table 19
BORAX-NaOH BUFFER*

x	pH
0.0	9.28
7.0	9.35
11.0	9.4
17.6	9.5
23.0	9.6
29.0	9.7
34.0	9.8
38.6	9.9
43.0	10.0
46.0	10.1

*Stock solutions

A:0.05 *M* solution of borax (19.05 g in 1,000 ml; 0.02 *M* in terms of sodium borate)
B:0.2 *M* NaOH 50 ml of A + *x* ml of B, diluted to a total of 200 ml

REFERENCE

1. **Clark and Lubs,** *J. Bacteriol.,* 2, 1 (1917).

Table 20
CARBONATE-BICARBONATE
BUFFER*

x	y	pH
4.0	46.0	9.2
7.5	42.5	9.3
9.5	40.5	9.4
13.0	37.0	9.5
16.0	34.0	9.6
19.5	30.5	9.7
22.0	28.0	9.8
25.0	25.0	9.9
27.5	22.5	10.0
30.0	20.0	10.1
33.0	17.0	10.2
35.5	14.5	10.3
38.5	11.5	10.4
40.5	9.5	10.5
42.5	7.5	10.6
45.0	5.0	10.7

*Stock solutions

A:0.2 M solution of anhydrous sodium carbonate (21.2 g in 1,000 ml)
B:0.2 M solution of sodium bicarbonate (16.8 g in 1,000 ml) x ml of A + y ml of B, diluted to a total of 200

REFERENCE

1. **Delory and King,** *Biochem. J.,* 39, 245 (1945).

INDICATORS FOR VOLUMETRIC WORK AND pH DETERMINATIONS

Indicator		Acid color	pH range	Basic color	Preparation
Methyl violet 6B	Tetra and pentamethylated p-rosaniline hydrochloride	Y	0.1–1.5	B	pH: 0.25% water
Metacresol purple (acid range)	m-Cresolsulfonphthalein	R	0.5–2.5	Y	pH: 0.10 g. in 13.6 ml. 0.02 N NaOH, diluted to 250 ml. with water
Metanil yellow	4-Phenylamino-azobenzene-3'-sulfonic acid	R	1.2–2.3	Y	pH: 0.25% in ethanol
p-Xylenol blue (acid range)	1,4-Dimethyl-5-hydroxybenzene-sulfonphthalein	R	1.2–2.8	Y	pH: 0.04% in ethanol
Thymol blue (acid range)	Thymolsulfonphthalein	R	1.2–2.8	Y	pH: 0.1 g. in 10.75 ml. 0.02 N NaOH, diluted to 250 ml. with water
Tropaeolin OO	Sodium p-diphenylaminoazo-benzenesulfonate	R	1.4–2.6	Y	pH: 0.1% in water Vol.: 1% in water
Quinaldine red	2-(p-Dimethylaminostyryl)-quinoline ethiodide	C	1.4–3.2	R	Vol.: 0.1% in ethanol
Benzopurpurine 4B	Ditolyl-diazo-bis-α-naphthyl-amine-4-sulfonic acid	B-V	1.3–4.0	R	pH, vol.: 0.1% in water
Methyl violet 6B	Tetra and pentamethylated p-rosaniline hydrochloride	B	1.5–3.2	V	pH, vol.: 0.25% in water
2,4-Dinitrophenol		C	2.6–4.0	Y	pH, vol.: 0.1 g. in 5 ml. ethanol, diluted to 100 ml. with water
Methyl yellow	p-Dimethylaminoazobenzene	R	2.9–4.0	Y	pH, vol.: 0.05% in ethanol
Bromphenol blue	Tetrabromophenolsulfon-phthalein	Y	3.0–4.6	B	pH: 0.1 g. in 7.45 ml. 0.02 N NaOH, diluted to 250 ml. with water
Tetrabromophenol blue	Tetrabromophenol-tetrabromo-sulfonphthalein	Y	3.0–4.6	B	pH: 0.1 g. in 5.00 ml. 0.02 N NaOH, diluted to 250 ml. with water
Direct purple	Disodium 4,4'-bis(2-amino-1-naphthylazo)-2,2'-stilbene-disulfonate	B-P	3.0–4.6	R	Vol.: 0.1 g. in 7.35 ml. 0.02 N NaOH, diluted to 100 ml. with water
Congo red	Diphenyl-diazo-bis-1-naphthyl-amine-4-sodium sulfonate	B	3.0–5.2	R	pH: 0.1% in water
Methyl orange	4'-Dimethylaminoazobenzene-4-sodium sulfonate	R	3.1–4.4	Y	Vol.: 0.1% in water
Brom-chlorphenol blue	Dibromodichlorophenolsulfon-phthalein	Y	3.2–4.8	B	pH: 0.1 g. in 8.6 ml. 0.02 N NaOH, diluted to 250 ml. with water Vol.: 0.04% in ethanol
p-Ethoxychrysoidine	4'-Ethoxy-2,4-diaminoazo-benzene	R	3.5–5.5	Y	Vol.: 0.1% in ethanol
α-Naphthyl red		R	3.7–5.0	Y	Vol.: 0.1% in ethanol
Sodium alizarinsulfonate	Dihydroxyanthraquinone sodium sulfonate	Y	3.7–5.2	V	pH, vol.: 1% in water
Bromcresol green	Tetrabromo-m-cresolsulfon-phthalein	Y	3.8–5.4	B	pH: 0.10 g. in 7.15 ml. 0.02 N NaOH, diluted to 250 ml. with water
2,5-Dinitrophenol		C	4.0–5.8	Y	pH, vol.: 0.10 g. in 20 ml. ethanol, then dilute to 100 ml. with water
Methyl red	4'-Dimethylaminoazobenzene-2-carboxylic acid	R	4.2–6.2	Y	pH: 0.10 g. in 18.6 ml. 0.02 N NaOH, diluted to 250 ml. with water Vol.: 0.1% in ethanol
Lacmoid		R	4.4–6.2	B	Vol.: 0.5% in ethanol
Azolitmin		R	4.5–8.3	B	Vol.: 0.5% in water
Litmus		R	4.5–8.3	B	Vol.: 0.5% in water

Note: The indicator colors are abbreviated as follows: B, blue; Br, brown; C, colorless; G, green; L, lilac; O, orange; P, pink; Pu, purple; R, red; V, violet; and Y, yellow.

INDICATORS FOR VOLUMETRIC WORK AND pH DETERMINATIONS (continued)

Indicator	Chemical name	Acid color	pH range	Basic color	Preparation
Cochineal	Complex hydroxyanthraquinone derivative	R	4.8–6.2	V	Vol.: Triturate 1 g. with 20 ml. ethanol and 60 ml. water, let stand 4 days, and filter
Hematoxylin		Y	5.0–6.0	V	Vol.: 0.5 % in ethanol.
Chlorphenol red	Dichlorophenolsulfonphthalein	Y	5.0–6.6	R	pH: 0.1 g. in 11.8 ml. 0.02 N NaOH, diluted to 250 ml. with water Vol.: 0.04 % in ethanol
Bromcresol purple	Dibromo-*o*-cresolsulfonphthalein	Y	5.2–6.8	Pu	pH: 0.1 g. in 9.25 ml. 0.02 N NaOH, diluted to 250 ml. with water Vol.: 0.02 % in ethanol
Bromphenol red	Dibromophenolsulfonphthalein	Y	5.2–7.0	R	pH: 0.1 g. in 9.75 ml. 0.02 N NaOH, diluted to 250 ml. with water Vol.: 0.04 % in ethanol
Alizarin	1.2-Dihydroxyanthraquinone	Y	5.5–6.8	R	Vol.: 0.1 % in ethanol
Dibromophenoltetrabromo-phenolsulfonphthalein		Y	5.6–7.2	Pu	pH: 0.1 g. in 1.21 ml. 0.1 N NaOH, diluted to 250 ml. with water
p-Nitrophenol		C	5.6–7.6	Y	pH, vol.: 0.25 % in water
Bromothymol blue	Dibromothymolsulfonphthalein	Y	6.0–7.6	B	pH: 0.1 g. in 8 ml. 0.02 N NaOH, diluted to 250 ml. with water Vol.: 0.1 % in 50 % ethanol
Indo-oxine	5,8-Quinolinequinone-8-hydroxy-5-quinolyl-5-imide	R	6.0–8.0	B	Vol.: 0.05 % in ethanol
Cucumin		Y	6.0–8.0	Br-R	Vol: saturated aq. soln.
Quinoline blue	Cyanine	C	6.6–8.6	B	Vol.: 1 % in ethanol
Phenol red	Phenolsulfonphthalein	Y	6.8–8.4	R	pH: 0.1 g. in 14.20 ml. 0.02 N NaOH, diluted to 250 ml. with water Vol.: 0.1 % in ethanol
Neutral red	2-Methyl-3-amino-6-dimethyl-aminophenazine	R	6.8–8.0	Y	pH, vol.: 0.1 g. in 70 ml. ethanol, diluted to 100 ml. with water
Rosolic acid aurin; corallin		Y	6.8–8.2	R	pH, vol.: 1 % in 50 % ethanol
Cresol red	*o*-Cresolsulfonphthalein	Y	7.2–8.8	R	pH: 0.1 g. in 13.1 ml. 0.02 N NaOH, diluted to 250 ml. with water Vol.: 0.1 % in ethanol
α-Naphtholphthalein		P	7.3–8.7	G	pH, vol.: 0.1 % in 50 % ethanol
Metacresol purple (alkaline range)	*m*-Cresolsulfonphthalein	Y	7.4–9.0	P	pH: 0.1 g. in 13.1 ml. 0.02 N NaOH, diluted to 250 ml. with water Vol.: 0.1 % in ethanol
Ethylbis-2,4-dinitrophenylacetate		C	7.5–9.1	B	Vol.: saturated soln. in equal volumes of acetone and ethanol
Tropaeolin OOO No. 1	Sodium α-naphtholazobenzene-sulfonate	Y	7.6–8.9	R	Vol.: 0.1 % in water
Thymol blue (alkaline range)	Thymolsulfonphthalein	Y	8.0–9.6	B	pH: 0.1 g. in 10.75 ml. 0.02 N NaOH, diluted to 250 ml. with water Vol.: 0.1 % in ethanol
p-Xylenol blue	1,4-Dimethyl-5-hydroxybenzene-sulfonphthalein	Y	8.0–9.6	B	pH, vol.: 0.04 % in ethanol
o-Cresolphthalein		C	8.2–9.8	R	pH, vol.: 0.04 % in ethanol
α-Naphtholbenzein		Y	8.5–9.8	G	pH, vol.: 1 % in ethanol

INDICATORS FOR VOLUMETRIC WORK AND pH DETERMINATIONS (continued)

Indicator	Chemical name	Acid color	pH range	Basic color	Preparation
Phenolphthalein	3,3-Bis(*p*-hydroxyphenyl)-phthalide	C	8.2–10	R	Vol.: 1% in ethanol
Thymolphthalein		C	9.3–10.5	B	pH, vol.: 0.1% in ethanol
Nile blue A	Aminonaphthodiethylamino-phenoxazine sulfate	B	10–11	P	Vol.: 0.1% in water
Alizarin yellow GG	3-Carboxy-4-hydroxy-3′-nitro-azobenzene	Y	10–12	L	pH, vol.: 0.1% in 50% ethanol
Alizarin yellow R	3-Carboxy-4-hydroxy-4′-nitro-azobenzene sodium salt	Y	10.2–12.0	R	pH, vol.: 0.1% in water
Poirrer's blue C4B		B	11–13	R	pH: 0.2% in water
Tropaeolin O	*p*-Benzenesulfonic acid-azo-resorcinol	Y	11–13	O	pH: 0.1% in water
Nitramine	Picrylnitromethylamine	C	10.8–13	Br	pH: 0.1% in 70% ethanol
1,3,5-Trinitrobenzene		C	11.5–14	O	pH: 0.1% in ethanol
Indigo carmine	Sodium indigodisulfonate	B	11.6–14	Y	pH: 0.25% in 50% ethanol

MIXED INDICATORS

Composition	Solvent	Transition pH	Acid color	Transition color	Basic color
Dimethyl yellow, 0.05% + Methylene blue, 0.05%	alc.	3.2	Blue–violet	—	Green
Methyl orange, 0.02% + Xylene cyanole FF, 0.28%	50% alc.	3.9	Red	Gray	Green
Methyl yellow, 0.08% + Methylene blue, 0.004%	alc.	3.9	Pink	Straw–pink	Yellow–green
Methyl orange, 0.1% + Indigocarmine, 0.25%	aq.	4.1	Violet	Gray	Yellow–green
Bromcresol green, 0.1% + Methyl orange, 0.02%	aq.	4.3	Orange	Light green	Dark green
Bromcresol green, 0.075% + Methyl red, 0.05%	alc.	5.1	Wine–red	—	Green
Methyl red, 0.1% + Methylene blue, 0.05%	alc.	5.4	Red–violet	Dirty blue	Green
Bromcresol green, 0.05% + Chlorphenol red, 0.05%	aq.	6.1	Yellow–green	—	Blue–violet
Bromcresol purple, 0.05% + Bromthymol blue, 0.05%	aq.	6.7	Yellow	Violet	Violet–blue
Neutral red, 0.05% + Methylene blue, 0.05%	alc.	7.0	Violet–blue	Violet–blue	Green
Bromthymol blue, 0.05% + Phenol red, 0.05%	aq.	7.5	Yellow	Violet	Dark violet
Cresol red, 0.025% + Thymol blue, 0.15%	aq.	8.3	Yellow	Rose	Violet
Phenolphthalein, 0.033% + Methyl green, 0.067%	alc.	8.9	Green	Gray–blue	Violet
Phenolphthalein, 0.075% + Thymol blue, 0.025%	50% alc.	9.0	Yellow	Green	Violet
Phenolphthalein, 0.067% + Naphtholphthalein, 0.033%	50% alc.	9.6	Pale rose	—	Violet
Phenolphthalein, 0.033% + Nile blue, 0.133%	alc.	10.0	Blue	Violet	Red
Alizarin yellow, 0.033% + Nile blue, 0.133%	alc.	10.8	Green	—	Red–brown

UNIVERSAL INDICATORS FOR APPROXIMATE pH DETERMINATIONS

No. 1. Dissolve 60 mg methyl yellow, 40 mg methyl red, 80 mg bromthymol blue, 100 mg thymol blue and 20 mg phenolphthalein in 100 ml of ethanol and add enough 0.1 N NaOH to produce a yellow color.

No. 2. Dissolve 18.5 mg methyl red, 60 mg bromthymol blue and 64 mg phenolphthalein in 100 ml of 50% ethanol and add enough 0.1 N NaOH to produce a green color.

pH	Color		pH	Color	
	No. 1	No. 2		No. 1	No. 2
1	Cherry–red	Red	7	Yellowish–green	Greenish–yellow
2	Rose	Red	8	Green	Green
3	Red–orange	Red	9	Bluish–green	Greenish–blue
4	Orange–red	Deeper red	10	Blue	Violet
5	Orange	Orange–red	11	—	Reddish–violet
6	Yellow	Orange–yellow			

ADSORPTION INDICATORS

Indicator	Color change	Indicator for
Chromotrope F4B	Red to gray–green (bromides), pink to green (iodides)	Bromides, iodides
Dichlorofluorescein	Yellow–green to pink	Bromides, chlorides
Diiodofluorescein	Yellow to pink	Iodides
Fluorescein	Green to pink	Chlorides
Phenosafranine	Pink to blue	Chlorides, bromides
Rhodamine 6G	Pink to violet	Silver with bromide
Rose bengal	Deep pink to bluish–pink	Iodides
p-Ethoxychrysoidine	Red to yellow	Chlorides, thiocyanates, silver with iodides or thiocyanates
Eosin	Changes to pink	Bromides, iodides, thiocyanates
Indo-oxine	Red to blue	Halides
Tartrazine	Colorless to yellowish–green	Silver with halides

OXIDATION-REDUCTION INDICATORS

Common name	Transition potential volts (N Hydrogen Electrode = 0.000)	Color	
		Reduced form	Oxidized form
p-Ethoxychrysoidin	0.76	Red	Yellow
Diphenylamine	0.776	Colorless	Purple
Diphenylbenzidine	0.776	Colorless	Purple
Diphenylamine-sulfonic acid or barium salt	0.84	Colorless	Purple
Naphthidine	—	Colorless	Red
Dimethylferroin	0.97	Red	Yellowish–green
Eriogreen B	0.99	Yellow	Orange
Erioglaucin A	1.0	Yellowish–green	Red
Xylene cyanole FF	1.0	—	—
2,2′-Dipyridyl ferrous ion	1.03	Red	Colorless
N-Phenylanthranilic acid	1.08	Colorless	Pink
Methylferroin	1.08	Red	Pale-blue
Ferroin (*o*-Phenanthrolineferrous ion)	1.12	Red	Pale-blue
Chloroferroin	1.17	Red	Pale-blue
Nitroferroin	1.31	Red	Pale greenish–blue
α-Naphtholflavone	—	Pale straw	Brownish–orange

Reprinted from *The Merck Index,* 7th ed., Merck and Co., Rahway, N.J., 1960, 1566-1570. With permission of the copyright owners.

CONCENTRATION OF ACIDS AND BASES

COMMON COMMERCIAL STRENGTHS

	Molecular weight	Moles per liter	Grams per liter	Per cent by weight	Specific gravity
Acetic acid, glacial	60.05	17.4	1,045	99.5	1.05
Acetic acid	60.05	6.27	376	36	1.045
Butyric acid	88.1	10.3	912	95	0.96
Formic acid	46.02	23.4	1,080	90	1.20
	—	5.75	264	25	1.06
Hydriodic acid	127.9	7.57	969	57	1.70
	—	5.51	705	47	1.50
	—	0.86	110	10	1.1
Hydrobromic acid	80.92	8.89	720	48	1.50
	—	6.82	552	40	1.38
Hydrochloric acid	36.5	11.6	424	36	1.18
	—	2.9	105	10	1.05
Hydrocyanic acid	27.03	25	676	97	0.697
	—	0.74	19.9	2	0.996
Hydrofluoric acid	20.01	32.1	642	55	1.167
	—	28.8	578	50	1.155
Hydrofluosilicic acid	144.1	2.65	382	30	1.27
Hypophosphorous acid	66.0	9.47	625	50	1.25
	—	5.14	339	30	1.13
	—	1.57	104	10	1.04
Lactic acid	90.1	11.3	1,020	85	1.2
Nitric acid	63.02	15.99	1,008	71	1.42
	—	14.9	938	67	1.40
	—	13.3	837	61	1.37
Perchloric acid	100.5	11.65	1,172	70	1.67
	—	9.2	923	60	1.54
Phosphoric acid	80	18.1	1,445	85	1.70
Sulfuric acid	98.1	18.0	1,766	96	1.84
Sulfurous acid	82.1	0.74	61.2	6	1.02
Ammonia water	17.0	14.8	252	28	0.898
Potassium hydroxide	56.1	13.5	757	50	1.52
	—	1.94	109	10	1.09
Sodium carbonate	106.0	1.04	110	10	1.10
Sodium hydroxide	40.0	19.1	763	50	1.53
	—	2.75	111	10	1.11

TRANSMISSION LIMITS OF COMMON IONS

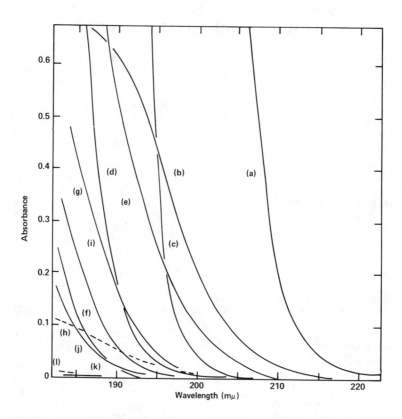

All spectra refer to aqueous solution in a 2-mm-path cell. Analytical grade reagents are used, without further purification: (a) sodium hydroxide, pH 11.9; (b) sodium hydroxide, pH 10.9; (c) 0.01 M sodium chloride; (d) 1 mM sodium chloride; (e) 0.01 M disodium phosphate; (f) 0.01 M monosodium phosphate; (g) 0.01 M sodium sulfate; (h) 1 mM disodium phosphate; (i) 0.01 M boric acid + sodium hydroxide, pH 9.1; (j) 0.1 M perchloric acid, (k) 0.1 M sodium fluoride; (l) 0.01 M boric acid. (Reprinted from Gratzer, W. B., in *Poly-α-Amino Acids*, Fasman, G., Ed., 1967, p. 232 by courtesy of Marcel Dekker, Inc., New York.)

Contributed by W. B. Gratzer.

SPECTROPHOTOMETRIC DETERMINATION OF PROTEIN CONCENTRATION IN THE SHORT-WAVELENGTH ULTRAVIOLET

W. B. Gratzer

Whereas the extinction coefficients of proteins in the aromatic absorption band at 280 nm vary widely, the spectrum at shorter wavelengths is dominated by the absorption of the peptide bond and, therefore, has only a secondary dependence on amino acid composition and conformation. Measurements in this region can therefore serve for approximate concentration determinations of any protein. The following relations are available:

1. Scopes, *Anal. Biochem.,* 59, 277 (1974):

    a. E(1 mg/ml; 1 cm) at 205 nm = 31 with a stated error of 5%.

    b. This can be improved by applying a correction for the relatively strongly absorbing aromatic residues, by measuring the absorbance also at 280 nm. Two forms of this correction are

$$E(1 \text{ mg/ml}; 1 \text{ cm}) \text{ at } 205 \text{ nm} = 27.0 + 120 \times (A^{280}/A^{205})$$

or

$$E(1 \text{ mg/ml}; 1 \text{ cm}) \text{ at } 205 \text{ nm} = \frac{27.0}{1 - 3.85(A^{280}/A^{205})}$$

where the bracket term is the ratio of the absorbances measured at 280 and 205 nm; stated error, 2%.

2. Tombs, Soutar, and Maclagan, *Biochem. J.,* 73, 167 (1959):

$$E(1 \text{ mg/ml}; 1 \text{ cm}) \text{ at } 210 \text{ nm} = 20.5$$

3. Waddell, *J. Lab. Clin. Med.,* 48, 311 (1956): To avoid wavelength-setting error on steeply sloping curves, measurements are made at two wavelengths 10 nm apart and the absorbance difference is used to give the concentration:

$$C(\text{mg/ml}) = 0.144(A^{215} - A^{225})$$

where A^{215} and A^{225} are the absorbances read in 1 cm at 215 and 225 nm.

Note that the longer the wavelength, the lower the sensitivity of the spectrophotometric method, but the hazard of interference from ultraviolet absorbing contaminants is less.

WEIGHTS OF CELLS AND CELL CONSTITUENTS

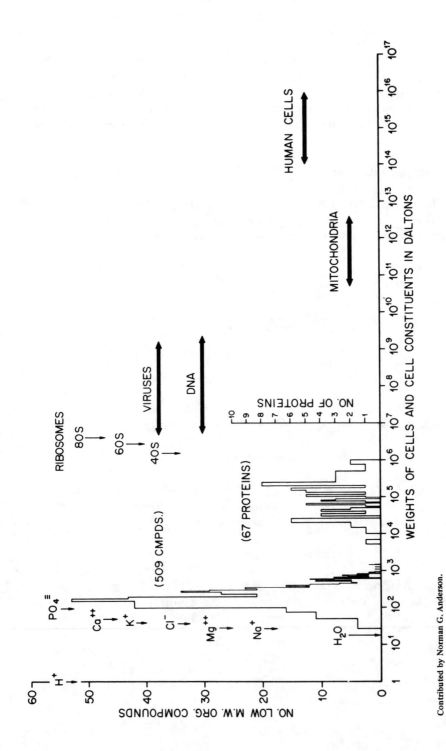

Contributed by Norman G. Anderson.

This figure originally appeared in Sober, Ed., *Handbook of Biochemistry and Selected Data for Molecular Biology*, 2nd ed., Chemical Rubber Co., Cleveland, 1970.

PARTICLE DIAMETER

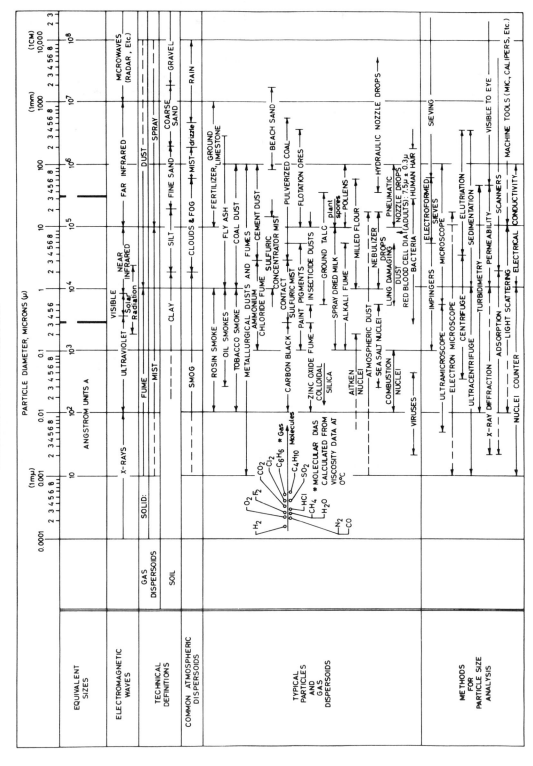

This figure originally appeared in Sober, Ed., *Handbook of Biochemistry and Selected Data for Molecular Biology*, 2nd ed., Chemical Rubber Co., Cleveland, 1970.

CHEMICAL AND PHYSICAL PROPERTIES OF
VARIOUS COMMERCIAL PLASTICS

The chart on the following page should be taken as a general guide only. It states that, for most compounds of the class indicated, the specific material is rated either Excellent, Good, Fair, or Not Recommended. Because so many factors can affect the chemical resistance, it is recommended that a test be run under specific conditions if any doubt exists. The combination of compounds of two or more classes may cause an undesirable effect. Other factors affecting chemical resistance include temperature, pressure and other stresses, length of exposure, and concentration of chemical. As the maximum useful temperature of the plastic is approached, resistance to attack decreases.

CHEMICAL AND PHYSICAL PROPERTIES OF VARIOUS COMMERCIAL PLASTICS

	Polyethylene conventional	Polyethylene linear	Polyallomer	Polypropylene	Polymethylpentene	Teflon FEP	Polycarbonate	Noryl	Polystyrene General purpose	Styrene-Acrylonitrile	Polyvinyl chloride
PHYSICAL PROPERTIES[a]											
Temperature Limit, °C	80	120	121	135	175	205	135	135	70	95	70
Specific Gravity	0.92	0.95	0.90	0.90	0.83	2.15	1.20	1.06	1.07	1.07	1.34
Tensile Strength, psi	2,000	4,000	2,900	5,000	4,000	3,000	8,000	9,600	6,000	11,000	6,500
Brittleness Temperature, °C	−100	−100	−40	0	—	−270	−135	—	Brittle[b]	−25	−30
Water Absorption, %	0.01	0.01	0.02	0.02	0.01	0.01	0.35	0.07	0.05	0.23	0.06
Flexibility	Excellent	Rigid	Slight	Rigid	Rigid	Excellent	Rigid	Rigid	Rigid	Rigid	Rigid
Transparency	Translucent	Opaque	Translucent	Translucent	Clear	Translucent	Clear	Opaque	Clear	Clear	Clear
Relative O_2 Permeability	0.40	0.08	0.20	0.11	2.0	0.59	0.15	—	0.11	0.03	0.01
Autoclavable	No	With caution	Yes	Yes	Yes	Yes	Yes	Yes	No	No	No
CHEMICAL RESISTANCE[c]											
Acids, inorganic	E	E	E	E	E	E	E	G	N	E	G
Acids, organic	E	E	E	E	E	E	G	E	N	E	G
Alcohols	E	E	E	E	E	E	G	E	N	G	G
Aldehydes	G	G	G	G	G	E	F	G	G	F	F
Amines	G	G	G	G	G	E	N	F	N	G	N
Bases	E	E	E	E	E	E	N	E	N	E	N
Dimethyl sulfoxide (DMSO)	E	E	E	E	E	E	N	E	N	E	E
Esters	G	G	G	G	G	E	F	F	F	N	F
Ethers	E	E	E	E	E	E	E	N	E	N	F
Foods	E	E	E	E	E	E	G	G	G	G	G
Glycols	G	G	G	G	G	E	F	E	N	E	F
Hydrocarbons, aliphatic	G	G	G	G	F	E	N	G	N	N	N
Hydrocarbons, aromatic	G	G	G	G	F	E	N	N	N	N	N
Hydrocarbons, halogenated	G	G	G	G	G	E	N	N	N	N	N
Ketones	G	G	G	G	G	E	N	N	N	N	E
Mineral oil	E	E	E	E	E	E	E	E	G	G	E
Oils, essential	G	G	G	G	G	E	G	F	G	F	N
Oils, lubricating	G	E	E	E	E	E	G	E	G	G	E
Oils, vegetable	E	E	E	E	E	E	E	E	G	E	E
Proteins, unhydrolyzed	E	E	E	E	E	E	E	G	G	G	G
Salts	E	E	E	E	E	E	E	E	E	E	E
Silicones	G	E	E	E	E	E	E	E	G	G	G
Water	E	E	E	E	E	E	E	E	E	E	E

[a] Many properties are measured on carefully controlled specimens. Data allow comparison between materials, but values given may not be realized on individual products.

[b] Normally somewhat brittle at room temperatures.

[c] E. Excellent. Long exposures (up to one year) at room temperatures usually have no effect. G. Good. Short exposures (less than 24 hours) at room temperature usually cause no damage. F. Fair. Short exposures at room temperature cause little or no damage under unstressed conditions. N. Not recommended. Short exposures may cause permanent damage.

By permission of Nalge Company, Rochester, New York.

CHEMICAL RESISTANCE AND OTHER PROPERTIES OF CENTERPIECE MATERIALS USED IN THE BECKMAN® MODEL-E ANALYTICAL ULTRACENTRIFUGE

The centerpieces of the cell assembly are offered in three types of materials: Kel-F® (registered trademark of 3M Company), Epon® (registered trademark of Shell Chemical Corporation), and aluminum. These materials vary in chemical resistance, temperature tolerence, and strength. The individual characteristics are given below.

Kel-F

Kel-F is a trifluorochlorethylene polymer. It has a durometer hardness of 80 (D scale), does not absorb water, and is not affected by sunlight. The chlorine constitutes exceptional rigidity of the material, while the fluorine is responsible for its chemical inertness and zero moisture absorption. Because Kel-F is chemically inert, it will withstand exposures to strong acids and alkalies and to most organic solvents.

Kel-F centerpieces are recommended for biological materials when nonaqueous solvents are to be used when the pH of the solution falls outside the range tolerated by the Epon centerpiece (see below). Notable exceptions are highly halogenated and aromatic compounds.

If Kel-F absorbs certain highly halogenated and aromatic compounds, it will swell slightly. Consequently, when using a Kel-F centerpiece with these materials, rinse it immediately after run with distilled water. The following is a partial list of materials that will cause a weight change in Kel-F of 1% or more in 7 days at 25°C.

			%
Chlorine	–	–	12.3
Diethylamine	–	–	1.9
Ethyl acetate	–	–	1.2
Ethyl ether	–	–	3.8
Ethyl propionate	–	–	1.0
Freon® 113	–	–	1.3
Furan	–	–	2.4
Methyl acetate	–	–	1.0
Methylal	–	–	1.3
Methyl ether	–	–	6.4
Methyl propionate	–	–	1.4
Trichloroethylene	–	–	2.3

High speeds may cause a slight distortion in Kel-F centerpieces; the distortion may lead to convection in the cell. Generally, this distortion is negligible but becomes permanent and significant after the centerpiece has been used in about 100 runs at maximum speed.

The maximum operating speed of Kel-F centerpiece varies as follows: for single sectors, maximum speed of rotor; for double sectors, 60,000 rpm; for 6-channel equilibrium, 20,000 rpm; and for fixed partition, 30,000 rpm. The maximum operating temperature for Kel-F centerpiece is 40° C.

Filled Epon

Filled-Epon centerpieces are made from an epoxy resin with a filler consisting of either powdered aluminum or powdered charcoal. The aluminum filler increases the strength and thermal conductivity of the Epon, while the powdered charcoal simply makes the epoxy opaque. After machining the aluminum-filled Epon centerpieces are subjected to a passivation process in which they are soaked in an 20% solution of sodium hydroxide. This process removes aluminum particles from the centerpiece surfaces. However, there is still a possibility of aluminum ions contaminating the sample, particularly after repeated use of the centerpiece. If this possibility can affect the results of an experiment or if results suggest that it is having an effect, charcoal-filled Epon centerpieces should be used.

Filled-Epon centerpieces are used for aqueous solutions in the pH range of 3 to 10. They are the most widely used and recommended for biological materials.

The water absorption of filled Epon is negligible. Some reagents, however, may penetrate the centerpiece and soften it slightly. Generally, any softening effect can be eliminated by rinsing the centerpiece with distilled water after run, regardless of what material was run. The following chemicals cause excess softening of filled Epon and should not be used with this centerpiece:

1. Acetic acid, glacial
2. Ammonium hydroxide (27%)
3. Chloroform
4. Diethylene triamine
5. Dimethyl sulphoxide
6. Ethylene diamine
7. Formaldehyde (40%)
8. *m*-Cresol
9. *M,N*-Dimethyl formamide
10. Nitric acid (10%)
11. Phosphoric acid (85%)
12. Sulphuric acid (70%)
13. Tetrahydrofuran

The maximum operating speeds are: for single sectors, maximum speed of rotor; for double sectors, 50,000 rpm; for single-sector capillary, 60,000 rpm; for 6-channel equilibrium, 48,000 rpm; and for interference wedges, 44,000 rpm. The maximum operating temperature is 40° C.

Aluminum

Aluminum is the strongest of the centerpiece materials, has the best resistance to temperature, and has the longest life expectancy. Aluminum centerpieces are recommended primarily for non-biological materials and for high temperature work. Solutions in aluminum centerpieces must be kept at the neutral pH range, and some of the organic solvents that plastics do not tolerate can be used. For protection against corrosion, each aluminum centerpiece is anodized. The hard anodized surface permits the use of aluminum centerpieces with a wide variety of materials. When handling these center-pieces, care must be taken not to scratch the anodized surface, for corrosion can begin in the scratches on the finish.

Table 1A classifies the groups of chemicals according to their effect on aluminum centerpieces while Table 1B is a partial list of individual chemicals corrosive to aluminum centerpiece.

Table 1A
EFFECTS OF VARIOUS CHEMICALS ON
ANODIZED ALUMINUM CENTERPIECES

Excellent resistance	Good resistance	Poor resistance
Most materials when absolutely dry	Solutions of weak acids including organic acids	Solutions of strong acids and strong alkalies
Solutions containing strongly oxidizing substances, such as dichromates	Solution of weak alkalies	Solutions of heavy metal salts such as compounds of iron, copper, lead, silver, and mercury
Solutions of non-electrolytes with access of air	Organic compounds containing chlorine, if free from moisture and heavy metals	Contact with above metals in presence of moisture
	Gases that form weak acids in water	
	Solutions of salts of active acids including chlorides	
	Solutions of alkali salts of weak acids	

Table 1B
PARTIAL LIST OF INDIVIDUAL CHEMICALS CORROSIVE TO
ALUMINUM CENTERPIECES

Acetyl chloride
Acetylene tetrachloride (wet)
Allyl chloride
Ammonium acid fluoride
Aniline (liquid)
Anthranilic acid
Antimony pentafluoride
Antimony trichloride
Barium chloride, hydroxide, peroxide
Bromine
Cadmium chloride
Calcium arsenide, hydroxide, oxalate
Chloric acid
Chlorine
Chloroform (boiling)
Chromic chloride
Cobaltous chloride
Codeine sulfate
Copper acetate, chloride, nitrate, sulfate
Diglycolic acid
Ferric chloride, nitrate, sulfate
Fluorine
Fluoroboric acid
Fluosilicic acid
Glycerophosphoric acid
Hydrobromic acid
Hydrochloric acid
Hydrofluoric acid
Hydrofluosilicic acid

Hypochlorous acid
Iodine
Lead chloride, nitrate, sulfate
Lithium hydroxide
Mercury salts
Meta-nitrobenzoyl chloride
Methyl bromide, chloride
Naphthylamine
Nickel chloride, nitrate, sulfate
Nitric acid (below 82%)
Nitrogen peroxide (wet)
Nitrosyl chloride
Nitrous oxide (wet)
Perchloric acid
Phosphoric acid
Phosphorous trichloride
Potassium cyanide, hydroxide, peroxide
Potassium persulfate, phosphate
Silver salts
Sodium cyanide, hydroxide, hypochlorite
Sodium lactate, persulfate, phenoxide
Sodium phosphate
Stannous and stannic salts
Sulfuric acid (other than fuming)
Thorium nitrate
Trichloroacetic acid
Tripotassium phosphate
Trisodium phosphate
Zinc chloride

CHEMICAL COMPATIBILITY OF MILLIPORE® FILTERS, O-RINGS, AND FILTER HOLDERS

In the following table chemical compatibilities are intended to serve as a guide in the selection of Millipore® filters, O-rings, and filter holders for use with any of a wide variety of solvents and solutions. Its recommendations are based on static tests involving immersion in the test fluid for 72 hr at 25°C.

Solvent/solution	O-Rings					Filters						Filter holders					
	Teflon®	Viton-A®	Buna-N®	Silicone	E-P®	MF-Millipore®	Duralon®	Polyvic®	Solvinert®	Celotate®	Mitex®	PVC-142®b	Sterifil®b	Swinnex®b	Microtube®	Lifegard® II	Millitube®
Acetic acid, glacial	R	N	N	N	N	N	N	R	R	N	R	R	N	R	N	N	N
5%	R	R	N	R	R	R	R	R	R	N	R	R	R	R	N	R	R
Acetone	R	N	R	N	R	N	R	N	R	N	R	N	N	R	R	N	N
Ammonium hydroxide	R	N	N	R	R	R	N	R	R	N	R	N	N	R	R	N	N
Amyl acetate	R	N	N	N	R	N	R	N	R	R	R	R	R	R	R	N	N
Amyl alcohol	R	N	N	N	R	N	R	R	R	R	R	R	R	R	R	N	N
Benzene	R	R	N	N	N	R	R	N	R	R	R	N	N	N	R	N	N
Boric Acid	R	R	R	R	R	R	R	R	R	R	R	R	R	R	R	R	R
Brine (sea water)	R	R	R	R	R	R	R	R	R	R	R	R	R	R	R	R	R
Butyl acetate	R	N	N	N	N	N	R	N	R	N	R	N	N	R	R	N	N
Butyl alcohol	R	R	R	N	N	R	N	R	R	N	R	R	R	R	R	N	N
Carbon tetrachloride	R	R	N	N	N	R	N	N	R	R	R	N	N	R	R	N	N
Cellosolve®	R	N	N	N	N	N	R	N	R	N	R	N	N	R	R	N	N
Chloroform	R	R	N	N	N	R	N	N	R	N	R	N	N	R	R	N	N
Chlorothene® NU	R	R	N	N	N	N	R	N	R	R	R	N	N	R	R	N	N
Cyclohexanone	R	N	N	N	N	N	R	N	R	R	R	N	N	R	R	N	N
Developers (photo)	R	R	R	N	N	N	R	R	R	N	R	R	N	R	R	N	N
Dioxane	R	N	N	N	N	N	N	N	R	N	R	N	N	R	R	N	N
Dowciene	R	R	N	N	N	N	R	N	R	R	R	N	N	R	R	N	N
Ethers	R	N	N	N	N	R	R	R	R	R	R	N	N	R	R	N	N
Ethyl acetate	R	N	N	N	N	N	R	N	R	N	R	N	N	R	R	N	N
Ethyl alcohol	R	N	R	R	R	N	R	R	R	N	R	R	R	R	R	N	N
Ethylene glycol	R	R	R	R	R	N	R	R	R	N	R	R	R	R	R	N	N
Formaldehyde	R	N	N	N	N	N	N	R	R	N	R	R	N	R	N	N	N
Freon® TF or PCA	R	N	R	N	N	R	R	R	R	R	R	R	N	R	R	R	R
Gasoline	R	R	R	N	N	R	R	R	R	R	R	R	N	R	R	R	R
Glycerine	R	R	R	R	R	R	R	R	R	R	R	R	R	R	R	R	R
Helium	R	R	R	R	R	R	R	R	R	R	R	–	–	R	R	R	R
Hexane	R	R	R	N	N	R	R	R	R	R	R	R	R	R	R	R	R
Hydrochloric acid	R	R	N	N	N	N	N	R	R	N	R	N	N	R	N	N	N
Hydrofluoric acid	R	R	N	N	N	N	N	R	R	N	R	N	N	N	N	N	N
Hydrogen (gas)	R	R	R	N	R	R	R	R	R	R	R	–	–	R	R	R	R
Hypo (photo)	R	R	R	R	R	R	R	R	R	R	R	R	R	R	R	R	R
Isobutyl alcohol	R	R	N	R	R	R	N	R	R	N	R	R	R	R	R	N	N
Isopropyl acetate	R	N	N	N	N	N	R	N	R	N	R	N	N	R	R	N	N
Isopropyl alcohol	R	R	N	R	R	N	N	R	R	N	R	R	R	R	R	N	N
JP-4	R	R	R	N	N	R	R	R	R	R	R	R	R	R	R	N	N

[a]R = recommended; N= not recommended.
[b]Recommendations apply to materials of construction other than O-rings. If the solvent of interest is compatible with the material of construction, select an O-ring which is also compatible.

CHEMICAL COMPATIBILITY OF MILLIPORE® FILTERS, O-RINGS, AND FILTER HOLDERS (continued)

Solvent/solution	O-Rings					Filters						Filter holders					
	Teflon®	Viton-A®	Buna-N®	Silicone	E-P®	MF-Millipore®	Duralon®	Polyvic®	Solvinert®	Celotate®	Mitex®	PVC-142®b	Sterifil®b	Swinnex®b	Microtube®	Lifegard® II	Millitube®
MEK	R	N	N	N	R	N	R	N	R	N	R	N	N	R	R	N	N
Methyl alcohol	R	N	R	R	R	N	N	R	R	N	R	R	R	R	R	N	N
Methylene chloride	R	N	N	N	N	N	N	N	R	N	R	N	N	R	R	N	N
MIBK	R	N	N	N	N	N	R	N	R	N	R	N	N	R	R	N	N
Mineral spirits	R	R	R	N	N	R	R	R	R	R	R	R	R	R	R	R	R
Nitric acid	R	R	N	N	N	N	N	R	R	N	R	N	N	R	N	N	N
Nitrobenzene	R	N	N	N	N	N	R	N	R	N	R	N	N	N	N	N	N
Nitrogen (gas)	R	R	R	R	R	R	R	R	R	R	R	–	–	R	R	R	R
Pentane	R	R	R	N	N	R	R	R	R	R	R	R	R	R	R	R	R
Perchloroethylene	R	R	R	N	N	R	N	N	R	N	R	N	N	R	R	N	N
Pet. base oils	R	R	R	N	N	R	R	R	R	R	R	R	R	R	R	R	R
Silicone oils	R	R	R	N	R	R	R	R	R	R	R	R	R	R	R	R	R
Sodium hydroxide	R	N	N	R	R	N	R	R	R	N	R	R	N	R	N	N	N
Sulfuric acid	R	R	N	N	N	N	N	R	R	N	R	N	N	R	N	N	N
Toluene	R	R	N	N	N	R	R	N	R	R	R	N	N	N	R	N	N
Trichloroethane	R	R	N	N	N	R	N	N	R	N	R	N	N	R	R	N	N
Trichloroethylene	R	R	N	N	N	R	N	N	R	N	R	N	N	R	R	N	N
Xylene	R	R	N	N	N	R	N	N	R	R	R	N	N	N	R	N	N

Compiled by V. S. Ananthanarayanan.

Index

INDEX